Bioprocessing, Bioengineering and Process Chemistry in the Biopharmaceutical Industry

Kumar Gadamasetti • Stephen A. Kolodziej
Editors

Bioprocessing, Bioengineering and Process Chemistry in the Biopharmaceutical Industry

Using Chemistry and Bioengineering to Improve the Performance of Biologics

Springer

Editors
Kumar Gadamasetti
Certum Bio
San Francisco, CA, USA

Stephen A. Kolodziej
Bioprocess Research and Development
Biotherapeutics Pharmaceutical Sciences
Pfizer, Inc.,
Chesterfield, MO, USA

ISBN 978-3-031-62006-5 ISBN 978-3-031-62007-2 (eBook)
https://doi.org/10.1007/978-3-031-62007-2

This Springer imprint is published by the registered company Springer Nature Switzerland AG
The registered company address is: Gewerbestrasse 11, 6330 Cham, Switzerland

KG would like to dedicate this volume to his parents, Rao and Devi, and children, Stuthi and Pratik, and to all those who work relentlessly towards finding solutions in alleviating the pain and suffering of patients.

SAK would like to dedicate this book to his wife Mary and his children Rebecca, Eric and Brian.

Foreword

Biotherapeutics are driving a remarkable healthcare revolution, propelling medicine into new frontiers and transforming the landscape of patient care. These advancements build on decades of success with monoclonal antibodies and are being further guided by technological innovations catalyzing novel therapeutic modalities. In parallel, important fundamental scientific discoveries are shedding new light on biochemical processes underlying human disease and illuminating new targets for therapeutic intervention. It is an exciting time for biotechnology, a spirit that the authors have captured in the current volume, which eloquently describes both the current state of affairs and the advancements in bioprocess engineering related to biotherapeutic development.

Biotherapeutics discovery, development, and manufacturing take a village, including academic researchers who make seminal scientific discoveries, entrepreneurs who start biotechnology companies based on this knowledge, venture capitalists who fund these companies, pharmaceutical companies who provide expertise and resources throughout, and regulatory agencies which oversee clinical trials and approvals. A thriving ecosystem relies on stakeholders throughout the value chain, where lines and roles are becoming blurred, particularly as contract organizations become more prolific and enable outsourcing of activities that have historically been infrastructure and resource heavy. However, despite all of the advancements made and partnerships forged, the cost of developing drugs is still prohibitively expensive and the timelines excruciatingly long. Researchers in academia and industry have created tools to accelerate early-stage drug screening and characterization efforts, yet significant bottlenecks and costs exist with manufacturing drug candidates at scale, especially for new therapeutic modalities.

Several chapters of this volume are devoted to innovations in biotherapeutics manufacturing including scalability of DNA synthesis, cell culture technologies, and considerations related to more complex therapeutics beyond monoclonal antibodies such as multi-specific and fusion proteins, drug conjugates, gene therapies, vaccines, and engineered cells. Complementing the discussion are chapters focused on data analytics, bioinformatics approaches for assessing developability, and regulatory considerations which are a key part of the manufacturing landscape. Also included are deep dives on some of the twenty-first century's most exciting topics under the broad theme of synthetic biology, which is enabling researchers to leverage engineering principles to construct new biological systems, molecules, and organisms with specific functionalities.

A convergence of scientific breakthroughs, technological advancements, and interdisciplinary collaboration has propelled biotherapeutics to the forefront of modern medicine. While innovations in biotherapeutics development and manufacturing discussed in this volume are a glimpse into the extraordinary progress being made in this dynamic field, as personalized medicine and targeted interventions continue to shape how we approach and conquer disease, it is imperative to foster continued investment in research, development, and regulatory frameworks that ensure the safety, efficacy, and accessibility of these transformative therapies. Thus, I highly recommend this book to those desiring education in these key areas.

Jennifer R. Cochran
Senior Associate Vice Provost for Research, Macovski Professor of Bioengineering and (by courtesy) Chemical Engineering, Stanford University, Stanford, CA, USA

Limitless possibilities are unleashed with continuing advances in biochemical discovery applied to human health. Within the context of the 2019 Covid pandemic, we witnessed an industry galvanized with a single purpose to pursue new medicines and vaccines to bring an end to the outbreak. While many of the approaches used were based on established technologies, we also saw disruptive solutions implemented with RNA technologies, scale-out strategies, and delivery solutions that have far-reaching applications.

Vaccines once again demonstrated the transformative nature of their application in providing a route out of the pandemic. Indeed, vaccination is considered one of the most successful public health interventions of all time and has undoubtedly resulted in preventing millions of deaths. But those applications are not limited to vaccines and have far-reaching potential in the field of therapeutics, including cell and gene therapies, with the promise of treating chronic or acquired diseases. To date, we have seen the licensure of several gene therapies, against beta-thalassemia, lymphoma and leukemia, inherited retinal disorders, and cerebral adrenoleukodystrophy. These are examples of early successes, but there are countless others that do not reach the end-point, whether for reasons of cell-specific targeting or durability of response; understanding this will be critical to advancing the field.

Our comprehension of cancer and novel mechanisms to identify, target, and kill these cells has also led to the development of bi- and tri- specific antibodies, and antibody-drug conjugates. The complexity of these new molecules, from their design, stability, expression, and purification, should not be underestimated. While antibodies have historically benefitted from platform processing, these new entities require a degree of customization not previously witnessed.

For the biochemical engineer or process scientist, the field has become a much more interesting world, both in the challenges and the tools they must utilize to deliver the promise of tomorrow's medicines. Tools such as synthetic biology and Crisper technologies allow a degree of cellular manipulation and editing for the identification of new drug targets, biomarkers, and identification of mechanisms leading to drug resistance. In all these tool sets unlock novel strategies for the treatment of infectious and non-infectious diseases. However, with such novel advancements, the regulatory framework will have to be negotiated and the role of our scientists will be to collate the data required to license these approaches, and a voice of advocacy in driving their adoption, thus the need of developing true expertise in the field.

Perhaps the most weighted consideration in any development plan is the cost of goods. Considering which market these products will launch and how to best achieve distribution to maximize utility is a key aspect of any target product profile. While first-generation vaccines may have achieved the lowest cost possible; multi-product, multi-valency vaccines increase these costs. The problem is further compounded in the therapeutic space, especially in cell therapies where the process is affected by high labor costs and low volume production. There are two possible strategies that could be executed in a complimentary manner—operating in a globalized manner to take advantage of localized distribution chains and labor costs, or to create manufacturing solutions that break the economies of scale model. It is possible that a little of both will be required, which means that tech transfer and regionalization will still be critical to successful execution.

I recommend this book to those in academia and industry in the translational sciences and process engineering. It covers a wide range of pertinent topics described above while giving the reader a focused context of current technologies and future trends, and I commend the authors on an excellent summary of an array of difficult topics.

Tarit K. Mukhopadhyay
Vice President, Head of Infectious Disease and Vaccines Discovery, Merck & Co., Rahway, NJ, USA

The biopharmaceutical industry has seen rapid advancements over the past four decades, dating back to the development of recombinant insulin. The significant change and progress in that time has been breathtaking, and the editors and authors of this volume have captured recent advances in the field which both capture our past and simultaneously look forward to future advancements.

This volume highlights the wide variety of modalities that are currently being developed within the industry. The authors begin with a discussion of the optimization and continued development of protein-based biotechnology products (mAbs, bi-specifics, etc.) and focus on concerns around cellular output and process intensification. This is critical, for as a more mature industry we need to deliver on the hope of biotechnology products in a global, equitable manner. Subsequent chapters review novel therapies including AAV-based gene therapies, CAR-T, ADCs, and mRNA-LNPs, emphasizing the continued evolution of modalities. This diversity of products demonstrates an expanding toolkit of potential therapeutics with which to address unmet medical needs, as well as it shows that, as a field, we can both continuously learn and apply expertise and talents to new compounds. In addition to highlighting developments in the creation of novel products, there are several chapters on areas which compliment the discussion of biotechnology product and process development, specifically oligonucleotide synthesis, advances in bioconjugation, informatics, regulatory strategies, continuous manufacturing, and technical transfer all of which are required for thoughtful development and implementation of biopharmaceutical products.

One aspect that this volume highlights is the continued intersection of protein chemistry, cell biology, process engineering, and analytical characterization that is required to develop these complex compounds and their scalable and controlled manufacturing processes. While these disciplines were combined at the initiation of the field of biotechnology, the level of sophistication and understanding of production cell biology (metabolism, genetics, optimal growth, and expression conditions) has expanded significantly providing us with confidence that these products can be produced at lab scale, as well as at production scale, in a manner that is predictable and sustainable. The principles and experiences associated with scale-up of monoclonal antibodies, developed over the past 30 years, can be applied to modalities which are still emerging such as potential CRISPR-based therapeutics for gene editing. Again, as technologies evolve, so do therapies, enabling patients to slowly migrate, for example, from the management and treatment of hemophilia-based disorders with Factor VIII and Factor IX to potential AAV-based cures. The modalities are remarkably different but the fundamentals of molecular design, cell culture, process design, and manufacture of aseptic injectables are the same.

Looking forward, we can be assured that there will be challenges with new modalities as well as increased scale and production of existing products. With these challenges, however, we can be confident that the foundation of biology, chemistry, and engineering described in this volume will enable continued success in the development and implementation of novel therapies for patients across the globe.

Nicholas W. Warne
Biotherapeutics Pharmaceutical Sciences, Pfizer, Inc., Cambridge, MA, USA

Preface

Drug development in the pharmaceutical and biopharmaceutical industry entails discovery, pre-clinical and clinical development and market launching of the drug under regulatory constraints. The previous two volumes (*Process Chemistry in the Pharmaceutical Industry*—1999 and 2008) by one of the authors (Gadamasetti) were dedicated to the development of the drug substance (active pharmaceutical ingredient), and in particular, the introduction of biologics (monoclonal antibodies) development to the process technologists and process development community in the pharmaceutical industry made a significant impact and has been an eye-opener for some of the readers. Given the explosive growth of biologics around the turn of the millennium on the need-basis of the technologies and the applications to address the patient population demands, it became relevant to discourse the need to understand and share the information on biologic drugs, bioprocessing and the bioengineering in developing the biologic drug to bring it to patients. The editors sense the compelling need to share the various aspects of technologies and the process of biologic drug development, CMC regulatory aspects and the pathways to bring the biologic drugs to patients to minimize the pain and suffering and maximize the outcomes.

The genesis and the devastating rapid spread of the pandemic in 2019 have challenged the global scientific community outlook to embrace and adapt the methods for the speedy development of vaccines to bring them to the patient population and the paradigm shift of speedy discovery and the development has once again proved the strength of unity among the researchers, developers and manufacturers alike, in addressing the challenges with sense of urgency, globally.

The work in this volume embraces the inclusivity of various disciplines to jointly undertake the problem-solving issues in discovering and developing the need-based biological medicines to market the drugs for alleviating the pain and suffering of the patients. The topics and subject material were deliberately chosen to cover a wide range of topics by global expertise: synthetic biology, oligo nucleotides and DNA synthesis, CAR-T and CRISPR technologies and applications, biosimilars, bioprocessing, process engineering, gene therapy and vaccines, fusion proteins, antibody-drug conjugates (ADCs), biopharmaceutical analytics, regulatory CMC and the tech transfer and emerging trends in biologics of biopharmaceuticals.

We are hoping to reach global readers across academia as well as the pharmaceutical and biopharmaceutical industries and look forward to their participation and input for future work.

San Francisco, CA, USA Kumar Gadamasetti
Chesterfield, MO, USA Stephen A. Kolodziej

Acknowledgements

We would like to thank the many contributors of this volume for their hard work, dedication, manuscripts, and their patience. In addition, we want to thank the reviewers of these manuscripts, as they were entrusted with assuring the quality of the content and strengthening the high level of the science presented. We believe the significant contributions of Rosalind Franklin laid the pathway to our modern understanding of the structure of DNA, and this led to the significant advances made in the twenty-first century in bioengineering and bioprocessing detailed in this book.

We would like to extend special thanks to Anita Lekhwani, Bo Arve, Aaron Sato and Dianne Shumay for working with us during the initial phases of the book and helping with valuable suggestions.

We thank the staff at Springer-Nature Publishers for their outstanding support throughout the journey and for cooperating with late-breaking changes.

Contents

Contributors

Erin Abbott Global Product Development and Supply, Bristol-Myers Squibb Company, East Syracuse, NY, USA

Adebola Adeniran Cell Therapy Process Development, The Center for Breakthrough Medicine, King of Prussia, PA, USA

Katherine Arch-Douglas Global Regulatory Sciences CMC, Pfizer Inc., Pearl River, NY, USA

Abdulrahman Baki Cell and Gene Therapy, Medicinal Science and Technology, GlaxoSmithKline, Stevenage, UK

Barak Barnoon Global Product Development and Supply, Bristol-Myers Squibb Company, Devens, MA, USA

Joschka Bauer Pharmaceutical Development Biologicals, Boehringer Ingelheim Pharma GmbH & Co. KG, Biberach/Riss, Germany

Melissa Bentley Global Product Development and Supply, Bristol Myers Squibb, Phoenix, AZ, USA

Ciaran Brady Global Product Development and Supply, Bristol Myers Squibb, Devens, MA, USA

Jennifer R. Cochran Schools of Medicine and Engineering, Stanford University, Shriram Center, Stanford, CA, USA

Hussain Dahodwala Institute for Bioscience and Biotechnology Research, University of Maryland, Rockville, USA

Yanhuai (Richard) Ding CMC Drug Substance/Drug Product, EvolveImmune Therapeutics, Branford, CT, USA

Nathalie Dubois Global Regulatory Sciences CMC, Pfizer Inc., Brussels, Belgium

Michael P. Dux Pfizer Global Supply, Pfizer Inc., Sanford, NC, USA

Kumar Gadamasetti Certum Bio, San Francisco, CA, USA

Izabela Gierach Protein Capture Science, Columbus, OH, USA

Pankaj Gupta In Silico Team, Biotherapeutics Discovery, Boehringer – Ingelheim Pharmaceutical Inc., Ridgefield, CT, USA

Salina Handy Manufacturing Science and Technology, Lonza, Houston, TX, USA

Gregory W. Hiller Bioprocess Research and Development, Biotherapeutics Pharmaceutical Sciences, Pfizer Inc., Andover, MA, USA

Kevin Holden Synthego Corporation, Redwood City, CA, USA

Brendan Hughes Global Product Development and Supply, Bristol Myers Squibb, Devens, MA, USA

Timothy Iskra Bioprocess Research and Development, Biotherapeutics Pharmaceutical Sciences, Pfizer Inc., Andover, MA, USA

David Y. Jackson DYJ Pharma, San Mateo, CA, USA

Arpana Khatri Rare Disease Research Unit, Pfizer, Inc., Cambridge, MA, USA

Rigel Kishton Moonwalk Biosciences, South San Francisco, CA, USA

Stephen A. Kolodziej Bioprocess Research and Development, Biotherapeutics Pharmaceutical Sciences, Pfizer Inc., Chesterfield, MO, USA

Frank Kotch Bioprocess Research and Development, Biotherapeutics Pharmaceutical Sciences, Pfizer Inc., Chesterfield, MO, USA

Sebastian Kube Pharmaceutical Development Biologicals, Boehringer Ingelheim Pharma GmbH & Co. KG, Biberach/Riss, Germany

Sandeep Kumar Computational Protein Design and Modeling Biopharmaceutical Informatics, Moderna, Cambridge, MA, USA

Sid Kundu Cell Culture and Bioprocess Operations, Genentech, South San Francisco, CA, USA

Conor Layden Global Product Development and Supply, Bristol Myers Squibb, Dublin, Ireland

Leo Letendre Leo Letendre Consulting, LLC, Oakdale, CT, USA

Amanda M. Lewis Global Product Development and Supply, Bristol Myers Squibb, Devens, MA, USA

Mairead Looby Global Product Development and Supply, Bristol Myers Squibb, Dublin, Ireland

Jun Luo Manufacturing Sciences, Genentech, Vacaville, CA, USA

Margaret (Peggy) Marino Downstream Process Development & Manufacturing, AnaptysBio, San Diego, CA, USA

Stephen Mayer Global Regulatory Sciences CMC, Pfizer Inc., Pearl River, NY, USA

Bobby Moon Synthego Corporation, Redwood City, CA, USA

Montse Morell EditCo Bio, Redwood City, CA, USA

Tarit K. Mukhopadhyay Merck Research Labs, West Point, PA, USA

Durgesh Nadkarni Bioprocess Research and Development, Biotherapeutics Pharmaceutical Sciences, Pfizer Inc., Chesterfield, MO, USA

Andrew Nelson Global Regulatory Sciences CMC, Pfizer Inc., Sanford, NC, USA

Rebecca L. Nugent Tessera Therapeutics, Sommerville, MA, USA

Kruti Patel Rare Disease Research Unit, Pfizer, Inc., Cambridge, MA, USA

Richard Pelt Global Regulatory Sciences CMC, Pfizer Inc., Sanford, NC, USA

Yanhuai (Richard) Ding Nonclinical Biostatistics, Genentech, South San Francisco, CA, USA

Meenakshi Prabhune The Scientist, Lab X Media Group, Midland, ON, Canada

Ranya Pranomphon School of Biotechnology, Institute of Agricultural Technology, Suranaree University of Technology, Nakhon Rachasima, Thailand

Lin Qiu Nonclinical Biostatistics, Genentech, South San Francisco, CA, USA

Stephen Riffle Twist Bioscience, South San Francisco, CA, USA

Rebecca Roberts Synthego Corporation, Redwood City, CA, USA

Jennifer A. S. Romine Bioprocess Research and Development, Biotherapeutics Pharmaceutical Sciences, Pfizer Inc., Chesterfield, MO, USA

Ashley Sacramo Solid Biosciences Inc., Charlestown, MA, USA

Aaron K. Sato Twist Bioscience, South San Francisco, CA, USA

John J. Scarcelli Sanofi, Framingham, MA, USA

Stefan R. Schmidt evitria AG, Zurich, Switzerland

Susan T. Sharfstein College of Nanoscale Science and Engineering, SUNY Polytechnic Institute, Albany, NY, USA

Suryanarayan Somanathan Rare Disease Research Unit, Pfizer, Inc., Cambridge, MA, USA

Grant Sumida Manufacturing Sciences, Genentech, Vacaville, CA, USA

Yang Tang Nonclinical Biostatistics, Roche, Mississauga, ON, Canada

Vijay Tejwani College of Nanoscale Science and Engineering, SUNY Polytechnic Institute, Albany, NY, USA

Tarl Vetter Novartis Pharmaceuticals Corporation, Durham, NC, USA

Nicholas W. Warne Biotherapeutics Pharmaceutical Sciences, Pfizer Inc., Andover, MA, USA

David W. Wood Department of Chemical and Biomolecular Engineering, Ohio State University, Columbus, OH, USA

Montarop Yamabhai College of Nanotechnology, Science and Engineering, State University of New York, Albany, NY, USA

About the Editors

Kumar Gadamasetti is an experienced pharmaceutical and biopharmaceutical executive with a strong operations mindset combined with integrating pragmatic strategic vision. Currently, he is the CEO of Certum Bio in San Francisco Bay area and has a joint appointment as the Member of the Board of Advisors with the University of California Berkeley Postdoctoral Entrepreneurial Program (BPEP). Over the past 30 years, his focus has been in developing small molecules and biologics and marketing programs, efficient operations, R&D involving drug substance (API) and drug product (formulations) and outsourcing management of both sides of business with CDMOs and CROs. Kumar actively consults with both pharma/biopharma and the academic institutions and enjoys being a liaison connecting academia to the industry and *vice versa*. The therapeutic areas include cancer, CV, CNS and infectious diseases. He worked in Fortune 500 companies (Bristol-Myers Squibb Co, NJ-USA and Amgen, CA-USA), mid-size and startups/CROs (Discovery Partners Intl. CA; X-Mine, CA; Delphian Pharma, CA and Certum LLC, CA). Of the several programs he worked at Pharma and Biotech companies, three culminated in making to the market (*Paclitaxel™*—Bristol-Myers Squibb; *Sensipar™*—Amgen and *Duexis™*—Horizon Therapeutics). As the Founder and Chairman, he spearheaded the ACS ProSpectives international conferences on "Process Chemistry in the Pharmaceutical Industry," and as the chair, speaker and moderator at numerous national and international conferences on API Process Chemistry and Drug Discovery and Development. He has been a Visiting Professor at Catholic U., Louvain, Belgium and U. Pittsburgh, Pittsburgh, PA and Visiting Speaker at Humboldt U. in Berlin, Germany. In addition to several peer-reviewed publications and patents, he published two volumes on *Process Chemistry in the Pharmaceutical Industry* [1997 & 2008] and the current volume entitled, *Bioprocessing, Bioengineering and Process Chemistry in the Biopharmaceutical Industry-Using Chemistry and Bioengineering to Improve the Performance of Biologics* [2024]. Kumar's current interests include cancer and immunotherapy—ADCs as cancer therapeutics, API process R&D, drug formulations and manufacturing, new paradigms and repurposing drugs, rapid DNA synthesis and DNA synthetic libraries. He earned his PhD from the University of Vermont and was a Post-doctoral Fellow at the University of Virginia.

Stephen A. Kolodziej is an Associate Research Fellow in the Biotherapeutics Pharmaceutical Sciences Division of Pfizer at the Chesterfield Missouri site. His group develops downstream manufacturing processes for biotherapeutic drug and vaccine candidates, specializing in protein conjugates. He made major contributions to process and product understanding on late-stage bacterial vaccine projects, including multivalent capsular polysaccharide conjugate vaccines against pneumococcus, staphylococcus and meningococcus. He started his professional career as a medicinal chemistry in Searle, Pharmacia and then Pfizer, developing novel small-molecule candidates for the treatment of inflammation, cancer and cardiovascular disease. He applied the tools of combinatorial chemistry to produce small libraries of candidates for evaluation in numerous preclinical programs and was a coinventor for Xalkori®, an ALK inhibitor approved for the treatment of non-small cell lung cancer. Steve has a BA and a PhD in Organic Chemistry from the University of Missouri-St. Louis with Professor Rudolph E. K. Winter on the synthesis of natural products.

Part I
Overview

Chapter 1
Bioprocessing, Bioengineering and Process Chemistry in the Biopharmaceutical Industry: Using Chemistry and Bioengineering to Improve the Performance of Biologics

Kumar Gadamasetti

Abstract The explosion and exploitation of biotechnology and bioengineering toward the end of the twentieth century led to the invention of valuable biologic drugs for the benefit of patients around the world. Opportunities continue to expand in dramatic and unforeseen ways through the present time. Advances in individual fields as well as multidisciplinary integration across science, technology, and bioengineering drive this innovation, emphasizing the importance of reducing timelines and cost for drug development and clinical studies, all under a regulatory and ethical framework that is vital for delivering safe and effective therapies to patients.

Medical disorders with associated genetic components and the understanding of single mutations in disorders like sickle-cell disease, cystic fibrosis, and progeria have led to possible solutions by synthetic biologists for such diseases. The latest advancements in synthetic biology and emerging new technologies manifested in uncovering the expression of monoclonal antibodies, antibody drug conjugates (ADCs), gene therapy and immunotherapies, cell therapies (e.g., CAR-T [Chimeric Antigen Receptor-engineered T-cells]), gene editing (e.g., CRISPR [Clustered Regularly Interspaced Short Palindromic Repeats]), and vaccines among other applications. Efficient processes for generating monoclonal antibodies, gene therapy, and cell therapy products require significant advances in upstream and downstream processes, analytical methods, as well as automation leading to process efficiencies and green processes.

The original version of the chapter has been revised. A correction to this chapter can be found at https://doi.org/10.1007/978-3-031-62007-2_20

K. Gadamasetti (✉)
Certum Bio, San Francisco, CA, USA
e-mail: kumar@certumus.com

K. Gadamasetti, S. A. Kolodziej (eds.), *Bioprocessing, Bioengineering and Process Chemistry in the Biopharmaceutical Industry*,
https://doi.org/10.1007/978-3-031-62007-2_1

The opening chapter discusses the differences and the relationship between small molecules and biologic drugs and topics representing the latest advances of bioprocessing, bioengineering, and process chemistry in biologics and biotechnology applications of interest to global readers from both industry and academia.

Keywords Building blocks of DNA · Mutations · Sickle cell disease (SCD) · Cystic fibrosis · Progeria · Crick's adaptor hypothesis · Process chemistry

> What is Life? The Physical Aspect of the Living Cell," Erwin Schrödinger in his book (1944) noted that "Organic chemistry, indeed, in investigating more and more complicated molecules, has come very much nearer to that 'aperiodic crystal' which, in my opinion, is the material carrier of life." Schrödinger continues to mention in Chap. III of his book: "The physicist would be inclined to call de Vries's mutation theory (1902) figuratively, the quantum theory of biology… The mutations are actually due to quantum jumps in the gene molecule" [1].

1.1 Introduction

The first non-small molecule drug that made a major impact in alleviating pain and suffering of patients over a hundred years ago was insulin, produced using *Escherichia coli* (*E. coli*) cells. Canadians Frederick Banting and his medical graduate student Charles Best had successfully isolated the hormone insulin in 1921 from dog pancreas [2], and Banting and John MacLeod were awarded the Nobel Prize in medicine in 1923 for mass producing insulin for the treatment of diabetes.

Human insulin is composed of 51 amino acids (MW 5808). While insulin meets the definition of a biologic or a biologic drug (a complex molecule made from *E. coli* cells), it has been treated both as a drug and a biologic, being on the market longer than the FDA rules for biologics were generated. Even though initially insulin was treated as a drug, during early 2020 it officially moved to biologic regulatory framework by the FDA [3]. There is a special discussion and definition dedicated to biologics in the related subheadings.

The premise in putting this volume together is not to show that either the biologics or small molecule drugs are superior to the other class of drugs. Typically, biologic drugs take longer lead times to develop, have stringent regulation requirements for clinical trials, and need relatively high capital investments as compared to the complementing small molecule drugs, but their aptness to certain drug targets and the therapeutic applications makes biologics subjugate in treating autoimmune disorders and the rare diseases like sickle-cell disease (SCD), cystic fibrosis (CF), and the genetic disease, progeria.

1.1.1 The Synopsis of the Volume

Biogenetic processes provide a pathway to biosynthesis of target molecules, leveraging the use and manipulation of the basic nucleotide building blocks, RNA, and eventually modifying DNA for constructive purposes. Thus, the fundamental aspects of genetic engineering allow the tweaking and maneuvering of building blocks for the biochemical and organic chemical processes to create novel and synthetic oligonucleotides, including plasmids, and relying on cellular machinery for applied functional outcomes. Extending the idea of utilizing the reconstruction processes of chemistry at the most fundamental level to generate significant end products has unlimited potential. The utilization of these fundamental building block maneuvering processes can facilitate the advancement of synthetic biology, in general, and the emerging new technologies and tools in biopharmaceuticals, including expression of monoclonal, bispecific, and single-domain antibodies, antibody drug conjugates (ADCs), fusion proteins, genetic alphabet engineering, gene therapy and immunotherapy, gene editing (e.g., CRISPR [Clustered Regularly Interspaced Short Palindromic Repeats]), cell therapies (e.g., CAR-T [chimeric antigen receptor-engineered T-cells]), vaccines, and chemical processes to generate biologics with desired properties, linking vaccines with enhanced immunogenicity and nanoparticles with targeting functions. Efficient processes for monoclonal antibodies (mAbs), gene therapy, and cell therapy products require significant advances in upstream and downstream processes, analytical methods, as well as automation leading to process efficiency, cost containment, and green processes, all topics for invaluable discussions. Unlike small molecule generics, biologics (biosimilars, biobetters, and proteins) cannot be precisely replicated in manufacturing. They demand precise processes to be put in place that meet regulatory constraints leading to significant manufacturing challenges. Certainly, the regulatory bodies for biologics are constantly challenged with continuous learning and evolution of regulatory perspectives and the fine-tuning of regulatory guidelines to support new technologies. The International Conference of Harmonization (ICH) is specially challenged with guiding the biopharmaceutical companies and allied CROs/CMOs to advance the biological products into global markets.

In visualizing this futuristic volume to greater reach, a deliberate and conscious attempt has been made by the editors to select the topics representing the latest advances of bioprocessing, bioengineering, and process chemistry in biologics and biotechnology applications of interest to global readers both from industry and academia. It is hoped that the readers will benefit from learning directly from top experts in the field to improve and streamline processes, problem-solve the complex challenges inherit in this effort, and further the vital and important work of advancing new drugs, vaccines, and therapies to alleviate the pain and suffering of humankind.

1.2 Synthetic Biology: A Perspective

As a chemist, it has always been intriguing and mindboggling to comprehend with the fact how the chemistry and the array of atoms and molecules in nature, nucleotides specifically, are essential to the storage, transmission, and evolution of genetic information. Understanding the stereoregularity of the building blocks of DNA (dA, dT, dG, dC; "d" refers to deoxy), and of RNA (A, U, G, C), the thermodynamic stability of the gene in the cell supporting the Darwinism of life is vital for every discipline and the faculties in science and technology to appreciate. Protein metabolism [12], the manifestation of gene expression to proteins, and functional proteins is a fascinating subject by itself. The captivating fact of mutations of DNA leading to evolution and the inevitable ugly truth of genetic disorders and diseases go hand in hand, so long as the existence of human beings and living organisms on the planet earth. The genetic mutations leading to medical disorders opens a plethora of opportunities for the fraternity of medical, scientific, bioengineers, and biotechnologists to look into finding the solutions to cure diseases and alleviate the human suffering. A selective group of the pharmaceutical and biotechnology companies conceptually have the mission, based on genetic modifications, to discover, develop, and provide the most advanced innovative therapies to patients and the medical community. An interesting description on how six commercially-available products that are changing our world by Voight is noteworthy [11].

1.2.1 Basic Alphabets and Expanded Alphabets of Genetic Building Blocks

The four natural deoxy nucleotides (dA, dT, dG, and dC) in DNA and nucleotides (A, U, G, and C) in RNA, respectively, as the genetic building blocks and corresponding matching nucleotide base pairs, are the central foundation of creating 20 amino acids to build proteins through the 64 codons, the proteins, and thus, the concept of the creation and the existence of the living organisms and the human race. Benner [6a], Romesberg [6b], and Hirao and others [6c] have been involved in experimenting with adding synthetic nucleotide-like base pairs to natural deoxy nucleotides to create, potentially, several synthetic proteins and unnatural mAbs.

This expanded genetic alphabet platform is expected to be a source for developing new class of synthetic proteins and possibly a differentiated therapeutic pipeline.

1.2.2 Mutations, Point Mutations, Chemistry, and Health Disorders

Mutation, in summary, is when a gene is altered through a modification in DNA structure due to any chemical or physical change [5]. Point mutation [4c] is a type of mutation in DNA or RNA in which a single nucleotide base is added, deleted, or changed. A single base pair mutation in the genome can result in many congenital disorders in humans and living organisms. The incidence of chemical and biochemical changes occurs that alters the genetic outcomes and amino acid sequences in proteins by the cell, thereby manifesting in medical disorders during the mutations. Point mutations contribute to about 62% of total genetic disorders. Base editing precisely edits DNA sequences in a specific locus without inducing harmful double-strand breaks (DSBs). Around 60% of pathogenic point mutations can potentially be corrected by base editors and can provide a cure for many blood diseases [4f]. Readers interested in the discussion are encouraged to read the work in the citations as it is beyond the scope of elaboration in this chapter. The content in Table 1.1 [4, 4a, 4c] lists the class and the type of mutation, description, and the human diseases associated with the specific mutation.

It is safe to say that nearly all medical disorders have a genetic component associated with the diseases. Some disorders like sickle-cell disease, cystic fibrosis, and progeria are caused by single gene mutations. Going into some details calls for special attention in understanding the root cause of these diseases how the synthetic biology and genetic engineering help with unraveling the intricacies of science involved in dealing with diseases like sickle-cell disease (SCD), cystic fibrosis (CF), and progeria, also known as Hutchison–Gilford progeria syndrome (HGPS).

1.2.2.1 Sickle-Cell Disease (SCD) and Potential Treatment

Sickle-cell anemia [7a] is a disease caused by the alteration in a single nucleotide in the gene (Fig. 1.1) for the beta chain of hemoglobin (the oxygen-carrying protein that makes blood red) that turns the normal hemoglobin gene into a sickle-cell hemoglobin gene. This change in turn alters only one amino acid protein chain reducing the ability to transport oxygen. The resulting protein with 147 amino acids sequence contains valine instead of the glutamic acid as the sixth amino acid. The impeded blood flow leads to recurrent pain and severe organ damage.

Recommended cure options include blood transfusions and stem cell transplantation. Clinical trials are ongoing to address stem cell transplantation in adults, and gene therapies and CRISPR-Cas9-mediated gene editing methods are underway to address the need to cure sickle-cell disease [7b]. Recent advances in the treatment of SCD have been reviewed by Cisneros and Thein [7c] and Fortenberry and co-authors [7d].

Table 1.1 Types of DNA mutations and the health disorder outcomes

Class of mutation	Type of mutation	Description	Human disease(s) linked to this mutation
Point mutation	Substitution	One base is incorrectly added during replication and replaces the pair in the corresponding position on the complementary strand	Sickle-cell anemia
	Insertion	One or more extra nucleotides are inserted into replicating DNA, often resulting in a frameshift	One form of beta-thalassemia
	Deletion	One or more nucleotides is "skipped" during replication or otherwise excised, often resulting in a frameshift	Cystic fibrosis
Chromosomal mutation	Inversion	One region of a chromosome is flipped and reinserted	Opitz-Kaveggia syndrome
	Deletion	A region of a chromosome is lost, resulting in the absence of all the genes in that area	Cri du chat syndrome
	Duplication	A region of a chromosome is repeated, resulting in an increase in dosage from the genes in that region	Some cancers
	Translocation	A region from one chromosome is aberrantly attached to another chromosome	One form of leukemia
Copy number variation	Gene amplification	The number of tandem copies of a locus is increased	Some breast cancers
	Expanding trinucleotide repeat	The normal number of repeated trinucleotide sequences is expanded	Fragile X syndrome, Huntington's disease

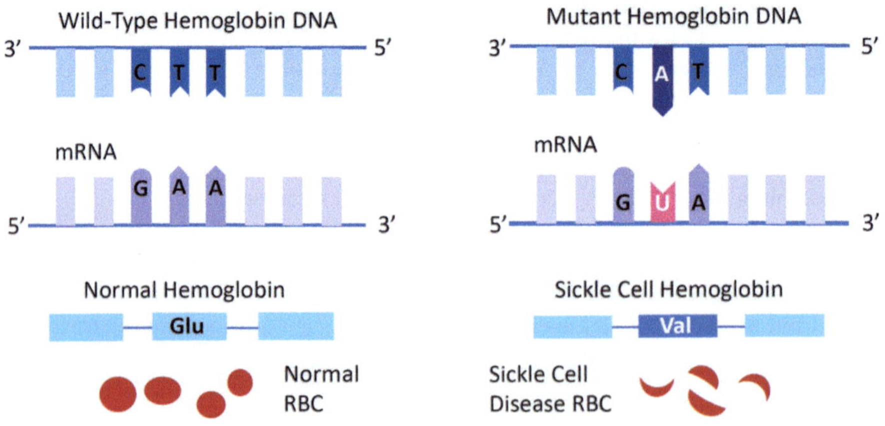

Fig. 1.1 Molecular basis of sickle-cell disease

1.2.2.2 Cystic Fibrosis

Cystic fibrosis (CF) is a progressive, genetic disease that causes persistent lung infections and limits the ability to breathe over time. CF is caused by mutations in the gene encoding the CF transmembrane conductance regulator (CFTR), a cyclic AMP-activated chloride channel [8a]. Most of these mutations either substitute one base pair (the building material for DNA) for another or delete a small number of DNA bases. This mutation is caused by the deletion of three base pairs of the CFTR gene leading to the loss of an amino acid, phenylalanine, in the CFTR protein. Everyone receives a copy of the CFTR from each parent. Mutations in CFTR lead to imbalanced water and ion movement across the airway epithelium, resulting in thickened mucus, chronic bacterial infection, and inflammation with progressive loss of pulmonary function.

A detailed discussion is presented in the article "Molecular Basis of CF: From Bench to Bedside" [8c]. The gene is located in chromosome 7 (a 27 exons structure), and the deletion of phenylalanine in position 508 of the polypeptide chain, known as Phe508del or F508del, is the most common *CFTR* mutation affecting from 50% to 90% of the chromosomes of CF patients along different geographical areas. It is a chloride (and iodide in some cases) ion transporter localized at the apical membrane of several polarized epithelia, although other small molecules seem to be transported by CFTR, including ATP.

There is no cure for cystic fibrosis, but treatment can ease symptoms, reduce complications, and improve quality of life [8b]. Promising gene therapies and therapies involving CFTR mRNA and gene editing with CRISPR as well as new viral and non-viral vector formulations are in development [8a].

1.2.2.3 Progeria

Progeria, also known as Hutchison–Gilford progeria syndrome (HGPS or HPS) [9], is a genetic condition characterized by the dramatic, rapid appearance of aging beginning in childhood. Mutations in the *LMNA gene* cause HGPS. The *LMNA* gene expresses to generate the protein, *lamin A* [4b], which plays an important role in determining the shape of the nucleus within the cells. The altered protein is accountable for progressively damaging the nucleus and the premature cell death.

There is no cure for progeria, but regular monitoring for heart and blood vessel (cardiovascular) disease may help with managing your child's condition [4d]. Recent studies by Fabrizio d'Adda di Fagagna and coworkers demonstrate an important role for telomeric DNA damage response (DDR) activation in HGPS progeroid detrimental phenotypes in vitro and in vivo [4e].

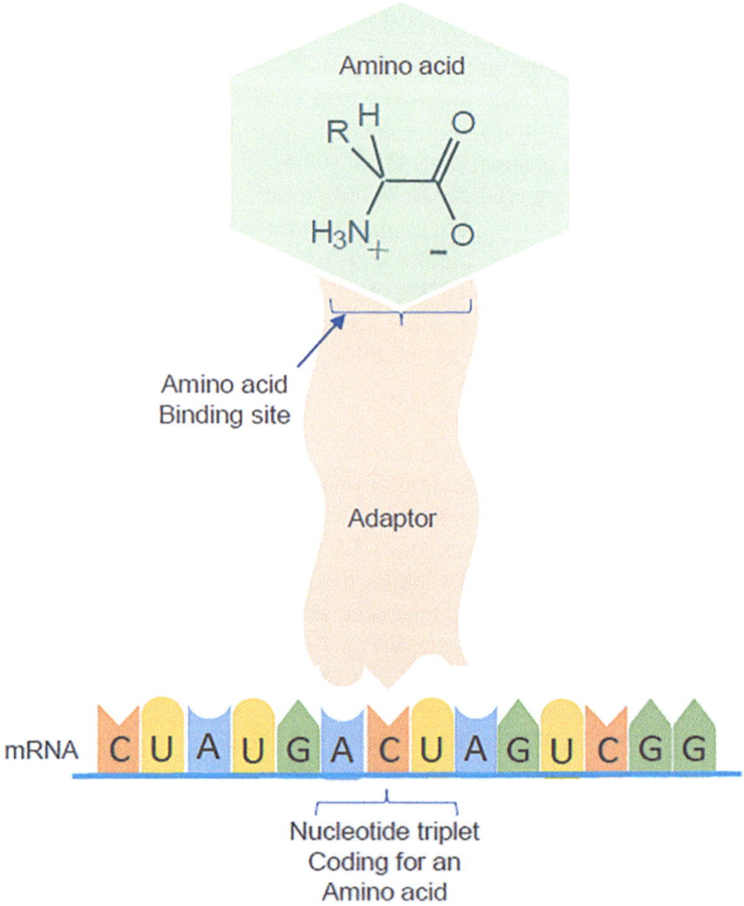

Fig. 1.2 Crick's adaptor hypothesis. The amino acid is covalently bound at the 3′ end of a tRNA molecule and that a specific nucleotide triplet elsewhere in the tRNA interacts with a particular triplet codon in mRNA through hydrogen bonding of complementary bases [12]

1.3 Synergy Between Biologics and Small Molecules

In nature, the small molecules interact with biological entities and vice versa to produce outstanding outcomes, manifestations of which make it a significant event. Among the following two examples, in the first one, the small molecular amino acid is selected via a covalent binding to the tRNA, a biological unit (Fig. 1.2). In the second example, the specific stereoselective chemical reaction takes place via binding of the substrate racemic mixture of **1** as the starting material to the enzyme-specific strains, *H. polymorpha SC 13865* and *H. fabianii SC 13894*, resulting in the transformation to the optically active single diastereomer **2,** which constitutes the side chain of the cancer drug, Paclitaxel™ (Fig. 1.3).

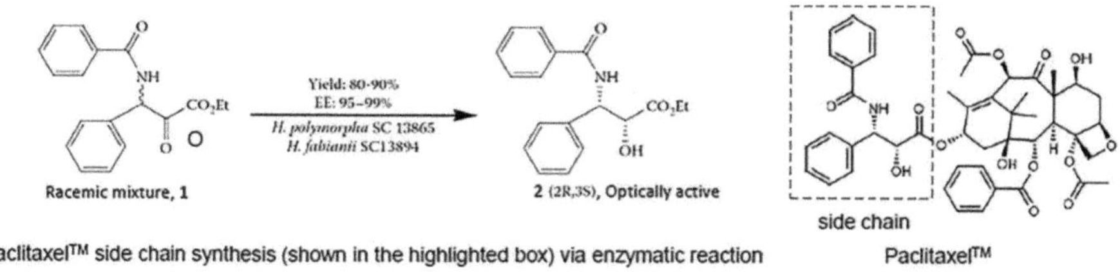

Fig. 1.3 Enzymatic intervention in generating optically active small molecules. Synthesis of side chain via enzymatic reaction in anticancer drug, Paclitaxel™

One can see, these manifestations (Fig. 1.2) known as Crick's adaptor hypothesis are represented in the DNA expression of the proteins where the amino acid building block is selected via the covalent binding of the amino acid with the 3′-end of a tRNA molecule and that a specific nucleotide triplet elsewhere in the tRNA interacts with a particular triplet codon in mRNA through hydrogen bonding of complementary bases [12]. The particular amino acid sequence of a protein is constructed through the translation of information encoded in mRNA. This process is carried out by ribosomes.

In synthetic organic chemistry, myriads of examples are known to use enzyme-mediated reactions to create new chiral sites in the molecule. Although the synthetic routes and processes developed to produce drug candidate compounds rely mainly on conventional chemistry, the usage of enzyme-mediated intervention appeared prudent for the production of chiral (optically active) intermediates [13]. Amino acylation, hydrolysis, reductive amination, and reduction, all using enzyme-mediated reactions, were used at industrial scale for generating the optically active chiral intermediate molecules. One such example is shown below for the semi-synthesis (side chain of Paclitaxel™) of the anticancer drug Paclitaxel™ at Bristol-Myers Squibb (Fig. 1.3).

1.3.1 Process Chemistry and Bioprocessing

The process or the event of "making or breaking of a covalent bond" can be brought forward between two atoms resulting in creating molecules or a molecule, respectively, using a traditional chemical reaction or an enzyme-mediated intervention. Process chemistry is the term used for efficiently carrying out the chemical transformations and the series of steps involved to reproduce the reaction or a series of reactions.

Traditionally, process development chemistry involves devising efficient and commercially viable synthetic routes in building small molecules. This process can advance an anticipative drug candidate through the preclinical and clinical development timeline, establishing and transferring a commercial synthetic chemistry route to the manufacturing division, and contributing to the CMC section of the new drug application (NDA) submission. Biologics, on the other hand, are large molecules such as therapeutic proteins, fusion proteins, monoclonal antibodies, and vaccines. Most biologic medicines are developed using recombinant DNA technology. Bioprocessing involves the processing of biologic molecules through genetically engineering whole living cells to generate the protein. Unlike the chemical synthetic process used to produce small molecules, bioprocessing involves upstream processing (upfront) starting with living cells and downstream processing involving larger material handling and eventual purification of the finished biologic material.

The role of the process chemist has evolved not only to accommodate advancements in regulatory affairs and to improve business efficiency but also to embrace at times the technical challenges to learn processes involving and understanding of synthetic biology and handling the biologics such as antibody drug conjugates (ADCs). Interestingly, ADCs opened the door to productive interactions between chemists and biologists. The FDA treats and regulates ADCs as biologics rather than as conventional chemically synthesized drugs. ADCs, therefore, are approved through a Biologics License Application (BLA).

1.3.2 Antibody Drug Conjugates (ADCs): A Conjugation of Small Molecules and Biomolecules

Antibody drug conjugates (ADCs) combine the cytotoxic potential of chemotherapeutic drugs (e.g., doxorubicin, auristatins, calicheamicins, maytansines, and other drugs with sub-nano molar activities) with the specificity of monoclonal antibodies (mAbs) bridged via the chemical linkers (Fig. 1.4). The site-specific conjugation of the cytotoxic drugs (e.g., doxorubicin, calicheamicin, auristatin, maytansine) involve specific cancer antigen targets (e.g., CD33, CD30, Her2) linking to the antibody sites via cysteine S- or lysine N-bonding to the cytotoxic drug. The mechanism of action (MoA) encompasses internalization of ADC into cancer cell, the release of the cytotoxic drug in the cell, and apoptosis (cell death). Some examples of selected ADCs include Adcetris™ from Seattle Genetics; Kadcyla™ from Genentech-Roche; and Mylotarg™ and Besponsa™ from Pfizer. A detailed discussion on the development of ADCs is presented in this volume in Chap. 12 by David Jackson and in Chap. 13 by the Pfizer group of scientific teams that provide the detailed work of discovery, development, manufacturing, and the regulatory challenges in launching Mylotarg™.

The author believes that the science and technology, as well as the development and manufacturing, involving process chemistry development tied to the development of specific monoclonal antibody for a successful ADC creation and the launching, embodies a genuine meaning of a "conjugation of small molecules with biomolecules." ADC technology and development challenges created an

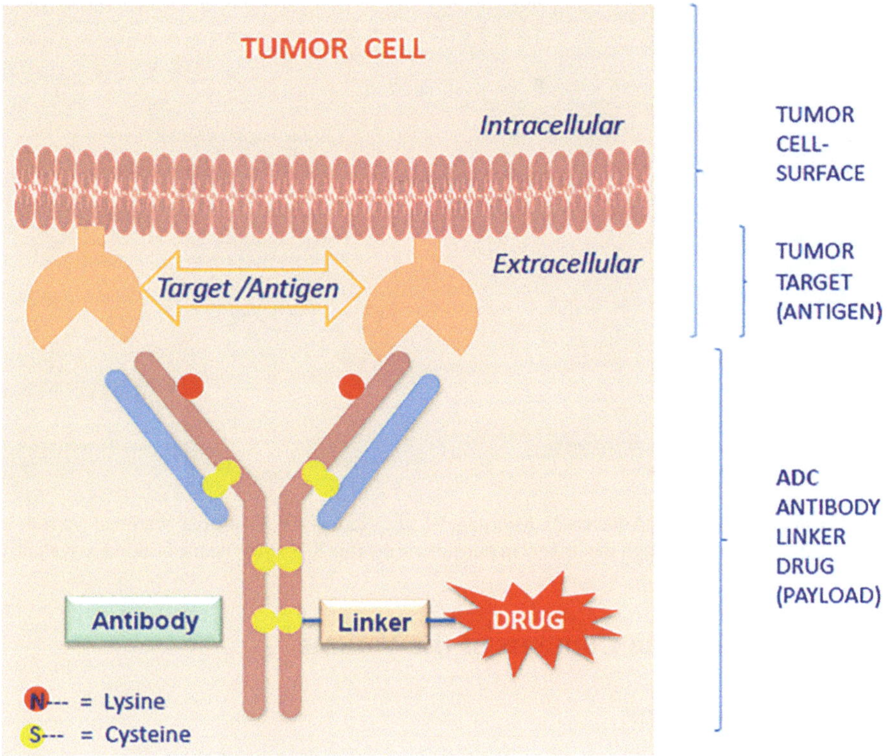

Fig. 1.4 Site-specific homogenous ADCs

opportunity for the biologists, chemists, and bioengineers to come together to successfully launch the specific anticancer ADCs to alleviate the pain and suffering of humankind.

1.3.2.1 ADC Drugs

Three of the marketed ADCs are shown in Fig. 1.5.

1.4 Top 25 Best-Selling Drugs

Drug sales hit records with the COVID-19 Comirnaty vaccine jointly developed by Pfizer and BioNTech [10]. Table 1.2 lists the top 25 best-selling drugs based on 2021 sales.

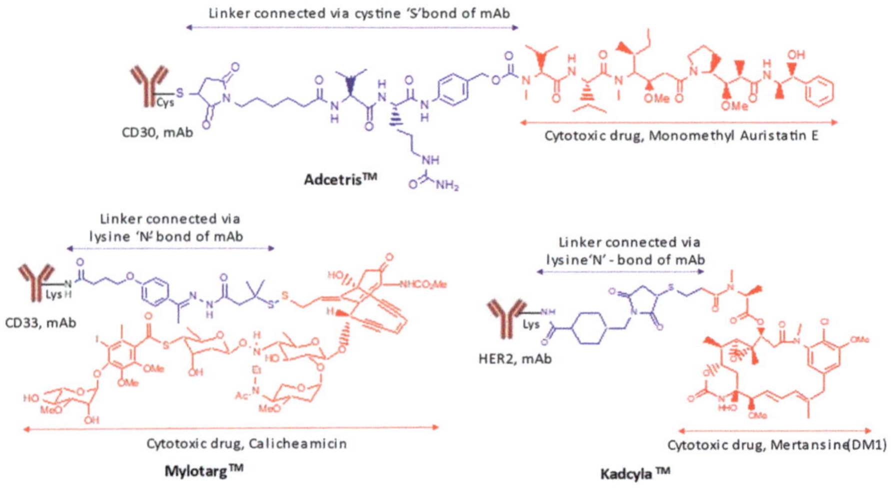

Fig. 1.5 Marketed ADCs, Adcetris™, Mylotarg™, and Kadcyla™. Cytotoxic warhead (in red) is connected to the mAb through the linker (in purple) via cystine S-bond or lysine N-bond, covalently

1.5 Outline of the Contents of the Volume

1.5.1 An Overview

The first chapter of this volume opens with a general introduction, and the author's perspective on synthetic biology, genetic mutations, and associated diseases. It attempts to give a general idea about the plot of the book, the purpose, the process, and the rational aspects behind putting the volume together. This chapter lists the top best-selling 25 drugs and also summarizes the outline of contents of the volume.

1.5.2 Synthetic Biology

In the overview, Gadamasetti in Chap. 1 opens the discussion on the outline of synthetic biology, genetic alphabets and expanded alphabets, point mutations, and specific examples of diseases related to mutations. Remarkably, the cure of sickle-cell therapy was through tweaking the specific genetic building block.

Table 1.2 Top 25 best-selling drugs

#	Drug name	Manufacturer(s)	2021 Sales (millions USD)	Indication(s)	SM/B[a]
1	Comirnaty COVID-19 vaccine	Pfizer/BioNTech	$59,109.00	Reduce risk of COVID-19 infection	B
2	Humira (adalimumab)	AbbVie	$20,694.00	Rheumatoid and psoriatic arthritis, ankylosing spondylitis, Crohn's disease, ulcerative colitis	SM
3	Spikevax COVID-19 vaccine	Moderna	$17,675.00	Reduce risk of COVID-19 infection	
4	Keytruda (pembrolizumab)	Merck	$17,186.00	Various cancers	SM
5	Eliquis (apixaban)	Bristol Myers Squibb and Pfizer	$16,732.00	Blood clots	B
6	Revlimid (lenalidomide)	Bristol Myers Squibb (Celgene)	$12,821.00	Myelodysplastic syndrome, multiple myeloma, and mantle cell lymphoma	SM
7	Imbruvica (ibrutinib)	Pharmacyclics (AbbVie) and Janssen (Johnson & Johnson)	$9777.00	Chronic lymphocytic leukemia/small lymphocytic lymphoma with 17p deletion, Waldenström's macroglobulinemia	SM
8	Eylea (aflibercept)	Regeneron pharmaceuticals, Bayer	$9244.00	Age-related macular degeneration, macular edema, and diabetic retinopathy	B
9	Stelara (ustekinumab)	Janssen (Johnson & Johnson)	$9134.00	Plaque psoriasis and psoriatic arthritis	B
10	Biktarvy (bictegravir, emtricitabine, and tenofovir alafenamide)	Gilead Sciences	$8624.00	HIV	SM
11	Xarelto (rivaroxaban)	Janssen (Johnson & Johnson)/Bayer	$8040.00	Reducing risk of stroke in non-valvular atrial fibrillation, deep vein thrombosis, pulmonary embolism, DVT prophylaxis following knee or hip replacement surgery	SM
12	Opdivo (nivolumab)	Bristol Myers Squibb	$7523.00	Various forms of cancer	B
13	Trulicity (dulaglutide)	Eli Lilly	$6472.00	Type 2 diabetes	B
14	Dupixent (dupilumab)	Sanofi Genzyme, Regeneron pharmaceuticals	$6210.00	Atopic dermatitis, asthma, chronic rhinosinusitis with nasal polyps	B

(continued)

Table 1.2 (continued)

#	Drug name	Manufacturer(s)	2021 Sales (millions USD)	Indication(s)	SM/B[a]
15	Darzalex (daratumumab)	(Janssen) Johnson & Johnson	$6023.00	Multiple myeloma	B
16	REGEN-COV (Casirivimab/ imdevimab)	Regeneron pharmaceuticals, Bayer	$5828.00	Reduce risk of COVID-19 infection	B
17	Trikafta/Kaftrio	Vertex pharmaceuticals	$5697.00	Cystic fibrosis	SM
18	Gardasil/Gardasil 9 (human papillomavirus 9-valent vaccine)	Merck	$5673.00	Various cancers caused by human papillomavirus	B
19	Veklury (remdesivir)	Gilead Sciences	$5565.00	COVID-19 requiring hospitalization	SM
20	Ibrance (palbociclib)	Pfizer	$5437.00	HR-positive and HER2-negative breast cancer	SM
21	Januvia/Janumet (sitagliptin)	Merck & Co.	$5288.00	Type 2 diabetes	SM
22	Prevnar family (pneumococcal vaccine)	Pfizer	$5272.00	Pneumococcal vaccine	B
23	Tagrisso (osimertinib)	AstraZeneca	$5015.00	Non-small-cell lung carcinomas	SM
24	Cosentyx (secukinumab)	Novartis	$4718.00	Plaque psoriasis, psoriatic arthritis, ankylosing spondylitis	B
25	Ocrevus (ocrelizumab)	Roche	$4622.00	Relapsing or primary progressive multiple sclerosis	B

Taken from Top 50 Best Selling Drugs [10]

B biologic drug, *COVID-19* Coronavirus Disease-2019, DVT, HER2, *HIV* human immunodeficiency virus, HR, *SM* small molecule drug

[a]It is interesting to note the significance of biologic drugs (13 out of 25) in the top 25 best-selling drugs from 2021.

1.5.2.1 Synthetic Biology in Drug Development and Beyond

In its simplest definition, Sato and Riffle in Chap. 2 note that synthetic biology is the creation of new biological entities for useful purposes. By manipulating an organism's genome, synthetic biologists can produce novel proteins for a wide range of applications, from the biosynthesis of industrial chemicals to the discovery and optimization of biotherapeutics. The field of synthetic biology has experienced a renaissance in recent years as technological advances have lowered the barrier to entry and increased the potential for innovation. Principal among these advances has been the development of highly precise and large-scale DNA synthesis platforms.

1.5.3 Oligonucleotide Synthetic Chemistry to DNA Synthesis, Bioprocessing, and Manufacturing

Outlined in this section are three chapters. In its simplest definition, Aaron Sato and Rebecca Nugent in Chap. 3 outline that creation of new biological entities defined by synthetic biology facilitates in manipulation for several useful purposes. By manipulating an organism's genome, synthetic biologists can produce novel proteins for a wide range of applications, from the biosynthesis of industrial chemicals to the discovery and optimization of biotherapeutics. The chapter outlines the evolution of gene synthesis starting with the humble beginnings of the oligonucleotide synthesis leading to gene synthesis, antibody synthesis, and the library creation.

In Chap. 4, Greg Hiller outlines the recent innovations and advances in cell culture technology enabling process intensification in detailing the upstream process development, the limitations of fed-batch processes, the recent resurgence of interest in traditional "classical steady-state" perfusion in which viable cell density is held nearly constant, and the development of non-conventional "dynamic" perfusion processes in which viable cell density is allowed to peak and decline. Other process intensification methodologies such as N-1 perfusion for high-density production reactor inoculation are also examined.

Timothy Iskra's Chap. 5 summarizes the implication of bispecific and trispecific antibodies on antibody process development paradigms. According to Iskra, in recent years the bispecific and trispecific (or multispecific) antibodies have become a large area of development for most companies, as the ability to bind two different antigens at a time opens new and unique therapeutic areas.

1.5.4 Process Engineering, Gene Therapy, and Vaccine

Under this title, three chapters are clustered. Sharfstein and coworkers, in Chap. 6, review the basics of glycoprotein and glycosaminoglycans biosynthesis with particular focus on the current understanding of how glycan structures are controlled in vivo. The studies in which glycan structures for recombinant proteins and glycosaminoglycans have been optimized by cell line metabolic engineering and bioprocess manipulations are also reviewed. It is also noted that CRISPR has permitted exquisite editing of host cells, allowing tailored production of glycan structures, facilitating the production of biosimilars and bioengineered glycosaminoglycans as well as setting the stage for "biobetters" in which improved functionality is obtained by glycoengineering. Chapter 7 by Somanathan and coworkers at Pfizer summarizes the host immune response activated by adeno-associated viral (AAV) gene therapy vectors and the approaches currently being pursued to mitigate immune recognition and activation. Kolodziej and coauthors, in Chap. 8, discuss the details of Coronavirus Disease-2019 (COVID-19) vaccine manufacturing processes in finding the solutions via making molecules to address the needs of eradicating the pandemic.

1.5.5 Special Topics: CAR-T and CRISPR Technologies and Applications

CAR-T and CRISPR technologies and the associated applications are addressed in Chaps. 9 and 10. In Chap. 9, the authors from Glaxo-Welcome Smith Kline (GSK) present a high-level overview of the general process for production of chimeric antigen receptor-engineered (CAR)-T cells. Also described are individual aspects of the workflow that ae utilized to generate and administer CAR-T cell products including production of lentiviral vector, acquisition of lymphocyte starting material, cell activation, transduction and expansion, and formulation and product release. The authors give some historical background about various technologies and how they have been implemented in production, along with recent changes that have been made in academia and industry to optimize CAR-T cell production. Chapter 10 is a comprehensive guide to Clustered Regularly Interspaced Short Palindromic Repeats (CRISPR) gene editing and its applications in biotherapeutics. It begins with an introduction to CRISPR-Cas9 technology, examining its discovery and adaptation, mode of action, advantages over previous tools, derivatives of the original CRISPR-Cas9 technology, and gene editing methods. This is followed by a review of the applications of CRISPR technology in biotherapeutics, including both gene therapies and gene-edited cell therapies, for the treatment of three disease categories: genetic disorders, infectious diseases, and cancer. The final section delves into the future of CRISPR therapeutics, including nascent applications, the industrialization and projected growth of the industry, safety and ethical considerations, and its transformative potential for human health.

1.5.6 Fusion Proteins, Antibody Drug Conjugates, and Process Chemistry

Stefan Schmidt from BioAtrium AG, focuses in Chap. 11 on fusion proteins engineered from two or more genes that code for separate proteins joined by genetic techniques. The result is a single polypeptide sharing functional properties of both parent proteins. Fusion proteins belong to the class of next-generation therapeutics as they are designed exclusively through human creativity and are not evolved from a natural process. Very often they represent the so called "biobetters," molecules with improved functionalities.

The subsequent two chapters are dedicated to ADCs. In Chap. 12, David Jackson summarizes the history of currently approved ADCs with emphasis on the challenges that were overcome during development and the new technologies that likely had contributed to their success. The safety and efficacy of each ADC are also discussed from a critical but honest perspective. The author's intention in writing this chapter is to encourage readers to educate themselves about the real benefits and risk of ADC therapeutics so that informed decisions can be made by cancer patients in collaboration with their doctors. And in Chap. 13, the team from Pfizer discusses the development and the challenges of the journey to FDA reapproval and broad international approval process of the ADC, Mylotarg™.

1.5.7 Biopharmaceutical Informatics and Analytics

Chapter 14 from Sandeep Kumar and coworkers at Boehringer Ingelheim provides a comprehensive overview of strategic vision of closer collaboration between drug discovery and development functions of biopharmaceutical industry by integrating the considerations of developability during early stages of drug discovery. Computational tools already available to enable Biopharmaceutical Informatics are reviewed in this work. While the focus is on monoclonal antibody-based biologics, the concepts discussed in this work are also applicable to novel formats such as multispecific biologics. Advanced data analytics (ADA) application in biomanufacturing processes is outlined in Chap. 15 by Jun Luo and coauthors from Genentech-Roche. The authors believe that using ADA may elucidate previously undetected relationships between process inputs and outputs, which holds promise as an additional tool to augment traditional design of experiments (DoE) as a means to gain actionable insights, on process and product knowledge.

1.5.8 Biopharmaceutical Regulatory CMC

Two chapters are presented in this volume dedicated to regulatory chemistry, manufacturing, and control (CMC) of biopharmaceuticals. Richard Pelt from Pfizer outlines the overview of complexities of global regulatory CMC aspects in Chap. 16. Noted in the overview is the evolution of the technologies within the biopharmaceutical industry and relating to the global CMC regulatory environment for regulatory approvals to benefit the patients. In Chap. 17, Ding and Marino describe the monoclonal antibody (mAb) continuous bioprocess (CBP) from design, implementation to manufacturing based on the scientific understanding of mAb physicochemical properties, proven bioprocessing principles, available technologies, CMC considerations, current industrial practices, regulatory guidelines, challenges, potential solutions, and future perspectives.

1.5.9 Technology Transfer

Hughes and Brady from Bristol-Myers Squibb, in Chap. 18, describe in detail the advancements in technical transfer of biologics. According to the authors, technology transfer of the manufacturing process and associated analytical methods is a critical process in the lifecycle of biotech products. It enables the production of clinical material to test a hypothesis developed in the drug discovery environment, expansion of clinical supplies, and establishment of a robust platform for commercial production. It can also provide flexibility in the product lifecycle by adding additional manufacturing nodes to supplement supply or provide business continuity.

1.5.10 Emerging Trends and Future of Biopharmaceuticals

The last chapter by Wood and Gierach from Department of Chemical and Biomolecular Engineering of The Ohio State University entitled, "Emerging Biopharmaceutical Technologies and Trends" introduces a few of the high-profile changes taking place in the industry, both in terms of new treatments and modalities and new manufacturing approaches.

1.6 Future of Biologics

Synthetic biology leading to biologics is here to stay and thrive for decades ahead. It will transform how we grow food, what we eat, and where we source materials and medicines from. Synthetic biology enabled in advancing metabolic engineering, directed evolution-automated strain engineering, metagenomic discovery, gene circuit design, and genome editing. Products from synthetic biology are rapidly growing and by 2030, it is highly likely that you will have eaten, worn, used, or been treated with one [11].

References

1. Schrödinger E (1944) What is life? The physical aspect of the living cell, 1st edn. Cambridge University Press, Cambridge. Schrödinger's contribution to the field of molecular biology from the perspective of quantum physics. 'Aperiodic crystal' is currently called the protein structure
2. Banting FG, Best CH, Collip JB, Campbell WR, Fletcher AA (1922) Pancreatic extracts in the treatment of diabetes mellitus. Can Med Assoc J 12(3):141–146. https://www.ncbi.nlm.nih.gov/pmc/articles/PMC1524425/
3. Insulin is a biologic—What does that mean? https://www.diabetes.org/blog/insulin-now-biologic-what-does-mean
4. The readers are encouraged to follow the citations 4a and 4c below especially, if new to this subject material of genetics:

4a. Genetics home reference: help me understand genetics. https://ghr.nlm.nih.gov/ and https://medlineplus.gov/genetics/

4b. The health conditions in https://medlineplus.gov/genetics/condition/. Lamins A and C are structural proteins called intermediate filament proteins. Intermediate filaments provide stability and strength to cells

4c. https://www.genome.gov/genetics-glossary/Point-Mutation
4d. Mayo Clinic. https://www.mayoclinic.org/diseases-conditions/progeria/diagnosis-treatment/drc-20356043
4e. Aguado J, d'Adda di Fagagna F et al (2019) Inhibition of DNA damage response at telomeres improves the detrimental phenotypes of Hutchinson–Gilford Progeria Syndrome. Nat Commun 10:4990. https://doi.org/10.1038/s41467-019-13018-3
4f. Rees HA, Leu DR (2018) Base editing: chemistry on the genome and the transcriptome in living cells. Nat Rev Genet 19:770–788
5. Clancy S (2008) Genetic mutation. Nat Educ 1(1):187. https://www.nature.com/scitable/topicpage/genetic-mutation-441/
6a. Benner SL et al (2019) Hachimoji DNA and RNA: a genetic system with eight building blocks. Science 363(6429):884–887 and the references therein
6b. Romesberg FE et al (2019) Eight-letter DNA. Biochem 58(22):2581–2583 and references therein; expanded genetic alphabet, Romesberg FL and coworkers, Acc Chem Res, 2018, 51, 2:394–403. https://pubs.acs.org/doi/full/10.1021/acs.accounts.7b00403, Angew Chem Int Ed Engl 2015, 54(41):11930–11944
6c. Hirao I et al (2021) Cognate base-pair selectivity of hydrophobic unnatural bases in DNA ligation by T4 DNA ligase. Biopolymers 112:e23407. https://doi.org/10.1002/bip.23407 and references therein; Genetic alphabet expansion transcription generating functional RNA molecules containing a five-letter alphabet including modified unnatural and natural base nucleotides by thermostable T7 RNA polymerase variants, Kimoto M, Meyer AJ, Hirao I, and Ellington AD (2017) Chem Commun, 53:12309–12312 and references therein
7a. Sickle cell disease, Gregory J. Kato and co-authors, Nat. Reviews, 2018, 4, Article No. 18010, 1; Ribeil J-A (2017) Gene therapy in a patient with sickle cell disease. New Engl J Med 376:848–855. Mayo Clinic Patient Care and Health Information. https://www.mayoclinic.org/diseases-conditions/sickle-cell-anemia/diagnosis-treatment/
7b. A CRISPR Focus into the Attitudes and Beliefs of Sickle Cell Patients, Parents, and Providers Towards Gene Editing, Vence L. Bonham and co-workers, NIH, National Human Genome Res Instt. 2018
7c. Gabriel Salinas Cisneros and Swee L. Thein, Front Physiol, 2020, 11, Article 435; Genome Editing Clinical Trials for Sickle Cell Disease, Stacy Desine and co-authors, AJOB, Empir Bioeth, 2020, 11(4), 195–207
7d. Current and Future Therapeutics for Treating Patients with Sickle Cell Disease by Fortenberry and co-authors, Cells 2024, 13, 848. https://doi.org/10.3390/cells13100848
8a. Hart SL et al (2020) New approaches to genetic therapies of cystic fibrosis. J Cystic Fibrosis 19:S54–S59. https://hopkinscf.org/knowledge/cftr/
8b. https://www.mayoclinic.org/diseases-conditions/cystic-fibrosis/diagnosis-treatment/drc-20353706
8c. Dechecchi MC et al (2018) Molecular basis of cystic fibrosis: from bench to bedside. Ann Transl Med 6(17):334 and references therein
9. Progeria. https://www.mayoclinic.org/diseases-conditions/progeria/symptoms-causes/syc-20356038#:~:text=Overview,generally%20appear%20normal%20at%20birth
10. 50 of 2021's best selling drugs, Brian Buntz, Drug Disc Dev, 29, 2022. https://www.drugdiscoverytrends.com/50-of-2021s-best-selling-pharmaceuticals/
11. Voigt CA (2020) Synthetic biology 2020–2030: six commercially-available products that are changing our world. Nat Commun 11:1–6
12. Protein metabolism, Lehninger, principles of biochemistry, Chapter 27, ISBN-13: 978-1-4641-2611-6
13. Hanson RL (2008) Process chemistry in the pharmaceutical industry.: Gadamasetti K, Braish T (eds) *Challenges in an ever-changing climate*. CRC Press, Taylor & Francis Group, pp 279–294

Part II
Synthetic Biology

Chapter 2
Synthetic Biology in Drug Development and Beyond

Aaron K. Sato and Stephen Riffle

Abstract In its simplest definition, synthetic biology is the creation of new biological entities for useful purposes. By manipulating an organism's genome, synthetic biologists can produce novel proteins for a wide range of applications, from the biosynthesis of industrial chemicals to the discovery and optimization of biotherapeutics. The field of synthetic biology has experienced a renaissance in recent years as technological advances have lowered the barrier to entry and increased the potential for innovation. Principal among these advances has been the development of highly precise and large-scale DNA synthesis platforms.

The synthesis of genetic material is a non-trivial, yet integral component of synthetic biology. After all, behind every novel protein is a novel DNA sequence. Synthetic oligonucleotides are also irreplaceable components in CRISPR (clustered regularly interspersed palindromic repeats) gene-editing, DNA sequencing, and the myriad tools that enable researchers to manipulate and interrogate genomes.

Historically, oligonucleotide synthesis was a slow, error-prone process that severely limited its usefulness beyond niche studies. However, the advent of phosphoramidite chemistry and solid-phase synthesis marked an inflection point after which the scale, efficiency, and precision of DNA synthesis markedly increased. And with this increase came an ever-growing list of applications for synthetic biology.

The field of synthetic biology is on a trajectory to play a pivotal role in addressing many of the world's most challenging and complex problems. Projects are underway to develop and apply engineered organisms in bioremediation, helping to clean polluted ecosystems. Crops, engineered to resist harsh weather conditions, have long been sought as a means to reduce starvation in drought-stricken environments. And with the ability to rapidly design and build DNA libraries, drug developers will be better equipped to discover and optimize novel therapeutic modalities.

A. K. Sato (✉) · S. Riffle
Twist Bioscience, South San Francisco, CA, USA
e-mail: asato@twistbioscience.com

K. Gadamasetti, S. A. Kolodziej (eds.), *Bioprocessing, Bioengineering and Process Chemistry in the Biopharmaceutical Industry*,
https://doi.org/10.1007/978-3-031-62007-2_2

Perhaps the most salient example of synthetic biology's ability to advance therapeutic development is the rapid expansion of cell therapies to include chimeric antigen receptor (CAR)-T and CAR-NK (natural killer) cells. These highly engineered cells sit at the vanguard of clinical oncology and, as the tools of synthetic biology continue to evolve, will likely play an ever-larger role in patient care.

In the following sections, we provide a brief overview of synthetic biology and how this nascent field is catalyzing rapid development of novel biotherapeutics.

Keywords Synthetic biology · Genetic circuits · Chassis organism · Hybridoma cells · Phage display

2.1 What Is Synthetic Biology?

In the throes of the industrial revolution, Mary Shelley sat down to write the now-famous story of Frankenstein. Having listened to galvanists argue over the mechanical and potentially electric nature of life, she was inspired by the possibility that "perhaps the component parts of a creature might be manufactured, brought together, and endued with vital warmth" [1]. Given the right knowledge set and tools, Shelley and many scientists believed that someday we may be able to engineer novel forms of life. More than 150 years later, the field of synthetic biology began.

Synthetic biology describes the general pursuit of solutions to the world's problems through the deliberate engineering of biological systems. It is a distant echo from what Shelley and her contemporaries envisioned, but synthetic biology retains the central notion that living organisms are a complex assemblage of parts whose activity can be both predicted and manipulated. As such, synthetic biology can modify—or in some cases, create—biological systems that perform specific, predictable functions.

For example, genes can be conceptualized as discrete "parts" that reliably produce proteins when in the right environment. It stands to reason that parts can be swapped and that inserting a specific gene into an organism's genome will enable that organism to start producing the corresponding protein. Such was the reasoning when researchers from UC Berkeley modified the *Escherichia coli* genome to contain a myriad of foreign genes sourced from plants, yeast, and bacteria [2]. The result was a novel bacterium whose synthetic genome gave it the power to convert a common metabolic byproduct (acetyl-CoA) into the antimalarial therapeutic artemisinin. In subsequent years, approximately 39 million treatments of artemisinin were produced through synthetic means [3].

As demonstrated in the example of artemisinin, synthetic organisms can be engineered to produce complex chemicals that may be completely novel, or else only exist in specific ecological niches. Artemisinin is naturally produced by sweet wormwood (also known as qinghao, *Artemisia annua*) in very low quantities [4, 5].

Extraction of the compound is a costly process and requires large quantities of the host plant. By producing artemisinin with a synthetic organism, researchers gain a stable therapeutic source that may increase the scale of production while reducing the process' costs and environmental impacts [5].

The potential applications of synthetic biology extend well beyond drug manufacturing. Already organisms have been engineered to improve crop yields, synthesize complex materials, and produce commercial chemicals on an industrial scale [6]. Yet, the field is still in its infancy, having only begun in earnest at the turn of the twenty-first century [7].

In these nascent stages, the field must not only grapple with technological and scientific challenges but also those concerning philosophical and ethical questions. As in the days of Mary Shelley, the notion that life can be re-engineered or otherwise artificially altered raises questions about what constitutes life and what boundaries should be placed on the engineering of living organisms. These questions drive much of Shelley's novel as well as modern discourse about the potential—both good and bad—of synthetic biology [8].

As these questions continue to be explored, the field of synthetic biology continues to mature and expand in its applications. While the full potential of synthetic biology is yet to be realized, it is clear that this field will dramatically influence—and indeed has already influenced—the drug discovery and development industry.

In this chapter, we explore synthetic biology and its ongoing influence in modern drug development. We begin by outlining the basic tools used in synthetic biology. We then provide a high-level overview of several ways in which these tools are used to advance drug discovery and development, from CRISPR screening to chimeric antigen receptor (CAR)-T cell therapies. While this is not an exhaustive overview, we hope that the reader will find it a useful introduction to the fast-expanding application of synthetic biology in drug development.

2.2 A Brief Overview of Synthetic Biology Tools

At this point, it is important to briefly review the tools that are commonly employed in synthetic biology, namely, synthetic DNA, gene assembly methods, gene circuits, and chassis organisms.

Synthetic DNA

Solid-phase DNA synthesis has had a transformative impact on the molecular sciences, including synthetic biology. This chapter will include a brief discussion of DNA synthesis technology, but for an in-depth exploration of synthesis chemistry, readers may find a recent review from Hughes et al., in *Cold Spring Harbor Perspectives in Biology*, helpful [9]. Modern advances in automation, solid-phase synthesis, and phosphoramidite chemistry have made it possible to precisely synthesize millions of oligonucleotides in a matter of days.

Synthesis companies can now reliably produce large-scale libraries consisting of distinct oligonucleotides up to 300 nucleotides in length. The added length and scale of synthesis has enabled rapid gene assembly, genome-wide single-cell CRISPR screening, and a multitude of other applications. Scale, precision, and uniformity are all important factors that will be discussed later in this chapter, but the important point is that technological advances have made it possible to routinely synthesize genetic sequences that can then serve as modular parts for synthetic biology applications.

Gene Assembly

Modularity is a core feature in synthetic biology as it allows for parts to be easily swapped when needed. Swapping segments of DNA with precision became possible in the 1970s with the advent of restriction enzyme digestion and ligation-based assembly methods [10]. Though useful, these original methods suffered from low fidelity and the need for unique restriction enzyme recognition sites—greatly limiting the scale and modularity of their application. Contemporary assembly methods include Biobrick Assembly, Golden-Gate Cloning, and Gibson Cloning [10, 11]. These methods make use of DNA scaring, sticky end ligation, and other techniques to achieve high-fidelity gene assembly. While none are perfect, these methods have been optimized to allow for many gene parts to be rapidly and reliably assembled into larger gene clusters. When these gene clusters interact with one another in a predictable and controlled manner, they may be referred to as circuits.

Genetic Circuits

Genetic circuits are biological networks designed to carry out logic functions akin to computational circuits. A more formal definition may describe them as interlinked transcriptional and posttranscriptional networks whose activity regulates molecular processes, such as gene expression. Genetic circuits can be designed to express molecules—such as reporters or specific enzymes—only when a specific stimulus is present. In the case of circuits following AND logic, the desired molecule is only expressed when multiple conditions are true, such as the presence of a synthetic transcription factor and an accessible target promoter [12–14]. Alternatively, an engineered circuit may express different proteins depending on the cell's environment, as in the case of NOR/NAND circuits [14].

The ability to program precise, conditional gene expression opens the door to many possible applications in drug development and target validation. Many natural therapeutics are built through a series of enzymatic relays that may involve negative and positive feedback loops and may require compartmentalization. Genetic circuits can be built to replicate these natural biosynthetic pathways in yeast, bacteria, and other chassis organisms, enabling the construction of complex chemicals en masse [12, 13].

Chasses

Finally, chassis organisms are the workhorses of synthetic biology. A chassis in synthetic biology is an organism that can host synthetic parts (genes, gene clusters, etc.) and provides the machinery for those parts to function, such as transcriptional and translational proteins. Two of the most commonly used chasses are *Escherichia coli* and *Saccharomyces cerevisiae*, primarily because these organisms have been extensively characterized and used in molecular research. Their genomes are well described and both tools and methods are already established for efficient engineering of their genomes. As such, these organisms can readily be engineered to host and express new genes and gene circuits in a laboratory environment [15–17].

Together, DNA synthesis and gene assembly technology, genetic circuits, and chassis organisms represent critical components in a synthetic biologist's toolkit. These enable the synthesis of novel DNA sequences, assembly of novel genes and gene clusters, and culturing of novel genetic sequences in a controlled laboratory setting—all of which are employed in the discovery and development of new therapeutics.

2.3 Synthetic Biology in Drug Discovery and Development

The synthesis of artemisinin is the most commonly cited example of how synthetic biology can be applied to drug discovery and development. And while it is an example worth delving into deeper, it is only one of many varied ways that synthetic biology has been advancing therapeutic development. Below, we provide select examples of how synthetic biology is currently reshaping this space. An exhaustive overview would likely warrant its own book. Therefore, we have selected examples to demonstrate both the core principles of synthetic biology and the breadth of its application. There are undoubtedly many more applications that are noteworthy but have been nonetheless excluded from this discussion. We encourage readers to follow their interest to the many citations provided throughout this text for further reading.

2.3.1 Gene Editing for the Identification of Drug Targets

The term synthetic biology often carries with it an air of artisanal craft, as though the application of this field were limited to niche projects. But as this chapter will emphasize, such a connotation is far from the truth. Perhaps the best example of this is the widespread application of CRISPR-Cas gene editing technology.

Multiple species of bacteria and archaea contain segments of DNA known as *clustered regularly interspersed palindromic repeats*, or CRISPR sequences. These sequences serve as memory centers wherein DNA that has been harvested from invading viruses can be stored. When these sequences are transcribed, they produce spacer RNA (spacerRNA) that is processed into a shorter crisprRNA (crRNA) that is complementary to a specific segment of viral DNA. When crRNA is joined to both a trans-activating CRISPR RNA (tracerRNA) and a CRISPR-associated (Cas) nuclease, they form a CRISPR-Cas complex that functions as an adaptive immune system. DNA sequences that are complementary to the spacerRNA will be targeted by the complex and, once bound, cleaved by the Cas nuclease (see Fig. 2.1) [18].

By storing fragments of foreign DNA in CRISPR sequences, these species are able to recognize and more effectively eliminate pathogens. And, in 2012, researchers showed that this system could be modified to enable highly precise gene editing.

Building on decades of work, Jennifer Doudna at UC Berkeley and Emmanuelle Charpentier at the University of Sweden collaborated to engineer a synthetic CRISPR DNA sequence that simplified the CRISPR-Cas complex (by combining spacerRNA and tracerRNA into a single transcript known as a single-guide RNA; Fig. 2.2a) and enabled it to target non-viral DNA sequences, opening the door to prescribed gene editing [19]. Further modification to the system by Feng Zhang of the Broad Institute made the heterologous expression of CRISPR-Cas systems in eukaryotic cells possible, further expanding its potential for engineering biological systems through precise gene editing [20].

Though it is not often discussed as such, CRISPR-Cas represents a powerful product of synthetic biology wherein modified bacterial genes are heterologously

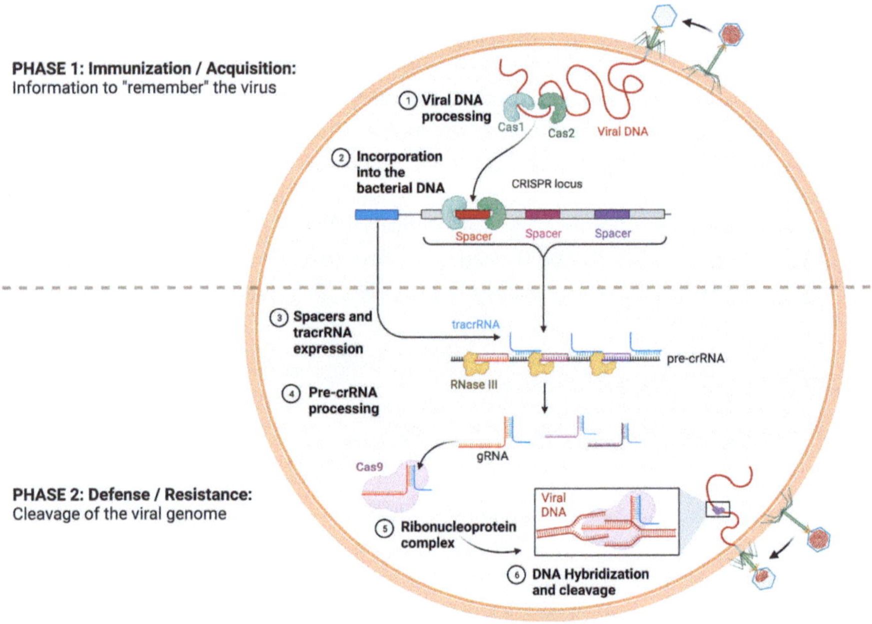

Fig. 2.1 CRISPR-Cas9 adaptive immune system of *Streptococcus pyogenes* against bacteriophages. Diagram showing how the CRISPR-Cas9 complex processes viral DNA within *Streptococcus pyogenes*. Upon entry into the cell, cas nucleases digest the DNA and insert it into the bacterial genome at CRISPR loci. These inserts are referred to as spacer sequences. Expression of the spacer sequence along with an inborn tracer sequence will generate a guide RNA comprised of both spacer and tracer RNAs. Together with the Cas9 nuclease, this complex targets and digests complimentary viral sequences. (Figure adapted from "CRISPR-Cas9 Adaptive Immune System of *Streptococcus pyogenes* Against Bacteriophages," by BioRender.com [2022]. Retrieved from https://app.biorender.com/biorender-templates)

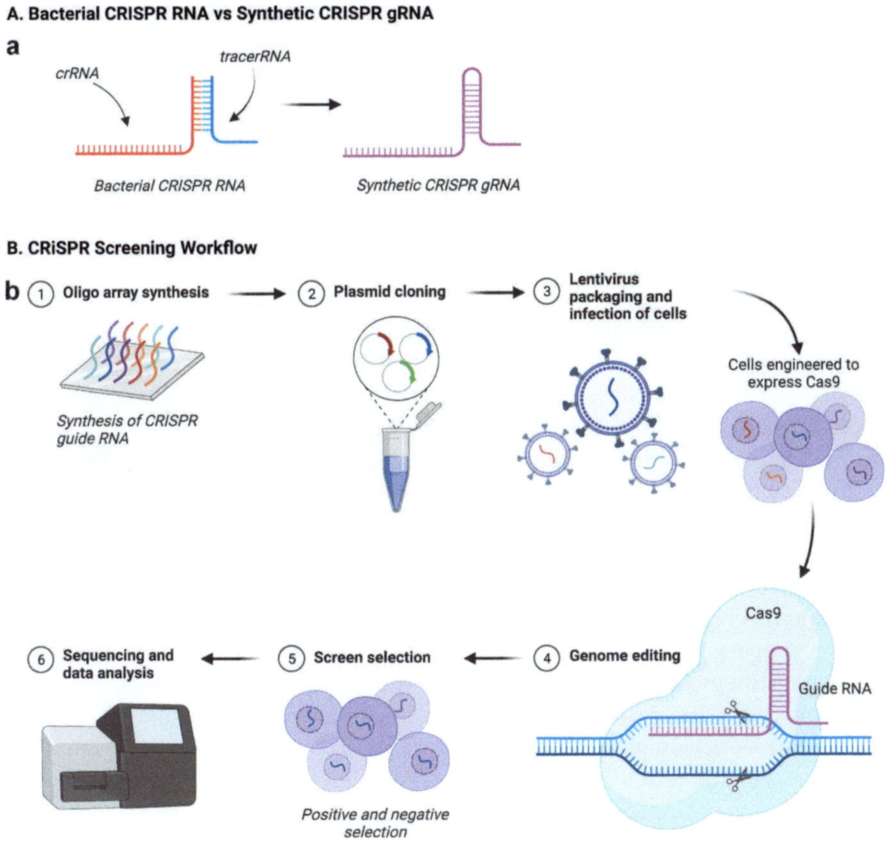

Fig. 2.2 CRISPR-Cas9 in Drug Discovery. (**a**) In bacteria, CRISPR guide RNAs are comprised of two distinct RNA sequences, known as the crRNA and tracerRNA. Use of CRISPR gene editing in drug discovery was enabled by the generation of a synthetic mimic of the natural guide RNA complex. This synthetic version is comprised of a single-guide RNA with sequences corresponding to the crRNA and tracerRNA. (**b**) CRISPR screening workflow. Large-scale CRISPR screening can be used to identify potential drug targets by perturbing individual genes and subsequently analyzing for desired effects. (Figure adapted from "CRISPR Screening Protocol," by BioRender.com [2022]. Retrieved from https://app.biorender.com/biorender-templates)

expressed in a wide range of hosts for productive means. Since its discovery, CRISPR-Cas systems have been developed for numerous applications, including the discovery of therapeutic drug targets [11].

Drug discovery can be carried out in one of two ways: target-forward approaches or phenotype-forward approaches. In the latter, libraries of compounds are screened for their ability to elicit a specific phenotype within a model system. This approach can be useful when the target disease pathology is not well characterized or when novel drug targets are desired. On the other hand, target-forward approaches involve the screening of libraries of compounds for their activity against specific drug targets that are known to hold therapeutic value.

CRISPR-Cas can support both approaches. Target-forward drug discovery starts with identifying targets of interest, a process greatly improved with CRISPR screening. Through large-scale CRISPR screens, researchers can methodically perturb gene expression across a large range of targets in an effort to elicit a desired phenotype (such as reduced proliferation). Results from these screens can produce new therapeutic targets by revealing the proteins and pathways that are critical to disease pathology. Once identified, libraries of compounds can be screened for their ability to target these specific proteins.

A recent example of this came from the Bassik lab at Stanford University, where a team of researchers performed a large-scale CRISPR screen to identify synthetic lethal drug combinations in a BCR-ABL expressing chronic myeloid leukemia cell line [21]. The team utilized a dual-guide approach so that each cell in the study would receive two sgRNAs, and thus have two genes simultaneously knocked out. Approximately 490,000 sgRNA pairs were created, enabling the team to simulate 21,321 drug combinations and identify numerous synthetic lethal combinations.

In addition to driving target identification and validation, CRISPR can also aid in the development of screening models. A significant challenge in drug discovery is ensuring that efficacy in the laboratory setting translates to the bedside. As many as 90% of therapeutics entering clinical trials fail due to a lack of efficacy or unforeseen toxicities [22]. Quantitative decision-theoretic modeling of the drug discovery pipeline suggests that better preclinical models, ones that more accurately reflect human physiology, could substantially improve drug development productivity [23].

To this end, CRISPR-Cas has been used to engineer both cell lines and animal models that better emulate human disease [24]. Having the ability to introduce disease-specific mutations into the desired model—be it human cells or rodents—enables the fine-scale study of disease pathogenesis, rapid screening of compounds in a disease-relevant context, and provides an opportunity for early go/no-go decision-making. All of this builds toward a more productive drug development pipeline.

Beyond its use as a tool in drug development, CRISPR-Cas is also being explored as a therapeutic in and of itself, largely for the treatment of monogenic disorders. A clinical trial initiated in 2019 by pharmaceutical company Editas Medicine is studying the ability of a CRISPR-Cas system to improve vision for patients harboring heterozygous or homozygous point mutations in the Centrosomal Protein 290 (CEP290) gene, which leads to Leber Congenital Amaurosis Type 10 [25, 26]. As of the writing of this book, the trial is ongoing and is expected to conclude in May of 2025. And, in December of 2023, the US FDA approved two of the first ever CRISPR-based therapeutics (Casgevy and Lyfgenia) for the treatment of sickle cell anemia [27, 28].

While CRISPR-Cas technology is proving to be an invaluable tool for drug development, there remain significant challenges that limit its application. Practical challenges include difficulties with effectively delivering multiple macromolecules (such as the vectors encoding both Cas nucleases and gRNAs) to primary cell lines and living tissues. More difficult is the challenge of anticipating and preventing off-target editing. Though the CRISPR-Cas system is capable of highly specific editing, promiscuous nuclease activity as well as imperfect gRNA design can lead to unintended editing. Here, imperfect gRNA designs are those that target non-unique DNA sequences that may be identical—in whole or in part—to other loci within the genome. These redundant loci can attract CRISPR-Cas activity. The many challenges associated with CRISPR-Cas systems and the ongoing efforts to overcome those challenges are covered in depth by Liu et al., in *Precision Clinical Medicine* [29].

2.4 Engineering Biosynthetic Gene Clusters for Drug Discovery

An analysis of 1881 approved drugs in the United States between January 01, 1981, and September 30, 2019, found that only 463 (24.6%) were produced through entirely synthetic means, meaning without inspiration from natural products [30]. Stated another way, nature remains a significant source of medicinal compounds, from biologics to small molecules.

Nature-derived therapeutics can often be traced back to secondary metabolites—biologically active small molecules whose activity is not critical to the growth of an organism but may contribute to signaling or defense systems. Secondary metabolites can be structurally complex and difficult, if not impossible, to synthesize in the laboratory. Therefore, extraction has been the primary approach to accessing secondary metabolites, which may then serve as the chemical skeleton for further modification and development [31].

However, extraction processes can put significant strain on natural resources and may be prohibitively expensive to operationalize at scale. Additionally, natural concentrations of these compounds can be very low, necessitating large quantities of starting material. Owing to these limitations, the sourcing of secondary metabolites has largely been focused on organisms that can be cultured in the laboratory setting or else farmed in significant quantities. Alternatively, some laboratories are attempting to synthesize chemical pathways that generate secondary metabolites in vitro.

The advent of both next-generation sequencing (NGS) and synthetic biology has greatly expanded opportunities for new drug discovery by enabling the production of secondary metabolites in chassis organisms [31–35].

This is done through the identification, synthesis, and cloning of biosynthetic gene clusters (BGCs). Secondary metabolites are frequently produced through an enzymatic relay that enables the formation of complex and pharmacologically active structures. Importantly, the DNA encoding these enzymes is often grouped together into clusters, or BGCs, that can range from several kilobases to >100 kb in length. These clusters can be identified by mining next-generation sequencing (NGS) data using advanced bioinformatics [31–36].

Though the specific enzymes used in this process can vary, enzymes like polyketide synthases (PKSs) and non-ribosomal peptide synthetases (NRPs) frequently appear in the production pathways for toxins, antibiotics, cytostatics, and immunosuppressants [36, 37]. Therefore, bioinformatic tools can be used to mine genomic databases for potential BGCs by searching for PKSs, NRPs, and their constituent elements. Once identified, these gene clusters can be synthesized and cloned into chassis organisms.

This approach has several advantages. Porting BGCs into chassis organisms obviates the need for the original host organism to be involved in compound screening, reducing ethical, financial, and operational concerns. Additionally, BGCs in chassis organisms can be optimized to produce large quantities of the candidate compounds, enabling testing of compounds that may have otherwise been overlooked due to low expression levels in their native host [38]. Further, the synthetic nature of this approach allows for combinatorial experimentation such that novel variants and unique gene combinations can be tested en masse.

A recent preprint from Liu et al. highlights this potential. Using metagenomic data from soil samples, Liu et al. analyzed >6.8 billion BGCs for the potential to produce inhibitors of human methionine aminopeptidase-1 (HsMetAP1)—a validated oncology target [39]. This search yielded 35 BGCs with enzymes that are likely to produce secondary metabolites against HsMetAP1. Heterologous expression of two of these BGCs in a bacterial chassis resulted in the discovery of several novel HsMetAP1 inhibitors, one of which demonstrated sub-micromolar potency and high selectivity in an in vitro enzymatic assay (specifically a methionine aminopeptidase colorimetric activity assay based on an R&D Systems commercial kit). This example demonstrates how heterologous expression of BGCs can be used to discover new small molecules with therapeutic potential.

Heterologous expression of BGCs is far from trivial, though. Gene regulation and cellular context can be important factors in determining the catalytic efficiency and end products of a BGC. When these enzymatic networks are ported into chassis organisms, the costly and time-consuming process of directed evolution may be required to optimize secondary metabolite production. Herein, directed evolution describes the creation of libraries in which BGC elements will undergo an iterative process that involves random mutation, screening of mutants for desired properties, and further mutation of selected mutants until optimal performance is attained. Such a process can be time and resource intensive for individual genes, much less gene clusters.

2.5 Unprecedented Antibody Discovery and Development Through Synthetic Biology

Over the past two decades, biologics have been an increasingly important source of new molecular entities [40]. Between the years 2000 and 2021, 723 new drugs were approved in the United States, 158 (21.9%) of which were biologics [41]. Within this category, monoclonal antibodies are by far the most common.

Owing to size and physiochemical differences, small molecules and antibodies display substantially different pharmacokinetics, which can be both beneficial and detrimental. For example, antibodies generally have longer half-lives in the body, enabling longer exposure to the therapeutic and intermittent dosing. However, antibodies also struggle to penetrate beyond tissue barriers (such as the blood-brain barrier), which can greatly limit their distribution [42]. As polymeric proteins, antibodies are also susceptible to deamidation, glycosylation, and oxidation events that can alter their biological properties.

Therefore, the development of antibodies and other biologics often includes extensive rounds of iterative optimization in an effort to tailor the antibody's pharmacological and pharmacokinetic properties. This process is widely referred to as a design, build, test, learn (DBTL) cycle. During the design phase, libraries of antibodies can be computationally designed such that each antibody clone contains a combination of missense mutations that are expected to affect the antibody's properties, such as binding affinity or stability.

Synthetic biology intersects with and drives this process in many ways. Much of this work relies on the same tools and high-level workflows already discussed: Synthetic DNA sequences representing whole or partial antibodies are heterologously expressed in a chassis organism—be it a bacteria cell, mammalian cell, or a transgenic mouse—in order to screen, optimize, and manufacture novel biologics.

In the following sections, we will briefly describe two of the primary approaches in biologics development—one based on B-cell technology, and one based on phage display—as well as the impact of synthetic biology in enabling these approaches.

2.5.1 B-Cells and Hybridoma Technology for Antibody Discovery and Development

Hybridoma technology advanced antibody discovery and development efforts considerably by enabling the long-term culture of B-cells collected from immunized animals. First described in 1975, hybridoma technology fuses mouse myeloma cells with antibody-producing B-cells to form an immortalized B-cell line from which antibodies can be collected [43]. Clonal expansion of B-cells, each expressing a different antibody variant, can then be used to produce antibody libraries (Fig. 2.3).

Hybridoma technology enables robust production of monoclonal antibodies in the laboratory and has been used to develop many currently approved biologics [43, 44]. However, antibodies derived from non-human species are very likely to be recognized as foreign proteins by the human immune system and elicit a life-threatening response. To mitigate this risk, therapeutic antibodies must be made to resemble human antibodies using tools from synthetic biology.

Antibodies are hetero-tetrameric proteins consisting of two identical heavy chains and two identical light chains joined together through disulfide bonds. At one end of the protein complex can be found variable regions for both the heavy and

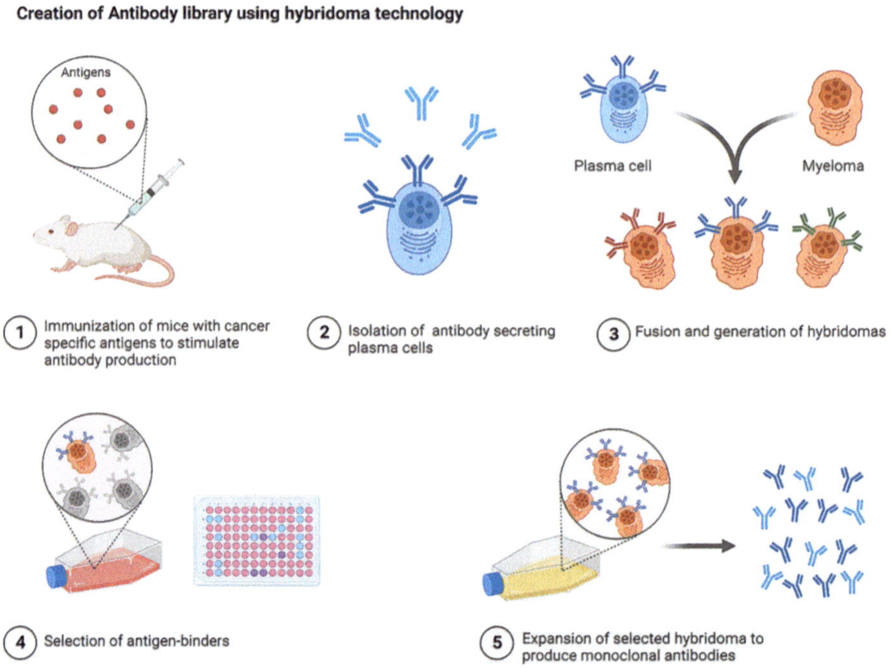

Fig. 2.3 B-cell hybridoma library creation. Workflow for the creation of antibody libraries using hybridoma technology. (Figure adapted from "Monoclonal Antibodies Production," by BioRender.com [2022]. Retrieved from https://app.biorender.com/biorender-templates)

light chains. As their name suggests, variable regions have highly varied amino acid sequences whose unique composition determines the antibody's unique antigen specificity and affinity. Within these variable regions, three amino acid loops (known as complementarity-determining regions, or CDRs) are critically influential in determining antigen recognition. The rest of the antibody complex consists of frameworks and constant regions whose amino acid sequences are largely conserved within species [45].

Reducing immunogenic potential requires careful modification of the candidate antibody to preserve antigen specificity while incorporating human frameworks wherever possible. Multiple approaches have been developed, all of which rely on creating novel antibody genes in which elements of the original antibody's DNA sequence are replaced with human antibody sequences [44].

For example, chimeric antibodies commonly fuse promising mouse antibodies with human antibody frameworks. In these synthetic antibodies, the DNA encoding the human antibody variable regions are replaced by the corresponding variable region in the mouse therapeutic antibody, such that the final product contains human constant regions and non-human variable regions. This is also sometimes referred to as CDR grafting because the CDR from one animal is transplanted onto a human antibody framework [44].

Novel chimeric antibodies are easily produced through the use of de novo DNA synthesis (discussed in more detail below), homology-based cloning methods, and chassis organisms. Though the majority of therapeutic antibodies used to treat humans are chimeras, the development of chimeric antibodies remains relatively risky compared to fully human antibodies [46].

Alternatively, transgenic mouse models, such as HuMabMouse and XenoMouse, have been created by replacing mouse immunoglobulin genes for those derived from the human genome [44]. Rather than expressing murine antibodies, these mice produce human antibodies and can thus be used to create fully human hybridoma cells. While undoubtedly useful, the application of transgenic mice is limited by their costly and slow nature, two attributes that make it difficult to apply to emergent cases, such as infectious diseases.

Rather than humanizing antibodies, it is also possible to *start* with human antibodies. Human B-cell screening is an approach whereby antibody libraries are sourced from the human B-cell repertoire of resear ch participants [44, 47]. Heterogeneous populations of plasma and memory B-lymphocytes can be collected from either lymphatic tissue or peripheral bone marrow cell populations and subsequently screened in cell-based or bead-based assays. As these cells naturally excrete antibodies, individual B-cells can be isolated and assessed for antigen-binding activity. Those that show binding activity can then be sequenced to identify the corresponding DNA sequence.

Such an approach provides several benefits. It is estimated that at least 10^7 antibody variants can be found among circulating human immune cells at any given time. Based on the mathematical possibility of different V(D)J recombinations, it is estimated that the true human B-cell repertoire size is between 10^{12} and 10^{18} distinct variants [47]. However, the human body has natural maturation processes that prevent B-cells containing autoreactive, truncated, or otherwise non-functional antibodies from entering circulation. This maturation process is advantageous for antibody development.

Additionally, B-cell screening can be a valuable tool when combating infectious diseases. Patients who have previously been infected with the target pathogen are very likely to have memory B-cells that produce high-affinity antibodies. Collection and screening of B-cells from these patients can help researchers rapidly develop therapeutic antibodies. This was made evident early in the SARS-CoV-2 pandemic when B-cells collected from previously infected patients were used to rapidly identify 49 different human anti-SARS-CoV-2 antibodies within a matter of weeks [48].

For a more in-depth review of B-cell technology, we refer you to Pedrioli and Oxenius' recent review in *Cell* [47].

Though a powerful method for antibody discovery, human B-cell screening has notable limitations related to throughput and scale. But perhaps the biggest limitation is that B-cell screening can only survey antibodies that nature has already produced. Sometimes, it is advantageous to go beyond what already exists and develop a de novo antibody sequence. For this, researchers turn to synthetic libraries.

2.5.2 Synthetic Libraries and Display Systems

As with small molecules, the discovery and development of novel biologics often begins with the screening of large libraries in which millions of antibody (or antibody fragment) sequences are represented. When these libraries are diverse, with uniform representation among variants, and are free of liability motifs that can lead to poor pharmacokinetics, hundreds of promising candidates can be found and carried forward for further characterization and optimization.

Historically, antibody libraries have been sourced from the serum of animals previously infected with target antigen [44]. With a natural, pre-existing repertoire of diverse antibodies, the animal's immune system is often readily able to produce polyclonal antibodies against a wide range of targets. However, this approach has many drawbacks related to variability, a general inability to scale, and ethical concerns surrounding animal welfare that preclude its use in the development of therapeutics.

Whereas hybridoma technology relies on animal immune systems to surface candidate antibodies, recent advances in both DNA synthesis technology and synthetic biology have given researchers the power to design, produce, and screen custom antibody libraries. Specifically, DNA synthesis can now be reliably performed on a large scale [9], such that millions of custom oligonucleotides can be generated per run on platforms such as Twist Bioscience's silicon-based DNA synthesis platform. This makes it possible to synthesize large libraries of deliberately designed antibodies.

There are many approaches to designing such a library that go beyond the scope of this chapter. However, whether starting with a known antibody sequence or taking a broader approach, computational tools can be used to identify key loci in an antibody sequence where missense mutations are likely to impact antigen recognition [49]. The library would subsequently consist of antibody variants containing combinations of missense mutations (a so-called combinatorial variant library). Such libraries may contain as many as 10^{10} unique antibodies for screening [50].

Once designed, antibody libraries may then be constructed using DNA synthesis technology. Modern synthesis platforms leverage a myriad of approaches to create variant libraries that can generally be divided into direct and degenerate approaches.

Degenerate approaches rely on randomly incorporating nucleotides at specified locations within the growing oligonucleotide. Most common among these methods is the NNK approach, wherein the first and second positions of a codon may be occupied by any of the four nucleotide possibilities (A, T, G, C). The final position denoted by a K, the so-called wobble position, is limited to just two potential nucleotides (G, T).

NNK synthesis can be extremely effective for directed evolution; however, its inefficiencies and bias-prone nature greatly limit its application in antibody development. Consider the redundancy of the genetic code, wherein multiple codons may apply to the same amino acid. Using NNK synthesis, you are likely to generate multiple distinct codons that correspond to the same amino acid, resulting in redundant variants that may not provide additional value to your study. This also means it is difficult to create uniform libraries—libraries where every variant antibody is equally represented—using NNK.

For example, because six different codons code for the amino acid arginine, it is far more likely to occur relative to amino acids like phenylalanine that only have two corresponding codons. Using NNK, then, a mutation in any given loci is more likely to substitute an arginine than phenylalanine, biasing the test pool away from antibody variants containing less common amino acids.

Finally, NNK synthesis can produce stop codons. When creating large libraries with multiple variants per oligo, premature stop codons are likely to affect a significant portion of the library, resulting in many truncated and non-functional antibodies [51].

Collectively, these drawbacks mean that creating combinatorial variant libraries with NNK synthesis is prohibitively inefficient. The combination of redundant antibody clones, non-uniform antibody representation, and truncated antibodies (produced by premature stop codons) generates an expensive library with few unique, viable antibody clones to screen. Put another way, NNK approaches reduce the functional diversity in antibody libraries and may require costly, time-consuming efforts to compensate for this, ultimately increasing the researcher's screening burden.

The use of libraries synthesized with trinucleotide phosphoramidites (trimers) can be a good alternative to NNK libraries because it gives the user more control. Unlike NNK libraries, which randomly incorporate individual nucleotides, trimer approaches incorporate blocks of three nucleotides at a time. The trinucleotide blocks can be synthesized through various means and are designed such that every desired codon is represented, whereas undesired codons—such as stop codons or those coding for cysteine—can be avoided.

While trimer synthesis affords the user with more control, it is still libel to produce biased oligonucleotide libraries where certain amino acids are over-represented. This is because each trimer phosphoramidite has a different reaction factor that dictates the rate of the coupling reaction during synthesis. For example, the trimer AAC (encoding asparagine) couples with the greatest efficiency—meaning it is very likely to be incorporated whenever given the opportunity—and thus has a reaction factor of 1.0. In contrast, the trimer CAG (coding for glutamine) has a lower coupling efficiency and a reaction factor of 2.0. In order to prevent over-representation of asparagine in the final product, the oligonucleotide synthesis reaction mixture will need to have two CAG trimers for every one AAC trimer [52, 53].

The coupling efficiency of trimers is also lower than individual nucleotides because three nucleotides are being coupled instead of one, which increases steric hindrance. If the coupling efficiency is high, we can expect to see fewer truncated oligos and a higher percentage of full-length material. Truncated material in the pool will impact downstream processes such as cloning because shorter fragments clone preferentially. For the same reasons as above, if trimers are used to construct a long domain, such as an antibody's third heavy chain complementarity-determining region (CDRH3), the percent full length can be very low.

There is currently significant interest in the development of single-domain heavy-chain (VHH) biologics owing to their antibody-like properties and smaller size (affording them distinct pharmacokinetics relative to full-length antibodies). Assembling VHH libraries can be challenging with trimers due to the potential for truncation. Truncated oligonucleotides can then lead to a diminished percentage of full-length VHH candidates and the need for increased screening to ensure that all full-length variants are tested.

With trimers, it is also difficult to avoid liability motifs that can result in instability of biologics following post-translational modifications or isomerization. In order to avoid motifs with trimers, certain amino acids would need to be avoided to eliminate the possibility of unwanted amino acid pairings that would create an undesirable motif.

Direct synthesis, on the other hand, has enabled a much higher degree of precision and efficiency. Through solid-phase synthesis along with miniaturized chemistry, companies like Twist Bioscience have made it possible to synthesize oligonucleotides without randomization. By deliberately adding every nucleotide in a growing strand, direct synthesis can be programmed to generate combinatorial variant libraries containing uniform variant representation and prespecified amino acid ratios (Fig. 2.4).

The improved accuracy and efficiency gained through direct synthesis has enabled researchers to focus their resources on generating and testing millions of rationally designed antibody variants. In so doing, direct synthesis has improved the scale of antibody discovery by allowing parallelization—rather than only creating a

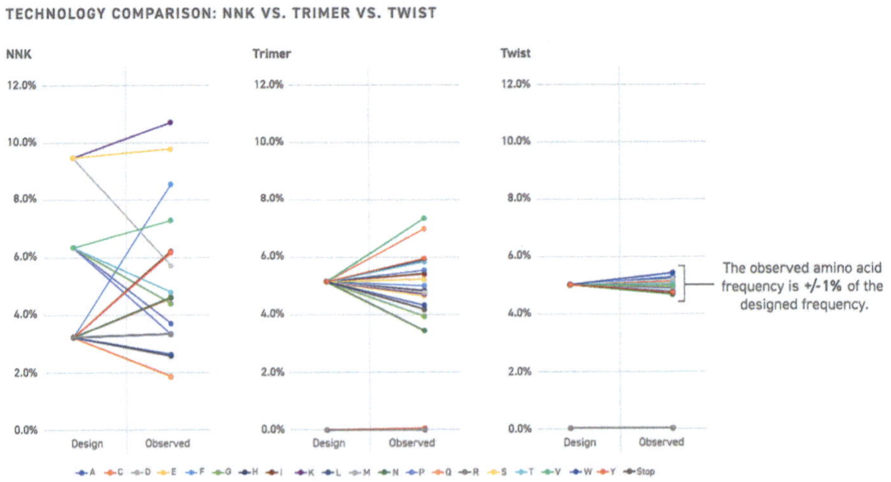

Fig. 2.4 Amino acid distribution after synthesis with NNK, Trimer, and direct synthesis. When compared to NNK and Trimer synthesis, libraries created with direct synthesis technology (Twist) showed less than 1% deviation from the designed amino acid frequency (y-axis). (Data produced by Twist Bioscience)

small handful of desired antibodies at a time, millions can be rapidly produced and screened in parallel.

Importantly, such libraries need not be full-length antibodies. As researchers are primarily interested in identifying antibodies with a specific and strong affinity for their target of interest, resources can be conserved by limiting synthesis to antibody fragments that contain CDR loops, such as VHH, single-chain variable fragments (scFv), and fragment antigen-binding regions (Fabs). This approach opens the door to high-throughput screening opportunities.

As with other synthesis methods, direct synthesis is not without its hurdles. Highly parallel direct oligonucleotide synthesis requires substantial infrastructure and automation. As such, these methods are often more expensive than degenerate methods and, like trimers, must contend with nucleotide coupling efficiencies to ensure uniform representation. Therefore, careful design and quality control (often through next-generation sequencing) is needed, further driving up costs. However, direct synthesis has greatly decreased the costs associated with DNA synthesis and will likely continue to do so as the technology advances [54].

Phage Display

Converting synthetic DNA libraries into screen-ready antibody libraries requires cellular machinery. Therefore, synthesized libraries must be inserted into a chassis organism whose internal machinery can be taken advantage of to produce antibody fragments.

Following DNA synthesis, antibody sequences can be assembled in plasmid or phagemid vectors and transduced into a display system. In these systems, antigen-binding fragments are heterologously expressed in a chassis organism such as a bacteriophage, yeast, or bacteria cell. Importantly, these antibody fragments are fused to a native surface protein within the chassis, enabling surface expression of the fragment [55, 56].

Phage display is the most commonly used display system in antibody discovery and development owing to its ease of use and scale. This system utilizes bacteriophages—typically M13 bacteriophage—that selectively infects *E. coli*. For M13 bacteriophage, antibody fragments can be assembled into phagemid vectors as fusion proteins with one of the phage's five surface proteins. Importantly, the phagemid vector contains genetic elements that enable it to be enclosed in the viral capsid during replication [55].

The resulting bacteriophage will express antibody fragments on its surface while also containing phagemid vectors within. Selection is then performed by first immobilizing purified antigens on solid surfaces, such as nitrocellulose membranes, polystyrene plates, magnetic beads, or column matrices. Then, a phage library corresponding to the synthesized antibody library can be exposed to the immobilized antigen. Successive cycles of washing will eliminate all unbound phages, leaving just those phages that express high-affinity antigen-binding fragments (Fig. 2.5).

High-affinity binders can then be recovered with the use of myriad elution buffers. Because the eluted phages each contain a phagemid, they can be transduced into *E. coli* and subsequently sequenced to determine which antibody sequences correspond to high-affinity binders.

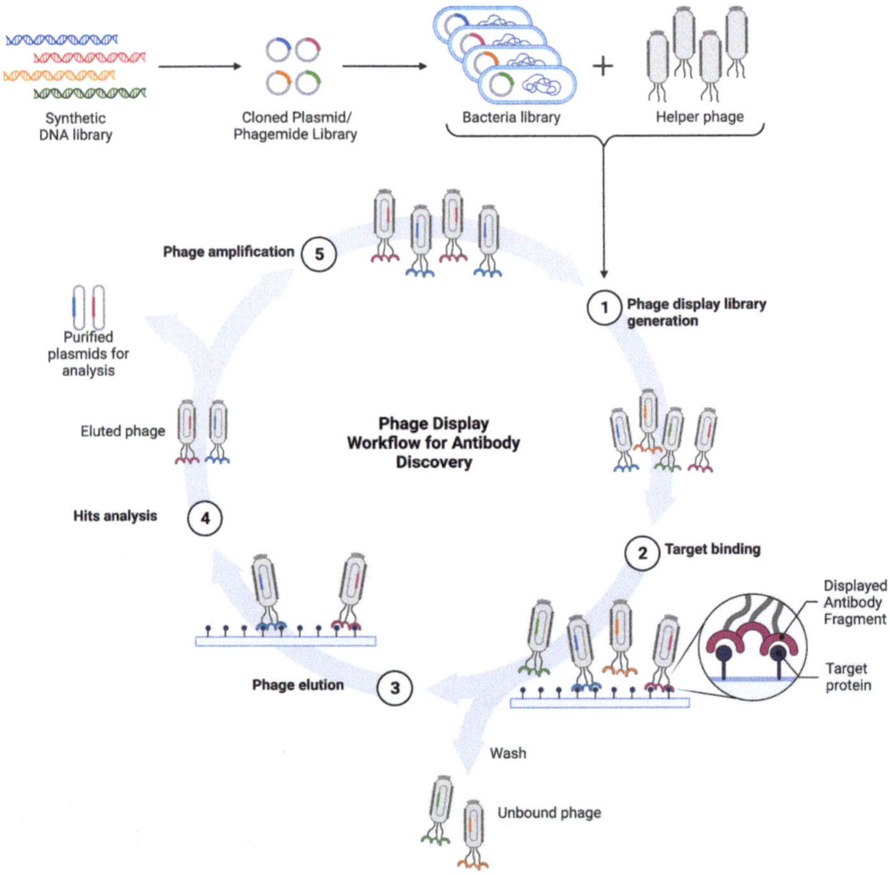

Fig. 2.5 Phage display for antibody discovery. Workflow for the discovery and optimization of antibodies using phage display. Helper phages may be needed to ensure production of engineered bacteriophages. (Figure adapted from "Phage Display Panning," by BioRender.com [2022]. Retrieved from https://app.biorender.com/biorender-templates)

Phage display screens can be used to test libraries as large as 10^{10} unique antibody fragments, enabling researchers to rapidly home in on antibody sequences with desired binding properties [55, 56]. In contrast to immunization approaches, in vitro display systems can be used to develop biologics against highly toxic and non-immunogenic targets.

Though phage display has clear advantages over hybridoma technology, antibodies discovered through phage display often require substantial optimization. Replicating in vitro the immune system's ability to rapidly optimize antibodies for given antigens is difficult. Without comprehensive optimization, phage-derived antibodies may be prone to in vivo aggregation, low levels of expression, and other issues that limit the antibody's clinical utility [57]. However, in vitro optimization systems are advancing and may reduce developability concerns stemming from phage display in the near future.

In the meantime, phage display remains a powerful discovery tool. As of August 2020, 70 phage-derived antibodies had advanced to clinical trials, 14 of which have already been approved [56].

Similar display systems have been developed with other chassis organisms, such as *S. cerevisiae* [55]. Herein, synthetic DNA encoding antibody fragments are transfected into the yeast where they integrate with the yeast's cell wall, akin to the integration that occurs in bacteriophages. These eukaryotic display systems can be useful as they replicate post-transcriptional modifications that may affect antibody performance in human cells, such as glycosylation. While yeast glycosylation patterns may differ from what is observed in humans, yeast cells can be engineered to express human glycosylation machinery. Useful as it is, yeast display suffers from decreased efficiencies that limit its scale to approximately 10^8 antibody fragments per experiment.

There are many more display systems worth diving into, and we encourage you to read Mahdavi et al.'s excellent overview in the *International Journal of Biological Macromolecules* to learn more [55].

2.6 CAR-T Cell Therapies

One of the most recognized therapeutic applications of synthetic biology is in the development of cell therapies broadly and chimeric antigen receptor (CAR) presenting T-cells specifically. Conceptually, cell therapies aim to endow an organism with enhanced immune cells, whether those are stem cells lacking pathogenic mutations or, in the case of CAR-T cells, differentiated cells with specific and potent pathogen targeting abilities. In contrast to the transient nature of antibodies and small molecules, cell therapies have the potential to help patients indefinitely, providing lasting protection against recurrent malignancies or viruses like HIV [58–60].

Among the many different cell therapies, CAR-T cell technology has seen remarkable advancement in recent years, driven largely by improvements in synthetic biology (Fig. 2.6).

T-cells are a class of lymphocytes that play a critical role in adaptive immunity [60]. Born in the thymus, naive T-cells roam the body awaiting activation. Unlike B-cells, which recognize pathogens through antibodies, T-cell activation is initiated when a unique cell surface protein—the T-cell receptor (TCR)—recognizes a major histocompatibility complex (MHC) protein displaying an antigenic peptide. This, combined with cascading signals from costimulatory factors, prompts T-cells to release a slew of cytokines that, depending on the type of T-cell, are often proinflammatory [61–63].

T-cells have the ability to direct adaptive immune cells against specific pathogens and thus have significant potential in therapeutic applications. Like antibodies, the human TCR repertoire is extremely diverse owing to V(d)J recombination. But

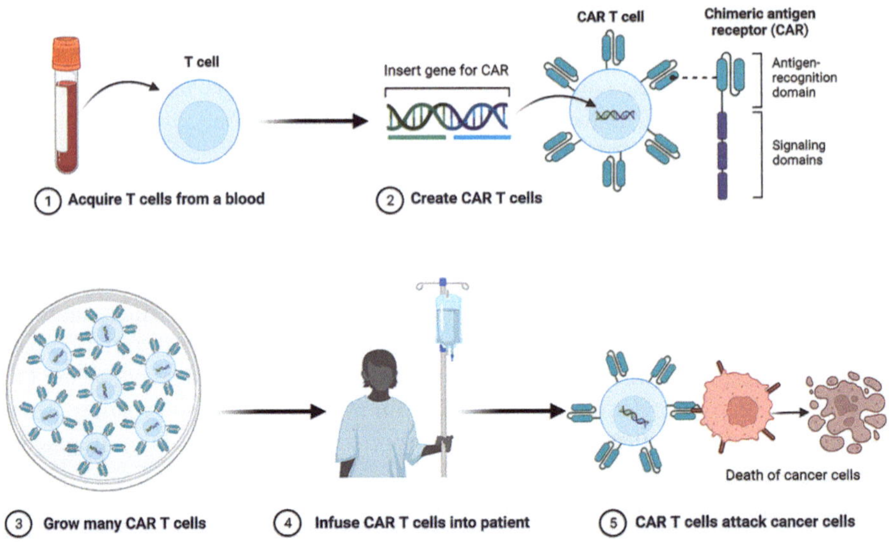

Fig. 2.6 CAR-T cell therapy overview. Workflow for CAR-T cell therapy. T-cells are isolated from donor or patient blood and engineered to express a chimeric antigen receptor targeting tumor-cell-specific antigens. After successive rounds of growth, the engineered T-cells can then be reinfused into the patient to affect therapeutic benefits. (Reprinted from "CAR T Cell Therapy Overview," by BioRender.com [2022]. Retrieved from https://app.biorender.com/biorender-templates)

while a T-cell may have the ability to recognize antigens, it must do so when the antigenic peptide is in complex with an MHC protein [63]. Because of this, researchers developing cell therapies based on T-cells would need to match highly antigen-specific T-cells against specific antigen-presenting cells together to enable a potent therapy. Or a simpler approach is to engineer therapeutic T-cells using synthetic biology.

In the late 1980s, researchers at multiple universities began engineering T-cells to express a synthetic fusion protein that combined antibody variable domains with T-cell receptor signaling domains [64, 65]. The resulting chimeric protein could thus recognize antigens with antibody-like specificity and, upon binding, elicit immunogenic activity from the host T-cell. These early experiments are broadly considered the first demonstrations of CAR-T technology [66].

Importantly, the inclusion of antibody variable domains in CARs enables antigen recognition without the involvement of MHC proteins; and, the addition of a costimulatory factor (CD3ζ) enables the chimeric protein to elicit T-cell activation without MHC complex recognition on antigen-presenting cells [66]. Collectively, these modifications make it possible to aim T-cells at specific antigens, such as proteins uniquely present on the surface of cancer cells, which can then independently mount an immune response.

The power of CAR-T cells was recognized around the world in 2012 when 6-year-old Emily Whitehead became the first child to receive CAR-T therapy [67].

Emily had undergone 16 months of intermittent chemotherapy, but the acute B-cell lymphoblastic leukemia that flowed through her veins resisted treatment. With no other options, Emily was enlisted in a clinical trial to test the effect of a new CAR-T therapy, known as tisagenlecleucel (also referred to as Kymriah). T-cells from Emily's body were removed and transduced with genes carrying a synthetic CAR that targeted the cell surface protein CD19. It has been more than 10 years since her treatment, and Emily remains cancer free.

While both promising and inspiring, it is also a warning: following treatment, Emily's body responded with a cytokine storm that nearly killed her. Cell therapies come with a substantial risk of inducing a potentially lethal condition known as cytokine release syndrome (CRS). The cause for critical side effects in cell therapies can be multifaceted. CRS can be induced through an overwhelming CAR-induced on-target, on-tumor inflammatory reaction. Careful design and validation of the CAR for a given target and therapeutic indication is key for development of safe and efficacious therapies.

Beyond designing the synthetic receptor, the cellular vehicle that the CAR is inserted into plays a major role in drug development. While autologous therapies focus on engineering a patient's own cells, allogeneic therapeutics offer the ability to build therapies with broader accessibility by using engineered cells from donors (discussed further below). However, such an option raises additional requirements in terms of design, engineering, and functional evaluation of cells. In contrast to autologous therapies, allogeneic therapies must take into consideration graft-versus-host disease (GVHD)—a condition wherein donor immune cells attack healthy host tissues [68]. Similarly, the engineered cells are at risk of being rejected by the host's body.

The development of safe and effective CARs thus requires careful and extensive optimization to ensure both safety, specificity, and efficacy. Fortunately, many of the tools that characterize synthetic biology can be leveraged during CAR development.

Antibody display systems, for example, enable large-scale and high-throughput screening of potential antigen-binding domains. Once identified as specific for the target antigen, DNA coding for binding domains can be synthesized and assembled into a CAR gene.

Subsequent optimization will be required, wherein variations in antigen-binding domains, hinge and transmembrane domains, and costimulatory factor combinations are tested. This is because cellular response to CAR-antigen binding is highly context specific, and alterations in the flexibility of transmembrane domains, as well as the intracellular environment, can alter downstream signaling. Therefore, researchers will need to generate and express combinatorial variant CAR libraries for iterative rounds of optimization [69, 70].

As with antibody discovery, the precise and large-scale production of synthetic DNA and the combinatorial assembly of variants can accelerate CAR-T development (Fig. 2.7).

In some allogeneic therapies, even when CARs are optimized to be highly specific, patients are still at risk for GVHD. This is because T-cells are trained to recognize foreign cells through mismatches in MHC proteins (the human equivalent of

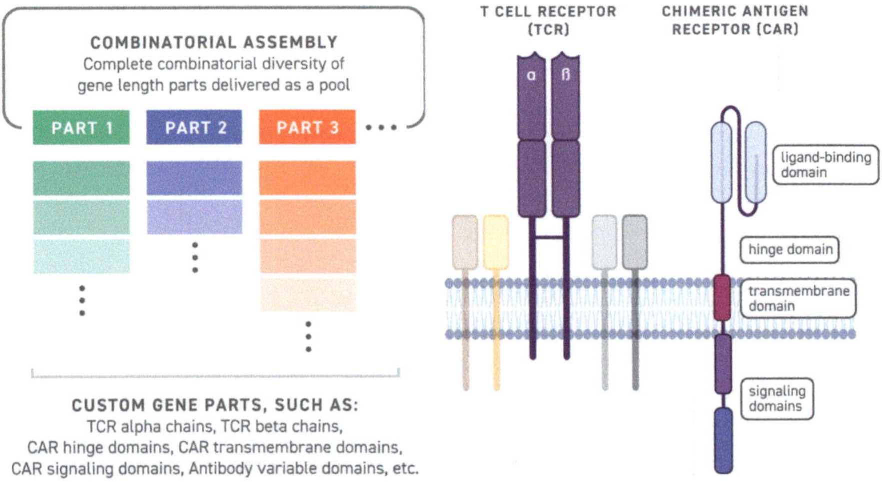

Fig. 2.7 Creation of chimeric antigen receptor and T-cell receptor libraries via combinatorial assembly. Through combinatorial assembly, user-defined combinations of gene fragments can be shuffled together to create highly uniform screening libraries that enable comprehensive screening of the variant space and the discovery of CAR formats with novel functionalities

HLA proteins). It is thus important for donors and patients to carry similar HLA profiles [71].

One approach to avoid GVHD is to use the patient's own cells, as in the case of Emily Whitehead. This is known as autologous cell therapy. However, the patient's immune cells may not be suitable for engineering due to the effects of previous therapeutics or the pathogen/malignancy affecting the patient.

In these situations, physicians may opt for allogeneic CAR-T cell therapy wherein T-cells are collected from a healthy donor, engineered to express the synthetic CAR, and then transfused into the patient. The risk of GVHD is significant in allogeneic CAR-T cell therapy. And while supplementary chemotherapies are being explored to mitigate the risks of GVHD, and HLA matching is possible, the risks remain.

Another option is to use gene editing technology to engineer T-cell stocks that do not express HLA proteins, or else express specific HLA profiles. These allogeneic T-cells could then serve as a repository, off-the-shelf option for researchers and physicians to access when needed. Research into this possibility is ongoing and holds significant promise. Currently, CAR-T cells are generated rapidly, providing little time for cells to be optimized. However, if T-cell stocks can be made, researchers may have ample time to engineer CAR-T cells that target multiple pathogenic antigens by expressing multiple different CARs [72].

The CAR-T therapy that saved Emily Whitehead's life was approved by the FDA for the autologous treatment of adult patients with relapsed or refractory follicular lymphoma after two or more lines of therapy [73]. While a substantial landmark for synthetic biology, much more research is needed to develop this technology. For instance, CAR-T cells are not yet durable enough to provide lifelong immunity, and the intense immunological side effects will need to be contended with.

To this end, the success of CAR-T cell therapy has spurred interest in similar cell therapies [74]. One such therapy leverages natural killer (NK) cells—a type of peripheral leukocyte with potent cytotoxic abilities. Unlike T-cells, NK cells do not become activated against targets based on T-cell receptor-antigen binding. Instead, NK cells are under a dynamic state of intracellular signaling that suppresses activation. However, when exposed to a myriad of contextual signals in the cell's microenvironment that decrease suppression and enable cell-killing, cells that downregulate self-identifying HLA proteins, for example, are more likely to be targeted by NK cells [75, 76].

Interest in the use of NK cells for CAR therapy stems from a growing body of evidence that shows CAR-NK cells are less likely to induce CRS and GVHD (owing to a different cytokine release profile relative to T-cells), are capable of potent on-tumor cell-killing, and may affect cytotoxicity independent of CAR-antigen binding (through natural activation by the tumor microenvironment) [76]. With a reduced risk of CRS and GVHD, CAR-NK cells may prove to be better suited for off-the-shelf, allogeneic cell therapy development.

As with CAR-T cells, CAR-NK cell therapy faces several challenges related to transduction efficiency, ex vivo expansion of engineered cells, tumor evolution toward the loss of CAR-targeted antigens, and more [76]. As such, development of CAR-NK will similarly require extensive preclinical optimizations in CAR design and engineering methods.

Nonetheless, development of both CAR-T and CAR-NK therapies represents shining examples of synthetic biology and its application in therapeutic development. It is likely that the number of clinical trials testing either CAR-T and CAR-NK will continue to increase as advances in synthetic biology continue to expand the scale and throughput of preclinical development.

Notably, there are many additional cell therapies that are worth exploring. The reader is encouraged to read more on cell therapies in a recent review by Wang et al. [74].

2.7 Conclusion

Synthetic biology is an expansive field, one that aims to solve many of the world's problems through the engineering of biological systems. Though relatively new, the field has already made a substantial impact on modern drug discovery and the advancement of biologics research. In the preceding sections, we have attempted to provide a brief overview of this impact, highlighting choice contributions where possible. We have undoubtedly left many details and areas of interest out of this discussion, not for a lack of importance but rather in service of brevity.

While there is still much to be covered, there are some key concepts that we have tried to emphasize. When precisely synthesized, synthetic DNA can be a versatile tool in the synthetic biologist's toolkit. Novel genes can be rapidly assembled and, through cloning, inserted into a chassis organism. This process represents a synthetic dogma that drives modern drug development in a myriad of ways.

CRISPR-Cas systems are an excellent example. The CRISPR-Cas complex is a synthetic creation, modified from a bacterial immune system and used extensively for gene editing. Among its many uses, CRISPR-Cas affords researchers the ability to carry out functional genomic screening on a large scale, such that individual genes can be systematically perturbed across the entire genome within a single screen [77]. Such a study can reveal key genes involved in disease progression and thus potential therapeutic targets. To do this, synthetic DNA coding the CRISPR-Cas complex will be synthesized, assembled, and inserted into the cell line of interest for high-resolution study.

Beyond target identification and validation, synthetic biology is greatly advancing our ability to discover therapeutic small molecules and biologics. Secondary metabolites, for example, are produced in myriad organisms and have proven to be powerful sources of therapeutic small molecules. However, screening of secondary metabolites has been severely limited. With complex structures and often low natural concentrations, secondary metabolites can be difficult to detect and harder to screen. As such, discovery efforts are largely limited to metabolites produced in large quantities or in organisms that can be easily cultured in the laboratory setting.

Synthetic biology offers a way to greatly expand the scope of these efforts. Once described, the molecular pathways—often identified in biosynthetic gene clusters—that produce secondary metabolites can be reconstituted in chassis organisms and leveraged for the large-scale production of diverse secondary metabolite libraries. Such libraries have already been used to surface promising small molecules.

DNA synthesis, cloning, and expression in chassis organisms drives therapeutic antibody discovery, as well. Advances in DNA synthesis technology have made it possible to precisely and rapidly engineer millions of unique DNA sequences. These in turn can be assembled into synthetic antibody libraries whose diversity extends beyond natural repertoires, providing researchers with a unique opportunity to design, discover, and optimize novel therapeutic antibodies.

In short, it is clear that synthetic biology is a versatile field and one that is integral to modern drug development.

References

1. Mary Wollstonecraft Shelley (1831) Frankenstein
2. Hale V et al (2007) Microbially derived Artemisinin: a biotechnology solution to the global problem of access to affordable antimalarial drugs. Am J Trop Med Hyg 77(6_Suppl):198–202. https://doi.org/10.4269/ajtmh.2007.77.198
3. Peplow M (2016) Synthetic biology's first malaria drug meets market resistance. Nature 530(7591):389–390. https://doi.org/10.1038/530390a
4. Septembre-Malaterre A et al (2020) Artemisia annua, a traditional plant brought to light. Int J Mol Sci 21(14):4986. https://doi.org/10.3390/ijms21144986

5. Paddon CJ, Keasling JD (2014) Semi-synthetic artemisinin: a model for the use of synthetic biology in pharmaceutical development. Nat Rev Microbiol 12(5):355–367. https://doi.org/10.1038/nrmicro3240
6. Voigt CA (2020) Synthetic biology 2020–2030: six commercially-available products that are changing our world. Nat. Commun 11(1):6379. https://doi.org/10.1038/s41467-020-20122-2
7. Cameron D, Ewen et al (2014) A brief history of synthetic biology. Nat Rev Microbiol 12(5):381–390. www.nature.com/articles/nrmicro3239/. https://doi.org/10.1038/nrmicro3239
8. van den Belt H (2009) Playing God in Frankenstein's footsteps: synthetic biology and the meaning of life. Nanoethics 3(3):257–268. link.springer.com/article/10.1007%2Fs11569-009-0079-6. https://doi.org/10.1007/s11569-009-0079-6
9. Hughes RA, Ellington AD (2017) Synthetic DNA synthesis and assembly: putting the synthetic in synthetic biology. Cold Spring Harb Perspect Biol 9(1):a023812. https://doi.org/10.1101/cshperspect.a023812
10. Chao R et al (2014) Recent advances in DNA assembly technologies. FEMS Yeast Res 15:1–9. https://doi.org/10.1111/1567-1364.12171
11. David F et al (2021) A perspective on synthetic biology in drug discovery and development—current impact and future opportunities. SLAS Discov 26(5):581–603. https://doi.org/10.1177/24725552211000669
12. Beitz AM et al (2022) Synthetic gene circuits as tools for drug discovery. Trends Biotechnol 40(2):210–225. https://doi.org/10.1016/j.tibtech.2021.06.007
13. Ausländer S, Fussenegger M (2013) From gene switches to mammalian designer cells: present and future prospects. Trends Biotechnol 31(3):155–168. https://doi.org/10.1016/j.tibtech.2012.11.006
14. Goñi-Moreno A, Amos M (2012) A reconfigurable NAND/nor genetic logic gate. BMC Syst Biol 6(1):126. https://doi.org/10.1186/1752-0509-6-126
15. Calero P, Nikel PI (2018) Chasing bacterial chassis for metabolic engineering: a perspective review from classical to non-traditional microorganisms. Microb Biotechnol 12(1):98–124. www.ncbi.nlm.nih.gov/pmc/articles/PMC6302729/. https://doi.org/10.1111/1751-7915.13292
16. Beites T, Mendes MV (2015) Chassis optimization as a cornerstone for the application of synthetic biology based strategies in microbial secondary metabolism. Front Microbiol 6:9. https://doi.org/10.3389/fmicb.2015.00906
17. Adams BL (2016) The next generation of synthetic biology chassis: moving synthetic biology from the laboratory to the field. ACS Synth Biol 5(12):1328–1330. https://doi.org/10.1021/acssynbio.6b00256
18. Ishino Y et al (2018) History of CRISPR-Cas from encounter with a mysterious repeated sequence to genome editing technology. J Bacteriol 200(7):10–1128. www.ncbi.nlm.nih.gov/pubmed/29358495. https://doi.org/10.1128/JB.00580-17
19. Jinek M et al (2012) A programmable dual-RNA-guided DNA endonuclease in adaptive bacterial immunity. Science 337(6096):816–821. https://doi.org/10.1126/science.1225829
20. Cong L et al (2013) Multiplex genome engineering using CRISPR/Cas systems. Science 339(6121):819–823. https://doi.org/10.1126/science.1231143
21. Han K et al (2017) Synergistic drug combinations for cancer identified in a CRISPR screen for pairwise genetic interactions. Nat Biotechnol 35(5):463–474. pubmed.ncbi.nlm.nih.gov/28319085/. https://doi.org/10.1038/nbt.3834
22. Sun D et al (2022) Why 90% of clinical drug development fails and how to improve it? Acta Pharm Sin B 12:3049–3062. https://doi.org/10.1016/j.apsb.2022.02.002
23. Scannell JW, Bosley J (2016) When quality beats quantity: decision theory, drug discovery, and the reproducibility crisis. PLoS One 11(2):e0147215. https://doi.org/10.1371/journal.pone.0147215
24. Fellmann C et al (2016) Cornerstones of CRISPR–Cas in drug discovery and therapy. Nat Rev Drug Discov 16(2):89–100. www.nature.com/articles/nrd.2016.238. https://doi.org/10.1038/nrd.2016.238

25. Single ascending dose study in participants with LCA10 – full text view. ClinicalTrials.gov. Clinicaltrials.gov, clinicaltrials.gov/ct2/show/NCT03872479
26. Editas Medicine. Corporate presentation for investors. Aug 2022. ir.editasmedicine.com/static-files/03350a40-33d7-413c-b2d6-c46f94ab273f. Accessed 23 Sept 2022
27. U.S. Food and Drug Administration (2024). FDA approves first gene therapies to treat patients with sickle cell disease. FDA Press Announcements. https://www.fda.gov/news-events/press-announcements/fda-approves-first-gene-therapies-treat-patients-sickle-cell-disease
28. A safety and efficacy study evaluating CTX001 in subjects with transfusion-dependent β-Thalassemia – full text view. ClinicalTrials.gov. Clinicaltrials.gov, clinicaltrials.gov/ct2/show/NCT03655678
29. Liu W et al (2021) Applications and challenges of CRISPR-Cas gene-editing to disease treatment in clinics. Precis. Clin. Med 4(3):179–191. https://doi.org/10.1093/pcmedi/pbab014
30. Newman DJ, Cragg GM (2020) Natural products as sources of new drugs over the nearly four decades from 01/1981 to 09/2019. J Nat Prod 83(3):12. https://doi.org/10.1021/acs.jnatprod.9b01285
31. Lin Z et al (2020) Bioprospecting through cloning of whole natural product biosynthetic gene clusters. Front Bioeng Biotechnol 8. https://doi.org/10.3389/fbioe.2020.00526
32. Li L et al (2021) Refactoring biosynthetic gene clusters for heterologous production of microbial natural products. Curr Opin Biotechnol 69:145–152. https://doi.org/10.1016/j.copbio.2020.12.011
33. Bharadwaj R et al (2021) Plant metabolic gene clusters: evolution, organization, and their applications in synthetic biology. Front. Plant Sci 12. https://doi.org/10.3389/fpls.2021.697318
34. Tran PN et al (2019) Detecting and prioritizing biosynthetic gene clusters for bioactive compounds in bacteria and fungi. Appl Microbiol Biotechnol 103(8):3277–3287. https://doi.org/10.1007/s00253-019-09708-z
35. Cimermancic P et al (2014) Insights into secondary metabolism from a global analysis of prokaryotic biosynthetic gene clusters. Cell 158(2):412–421. https://doi.org/10.1016/j.cell.2014.06.034
36. Chavali AK, Rhee SY (2018) Bioinformatics tools for the identification of gene clusters that biosynthesize specialized metabolites. Brief Bioinform 19(5):1022–1034. www.ncbi.nlm.nih.gov/pubmed/28398567. https://doi.org/10.1093/bib/bbx020
37. Wang H et al (2014) Atlas of nonribosomal peptide and polyketide biosynthetic pathways reveals common occurrence of nonmodular enzymes. Proc Natl Acad Sci U S A 111(25):9259–9264. www.ncbi.nlm.nih.gov/pmc/articles/PMC4078802/. https://doi.org/10.1073/pnas.1401734111
38. Raman S et al (2014) Evolution-guided optimization of biosynthetic pathways. Proc Natl Acad Sci U S A 111(50):17803–17808. https://doi.org/10.1073/pnas.1409523111
39. Liu OW et al (2022) Precision discovery of novel inhibitors of human cancer target HsMetAP1 from vast unexplored metagenomic diversity. https://doi.org/10.1101/2022.06.11.495772
40. Batta A et al (2020) Trends in FDA drug approvals over last 2 decades: an observational study. J Family Med Prim Care 9(1):105–114. www.ncbi.nlm.nih.gov/pmc/articles/PMC7014862/. https://doi.org/10.4103/jfmpc.jfmpc_578_19
41. de la Torre BG, Albericio F (2022) The pharmaceutical industry in 2021. An analysis of FDA drug approvals from the perspective of molecules. Molecules (Basel, Switzerland) 27(3):1075. pubmed.ncbi.nlm.nih.gov/35164339/. https://doi.org/10.3390/molecules27031075
42. Ryman JT, Meibohm B (2017) Pharmacokinetics of monoclonal antibodies. CPT Pharmacometrics Syst Pharmacol 6(9):576–588. www.ncbi.nlm.nih.gov/pmc/articles/PMC5613179/. https://doi.org/10.1002/psp4.12224
43. Mitra S, Tomar PC (2021) Hybridoma technology; advancements, clinical significance, and future aspects. J Genet Eng Biotechnol 19(18):159. www.ncbi.nlm.nih.gov/pmc/articles/PMC8521504/. https://doi.org/10.1186/s43141-021-00264-6
44. Lu R-M et al (2020) Development of therapeutic antibodies for the treatment of diseases. J Biomed Sci 27(1) jbiomedsci.biomedcentral.com/articles/10.1186/s12929-019-0592-z. https://doi.org/10.1186/s12929-019-0592-z

45. Charles A et al (2017) The structure of a typical antibody molecule. Nih.gov, Garland Science. www.ncbi.nlm.nih.gov/books/NBK27144/
46. The mouse that roared for human advantage. Www.nature.com, www.nature.com/articles/d42473-018-00011-5
47. Pedrioli A, Oxenius A (2021) Single B cell technologies for monoclonal antibody discovery. Trends Immunol 42(12):1143–1158. www.cell.com/trends/immunology/fulltext/S1471-4906(21)00213-1. https://doi.org/10.1016/j.it.2021.10.008
48. Zost SJ et al (2020) Rapid isolation and profiling of a diverse panel of human monoclonal antibodies targeting the SARS-CoV-2 spike protein. Nat Med 26(9):1422–1427. https://doi.org/10.1038/s41591-020-0998-x
49. Krawczyk K, Dunbar J, Deane CM (2017) Computational tools for aiding rational antibody design. In: Samish I, Rehovot I, Samish I (eds) Computational protein design. Methods in molecular biology, vol 1529. Humana Press, New York. https://doi.org/10.1007/978-1-4939-6637-0_21
50. Library of Libraries | Twist Bioscience. Www.twistbioscience.com, www.twistbioscience.com/resources/product-sheet/library-libraries. Accessed 19 Aug. 2022
51. vbpinheiro. DNA Library Synthesis for Directed Evolution. Synthetic Biology through Directed Evolution, 28 Apr 2020. pinheirolab.com/2020/04/28/dna-library-synthesis-for-directed-evolution/. Accessed 19 Aug 2022
52. Kayushin A (1996) A convenient approach to the synthesis of trinucleotide phosphoramidites—synthons for the generation of oligonucleotide/peptide libraries. Nucleic Acids Res 24(19):3748–3755. https://doi.org/10.1093/nar/24.19.3748
53. Yagodkin A et al (2007) Improved synthesis of trinucleotide phosphoramidites and generation of randomized oligonucleotide libraries. Nucleosides Nucleotides Nucleic Acids 26(5):473–497. https://doi.org/10.1080/15257770701426260
54. Meiser LC et al (2020) Reading and writing digital data in DNA. Nat Protoc 15(1):86–101. www.nature.com/articles/s41596-019-0244-5. https://doi.org/10.1038/s41596-019-0244-5
55. Mahdavi SZB et al (2022) An overview on display systems (phage, bacterial, and yeast display) for production of anticancer antibodies; advantages and disadvantages. Int J Biol Macromol 208:421–442. https://doi.org/10.1016/j.ijbiomac.2022.03.113
56. Alfaleh MA et al (2020) Phage display derived monoclonal antibodies: from bench to bedside. Front Immunol 11. https://doi.org/10.3389/fimmu.2020.01986
57. Kaleli NE et al (2019) Phage display derived therapeutic antibodies have enriched aliphatic content: insights for developability issues. Proteins Struct Funct Bioinf 87(7):607–618. https://doi.org/10.1002/prot.25685
58. Finck A (2022) Engineered cellular immunotherapies in cancer and beyond. Nat Med 28(19):678–689. https://doi.org/10.1038/s41591-022-01765-8
59. June CH et al (2018) CAR T cell immunotherapy for human cancer. Science 359(6382):1361–1365. https://doi.org/10.1126/science.aar6711
60. Mu W et al (2020) Engineering CAR T cells to target the HIV reservoir. Front Cell Infect Microbiol 10. https://doi.org/10.3389/fcimb.2020.00410
61. Wan YY, Flavell RA (2009) How diverse—CD4 effector T cells and their functions. J Mol Cell Biol 1(1):20–36. https://doi.org/10.1093/jmcb/mjp001
62. Janeway C (1999) Immunobiology: the immune system in health and disease, 5th edn. Harcourt Brace & Company, London
63. Pennock ND et al (2013) T cell responses: naïve to memory and everything in between. Adv Physiol Educ 37(4):273–283
64. Kuwana Y et al (1987) Expression of chimeric receptor composed of immunoglobulin-derived v resions and T-cell receptor-derived c regions. Biochem Biophys Res Commun 149(3):960–968. https://doi.org/10.1016/0006-291X(87)90502-X
65. Gross G et al (1989) Expression of immunoglobulin-T-cell receptor chimeric molecules as functional receptors with antibody-type specificity. Proc Natl Acad Sci 86(24):10024–10028.

www.ncbi.nlm.nih.gov/pmc/articles/PMC298636/pdf/pnas00291-0398.pdf. https://doi.org/10.1073/pnas.86.24.10024
66. Feins S et al (2019) An introduction to chimeric antigen receptor (CAR) T-cell immunotherapy for human cancer. Am J Hematol 94(S1):S3–S9. https://doi.org/10.1002/ajh.25418
67. Valerie NN (2022) The incredible story of Emily Whitehead & CAR T-cell therapy. Oncology Times 44(6):19–21. https://doi.org/10.1097/01.cot.0000824668.24475.b0
68. Sanber K et al (2021) Graft-versus-host disease risk after chimeric antigen receptor T-cell therapy: the diametric opposition of T cells. Br J Haematol 195(5):660–668. https://doi.org/10.1111/bjh.17544
69. Stock S et al (2022) Chimeric antigen receptor T cells engineered to recognize the P329G-mutated Fc part of effector-silenced tumor antigen-targeting human IgG1 antibodies enable modular targeting of solid tumors. J ImmunoTher Cancer 10(7):e005054. jitc.bmj.com/content/10/7/e005054. https://doi.org/10.1136/jitc-2022-005054
70. Rafiq S et al (2020) Engineering strategies to overcome the current roadblocks in CAR T cell therapy. Nat Rev Clin Oncol 17(3):147–167. www.ncbi.nlm.nih.gov/pmc/articles/PMC7223338/. https://doi.org/10.1038/s41571-019-0297-y
71. Graham C (2018) Allogeneic CAR-T cells: more than ease of access? Cells 7(10):155. https://doi.org/10.3390/cells7100155
72. Torikai H (2016) Translational implications for off-the-shelf immune cells expressing chimeric antigen receptors. Mol Ther 24(7):1178–1186. https://doi.org/10.1038/mt.2016.106
73. Research, Center for Biologics Evaluation and KYMRIAH (Tisagenlecleucel). FDA, 5 Apr 2019. www.fda.gov/vaccines-blood-biologics/cellular-gene-therapy-products/kymriah-tisagenlecleucel
74. Wang LL-W et al (2021) Cell therapies in the clinic. Bioeng Transl Med 6(2). https://doi.org/10.1002/btm2.10214
75. Marofi F et al (2021) CAR-NK cell: a new paradigm in tumor immunotherapy. Front Oncol 11. https://doi.org/10.3389/fonc.2021.673276
76. Xie G et al (2020) CAR-NK cells: a promising cellular immunotherapy for cancer. EBioMedicine 59., www.thelancet.com/journals/ebiom/article/PIIS2352-3964(20)30351-0/fulltext. https://doi.org/10.1016/j.ebiom.2020.102975
77. Replogle JM et al (2022) Mapping information-rich genotype-phenotype landscapes with genome-scale perturb-Seq. Cell. https://doi.org/10.1016/j.cell.2022.05.013

Part III
Oligonucleotide Synthetic Chemistry to DNA Synthesis, Bioprocessing and Manufacturing

Chapter 3
Increasing the Scalability of DNA Synthesis and Its Key Role in Expanding the Biopharmaceutical Discovery Process

Rebecca L. Nugent and Aaron K. Sato

Abstract DNA synthesis is a naturally occurring process that happens during DNA replication. In the mid-twentieth century, a chemical process was developed to synthesize single-stranded oligonucleotides. By the early 1980s, multiple commercial companies were selling synthetic DNA. Gene synthesis (i.e., DNA assembly) methods were developed in parallel to generate long double-stranded DNA (dsDNA) fragments from many short, synthetic oligonucleotides. These classical methods have been improved upon and miniaturized over the past few decades, culminating in modern gene synthesis platforms that can simultaneously synthesize ~10,000 genes where legacy gene synthesizers could synthesize only one. In this chapter, we review the key advances that enabled modern-scale gene synthesis and how they set the stage for an expanded role for synthetic gene fragments in antibody biopharmaceutical development. We conclude that, in this world of nearly unlimited gene synthesis capacity, the test and learn phases of the design-build-test-learn cycle have supplanted the build phase as the new bottlenecks in biopharmaceutical development.

Keywords Solid-phase phosphoramidite chemistry · Oligonucleotide · Gene synthesis · Semi-synthetic antibody library · Fully synthetic antibody library

3.1 Introduction

> Every step of this synthesis has obviously been executed with consummate skill and the whole constitutes perhaps the greatest *tour de force* organic and biochemists have yet achieved. Like NASA with its Apollo programme, Khorana's group has shown it can be done, and both feats may well never be repeated.
>
> – An anonymous cell biology correspondent on the first synthetic genes [1]

R. L. Nugent
Tessera Therapeutics, Somerville, MA, USA

A. K. Sato (✉)
Twist Bioscience, South San Francisco, CA, USA
e-mail: asato@twistbioscience.com

K. Gadamasetti, S. A. Kolodziej (eds.), *Bioprocessing, Bioengineering and Process Chemistry in the Biopharmaceutical Industry*,
https://doi.org/10.1007/978-3-031-62007-2_3

Writing for *Nature New Biology* in the seventies, the anonymous correspondent captured a common sentiment of the time [2]: the first synthetic genes, like the Apollo missions, were outstanding achievements in and of themselves, but little more. Thankfully, the chemists persisted, inventing the foundational methods that underlie modern gene synthesis. The invention of the phosphoramidite method of oligo synthesis came at a time of explosive innovation in molecular biology that also included the invention of the polymerase chain reaction (PCR) and recombinant DNA technology. Gene synthesis—or the assembly of chemically synthesized oligos into double-stranded DNA (dsDNA) fragments—forms a substantial part of the bedrock of modern biopharmaceutical discovery processes, as exemplified by the advent of massive, fully synthetic antibody libraries.

In this chapter, we review the progress that has been made toward realizing the promise of gene synthesis in the biopharmaceutical industry. More specifically, we discuss how the foundational methods that were developed and commercialized in the twentieth-century spawned modern, large-scale gene synthesis platforms. We also outline how the overall scaling up and automation of gene synthesis over the years has transformed the biopharmaceutical discovery process. We conclude that these high-throughput technologies have shifted the biopharmaceutical discovery bottleneck from the fabrication of building blocks (gene synthesis) to the expression, purification, and testing of biopharmaceuticals, enabling a new era in antibody discovery.

3.2 Evolution of Gene Synthesis

Gene synthesis encapsulates two discrete processes: the de novo synthesis of short single-stranded DNA oligonucleotides (oligos) and their subsequent assembly into gene-length dsDNA fragments. The technologies for each process were pioneered concurrently; in fact, the first synthetic genes were reported before the development of the solid-phase phosphoramidite chemistry that dominates modern oligo synthesis platforms. In this section, we review the key advances that inspired the gene synthesis technologies used today.

3.2.1 Oligonucleotide Synthesis

Despite being developed by chemists, oligo synthesis technologies largely benefit biologists. The evolution of oligo synthesis largely reflects an effort by chemists to make synthetic DNA accessible to biologists. Whereas the earliest oligos were manually synthesized by skilled chemists, modern oligos are produced by highly automated machines. The automation of modern oligo synthesis was made possible by the development of a solid-phase phosphoramidite method.

3.2.1.1 Early Synthetic Chemistries

Sir Alexander Todd described the first chemical dinucleotide synthesis, using a phosphodiester approach, in 1955 [3]. The phosphodiester approach became the dominant strategy for the next two decades even as additional chemistries became available, including the phosphotriester and H-phosphonate chemistries [4]. Indeed, Khorana and colleagues employed phosphodiester chemistry to elucidate the genetic code in the 1960s and to synthesize oligo building blocks for the first synthetic genes in the 1970s [5, 6]. Despite its early success, the approach was only viable in the hands of skilled chemists, precluding its automation [2].

In the mid-1970s, Letsinger and colleagues debuted a new, highly efficient oligo synthesis approach based on phosphite triester chemistry [7]. This strategy enabled rapid oligo synthesis through the use of highly reactive nucleosides (3′-O-chlorophosphites). Although these intermediates hastened the synthesis reaction, they were also difficult to handle [4]. Beaucage and Caruthers [8] addressed this problem by developing the phosphoramidite chemistry that forms the basis of modern chemical oligo synthesis platforms.

3.2.2 Solid Supports

Oligos can be synthesized in solution or covalently attached to solid supports (i.e., solid-phase synthesis). These methods differ substantially in the time and resources they require [2]. Whereas solution-based synthesis requires extensive purification at multiple steps in the synthesis cycle, solid-phase synthesis eliminates these time-consuming steps entirely by providing a simple means of separating oligos from reagents, solvents, and by-products. In addition, solid-phase synthesis supports the synthesis of the microgram amounts of DNA—an adequate, yet not excessive, amount for a typical biological experiment.

The concept of solid-phase synthesis originated with Merrifield, who applied it to the synthesis of peptides [9]. The idea—and the same exact solid support—was soon adapted to oligo synthesis by Letsinger and Mahadevan [10]. In this early iteration, a nonpolar "popcorn" polystyrene resin provided a solid support matrix that was insoluble in the solvents and solutions commonly used for oligo synthesis. Another organic polymer, polyamide resin, was later found to be compatible with the polar solvents used in phosphodiester and phosphotriester chemistries [11]. Inorganic matrices, including silica gel, were attempted by Koster and colleagues in 1972 [12], albeit with minimal success initially [2]. These early solid support matrices were nonideal, as they tended to impede the diffusion of reagents either via swelling (organic polymers) or fracturing (silica gel) [11].

Interest in silica-based supports—namely, controlled-pore glass (CPG)—renewed in the 1980s alongside the development of phosphite triester and phosphoramidite chemistries [4]. Unlike earlier supports, CPG was not prone to compaction, making it ideal for use under the continuous flow conditions used in automated synthesizers.

3.2.3 Modern Oligo Synthesis Platforms

Modern oligo synthesis platforms invariably use solid-phase phosphoramidite chemistry due to its amenability to automation [2]. Synthesizing DNA with phosphoramidite reaction cycles starts when a phosphoramidite is attached to a solid surface in a catalyzed condensation reaction. An oxidation step follows whereby the unstable phosphite triester is converted to a phosphate with the goal of improving sequence integrity (Fig. 3.1). Next, deblocking occurs such that the 5′ protecting group is removed (using trichloroacetic acid). Then coupling and capping occur when the next phosphoramidite in the chain is coupled to the available 5′-hydroxyl group on the previously deblocked molecule in a catalyzed reaction. Coupling is not 100% efficient, and sometimes the coupling fails. Because uncoupled sequences could create deletion errors in the synthesized molecule, an unreactive group is added to prevent the further extension of these sequences. Oligos are repeatedly cycled through these oxidation, deprotection, and coupling steps until the desired length(s) and sequence(s) is (are) achieved.

The first automated synthesizers used columns packed with CPG beads. Each column supports the synthesis of a single sequence, but high-throughput plate synthesizers can operate up to 1536 columns at once (although 96–384 columns is more common) [13]. Such synthesizers are limited in the length (<100 nucleotide [nt]) and sequence fidelity (0.5% error rate) they can produce as a result of even minor

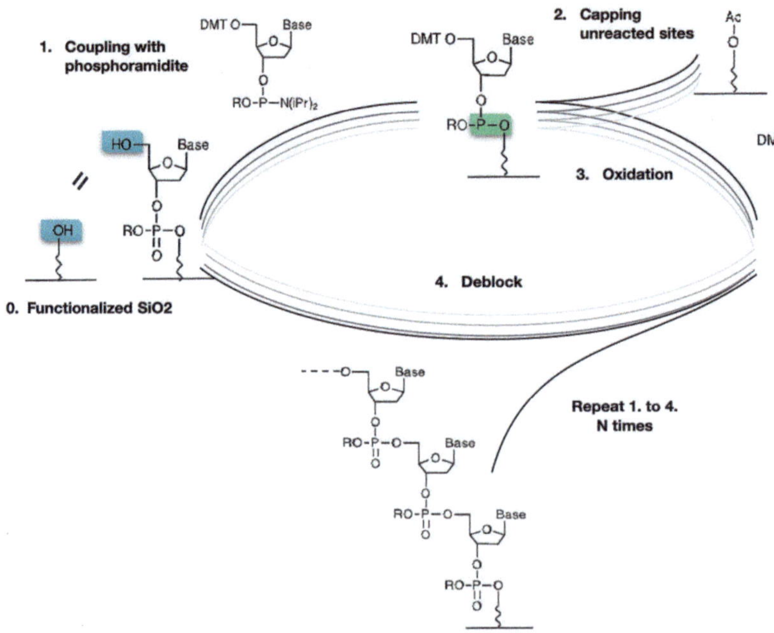

Fig. 3.1 Phosphoramidite chemistry for oligonucleotide synthesis

inefficiencies in the process [14]. Even near-perfect (99%) yield at each cycle can culminate in a poor overall yield for longer synthesis lengths.

The throughput of oligo synthesis dramatically increased with the advent and commercialization of DNA microarrays beginning in the nineties [15]. Oligo synthesis by microarray was made possible by the development of technologies that enabled spatial control over the synthesis reaction. Affymetrix was the first to develop light-activated chemistries for the synthesis of oligos on microchip surfaces [16, 17]. NimbleGen and LC Sciences simplified these light-activated chemistries by using programmable micromirror devices to direct light in lieu of photolithography masks [18, 19]. CombiMatrix (subsequently CustomArray and now Genscript) and Agilent commercialized semiconductor-based [20] and inkjet printing-based [21] oligo synthesis technologies, respectively, abrogating the need for expensive processes associated with light-activated chemistries [13]. These later high-throughput technologies effectively reduced the cost of oligo synthesis from approximately $0.05–0.15 per nucleotide (using columns) to a mere $0.00001–0.0001 per nucleotide.

Microarray-based synthesis technologies continue to be innovated to this day. One area of innovation has been in the control of the depurination side reactions, as demonstrated by LeProust and colleagues [22]. Depurination occurs when adenine and guanine bases are hydrolyzed as a result of prolonged exposure to deprotecting/detritylation agents during synthesis [23]. This side reaction effectively incorporates an apurinic base without interfering with chain elongation. Deprotection of the final base also triggers the cleavage of abasic sites created by depurination, generating truncated sequences. LeProust and colleagues solved this problem by optimizing the flow of reagents within the Agilent flow cell [22].

Increasing the efficiency of the stepwise, phosphoramidite-based oligo synthesis reaction represents a second area of innovation. Patterned substrates (for light-based chemistries) [24, 25] and highly patterned chips (for inkjet synthesizers) [26] have provided better control over spatially controlled steps in synthesis reaction, reducing the number of errors caused by misalignment. The efficiency of the subsequent coupling and capping steps has also improved. A coupling efficiency of >99.5% is a requirement for the synthesis of long oligos, and even small (~0.1%) improvements to this metric can enable synthesis of longer oligos with fewer errors. (Conversely, even small reductions in coupling efficiency can dramatically decrease the length at which quality oligos can be synthesized.) These outsized effects reflect the compounding of miniscule inefficiencies across the stepwise synthesis of long oligos. In other words, as the length of the synthesized oligo increases, the more apparent small differences in reaction efficiencies become. Today's high-throughput synthesizers are highly efficient, reaching coupling efficiencies as high as 99.9% (Twist Bioscience).

As of 2024, multiple companies offer oligo pool products synthesized by microarray technologies; notable examples include Integrated DNA Technologies, Genscript, Agilent, and Twist Bioscience. Across these commercial synthesis platforms, the synthesis of hundreds of thousands of unique oligos on a single chip has become routine. Although the product specifications of the oligos synthesized by these commercial microarray-based platforms vary widely, lengths of up to 350 nt have been attained, more than doubling the ~150 nt achieved a decade prior [22]. Error rates as low as 1 in 2000 nt are also possible with today's synthesizers. The oligo lengths achieved by modern oligo synthesis platforms, however, will inevitably plateau as phosphoramidite reaction efficiencies approach 100%. Nevertheless, synthesis density—how many oligos can be synthesized per chip—is expected to grow with the continued miniaturization of synthesis platforms.

3.3 Gene Synthesis

As discussed above, oligo synthesis methods, although substantially improved over the years, still struggle to synthesize oligos longer than a few hundred nucleotides. As a result, methods were developed to stitch short, chemically synthesized oligos into dsDNA gene fragments, called gene synthesis. Although the first successful attempts at gene synthesis utilized T4 DNA ligase to stitch oligos together sequentially, the discovery of new enzymes (thermostable ligases) and technologies (PCR) paved the way for one-pot gene syntheses in the subsequent decades. These strategies, subject to continued innovation within the context of array-based oligo synthesis, form the basis of modern commercial gene synthesis platforms.

3.3.1 Early DNA Assembly Methods

The first synthetic gene took more than 5 years to synthesize. Working with limiting synthetic chemistries, Khorana and colleagues generated the first synthetic genes by chemically synthesizing, annealing, phosphorylating, and ultimately ligating a series of short (8–12 nt) oligos [27, 28]. Without thermostable ligases, Khorana's group resorted to sequential ligation reactions to tack on each chemically synthesized oligo. One-pot, ligation-dependent gene synthesis was not demonstrated until 1998 [29]. Ligation-based DNA assembly methods have inherently low error rates, unlike the PCR-based methods that have largely, but not completely, superseded them. At least two commercial options for ligation-based gene synthesis exist today: Blue Heron Biotechnology (acquired by Eurofins Genomics in 2019) and Sloning BioTechnology GmbH (acquired by MorphoSys AG in 2010).

The invention of PCR in the 1980s spawned multiple PCR-based methods for DNA assembly. Yet, to this day, Willem Stemmer's polymerase chain assembly (PCA) method [30] remains the most common among them (Fig. 3.2). PCA, also called oligo shuffling, harnesses the ability of DNA polymerases to fill in gaps in complementary single-stranded DNA by PCR extension. DNA polymerases randomly form larger and larger assemblies during thermocycling, and the final assembly, once formed, is ultimately amplified by PCR using primers unique to each end of the fully assembled DNA fragment. Stemmer and colleagues leveraged this

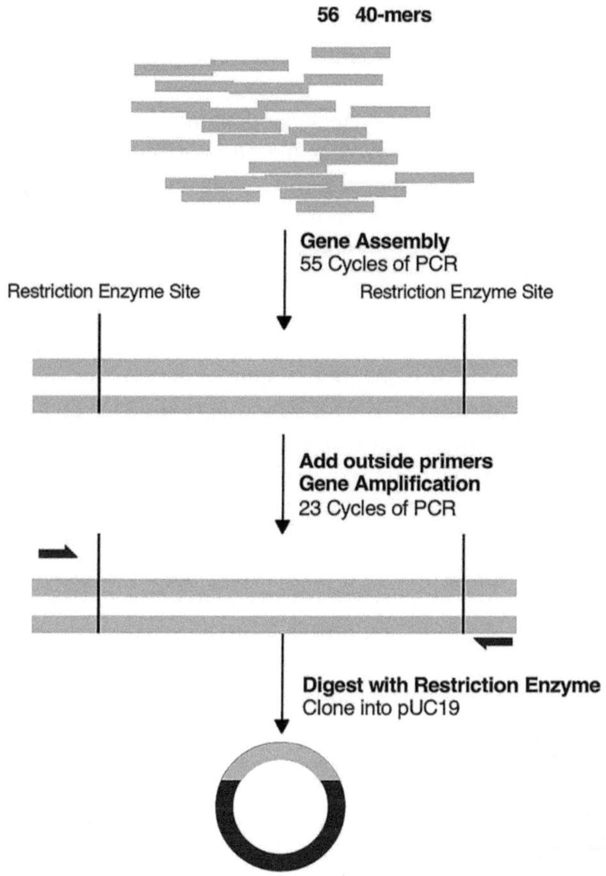

Fig. 3.2 Classic gene synthesis. (Adapted from Stemmer et al. [30])

method to assemble as many as 134 chemically synthesized oligos into DNA fragments up to 2.7 kb in length [30]. The numerous other PCR-based assembly methods have been reviewed in detail elsewhere [31] and will not be discussed here, except to note that they largely involve a subassembly approach whereby smaller fragments are joined first into subassemblies, which are subsequently stitched together into full-length gene fragments.

3.3.2 Array-Based Gene Synthesis

By necessity, early gene synthesis attempts used column-synthesized oligos as building blocks. The advent of microarray-based oligo synthesizers promised increased throughput while posing new challenges for gene synthesis. Compared to column-synthesized oligos, array-synthesized oligos are produced as a complex mixture of low-quantity (femtomolar) and relatively low-quality oligos. These features—complexity, quantity, and quality—have complicated gene synthesis using array-synthesized oligos.

Current microarray oligo synthesizers can generate one million oligos on a single chip, with each oligo synthesized at the atto- or femtomolar scale. Although these minute quantities reduce the overall cost of oligo synthesis by minimizing reagent use, they are insufficient for most DNA assembly methods. Moreover, the complexity of the synthesized oligo pools can lead to unintended hybridization of oligos during assembly. Two general strategies have been deployed to solve these concentration and complexity problems: amplifying oligos by PCR and miniaturizing the volume of the assembly reaction.

Amplifying oligos by PCR is the more common method for increasing the concentration of array-synthesized oligos for gene synthesis. Tian et al. [32] solved this problem by attaching a common primer sequence and a nicking endonuclease recognition site to the 3′-end of synthesized oligos. Once synthesized, oligos could be amplified from the 3′-end by primer extension and the resulting complementary strands cleaved by a nicking endonuclease, effectively amplifying the array-synthesized pool. The strategy, however, fails when applied to more complex (>1000 oligo) pools [33]. Kosuri et al. [34] conceived a barcoding strategy whereby primer sites were appended to each end of a subset of oligos to enable their selective amplification. This strategy enabled multiple subpools to be amplified from the master oligo pool. Although these PCR-based methods are straightforward, they are also error-prone. Not only is PCR inherently error-prone, but extending the length of oligo synthesis to incorporate primer sites can increase the error rate of the synthesis process, too.

Miniaturizing the assembly reaction is a less common, but nonetheless effective, solution to the concentration problem. It also addresses the complexity problem. Kong et al. [35] demonstrated the feasibility of this approach by partitioning oligo synthesis and assembly into 500 nL microfluidic reactors. Quan et al. [26] utilized an inkjet printing approach to spatially control the synthesis of oligos into physically separated microwells for subsequent gene synthesis.

Overall, the success of these strategies has led to their commercialization, in principle, by Gen9 (now Ginkgo Bioworks) and Twist Bioscience.

3.3.3 Error Correction and Sequence Validation

If oligo synthesis were a perfect process, gene synthesis would be straightforward and relatively quick. Unfortunately, it is not. Errors in oligo synthesis lead to errors in gene synthesis, as error-containing oligos are assembled along with perfect (error-free) ones. Traditionally, error-free gene fragments were identified by cloning and sequencing individual clones. This laborious and expensive process serves as a bottleneck in the discovery workflow. Because biological features are encoded by DNA, even minor errors can have an outsized impact on the functionality of the final product. The quality of array-synthesized oligos is generally lower than that of their column-synthesized counterparts, necessitating methods for error filtration and correction.

The process of array-based gene synthesis spans multiple error-prone steps. Array-based oligo synthesizers can generate errors as a result of inefficiencies in the synthesis cycle, side reactions (i.e., depurination), and deviations in spatially controlled steps (as discussed in Sect. 3.2.3 Modern Oligo Synthesis Platforms). DNA assembly methods, particularly the PCR-based ones in common use, are as error-prone as the DNA polymerases they use. To mitigate the impact of these error sources on gene synthesis, two general strategies have been developed to filter error-free oligos from those containing errors: those that use mismatch-binding/cleaving proteins to remove errors and those that enrich for error-free DNA. These strategies are discussed briefly below; see Ma et al. [23] for an in-depth review on the topic.

Mismatch detection is commonly used to filter errors in synthetic genes (i.e., duplexes). This strategy involves using a mismatch-binding/cleaving protein to either filter out or degrade, respectively, DNA assemblies containing errors. Mismatches are generated by denaturing the DNA duplexes and reannealing them; this process causes error-containing strands to randomly pair with error-free strands. Once this is done, mismatch-binding proteins (e.g., MutS) [36] or mismatch-cleaving enzymes (e.g., T7 endonuclease 1) [37, 38] can be used to selectively isolate or degrade error-containing duplexes, respectively. The choice of strategy depends on the length of the gene fragments and how many errors they contain; mismatch cleavage is generally preferred for both long and error-rich fragments [39].

Several groups have used next-generation sequencing (NGS) to filter error-containing gene fragments. Among them, the most accessible is dial-out PCR [40, 41], which involves appending primer sites to the ends of each gene fragment, identifying error-free fragments by sequencing, and selectively amplifying them by PCR. This approach is highly cost-efficient but time-consuming. Other sequencing-based methods have been proposed, but they are either low-throughput, incompatible with some NGS platforms, or require specialized equipment [15].

3.4 New Discovery Bottleneck

Modern DNA reading (sequencing) and writing (synthesis) technologies allow bioengineering to be approached more systematically. This systematic approach, referred to as the design-build-test-learn (DBTL) cycle, starts by modeling a bioprocess, including all the parts (e.g., oligos and gene fragments) and processes that will be used to produce and evaluate a bioengineered product; proceeds through the building and testing of the bioengineered product; and ends with a learning phase, where insights gained are used to inform the next iteration of the cycle. Traditionally, low-throughput oligo synthesis has bottlenecked the cycle at the build phase. Now, with modern gene synthesis platforms like Twist Bioscience's, which can synthesize nearly 10,000 gene fragments (or >one million oligos) on a silicon chip the size of a 96-well plate, the bottleneck has been shifted to the testing and learning phases of the cycle.

In the next section, we discuss how modern gene synthesis technologies have transformed the antibody discovery process by enabling the fabrication of massive synthetic antibody libraries.

3.4.1 Evolution of Antibody Discovery

Antibodies represent the largest class of biopharmaceuticals on the market. The first therapeutic monoclonal antibody (mAb), muromonab-CD3 (Orthoclone OKT3; Janssen-Cilag), entered the market in 1986 and was developed using the traditional hybridoma method [42]. The first therapeutic *human* mAb came a decade and a half later in 2002 with the approval of adalimumab (Humira; Abbott), a feat that was made possible by the invention of phage display technology [43]. Spurred by the concurrent development of molecular biology methods (e.g., molecular cloning, oligo synthesis, and gene synthesis), mAb-generating technologies have matured over the past half century. In this section, we discuss how modern gene synthesis platforms have expanded and hastened the therapeutic antibody discovery process.

3.4.1.1 Hybridoma Technology

The year 1975 saw the emergence of a breakthrough technology for the facile production of mAbs: hybridomas [44]. George Köhler invented the technology in César Milstein's lab by fusing mortal B-cells with transformed ones (i.e., myelomas) [45]. Unlike the myeloma–myeloma fusions attempted by Milstein previously, Köhler's hybridomas produced highly specific antibodies of a single specificity. Hybridomas unleashed mAbs onto the world, contributing to their pervasiveness in the laboratory and clinical settings.

Hybridomas continue to be used to this day. As of 2020, greater than 90% of therapeutic mAbs that have received regulatory approval were generated using hybridoma technology. Despite this clear success, using hybridomas to produce therapeutic mAbs can be time-consuming and costly. Also, the technology cannot be used to generate mAbs against toxic or nonimmunogenic antigens. Cellular stability and lack of direct access to antibody genes are additional challenges faced by hybridoma technology [46].

3.4.1.2 Phage Display Technology

The invention of molecular technologies such as PCR and recombinant DNA gave rise to the field of antibody engineering in the late 1980s. Skerra and Pluckthun [47] were the first to express recombinant antibody fragments in *Escherichia coli*.

Shortly after, Orlandi et al. [48] demonstrated that antibody domains could be isolated using PCR without a priori knowledge of their coding sequences. The first antibody libraries followed, generated via PCR by Ward et al. [49] and then by Huse et al. [50]. These studies demonstrated that antibody genes could be isolated and screened for antigen binding in a library format, laying the foundation for the invention of phage display technology by McCafferty et al. [51] a year later. The invention spurred the development of hundreds of antibody libraries over the next decade and a half. By 2006, over 30 companies were using phage display in their discovery platforms [52].

The invention of phage displays substantially increased the throughput of antibody discovery. Unlike the laborious process of screening individual hybridoma clones, phage display enables the screening of naïve, immunized, or synthetic repertoires en masse. Phage display involves fusing the genes encoding antibody fragments and phage coat proteins together. This process results in the display of antibody fragments on the coat protein surface of phages. The resulting phages can be subsequently screened for antigen binding by iteratively enriching them on antigen-immobilized plates. The process is highly amenable to automation, much more so than hybridoma technology [53]. It is also easily interfaced with high-throughput gene synthesis technologies.

3.4.1.3 Synthetic Antibody Library Construction

The diversification processes used to generate diverse antibody libraries mimic the natural processes that occur in situ [54]. The natural process starts with the combinatorial recombination of V, D, and J gene segments in B-lymphocytes and ends with affinity maturation via somatic hypermutation in response to antigenic stimulation. Similarly, the recombinant process involves combinatorial shuffling of cloned, semi-synthetic, or synthetic gene fragments within established antibody frameworks to obtain an antibody library. Subsequent affinity maturation occurs during phage display. The affinity of antibody candidates can be matured further by generating a secondary library based on candidates identified from an initial library, although this is not necessary. Indeed, antibody fragments with picomolar affinities are routinely isolated from antibody libraries by phage display without the need for additional affinity maturation [53].

Unlike natural and immunized antibody libraries, synthetic antibody libraries incorporate synthetic sources of diversity in their complementarity-determining regions (CDRs). Synthetic diversity can be sourced from degenerate primers, synthetic trinucleotides, and oligo pools of discrete sequences. The application of these methods to the construction of semi- and fully synthetic antibody libraries is outlined below.

Semi-Synthetic Libraries

The first semi-synthetic antibody libraries obtained their CDR diversity from degenerate primers [55, 56]. These libraries were constructed by shuffling a bank of germline heavy chain variable regions (V_H), which spanned CDRs 1 and 2; adding a third, diversified CDR (CDR3) by PCR amplification using NNK/NNS degenerate primers (N = any nucleotide; $K = G/T$; $S = G/C$); and cloning the resulting V_H and a human light chain (V_L) as scFv fragments (Fig. 3.3). Later degenerate antibody libraries improved upon these initial libraries by minimizing the diversity of certain residues to that found in natural antibodies [57] or by leveraging favorable antibody frameworks [58]. These early libraries were relatively modest in size, at least by present-day standards, at ~10^8 clones each. Although degenerate primers provide a straightforward and cost-effective method for diversifying CDRs, the use of degenerate codons limited the diversity that could be obtained at each position within the CDR and this diversity often did not exactly match the diversity seen in natural human sequences [46].

Fully Synthetic Libraries

The first *fully* synthetic antibody (and fully human) library was actually a set of libraries called the Human Combinatorial Antibody Libraries (HuCAL), constructed by Knappik et al. [59] using the trinucleotide phosphoramidite method (also known

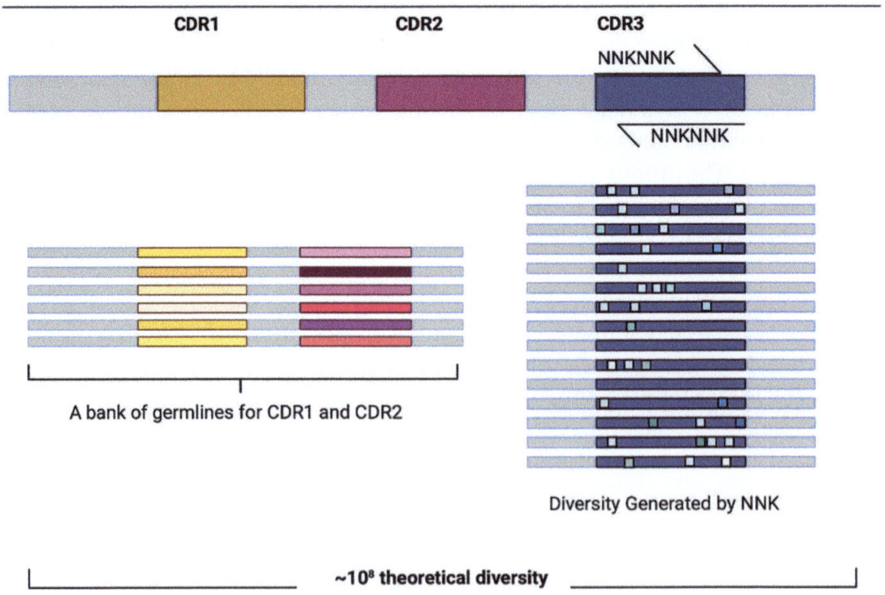

Fig. 3.3 Semi-synthetic libraries created by NNK

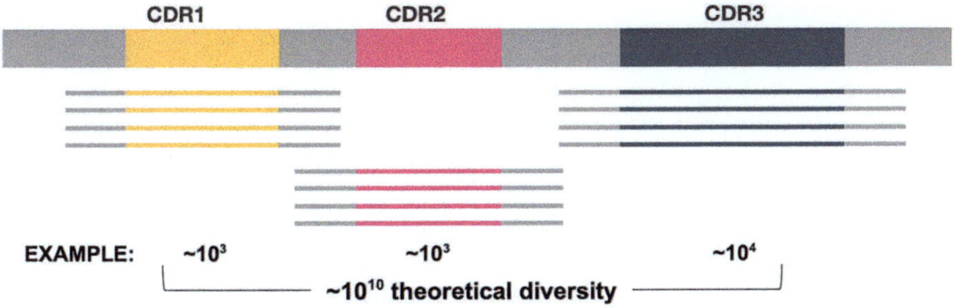

Fig. 3.4 High-diversity antibody libraries created by overlap PCR

as trinucleotide-directed mutagenesis [TRIM]). Instead of using PCR to diversify CDR3, Knappik et al. [59] sourced diversity from oligos synthesized from trinucleotide phosphoramidites. The final libraries were generated by stitching together synthetic oligos by overlap-extension PCR (Fig. 3.4). Although the original HuCAL design diversified only HCDR3 and LCDR3, a later iteration (HuCAL GOLD) synthetically diversified all six CDRs [60]. HuCAL and HuCAL GOLD comprised 2×10^9 and 3.6×10^{10} clones, respectively. HuCAL and HuCAL GOLD are commercially available for antibody discovery via MorphoSys AG, who developed them.

Sloning BioTechnology GmbH (since acquired by MorphoSys AG) debuted a new gene synthesis strategy—a ligation-based technology called Slonomics—in 2008 [61], providing a new method for antibody library construction. The method uses a set of common building blocks to synthesize DNA sequences three nucleotides at a time by iterating a series of ligation, immobilization, and cleavage steps. The first library generated by this approach, a Fab library, was as massive as HuCAL GOLD (3.6×10^{10}) and exhibited more favorable expression in phages than other, similarly diverse synthetic libraries [62]. Compared to the synthetic diversification methods that preceded it, Slonomics offers more control over the amino acid composition at each residue position while avoiding the introduction of PCR-related errors. The length of gene fragments synthesized by this method, however, is capped at 462 bp. Gene synthesis occurs in parallel using 96-well plates.

Innovations in microarray-based gene synthesis and antibody library technologies have converged to enable the construction of fully synthetic antibody libraries at unprecedented scales. Twist Biopharma, a division of Twist Bioscience, has capitalized on the massive throughput of the Twist synthesis platform to create the largest collection of fully synthetic antibody libraries to date, totaling 15 distinct libraries of ~10^{10} size each, with more libraries on the horizon. Collectively, these libraries span multiple antibody formats (scFv, Fab, VHH), diversification strategies, and target classes, providing rich sources of unique antibodies for antibody drug development.

3.5 Perspectives

Modern gene synthesis platforms have ushered in a new era of biopharmaceutical development by enabling the construction of unprecedented antibody libraries. Our goal in writing this chapter was not to comprehensively review the development of gene synthesis and antibody screening technologies. For those interested, many such reviews have been cited within this chapter. Rather, our goal was to describe how select technological achievements led to the expansion of the biopharmaceutical discovery process to its current scale. The methods that contributed to the first successes in biopharmaceutical development have been superseded by arguably superior and definitively more expansive technologies.

Recently, the combination of gene synthesis, automated experimentation, and faster and cheaper sequencing technologies has led many in the field to take advantage of machine learning to generate discoveries from ever-growing datasets. In one example, Mason and colleagues [63] developed a neural network from library screening and site-directed deep-sequencing data obtained from mammalian cells. They subsequently used this model to predict the specificity of trastuzumab variants to human epidermal growth factor receptor 2 (HER2). By narrowing down the original unfiltered antibody library to HER2-specific binders, Mason et al. were able to greatly reduce the cost of antibody screening while generating a more successful library. The success of this approach bodes well for the marriage of computational and gene synthesis techniques in antibody discovery, a combination we believe offers nearly limitless potential. We believe the continued co-development of these technologies will spur a renaissance in biopharmaceutical discovery.

Still, there is much to learn when it comes to gene synthesis and antibody discovery. High secondary structure, repetitive sequences, and unknowingly toxic sequences continue to complicate the fabrication of genetic parts for discovery workflows [13]. Moreover, our ability to generate antibodies of diverse formats, specificities, and naturalnesses has outstripped our ability to understand how these new engineered biomolecules behave in complex biological systems such as the human body. Developability—ensuring that antibody candidates meet basic requirements for clinical utility, including expression, stability, and solubility—also remains a challenge, albeit one that has been partially addressed by the advent of synthetic antibody libraries. Moving forward, biopharma will need to leverage insights gained from the latest synthetic antibody libraries to inform the development of the next generation of antibody therapeutics.

References

1. Anonymous (1973) Is enough too much? Nat New Biol 241:33
2. Caruthers MH (2011) A brief review of DNA and RNA chemical synthesis. Biochem Soc Trans 39:575–580
3. Michelson AM, Todd AR (1955) Nucleotides part XXXII. Synthesis of a dithymidine dinucleotide containing a 3′: 5′-internucleotidic linkage. J Chem Soc:2632–2638

4. Beaucage SL, Iyer RP (1992) Advances in the synthesis of oligonucleotides by the phosphoramidite approach. Tetrahedron 48:2223–2311
5. Khorana HG (1971) Total synthesis of the gene for an alanine transfer ribonucleic acid from yeast. Pure Appl Chem 25:91–118
6. Khorana HG, Büchi H, Caruthers MH et al (1968) Progress in the total synthesis of the gene for ala-tRNA. Cold Spring Harb Symp Quant Biol 33:35–44
7. Letsinger RL, Finnan JL, Heavner GA, Lunsford WB (1975) Nucleotide chemistry. XX. Phosphite coupling procedure for generating internucleotide links. J Am Chem Soc 97:3278–3279
8. Beaucage SL, Caruthers MH (1981) Deoxynucleoside phosphoramidites—a new class of key intermediates for deoxypolynucleotide synthesis. Tetrahedron Lett 22:1859–1862
9. Merrifield RB (1962) Peptide synthesis on a solid polymer. Fed Proc 21:412
10. Letsinger RL, Mahadevan V (1965) Oligonucleotide synthesis on a polymer support. J Am Chem Soc 87:3526–3527
11. Pon RT (1993) Solid-phase supports for oligonucleotide synthesis. In: Agrawal S (ed) Protocols for oligonucleotides and analogs: synthesis and properties. Humana Press, Totowa, pp 465–496
12. Köster H (1972) Polymer support oligonucleotide synthesis VI use of inorganic carriers. Tetrahedron Lett 13:1527–1530
13. Hughes RA, Ellington AD (2017) Synthetic DNA synthesis and assembly: putting the synthetic in synthetic biology. Cold Spring Harb Perspect Biol 9. https://doi.org/10.1101/cshperspect.a023812
14. Kosuri S, Church GM (2014) Large-scale de novo DNA synthesis: technologies and applications. Nat Methods 11:499–507
15. Kuhn P, Wagner K, Heil K et al (2017) Next generation gene synthesis: from microarrays to genomes. Eng Life Sci 17:6–13
16. Fodor SP, Read JL, Pirrung MC et al (1991) Light-directed, spatially addressable parallel chemical synthesis. Science 251:767–773
17. Pease AC, Solas D, Sullivan EJ et al (1994) Light-generated oligonucleotide arrays for rapid DNA sequence analysis. Proc Natl Acad Sci USA 91:5022–5026
18. Singh-Gasson S, Green RD, Yue Y et al (1999) Maskless fabrication of light-directed oligonucleotide microarrays using a digital micromirror array. Nat Biotechnol 17:974–978
19. Gao X, LeProust E, Zhang H et al (2001) A flexible light-directed DNA chip synthesis gated by deprotection using solution photogenerated acids. Nucleic Acids Res 29:4744–4750
20. Ghindilis AL, Smith MW, Schwarzkopf KR et al (2007) CombiMatrix oligonucleotide arrays: genotyping and gene expression assays employing electrochemical detection. Biosens Bioelectron 22:1853–1860
21. Hughes TR, Mao M, Jones AR et al (2001) Expression profiling using microarrays fabricated by an ink-jet oligonucleotide synthesizer. Nat Biotechnol 19:342–347
22. LeProust EM, Peck BJ, Spirin K et al (2010) Synthesis of high-quality libraries of long (150mer) oligonucleotides by a novel depurination controlled process. Nucleic Acids Res 38:2522–2540
23. Ma S, Saaem I, Tian J (2012) Error correction in gene synthesis technology. Trends Biotechnol 30:147–154
24. Saaem I, Ma K-S, Marchi AN et al (2010) In situ synthesis of DNA microarray on functionalized cyclic olefin copolymer substrate. ACS Appl Mater Interfaces 2:491–497
25. Chow BY, Emig CJ, Jacobson JM (2009) Photoelectrochemical synthesis of DNA microarrays. Proc Natl Acad Sci USA 106:15219–15224
26. Quan J, Saaem I, Tang N et al (2011) Parallel on-chip gene synthesis and application to optimization of protein expression. Nat Biotechnol 29:449–452
27. Khorana HG, Agarwal KL, Büchi H et al (1972) CIII. Total synthesis of the structural gene for an alanine transfer ribonucleic acid from yeast. J Mol Biol 72:209–217
28. Gupta NK, Ohtsuka E, Sgaramella V et al (1968) Studies on polynucleotides, 88. Enzymatic joining of chemically synthesized segments corresponding to the gene for alanine-tRNA. Proc Natl Acad Sci USA 60:1338–1344

29. Au L-C, Yang F-Y, Yang W-J et al (1998) Gene synthesis by a LCR-based approach: high-level production of leptin-L54 using synthetic gene in Escherichia coli. Biochem Biophys Res Commun 248:200–203
30. Stemmer WP, Crameri A, Ha KD et al (1995) Single-step assembly of a gene and entire plasmid from large numbers of oligodeoxyribonucleotides. Gene 164:49–53
31. Hughes RA, Miklos AE, Ellington AD (2011) Gene synthesis: methods and applications. Methods Enzymol 498:277–309
32. Tian J, Gong H, Sheng N et al (2004) Accurate multiplex gene synthesis from programmable DNA microchips. Nature 432:1050–1054
33. Borovkov AY, Loskutov AV, Robida MD et al (2010) High-quality gene assembly directly from unpurified mixtures of microarray-synthesized oligonucleotides. Nucleic Acids Res 38:e180
34. Kosuri S, Eroshenko N, Leproust EM et al (2010) Scalable gene synthesis by selective amplification of DNA pools from high-fidelity microchips. Nat Biotechnol 28:1295–1299
35. Kong DS, Carr PA, Chen L et al (2007) Parallel gene synthesis in a microfluidic device. Nucleic Acids Res 35:e61
36. Smith J, Modrich P (1997) Removal of polymerase-produced mutant sequences from PCR products. Proc Natl Acad Sci USA 94:6847–6850
37. Fuhrmann M, Oertel W, Berthold P, Hegemann P (2005) Removal of mismatched bases from synthetic genes by enzymatic mismatch cleavage. Nucleic Acids Res 33:e58
38. Saaem I, Ma S, Quan J, Tian J (2012) Error correction of microchip synthesized genes using surveyor nuclease. Nucleic Acids Res 40:e23
39. Binkowski BF, Richmond KE, Kaysen J et al (2005) Correcting errors in synthetic DNA through consensus shuffling. Nucleic Acids Res 33:e55
40. Schwartz JJ, Lee C, Shendure J (2012) Accurate gene synthesis with tag-directed retrieval of sequence-verified DNA molecules. Nat Methods 9:913–915
41. Kim H, Han H, Ahn J et al (2012) "Shotgun DNA synthesis" for the high-throughput construction of large DNA molecules. Nucleic Acids Res 40:e140
42. Kung P, Goldstein G, Reinherz EL, Schlossman SF (1979) Monoclonal antibodies defining distinctive human T cell surface antigens. Science 206:347–349
43. Jespers LS, Roberts A, Mahler SM et al (1994) Guiding the selection of human antibodies from phage display repertoires to a single epitope of an antigen. Biotechnology 12:899–903
44. Köhler G, Milstein C (1975) Continuous cultures of fused cells secreting antibody of predefined specificity. Nature 256:495–497
45. Alkan SS (2004) Monoclonal antibodies: the story of a discovery that revolutionized science and medicine. Nat Rev Immunol 4:153–156
46. Shim H (2015) Synthetic approach to the generation of antibody diversity. BMB Rep 48:489–494
47. Skerra A, Plückthun A (1988) Assembly of a functional immunoglobulin Fv fragment in Escherichia coli. Science 240:1038–1041
48. Orlandi R, Gussow DH, Jones PT, Winter G (1989) Cloning immunoglobulin variable domains for expression by the polymerase chain reaction. Proc Natl Acad Sci 86:3833–3837
49. Ward ES, Güssow D, Griffiths AD et al (1989) Binding activities of a repertoire of single immunoglobulin variable domains secreted from Escherichia coli. Nature 341:544–546
50. Huse W, Sastry L, Iverson S et al (1989) Generation of a large combinatorial library of the immunoglobulin repertoire in phage lambda. Science 246:1275–1281
51. McCafferty J, Griffiths AD, Winter G, Chiswell DJ (1990) Phage antibodies: filamentous phage displaying antibody variable domains. Nature 348:552–554
52. Rothe A, Hosse RJ, Power BE (2006) In vitro display technologies reveal novel biopharmaceutics. FASEB J 20:1599–1610
53. Bradbury ARM, Sidhu S, Dübel S, McCafferty J (2011) Beyond natural antibodies: the power of in vitro display technologies. Nat Biotechnol 29:245–254
54. Benhar I (2007) Design of synthetic antibody libraries. Expert Opin Biol Ther 7:763–779

55. Hoogenboom HR, Winter G (1992) By-passing immunisation. Human antibodies from synthetic repertoires of germline VH gene segments rearranged in vitro. J Mol Biol 227:381–388
56. Nissim A, Hoogenboom HR, Tomlinson IM et al (1994) Antibody fragments from a "single pot" phage display library as immunochemical reagents. EMBO J 13:692–698
57. de Kruif J, Boel E, Logtenberg T (1995) Selection and application of human single chain Fv antibody fragments from a semi-synthetic phage antibody display library with designed CDR3 regions. J Mol Biol 248:97–105
58. Pini A, Viti F, Santucci A et al (1998) Design and use of a phage display library. Human antibodies with subnanomolar affinity against a marker of angiogenesis eluted from a two-dimensional gel. J Biol Chem 273:21769–21776
59. Knappik A, Ge L, Honegger A et al (2000) Fully synthetic human combinatorial antibody libraries (HuCAL) based on modular consensus frameworks and CDRs randomized with trinucleotides. J Mol Biol 296:57–86
60. Rothe C, Urlinger S, Löhning C et al (2008) The human combinatorial antibody library HuCAL GOLD combines diversification of all six CDRs according to the natural immune system with a novel display method for efficient selection of high-affinity antibodies. J Mol Biol 376:1182–1200
61. Van den Brulle J, Fischer M, Langmann T et al (2008) A novel solid phase technology for high-throughput gene synthesis. Biotechniques 45:340–343
62. Zhai W, Glanville J, Fuhrmann M et al (2011) Synthetic antibodies designed on natural sequence landscapes. J Mol Biol 412:55–71
63. Mason DM, Friedensohn S, Weber CR et al (2021) Optimization of therapeutic antibodies by predicting antigen specificity from antibody sequence via deep learning. Nat Biomed Eng. https://doi.org/10.1038/s41551-021-00699-9

Chapter 4
Advancements in the Manufacture of Monoclonal Antibodies and Other Large Molecule Protein Therapeutics: Recent Innovations in Cell Culture Technology Enabling Process Intensification

Gregory W. Hiller

Abstract This chapter on recent advances in upstream process development will discuss the limitations of fed-batch processes, the recent resurgence of interest in traditional "classical steady-state" perfusion in which viable cell density is held nearly constant, and the development of non-conventional "dynamic" perfusion processes in which viable cell density is allowed to peak and decline. Many variants and hybrid processes that combine elements of both perfusion and fed-batch, or even link bioreactors together for a unique, isolated control of the growth phase and the more quiescent production phase will be explored. Other process intensification methodologies such as N-1 perfusion for high-density production reactor inoculation will also be examined.

A section on the pragmatic control of mammalian cell metabolism will explain recent advances designed to precisely control lactic acid formation and limit ammonium ion accumulation. Genetic engineering approaches such as enzyme knock outs and catabolic pathway reconstitution to limit the formation of previously unknown growth-inhibitory by-products of metabolism will also be delineated.

Keywords Perfusion · Glucose limitation · N-1 perfusion · Batch · Fed-batch · Linked bioreactors · High-intensity low-volume perfusion

---

G. W. Hiller (✉)
Bioprocess Research and Development, Biotherapeutics Pharmaceutical Sciences, Pfizer Inc., Andover, MA, USA
e-mail: greg@greghiller.com

K. Gadamasetti, S. A. Kolodziej (eds.), *Bioprocessing, Bioengineering and Process Chemistry in the Biopharmaceutical Industry*,
https://doi.org/10.1007/978-3-031-62007-2_4

4.1 Introduction

The market for monoclonal antibodies (mAbs) and large molecule protein therapeutics continues to grow at a vigorous pace partly because such medicines are now more affordable as the economic prosperity of enormous populations around the world has increased over the past several decades [1]. The incredible specificity, and lack of off-target effects and therefore adverse therapeutic events, continues to make protein therapeutics an essential tool to treat an enormous array of human diseases. When using animal cell culture as the production methodology, the vast majority of large molecule protein therapeutics are produced as secreted, extracellular products. The same cell culture principles that will be discussed in this chapter also apply to a great extent to the newly developing fields of autologous or allogenic cell therapy and the production of recombinant adeno-associated virus (rAAV) for gene therapy (though rAAV production is intracellular).

In recent years, some have questioned the need for increased productivity at the large scale [2] as the number of blockbuster monoclonal antibodies may be decreasing as we move into an age of more personalized medicine, with genetic variant testing of individual forms of disease determining a highly specific and potentially more efficacious treatment plan. This potentially logical argument for less interest in high-capacity processes, however, did not envision the massive and sudden global need for effective antibody therapeutics to treat coronavirus disease 2019 (COVID-19) in the face of the first global pandemic to strike the world for more than a century in 2020.

4.2 Modes of Bioreactor Operation

4.2.1 Batch

The simplest mode of bioreactor operation is batch. If repeated multiple times in succession, these processes are known as repeated batch or batch re-feed operations [3]. Repeated batch operations are still performed today in some commercial processes for labile molecules or proteins that are primarily produced only during the growth phase (growth-associated). Due to the limited solubility of many media components, those that are quickly consumed may be depleted from a batch culture and limit growth or productivity. Additionally, high concentrations of even benign nutrients such as certain amino acids and even glucose can inhibit initial cell growth, particularly when inoculation densities are low. Furthermore, as each component is added to the media at ever higher concentrations, the increasing osmolality of the medium also becomes a constraining factor since mammalian cells will only grow in a narrow range of osmotic strength [4] and will produce proteins only in a slightly wider range of osmotic strength than they will grow [5].

Innovative cell culture scientists realized decades ago that while cells consume certain nutrients, and these must be provided to continue growth and protein expression, components such as salts and other ions in many cases are not appreciably depleted during the course of growth, even when reaching very high cell densities. As mammalian cells are limited in their tolerance of osmotic strength, the ability to further fortify media is also limited. It was realized that much of the excess salts could be removed from the "classical" cell culture media that were designed for low density cell culture back in the 1960s and 1970s. Somewhat surprisingly, some mammalian cells seem to be able to adapt to ion concentrations of sodium and potassium well outside of the physiological concentration range observed within the body of any mammal in existence, despite what one might have learned in a course on cellular physiology!

The removal of these salts supplied significant osmotic "space" for the supplementation of amino acids and additional glucose. Provided such parameters as pH and dissolved oxygen (DO) were controlled, adding these nutrients "up front" in the cell culture medium allowed cell cultures to reach much higher cell densities without any feeds being required. Of course, cell culture scientists were never satisfied with those cell densities either and after optimizing cell culture processes with high nutrient content basal media the scientists also began again to include concentrated nutrient feeds in their processes.

4.2.2 Fed-Batch

Fed-batch cell culture processes add the key components that cells need to continue to grow and produce protein, principally glucose and amino acids. It is sometimes useful to think of a mammalian cell in culture as a polymerization catalyst. In essence, the only important purpose that all mammalian cells serve in an industrial protein production process is to string together amino acids (the monomers) into long chains that fold into proteins (the polymer). Of course, N- and O-linked glycosylation also occurs, but as a fraction of the mass of a therapeutic polypeptide, these are minor reactions by comparison for most proteins. For an efficient fed-batch process with discrete occasional bolus feeds and little waste products being formed, it is quite possible over the course of only 5–10 hours to detect a significant decline of osmotic strength. This can easily be observed by running frequent samples on a freezing-point osmometer. The phenomenon occurs because the monomer amino acids are being added to large polypeptides with such high molecular weights that their accumulation to even multi-gram/liter concentrations has a negligible influence on the osmotic strength of the culture. As cultures become ever more productive with very high cell densities, it is sometimes no longer practical to add nutrients to the culture as discrete bolus feeds because even once daily feeds can start to impart a significant osmotic shock of more than 30–70 mOsm/kg. Frequent osmotic shocks of such magnitude may become detrimental to the health of the culture and may contribute to amino acid misincorporation should the level of any particular amino acid fall too low between bolus feeds [6]. In such cases, a continuous or semi-continuous feeding system may be required.

With a highly efficient fed-batch cell culture process, it is quite feasible to feed the equivalent of 400–700 mOsm/kg of feeds over the course of 10–18 days. With such a process, feeds must be quite concentrated in order to minimize the total volume of liquid added to the culture. Many industrial-scale bioreactors have limits on the total volume that can be fed due to the placement of mixing impellers and the desire not to transition the liquid level through this zone of increased shear with its ensuing potential for increased foam formation. The science (or art!) of preparing very concentrated feed mixtures often relies on the careful order of addition of chemicals, manipulation of pH (up and down) and temperature, and separate preparation of certain concentrated solutions before addition to the bulk feed mixture. In some cases, it may even be necessary or efficient to separate the components into multiple distinct feeds that might not be sufficiently stable together in a single solution. Furthermore, the pH manipulations often necessary to get all the components into solution will in turn introduce additional ions to the solution that are not consumed by the cells. Such ions, sodium and chloride being the most common of these, will ultimately accumulate in the culture. Some fed-batch culture processes then also need to manage the accumulation of these ions and other cell-produced waste products that might not be toxic in and of themselves, but do contribute to the continuous slow increase in the osmotic strength of the culture until the high osmotic strength itself becomes detrimental to cell health and productivity. While probably not yet necessary for the vast majority of fed-batch cultures, our laboratory, in consultation with drug product formulation experts, has also experimented with the lyophilization and even sterile spray-drying of cell culture feed media. Lyophilization was found to be largely impractical due to the amounts of high osmotic strength excipients that need to be added to facilitate stable cake formation. Aseptic spray drying, however, was found to produce a very stable and incredibly fast dissolving, uniform particle size powder that can be added directly to the surface of a production bioreactor. Naturally, methods of addition of an aseptic dry powder to the liquid surface of a bioreactor would then need to be worked out before practical implementation. Should much higher efficiency cell lines and cell culture processes be developed in the future, dry feeds might inevitably increase overall bioreactor productivity by minimizing culture volume increases.

Thus far we have only discussed the requirement to add nutrients to the cell culture. Simultaneously, of course, cells in culture are producing certain waste products. Cells in culture undoubtedly lack the life-supporting systems that a whole organism enjoys. Individual cells in culture cannot control glucose levels, remove and process most metabolic waste products, or manage pH, dissolved oxygen, carbon dioxide, or temperature. Most of these parameters are reasonably easily controlled to the degree necessary by basic bioreactor operations that have been in existence for decades. The removal of waste products, however, is not so simple; think of the complexity of chemical processes occurring in the kidney or liver of an intact mammal. We will discuss in more detail methods to reduce the formation of waste products in a later section on control of cellular metabolism. Clearly however, for the state-of-the-art fed-batch process today, the buildup of waste products is the most important factor that ultimately limits the productivity of the fed-batch mode of operations.

4.2.3 *Perfusion—Or the Solution to Pollution Is Dilution!*

When it becomes difficult to either add sufficient nutrients to a culture, or control the level of inhibitory metabolic by-products, or merely control the ever-increasing osmotic strength of the culture, many cell culture scientists have resorted to perfusion. Perfusion culture solves all of these problems simultaneously, depending upon how it is implemented and is somewhat analogous to the processes that occur for cells within an organ in the body. In an animal's body, the cells in an organ do not exit the organ but are continuously supplied with nutrients and have waste products washed away by virtue of the blood supply.

In perhaps its most widely used and recognized form, perfusion cell culture utilizes a device to maintain all, or nearly all cells within the bioreactor (a filter or other cell retention device), while nutrient medium (perfusion medium) flows in and out of the bioreactor and thus the bioreactor maintains a constant volume.

Many early cell culture processes produced key enzymes or blood factors missing in small populations of humans due to genetic abnormalities. These proteins tended to be susceptible to biochemical or enzymatic breakdown or form high molecular weight aggregates under cell culture conditions, and therefore it was advantageous to produce them by perfusion culture. In perfusion culture the average residence time in the bioreactor of the newly secreted protein could be controlled and reduced to a few days or even a few hours if necessary. Early perfusion mammalian cultures often used cells grown on microcarriers or immobilized between the fibers of a microfiltration hollow-fiber bioreactor. Freely suspended cells are inherently easier to scale up but are more difficult to separate from fluid leaving a bioreactor. After the "low hanging fruit" products of the nascent biotechnology industry were developed and supplied for clinical use (e.g., enzymes and blood component replacements), many companies started to produce protein therapeutics (antibodies and fusion proteins) that were more stable in the cell culture bioreactor environment. Additionally, scientists became better at providing nutrients or controlling waste product formation in a batch or fed-batch format, and perfusion processes seemed to fall out of favor for a time in the 1990s and early 2000s. Eventually however, as the inherent limitations of fed-batch culture mentioned before were reached, and better designed methods for cell retention became commercially available, the bioprocessing community exhibited a resurgence of interest in the process intensification capabilities afforded by perfusion.

There have been a large number of clever devices designed as cell retention systems [7], but the more commonly utilized today consist of external-to-the-bioreactor microfiltration membranes, principally hollow-fiber filtration devices for scales up to about 1000 L, and tangential flow filtration devices in a stacked plate configuration (Prostak™) for larger bioreactors. Cells are recirculated at considerable flowrates through the tube side of these filtration cartridges, or through the open channels of the stacked plate systems, by several different forms of low-shear pumps (tangential flow filtration). Compared with batch or fed-batch processes, perfusion processes can reach much higher cell densities. Provided sufficient oxygen transfer into and carbon dioxide removal from the culture can be maintained, it is possible in perfusion cultures to come very close to the theoretical maximum cell density which would approximate close packed spheres [8]. Such very high cell densities, and the corresponding productivities that can be achieved, are primarily possible because of the capacity of waste product removal that perfusion can provide. Even with a poorly designed perfusion process, using low nutrient content media and conditions that do not reduce the formation of cell-produced inhibitory metabolic by-products, extremely high cell densities are possible if sufficient cell retention capacity is available (i.e., high filtration area). As mentioned earlier, near close packed spheres or tissue density is possible provided one is willing to use vast quantities of perfusion media and the viscosities of the resulting culture do not overly inhibit the recirculation of the fluid though any cell retention device. Perfusion systems that recirculate culture using diaphragm pumps may be constrained in this regard because vacuum (negative 14.7 psig, or ~1 bar) is the maximum driving force they can provide for moving the culture into the hollow-fiber filtration device unless the bioreactor itself was under pressure (difficult for single-use vessels) [8].

It should be mentioned that reliable sterilization of large external microfiltration devices for cell retention can sometimes be problematic. Precise protocols must be established and strictly adhered to for cleaning and steam sterilization of stacked plate systems. The advent of gamma-irradiated, pre-sterilized hollow fiber devices has simplified and increased the success rate of moderate scale perfusion, but sterilization by radiation may not be feasible for very large hollow fiber cartridges.

4.2.4 Hybrid Processes

Our laboratory introduced the concept of "hybrid perfusion fed-batch" for CHO cell culture several years back [9]. This mode of operation uses several days of 100% cell retention perfusion (the perfusate is discarded) followed by several days of conventional fed-batch operations culminating in a single drug substance harvest. Utilizing perfusion at the optimal time and performing it in a manner that minimized the volumes of media required, we were able to boost the overall productivity of a number of cell lines/processes to more than double that achieved in optimized fed-batch processes without increasing the length of the processes. Additionally, since the perfusion only occurs for a short time at the beginning of the culture when the cells are at very high viability, the surface area of the cell retention membrane can be minimized. Because the early perfusate was discarded in such a process, and the cell densities reached could be very high, we hypothesized at the time that such a process could also theoretically be used for moderately labile proteins since the fed-batch portion of the process could be as short as 2–3 days, yet produce product titers in the 3–6 g/L range. Finally, the perfusion volumes needed for such a process were quite reasonable at between 1 and 2 reactor volumes total for the perfusion portion of the operation.

One of the essential features of this hybrid process, and many intensified cell culture processes, is the concept of maximizing integrated viable cell density (IVCD). IVCD in units of cells * time/volume is a term that is defined very simply as the area under a graph of the viable cell density versus time for any process. If we assume for a moment that the per cell, or specific productivity (picograms/cell/day) of any cell line remains relatively constant if we can provide healthy culture conditions, we can quickly see that any increase in the IVCD directly leads to a proportional increase in overall culture productivity.

Part of the reason the hybrid perfusion fed-batch process worked well with minimal development effort is likely due to a phenomenon of cell division that might be called "division momentum." If at the beginning of the culture conditions are maintained sufficiently favorable, for instance with a nearly constant cell-specific perfusion rate (commonly in picoliters/cell/day), cells will continue to divide near their exponential growth rate. When the perfusion is suddenly stopped, the cell division does not stop immediately, likely because a fraction of cells at any point are already "committed" to cell division due to their instantaneous position in the cell cycle. Additionally, it likely takes a day or so of fed-batch conditions to accumulate the metabolic by-products to levels that inhibit further growth. Because of this phenomenon, if sufficient nutrients are made available through appropriate timed additions of highly concentrated feed media, an additional near doubling of cell density can occur on the days immediately following the termination of perfusion.

While generally the loss of product to the perfusion stream at the beginning of the hybrid perfusion fed-batch process is a small fraction of the total produced over the time in culture, an ultra-filtration membrane of the appropriate molecular weight cutoff to retain protein in the bioreactor could be used in place of the more commonly used micro-filtration membranes for cell retention. Such a replacement would result in near 100% capture of all protein produced during the process [10].

4.2.5 *Concentrated Fed-Batch or Perfusion with Ultrafiltration*

A fed-batch process or the hybrid process described above inherently terminates with a discrete, single harvest. There are of course advantages and disadvantages to a single harvest when it comes to sizing and efficiency of the downstream purification process. Perfusion processes, as mentioned previously, are oftentimes used with labile proteins because of the necessity of quickly purifying the protein away from the cell culture fluid. Fed-batch processes are limited by the volume of feeds possible before the bioreactor reaches its maximum working volume, and there are significant disadvantages in taking a partial harvest prior to the end of the culture. In instances of lower productivity of stable proteins, there may be value in performing

100% cell retention perfusion with the use of an ultrafiltration (UF) membrane of the appropriate molecular weight cutoff. In such a process (also called concentrated fed-batch) no, or very little, protein product or cells leave in the permeate that exits the system. Oftentimes such a process will operate at around 1 vessel volume per day (VVD) rate of continuous or semi-continuous perfusion. Since waste products are removed and nutrients are easily provided in the perfusion media, very high cell densities can be achieved, and culture duration can be longer than a typical fed-batch. Because of the very high cell densities achieved (commonly greater than one-quarter by volume cells), care must be taken in considering the volume of the cells in the final calculation of productivity of such a culture as there is no significant product typically recovered from the cell's cytoplasm. Such a concentrated fed-batch process typically ends in a single discrete harvest, although there may be value in minimizing the size of the downstream process by taking several smaller, discrete harvests towards the end of the culture.

Other variants on this process include the alternating use of two different forms of cell retention device for the perfusion: one being microfiltration (MF) and the other ultrafiltration (UF). When the UF filter is used, no product comes out of the bioreactor with the exiting perfusion medium and protein product builds up in the bioreactor. The MF filter is then used for the perfusion when a harvest is desired as it allows the protein to flow across the membrane. Thus, it is possible to switch back and forth between the two forms of cell retention device and even use different perfusion rates, on alternate days as desired, taking harvests whenever might be most appropriate for the downstream process that is designed [11].

4.2.6 *Intensification of Classical Perfusion*

Many of the continuous perfusion processes developed over the past several decades have attempted to achieve a steady-state operation. Possibly because many of the cell culture scientists working in the industry were originally classically trained chemical engineers, there has long been a fondness for the type of steady-state continuous (and highly efficient) process that is very common in the petroleum refining industry. The problem with this approach, however, is that cells have a certain amount of "memory" for what they have seen before [12]. Unlike hydrocarbons, they do not necessarily immediately revert to the same behavior when placed under the same conditions as had previously been encountered. Minor perturbations in culture can sometimes lead to instabilities in the process that can be difficult to predict, reliably prevent, or easily recover from. Additionally, increasing metabolic efficiency as nutrient levels drop below certain critical levels can cause cell densities to slowly increase and also slow the progress towards a steady-state condition.

Ideally, perfusion culture process development is performed in a fairly standard manner. There is an initial decision made perhaps on ideal maximal cell density that can be sustained by the bioreactor system, or some productivity target is used. A decision is also commonly made as to the practical perfusion rate that can be maintained. That perfusion rate is typically in the 1–2 VVD range, but ideally should always try to be minimized due to the costs and logistics of supplying the massive amounts of liquid media that will need to be continuously available during the process.

A somewhat conventional cell culture perfusion process then starts usually with a day or so of batch growth, followed by a slow ramp up of the perfusion rate to match the rate of growth of cells. This is the concept of a fixed cell-specific perfusion rate used at least during the expansion phase to maintain cells in exponential growth in order to increase the cell density in the bioreactor as quickly as possible (maximize the IVCD). Once the peak perfusion rate chosen is reached, if no other interventions are taken the cell density will typically peak at a very high and potentially unsustainable cell density and subsequently significant cell death will occur. Provided perfusion media with a sufficient nutrient concentration is used, and other standard cell culture parameters are controlled (pH, DO, etc.), the cell death is usually due to waste product accumulation. If the perfusion continues long enough in this theoretical perfusion process, eventually the bioreactor would be a solid mass of dead or dying cells.

Just as a bleed of inert materials is required in many continuous industrial chemical processes, so too is a cell bleed required in most cell culture perfusion processes to reach a "steady-state." Perfusion processes commonly use cell bleeds of 5–20% of the bioreactor volume per day. The cell bleed does not usually distinguish between live and dead cells and is taken directly from the bulk of the cell culture bioreactor. Sometimes the cell bleed rate will be varied and used to directly control the sustained viable cell density of the bioreactor. This can be achieved with offline cell counts and manual manipulations of the cell bleed, or some method of continuous online measurement or approximation of cell mass (commonly using capacitance probes) and a continuous computer-controlled variable cell bleed.

Because different downstream operating conditions/unit operations and potentially an entirely different downstream process would be required, most perfusion processes do not attempt to purify protein product from the cell bleed as this cell-containing stream is different from that in the cell-free permeate coming through the main cell-retention filter. Thus, the cell bleed represents a continuous loss of product and attempts to minimize the necessary cell bleed are typically a significant effort during process development [13]. As an example, if the perfusion rate was 1 VVD and the cell bleed rate was 20% per day (or 0.2 VVD), the loss of product would be close to 20% (potentially less since the cells will occupy a fraction of the volume). However, there is another phenomenon at play that very often significantly increases the product loss in the cell bleed far above what would be estimated merely using volume ratios. When MF cell retention membranes begin to plug with cellular debris, they start to slowly transform into UF membranes with a continuously changing (decreasing) molecular weight cutoff. When this molecular weight cutoff gets low enough, product protein is concentrated in the bioreactor and the cell bleed can become a much larger source of product loss.

If a therapeutic protein-producing cell line is sufficiently stable both with respect to its cellular productivity and its phenotypic nature, a steady-state perfusion process with an appropriate cell bleed might be able to operate continuously for weeks or even months [14]. The efficiency of such long-term operations can be very high and the flexibility of running a process as long as necessary to produce the required amount of protein is desirable. However, proving the stability and consistency of a very long duration process for validation or process characterization purposes can be daunting. As mentioned in the previous paragraph, it is desirable to minimize the required cell bleed in a continuous steady-state, or one might say "sustainable," perfusion process. The cell bleed is usually used to control the viable cell density at some predetermined value. In such a process where the viable cell density is not changing, there is a balance between the fraction of viable cells continuously dying, the new viable cells being created by division, and the viable cells that are leaving the system in the cell bleed. The number of cells dying is usually small compared to the division rate and thus the cell bleed rate is nearly equivalent to the growth rate.

While not rigidly true for all mammalian cell processes, it is generally accepted and observed that high growth rates are usually linked to lower per cell, or specific productivity rates [15]. When cells are focusing their metabolic precursors and cellular energy on biomass production (division), rates of recombinant protein production are generally lower. This is part of the reason that in fed-batch cultures the majority of protein production occurs in the last several days of the batch after most division is complete, and the cell density is much higher than earlier in the culture. So, these two constraints are actually somewhat synergistic. Minimizing the cell bleed once the target steady-state cell density is reached can minimize product losses, and simultaneously maintaining a lower cellular growth rate can help keep per cell productivity high. Low cell growth rate = low cell bleed rate. By increasing the nutrient content of the perfusion media and checking the culture for residual nutrient levels occasionally, one can ensure that sufficient nutrients are always available to the cells [16]. In this case then the growth rate of cells in a perfusion culture is directly controlled by the rate of removal of metabolic wastes and other inhibitors of cell division. Lower perfusion rates lead to more accumulation of inhibitors of cell growth which ultimately leads to lower cellular growth rates. A slow reduction in perfusion rate during process development can then lead to a slow reduction in growth rate, increased productivity and reduced medium usage. However, the difficulty of continuously decreasing the perfusion rate (while increasing the available nutrients in that perfusion medium) is that the most significant benefits in per cell productivity tend to come at very low growth rates which may be difficult to maintain [17]. A minor process deviation, even one that might not be readily detectable, might be sufficient to push cells into a no-growth state, out of the steady-state condition, and result in a culture with a changing and decreasing cell density. Additionally, mammalian cells in culture are continuously undergoing slight epigenetic changes as they continuously adapt to environmental factors [18], making their long-term responses to culture conditions slightly unpredictable.

For many of the reasons stated in the previous paragraph, while a sustainable, or true steady-state perfusion process has many advantages, there are considerable challenges in the design and operation of such a process with a high cell-specific productivity. To add to the difficulty, the low cellular growth rates that are usually encountered in such a process also result in more fragile cells which are oftentimes less capable of withstanding the shear stresses that may be encountered in the cell retention system [19]. The additional shear death results in additional cellular debris/small particles which more quickly plug the small pores of most membrane-based cell retention devices, complicating scale up since ever larger surface areas, or frequent change outs of the cell retention membrane are ultimately required.

4.2.7 Dynamic Perfusion Processes

In an effort to solve some of the undesirable features of "sustainable" or steady-state perfusion, various groups in recent years have been experimenting with a more dynamic perfusion process in which cell densities, perfusion rates, and most culture conditions are expected to change during the process [8, 17, 20]. Since the goal of attaining steady state is abandoned, the cell bleed is entirely eliminated. Cell density, however, still reaches some controlled peak, but that peak is typically determined by the timing of the start of perfusion, how long and at what rates the perfusion is performed, and the nature of the medium or solutions used for the perfusion. Again, the goal is to never limit the culture for nutrient availability, but also accept that growth inhibitory metabolic by-products are accumulating and will eventually slow and ultimately halt cell division in a manner analogous to that which occurs towards the end of a fed-batch process.

The initial goal of such a dynamic perfusion process is identical to that of a conventional perfusion process: grow the cells to high cell density as quickly as possible by using perfusion to remove waste products and supply nutrients. The approximate linear timeline of one such dynamic perfusion process developed in our laboratory, the high-intensity low-volume perfusion process (HILVOP) [17] is depicted in the figure below (Fig. 4.1). For the HILVOP dynamic perfusion process, perfusion starts within a day or so of inoculation depending upon the inoculation density. This initial perfusion phase (the first stage) is generally conducted as a continuous increase in perfusion rate that closely matches the increase in cell density (therefore a nearly constant cell-specific perfusion rate). Since the product concentration in the permeate at this stage is quite low, generally the perfusate is sent to

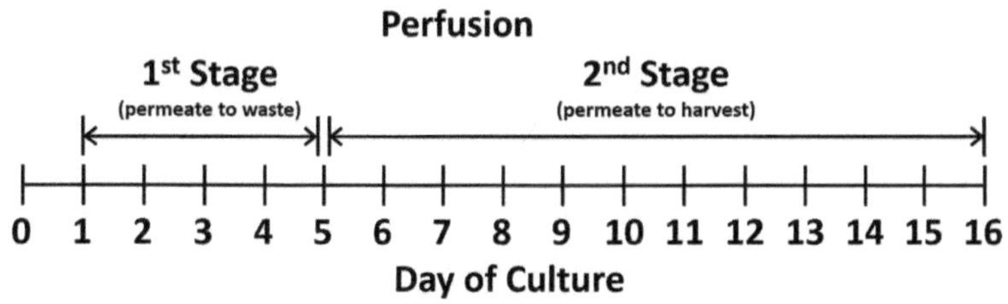

Fig. 4.1 Stages of perfusion

waste. Our laboratory developed a unique method of "cell-controlled" perfusion that can be utilized in this first stage of perfusion that minimizes the overall requirements for perfusion volumes by eliminating the net production of lactic acid [9]. The perfusion control method also does not require constant counting or estimation of the viable cell density, but still maintains a nearly constant cell-specific perfusion rate (picoliters/cell/day) allowing for the maintenance of a nearly exponential growth rate. More details on cell-controlled perfusion will be discussed in the section on metabolism.

During the second stage of perfusion, the goal of the perfusion is no longer to maintain exponential growth, but mainly to supply the cells with nutrients for the production of the recombinant protein and for any additional cellular division. Since we are no longer attempting to dilute out inhibitory compounds, the perfusion media used during the first part of the second stage of perfusion is extremely concentrated, perhaps over 700 mM amino acids combined with additional nutrients. However, since the cell density is so high, and the cells so metabolically active, the osmotic strength in the bioreactor does not rise much above the normal physiological range even while delivering the perfusion media at the relatively low rate of around 0.2–0.3 VVDs.

Even if lactic acid is well controlled at this point in the culture (during the beginning of the second stage), other cell-produced metabolic by-products quickly accumulate, and division slows as the cells reach their peak cell density. The length of time the perfusion continues in the first stage controls the ultimate trajectory and peak viable cell density eventually achieved later in the culture. In approximately the second half of the second stage of perfusion, the perfusion rate is slowly ramped up by adding an ever-larger fraction of a saline diluent to the perfusion medium. Cellular division does not resume, but the large volumes of saline flushing the bioreactor facilitate the transport of the majority of the produced protein out of the bioreactor/cell retention system and into the downstream process. If a continuous downstream process is integrated directly with the upstream process, then the ramp up of saline delivery to the bioreactor can be used to help control the mass per day of product delivered to the downstream operation. Keeping the mass/day within a fairly narrow range allows for the protein A capture chromatography columns and the other associated columns and equipment throughout the purification train to be much smaller than would typically be needed with a fed-batch process. Such an integrated downstream and upstream process utilizing a pair of protein A columns operated in a simple bind and elute manner allows for highly efficient column utilization, achieving two to six column cycles per day and 40–80 cycles over the course of a 14–18-day dynamic perfusion process.

Our laboratory has found that this dynamic perfusion/HILVOP process achieves specific productivity (per cell) rates near that of an optimized fed-batch process and significantly higher than sustainable perfusion processes that we have explored for several cell lines [17]. The overall volumes of perfusion media used are moderate and significantly lower than most sustainable "steady-state" perfusion processes. Because the cell density is much higher than fed-batch and the specific productivity (picograms/cell/day) is usually similar to a fed batch, the overall volumetric productivity of the process is also increased proportionally, typically three to four-fold higher than an optimized fed-batch process.

4.3 Pragmatic Control of Cellular Metabolism

When in a highly proliferative state and therefore dividing quickly, most mammalian cell lines in ex vivo culture tend to exhibit some degree of "aerobic glycolysis," meaning that in spite of readily available oxygen, cells shift to metabolize significant amounts of glucose directly to lactic acid [21]. This metabolic state may have advantages during fast growth as large amounts of various biochemicals may be drawn off from the glycolytic pathway for the production of needed precursors of DNA, RNA, lipids or other cellular building blocks required for accelerated biomass production. One can imagine that in tumorigenic growth, there would be significant competitive advantages to fast growth in that the tumor cells themselves do not have to trouble with the lactic acid produced as it is merely carried away to other tissues, including liver and muscle, by the nearly endless supply of fresh blood provided by the host organism.

In flask or bioreactor culture however, without some form of pH control, the rapid buildup of lactic acid oftentimes drives pH quickly below that at which the cells can continue fast division (6.3–6.8) [22]. Employing pH control via the addition of a base titrant to the culture tied to a continuous pH measurement effectively restores growth. In some cultures, however, even with pH control the production of lactic acid is so significant that the lactate ion accumulates to an extent that eventually the culture growth stops due to both the level of lactate ion and the accumulation of osmolality from the lactate ion in addition to the sodium or potassium counter ions from the base titrant [23].

4.3.1 The Development of pH or Temperature Shifts

In a batch or fed-batch process, any ion added to the culture or chemical that is produced by the cells accumulates in the culture unless it is later taken up by the cells or degraded via some extracellular chemical or enzymatic process. Within the normal pH operating ranges of a bioreactor (6.9–7.2), lactic acid output by cells in culture is usually directly proportional to the pH of the culture [22]. To an extent, higher pH (up to about 7.3) may result in faster cell growth, but also result in higher rates of lactic acid production, and therefore lactate accumulation. To control lactic acid production rates, some researchers have found it advantageous early in the culture to reduce the pH control set point closer to neutral or slightly lower pH [22, 24, 25]. While such a practice might slow cell growth, the slower accumulation of lactate sometimes allows the growth phase to be extended longer, resulting ultimately in a higher IVCD, or at a minimum a healthier biomass than would have been achieved with a higher pH set point. It was also discovered that while culture temperatures below the normal mammalian body temperature of about 37 °C slow cellular division, sometimes a decrease in temperature several degrees below 37 °C (30–35 °C) at the optimum time will also slow or even completely stop the net production of lactic acid [24, 26].

When culture conditions are sufficiently conducive, it is even possible for cells to begin to take up lactate from the bulk culture fluid. This usually occurs several days into a batch or fed-batch culture when cell growth has slowed and some significant amount of lactate has formed [27]. Sometimes facilitated by decreasing culture temperature below 37 °C, when this "metabolic shift" occurs the lactate is taken up from the culture in the form of lactic acid, and the uptake rate can be rapid enough to be detected by an increase in bulk culture pH by conventional pH probes. Under such conditions lactate in the culture can sometimes be consumed down to nearly undetectable levels. It should be noted, however, that while the cells can effectively completely consume the lactate ion from the culture, sodium or potassium ions also were entering the culture as a result of the use of the base titrant to control the pH near neutrality early in the culture. These sodium or potassium ions are not appreciably consumed by the cells and therefore contribute to the osmotic strength (the gift that keeps on giving!) of the culture in what might be called the "base titrant ion penalty." In some cases, so much lactate is produced early in culture that when the lactate is consumed as lactic acid, it can actually drive the pH so high (>7.4) that some processes add pH control on the high end by sparging gaseous carbon dioxide (sometimes leading to its own problems) [28] or other forms of acid (phosphoric, etc.).

4.3.2 *Glucose Limitation*

Uptake of lactic acid by mammalian cells, or at a minimum a lactic acid production rate of zero, also can occur when the level of glucose in the culture drops below about 2–3 mM, but the exact threshold where this occurs might be quite variable. This phenomenon was recognized several decades ago, oftentimes in continuous cultures (chemostats) [22], or perfusion cultures [29] in which the glucose concentration would slowly decrease but cells had time to adapt to the conditions, and there was always a low, but continuous rate of glucose addition. Some researchers attempted to determine the precise, optimal glucose level to maintain during culture in order to prevent lactate production, but also to not overly starve the cells and slow proliferation [30, 31]. Others resorted to very frequent glucose measurement via offline analytical instruments and then sophisticated control algorithms to attempt to keep the glucose at the precise level thought to be ideal [32]. Another experimental feedback control method for glucose delivery was based on the continuous measurement of oxygen uptake rate [33]. If the oxygen uptake rate dropped, additional glucose containing feed was added as it was assumed cells had depleted the available glucose.

While these techniques proved to be moderately successful in controlling the lactate production in a research environment at the bench scale, frequent sampling and accurate glucose quantitation at very low levels, combined with precise feeding of glucose, were generally considered to be too difficult to implement in large-scale commercial bioreactor operations [34]. Additionally, while cultures were actively growing, the length of time between sampling, accurate determination of the glucose level, and then a change in the glucose feeding rate may have been too long. Likely this resulted in some level of continuous glucose starvation which resulted in sub-optimal growth rates when compared with non-glucose limited cells.

Other groups have experimented with alternative carbon sources to glucose such as galactose, fructose, or mannose, but while those substitutions often do reduce the formation of lactate it is usually at the expense of fast growth rates and results in considerably reduced peak cell densities [35–38]. Additional researchers tried to slow the rate of six-carbon sugar membrane transport with chemical inhibitors or genetic engineering techniques such as amplification, suppression, addition, or silencing of enzymes or substrate membrane transporters [39–41]. Because in most cases the cells eventually reverted to their original wild-type metabolism, practical success with these techniques was not achieved.

Experimentation in our laboratory with additions of galactose to CHO cell cultures in the presence and absence of glucose lead us to the conclusion that in the presence of glucose, galactose might be competing for available membrane transporters and potentially slowing the transport of glucose across the cell membrane [23]. At that point we decided to investigate other methodologies to achieve a slow glucose uptake rate without slowing cellular growth rates.

4.3.3 The Development of HiPDOG Control Methodology

After experiments in our laboratory demonstrated that the accumulation of lactate in our CHO cell fed-batch cultures was the primary factor suppressing continued cell division and productivity [23], we built on controlled glucose feeding experiments conducted earlier [25]. After realizing that the uptake of lactic acid from the bulk culture fluid was sufficient to result in a significant rise in pH detectable by conventional in-situ pH probes, we developed a feedback control system for lactate that would prove to be easy to manage and implement even in the largest bioreactors in common use (13,000 L) in our production facilities. It was discovered that cells in culture would "signal" a low level of glucose in the bulk culture by taking up lactic acid from the surrounding culture fluid. The switch to consumption of lactic acid was fast enough and of sufficient magnitude that the bulk culture increase in pH was enough to trigger a feed pump to add a small amount of glucose to the culture before the cells were overly starved for glucose. Starting with a low level of glucose in the initial medium in the production bioreactor, this control loop strategy, later given the acronym HiPDOG, for Hi-end pH delivery of glucose (shown as a hypothesized sequence of events in Fig. 4.2 below) would then turn on and off the glucose feed

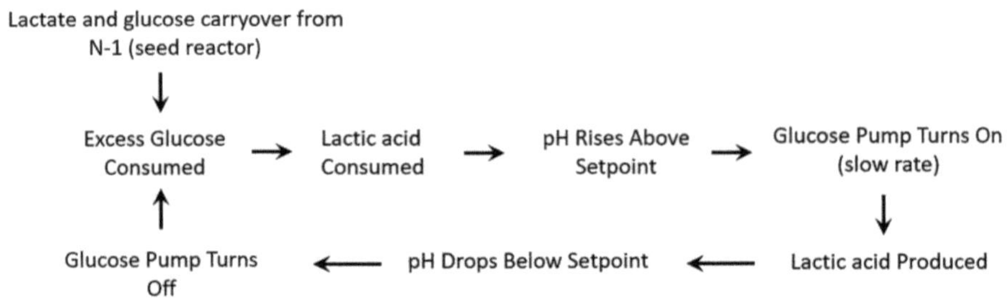

Fig. 4.2 HiPDOG sequence of events

pump periodically throughout the growth phase of the culture. Somewhat surprisingly, with a sufficiently tuned system, the growth rate of a fed-batch culture utilizing the HiPDOG control strategy was found to be nearly identical to that of parallel cultures which never experienced glucose limitation. The net effect of the glucose/lactic acid control scheme resulted in zero net lactic acid formation during the growth phase once the control system first activated as the glucose concentration first fell into a limiting range. Additionally, since little to no titrant is added in such a process, the "base titrant ion penalty" mentioned above never occurs.

One would expect that if the HiPDOG control strategy worked as simply as explained above (lactic acid produced, then almost immediately consumed again, ion for ion), when operating in a fed-batch culture there would be a very slow decrease in the lactate level of the bioreactor as the culture was slightly diluted by the glucose solution, or other feeds. However, it was found during experimentation that the rate of decrease of the culture's lactate concentration was actually significantly faster than could be accounted for by dilution alone. It was hypothesized and later confirmed by additional experiments that CHO cells in a highly proliferative state are producing many other by-products of metabolism, albeit at lower levels, a significant fraction of which are acidic in nature [42]. As these acids are secreted into the culture medium, they depress the culture pH slightly and result in a partial decoupling of the predominant glucose/lactic acid pH control and thereby slightly over-limit the cells for glucose, resulting in a slow reduction in the concentration of lactate in the culture.

Because a rapidly expanding fed-batch culture requires many other nutrients, it was found to be useful to also include these nutrients in the glucose feed being delivered by the HiPDOG control scheme. In this way, cells in a fed-batch culture would precisely control their own rate of feeding without any operator intervention. The amount of nutrients delivered would of course depend upon the glucose concentration in the feed since only the delivery of glucose was actually being controlled by the culture. The higher the concentration of glucose in the feed, the less volume of feed that would be delivered, and vice-versa. Finding the right ratio of glucose to other nutrients allowed for an optimum delivery rate to the culture, maintaining all amino acids and other components above growth limiting concentrations, yet not leading to excess accumulation which could lead to toxicity, excess ammonium production if amino acids were used as a carbon source, or merely high osmotic strength.

The HiPDOG control scheme was found to be widely applicable to most CHO cell lines that produced even modest amounts of lactic acid, and particularly valuable for those with inherent high levels of lactic acid production. Many cell lines grew to much higher peak cell densities and maintained higher specific productivities even late in culture because of the reduced level of lactate that had accumulated in the culture. Additionally, when utilizing HiPDOG, cell lines in which the previous practice had been to decrease the pH set point during the growth phase to minimize lactate formation could now be grown at whatever higher pH was found to be optimum to allow for faster accumulation of IVCD.

While accumulations of ammonium ion can sometimes limit growth in CHO fed-batch cultures, oftentimes with HiPDOG controlling lactate (particularly with cell lines utilizing the glutamine synthetase expression system), growth inhibitory concentrations of ammonium are not reached [42]. For expression systems that utilize glutamine in the cell culture media, limiting its availability in the production bioreactor with a lower starting concentration, or substituting the majority of it with a glutamine-alanine dipeptide, or asparagine, glutamate, or aspartate can help to limit the accumulation of ammonia [43–45]. Glutamate and aspartate may be preferred in some cases over asparagine as these two amino acids only contribute one amine group if the carbon skeleton of the amino acid is utilized by the cells as an energy source.

Even when lactic acid and ammonia production are well controlled, cells still eventually cease division even if all necessary nutrients are available and the osmotic strength of the culture is not in a growth-inhibitory range. Other metabolic by-products accumulate in these cultures and eventually reach growth inhibitory levels after many days in culture [42]. As growth slows in this environment, it is generally no longer necessary to utilize the HiPDOG control scheme and the culture can be allowed to return to a non-glucose limited state for the remainder (usually the last half) of the culture. After working with dozens of cell lines, our lab was never able to detect any problematic features of the product quality (of particular concern was glycosylation) of protein produced during the HIPDOG phase of a culture. Additionally, since the vast majority of the recombinant protein product in a fed-batch process is produced late in culture when glucose was no longer limiting, this theoretical concern was further minimized. In many cases, the improved health of the culture as a whole when using HiPDOG resulted in superior product quality for many programs.

4.3.4 Extension of Controlled Glucose Limitation to Perfusion Cultures

As mentioned in the above sections, even in a very carefully fed culture with methods to minimize lactic acid and ammonia production being utilized, eventually a variety of other growth-inhibitory metabolic by-products will accumulate which slow and eventually stop cellular division. The logical progression of CHO cell culture science would then be to determine the other by-products and devise methodologies to limit or control their formation. Easier said than done! We will touch on such efforts later in this chapter, but for now remind the reader that "the solution to pollution is dilution," or in the case of cell culture...perfusion. Perfusion is the brute force method that will solve nearly any waste product accumulation problem. However, if the waste product is merely lactate, perfusion is a very inefficient method of keeping lactate below a growth inhibitory concentration [29, 46]. A much more efficient way to implement perfusion is to allow cells to control their own perfusion rate and simultaneously reduce net lactate production to near zero. This can be accomplished by modifying and extending the HiPDOG control scheme to work with perfusion [9]. The technique can reduce by several fold the rate of perfusion required to maintain a culture in an exponential growth state and allow us to coin yet another cute acronym, HiPCOP, or Hi-end pH Control Of Perfusion. Instead of controlling the rate of nutrient addition to a fed-batch culture, now when the pH reaches a high-end set point (indicating the culture is limited for glucose) a pump is activated that delivers perfusion media containing glucose to the culture, while a second pump maintains the volume of the culture constant by removing an equal amount of liquid from the culture through a cell retention device. In a similar manner to the HiPDOG control, the addition of perfusion media via HiPCOP stops when the pH drops back below the high-end set point as a small amount of lactic acid is produced. However, at this point the astute cell culture scientist would realize the fundamental problem with this this approach. The control of lactic acid formation would almost be *too* good in this case. Recall that when HiPDOG was used with fed-batch cultures the lactate concentration would slowly fall throughout the course of a culture since other acidic metabolic by-products partially interfered with the control. Even with many dozens of CHO cell lines tested with HiPDOG in fed batch, provided the initial glucose level of the culture was high enough, the level of lactate never fell too low for the HiPDOG control scheme to continue to operate effectively during the growth phase. However, in the case of perfusion, with the simultaneous reduction in lactate concentrations just due to the perfusion operation itself (the flushing of the pollution, remember?), reduction in lactate to near zero could be a real possibility and might cause the HiPCOP control scheme to fail, overly restricting the perfusion rate and slowing cell growth.

Upon realizing this complication, in our laboratory we began adding sodium-L-lactate to the incoming perfusion medium at low levels to ensure that there would always be a low level of lactate in the perfusion bioreactor culture so that the HiPCOP control scheme would not fail. While almost a blasphemous proposal to add lactate to a culture when for the last few decades we were trying to minimize lactate formation in mammalian cell culture, it seemed pragmatic initially. We found, however, that pure sodium-L-lactate is not an inexpensive raw material. Far less expensive is the racemic mixture of D/L-lactic acid which is a by-product of certain industrial processes, and seems to be well tolerated when added to cell cultures. However, since the D form of lactic acid is not readily consumed by mammalian cultures, it would be an extra ion contributing to osmotic strength and the chemical mixture itself would be another raw material requiring suitable sourcing. In continued experimentation and understanding of the chemical and biochemical processes occurring in these cell cultures, it was realized that just as in fed-batch cultures where HiPDOG control was partially decoupled due to the formation of other acidic species, a perfusion culture with HiPCOP control could be continuously maintained at a higher (but still relatively low and controlled) lactate level merely by continuously forcing the pH up independent of the HiPCOP control scheme. This could be accomplished with a continuous slow addition of sodium carbonate titrant or by the continuous flashing-off of carbon dioxide (a removal of carbonic acid if you will) coming from perfusion media entering the bioreactor. The perfusion media was modified to have slightly higher than normal levels of sodium bicarbonate [47]. While carbon dioxide removal can sometimes be an issue with scale up of cell culture processes, a simple calculation reveals that the amount of additional carbon dioxide necessary to be removed from a perfusion culture using extra sodium bicarbonate in the perfusion media is a small fraction of that needed to be removed simply due to the metabolism of glucose to carbon dioxide in the tricarboxylic acid cycle.

4.3.5 "Whack a Mole" with Inhibitory By-products

The techniques discussed in the above sections describe methods to control lactate and ammonium accumulation, and, at least with perfusion culture, flush out everything else that might be slowing growth. Ideally, however, to minimize process complexity, we'd prefer not to use perfusion at all and just keep cells growing to ever higher peak densities in simple fed-batch culture. We determined early on that lactate accumulation was the primary factor limiting cell growth in our cultures [23], but once that was controlled, we could start to see the "forest for the trees," that is, we began to see all the other inhibitors being formed when previously all we could see was lactate. We conducted a series of experiments which ruled out a number of other potential reasons and confirmed that the primary factor causing cessation of CHO cell growth was the accumulation of small molecular weight chemical compounds that were by-products of cellular metabolism [42]. A wide range of metabolism by-products were identified and eventually quantitated by NMR (nuclear magnetic resonance) and LC/GC-MS (liquid chromatography/gas chromatography—mass spectroscopy). The levels of their accumulation in fed-batch cultures as cell growth slowed were determined, and a painstaking analysis directed by conventional understandings of mammalian cell metabolism pathways was performed to determine the growth inhibitory potential of each compound.

While most of these compounds accumulate in fed-batch cultures to only sub-millimolar concentrations, or at most single to low double digit millimolar concentrations, the compounds had independent and synergistic negative effects on cell growth. When added back to CHO cell cultures as pure chemicals in the ratios and levels to which they accumulate when growth ceases in CHO cell fed-batch culture experiments, the inhibitors could be shown to completely shut down all cell division. The biochemical pathways which produce the various inhibitors were then investigated and potential strategies for dealing with the most potent inhibitors were devised.

Many of the more potent inhibitors were found to be by-products of amino acid catabolism. It was found that simply keeping the levels of a number of amino acids below a certain upper threshold level, while simultaneously not allowing them to be completely depleted, could minimize the formation of some inhibitors and enable additional cell growth [42]. However, as with early efforts of cell culture scientists to measure and precisely feed glucose and keep it in a lower range to limit lactate formation, it was realized that other metabolic engineering approaches might be more straightforward and could be implemented across the board for CHO host cell lines. It was realized that three of the inhibitors: isovalerate, 2-methylbutyrate, and isobutyrate, which were primarily formed from leucine, isoleucine, and valine catabolism, respectively, shared the same enzyme at the beginning of their catabolic pathway. Efforts were made to knock out the enzyme—branched chain amino acid transaminase-1, or BCAT1. The resulting CHO cell lines grew to higher cell densities, maintained higher cell viability longer, and when transfected with the appropriate genetic elements also produced higher levels of recombinant proteins in fed-batch and perfusion cultures than CHO cells without the knockout [48].

Of course, even fed-batch cultures of these BCAT1 knock out cells eventually reach a peak cell density and cease division. Provided the cultures were appropriately fed required nutrients, logically these cultures also stopped division due to some other inhibitory by-products. Further investigation suggested that in CHO cells, missing or low levels of certain enzymes in the catabolic pathway of phenylalanine and tyrosine might be leading to the formation of other inhibitory compounds. In this case increasing the level of several of these enzymes (and even adding back a co-factor pathway!) reduced the level of the inhibitor production and had the added benefit of making the cells no longer require exogenous additions of tyrosine [48].

Our laboratory continues in this effort to determine each inhibitory factor that might be appreciably limiting the maximum cell density reachable in fed-batch cultures, or that might require the use of higher levels of perfusion in processes that resort to perfusion to maintain exponential growth. We are currently investigating the implementation of the most beneficial and pragmatic metabolism modifications directly into the CHO host cell lines used for expression of therapeutic proteins in large-scale production.

4.4 Other Methods of Process Intensification

4.4.1 N-1 Perfusion

As was mentioned earlier in this chapter, there are significant advantages to increasing the IVCD, or the area under the curve of a viable cell density versus time graph. This is because cells act as the workhorses, catalyzing the conversion of the amino acids into the polymer, which is the protein therapeutic. Provided that specific productivity is maintained appropriately, more cells equal more stuff. The working volume and efficient utilization of the production bioreactor are typically the principal limiting factor that determines the overall annual productivity of any manufacturing facility. Therefore, there is incentive to have that vessel filled with the highest density of healthy cells at all times. Any time used during the expansion of the cell mass in the production bioreactor is potentially wasted time. With this in mind it is logical to minimize the time in the production bioreactor during which the cells are dividing. N-1 (or seed bioreactor) perfusion can help to achieve this.

For reasons discussed earlier, perfusion allows for the maintenance of a much higher number of cells in a bioreactor when compared with simple batch or fed batch. When large volumes of media are perfused through a cell mass that is retained within the bioreactor by a cell retention device, very high cell densities can be achieved. However, there are limits to the practicality of N-1 perfusion. At cell densities above 100×10^6 cell/mL the properties of the fluid can begin to change merely due to the fact that a significant fraction of the fluid is occupied by cells [8]. Viscosities can significantly increase making it difficult to move cells in and out of a cell retention filter, particularly if a vacuum diaphragm pump is being utilized for generating such flow. The cell retention filters can also reach the limits of filtration capacity, though this is mitigated to some extent by the fact that the rapidly dividing cell mass of an N-1 bioreactor is typically at very high viability. At extremely high viable cell densities, oxygen transfer in, and carbon dioxide removal from the bioreactor can become problematic, particularly when foaming starts to occur. Interruptions to gas supply in such cultures can drop the dissolved oxygen to near undetectable levels within less than a minute.

Most large-scale mammalian cell culture manufacturing facilities use a 1:5 volume ratio for each step of their scale up train, and this also typically includes the final seed bioreactor. If we assume for the moment that 100×10^6 cell/mL is a reasonable practical limit for the maximum density achievable in the N-1 seed bioreactor, that means that the maximum inoculation density in the production bioreactor would be about 20×10^6 cell/mL, or slightly higher as 20–30% of the volume of the production bioreactor must be left available for the fed-batch feeds that will occur later. If the N-1 bioreactor were instead merely a batch or fed-batch bioreactor the maximum inoculation density of the production bioreactor might be closer to 1 or 2 $\times 10^6$ cell/mL which is more typical of platform cell culture processes today. As exponential growth doubling times are only around 24 hours for mammalian cell, a 20×10^6 cell/mL inoculation coming from an N-1 culture using perfusion could potentially shave three to 4 days off the typical 12–18-day length of a production bioreactor operating in fed-batch mode and still yield equivalent or higher titers [49, 50].

4.4.2 N-1 Intensification

N-1 perfusion is the extreme example of N-1 intensification to enable higher production bioreactor inoculations, but a well-managed N-1 fed-batch culture can also provide sufficient cells to accelerate the production bioreactor. The key here is to provide cells that have not overgrown and are therefore not likely to suffer a lag in the production bioreactor. The HiPDOG glucose limiting strategy has been shown to accomplish this for N-1 cultures [23], but any technique that reduces the formation of growth inhibitors, such as those that were discussed in the earlier sections of this chapter, will also serve the same purpose provided the method does not overly extend the length of the N-1 culture.

4.4.3 Linked Bioreactors

In the last section we investigated and demonstrated the logic of increasing the inoculation density of the production bioreactor. If one follows this to the extreme, we see the value in having the production bioreactor at the highest viable cell density possible all the time. There are practical limits to the length of a fed-batch bioreactor operation as we explored in the section on fed batch. Briefly revisiting, anything the cells produce (including all waste products) remains in the culture, as does anything fed to the culture but not consumed by the cells. In addition, dying cells can release the contents of their cytoplasm and various organelles to the culture milieu. Depending upon the stability of the protein product, 10–18 days in such a mixture of various enzymes and cellular components might be problematic to product quality. Quiescent cells also have limits to their ability to withstand deteriorating culture conditions as waste products accumulate and the osmotic strength of the culture continuously increases. Late in a deteriorating culture, cells also sometimes revert to a more glycolytic metabolism leading to increases in lactate and sometimes also in ammonium. Lower viability cells are also more likely to truncate the glycosylation process resulting in product material with lower levels of sialylation, less terminal galactosylation, and increased levels of high mannose species. In short…things go bad when the cells get tuckered out!

We noted earlier that the commonly observed phenomenon that cells with low or near zero growth rates also tend to have higher per cell, or specific productivities. Also, the optimum conditions for cell division might not be the optimum conditions for the production of the protein, especially if it is found that some particular product quality parameter can best be optimized at an extreme of the normal growth range (pH or temperature for instance). While we can certainly change conditions post-growth phase in a fed-batch production bioreactor, it may take cells time to adapt to the new conditions and some significant fraction of product will be produced during the growth phase under the unfavorable environmental parameter for product quality.

Some of the above, long recognized limitations of fed-batch culture might be mitigated by completely separating the growth and production phases that usually occur in the production bioreactor. This can be achieved by linking two bioreactors together in continuous operation. The N-1 bioreactor continuously supplies cells dividing near the maximum growth rate to a production bioreactor where conditions are more conducive to a low-growth but highly productive, somewhat quiescent state. The cells entering the production CSTR (continuous-flow stirred-tank reactor) perhaps undergo one more division (recall the "division momentum" mentioned in an earlier section) before adapting to the production bioreactor conditions which are optimal for protein production. While certainly a more complex operation, such a linked bioreactor system can be very productive.

The simplest version of such a linked bioreactor system would be a pair of cascading CSTR's or chemostats [51]. If one were to make use of most existing large-scale cell culture facilities this would generally mean a 1–5, or 1–10 volume ratio of the N-1 to production bioreactor. The cascading CSTR system, however, suffers from the fact that the maximum dilution rate of the N-1 bioreactor is limited to the washout dilution rate, which for mammalian cells is not much above 1/day or 1 VVD. Even near that washout dilution rate the cell density in the N-1 would be low and the overall number of cells per unit time delivered to the production bioreactor would be low. Cells in an N-1 bioreactor operating at dilution rates much below the washout rate would by default be growing slower, and therefore on average less likely to continue division when transferred to the production bioreactor. So, we see that the N-1 operating as a CSTR significantly restricts the operating parameters and therefore the number of cells the bioreactor can produce per day. Introducing perfusion/cell retention in the N-1 removes this constraint and allows for high perfusion rates, very high growth rates, and enormous quantities of cells for transfer to the production reactor. If the production bioreactor then operates as a simple CSTR or chemostat (no cell retention), with additional nutrients added directly to it, volumetric productivities approaching or exceeding those in a stand-alone continuous perfusion bioreactor, ~1 g/L/d, may be achieved [52]. In such a system, the additional complexity of requiring a cell retention system on the largest vessel, the production bioreactor, is eliminated. Such a linked-bioreactor system could deliver harvest (containing whole cell culture) continuously to a suitably designed continuous downstream operation, or could even be cycled slightly in working volume, allowing bolus harvests to be taken once a day or once per 2 days.

To be efficient, a linked-bioreactor system must reach a high-productivity steady state quickly. It is therefore important in this reactor configuration not to overshoot the sustainable cell density as this could lead to a loss of cell viability. Getting to the highest productivity state might take additional time when compared to a standard fed-batch culture as the cells in the production bioreactor need to adapt to the conditions that are ultimately controlled by the dilution rate and rate of addition of cells and nutrients. To be a practical way to produce protein, and to demonstrate the advantage over fed-batch, a linked-bioreactor system would need to be operated for a comparatively long period of time; at least several weeks and perhaps ideally continuously for several months. At first glance, this may present a problem as genetically engineered cells have varying degrees of genotypic stability for production of the recombinant protein. The limit of such genotypic stability is generally demonstrated for any industrial cell culture process when the "limit of *in vitro* cell age" or LIVCA study is performed to determine how often a production facility needs to return to the working or master cell banks to expand cells for additional production batches.

The potential longevity of the linked culture system was demonstrated in our labs with the continuous operation of a linked-bioreactor system for over 80 days [52]. Additionally, it was established that the N-1 bioreactor (and ultimately the production bioreactor) could be "refreshed" by expanding again from frozen vials using a second parallel or twin N-1 bioreactor and switching the stream of cells entering the production bioreactor from the N-1 bioreactor containing the "old" cells to that containing the "new" cells. Performed correctly, it was shown that the viable cell density and volumetric productivity of the production bioreactor can remain nearly constant through this transition. This allows for the production bioreactor to operate nearly indefinitely in a very high volumetric productivity steady state, with a regularly "rejuvenated" set of cells (if only this could be replicated for humans!) producing product continuously with nearly identical product quality parameters. Since cells in the production bioreactor CSTR divide much slower than those in the N-1 culture, cells in the production CSTR that have accrued a larger number of generations are washed out of the system over time at a rate that depends upon the dilution rate of the production bioreactor among other factors.

The linked-bioreactor system just described provides volumetric productivities approximately double that of an optimized fed-batch process, and nearly as high as a stand-alone steady-state perfusion bioreactor, without the need for perfusion *in* the production bioreactor. This may decrease the size of the cell retention system by a factor of five or perhaps ten. For high demand products, production bioreactors in mammalian cell culture facilities are often in the 13,000-L working volume range, meaning that the N-1 reactor might be 2500 or perhaps as small as 1000-L working volume. Still, performing long term consistent perfusion using conventional microfiltration methods at such a scale might give some cell culture engineers nightmares.

4.4.4 Hydrocyclone Use with Linked Bioreactors

As mentioned earlier in this chapter in the section on perfusion, a great many clever cell retention devices have been developed over the past few decades but scale up for use on mammalian cells in suspension in bioreactors in the 1000 L range of any but the microfiltration methods have been particular problematic. Long used in microbial culture harvest and in mining and oil extraction industries, one cell retention system that seemed particularly promising for scale up was the hydrocyclone [53, 54]. While early experiments suggested that mammalian cells could withstand the very high shear forces generated as cells passed through a hydrocyclone [55], generally their use for cell retention in mammalian cell culture did not catch on widely, with very few mentions in the literature throughout 2005–2020. This may be because while highly viable cells dividing at near exponential growth rates might be able to withstand the shear forces, cells in perfusion production bioreactors are often not growing quickly because conditions that favor fast growth often cause per cell productivity to be lower. Alternatively, if used for an N-1 perfusion culture for the generation of large amounts of cells for a single inoculation of a production bioreactor, hydrocyclones also suffer when compared with membrane cell retention devices due to their somewhat low separation efficiency. As a result, it might take several additional days for an N-1 perfusion bioreactor using a hydrocyclone to reach a sufficiently high cell density to inoculate a production bioreactor at high density. Finally, while running multiple hydrocyclones in series might seem like a method to increase the efficiency of the cell separation, the additional pressure drop required might prove too extreme of a shear environment for mammalian cells to survive.

While hydrocyclones have the benefit of being very unlikely to plug with cell debris, even fairly optimized geometry hydrocyclones lose too many cells to the perfusion stream. The additional significant advantage of the hydrocyclone, its scale-up capacity, could even be considered a disadvantage as scaling down a hydrocyclone for use in bench scale bioreactors is problematic. The best way to scale down a hydrocyclone is merely to turn it on and off periodically by stopping the pump that delivers cell culture fluid to it. When running continuously, a hydrocyclone the size of a 50-mL centrifuge tube is sufficient to process ~500 L of cell culture fluid a day.

Understanding the limitations of hydrocyclones—low efficiency and high shear forces making them suitable only for high viability and shear resistant cells—our laboratory experimented with their use as a cell retention device in the linked bioreactor system [56]. In this system, cells in the N-1 bioreactor were typically dividing at near exponential growth rates, and the overflow fluid from the hydrocyclone (the permeate fluid that had reduced numbers of cells) could be used as the cell source for the continuously linked production bioreactor operating as a CSTR. In this highly specialized use, the hydrocyclone was nearly ideal as it would not plug, and a pair of 3-D printed then autoclaved hydrocyclones operating in parallel, fed by a single high-flow pump, could be shown to be sufficient to perfuse a 1000-L N-1 bioreactor at 1 VVD. Coupled with a production bioreactor operating as a CSTR with a working volume of 10,000 to possibly 13,000 L the system could achieve volumetric productivities near those of the linked system using a microfiltration membrane for N-1 perfusion, ~1 g/L/day [52].

4.5 Process Analytical Technology

Process analytical technology is fancy terminology that merely describes sensors for the continuous monitoring and control of parameters such as dissolved oxygen, pH, and temperature that have been in place on bioreactors for many decades. New sensors are always being developed, some of which may have pragmatic applications, while others seem to be searching about for practical value. Dissolved carbon dioxide probes, conductivity or capacitance probes for cell density approximation, and foam sensors are certainly of increased value as mammalian cell culture processes continue to be intensified.

The reliability, accuracy, and durability of dissolved carbon dioxide probes have certainly improved in recent years; however, it should be mentioned that extreme accuracy is not necessary for a carbon dioxide probe to yield valuable information. High intensity cultures with very high oxygen demands that often require use of pure oxygen and small bubble generating sparge elements can frequently suffer from high levels of dissolved carbon dioxide. Allowing the level of dissolved carbon dioxide to rise above approximately 12% of saturation can slow cell growth, and in some cases affect product quality parameters [28, 57]. While the two parameters, oxygen delivery and carbon dioxide removal, are certainly closely interlinked, large scale operations with highly intensified processes at our facilities have found it practical to independently control these parameters. Dissolved oxygen is controlled via the delivery of tiny pure oxygen bubbles from suitable spargers. Dissolved carbon dioxide is continuously measured using an in situ probe and is maintained below a target high-end setpoint by sparging either air or pure oxygen through sparge elements that deliver large volumes of gas and generate larger bubbles than those used exclusively for oxygen delivery.

If they make it to the surface without completely dissolving, the high efficiency oxygen transfer afforded by small bubbles of pure oxygen also can create a very stable foam layer, particularly in a culture with high cell densities and high levels of dissolved protein. Cell entrapment into foam has long been identified as potential source of high shear forces capable of killing cells as the bubbles burst [58]. Additionally, if allowed to reach the vent filter such foam can quickly plug vent filters, increase bioreactor hydrostatic pressure, and greatly restrict the ability to continue to sparge the bioreactor. Many forms of foam sensors are available from various industrial processes, but some may be difficult to implement in an aseptic fashion, particularly on the many single-use bioreactor designs being adopted today. Our pilot and large-scale facilities are currently adopting automated control systems with cameras and light strobes that can accurately estimate the foam layer and deliver antifoam as necessary to reduce the likelihood of vent plugging. Such a non-invasive foam control system merely requires a modestly clear window for a view down onto the culture surface.

Conductivity or capacitance probes measure the total volume of cells with intact cell membranes. Generally, this can be closely correlated with the number of viable cells in the culture, at least while the cells are growing quickly [46]. As manual or even computer assisted cell counts performed via automated microscopic methods are subject to significant variability due sample handling and dilution errors, a capacitance probe can better indicate the true growth rate of the culture. Since even slight changes in growth rate can over time result in significant differences in cell density, a continuous, precise knowledge of the biomass of a culture can assist in delivering concentrated feeds at appropriate rates to a culture, or assist in controlling perfusion rates to maintain a constant cell-specific perfusion rate. Unfortunately, later in culture the capacitance probe readings and the actual viable cell density can start to deviate from the strong correlation earlier in the process. This may occur because late in culture sometimes cell size increases without additional division. We have found that oxygen consumption rates, or even merely the value of the oxygen delivery rate of a culture that has controlled dissolved oxygen levels can sometimes be used in conjunction with capacitance probe values to estimate the viable cell mass and adjust feeding rates.

In situ, autoclavable probes for Raman spectroscopy can simultaneously collect spectra that can be used to quantitate a whole range of cell culture parameters. Factors such as glucose, lactate, and even certain amino acid concentrations can be correlated to the appropriately analyzed Raman signal; however, the ability to accurately determine any of these metabolites at low concentrations (below 3–5 mM) is extremely limited [59–61]. Also, unfortunately, the Raman calculated value for any of these parameters can be affected by a large number of factors that change throughout the culture. As a result of this, to accurately predict the concentration of a substrate or by-product in a culture it may be necessary to go through the somewhat laborious process of creating a large data set of Raman spectra on many bioreactor runs with minor process differences. This "teaching" spectra can be used with an orthogonal method of accurately quantitating the parameter, ultimately generating an algorithm which in the future can be used to calculate the parameter directly from the Raman spectra.

Through the process described above, our laboratory was able to continuously approximate the concentration of both glucose and lactate using Raman spectroscopy [60]. We realized that there could be some value in the number, even if only approximate. We surmised that even knowing the concentration of the value of glucose and lactate combined within the range of 1–2 g/L could allow us to control the rate of glucose feeding. As mentioned earlier in the section on control of cellular metabolism, during the growth phase if a cell culture becomes limited for glucose it will often begin to consume lactic acid from the culture. So, monitoring the combined concentration of glucose and lactate and feeding glucose in an attempt to keep the combined value nearly constant should minimize the formation of lactate. Of course, the same result could be achieved merely using the HiPDOG control strategy without all the added Raman correlation experiments; and certainly an autoclavable pH probe is far less expensive (and nearly ubiquitously implemented) than a Raman probe for every bioreactor. In certain circumstances when the HiPDOG control strategy might be undesirable Raman might prove useful.

Raman might also be utilized to better control in an automated fashion the mass load of antibody to the first column chromatograph step (e.g., protein A affinity). As part of an integrated downstream purification process, our laboratory is currently evaluating Raman to determine the antibody concentration of permeate leaving the microfiltration cell retention system of a perfusion bioreactor. Unfortunately, as with other uses of Raman spectroscopy, we have found it may be necessary to generate a complete set of "learning spectra" for each new antibody molecule despite the assumed similar properties that antibodies with only changes to the antigen binding site would have.

4.6 Single-Use Bioreactors (SUBs)

Advancements in multilayer polymer films and clever designs for structural support of the inflatable bioreactor bags over the past decade have culminated with single-use or disposable bioreactors with volumes approaching 6000 L [62]. SUBs eliminate the need for cleaning solutions, and cleaning and sterilization protocols and associated testing, saving considerable resources and changeover time. As many can attest, however, with the many pools of cell culture fluid on the floor of the bioreactor suite, leaks are still quite common and significant strides are required to improve the quality control of the manufacturing of single-use bioreactor inserts. Compared with their stainless-steel counterparts, SUBs also typically suffer with respect to their capability to handle highly intensified cell culture processes due to longer mixing times and reduced oxygen transfer capabilities due to substandard agitation and aeration equipment.

However, even with the mentioned shortcomings, the time and short-term cost to set up a facility for clinical production is greatly reduced using SUBs. If additional scale up is required, in the short run the "number up" methodology of scale up can be used, meaning just acquire additional SUBs of the same size. Since such bioreactors typically require little in the way of utilities, they can usually be placed in (merely rolled into) a simple "ballroom" type facility. Since the same scale and presumably the identical bioreactor design can be used, efforts to prove equivalent product quality with a change in scale are eliminated, and even concerns regarding process transfer to new manufacturing sites can also be reduced. While the efficiencies of scale are certainly forfeited with such a strategy, the accelerated time to market and ease of regulatory concerns can make a single-use bioreactor facility very attractive, particularly for a small company with an independent mindset but little existing infrastructure.

4.7 Conclusions

Significant advances in intensification of cell culture processes have occurred over the last decade. Higher productivity and a better understanding of the science of methods to manipulate product quality have been achieved. The versatility of biological products to treat disease never before considered has led numerous biopharmaceutical companies recently to announce expansions of production bioreactor capacity in the hundreds of thousands of liters. If this much additional capacity is necessary, clearly there remains a need for significant increases in efficiency and productivity from upstream operations. Within this chapter, we've described multiple new technical innovations such as hybrid cell culture processes and a renewed interest in continuous perfusion processes, particularly with the potential to link directly to continuous or nearly continuous downstream recovery operations. New molecule modalities and fields such as cell and gene therapy, with minor modifications, can also benefit from the innovations described in this chapter.

References

1. Walsh G (2018) Biopharmaceutical benchmarks 2018. Nat Biotechnol 36:1136–1145
2. Kelley B, Kiss R, Laird M (2018) A different perspective: how much innovation is really needed for monoclonal antibody production using mammalian cell technology? In: In GU, Kiss B (eds) New bioprocessing strategies: development and manufacturing of recombinant antibodies and proteins. Springer, Cham, pp 443–462
3. Drapeau D, Luan Y, Popoloski J, Richard D (1994) Extracellular insulin degrading activity creates instability. Cytotechnology 15:103–109
4. Waymouth C (1970) Osmolality of mammalian blood and of media for culture of mammalian cells. In Vitro 6:109–127
5. Takagi M, Moriyama T, Yoshida T (2001) Effects of shifts up and down in osmotic pressure on production of tissue plasminogen activator by Chinese hamster ovary cells in suspension. J Biosci Bioeng 91:509–514
6. Wong HE, Huang C-J, Zhang Z (2018) Amino acid misincorporation in recombinant proteins. Biotechnol Adv 36:168–181
7. Chotteau V (2015) Perfusion processes. In: Al-Rubeai M (ed) Animal cell culture. Cell engineering. Springer, pp 407–443
8. Clincke M, Mölleryd C, Zhang Y, Lindskog E, Walsh K, Chotteau V (2013) Very high density of CHO cells in perfusion by ATF or TFF in WAVE bioreactor™. Part I. Effect of the cell density on the process. Biotechnol Prog 29:754–767
9. Hiller GW, Ovalle AM, Gagnon MP, Curran ML, Wang W (2017) Cell-controlled hybrid perfusion fed-batch CHO cell process provides significant productivity improvement over conventional fed-batch cultures. Biotechnol Bioeng 114:1438–1447
10. Hiller GW (2007) Patent No. US20090042253A1. US
11. Tharmalingam CG (2015) Patent No. PCT/US2015/034297. US
12. Mulukutla BC, Yongky A, Grimm S, Daoutidis P, Hu W-S (2015) Multiplicity of steady states in glycolysis and shift of metabolic state in cultured mammalian cells. PLoS One 10(3):e0121561
13. Moritz KF, Closet A, Bzowska M, Bielser J-M, Souquet J, Broly H, Massimo M (2018) Improved performance in mammalian cell perfusion cultures by growth inhibition. Biotechnol J 14:1700722
14. Chong L, Saghafi M, Knappe C, Steigmiller S, Matanguihan C, Goudar CT (2013) Robust on-line sampling and analysis during long-term perfusion cultivation of mammalian cells. J Biotechnol 165:133–137
15. Kumar N, Gammell P, Clynes M (2007) Proliferation control strategies to improve productivity and survival during CHO based production culture. Cytotechnology 53:33–46
16. Konstantinov K et al (2006) The "push-to-low" approach for optimization of high-density perfusion cultures of animal cells. Adv Biochem Eng Biotechnol 101:75–98
17. Gagnon M, Nagre S, Wang W, Hiller G (2018) Shift to high-intensity, low-volume perfusion cell culture enabling a continuous, integrated bioprocess. Biotechnol Prog 34:1472–1481
18. Hernandez-Bort J, Hackl M, Höflmayer H, Jadhav V, Ernst W, Grillari J, Borth N (2012) Dynamic mRNA and miRNA profiling of CHO-K1 suspension cell cultures. Biotechnol J 7:1–16
19. Vickroy B, Lorenz K, Kelly W (2007) Modeling shear damage to suspended CHO cells during cross-flow filtration. Biotechnol Prog 23(1):194–199
20. Yang WC, Minkler DF, Kshirsagar R, Ryll T, Huang Y-M (2016) Concentrated fed-batch cell culture increases manufacturing capacity without additional volumetric capacity. J Biotechnol 217:1–11
21. Warburg O (1956) On the origin of cancer cells. Science 123:309
22. Miller W, Blanch H, Wilke C (1988) A kinetic analysis of hybridoma growth and metabolism in batch and continuous suspension culture: effect of nutrient concentration, dilution rate, and pH. Biotechnol Bioeng 32:947–965

23. Gagnon M, Hiller G (2011) High-end pH-controlled delivery of glucose effectively suppresses lactate accumulation in CHO fed-batch cultures. Biotechnol Bioeng 108(6):1328–1337
24. Yoon SK, Choi S, Song J-J (2005) Effect of culture pH on erythropoietin production by Chinese hamster ovary cells grown in suspension at 32.5 and 37.0 degrees C, vol 89. Biotechnol Bioeng, pp 345–356
25. Luan YM (1987) Factors governing lactic acid formation in long term cultivation of hybridoma cells. Biotechnol Lett 9:751–756
26. Bollati-Fogolin M, Forno G, Nimtz M, Conradt H, Etcheverrigaray M, Kratje R (2005) Temperature reduction in cultures of GM-CSF-expressing CHO cells: effect on productivity and product quality. Biotechnol Prog 21:17–21
27. Mulukutla BC, Gramer M, Hu W-S (2012) On metabolic shift to lactate consumption in fed-batch culture of mammalian cells. Metab Eng 14(2):138–149
28. Borys MD-A (2010) Effects of culture conditions on N-glycolylneuraminic acid (Neu5Gc) content of a recombinant fusion protein produced in CHO cells. Biotechnol Bioeng 105:1048–1057
29. Hiller G, Clark D, Blanch H (1993) Cell retention-chemostat studies of hybridoma cells-analysis of hybridoma growth and metabolism in continuous suspension culture in serum-free medium. Biotechnol Bioeng 42(2):185–195
30. Kurokawa H, Park Y, Iijima S, Kobayashi T (1994) Growth characteristics in fed-batch culture of hybridoma. Biotechnol Bioeng 44:95–103
31. Ljunggren J, Haggstrom L (1994) Catabolic control of hybridoma cells by glucose and glutamine limited fed. Biotechnol Bioeng 44:808–818
32. Fleischaker R (1982) An experimental study in the use of instrumentation to analyze metabolism and product. Ph.D. thesis. Cambridge, MA, USA
33. Zhou W, Rehm J, Hu W-S (1995) High viable cell concentration fed-batch cultures of hybridoma cells through on-line nutrient feeding. Biotechnol Bioeng 46:579–587
34. Freund NW, Croughan MS (2018) A simple method to reduce both lactic acid and ammonium production in industrial animal cell culture. Int J Mol Sci 19(2):385
35. Altamirano C, Paredes C, Cairó JJ, Gòdia F (2000) Improvement of CHO cell culture medium formulation: simultaneous substitution of glucose and glutamine. Biotechnol Prog 16:69–75
36. Anderson DC, Goochee CF, Cooper G, Weitzhandler M (1994) Monosaccharide and oligosaccharide analysis of isoelectric focusing-separated and blotted granulocyte-colony-stimulating factor glycoforms using high-pH anion-exchange chromatography with pulsed amperometric detection. Glycobiology 4:459–467
37. Reitzer LJ, Wice BM, Kennell D (1979) Evidence that glutamine, not sugar is the major source for cultured HeLa cells. J Biol Chem 254:2669–2676
38. Wagner A, Marc A, Engasser JM, Einsele A (1991) Growth and metabolism of human tumor kidney cells on galactose and glucose. Cytotechnology 7:7–13
39. Chen K, Liu Q, Xie L, Sharp PA, Wang DI (2001) Engineering of a mammalian cell line for reduction of lactate formation and high monoclonal antibody production. Biotechnol Bioeng 72:55–61
40. Irani N, Beccaria AJ, Wagner R (2002) Expression of recombinant cytoplasmic yeast pyruvate carboxylase for the improvement of the production of human erythropoietin by recombinant BHK-21 cells. J Biotechnol 93:269–282
41. Wlaschin KF, Hu W-S (2007) Engineering cell metabolism for high-density cell culture via manipulation of sugar transport. J Biotechnol 131:168–176
42. Mulukutla B, Kale J, Kalomeris T, Jacobs M, Hiller G (2017) Identification and control of novel growth inhibitors in fed-batch cultures of Chinese hamster ovary cells. Biotechnol Bioeng 114(8):1779–1790
43. Kim DY et al (2013) Fed-batch CHO cell t-PA production and feed glutamine replacement to reduce ammonia production. Biotechnol Prog 29(1):165–175

44. Hong J, Cho S, Yoon S (2010) Substitution of glutamine by glutamate enhances production and galactosylation of recombinant UgG in Chinese hamster ovary cells. Appl Microbiol Biotechnol 88:869–876
45. Kurano N, Leist C, Messi F, Kurano S, Fiechter A (1990) Growth behavior of Chinese hamster ovary cells in a compact loop bioreactor. J Biotechnol 15:113–128
46. Dowd J, Jubb A, Kwok K, Piret J (2003) Optimization and control of perfusion cultures using a viable cell probe and cell specific perfusion rates. Cytotechnology 42:35–45
47. Hiller GW, Gagnon MP, Ovalle AM (2016) Patent No. PCT/US2016/034570. US
48. Mulukutla BC et al (2019) Metabolic engineering of Chinese hamster ovary cells towards reduced biosynthesis and accumulation of novel growth inhibitors in fed-batch cultures. Metab Eng 54:54–68
49. Yang W, Lu J, Kwiatkowski C, Yuan H, Kshirsagar R, Ryll T, Huang Y-M (2014) Perfusion seed cultures improve biopharmaceutical fed-batch production capacity and product quality. Biotechnol Prog 30:616–625
50. Padawer W, Ling YB (2013) Case study: an accelerated 8-day monoclonal antibody production process based on high seeding densities. Biotechnol Prog 29(3):829–832
51. Bakker WA, Schäfer T, Beeftink HH, Tramper J, De Gooijer CD (1996) Hybridomas in a bioreactor cascade: modeling and determination of growth and death kinetics. Cytotechnology 21:263–277
52. Gagnon M, Nagre S, Wang W, Coffman J, Hiller GW (2019) Novel, linked bioreactor system for continuous production of biologics. Biotechnol Bioeng 116:1946–1195
53. Bicalho IC, Mognon JL, Shimoyama J, Ataíde CH, Duarte CR (2012) Separation of yeast from alcoholic fermentation in small hydrocyclones. Sep Purif Technol 87:62–70
54. Dikov VA, Sukhanov DE (2014) Feasibility of hydrocyclones for combined cleaning of circulation and waste water. Chem Pet Eng 49:619–623
55. Elsayed EA, Medronho RA, Wagner R, Deckwer WD (2006) Use of hydrocyclones for mammalian cell retention: separation efficiency and cell viability (part 1). Eng Life Sci 6(4):347–354
56. Kundu AM, Hiller GW (2021) Hydrocyclones as a cell retention device for an N-1 perfusion bioreactor linked to a continuous-flow stirred tank production bioreactor. Biotechnol Bioeng 118:1973–1986
57. Kimura R, Miller W (1996) Effects of elevated pCO2 and/or osmolality on the growth and recombinant tPA production of CHO cells. Biotechnol Bioeng 52:152–160
58. Chalmers JA (1991) Microscopic visualization of insect cell-bubble interactions. II: The bubble film and bubble rupture. Biotechnol Prog 7:151–158
59. Moretto JS, Cuellar M, Doane A, Ryll T, Berry B, Wiltberger K (2011) Raman spectroscopy for in-line CHO cell culture monitoring. Am Pharm Rev 14(3):18
60. Mehdizadeh H, Lauri D, Karry KM, Moshgbar M, Procopio-Melino R, Drapeau D (2015) Generic R aman-based calibration models enabling real-time monitoring of cell culture bioreactors. Biotechnol Prog 31(4):1004–1013
61. Bhatia H, Mehdizadeh H, Drapeau D, Yoon S (2018) In-line monitoring of amino acids in mammalian cell cultures using raman spectroscopy and multivariate chemometrics models. Eng Life Sci 18:55–61
62. Barba V (2019) ABEC launches 6,000 L single-use bioreactor to answer scalability needs. BioPharma-Reporter.com

Chapter 5
Process Development and Manufacturing Considerations for Multispecific (Bispecific and Trispecific) Antibodies: Case Study

Timothy Iskra, Ashley Sacramo, and John J. Scarcelli

Abstract Antibodies as a therapeutic treatment have been the focus of numerous companies for many years, resulting in over 100 currently on the market. This has led to the creation of a rich understanding of how to develop and manufacture these molecules. The result has been the creation of platforms consisting of high productivity cell lines and optimized culture conditions that can generate titers as high as 10 g/L. These platforms have also seen the introduction of streamlined downstream processes which typically consist of two to three chromatography steps.

In recent years, the bispecific and trispecific (or multispecific) antibodies have become a large area of development for most companies, as the ability to bind two different antigens at a time opens new and unique therapeutic areas. Although these are antibody-like, the complexity created due to the structurally diverse molecular formats and engineering adjustments presents a challenge to the current antibody development and manufacturing paradigm. New steps such as reactions to bring these molecules together as well as new impurities and stability issues have meant that these platforms have needed to be adjusted to enable production of suitable quantities of high-quality product.

Expression and production of multispecific antibodies have brought new challenges and considerations to cell line generation. Depending upon molecular

T. Iskra (✉)
Bioprocess Research and Development, Biotherapeutics Pharmaceutical Sciences, Pfizer Inc., Andover, MA, USA
e-mail: Timothy.iskra@pfizer.com

A. Sacramo
Pfizer Inc., Andover, MA, USA

Solid Biosciences Inc., Charlestown, MA, USA
e-mail: asacramo@solidbio.com

J. J. Scarcelli
Pfizer Inc., Andover, MA, USA

Sanofi, Framingham, MA, USA
e-mail: Jack.Scarcelli@sanofi.com

K. Gadamasetti, S. A. Kolodziej (eds.), *Bioprocessing, Bioengineering and Process Chemistry in the Biopharmaceutical Industry*,
https://doi.org/10.1007/978-3-031-62007-2_5

format, a strategy needs to be implemented encompassing choice of cell host organism (microbial vs. mammalian), and in the case of mammalian expression, number of cell lines generated (single vs. dual cell lines), as well as random versus targeted integration of transgenes. These initial choices can have far reaching implications, often necessitating expanded cell line screening efforts when compared to traditional monoclonals.

Once a suitable cell line is created, the production of multispecific molecules uses culture conditions similar to typical antibodies. Some adjustments are required however as the engineering of these molecules may result in a higher occurrence of clip species formation or presence of different molecule fragments or impurities. Attention has been paid to ways to reduce these impurities while increasing titers through culture growth conditions and setpoints.

After production there are several considerations that must be focused on to achieve a final high-quality product. These complex molecules can be less stable than traditional antibodies and they tend to contain new and unique impurities that must be removed. Due to the engineering of the molecules, there can be issues with stability that may prevent traditional operations from being conducted such as low pH viral inactivation. This results in the need to design or find additional means to achieve sufficient and robust viral safety. Also, most of the molecules have some level of homodimers present, and due to the similarity of these impurities to the heterodimer, the separation and removal can be difficult. This often results in either the addition of new steps or the use of less traditional steps being employed.

Keywords Multispecific antibodies · Bispecific antibodies · Trispecific antibodies · Expression · Purification · Knob into hole · Charge-based multispecific

5.1 Introduction

Monoclonal antibodies (mAbs) have had a significant impact over the last few decades in several therapeutic areas such as cancer, inflammation, autoimmune, cardiovascular, and infectious diseases [1, 2]. There are now more than 100 approved mAb products marketed worldwide [3]. The focus on mAbs has created a rich understanding of how to develop and manufacture these molecules. The result has been the creation of mAb platform processes consisting of high productivity cell lines and optimized culture conditions that commonly generate titers as high as 10 g/L. These mAb platforms have also seen the introduction of streamlined downstream processes which typically consist of three or in some cases two chromatography steps [1] that have yields typically greater than 65% [4].

In recent years, improvements in molecular design have led to the creation of bispecific and trispecific (or multispecific) mAb-based modalities as new candidate therapeutics. The ability to simultaneously bind multiple antigens has opened up new and exciting therapeutic areas. Multispecific antibodies have the ability to target specific cancer cells and coupled with the body's own immune cells, can enable

simultaneous mediators or pathways, or potentially increase binding specificity by interacting with two different antigens instead of only one [5]. As of 2021 there have been four bispecific molecules approved, with only three of them currently still on the market (amivantamab in the United States and European Union (EU); emicizumab and blinatumomab in the USA, EU, and Japan, and catumaxomab approved originally in EU and USA and later withdrawn from the market), with many more in either clinical studies or early phase development.

Dozens of structurally diverse, multispecific molecular formats are currently available, which can lead to differences in properties such as valency, pharmacokinetics, and mechanism of action [6, 7]. The decision to select a particular format can impact development and manufacturing. One example of this is the choice of host cell organism. If a particular molecular format does not require glycosylation or other complex post-translational processing, then a microbial host, such as *Escherichia coli*, having simple, scalable fermentations and shorter development timelines, may be the optimal choice. On the other hand, if glycosylation is central to the molecule's mechanism of action (e.g., Fc-mediated effector function), then a mammalian cell host is required. This chapter will focus exclusively on immunoglobulin G (IgG)-like multispecifics, which are most suitably expressed using mammalian cell culture.

Multispecific antibody expression has brought new challenges and considerations to cell line generation. Molecular format can dictate the number of cell lines that require generation (single vs. dual cell lines), as well as transgene integration strategy, and clone screening approach. These initial choices can have far reaching implications, and often necessitate expanded cell line screening efforts when compared to traditional mAbs.

Once a suitable cell line is created, the production of multispecific molecules typically uses culture conditions traditionally employed for mAbs. Some adjustments are required as the engineering of these molecules may result in a higher occurrence of clipped species or the presence of different molecule fragments or product-related impurities. Development focus must be on ways to reduce these impurities, while at the same time increasing process yields through cell culture growth and downstream process optimization.

Beyond production of these molecules, there are several considerations to achieve a high-quality final product. These complex molecules can be less stable than traditional mAbs and they tend to have an increased level of product-related impurities such as high molecular mass species (HMMS), residual homodimer, half antibodies, heavy chain-light chain mispairing, and/or fragments depending upon the molecular architecture [8, 9]. Some engineered multispecifics may be less stable than traditional mAbs thus limiting the use of certain production technologies. One example is an increased sensitivity to low pH resulting in the need to find alternative means to achieve an acceptable level of viral clearance. In addition, many of the techniques used to create these molecules result in low levels of product-related impurities, such as homodimers. Since homodimers are similar to the desired heterodimer, separation of these impurities is often difficult. This may result in the addition of new and sometime less traditional downstream process steps.

5.2 Strategies for Molecule and Cell Engineering to Produce IgG-Like Multispecifics

5.2.1 Molecular Format Considerations

The multivalent nature of IgG-like multispecific molecules is facilitated by sequence differences in the fragment variable (Fv) domain between at least one of either the heavy chain (HC) or light chain (LC). Depending on the heterodimerization strategy, differences in the constant domains may exist as well. Once a decision is made to develop an IgG-like multispecific molecule, the challenges and constraints associated with development of a robust expression cell line and process must also be considered. As these molecules are often comprised of at least four distinct peptide chains (two LCs and two HCs), a strategy to ensure proper pairing of each chain within a given molecule becomes paramount. Without such a strategy, there exists the possibility of producing several different product-related impurities such as monovalent structures and other nonfunctional IgG-like structures, all resulting from mispairing between either a LC and HC (the LC pairing problem), or the two HCs (the HC pairing problem, see Fig. 5.1) [10].

Catumaxomab is one of the first examples of a multispecific molecule approved as a therapeutic, having affinity for both human CD3 and human EpCAM, and targets T-cells to EpCAM-expressing tumor cells. Catumaxomab production

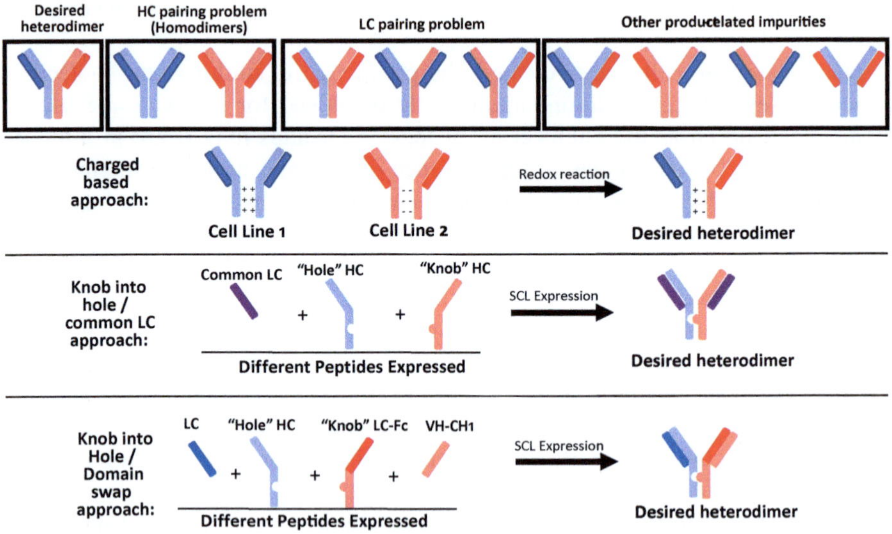

Fig. 5.1 Possible product-related impurities encountered during generation of IgG-like multispecific molecules, and strategies to reduce/eliminate formation of product-related impurities. HC—heavy chain; LC—light chain; Redox—reduction/oxidation reaction; SCL—single cell line; LC-Fc—light chain fused to HC Fc domain; VH-CH1—peptide containing only the variable domain and first constant region of HC

overcomes chain pairing challenges by using isotype combinations from orthologous species. The molecule is comprised of four total chains and two different LC/HC pairs: one LC/HC pair of a mouse IgG2b isotype and the other being a rat IgG2a isotype [11]. The molecule is expressed in a single cell line, generated using quadroma technology [11]. These cell lines are created via somatic hybridization of two hybridoma cell lines (a hybrid hybridoma), each expressing either the mouse IgG2b or the rat IgG2a. This strategy can accomplish high levels of correctly paired heterodimer via (a) favorable LC-HC binding between homologous species chains, and (b) specific purification of properly heterodimerized Fab molecules [7]. Although this strategy results in properly paired chains, this format is highly immunogenic in humans and results in the development of antidrug antibodies in patients, which in theory limits its dosing [12].

Advances in both protein and cellular engineering have made alternative expression and production strategies possible. First, the more commonly used Chinese Hamster Ovary (CHO) cell expression systems are suitable for the production of these molecules and provide distinct advantages over hybridoma lines [13]. CHO cells are well-known for the expression of large molecule biotherapeutics due to their ability to correctly assemble, fold, and properly modify recombinant proteins. CHO cells are also capable of rapid growth and reaching high cell densities to facilitate high recombinant protein yield and have been adapted to suspension growth in serum-free, chemically defined media. Despite this, methods used to produce IgG-like multispecifics must still overcome pairing problems with both the HCs and LCs. Here, we examine two common approaches employed to overcome both challenges—the charge-based electrostatic approach, and the steric-based knob into hole (KiH) approach (Fig. 5.1), and cell line generation considerations for each strategy.

5.2.1.1 The Charge-Based Electrostatic Approach

Under certain conditions, HC disulfide bonds of IgG4 molecules reduce and subsequently reform new interchain disulfide bonds both in vivo and in vitro, resulting in multispecific molecules [14]. This observation is unique to the IgG4 isotype and is facilitated by specific residues within the hinge and CH3 regions. Mutation of corresponding residues in IgG1 and IgG2 molecules renders them competent for exchange as well. Further mutation of these residues to have opposing charges on each HC (glutamic acid for the negatively charged HC, arginine for the positively charged HC) favors formation of a stable heterodimeric molecule when the differently charged HCs are mixed and mildly reduced [15].

Production of these molecules is facilitated by a two-cell line approach, in which each cell line produces a different parental homodimer possessing unique variable domains and opposing charges engineered into the HCs (either glutamic acid or arginine residues). After separate upstream manufacturing processes, purified homodimers are mixed together under appropriate reduction and oxidation (redox) conditions to create the heterodimeric molecule. An example of such a

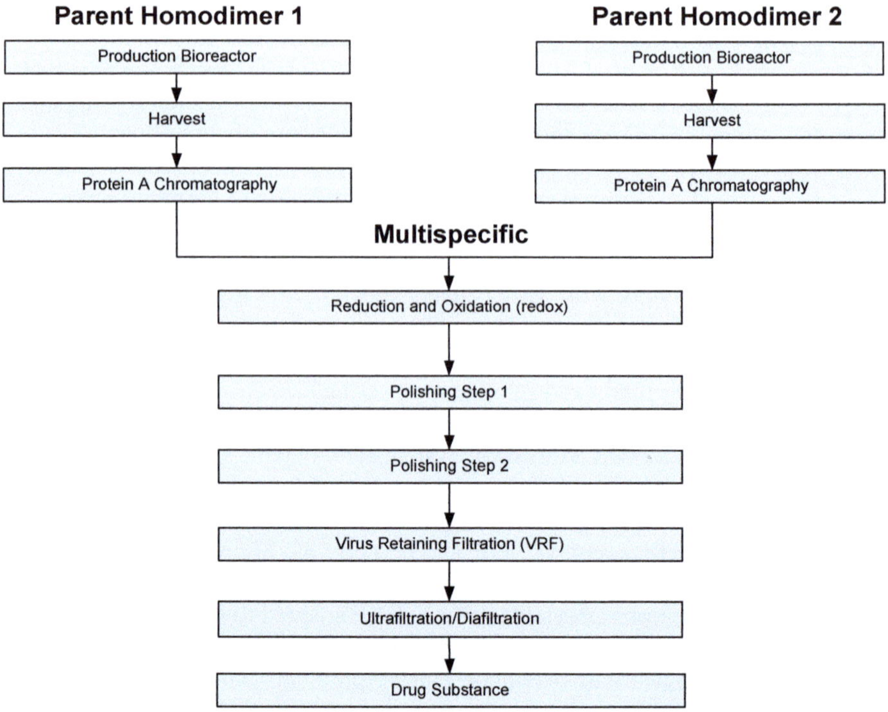

Fig. 5.2 Example of a manufacturing process using the two-cell line approach

manufacturing process, including a potential position for the redox step, is presented in Fig. 5.2. This approach solves the LC pairing problem by simply restricting LCs to one cell line and one process. While the reduction step likely reduces the disulfide bond formed between the HC-LC pair, the Fab does not dissociate, due to the tight molecular interactions between the Fv, as well as the light chain constant domain and CH1. Cell line generation for this type of multispecific is much like a standard monoclonal antibody; vector configurations, CHO host cells, and clone screening strategies used for the generation of the homodimer cell lines are compatible with those used for standard mAbs. The only major difference between generation of cell lines for mAbs and charged based multispecifics is that two cell lines are established. The need for multiple cell lines and at least separate upstream manufacturing processes/Protein A chromatography steps can pose a challenge for at scale manufacturing, which will be addressed later in this chapter. It should also be noted that if observed, a half-antibody impurity after Protein A purification is acceptable, as this species is competent for incorporation into a heterodimeric molecule during the redox step. Since the cell line generation approach used for two cell line multispecifics is highly similar to well-established approaches for mAbs, the rest of this section will focus on cell line approaches for single cell line multispecific molecules.

5.2.1.2 The Knob into Hole Approach

Based in part on a model proposed by Francis Crick for packing of amino acid side chains, the concept of the knob into hole approach is to facilitate HC heterodimerization by mutation of an amino acid residue within the CH3 domain in one HC from a large to a small amino acid (the "hole" chain) and then from a small amino acid to a larger one (the "knob" chain) [16]. In the first iteration of this strategy, the hole chain harbored a tyrosine to threonine (Y407T) mutation, while the knob chain bore a threonine to tyrosine (T366Y) mutation. Following co-expression from a single cell line, the resulting Protein A purified pool was comprised of greater than 90% of the desired heterodimer [16]. Further protein engineering efforts obtained via phage display screening yielded a change to the knob mutation, from tyrosine to tryptophan (T366W), as well as two additional mutations in the hole heterodimer, T366S and L368A, which improved molecule stability following cellular expression [17].

While this heterodimerization strategy is quite robust, it does not account for the LC pairing problem. One simple way to alleviate this concern is through use of a common LC for both antigen binding arms; however, not all targets may be easily amenable to this approach. Another potential strategy to overcome LC pairing challenges is through domain crossovers [18]. In this approach, one Fab (fragment antigen binding) arm would retain its canonical LC/HC structure, while the other Fab HC variable domain would be expressed as part of a "new" LC, either fused to CH1 (HC constant domain one) or CL (LC constant domain), and the LC variable domain would be expressed as a "new" HC, and be fused to either CH1-Fc or CL-Fc (e.g., see Fig. 5.1). This approach solves the LC pairing problem as the "new" LC would not be competent for binding to the standard HC on the first Fab arm and likewise the same for the new HC's ability to bind to the standard LC. However, depending on the molecular format selected, new impurities can be generated [18].

Another strategy to overcome the LC pairing problem is through genetic engineering based upon either charge or hydrophobicity of the CH1-CL interface to favor correct pairing [19]. In a previously described charged-based approach, a negatively charged amino acid was substituted in the CH1 domain (T192E), while a positively charged amino acid was introduced into the CL domain (N137K). These mutated residues form a salt bridge, promoting specificity between this HC/LC pair while helping to avoid improper pairing. A similar strategy can be used by swapping out a pair of amino acids on each chain, one apolar pair of residues for a polar pair (L143Q in CH1 and V133T in CL), and vice-versa (S188V on CH1 and S176V on CL).

In terms of cell line generation, when comparing HC heterodimerization approaches (charged-based vs. knob into hole), it may seem as though the knob into hole approach would be favorable, as it only requires development of one cell line per molecule, as opposed to two. This may be true in some circumstances, such as in cases where multiple targets are being considered for a particular Fab, or multiple constructs are being considered for a single target. However, the single cell line-based knob into hole approach may present more technical challenges that are difficult to overcome. These include the ability to screen for high levels of product titer, as well as low levels of product-based impurities (e.g., homodimers and mispaired LCs), and difficulties in cellular genetic characterization. Cell line stability may also offer a challenge with the knob into hole approach, as a greater number of peptide chains (which can also be more unique than LC/HC) are being expressed in a single cell. Proper expression ratios must be maintained over generational age, which may be more challenging for the cell, when compared to expressing homodimers in separate cell lines. The following sections offer considerations and strategies for overcoming challenges faced when generating single cell lines for the expression of knob into hole multispecifics.

5.2.2 Cell Line Generation Considerations for Single Cell Line (SCL) Multispecifics

5.2.2.1 Stable CHO Host Cell Integration System—Random or Targeted?

Currently two prevailing stable CHO host cell expression strategies are used for the production of biotherapeutics. One strategy, the random integration platform, results in transgene sequences that are integrated within the CHO host cell genome at random numbers of copies and chromosomal locations. An alternative to this approach is the targeted integration platform, wherein the transgene sequences are enzymatically inserted into pre-defined genomic loci [20]. Both systems have proven to be efficient vehicles for the production of SCL multispecifics [21, 22].

The indiscriminate nature of the random integration system, which is sometimes viewed as a negative for the expression of standard monoclonal antibodies, can be useful for the expression of SCL multispecifics. The differing chromosomal locations of transgene vector integration can vary with respect to the amount of transcriptional activity they provide transgenes [23]; this variability can be used as an advantage. Certain chains within the multispecific may exist at a higher molar ratio within the molecule relative to other distinct chains (e.g., common LC multispecifics), which would necessitate higher expression of these genes. Further, some chains that comprise the multispecific may not be transcribed and/or translated as efficiently as others. By virtue of transgene insertion at arbitrary genomic locations, random integration can provide a range of transcriptional efficiency for transgenes that can easily be harnessed through single cell cloning and screening for proper product formation. In other words, it is possible to leverage the "random" nature of the random integration system for the production of multispecifics, but appropriate analytical screening strategies are required; this will be discussed further below.

For targeted integration, typically the pre-defined loci for insertion of transgenes are ones that provide high and stable transgene expression [24, 25]. Unless a battery of insertion sites is identified, characterized, and developed, targeted integration does not offer the benefit of transcriptional differences that random integration has the potential to provide. One potential benefit of targeted integration is that determination of the likelihood of success/failure happens more quickly; analysis of material generated from targeted integration transfection pools can generate meaningful results with respect to both product titer and product quality, whereas with random integration a cloning step is more likely to be required. If targeted integration is found to be suitable for the expression of SCL multispecifics, cell line stability is also less of a concern compared to random integration due to the predictable stability of the integration site.

5.2.2.2 Expression Vector Considerations

The expression vector and transgene topologies serve as another potential source of variability of transcriptional efficiency; if appropriately leveraged, knowledge of this variability can lead to higher total multispecific yields and lower product-related impurities. Governing rules from expression of standard mAbs can serve as a foundation for the production of SCL multispecifics. First, transcriptional analysis of several standard mAb expressing cell lines has established that of the two chains, the HC is limiting in its expression. In other words, there is a direct correlation between HC transcript/peptide and mAb product titer [26]. In addition to this relationship, it has also been demonstrated that LC needs to be in excess to HC, as mathematical modeling of the mAb synthesis pathway suggested that IgG assembly time and LC-HC ratio are inversely correlated [27]. Cell lines which both naturally and have been engineered to express mAbs typically express LC in excess [28, 29]. Scenarios in which the HC is forced to higher expression than LC result in HC aggregation, as well as a decrease in cell viability [30]. The simplest way to account for these rules in developing expression vectors for mAbs is to construct a vector positioning the LC upstream of the HC, using identical promoters and polyadenylation signals that confer very high expression. In this topology, the LC would theoretically not be subjected to transcriptional interference in the same way that the HC would (from the LC), thus providing a boost in expression relative to HC such that the LC would not be limiting. Other options for optimizing LC-HC expression ratio include varying promoter/terminator strength, and varying gene copy number within the vector.

If opting for a single vector approach, the position of each chain within the vector must be carefully considered. Chain position within the vector has the potential to impact transcriptional activity within the cell, which can have downstream effects on overall product yield, product quality, and therefore even efficacy and safety. To compensate for any transcriptional interference effects on chains positioned further downstream within the vector, the use of stronger promoters/terminators, or inclusion of multiple copies of the same gene should be considered such that desired molar balance in the final product can be achieved. The use of different promoters/terminators may be preferable in this instance, as they can aid in ease of vector construction, sequencing, and genetic characterization of production cell lines.

Multivector strategies could also be employed. In this scenario, several vectors could be constructed having either some or all transgenes, in different orientations.

Selection of cells that have taken up the expression vector is usually accomplished through the vector harboring a gene expressing either (a) a protein that confers resistance to an antibiotic such as geneticin or hygromycin or (b) a protein that complements an auxotrophic phenotype, such as glutamine synthetase. In the random integration host system, cells are amenable to integrating several copies of the vectors harboring these genes as a means of surviving the selection. To further enhance clonal diversity, cell transfections could be performed using different ratios of vectors. A single selection scheme could be employed, such that all vectors have the same selectable marker. This approach would require intensive screening for clones producing the desired product quality, and cell line stability could also become an issue. To limit these issues, multiple selection schemes could be used (e.g., one vector harboring a geneticin resistance gene, another harboring a hygromycin resistance gene); however, this approach may limit clonal diversity, as cell selection would be more stringent.

For a targeted integration approach to the production of SCL multispecifics, variations in vector topology serve as the only sure way to influence expression differences between genes coding for the different molecule chains. Due to the genetic loci of insertion being controlled across the entire transfection pool cell population, if a vector topology is chosen that results in poor expression/molecule product quality, it will be very difficult/impossible to screen for a high-expressing clone. Therefore, an upfront evaluation of multiple vector configurations is paramount for the expression of SCL multispecifics in a targeted integration system [21]. Fortunately, this can be facilitated by the targeted integration platform, as these types of screens can be performed prior to single cell cloning, on material generated from the cellular transfection pools. Depending upon the exact mechanism/genetics of the targeted integration host cell system, pool screening results should not be expected to exactly mirror results from the clones; however, general trends should remain consistent; pools with higher expression and better product quality should yield better clones. Once cellular pools harboring a certain vector topology yield adequate results, clones can be generated, which can then be subjected to further screening. Thus, in the absence of a clear understanding of the relative expression of the individual polypeptide chains, and a similar understanding of vector positional effects, early vector screening is essential.

5.2.2.3 Cell Line Screening Strategy Considerations

When compared to cell line screening for the expression of standard monoclonal antibodies, clone screening for SCL multispecifics typically requires (a) screening a larger number of clones, and (b) different clone screening analytical tools (for illustrative example, see Fig. 5.3). The natural existence of monoclonal antibodies coupled with decades of experience and learnings can facilitate development of standard monoclonal antibodies without much upfront analytical screening. In the absence of these benefits, SCL multispecifics need tools that can provide both meaningful information and rapid results such that clone culling can happen as

Monoclonal antibody

Screening Criteria	Illustrative Result
Protein A titer measurement	~95% of clones analyzed positive for protein A titer
Cell growth	~95% of clones successfully scaled
Cell growth, protein A titer	>95% of clones could be progressed
Cell growth, protein A titer, intact product of interest via SEC / gel electrophoresis	95% of clones could be progressed

Screening step:

Cloning plate outgrowth
↓
Static well plate screening
↓
Shake flask screening
↓
Small-scale reactor screening

Single Cell Line (SCL) Multispecific

Screening Criteria	Illustrative Result
Protein A titer measurement	~80% of clones analyzed positive for protein A titer
Cell growth and ELISA screening for binding to multiple antigens	~25% of remaining clones positive for binding to all antigens
Cell growth, protein A titer, intact product of interest via SEC / gel electrophoresis	~10% of remaining clones expressing significant amounts of product of interest
Cell growth, protein A titer, intact product of interest via SEC / gel electrophoresis	~60% of remaining clones could be progressed

Fig. 5.3 Comparison of example cell line screening strategies and potential results for both a traditional monoclonal antibody and a single cell line (SCL) multispecific. Since SCL Multispecific cell line screening strategies need to consider product quality results more than traditional monoclonal antibody cell line screening, these typically employ more screens with lower success rates, resulting in the need to screen significantly more clones

efficiently as possible. As the cell line generation process progresses and the number of clones that are being analyzed decreases, different, more intensive molecular screening can take place. Because of this, an upfront assessment of the possible impurities that may be encountered, development of a robust screening strategy, and inventory of other cell screening reagents (e.g., homodimer impurities and target antigens) must be completed to ensure a successful cell line generation campaign. Creation of cell lines intended to exclusively express the homodimers should be strongly considered, as this reagent is not only a critical component of cell line screening, but may be useful for other facets of development, such as assay development for characterization and release testing. For these, cell line screening and generation will likely not need to be as extensive, and dependent upon the expression system used, transfection pools are often sufficient for their production.

One of the first screens that should be employed while clone screening for IgG-like multispecifics is the ability to make material capable of binding to Protein A (Fig. 5.3). This screen should be implemented shortly after single cell cloning, perhaps even as quickly as upon outgrowth of the single cell cloning vessel. While this assay will give a general idea as to whether a given clone is expressing IgG-like molecules, it does not discriminate between clones expressing high amounts of HC homodimers, or mispaired LCs. Therefore, results from this screen should be treated as preliminary. If feasible, taking forward all clones that can produce Protein A bound material should be considered, allowing for the highest amount of clonal diversity. If utilizing the domain crossover approach [18], the framework of the VH-CH1 peptide should be considered. Binding of HC to Protein A can be accomplished through interactions other than the HC Fc domain, in particular for HCs with the VH3 framework domain [31, 32]. If VH3 is used for the VH-CH1, the resulting peptide will be independently capable of binding to certain Protein A resins, resulting in artificially high titer values. In this case, preliminary titer determinations that do not rely solely on Protein A binding may be used.

Either simultaneous to, or shortly after the initial Protein A screen, the ability of clones to express molecules that can bind to both/all target antigens should be considered (Fig. 5.3). For this assay, we typically employ ELISA (enzyme-linked immunosorbent assay) screening, which offers a robust combination of information and speed to complement a clone screen strategy, as hundreds to even thousands of clones can be efficiently subjected to this screening. In theory, two ELISA screens, one for each antigen, could be conducted, although it would be preferable to design a format which can provide information on the ability to simultaneously engage both antigens in the same assay, as this has the potential to eliminate clones producing only mixtures of incorrect homodimers. Much like a Protein A titer assay, the ELISA assay can be used to determine protein quantities. For those clones that can bind to both antigens, ratios between material bound to one antigen versus another or versus total Protein A bound material can be established and used as screening criteria. It should be noted, however, that these ratios represent a snapshot in time of an early screening stage. If the cell line experiences genetic drift, these values may not remain static throughout cell line development. Results from these screens should be treated semi-quantitatively, and ratio ranges to determine progression to the next round of screening should be kept large.

Once clones producing material capable of simultaneously binding to both target antigens have been identified, product quality must be examined (Fig. 5.3). One such screening assay to be considered for this purpose is analytical size exclusion chromatography (SEC). This method is useful for obtaining high-resolution separations of desired protein molecules from other product related species, such as HMMS and degraded forms. For detection of smaller fragments, methods based on gel electrophoresis should be considered. These methods can either utilize standard polyacrylamide gel electrophoresis or capillary electrophoresis, the latter of which may lend itself better to high-throughput screening. The electrophoretic methods are suitable for identifying clones producing material with significant size differences, for example, missing one or more of the desired polypeptide chains, but may not be suitable for identifying clones producing product-related impurities that are similar in size to the desired product (e.g., homodimers). If better resolution is required than what can be provided by electrophoretic methods, mass spectrometry may be needed, although screening material from large numbers of clones by this method can prove cumbersome.

For some molecules, the size of homodimer impurities may be too close in size to the desired heterodimer to be able to differentiate for screening purposes. In these cases, other methods of separation must be explored. This can include separation based on attributes such as charge, hydrophilicity, and hydrophobicity. These types of screens will require sufficient amounts of homodimer impurities to serve as controls, which can be generated via their own cell line generation efforts, as mentioned above.

Once clones have been identified that can produce high amounts of the properly assembled molecule, they should continue to be assessed using more traditional clone screening strategies. This includes phenotypic assays such as performance in a small-scale bioreactor process, using cells that have been passaged to an appropriate generational age. Upon harvest, it is prudent to confirm molecular structure with one of the previously established assays to determine the impact (if any) that the generational age and bioreactor environment have on molecular assembly. It has been our experience thus far that analytical results from material harvested from a small-scale vessel (e.g., plate well or shake flask) have been comparable to material from a more representative process, such as a fed-batch bioreactor.

Just as a standard mAb, stability of the expression construct to the limit of in vitro cell age (LIVCA) needs to be confirmed. In addition to loss of product yield, loss of transgene(s) expression can also have significant impact on product quality for multispecifics. If screening clones transfected with different vector configurations, it is good practice to maintain clone vector diversity at least through the cell line stability assessment. Certain vector configurations may be able to yield high/optimal expression and product quality, but unable to maintain this phenotype over cellular generational age. Traditional methods of ensuring transgene stability (quantitative PCR, Southern/northern blotting, etc.) may have difficulty in discerning between the genes coding for the different chains of a multispecific molecule. For this reason, variant codons that can provide enough heterogeneity between similar chains should be considered and incorporated where possible.

5.3 Process Development of Multispecifics

5.3.1 Upstream Process Development

The production of multispecific antibodies is similar to traditional antibodies in terms of upstream development and optimization. Disulfide bonds between LC and HC as well as between the two HCs comprising an antibody molecule play an essential role in maintaining structure. As a result, more attention may be required to ensure parameters that might affect the oxidative or reductive capacity of the cell culture are carefully determined and well controlled.

The production bioreactor can be operated in a wide variety of ways including batch, fed batch, perfusion (either more traditional steady-state, or with fluctuating viable cell densities [dynamic]), or even hybrid modes that combine perfusion and fed batch [33]. The choice of the bioreactor operational mode as well as the operating conditions is optimized for each unique molecule. Commonly manipulated and important parameters include seeding density, pH, temperature, and the composition of media and feeds as well as the volume and timing of addition of feeds, and cell culture duration. The adjustment of these and other parameters enables not only the ability to produce suitable quantities of multispecific antibodies but also ensue the product is of high quality and potential issues such as covalent dissociation, clips, and aggregation are minimized or eliminated (Chap. 4 in this volume by Gregory W. Hiller).

5.3.2 Downstream Process Development Considerations

The main objective when developing a downstream process for any protein is to develop a robust, high-yielding, and scalable process that can reproducibly deliver high-quality product. Downstream processes normally start with the removal of cells and cellular debris followed by purification of the product of interest through a series of chromatography and filtration steps. Once fully purified, the product is concentrated and exchanged into a stable buffer solution for long term storage.

Downstream development for multispecific antibodies shares much with typical monoclonal antibodies, but there are a few unique challenges that these molecules provide that require additional attention. Using mammalian systems, the product along with impurities is secreted from the cells into the culture medium. Similar to mAbs, downstream processes for multispecifics are designed to remove host cell DNA, host cell proteins (HCPs), adventitious viruses, endogenous noninfective retroviral particles, aggregated product, endotoxin, process additives, and residual media components.

Multispecific antibodies often contain additional impurities not found in typical mAb processes, such as homodimers, fragments, or product with unwanted subunit pairings/mis-parings [8, 34, 35]. As detailed earlier, chain mispairing can result in several different impurities that make removal challenging (Fig. 5.1). The separation of the product of interest from these impurities can be approached through traditional purification strategies. These include optimization of existing or traditional process steps through changes such as collection criteria, pH, buffer composition, or inclusion of multiple buffer/wash steps. In some cases, optimization of existing/traditional unit operations may not be sufficient and additional or less commonly employed steps would be required. For example, additional chromatography modes may be required that exploit different separation mechanisms based on charge, size exclusion, metal affinity, or hydrophobicity. The order of these steps can also have an impact, so assessing the order of the various process steps should be considered. For example, removal of a specific impurity in an early stage of the process might lead to an overall downstream process that is simpler and more robust.

The following sections describe some of the purification challenges as well as potential product stability issues that may be encountered when working with multispecific antibodies. In addition, a case study is presented to further highlight the complexity of multispecific antibodies as compared to more traditional mAbs.

5.3.2.1 Unique Impurity Challenges

Unique impurities are common when producing multispecific antibodies. The type and amount of these impurities can depend upon both the molecular architecture as well as host cell line used (some example impurities are shown in Fig. 5.1). Common impurities consist of knob/knob (K/K) and hole/hole (H/H) homodimers for the knob into hole (KiH) approach or full antibody or half antibody homodimers for the dual cell line approach. Other impurities that can be found consist of mispaired HC and LC combinations as well as fragments (HC/LC variants) [8, 9]. The type of fragments found can consist of heterodimer molecules where one or more LCs is absent. Another common impurity can occur when LCs are swapped resulting in undesirable molecular structures.

Product-related impurities add to the difficulty in achieving multispecific drug substance that meets an acceptable purity specification. Homodimers, for instance, are often very similar to the desired product in terms of size, shape, charge profile, and hydrophobicity. Thus, multiple purification approaches using different modes to achieve drug substance of acceptable quality may be required. Although homodimers are typically similar to the desired product, some are different enough that they can be more easily separated from the heterodimer. As an example, examining the crystal structure of K/K and H/H homodimers compared to the desired heterodimer showed that the K/K homodimer much more closely resembled the heterodimer than the H/H homodimer [36]. The H/H homodimer has a slightly more open structure than either the K/K or the multispecific molecule which exposes different patches of amino acids and consequently provides a means to separate this impurity from the heterodimer [36]. Due to the similar structural nature of the K/K homodimer to the heterodimer, it can be more difficult to remove than the H/H homodimer. To avoid more complex downstream processes, it is important to use a cell line and cell culture process that maximizes the output of desired heterodimer while minimizing product-related impurities. As previously discussed, by optimizing the expression vector and cell line, different transgene transcript ratios can be achieved to limit formation of unwanted species.

Depending upon the format chosen, product-related impurities such as multispecific antibodies with LCs swapped or the loss of a LC(s) can also challenge the downstream process. These impurities, sometimes referred to as fragments, not only pose challenges for the downstream process but can sometimes form HMMS thus presenting an additional purification challenge [8, 9, 37]. Depending on the level and nature of fragment(s) during the production of a multispecific, there might be one or more additional downstream process steps needed to reduce levels to acceptable limits. mAb-like impurities that contain swapped lights chains are the most problematic as these molecules are very similar to the intended multispecific antibody. Removing these mispaired molecules downstream is a significant challenge and relies heavily on exploiting as many minor differences as possible, often leading to more complicated purification processes [8]. As mentioned earlier, an appropriate engineering of the molecule is the best way to prevent mispairing of the HC-LC.

5.3.2.2 Stability Concerns

Through a combination of molecular design and genetic engineering, many of the initial issues encountered in the development of multispecific antibodies can be resolved. However, depending upon the molecular architecture used, product stability issues can still be encountered. One example is low pH instability that is sometimes observed in the charge-based approach. The mutations incorporated in the homodimer molecules to enable correct formation of the heterodimer can also have the potential to adversely impact product stability. As described previously, these mutations consist of charged amino acids that work to destabilize the homodimers when produced as full-length antibodies (parents) as the charges repel each other [15]. In some cases, this destabilization is great enough that the full antibodies can separate into half antibodies under normal conditions without the addition of reductant. Additional molecular design can minimize or even prevent these issues altogether, but in some cases, there may be no solution. In this circumstance, limiting exposure to the low pH condition required for elution from the Protein A capture column, such as raising the elution buffer pH and/or neutralizing the product pool quickly post elution may be necessary. Depending upon how sensitive the molecules are a Protein A capture step may not even be possible. Alternative capture steps may be required; however, this will likely require additional development to effectively design a process capable of delivering a product that meets acceptable quality. This possibility further highlights the importance of performing molecular assessment upfront on candidate molecules and, as needed, providing the necessary time to potentially re-engineer a candidate molecule prior to moving forward with development.

5.3.2.3 Reduction and Oxidation (REDOX) Step

There are two strategies for producing multispecifics using a two cell line approach: the homodimers can be produced as either half antibodies or full antibodies. In the case of the half antibody approach, each half antibody contains a different binding epitope. In order to ensure that these combine in the correct configuration to form the desired multispecific, knob into hole technology is employed such that one half antibody contains amino acid mutations to create the knob, while the other half antibody contains amino acid mutations to create the corresponding hole in the Fc region. These half antibodies are produced with the inter disulfide bond cysteines being cyclized or capped. For the full antibodies two completely intact antibodies are produced such that each contains a different epitope. These two antibodies also contain mutations in the hinge and CH3 region that enable reduction and ultimately oxidation creating the desired multispecific. In both cases, a redox reaction is needed, consisting of a reduction and oxidation of the disulfides for full antibodies [15] or the removal of the caps or recyclization of the cysteines for half antibodies [9]. Traditional reducing agents such as dithiothreitol (DTT), 2-mercaptoethanol (BME), and tris(2-carboxyethyl)phosphine (TCEP) are very common reducing agents for refolding or analytical analysis of disulfide bonds [38]; however, these are considered strong reducing agents and tend to reduce all of the disulfide bonds within an antibody, not just the inter-HC disulfide bonds that must be reduced and then reoxidized to form the desired multispecific. Mild reducing agents such as reduced glutathione, 2-mercaptoethylamine, or cysteine have proven successful in these cases as they primarily impact the intra-HC disulfide bonds or the caps or cyclized cysteines without significantly reducing the LC-HC or intrachain disulfide bonds [9, 15, 39]. The choice of reducing agent depends on several factors including the ability to successfully reduce the molecule, availability, cost, toxicity, and the ability to remove the reducing agent following the redox reaction.

There are several key parameters that should be considered when developing the redox reaction step including buffering capacity/pH, temperature, reductant stoichiometry, molar ratio of proteins, mixing, and oxygen levels. Optimizing the reaction timing or kinetics and the interplay of these parameters on timing/kinetics must be understood when optimizing formation of the desired multispecific. Reduction and oxidation occur under alkaline conditions and sufficient buffering capacity to control the pH above neutral is important. Higher pH conditions tend to drive the reaction faster [40]; however, elevated pH can destabilize the proteins and potentially impact the charged species profile (e.g., asparagine deamidation).

Similar to pH optimization, the impact of temperature must also be well understood. Operating at elevated temperatures (30–37 °C) will drive faster reaction kinetics than operating at room temperature but exposing the molecule to elevated temperatures may also lead to instability. In most cases, reactions operating at room temperature provides acceptable reaction kinetics and is more stable. This is more amenable when considering manufacturing facilities as warmer temperatures require jacketed or temperature controlled rooms or tanks [41].

In addition to the reductant employed, the ratio of reductant to protein is also a critical parameter. There is a strong correlation between increased reductant and more rapid reaction kinetics. This increase is seen until the reductant level is saturated [42]. There is also the potential to add too much reductant leading to degradation of the molecule due to reduction of additional disulfide bonds. Beyond the ratio of reductant to protein, the molar ratio of each half antibody or full antibody is also a parameter that should be considered. A 1:1 molar ratio of either the full antibodies or half antibodies to each other is typically employed during the reaction as this will lead to full conversion to the multispecific, eliminating half antibodies or homodimers, thus simplifying subsequent purification steps [9, 43]. This ratio can be adjusted in certain cases if there is a driver such as enabling a more efficient and robust downstream process.

Once the molecules are reduced, the oxidation can occur naturally over time by ensuring sufficient oxygen is present. Oxidation can be accelerated through the addition of either the oxidized form of the reducing agent and/or metals that are known to promote oxidation. Ensuring sufficient levels of oxygen may consist of either sparging with oxygen or providing an air overlay while continuously mixing. If there is insufficient oxygen, then other means such as chemical oxidation or buffer exchange to remove the reductant can be considered.

Another consideration is the potential for undesired reduction of the molecule as has been observed during harvest and clarification operations using depth filtration [44]. Although not a part of the desired process, it has been observed that in harvest and clarification using a depth filter there is the potential of reduction of disulfide bonds. Depending on the format chosen, this might have significant impact on the molecule. This may result in undesirable impurities such as fragments and should be carefully considered and or evaluated for each molecule and appropriate corrections made to prevent reduction during harvest and clarification.

5.4 Manufacturing of Multispecifics

The production of multispecific antibodies is complex; although extensive development work is done to ensure that processes are robust, efficient, consistent, and scalable there are several important considerations when manufacturing these molecules. The need for additional process operations, stability data to support hold times and manufacture scheduling considerations are just a few. An evaluation of both multispecific approaches, KiH and electrostatic, may initially favor the single cell line KiH approach, as this eliminates the need for a second series of upstream process operations, but other aspects of each process need to be considered in determining which process to use.

As such, one aspect to consider is protein stability and the potential need for additional hold times. In the case of a two-cell line multispecific process, the operation of the production bioreactors may require additional hold time if the bioreactors are operated sequentially. This challenge may be circumvented by operating homodimer production bioreactors in parallel. Considering the risk mentioned above, if operating two bioreactors and one fails and hold times are not sufficient to cover the time needed to generate a back-up reactor, this would have significant impact on both manufacturing time and cost. Another consideration with the two-cell line approach is that one homodimer might be used for other multispecific molecules. Thus, it is extremely important to carefully consider homodimer stability in the early design of the manufacturing process for these types of multispecifics.

Another consideration for manufacturing of multispecifics is cost. If total process yield is equal, the single cell line approach would likely be favored in terms of cost, as two bioreactors run either in parallel or sequentially will have higher overall costs due to an increased need for more raw materials, equipment, and personnel. Extending this, the cost of failure for a two-cell line approach would also be higher. In our experience, however, total process yield between the single cell line and dual cell line approach is not equal, with the dual cell line approach typically having significantly higher yield. This demonstrates the need for a careful risk-benefit analysis when considering cost of the multispecific heterodimerization strategy. Other variables that may need consideration here are anticipated product demand and drug substance stability.

Considering the complexities associated with the production of multispecifics, scheduling and equipment utilization should not be ignored. Production using two cell lines either means that the manufacturing suite is occupied longer if the bioreactors are run sequentially or more bioreactors are occupied if the homodimers or half antibodies are produced in parallel.

In addition to these considerations, when manufacturing multispecific proteins thoughtful handling of product contact waste streams is necessary for multispecific molecules with high potency designations. Multispecific oncology therapeutics traditionally have very high potency and therefore can reach potentially toxic concentrations within the bioreactor. As a result, special considerations are needed in treating product contact waste streams and properly protecting personnel during the handling of such potent biologics.

As the realm of biotherapeutics expands from the standard monoclonal antibody to multispecific modalities, complexities and challenges are introduced to nearly every step of development. Early development choices, such as molecular format and heterodimerization strategy, can have a profound impact on subsequent development steps. To make the optimal choice, one needs to consider all program aspects, including but not limited to cell line generation, process development, manufacturing, as well as product safety profiles, and material demands.

5.5 Case Study: Challenges in Removal of Bispecific Product Related Impurities

A case study is presented to highlight some of the challenges that multispecifics (in this case a bispecific) may encounter compared to traditional mAbs. This case study highlights development strategies and process adjustments that were made to enable manufacture of a product with acceptable quality. The case study focuses on using a single cell line to produce the bispecific, utilizing the knob into hole technology.

The molecule presented in this case study is a diabody fused to an Fc domain (diabody-Fc) [45]. The diabody portion of the molecule is comprised of two peptide chains, each chain consisting of an scFv (single chain fragment variable) region with VL (variable light)/VH (variable heavy) domains having affinity for different targets. In the final molecule, each Fv is comprised of a VL from one chain and a VH from the other chain (Fig. 5.4). Compact packaging of the properly paired VL-VH is facilitated by short linkers in the scFv peptides; these structural constraints favor the formation of the diabody without domain exchange [46]. To favor

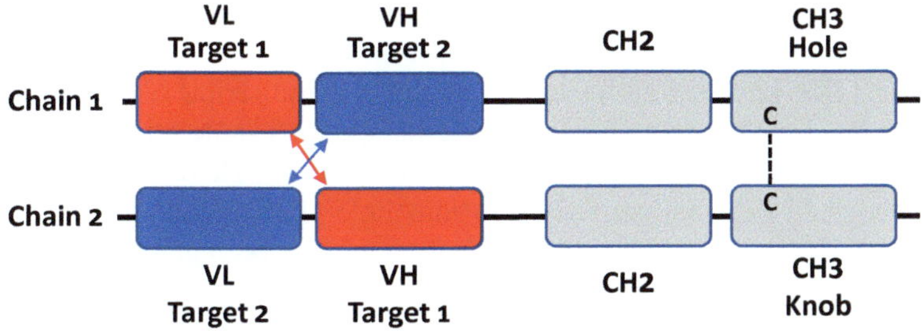

Fig. 5.4 Molecular architecture of the molecule described in the case study

high yields of the heterodimer, knob into hole technology was engineered into the Fc domains [47]. The choice of the knob into hole molecular design was in part facilitated by the relatively low number of theoretical product related impurities that would be predominantly present (two—the K/K homodimer and the H/H homodimer) in the expression of the two chain diabody-Fc format.

While molecular design inhibits biological activity of the K/K and H/H homodimers, a downstream process capable of generating drug substance with low homodimer impurity is desired. This is especially important in this case study as drug substance stability was impacted by homodimer levels of approximately 4% and HMMS impurity levels of approximately 3%. This impact was eliminated when these two impurities were reduced to ≤1% for each impurity. As outlined in the case study below to robustly deliver stable drug substance, strategic decisions had to be made during both the design of the cell line and throughout biomanufacturing process development. Consideration and understanding of the biophysical characteristics of the homodimer impurities helped to drive several project decisions which resulted in a robust process capable of delivering drug substance with acceptable quality.

5.5.1 Vector Orientation and Cell Line

The host cell line selected for the expression of this diabody-Fc molecule utilizes targeted integration, which harbors one landing pad previously engineered into the host cell at a defined chromosomal location, competent for receipt of the vector payload [20, 24]. Given the use of this host, cell line generation efforts were likely to yield clonal cell lines harboring one integration of the expression vector.

To express the molecule, the vector contained a single copy of both the gene coding for the hole and knob transcripts, under the control of identical transcriptional promoters and terminators. With chromosomal integration and transcriptional promoting signals being equal, the configuration of the cell line vector can be manipulated to differentially impact expression levels of hole and knob transcripts.

Since recombinase-mediated cassette exchange in this host cell is directional, vector configuration will dictate which chain is upstream of the other in the host cell genome. In our experience, when comparing similar genes/transcripts at a fixed chromosomal location, the upstream chain is slightly more transcriptionally active than the downstream gene. For example, in Fig. 5.5, a vector configuration using orientation 1 (the knob transcript preceding the hole transcript) will bias for higher expression levels of knob protein while orientation 2 (the hole transcript preceding the knob transcript) will bias higher expression levels of hole protein.

Knowledge of heterodimerized knob into hole (KiH) Fc domains and corresponding homodimerized H/H and K/K Fc domains has been gained through X-ray crystal structures [36]. As previously discussed, homodimerized H/H Fc domains are more open than KiH and K/K Fc domains. The differentially exposed surface

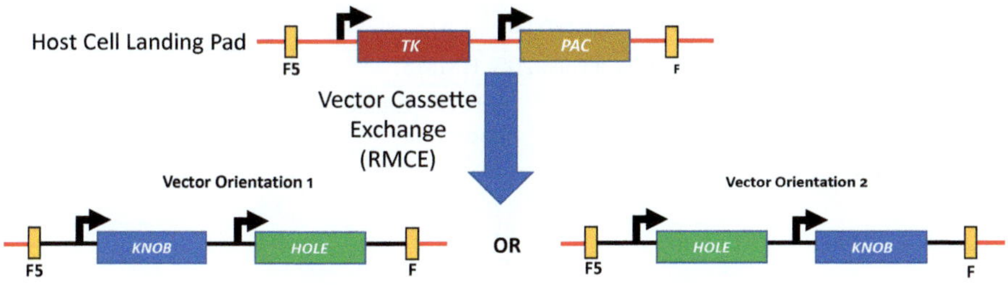

Vector Orientation	Level of HH Homodimer Observed	Level of KK Homodimer Observed
1	Trace	Moderate
2	Moderate	Trace

Fig. 5.5 For targeted integration systems, vector orientation impacts the types of product-related impurities observed. Host cell landing pad diagram adopted from Zhang et al. (2015). RMCE—recombinase mediated cassette exchange; HH—hole-hole; KK—knob-knob; TK—thymidine kinase gene; PAC—puromycin N-acetyltransferase gene

areas of H/H homodimers provide unique handles for removal through various protein purification techniques.

In this case study, orientation 2 was selected for the vector design to bias higher expression levels of the hole protein. In doing so, the H/H homodimer impurity, an impurity easier to separate from the heterodimer product than K/K homodimer, had levels of approximately 20% H/H in the Protein A pools (Fig. 5.5) while the K/K homodimer was undetected.

5.5.2 Downstream Process Design: Removal of Impurities

5.5.2.1 H/H Removal

By intentionally designing the single cell line to preferentially produce the hole HC, H/H homodimer was observed in the Protein A captured pool. H/H levels of over 20% were observed after Protein A capture. Although it is easier to remove the H/H homodimer, due to the significant presence of this impurity, significant downstream removal capacity is required.

Through high-throughput resin screening and additional development experiments, two differentiating biophysical properties between H/H homodimer and the heterodimer product were identified and leveraged to design a robust downstream process:

1. H/H bound more tightly to anion exchange (AEX) resin than heterodimer (Table 5.1).
2. H/H had reduced low pH stability than heterodimer with H/H converting to HMMS during low pH exposure (Fig. 5.6).

Table 5.1 Impurity data from AEX resin screen

Conditions		Load impurity levels: 20% HMMS, 23% H/H		
pH	Anion level	% HMMS—Eluate pool	%H/H—Eluate pool	% heterodimer—Eluate pool
Lower	Low	6	16	78
Medium		5	7	88
Higher		3	5	92

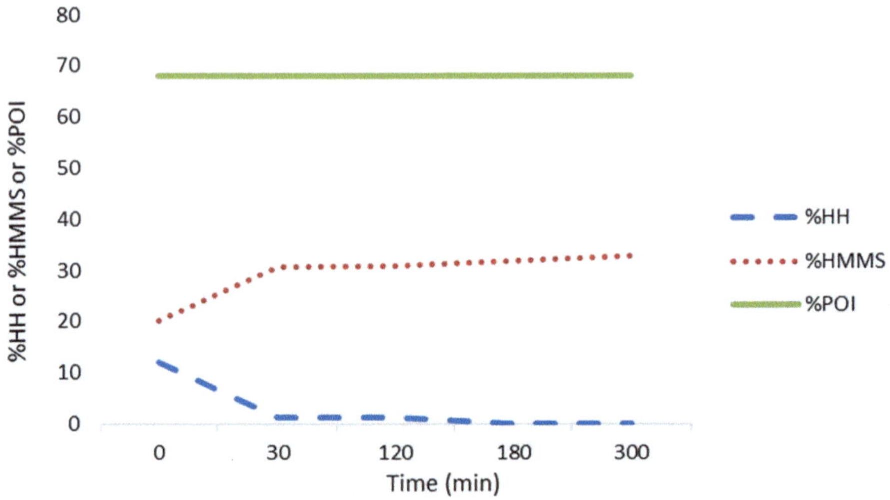

Fig. 5.6 Impact of low pH on product of interest (POI), hole/hole homodimer (H/H), and high molecular mass species (HMMS) levels for knob into hole (KiH) demonstrating conversion of H/H homodimer to HMMS under low pH conditions

5.5.2.2 HMMS Removal

HMMS is a very common product-related impurity observed during mAb manufacture. Several different chromatography modalities (such as ion exchange, hydrophobic, multi-modal) under optimized conditions have proven capable of separating HMMS from the monomer mAbs. Similarly, these modalities can also be applied to bispecific antibodies.

In this case study, HMMS levels of up to 20% were observed following Protein A capture, thus requiring significant downstream removal capacity for this product-related impurity.

Both anion exchange (AEX) and cation exchange (CEX) chromatography provided robust HMMS clearance (Tables 5.1 and 5.2). CEX was observed to have high HMMS capacity, but unlike AEX, had very limited H/H capacity. Identifying more than one step capable of removing HMMS was necessary given the high level of HMMS observed in the Protein A capture pool.

Table 5.2 CEX impurity results from resin screen

Conditions		Load impurity levels: 5% HMMS, 5% H/H		
pH	Anion level	% HMMS—pool	%H/H—pool	%Heterodimer—pool
Lower	Low	NMT 0.2	4	96
Lower	Medium	NMT 0.2	4	96
Higher	Low	1	5	94
Higher	Medium	1	5	94

NMT not more than

5.5.2.3 Balancing H/H and HMMS Removal

Due to the high level of H/H homodimer and HMMS present significant removal capacity for these impurities was needed to build a robust downstream process capable of delivering >99% product purity. To achieve this, several strategies were employed during the design of the downstream process based on previously described experimental observations.

The AEX resin demonstrated limited capacity for HMMS removal, but reasonable capacity to remove H/H homodimer. Although the reduction was significant, the AEX was still limited and could only decrease H/H levels to approximately 5%. The CEX resin demonstrated capacity to remove all the HMMS but could not remove any of the H/H homodimer. The combination of these two process steps (AEX and CEX) demonstrated that the process was capable to clearing all the HMMS, but additional capacity to remove the H/H homodimer was required to achieve the desired drug substance H/H homodimer purity targets. To increase the capacity of the process to remove H/H homodimer, a strategic placement of a low pH step within the downstream process was used to drive a conversion of H/H homodimer to HMMS. In doing so, H/H homodimer, a difficult to remove impurity, would be converted to HMMS, which is an easier to remove impurity from the process stream.

The placement of the low pH step was influenced by the anion concentration ranges that were in use for the AEX and CEX steps. Robust operating conditions for the AEX resin were achieved at high pH, low anion concentrations (Table 5.1), and robust operating conditions for the CEX resin were achieved at low pH and low-to-medium anion concentrations (Table 5.2). Placement of the low pH step following Protein A capture, a traditional location for mAb processes, would have had a negative impact on AEX step performance. The introduction of anions into the AEX load through the low pH titration of the Protein A pool would shift the AEX into a medium anion operating space, impacting impurity removal over this step. Placement of the low pH titration after the AEX step introduced additional anions into the CEX load but well within the anion concentration range and robust operating space for impurity removal over the CEX.

By rearranging the placement of the low pH hold from a traditional mAb process, the capacity for unique bispecific process impurity removal increased while negating the need for in-process pool dilution or an intermediate diafiltration step.

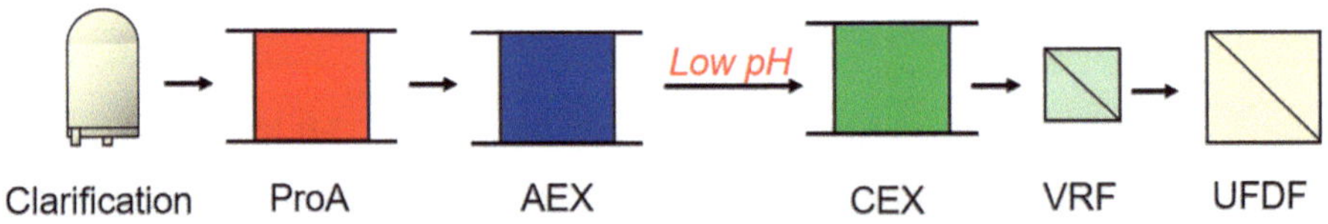

Fig. 5.7 Downstream process overview for knob into hole molecule case study

Table 5.3 In-process impurity levels and yields for KiH case study

Step	Yield (%)	HCP (ppm)	DNA (ppb)	rProA (ppm)	H/H (%)	HMMS (%)
Harvest	90	230,000	2,000,000	N/A	N/A	N/A
ProA pool	90	500	10	10	20	20
AEX pool	65	BLOQ	BLOQ	BLOQ	3	4
Low pH pool	98				BLOQ	7
CEX pool	85					BLOQ
VRF	98					
UF/DF	98					

BLOQ below limit of quantitation, *VRF* virus retentive filtration, *UF/DF* ultrafiltration/diafiltration, *N/A* not applicable

The overall downstream process is illustrated in Fig. 5.7 with typical in-process impurity levels highlighted in Table 5.3.

References

1. Shukla AA, Wolfe LS, Mostafa SS, Norman C (2017) Evolving trends in mAb production processes. Bioeng Transl Med 2:58–69. https://doi.org/10.1002/btm2.10061
2. Foltz IN, Karow M, Wasserman SM (2013) Evolution and emergence of therapeutic monoclonal antibodies: what cardiologists need to know. Circulation 127:2222–2230. https://doi.org/10.1161/CIRCULATIONAHA.113.002033
3. Mullard A (2021) FDA approves 100th monoclonal antibody product. Nat Rev Drug Discov 20:491–495. https://doi.org/10.1038/d41573-021-00079-7
4. Fahrner RL, Knudsen HL, Basey CD, Galan W, Feuerhelm D, Vanderlaan M, Blank GS (2001) Industrial purification of pharmaceutical antibodies: development, operation, and validation of chromatography processes. Biotechnol Genet Eng Rev 18:301–327. https://doi.org/10.1080/02648725.2001.10648017
5. Fan G, Wang Z, Hao M, Li J (2015) Bispecific antibodies and their applications. J Hematol Oncol 8:130. https://doi.org/10.1186/s13045-015-0227-0
6. Dickopf S, Georges GJ, Brinkmann U (2020) Format and geometries matter: structure-based design defines the functionality of bispecific antibodies. Comput Struct Biotechnol J 18:1221–1227. https://doi.org/10.1016/j.csbj.2020.05.006
7. Spiess C, Zhai Q, Carter PJ (2015) Alternative molecular formats and therapeutic applications for bispecific antibodies. Mol Immunol 67:95–106. https://doi.org/10.1016/j.molimm.2015.01.003
8. Li Y (2019) A brief introduction of IgG-like bispecific antibody purification: methods for removing product-related impurities. Protein Expr Purif 155:112–119. https://doi.org/10.1016/j.pep.2018.11.011

9. Giese G, Williams A, Rodriguez M, Persson J (2018) Bispecific antibody process development: assembly and purification of knob and hole bispecific antibodies. Biotechnol Prog 34:397–404. https://doi.org/10.1002/btpr.2590
10. Brinkmann U, Kontermann RE (2017) The making of bispecific antibodies. MAbs 9:182–212. https://doi.org/10.1080/19420862.2016.1268307
11. Chelius D, Ruf P, Gruber P, Ploscher M, Liedtke R, Gansberger E, Hess J, Wasiliu M, Lindhofer H (2010) Structural and functional characterization of the trifunctional antibody catumaxomab. MAbs 2:309–319. https://doi.org/10.4161/mabs.2.3.11791
12. Ott MG, Marme F, Moldenhauer G, Lindhofer H, Hennig M, Spannagl R, Essing MM, Linke R, Seimetz D (2012) Humoral response to catumaxomab correlates with clinical outcome: results of the pivotal phase II/III study in patients with malignant ascites. Int J Cancer 130:2195–2203. https://doi.org/10.1002/ijc.26258
13. Dhara VG, Naik HM, Majewska NI, Betenbaugh MJ (2018) Recombinant antibody production in CHO and NS0 cells: differences and similarities. BioDrugs 32:571–584. https://doi.org/10.1007/s40259-018-0319-9
14. van der Neut KM, Schuurman J, Losen M, Bleeker WK, Martinez-Martinez P, Vermeulen E, den Bleker TH, Wiegman L, Vink T, Aarden LA, De Baets MH, van de Winkel JG, Aalberse RC, Parren PW (2007) Anti-inflammatory activity of human IgG4 antibodies by dynamic Fab arm exchange. Science 317:1554–1557. https://doi.org/10.1126/science.1144603
15. Strop P, Ho WH, Boustany LM, Abdiche YN, Lindquist KC, Farias SE, Rickert M, Appah CT, Pascua E, Radcliffe T, Sutton J, Chaparro-Riggers J, Chen W, Casas MG, Chin SM, Wong OK, Liu SH, Vergara G, Shelton D, Rajpal A, Pons J (2012) Generating bispecific human IgG1 and IgG2 antibodies from any antibody pair. J Mol Biol 420:204–219. https://doi.org/10.1016/j.jmb.2012.04.020
16. Ridgway JB, Presta LG, Carter P (1996) 'Knobs-into-holes' engineering of antibody CH3 domains for heavy chain heterodimerization. Protein Eng 9:617–621. https://doi.org/10.1093/protein/9.7.617
17. Atwell S, Ridgway JB, Wells JA, Carter P (1997) Stable heterodimers from remodeling the domain interface of a homodimer using a phage display library. J Mol Biol 270:26–35. https://doi.org/10.1006/jmbi.1997.1116
18. Schaefer W, Regula JT, Bahner M, Schanzer J, Croasdale R, Durr H, Gassner C, Georges G, Kettenberger H, Imhof-Jung S, Schwaiger M, Stubenrauch KG, Sustmann C, Thomas M, Scheuer W, Klein C (2011) Immunoglobulin domain crossover as a generic approach for the production of bispecific IgG antibodies. Proc Natl Acad Sci USA 108:11187–11192. https://doi.org/10.1073/pnas.1019002108
19. Golay J, Choblet S, Iwaszkiewicz J, Cerutti P, Ozil A, Loisel S, Pugniere M, Ubiali G, Zoete V, Michielin O, Berthou C, Kadouche J, Mach JP, Duonor-Cerutti M (2016) Design and validation of a novel generic platform for the production of tetravalent IgG1-like bispecific antibodies. J Immunol 196:3199–3211. https://doi.org/10.4049/jimmunol.1501592
20. Scarcelli JJ, Shang TQ, Iskra T, Allen MJ, Zhang L (2017) Strategic deployment of CHO expression platforms to deliver Pfizer's monoclonal antibody portfolio. Biotechnol Prog 33:1463–1467. https://doi.org/10.1002/btpr.2493
21. Tevelev B, Patel H, Shields K, Wei W, Cooley C, Zhang S, Bitzas G, Duan W, Khetemenee L, Jackobek R, D'Antona A, Sievers A, King A, Tam A, Zhang Y, Sousa E, Cohen J, Wroblewska L, Marshall J, Jackson M, Scarcelli JJ (2021) Genetic rearrangement during site specific integration event facilitates cell line development of a bispecific molecule. Biotechnol Prog:e3158. https://doi.org/10.1002/btpr.3158
22. Mistry RK, Kelsall E, Sou SN, Barker H, Jenns M, Willis K, Zurlo F, Hatton D, Gibson SJ (2021) A novel hydrogen peroxide evolved CHO host can improve the expression of difficult to express bispecific antibodies. Biotechnol Bioeng 118:2326–2337. https://doi.org/10.1002/bit.27744
23. Wurm FM (2004) Production of recombinant protein therapeutics in cultivated mammalian cells. Nat Biotechnol 22:1393–1398. https://doi.org/10.1038/nbt1026

24. Zhang L, Inniss MC, Han S, Moffat M, Jones H, Zhang B, Cox WL, Rance JR, Young RJ (2015) Recombinase-mediated cassette exchange (RMCE) for monoclonal antibody expression in the commercially relevant CHOK1SV cell line. Biotechnol Prog 31:1645–1656. https://doi.org/10.1002/btpr.2175
25. Inniss MC, Bandara K, Jusiak B, Lu TK, Weiss R, Wroblewska L, Zhang L (2017) A novel Bxb1 integrase RMCE system for high fidelity site-specific integration of mAb expression cassette in CHO cells. Biotechnol Bioeng 114:1837–1846. https://doi.org/10.1002/bit.26268
26. Schlatter S, Stansfield SH, Dinnis DM, Racher AJ, Birch JR, James DC (2005) On the optimal ratio of heavy to light chain genes for efficient recombinant antibody production by CHO cells. Biotechnol Prog 21:122–133. https://doi.org/10.1021/bp049780w
27. Gonz Lez R, Andrews BA, Asenjo JA (2002) Kinetic model of BiP- and PDI-mediated protein folding and assembly. J Theor Biol 214:529–537. https://doi.org/10.1006/jtbi.2001.2478
28. O'Callaghan PM, McLeod J, Pybus LP, Lovelady CS, Wilkinson SJ, Racher AJ, Porter A, James DC (2010) Cell line-specific control of recombinant monoclonal antibody production by CHO cells. Biotechnol Bioeng 106:938–951. https://doi.org/10.1002/bit.22769
29. Baumal R, Scharff MD (1973) Synthesis, assembly and secretion of gamma-globulin by mouse myeloma cells. V. Balanced and unbalanced synthesis of heavy and light chains by IgG-producing tumors and cell lines. J Immunol 111:448–456
30. Ho SC, Koh EY, van Beers M, Mueller M, Wan C, Teo G, Song Z, Tong YW, Bardor M, Yang Y (2013) Control of IgG LC:HC ratio in stably transfected CHO cells and study of the impact on expression, aggregation, glycosylation and conformational stability. J Biotechnol 165:157–166. https://doi.org/10.1016/j.jbiotec.2013.03.019
31. Ghose S, Allen M, Hubbard B, Brooks C, Cramer SM (2005) Antibody variable region interactions with Protein A: implications for the development of generic purification processes. Biotechnol Bioeng 92:665–673. https://doi.org/10.1002/bit.20729
32. Sasso EH, Silverman GJ, Mannik M (1991) Human IgA and IgG F(ab')2 that bind to staphylococcal protein A belong to the VHIII subgroup. J Immunol 147:1877–1883
33. Hiller GW, Ovalle AM, Gagnon MP, Curran ML, Wang W (2017) Cell-controlled hybrid perfusion fed-batch CHO cell process provides significant productivity improvement over conventional fed-batch cultures. Biotechnol Bioeng 114:1438–1447. https://doi.org/10.1002/bit.26259
34. Egan TJ, Diem D, Weldon R, Neumann T, Meyer S, Urech DM (2017) Novel multispecific heterodimeric antibody format allowing modular assembly of variable domain fragments. MAbs 9:68–84. https://doi.org/10.1080/19420862.2016.1248012
35. Chen T, Han J, Guo G, Wang Q, Wang Y, Li Y (2019) Monitoring removal of hole-hole homodimer by analytical hydrophobic interaction chromatography in purifying a bispecific antibody. Protein Expr Purif 164:105457. https://doi.org/10.1016/j.pep.2019.105457
36. Kuglstatter A, Stihle M, Neumann C, Muller C, Schaefer W, Klein C, Benz J, Roche Pharmaceutical R, Early D (2017) Structural differences between glycosylated, disulfide-linked heterodimeric Knob-into-Hole Fc fragment and its homodimeric Knob-Knob and Hole-Hole side products. Protein Eng Des Sel 30:649–656. https://doi.org/10.1093/protein/gzx041
37. Chen SW, Zhang W (2021) Current trends and challenges in the downstream purification of bispecific antibodies. Antib Ther 4:73–88. https://doi.org/10.1093/abt/tbab007
38. Zhao DS, Gregorich ZR, Ge Y (2013) High throughput screening of disulfide-containing proteins in a complex mixture. Proteomics 13:3256–3260. https://doi.org/10.1002/pmic.201300242
39. Makaraviciute A, Jackson CD, Millner PA, Ramanaviciene A (2016) Considerations in producing preferentially reduced half-antibody fragments. J Immunol Methods 429:50–56. https://doi.org/10.1016/j.jim.2016.01.001
40. Goulet DR, Orcutt SJ, Zwolak A, Rispens T, Labrijn AF, de Jong RN, Atkins WM, Chiu ML (2018) Kinetic mechanism of controlled Fab-arm exchange for the formation of bispecific immunoglobulin G1 antibodies. J Biol Chem 293:651–661. https://doi.org/10.1074/jbc.RA117.000303

41. Carlring J, De Leenheer E, Heath AW (2011) A novel redox method for rapid production of functional bi-specific antibodies for use in early pilot studies. PLoS One 6:e22533. https://doi.org/10.1371/journal.pone.0022533
42. Goulet DR, Zwolak A, Chiu ML, Nath A, Atkins WM (2017) Diffusion of soluble aggregates of THIOMABs and bispecific antibodies in serum. Biochemistry 56:2251–2260. https://doi.org/10.1021/acs.biochem.6b01097
43. Paul S, Connor J, Nesspor T, Haytko P, Boakye K, Chiu ML, Jiang H (2016) An efficient process of generating bispecific antibodies via controlled Fab-arm exchange using culture supernatants. Protein Expr Purif 121:133–140. https://doi.org/10.1016/j.pep.2016.01.014
44. O'Mara B, Gao ZH, Kuruganti M, Mallett R, Nayar G, Smith L, Meyer JD, Therriault J, Miller C, Cisney J, Fann J (2019) Impact of depth filtration on disulfide bond reduction during downstream processing of monoclonal antibodies from CHO cell cultures. Biotechnol Bioeng 116:1669–1683. https://doi.org/10.1002/bit.26964
45. Chen W, Yang F, Wang C, Narula J, Pascua E, Ni I, Ding S, Deng X, Chu ML, Pham A, Jiang X, Lindquist KC, Doonan PJ, Van Blarcom T, Yeung YA, Chaparro-Riggers J (2021) One size does not fit all: navigating the multi-dimensional space to optimize T-cell engaging protein therapeutics. MAbs 13:1871171. https://doi.org/10.1080/19420862.2020.1871171
46. Moore PA, Zhang W, Rainey GJ, Burke S, Li H, Huang L, Gorlatov S, Veri MC, Aggarwal S, Yang Y, Shah K, Jin L, Zhang S, He L, Zhang T, Ciccarone V, Koenig S, Bonvini E, Johnson S (2011) Application of dual affinity retargeting molecules to achieve optimal redirected T-cell killing of B-cell lymphoma. Blood 117:4542–4551. https://doi.org/10.1182/blood-2010-09-306449
47. Gupta VR, Root A, Fisher T, Norberg R, David J, Clark T, Cohen J, May C, Giddabasappa A (2020) Molecular imaging reveals biodistribution of P-cadherin LP-DART bispecific and trafficking of adoptively transferred T cells in mouse xenograft model. Oncotarget 11:1344–1357. https://doi.org/10.18632/oncotarget.27544

Part IV
Process Engineering, Gene Therapy and Vaccines

Chapter 6
Metabolic and Process Engineering to Control Glycan Structures for Biopharmaceuticals Produced in Cultured Mammalian Cells

Ranya Pranomphon, Vijay Tejwani, Hussain Dahodwala, Montarop Yamabhai, and Susan T. Sharfstein

Abstract Biopharmaceuticals (also known as biologics) play an increasing role in the treatment of a wide range of diseases, particularly cancer, autoimmune diseases, and infectious diseases. In 2017 and 2018, 11 of the 15 best-selling drugs worldwide were biopharmaceuticals, generally produced in cultured mammalian cells. The vast majority of biopharmaceuticals are glycoproteins, in which the attached glycan moieties play important and often critical roles in controlling activity, clearance, and immunogenicity. In addition to glycoproteins, carbohydrates, particularly glycosaminoglycans (GAGs) such as heparin, the most widely used anticoagulant drug in the world, are critically important biopharmaceutical products.

Many blockbuster biopharmaceuticals such as adalimumab (Humira), trastuzumab (Herceptin), and bevacizumab (Avastin) have recently come off patent, providing an opportunity for production of biosimilar versions by companies other than the innovator. In addition, there is increasing interest in producing

R. Pranomphon
College of Nanotechnology, Science and Engineering, University at Albany, State University of New York, Albany, NY, USA

School of Biotechnology, Institute of Agricultural Technology, Suranaree University of Technology, Nakhon Rachasima, Thailand

V. Tejwani · S. T. Sharfstein (✉)
College of Nantechnology, Science and Engineering, Albany, NY, USA
e-mail: ssharfstein@albany.edu

H. Dahodwala
Institute for Bioscience and Biotechnology Research, University of Maryland, Rockville, MD, USA

M. Yamabhai
College of Nanotechnology, Science and Engineering, State University of New York, Albany, NY, USA

K. Gadamasetti, S. A. Kolodziej (eds.), *Bioprocessing, Bioengineering and Process Chemistry in the Biopharmaceutical Industry*,
https://doi.org/10.1007/978-3-031-62007-2_6

glycosaminoglycans from cultured mammalian cells, rather than the current purification from animal tissues with the attendant risks of contamination by adventitious agents and adulteration due to isolation under non-cGMP conditions. To successfully produce biosimilars and bioengineered GAGs, control of glycan composition and structures is critical. This control is challenging as glycan synthesis is a non-templated process, which is controlled by a complex collection of factors including the glycoprotein being synthesized or the core protein to which the GAG is attached, production host, enzyme activities, and bioprocess conditions.

In this chapter, we review some basics of glycoprotein and GAG biosynthesis with a particular focus on our current understanding of how glycan structures are controlled in vivo. We then review studies in which glycan structures for recombinant proteins and glycosaminoglycans have been optimized by cell line metabolic engineering and bioprocess manipulations. In particular, CRISPR has permitted exquisite editing of host cells, allowing tailored production of glycan structures, facilitating the production of biosimilars and bioengineered GAGs as well as setting the stage for "biobetters" in which improved functionality is obtained by glycoengineering. However, complicated new products including bispecific antibodies, intricate Fc-fusion proteins and molecules not yet envisioned may necessitate new hosts and further advances in glycoengineering.

Keywords Glycosylation · Glycoengineering · Bioprocessing · Chinese hamster ovary cells · Monoclonal antibodies · Biosimilars

6.1 Introduction

Biological therapeutics, including monoclonal antibodies (mAbs), vaccines, hormones, and other proteins, play an increasing role in the global pharmaceutical market, providing novel treatments for a wide range of conditions and indications. Since 2002 there have been more than 300 biopharmaceutical approvals by the FDA (U.S. Food and Drug Administration), and this number is continuously growing. The vast majority of these biologics are produced in mammalian cell culture with Chinese hamster ovary (CHO) cell lines serving as the preferred host, but other cell lines including human embryonic kidney cells (HEK293), human retinal cells (PERC.6), baby hamster kidney (BHK), murine myeloma (NSO) cells, and murine (SP2/0) hybridoma cells have also been employed [1, 2]. Mammalian cells are utilized due to their ability to produce the appropriate posttranslational modifications including proper folding, glycosylation, phosphorylation, and disulfide bond formation, resulting in high-quality proteins [3, 4].

Protein glycosylation is the most common posttranslational modification in eukaryotes, formed by the attachment of glycans onto proteins [5]. Glycosylation occurs in the endoplasmic reticulum (ER) and/or the Golgi apparatus including

N- and *O*-linked glycosylation, where glycan chains are added to the amide nitrogen of the asparagine residue and hydroxyl group of the serine/threonine residues, respectively [6–8]. Glycosylation is a complex, nontemplated process, resulting in heterogeneity in the glycan profile (microheterogeneity) and in glycan site-occupancy (macroheterogeneity) [9, 10]. Many factors including cell type and cell engineering, media-feed compositions, and cell culture conditions such as temperature, pH, and dissolved oxygen and ammonia concentrations can affect glycan patterns [1, 11, 12]. Glycosylation patterns on therapeutic glycoproteins significantly affect biological and physiochemical properties including efficacy, immunogenicity, in vivo half-life, protein conformation, stability, and solubility [13–15]. Therefore, glycosylation is a critical quality attribute (CQA) of therapeutic proteins [16, 17].

In addition to glycoprotein therapeutics, there is increased interest in the recombinant or bioengineered production of carbohydrate drugs, particularly glycosaminoglycans (GAGs) such as heparin, the most widely used anticoagulant drug worldwide [18]. GAGs are synthesized and attached to a core protein, and the structure of the GAG is controlled by an extensive biosynthetic pathway consisting of more than 20 enzymes. Consequently, GAGs are highly heterogeneous, and their properties are strongly influenced by the cell line and culture conditions used for production. The CQAs for GAGs include their sulfation pattern and for heparin, the anticoagulant activity.

CQAs are defined as physical, chemical, biological, or microbiological properties or characteristics that must be monitored during manufacturing processes and controlled within an appropriate limit, range, or distribution to ensure the desired product quality and hence, the safety and efficacy of the drug products [19–21]. Due to the natural glycan heterogeneity, there is not "one correct glycan," though there are certainly undesirable glycans including immunogenic forms described in Sects. 6.4 and 6.5. Moreover, the glycosylation patterns, such as degree of branching, sialylation and fucosylation have significant effects on biological activity as described in Sect. 6.6.3. Consequently, glycoengineering and process engineering may be performed to achieve a specific biological goal for the therapeutic protein. Most critical is the need to maintain consistency throughout the product lifetime, which may include a variety of process changes and manufacturing at many different sites.

In addition to the need to maintain quality attributes in the production of novel therapeutics, the end of patent protection for many biotherapeutics (e.g., adalimumab (Humira), trastuzumab (Herceptin), and bevacizumab (Avastin)) provides an opportunity for the production of biosimilars. Biosimilars are off-patent biologic products that are highly similar but not identical to the originator biologic product (Reference Biologic) in terms of structure, function, biological activity, immunogenicity, efficacy, and safety due to differences in manufacturing processes [22, 23]. However, in biosimilar production, it is vitally important to *approximate* the CQAs of the innovator molecule to ensure efficacy. Significant effort is often employed during the process development of biosimilars to match the reference product including host and cell line selection as well as optimization of bioprocess conditions [24]. Furthermore, CQA identification is a key step in the quality by design (QbD) approach to product development, a systematic approach that begins with predefined objectives and emphasizes product and process understanding and process control, based on sound science and quality risk management. QbD has been implemented for developing and manufacturing biologics, including biosimilars [25–27].

To understand how these steps are being implemented, this chapter provides an overview of protein glycosylation, strategies for glycoengineering cell lines to control glycan patterns, and a discussion of the effects of process conditions on the glycosylation profile of therapeutic protein products.

6.2 Biology of Glycosylation

The addition of oligosaccharides onto a protein is a complex metabolic pathway, characterized by the en-bloc transfer of polysaccharide chains, as well as the stepwise addition and removal of individual monosaccharides. The number of pathways traversed is dependent on reaction site accessibility, the enzymatic substrate specificities, as well as spatial localization of the various enzymes and nucleotide-sugar substrates that are necessary for the reactions to proceed in a particular order [28, 29].

6.2.1 N-Linked Glycosylation

N-linked glycosylation initiates primarily in the ER of mammalian cells. The entire pathway in mammalian cells involves a highly complex and interconnected reaction network, catalyzed by glycosidases and glycosyltransferases contained within different compartments of the ER and Golgi apparatus, as depicted in the schematic in Fig. 6.1. The biosynthesis of mammalian *N*-glycans initiates at the cytoplasmic face of the ER membrane with the transfer of a phosphorylated *N*-acetylglucosamine (GlcNAc-P) from uridine diphosphate (UDP-GlcNAc) to the dolichol phosphate (Dol-P) lipid carrier to generate dolichol pyrophosphate *N*-acetylglucosamine (Dol-P-P-GlcNAc) [30]. Fourteen sugars are then sequentially added to Dol-P-P-GlcNAc to form an oligosaccharide precursor (Glc3Man9GlcNAc2) [10]. Next, oligosaccharyltransferase identifies Asn-X-Ser/Thr sequons in a nascent polypeptide and proceeds with an en-bloc transfer of Glc3Man9GlcNAc2 to the side chain amide of asparagine, releasing Dol-P-P in the process [31]. The glucose (Glc) residues on the precursor are sequentially trimmed by ER α-glucosidase I and II to form a monoglucosylated glycan, which is a key intermediate in the calnexin/calreticulin-associated glycoprotein folding control cycle, an ER protein-quality assurance mechanism [32].

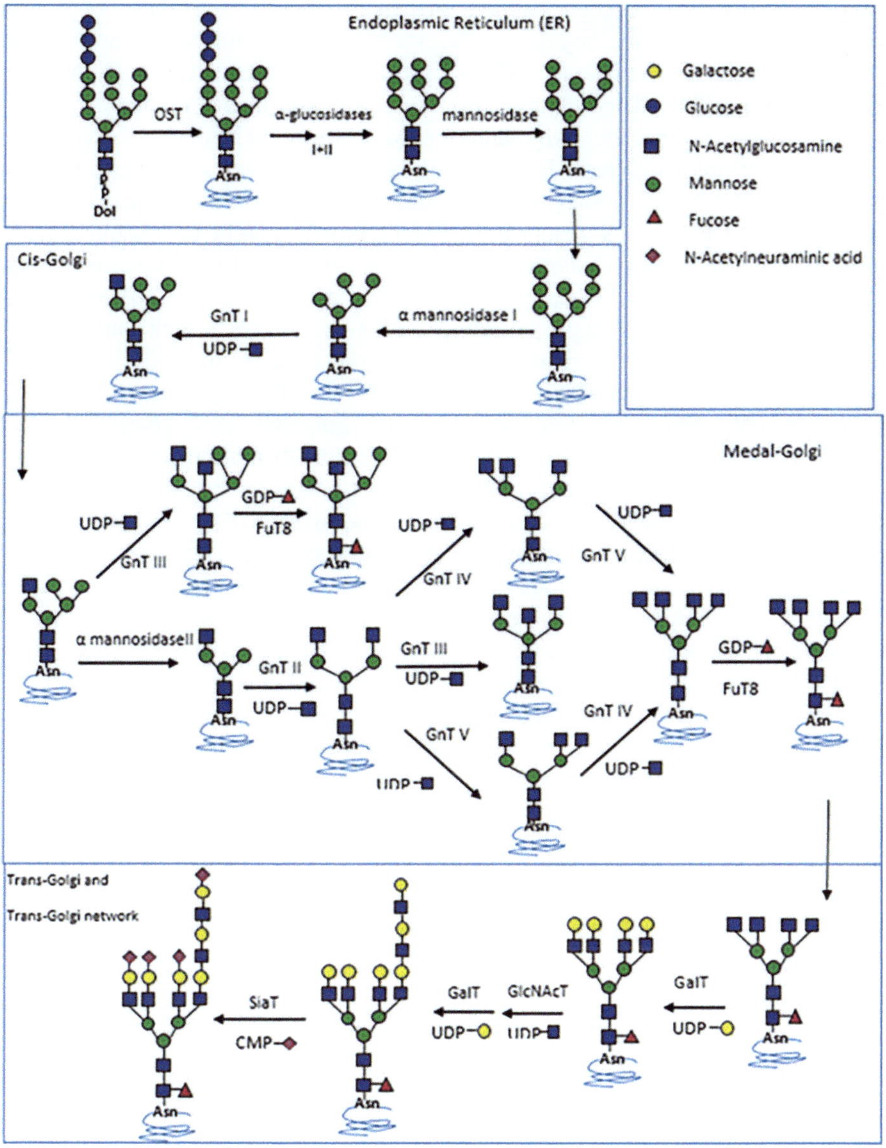

Fig. 6.1 An overview of glycosylation pathway in most eukaryotic cell systems. (Adapted from [33] with permission from Springer Science+Business Media and from [1] with permission from Oxford University Press)

Once correctly folded, the precursor is trimmed by ER α-mannosidase I to yield Man8GlcNAc2-protein before exiting the ER. After translocation into the cis-Golgi, the Man8GlcNAc2 glycoform is further trimmed by Golgi α-mannosidases I to give Man5GlcNAc2, a key intermediate along the pathway to form hybrid and complex *N*-glycans that occur as final glycan products. Biosynthesis of hybrid and complex *N*-glycans begins in the medial-Golgi with the action of an *N*-acetylglucosaminyltransferase (GnT-1 or mgat1), which adds a GlcNAc to Man5GlcNAc2 [34]. The majority of *N*-glycans are then trimmed by Golgi α-mannosidase II, removing two mannoses (Man) from GlcNAcMan5GlcNAc2 to yield GlcNAcMan3GlcNAc2. Hybrid *N*-glycans result when a structure such as GlcNAcMan3GlcNAc2 undergoes no further extension or trimming to remove exposed mannose residues, resulting in structures with one or two terminal mannose residues. In addition, another GlcNAc can be added to the innermost Man group by the enzyme β1,4-N-acetylglucosaminyltransferase III (GnT-III or mgat3) in the medial Golgi, resulting in bisecting GlcNAc structures, which can also alter the capacity for other downstream enzymes to act on the glycan structure. Next, the enzyme β-1,2-N-acetylglucosaminyltransferase II (GnT-II or mgat2) adds a GlcNAc to the GlcNAcMan3GlcNAc2 structure to generate the glycan product GlcNAc2Man3GlcNAc2, which is the precursor for all multiantennary complex *N*-glycans. Tri- and tetra-antennary branches can be achieved by adding GlcNAc at the α(1,3)-mannose site by *N*-acetylglucosaminyltransferase IV (GnT-IV or mgat 4) and/or at the α(1,6)-mannose site by *N*-acetylglucosaminyltransferase V (GnT-V or mgat 5). Additional modifications of complex and hybrid *N*-glycans can occur in the trans-Golgi including the addition of core α(1,6)-fucose (Fuc) to the GlcNAc adjacent to Asn at the *N*-glycan sites by α-(1,6)-fucosyltransferase and branch elongation by the addition of a β-linked galactose residue to GlcNAc by galactosyltransferase to produce Galβ1-4GlcNAc, or poly-acetyllactosamine (poly-LacNAc) sequences. Finally, these terminal Gal residues can serve as acceptors for several sialyltransferases, leading to even more complexity.

6.2.2 O-*Linked Glycosylation*

O-linked glycosylation occurs in either the ER or Golgi apparatus of animal cells. *O*-glycosylation begins with the attachment of a single monosaccharide (*N*-acetylgalactosamine, GalNAc) onto the hydroxyl group of the amino acids Ser or Thr on the polypeptide [35]. As *O*-glycosylation lacks consensus sequences, glycosylated Ser/Thr residues are often located in the "proline-glutamate-serine-threonine (PEST)" region of the protein [36]. This processing step is further modified by the addition of other monosaccharides including GlcNAc, GalNAc, sialic acid, fucose, galactose, xylose, and/or polylactosamine [14, 35]. *O*-glycosylation does not share a common core structure; it is believed to be primarily involved in protein stability and degradation [36].

6.2.3 *Glycosaminoglycan Synthesis*

Glycosaminoglycans are generally synthesized as proteoglycans attached to core proteins such as serglycin, syndecan, and glypican. Biosynthesis occurs in the ER or Golgi apparatus. GAGs are attached to the core proteins through a β-linkage between xylose and either serine or threonine on the core protein. GAGs are long (up to ~1 Mda) linear chains of repeating disaccharide units, consisting of either GlcNAc or GalNAc alternating with glucuronic acid and/or iduronic acid or Gal. Extensive chain modifications including *N*-deacetylation and *N*-sulfonation, *O*-sulfonations, and epimerization then occur under the actions of specific enzymes, many of which have a variety of isozymes whose expression is cell-type dependent.

6.3 Effects of Glycosylation on Biological Activity

The importance of glycosylation and the question of how glycosylation patterns affect different properties of recombinant proteins have attracted much attention over the last decade. For potential therapeutic use, large, complex proteins need to have human-like posttranslational modifications to be functional and nonimmunogenic. Proper glycosylation profiles promote biological activity and stability, increase half-life, and reduce the immunogenicity of protein therapeutics [37, 38]. Obtaining a consistent glycoform profile in production is desired due to regulatory concerns because a molecule can be defined, in part, by its carbohydrate structures. An optimal profile may involve a spectrum of product glycans that confers a desired therapeutic efficacy, or a homogeneous glycoform profile that can be systemically monitored [1]. As monoclonal antibodies have dominated protein therapeutics, much work has focused on glycosylation effects on mAbs, though certainly other proteins, particularly those containing sialic acids can be substantially affected. *N*-linked glycosylation is the most prevalent glycan form seen in recombinant mAbs, and IgG antibodies contain one conserved glycosylation site on each of the two Asn–297 in the constant heavy 2 (CH2) domains. This site has gained recent importance as it has been demonstrated that the Fc domain of the IgG can be engineered for improved functionality [39].

6.3.1 *Mannosylation*

In general, human IgG contains low levels (<5%) of high mannose (HM) glycoforms (Man5-9) [5, 40]. The HM glycans increase antibody-dependent cellular cytotoxicity (ADCC), which is likely due to the lack of core fucose, but significantly increase serum clearance, thereby impacting the therapeutic antibody efficacy [41–43]. For this reason, antibodies with HM glycans must be controlled. Cell type and cell culture parameters modulate the level of high mannose glycans [44].

6.3.2 Fucosylation

Fucosylation involves the addition of fucose from GDP-Fuc to the GlcNAc residue of Fc oligosaccharides via an α1,6-linkage catalyzed by fucosyltransferase 8 (fut8) [45]. Almost all therapeutic antibodies produced in CHO cells, mouse myeloma NS0, and SP2/0 are fucosylated [46, 47]. Removal of the core fucose residue from Fc oligosaccharides enhances ADCC by improving the binding affinity of the IgG Fc domain to Fcγ receptor IIIa [48–50]. For example, non-fucosylated therapeutic antibodies show up to 100-fold increased ADCC [51].

6.3.3 Galactosylation

Galactosylation involves the addition of Gal residues from UDP-Gal to the nonreducing terminal of the Man α1-3 and Man α1-6 arms of the bi-antennary core glycan structure [52, 53]. Terminal Gal residues of Fc glycans affect complement-dependent cytotoxicity (CDC) [54, 55], protein folding, stability, and aggregation [56, 57].

6.3.4 Sialylation

Terminal sialic acids in *N*-linked complex glycans play a major role in the circulatory half-life of glycoproteins as sialic acid masks the penultimate sugar, Gal, and prevents recognition and uptake of the glycoprotein by asialoglycoprotein receptors (ASGPR) on hepatocytes, resulting in a longer serum half-life of the therapeutic protein [58–60].

A variety of factors, both extrinsic and intrinsic, limit our abilities to produce the desired glycans in cultured mammalian cells. A summary of commonly understood cellular bottlenecks influencing glycosylation in CHO cells is illustrated in Fig. 6.2. The remaining sections in the chapter address strategies to overcome these limitations, including choice of host cell line, glycoengineering, and bioprocess optimization.

6.4 Choice of Host Cell Line

While CHO cell lines are the dominant mammalian cell host for producing therapeutic proteins, other rodent and human cell lines have been employed, with varying success at obtaining desired glycan profiles. Enzymatic differences between different host cell types and the implications are shown in Table 6.1 (adapted from

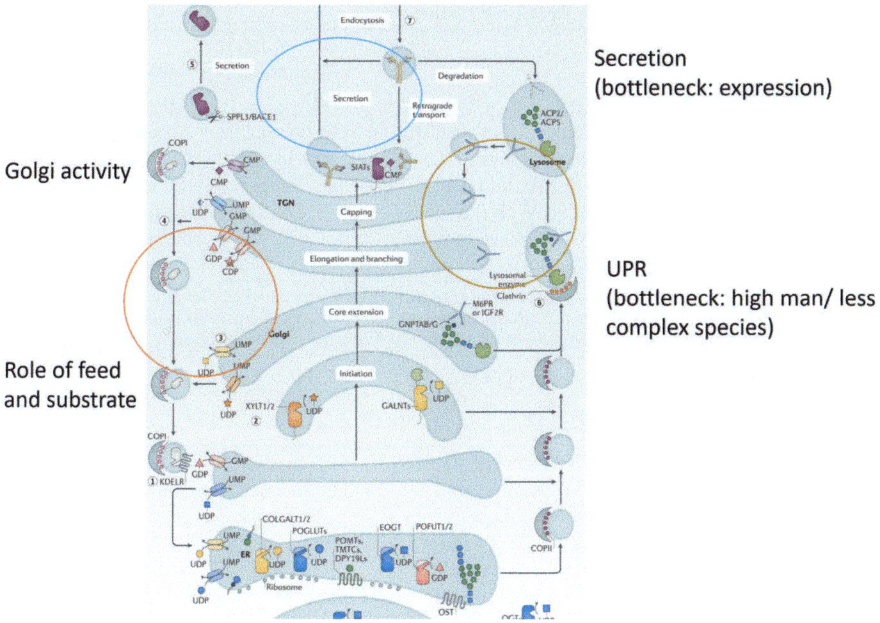

Fig. 6.2 Various experimentally observed bottlenecks in the glycosylation pathway in cells. (Adapted from [61] with permission from Springer-Nature)

[62, 63]). In addition, different CHO cell hosts (e.g., CHO-K1, CHO DG44) exhibit different glycosylation patterns, an important consideration when developing a biosimilar product in a different host cell line.

Goh and Ng [63] provided a detailed comparison of the glycosylation profiles of four recombinant glycoproteins: immunoglobulin G (IgG), coagulation factor VII (FVII), erythropoietin (EPO) and alpha-1 antitrypsin (A1AT) produced from different sources including human urinary- or plasma-derived proteins, different CHO cell hosts, NS0 and J558L murine myeloma cells, baby hamster kidney (BHK) cells, and the human cell lines HEK293/T (human embryonic kidney with or without Epstein Barr transformation), PER.C6 (retinoblastoma), HT-1080 (fibrosarcoma), and AGE1.HN (human neural tissue derived). They noted that for IgG the NS0-derived structures had significantly lower galactose levels than the plasma- or CHO-derived structures, with increased high mannose structures in the NS0-derived IgG. Interestingly, while terminal sialylation was under 5% on IgG derived from human plasma, CHO-K1, J558L, and HEK293 cells, 11.8% of glycans on IgG produced in NS0 cells were sialylated. For coagulation factor VII (FVII), which has two N glycosylation sites, N145 and N322, glycans on plasma-derived FVII were predominantly complex-type bi- and tri-antennary structures without core fucosylation. CHO- and BHK-derived FVII showed primarily complex-type biantennary di-sialylated core-fucosylated structures, while HEK293-derived FVII demonstrated the most heterogeneous *N*-linked glycans with about 20 different glycostructures. Native EPO is *N*-glycosylated at sites N24, N38, and N83 and is primarily

Table 6.1 Enzymatic differences between host species

Glycosylation enzyme	Function	Expression in mammalian cells	Impact on recombinant proteins
β-1,4 *N*-acetylglucosaminyl-transferase III (GnT- III)	Catalyzes the transfer of GlcNAc from UDP-GlcNAc to a core-β-Man of an *N*-glycan via β1,4-linkage to form a bisecting GlcNAc residue	Expressed in human cells; absent in CHO cells	Absence of bisecting GlcNAc in CHO derived proteins, leading to reduced biological activity of mAbs
α1,3/4 fucosyltransferase	Catalyzes the transfer of L-fucose from GDP-Fuc to *N*-acetyllactosamine, generating Lewis^x, sialyl Lewis^x structures and their combinations	Expressed in human cells; absent in CHO cells	May perturb proper biodistribution and pharmacokinetics
α1,3 galactosyltransferase (α1,3-GT)	Catalyzes the transfer of Gal onto terminal lactosaminide residues via an α1,3 linkage, to synthesize a Gal-α 1,3-Gal group (alpha-Gal)	Inactive in human and CHO cells; functional in BHK and murine cells	Possibility of immunogenic glycoforms in BHK- and murine-derived proteins
β-galactoside α2,6-sialyltransferase (Gal α2,6-ST)	Attaches sialic acid with an alpha-2,6 linkage onto Gal or GalNAc residues	Expressed in human and murine cells; absent in CHO and BHK cells	Undersialylation, reduced serum half-life of CHO- and BHK-derived proteins
CMP-Neu5Ac hydroxylase (CMAH)	Catalyzes the conversion of the precursor molecule CMP-*N*-acetylneuraminic acid (Neu5Ac) to CMP-*N*-glycolylneuraminic acid (Neu5Gc)	Inactive in humans; functional in animal cells	Presence of immunogenic form of sialic acid particularly in murine-derived proteins

decorated with tetra-antennary complex-type glycostructures. Recombinant EPO expressed in BHK-21, CHO, and HT-1080 cells demonstrated predominantly tetra-antennary structures for ~60–80% of the total glycoforms observed. The structures were largely sialylated in both the native and recombinant forms although the fraction of antennae that were sialylated varied between host cells, leading to charge variation in the proteins depending on host. Alpha-1 antitrypsin (A1AT) has three main *N*-linked glycosylation sites at positions N46, N83, and N247. In the native A1AT, the glycans are typically disialylated biantennary complex-type structures. The predominant glycan in CHO-derived, recombinant A1AT was also a disialylated, complex-type biantennary structure but with a lower fraction of 47.9% compared with the 74–77% found in two plasma studies. Studies in human cell lines

AGE1.HN, HEK293, HEK293T, and PER.C6 also observed that the predominant glycan in recombinant A1AT was a biantennary complex-type structure; however, the terminal sialylation on the predominant glycan in the AGE1.HN, HEK293, and HEK293T cell lines was significantly lower than in the native or CHO-derived glycans. In addition, while minimal fucosylation was observed in the plasma-derived A1AT, recombinant A1AT produced in CHO and human cell lines predominantly contained core α(1,6) fucose and to a certain degree, Lewis^x fucose.

Even among different host CHO cell lines (e.g., CHO-K1 and CHO DG44) and individual clones derived from the same host, different glycosylation patterns occur. Yeo and colleagues compared the glycan profiles for a recombinant mAb produced in CHO-K1 and CHO DG44 cell lines [64]. The mAb produced in CHO-K1 cells had predominantly G0F glycans (~60%) with approximately 25% G1F glycans and a much smaller fraction of G2F glycans. In contrast, the DG44-derived mAbs had a much lower fraction of G0F glycans (typically 20–30%) with 40–50% G1F glycans and ~20% G2F glycans. In both cell lines, the fraction of sialylated and high mannose glycans was relatively small, but it was significantly greater in the DG44-derived mAbs than in the mAb produced in CHO-K1. A similar study was performed by Könitzer and coworkers at Boehringer Ingelheim in which they compared mAbs produced in two DG44-derived cell lines and one CHO-K1-derived cell line. In addition to comparing the glycan patterns, they performed RNA-Seq to compare the gene expression patterns between the different host cell lines. The CHO-K1-derived cell line produced a higher fraction of nonfucosylated antibodies compared to either of the DG44 cell lines. When looking at high mannose sugar structures, antibodies produced in DG44 cell line 2 demonstrated the lowest amount (median: 3.7%; range: 0.1–35.0%), followed by DG44 cell line 1 (median: 7.9%; range: 1.1–37.0%) and the CHO-K1 cell line (median: 9.7%; range:1.7–24.4%). Terminal sialylation was also increased in DG44 cell line 2 compared to the other two cell lines. Lastly, antibody galactosylation was consistently lower in the DG44 cell line 1 and CHO-K1 cell line than in DG44 cell line 2. In general, the glycan profiles from DG44 cell line 1 and the CHO-K1 cell line show more similarity to each other than to DG44 cell line 2. Using RNA-Seq, they compared the transcriptome of the three cell lines and identified 188 glycosylation-related genes that were expressed in all 3 lines, 4 genes that were unique to the CHO-K1 cell line, 8 genes that were only expressed in the DG44 cell lines, and 2 genes expressed uniquely in either of the 2 DG44 cell lines. They further characterized relative expression levels of several key enzymes and noted that expression of a recombinant mAb led to differential expression of several glycosylation-related genes.

For production of recombinant or bioengineered glycosaminoglycans such as heparin, the choice of host cell lines appears even more critical. Baik and coworkers attempted to produce a bioengineered heparin in CHO cells by engineering the GAG biosynthesis pathways [65]. They overexpressed two critical enzymes in the pathway, NDST2, which removes the acetyl group from GlcNAc replacing it with a sulfate group, and 3OST, which sulfonates the critical 3-hydroxyl position on

GlcNAc responsible for anticoagulant activity. While they were able to dramatically increase sulfation and significantly increase anticoagulant activity, the resulting product was very different from pharmaceutical heparin both in terms of structure and anticoagulant activity. In contrast, Thacker and colleagues performed similar metabolic engineering manipulations on murine mastocytoma cells (a tumorigenic form of mast cells, which are the source of heparin in vivo). They were able to obtain heparin-like GAGs with an anticoagulant potency that exceeds porcine-derived heparin [66].

6.5 Glycoengineering

Genetic approaches for enhancing *N*-glycosylation are the most common strategies and employ gene editing or transient expression techniques to change the activity of glycosyltransferases and increase or decrease the precursors involved in the *N*-glycosylation process. Genetic approaches altering heterogeneity, sialylation, fucosylation, and branching in *N*-glycan structures are described below and shown in Fig. 6.3.

6.5.1 Manipulating Heterogeneity

Although glycoprotein biologics produced by CHO cells containing heterogeneous mixtures of *N*-glycans are considered safe as human therapeutics, excess heterogeneity can be an issue. This excess heterogeneity arises due to the variability of *N*-glycan processing and can compromise the safety and activity of such glycotherapeutics. Having homogenous glycoforms allows comparative studies of their biological effects, which can be advantageous in the development of therapeutic candidates [67]. Yang et al. showed that CHO cells can be genetically engineered to produce glycoproteins in a nearly homogenous form without any deleterious effect on their growth or other compensatory changes [68]. To achieve this goal, the in vivo function of each of the 19 glycosyltransferases potentially participating in *N*-glycan formation and processing was determined in CHO cells by individual and/or multiple glycosyltransferase gene knockouts. The effects of knocking out each of the 19 genes involved in *N*-glycan branching (mgat1/2/3/4A/4B/4C/5/5B), galactosylation (B4galt1/2/3/4), *N*-acetyllactosamine (LacNAc) elongation (B3gnt1/2/8), terminal capping by sialylation (st3gal3/4/6), and core α-6-fucosylation (fut8) were determined using a stably expressed model protein, a recombinant human erythropoietin (rhEPO) containing in its structure three *N*-glycans with heterogeneous tetra-antennary structures, low poly-LacNAc and terminal α-2,3-linked sialic acid.

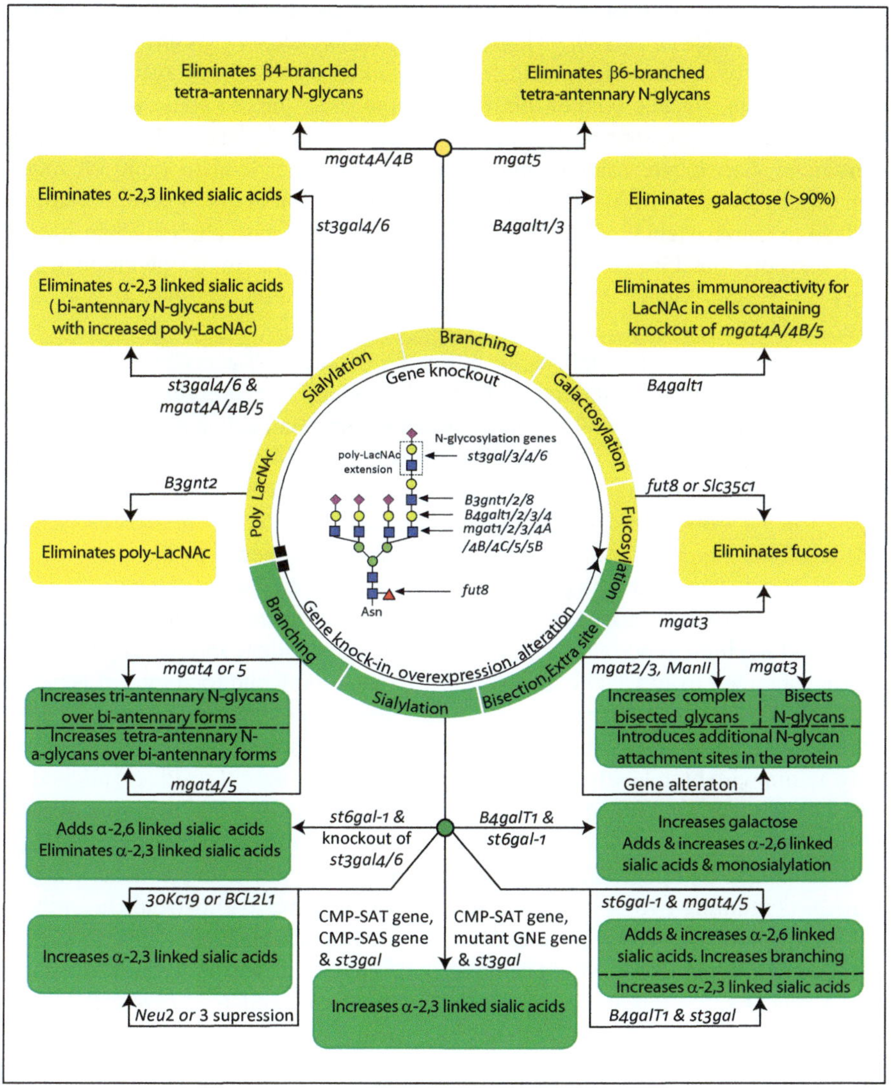

Fig. 6.3 Summary of genetic approaches for manipulating glycosylation in CHO cells described in this chapter. Gene knockouts are represented by yellow segments and gene knock-ins, overexpression and alteration are represented by green segments. For convenience, tetra-antennary *N*-glycans and genes participating in *N*-glycosylation are also shown

6.5.2 *Manipulating Sialylation*

Sialylation refers to the glycosidic addition of a negatively charged monosaccharide, a sialic acid, by sialyltransferases, generally to terminal Gal or GalNAc, though occasionally to GlcNAc or sialic acid itself in complex *N*-glycans [69]. CHO cells contain α-2,3-sialyltransferases (α-2,3-SiaT), whereas human cells have α-2,6-SiaT in addition to α-2,3-SiaT [70, 71]. Due to this difference, *N*-glycans produced by CHO cells only contain sialic acid residues linked by α-2,3-glycosidic linkages, whereas human glycans contain both α-2,3- and α-2,6-linked sialic acid residues [72]. Sialic acid at the termini in complex *N*-glycans masks terminal Gal from recognition by hepatocyte asialoglycoprotein receptors that lead to rapid clearance from the circulation [73]. Insufficient or lack of sialylation in glycoprotein biologics can lead to inconsistency in the pharmacodynamics and cause challenges in formulating reproducible dosages. Thus, correct and generally, maximal sialylation is necessary to ensure longer plasma half-lives and maximum in vivo activity and therapeutic efficacy [74].

Of the numerous (>50) sialic acids occurring in nature, *N*-acetylneuraminic acid (Neu5Ac) is the most abundant [69]. Its derivative, *N*-glycolylneuraminic acid (Neu5Gc), is also a major sialic acid found on mammalian cell surfaces, formed by Neu5Ac hydroxylation by cytidine monophosphate (CMP)-Neu5Ac hydroxylase. Humans lack CMP-Neu5Ac hydroxylase and are unable to synthesize Neu5Gc. Glycoproteins synthesized by CHO cells occasionally contain *N*-glycans capped with Neu5Gc. Neu5Gc in glycans can be a cause for concern as Neu5Gc-capped glycans act as "xeno-autoantigens" in humans and are a cause of "xenosialitis," an inflammatory process initiated by binding of naturally occurring antibodies against Neu5Gc in the human body [75–77]. The Neu5Gc-dependent antigenicity of glycoproteins obtained from CHO cells depends on the amount and locations of Neu5Gc in the glycan structure. For example, mAbs and rhEPO produced by CHO cells containing 1–2% Neu5Gc did not elicit an immune response, whereas, in the same study, fetuin with high levels of Neu5Gc (7% of total sialic acid residues) elicited an immune response in chickens [78]. A recent study by Yu et al. on different clinical mAbs containing Neu5Gc residues concluded that mAbs containing a single Neu5Gc residue do not bind to anti-Neu5Gc antibodies, while only a minor fraction of mAbs containing two or more Neu5Gc showed binding to anti-Neu5Gc antibodies [79]. Further, this study suggested that the binding of anti-Neu5Gc antibodies to mAbs containing multiple Neu5Gc depends on Neu5Gc location in the mAb structure. A recent study has implicated a newly discovered miRNA, *cgr*-miR-111, in controlling the expression of CMP-Neu5Ac hydroxylase, suggesting another strategy for control of Neu5Gc on recombinant proteins produced in CHO cells [80].

6.5.2.1 Increasing α-2,6 Sialylation

Glycoprotein biologics that contain α-2,6-linked sialic acid residues are suggested to be more "human-like" and perform better in vivo [81, 82]. In two recent studies, chemo-enzymatic modification (in vitro glycosylation) of two IgGs (anti-Her2 antibody and rituximab) creating homogenous glycans containing α-2,6-sialic acid residues enhanced ADCC due to stronger interaction of α-2,6-sialylated glycans with FcγRIIIa receptors on natural killer cells [81, 83]. Similarly, in two previous studies involving in vitro glycosylation, α-2,6-sialylated versions of IgGs produced superior anti-inflammatory responses compared to asialylated or α-2,3-sialylated versions [82, 84].

CHO cells were successfully engineered to produce rhEPO containing almost exclusively α-2,6-sialylation by knockout of st3gal4/6 genes (encoding α-2,3-SiaTs) and knock-in of st6gal-I gene (encoding an α-2,6-SiaT) [68]. Furthermore, in the same study, homogeneous bi-antennary *N*-glycans capped by α-2,6-NeuA were produced by additional knockout of mgat4A/4B/5 genes.

6.5.2.2 Increasing the Sialic Acid Content

Besides knockout and knock-in of sialyltransferase genes, increases in sialic acid content can be obtained by overexpressing genes participating in steps prior to sialylation or inhibiting genes encoding for sialidases that remove the sialic acid after the sialylation step. The addition of sialic acid residues to a growing *N*-glycan chain can be limited by the absence of Gal, which acts as an acceptor substrate for sialyltransferases. Increasing the Gal content in *N*-glycans by overexpressing GalT produces a corresponding increase in the sialic acid content in CHO cells overexpressing sialyltransferase enzymes. Raymond et al. transiently coexpressed genes encoding for GalT, α-2,6-SiaT, and an IgG1 antibody (trastuzumab/Herceptin®) and produced efficient α-2,6-sialylation in trastuzumab's Fc region. The glycans under investigation were monosialylated, a physiologically relevant form found in circulating human IgGs. Over 85% of sialic acids on trastuzumab showed α-2,6-sialylation due to overexpression of human GalT, which inserts Gal residues in the glycan that are preferentially used by α-2,6-SiaT rather than α-2,3-SiaT. Overexpression of GalT alone increased the Gal content of the Fc glycans, but had no effect on increasing the sialylation, whereas overexpression of α-2,6-SiaT only increased the sialylation moderately [85]. In a similar study, coexpression of GalT was beneficial in CHO cells overexpressing α-2,3-SiaT to increase the sialylation. Trisialylated glycans on rhEPO increased f*rom* 17.3 to 35.5% when expressed in CHO EC1 cells co-overexpressing both human α-2,3-SiaT and GalT [86].

Sialic acid content can also be enhanced by increasing the concentration and availability of the donor sugar nucleotide CMP-sialic acid (CMP-SA) present in the Golgi apparatus. Inside the nucleus, CMP-SA is generated from sialic acid by CMP sialic acid synthetase (CMP-SAS) and later transported to the Golgi by CMP-sialic acid transporter (CMP-SAT). In eukaryotes, sialic acid is synthesized in the cytoplasm by three enzymes in a four-step process. The two first steps are catalyzed by a bifunctional enzyme, GNE (UDP-N-acetylglucosamine-2-epimerase/N--acetylmannosaminekinase), having kinase and epimerase activity. The epimerase activity of GNE converts UDP-GlcNAc to *N*-acetylmannosamine (ManNAc), which is then converted to ManNAc-6-phosphate by the kinase activity of GNE. The last two steps are catalyzed by two enzymes, Neu5Ac-9-phosphate synthase and Neu5Ac-9-phosphate phosphatase, which produce Neu5Ac from ManNAc-6-phosphate by condensation and dephosphorylation reactions, respectively. GNE is a rate-limiting enzyme for the synthesis of sialic acid in the cytoplasm [71, 87].

Increasing sialic acid content in the cytoplasm or CMP-SA in the nucleus by different strategies increases sialylation by varying amounts. Overexpression of CMP-SAT in CHO cells expressing recombinant human interferon gamma (IFN-γ) increased IFN-γ sialylation by 4–16% [88]. Similarly, a modest increase in sialylation (10–20%) was observed upon supplementation with ManNAc, which led to a 12-fold increase in the intracellular pool of CMP-sialic acid [89]. In another study, supplementation with 1,3,4-*O*-Bu_3ManNAc, a chemical analog of the sialic acid precursor ManNAc increased the final sialic acid content of rhEPO >40% in CHO cells at a 100-fold lower concentration than natural ManNAc [90].

Co-overexpressing several genes together enhanced sialylation more significantly than overexpressing single genes individually. CHO cells expressing CMP-SAS in combination with CMP-SAT and human α-2,3-SiaT exhibited greater rhEPO sialylation compared with CHO cells overexpressing α-2,3-SiaT or CMP-SAS individually [91]. GNE, catalyzing the first two steps in the synthesis of sialic acid, is a rate-limiting enzyme, and its activity is regulated by feedback inhibition from free cytoplasmic CMP-Neu5Ac [71, 87]. Coexpressing a mutant version of GNE lacking feedback regulation with CMP-SAS enhanced the sialylation modestly, but when the mutant version of GNE was coexpressed with CMP-SAT and human α-2,3-SiaT, CHO cells produced rhEPO having 43% increased sialylation. Coexpression experiments involving CMP-SAT clearly indicated that endogenous CMP-SAT is insufficient, and its overexpression was essential for increasing the sialylation [92]. Recently, epigenetic modulators have been employed to turn on the silenced α-2,6-SiaT in CHO cells by demethylating its promoter using a CRISPR-based approach [93].

Sialidases are enzymes that catalyze the removal of sialic acid residues from glycoproteins and glycolipids. CHO cells contain four different sialidases (Neu1-4) distributed in the lysosome (Neu1 and Neu4), cytosol (Neu2), and plasma membrane (Neu3) [94]. During glycoprotein biologic manufacture, lowering the activity of these enzymes is desirable, but not completely, due to their important roles in crucial biological functions [95–97]. The cytosolic sialidase, Neu2, is released into the supernatant during cell lysis and preferentially removes α-2,3-linked sialic acids from the glycoprotein products [98]. RNA-mediated suppression of Neu2 [99] and plasma membrane-bound Neu3 [74] activities by 40 and 98%, respectively, increased sialic acid content (up to 33% in model proteins), but the effect of Neu2 suppression was observed in the death phase only.

6.5.3 Manipulating Fucosylation

Core Fuc residues on IgG antibodies have a negative effect on their effector function [51]. Effector function is essential in IgGs designed for use in tumor therapy; after binding to antigens on cancer cells, the IgG Fc region binds strongly to FcγRIIIa receptors present on natural killer cells, causing cancer cell death by lysis via the ADCC mechanism [48, 51]. Fucose-containing *N*-glycans present at Asn-297 in the

IgG adversely affect the Fc-FcγRIII interaction [100]. Multiple studies have successfully demonstrated that removal of the core Fuc residue from the *N*-glycan in human IgG1 increases the binding affinity of Fc toward FcγRIII, which, in turn, enhances the in vivo ADCC significantly [48, 51]. Such fucose-free antibodies can be beneficial to patients as their higher potency enables lower dosage administration [101, 102].

In mammals, fut8 is the only gene encoding for a fucosyltransferase capable of adding Fuc to *N*-glycans [103]. Inhibition or knockout of fut8 and interference with transport and synthesis of donor substrate GDP-Fuc are two approaches for reducing or inhibiting Fuc addition to *N*-glycans in CHO cells [33]. Knockout of fut8 in CHO cells using zinc finger nucleases and homologous recombination produced fully afucosylated antibodies, showing enhanced ADCC [104, 105]. Further, Chan et al. showed that inactivating Slc35c1, encoding for GDP-fucose transporter, also generated CHO cells producing fucose-free glycans [106].

Alternatively, antibodies devoid of core Fuc can also be produced by CHO cells overexpressing the GnT-III enzyme, which catalyzes bisecting GlcNAc addition onto the common core structure (Man3GlcNAc2) in *N*-glycans [107]. Upon addition of bisecting GlcNAc, the oligosaccharide cannot act as a suitable substrate for subsequent glycosylation enzymes, especially Golgi-mannosidase II (Man-II), GalT, and FucT [108], leading to the production of fucose-free mAbs showing increased ADCC, but decreased complement-dependent cytotoxicity. The decreased CDC is due to hybrid *N*-linked oligosaccharide structures, resulting from incomplete mannose cleavage [109].

6.5.4 Manipulating Branching

In mammals, branching in *N*-glycans occurs in the medial Golgi and is carried out by GnT-I, -II, -IV, and –V enzymes, which control GlcNAc addition at the branch point in a stepwise manner [108]. Branching produces bi-, tri-, and tetra-antennary structures, which can be extended by enzymes in the trans-Golgi capable of adding Gal, Fuc, and sialic acids (Fig. 6.2) [110]. GnT-I and GnT-II control the formation of bi-antennary structures; tri-antennary structure formation is controlled by GnT-IV or GnT-V, and the tetra-antennary structure formation is controlled by the combined action of GnT-IV and GnT-V (Fig. 6.2). Higher branching (tri and tetra) provides additional sites for the attachment of sialic acid residues, which enhance biological activity and circulatory lifetime. Increased branching of therapeutic glycoproteins is thus of significant clinical as well as commercial interest.

Multiple studies have shown that branching can be increased by overexpression of GnT-IV and/or GnT-V enzymes. Tri-antennary structures were significantly increased (over 50%) by overexpression of GnT-IV or GnT-V individually in CHO cells producing IFN-γ with predominantly bi-antennary sugar chains [111]. In the same study, tetra-antennary structures were increased up to 56% of the total sugar chains when GnT-IV and GnT-V enzymes were coexpressed. However, compared to observed increases in tri- and tetra-antennary structures, the corresponding increase in sialylation was insignificant. Inadequate sialylation was attributed to insufficient intracellular sialyltransferase or CMP-SAS activity. The reduced sialic acid capping on tri- and tetra-antennary structures in rhEPO was solved by Yin and coworkers by additionally expressing human α-2,6-SiaT [112]. They showed that coexpression of α-2,6-SiaT with GnT-IV and GnT-V produced rhEPO containing approximately 92% tri- and tetra-antennary *N*-glycans with a 45% increase in the sialic acid content compared with rhEPO obtained from wild-type CHO-K1 cells.

6.6 Effects of Bioprocess Conditions

A variety of bioprocess strategies have been employed to alter glycosylation profiles, sometimes individually, but often in combination. Here, we group the studies based on the predominant manipulation while recognizing that more than one parameter may have been varied in these studies.

6.6.1 Temperature

Mellahi and colleagues investigated the effect of various feeding regimes and temperature shift strategies to improve antibody productivity and ensure product quality of an inducible CHO cell line (cumate gene switch) expressing rituximab [113, 114]. Experiments were performed in both shake flasks and bioreactors, with induction performed at cell densities ranging from 1×10^6 to 10×10^6 cells/mL. While the cell density at induction had a significant impact on product titer, it had little effect on the glycan distribution. In contrast, the culture duration significantly affected glycosylation with a significant decrease in galactosylation observed as the culture proceeded. They further explored temperature shifts from 37 °C to either 34 °C or 30 °C and from 34 °C to 30 °C, coinciding with induction of production at either 4×10^6 or 10×10^6 cells/mL. Induction at high cell density combined with a temperature shift from 37 °C to 30 °C led to the highest antibody concentrations. Growth at 37 °C before induction yielded slightly higher percentages of galactosylation, sialylation, and fucosylation, compared to cultures started at 34 °C, but the differences were not statistically significant. Slightly lower galactosylation was seen later in culture, an effect that was less pronounced for cultures grown at 34 °C before induction.

McHugh and colleagues examined the effects of a temperature shift on therapeutic protein production, charge variants, *N*-linked glycosylation, and protein aggregation [115]. In this study, two CHO-K1 GS cell lines expressing different human mAbs (IgG1 mAb1 and IgG4 mAb2, respectively) exhibited differential responses to a temperature shift. They observed that the final mAb1 titer with a temperature shift to 32 °C was about 25% greater than unshifted cultures. Shift temperatures of

Δ1.5 °C increments significantly affected final titer and charge variants. Furthermore, lower shift temperature decreased acidic charge variants for both cell lines. Significant differences in G0 and Man5 of about 1% and a slight decrease in galactosylation were observed between shifted and unshifted conditions for mAb1 production. In addition, an early shift temperature on Day 3 significantly reduced mAb2 galactosylation similarly to mAb1 such that G1F was reduced by 5% and G2F was reduced by 0.8 with a 3% increase in G0F.

6.6.2 pH

Ivarsson and colleagues investigated the impact of chemical stress parameters (pH, dissolved oxygen tension (DOT) and osmolarity) and mechanical stress parameters (sparging) on cell growth, productivity, and *N*-linked glycosylation of a murine hybridoma cell line producing an IgG1 antibody, using a shift-experiment methodology in batch cultures [116]. A shift in one or more of the process parameters was performed during the early exponential growth phase. They observed that cell growth was dependent on pH, followed by osmolarity. mAb productivity was dependent on DOT, followed by pH. In contrast, sparging had little effect on the specific growth rate or production rate in a range of 0.05–0.2 vvm. Increasing pH between 6.8 and 8.0 monotonically decreased galactosylation and sialylation indices, with a maximum decrease of 50% between pH 6.8 and 8.0. The reduction in galactosylation was due to decreases in both G1F and G2F glycans with an increase in G0F glycans from 16 to 35% of total glycans. A slight decrease in fucosylation (from 99% to 94%) was also observed over that pH range. Compared with 50% of air saturation, DOT of 10% and 90% air saturation showed small increases in both galactosylation and sialylation. Increases in osmolarity from 320 to 420 mOsm/kg at the control pH of 7.2 did not affect glycosylation; however, at elevated pH (>7.2), osmotically stressed cultures showed increased galactosylation and sialylation, offsetting the decreases caused by increased pH.

Seo and coworkers evaluated the effects of different cultivation pH (6.8, 7.0, 7.2, 7.4, and 7.6) and temperature (33.0 °C and 37.0 °C) on a novel human cell line, F2N78, established by somatic fusion of HEK293 and Namalwa lymphoma cells, producing an antibody against rabies virus. The maximum antibody concentration was observed at 37 °C and pH 6.8. Regardless of temperature, the highest specific growth rate occurred at a pH range of 7.0–7.4. The dominant glycan form was G1F, followed by G2F and then G0F, a significant increase in galactosylation compared with antibody production in CHO cells. Increasing pH decreased G2F glycans and increased G0F glycans while culture temperature had little effect on the glycan distribution. Notably, a significant fraction of the antibody obtained from the stationary phase and later points in culture was aglycosylated, particularly at pH values between 7.0 and 7.4, which they attributed to glucose depletion [117].

Jiang and colleagues investigated the impact of pH excursions on cell culture performance and *N*-linked glycosylation for three mAb-producing Chinese hamster ovary cell lines [118]. Antibody galactosylation (G1F + G2F) was greater at higher pH values in the range of 6.90–7.10 while cell-specific productivity decreased with the increased pH for one cell line, suggesting that the impact of pH excursions on antibody *N*-linked glycosylation is cell-line specific.

Zheng and coworkers investigated the effect of a pH shift during the expression phase on a CHO cell line producing anti-CD52 mAbs in perfusion culture and compared the production process, CQAs and the biological potency in vitro [119]. In this study, they shifted the culture pH from 7.15 ± 0.05 to 6.85 ± 0.05 at day 9 and maintained a control group at 7.15 ± 0.05 in a 15-L bioreactor. Under pH shift conditions, the cell density and viability were stable and high throughout the perfusion culture. In addition, the qGlc and qLac were higher in the pH-shifted conditions than in the pH control conditions. Even though the specific productivity was lower under pH shift, the volumetric productivity remained high. Conversely, the charge and size variants of mAbs were similar in both pH conditions. Interestingly, six major glycoforms (G0, G0F, G1, G1F, G2F, and Man5) were observed under the two pH conditions. All forms were significantly different except Man5 and G1. The galactosylation (G1 + G1F + G2F) increased significantly upon pH shift, due primarily to a decrease in G0F and increases in G1F and G2F. However, a significant increase in G0 was also observed, leading to an overall decrease in fucosylation. CDC and ADCC efficacy were improved significantly in the antibodies derived from the pH-shifted cultures, which can be attributed to the increased galactosylation and decreased fucosylation, respectively.

6.6.3 Feeding Strategies and Other Bioprocess Manipulations

Reinhart and coworkers examined cell growth and product formation in batch, fed-batch, and semicontinuous perfusion cultures of three CHO cell lines (CHO-K1, CHO-S, and CHO DG44) producing the same monoclonal antibody in an isogenic format by transfection with the same bacterial artificial chromosome (BAC) constructs [120]. Additionally, two different cell culture media were used to investigate the effects on the bioprocess and mAb quality. Noticeably, fucosylated, mannosylated and aglycosylated glycans were less influenced by the media but were primarily influenced by the host cell line. Generally, there were core-fucosylated, complex biantennary Fc glycans with predominantly G0F, G1F, and G2F glycoforms present in all three CHO cell lines. Mannosylation was highest in CHO DG44 (11–13%) followed by CHO-K1 (5–9%), and then CHO-S (2–3%). Fucosylation was lowest in CHO DG44 (71–83%) followed by CHO-K1 (82–84%) and CHO-S (94–96%). The light chain region also had a glycosylation site containing sialic acid residues. CHO-K1 cells had the highest amount of sialylated mAb (21–36%) followed by CHO DG44 (15–19%) and CHO-S cultures (6–15%). Generally, CHO-S cells showed greater biomass synthesis while CHO-K1 showed the highest mAb production.

Wang and coworkers investigated the effects of different media compositions, EX-CELL (EX) medium and immediate advantage (IA) medium from Millipore-Sigma-Aldrich, on the glycan profile of a recombinant protein, EPO-Fc, secreted from a CHO-GS cell line using an intact glycopeptide analysis method [121]. In this study, the culture supernatant containing recombinant EPO-Fc protein was harvested on day 5 and day 8 of a batch culture. The three EPO *N*-glycosylation sites exhibited bi-, tri-, and tetra-branched glycan structures with various fucosylation and sialylation levels at each site, while the Fc *N*-glycosylation site exhibited mainly fully fucosylated, asialylated bi-antennary *N*-glycans as expected. Noticeably, EPO-Fc protein in the EX medium produced a more complex tetra-antennary *N*-glycan profile on the three EPO *N*-glycosylation sites than IA medium, while IA medium produced more bi- and tri-antennary structures at these same sites. The EPO-Fc exhibited about 25% sialylation, predominantly monosialylation on the EPO *N*-glycosylation sites. No sialylation was seen on the Fc site, as expected. While sialylation remained relatively constant between day 5 and day 8 in the EX medium, it decreased somewhat in the IA medium, particularly at the first *N*-glycosylation site. Furthermore, small amounts of the immunogenic Neu5Gc (2–6%) were detected in this study, and the content significantly increased on day 8 compared to cultures on day 5 in both media.

Wang and coworkers investigated the effect of ultralow carbon dioxide partial pressure (pCO_2) on mAb *N*-glycosylation in CHO cell culture [122]. In fed-batch bioreactor cultures, they decreased the medium pCO2 to an ultralow value (<25 mmHg) by increasing the headspace aeration rate and observed that cell viability and mAb production decreased. In addition, low pCO_2 increased the percentage of galactosylation to 27.45 ± 2.13% compared to the normal pCO_2 condition (21.36 ± 1.66%), while the fucosylation percentage was not affected. They suggested that the increased mAb galactosylation under low pCO_2 conditions may be caused by the upregulated transcription levels of relevant glycosyltransferases (GnT-1 and B4galt1) and nucleotide sugar transporters (UDP-GlcNAcT and UDP-GalT).

6.6.3.1 Glucose and Other Glycosylation Precursors

Zhang and coworkers investigated the effect of glycosylation precursors on the glycosylation profiles of IgG expressed in CHO cells [123]. In this study, when glucose was depleted, leaving lactate as a complementary carbon source, different sugar combinations of fructose, fucose, mannose, and galactose were fed into the culture to study their effects on glycosylation. They observed that feeding with mannose or glucose lowered HM glycans by 3–7% compared to feeding without mannose or glucose. Feeding galactose increased G1 glycans by 8–20% with a 2–6% increase in G2 glycans compared to feeding without galactose or glucose. Furthermore, feeding fucose significantly increased concentrations of intracellular GDP-Fuc.

However, as the afucosylated fraction (other than high mannose glycans) was very small, fucose feeding had no effect on fucosylation.

6.6.3.2 Amino Acids

Ghaffari and coworkers investigated the effect of limiting cysteine, asparagine, or glutamine on batch and fed-batch cultures for three antibody-producing Chinese hamster ovary cell lines (CHO-DXB11, CHO-K1SV, and CHO-S) [124]. They found that cysteine limitation had a negative impact on CHO cell growth, mAb productivity, and the glycosylation pattern of all three cell lines. Glutamine limitation reduced cell growth but not specific productivity. In contrast to glutamine, asparagine limitation did not significantly affect either specific productivity or cell growth. Under the limiting conditions, there were minor differences in the glycosylation profiles of the EG2-hFc antibody produced in the CHO-DXB11 cell line. In the case of cysteine limitation in the CHO-K1SV cells, the fucosylated glycans were reduced by 16% and the antennarity index by 15%. For the CHO-S cells, the fucosylated glycans were reduced by 12% and in the antennarity index by 16%. Restoring the cysteine levels after 1 day of cysteine limitation resulted in partial recovery of cell growth and productivity in CHO-DXB11 culture. Interestingly, they observed that supplementing cultures with cysteine resulted in similar cell-specific productivity as using commercial feeds.

6.6.3.3 Glycosaminoglycan Production

In an effort to improve the glycan structures for the bioengineered heparin described above, Baik et al. evaluated the effects of bioprocess conditions on productivity, structure, and anticoagulant activity. Supplementing the cultures with a chemically defined feed medium increased IVCD approximately twofold and product titer nearly threefold. However, the glycan compositions were largely unchanged. Adding a bolus of cysteine increased the anticoagulant activity approximately twofold, but this was not sufficient to obtain pharmacological anticoagulant activity [125].

6.6.4 Culture Additives

Dang and coworkers investigated the effects of carbon dioxide partial pressure, media hold duration (at 37 °C), manganese, and high-temperature short-time (HTST) treatment of culture media on afucosylation of mAb (IgG1) produced in CHO cells [126]. Afucosylation level was highest (~7.1%) in the low pCO_2 model employing the other process parameters (i.e., media HTST treatment, Mn, and media hold), while the lowest afucosylation level (~2.5%) was observed for the condition without these three factors. Moreover, they observed that HTST treatment may not be associated with the higher afucosylation in the condition with Mn supplementation and media HTST. Noticeably, G0F was reduced in conditions with Mn supplementation, whereas Man5 was ~1% lower in all the conditions tested. Interestingly, there was a synergistic effect of these three independent parameters (pCO_2, Mn, and media hold) on afucosylation. Afucosylation increased (~4%) with increasing these process parameters. In addition, they investigated the effects of high pCO_2, Na^+, and osmolality on afucosylation. They found that afucosylation increased with increasing Na^+ resulting from Na_2CO_3 addition to maintain pH in the high pCO_2 model.

Zhou and coworkers investigated the effects of different components, in particular bovine serum albumin (BSA), on IgG *N*-glycoforms in hybridoma culture media [15]. They found that small molecule additives including glucose, glutamine, pyruvate, and sodium bicarbonate caused minor changes in glycan species. Noticeably, glutamine addition decreased galactosylation and sialylation, presumably due to ammonia accumulation. On the other hand, BSA addition increased galactosylation and sialylation levels, with a decrease in fucosylation level. Furthermore, two new species of IgG *N*-glycans in the BSA+ conditions were discovered, afucosylated sialylated *N*-glycans and Neu5Ac-containing *N*-glycans, rather than the Neu5Gc-containing glycans commonly seen in rodent species.

The effects of deoxymannojirimycin (DMJ), an α-D-mannosidase inhibitor, on high mannose and core fucose levels were studied during IgG1 production in two different CHO cell lines (CHO-DXB-11 and CHO-K1SV) by Schwarz et al. (2020) [127]. DMJ at concentrations of 500 μM significantly decreased the levels of complex-type glycans and core-fucosylated complex glycans by 15- and 14-fold, respectively, and significantly increased the HM glycan level by 21-fold. However, they observed that the relative fucosylated complex and fucosylated bi-antennary levels were not affected by DMJ, demonstrating that DMJ had no effect on fucosyltransferase activity. Therefore, they suggest that the decrease in core-fucosylated glycans upon DMJ feeding resulted from increased HM glycans rather than an increase in afucosylated complex glycoforms.

The effect of raffinose supplementation on high mannose species was studied in high-throughput systems using two cell lines (CHO-K1 and CHO-S) expressing humanized and human IgG1 antibodies (cell line 1 and cell line 2) by Bruhlmann et al. [128]. Cultures supplemented with raffinose at constant medium osmolality (315 mOsm/kg) exhibited up to sixfold increased HM species with increasing raffinose. Interestingly, cell line 1 predominantly expressed Man5 and Man6 glycans, while Man7 was not detected, and small amounts of Man8 were observed. In addition, cell line 2 exhibited mainly Man5 glycans. They highlighted that raffinose mainly favored Man5 glycans. Supplementing the culture with 30 mM raffinose at three different osmolalities significantly increased HM glycans. Notably, culture supplementation with raffinose influenced the expression levels of glycosylation-related genes. In particular, sialyltransferase gene expression was highly upregulated, whereas galactosyltransferase gene expression was downregulated.

Ehret and coworkers investigated the impact of cell culture media additives on glycosylation profiles of recombinant IgG1 produced in Chinese hamster ovary cells (CHO-K1 and CHO DG44) [129]{Prabhu, 2019 #157}. Addition of 15 μM kifunensine into the feed medium increased HM species by 85.8%. Moreover, supplementing the culture with 800 μM 2-F-peracetyl fucose lowered fucosylation by 76.1%. Notably, the combination of 24 μM uridine, 48 μM manganese, and 120 mM galactose increased total galactosylation by 40.9%. In addition, the presence of 30 μM dexamethasone in combination with galactose (120 mM), manganese (48 μM), and uridine (24 μM) increased sialylation by 6.9%.

Yin and coworkers examined the impact of the butyrated *N*-acetylmannosamine analog, 1,3,4-*O*-Bu_3ManNAc on recombinant protein expression in CHO cell culture [90, 130]. Addition of 1,3,4-*O*-Bu_3ManNAc increased protein expression of EPO- and human IgG-expressing CHO cell lines with negligible impact on cell growth, viability, glucose consumption, and lactate production. Conversely, sodium butyrate (NaBu) addition decreased viable cell density and viability by ~20 and ~30%, respectively, compared to untreated or 1,3,4-*O*-Bu_3ManNAc treated cells for both products. NaBu-treated cells exhibited lower sialylation levels on purified EPO compared to the untreated control while the addition of 1,3,4-*O*-Bu_3ManNAc increased sialylation levels compared to both the NaBu-treated and untreated control cultures. They concluded that, at the same butyrate concentration, supplementation with 1,3,4-*O*-Bu_3ManNAc has a negligible impact on cell viability, cytotoxicity, and apoptosis while also enhancing protein sialylation, compared to NaBu in CHO cell culture.

Wells and colleagues investigated the effects of galactose, uridine, and 2-F-peracetyl fucose (2FP) on cell growth, titer, and gene expression of a CHO cell line producing an IgG1 antibody in a batch culture [131]. Supplementation with 200 μM uridine increased galactosylation by 2% and decreased fucosylation by 6% compared with unsupplemented controls. Furthermore, the mRNA levels of mgat1–2, B4galt1–6, and the sugar transporter Slc35a3 were upregulated compared with controls on day 4. Cells grown in the presence of 100 mM galactose exhibited up to 21% decreased G0 glycans with increased G1 or G2 glycans compared to unsupplemented cultures on day 7. In addition, this supplementation increased fucosylation by about 5%. Interestingly, galactose addition upregulated mgat2, mgat5, and B4galt3–4 by Day 4 of the cultures. Supplementing cultures with 50 μM 2FP decreased relative fucosylation by 48%. Moreover, 2FP supplementation significantly decreased the expression of fut8 and nucleotide sugar transporter gene expression (Slc35a3, Slc35a1, and Slc35c1). They observed that galactose and 2FP had no negative effects on cell growth, metabolism, or titer. In addition, the uridine-supplemented cells demonstrated ~23% increased viable cell density and ~30% decreased mAb titer.

Prabhu and coworkers investigated the effects of nickel and cobalt on glycosylation of a recombinant IgG produced in CHO cells [132]. Nickel concentrations greater than 500 μM significantly reduced galactosylation, while concentrations lower than 500 μM had no significant impact on glycosylation. In addition, fucosylation was not impacted at any nickel concentration tested. 50 μM cobalt slightly increased G1F glycans, whereas higher concentrations increased G0F glycans. Interestingly, uridine and galactose supplementation into cultures containing either Ni^{2+} or Co^{2+} may alleviate the effect of Ni^{2+} and Co^{2+} on galactosylation due to the increased availability of UDP-Gal.

Pande and colleagues investigated the effect of a small molecule ionophore, monensin, on high mannose glycans in different IgG-producing CHO cell lines [133]. Noticeably, monensin addition increased HM glycans, particularly Man5. Furthermore, no negative impacts on HM or cell culture parameters were observed at monensin concentrations between 0.1 to 10 nM. However, 50 nM monensin greatly increased HM glycans with negative effects on cell growth, viability, and titer. Interestingly, when monensin was added as a 25 nM bolus into the culture, total HM glycans on IgG increased without negative impacts on cell culture parameters. They next tested the effect of monensin on different production cell lines in a 10-day mock perfusion assay designed to mimic the conditions in a perfusion bioreactor. Three production cell lines producing mAbs with low levels of HM glycans (less than 10%) and one production cell line that consistently produced mAbs with high levels of HM glycans (more than 10%) were used. Monensin was added at 25 nM to all cultures on day 3. After that, cultures were subjected to daily partial medium exchanges with perfusion medium containing either 25 nM monensin or increasing monensin concentrations. They observed that monensin increased HM glycans on all four antibodies in a dose-dependent manner though the magnitude of the response was cell-line specific.

6.7 Perspectives and Future Directions

CHO cells have been successfully used to express complex and efficacious biotherapeutics at significant scales. One cannot overstate the progress in the field made possible by complex cellular engineering [134] and process engineering [1], which has been vital to large-scale manufacturing. Indeed, the examples of the highly complicated glycoproteins erythropoietin [135] and abatacept (Orencia®) [136] produced and marketed as therapeutics are an indication of how CHO cells can achieve reproducible CQAs without immunogenic glycans. While significant process development is required for successful production of a biosimilar, including cell line screening, manipulation of process conditions, and possibly cell line glycoengineering, the regulatory approval of over 60 biosimilars in Europe and over 30 in the United States demonstrates that matching CQAs without the innovator cell line or knowledge of proprietary bioprocess conditions is achievable. This success is a testament to the plasticity of CHO cells and a growing understanding of the effects of process conditions on glycan distributions.

Still, monoclonal antibodies, the dominant class of recombinant therapeutic proteins, have relatively straightforward glycosylation patterns with typically one glycosylation site on each heavy chain and biantennary, asialylated glycans [137]. As more challenging structures such as heparin, bispecific antibodies, complicated fusions of receptors with Fc domains, and more complex glycans on mAbs emerge, new challenges may arise. For example, while mAbs are generally not sialylated, new studies indicate that increased sialyation increases half-life, impacts stability, and may reduce dosing frequency [138, 139]. These new demands may require alternative hosts including human cell lines or highly engineered lower eukaryotes such as yeast or modifications to CHO using targeted gene and process manipulation to maximize their post-translational potential.

References

1. Hossler P, Khattak SF, Li ZJ (2009) Optimal and consistent protein glycosylation in mammalian cell culture. Glycobiology 19(9):936–949. https://doi.org/10.1093/glycob/cwp079
2. Wurm F (2004) Production of recombinant protein therapeutics in cultivated mammalian cells. Nat Biotechnol 22:1393–1398. https://doi.org/10.1038/nbt1026
3. Gupta SK, Shukla P (2018) Glycosylation control technologies for recombinant therapeutic proteins. Appl Microbiol Biotechnol 102(24):10457–10468. https://doi.org/10.1007/s00253-018-9430-6
4. Dalton AC, Barton WA (2014) Over-expression of secreted proteins from mammalian cell lines. Protein Sci 23(5):517–525. https://doi.org/10.1002/pro.2439
5. Walsh G, Jefferis R (2006) Post-translational modifications in the context of therapeutic proteins. Nat Biotechnol 24(10):1241–1252. https://doi.org/10.1038/nbt1252
6. Krasnova L, Wong C-H (2016) Exploring human glycosylation for better therapies. Mol Asp Med 51:125–143. https://doi.org/10.1016/j.mam.2016.05.003
7. Liu L (2015) Antibody glycosylation and its impact on the pharmacokinetics and pharmacodynamics of monoclonal antibodies and fc-fusion proteins. J Pharm Sci 104(6):1866–1884. https://doi.org/10.1002/jps.24444
8. Steentoft C, Bennett EP, Schjoldager KT-BG, Vakhrushev SY, Wandall HH, Clausen H (2014) Precision genome editing: a small revolution for glycobiology. Glycobiology 24(8):663–680. https://doi.org/10.1093/glycob/cwu046
9. Andersen DC, Bridges T, Gawlitzek M, Hoy C (2000) Multiple cell culture factors can affect the glycosylation of Asn-184 in CHO-produced tissue-type plasminogen activator. Biotechnol Bioeng 70(1):25–31. https://doi.org/10.1002/1097-0290(20001005)70:1<25::aid-bit4>3.0.co;2-q
10. Liu B, Spearman M, Doering J, Lattova E, Perreault H, Butler M (2014) The availability of glucose to CHO cells affects the intracellular lipid-linked oligosaccharide distribution, site occupancy and the N-glycosylation profile of a monoclonal antibody. J Biotechnol 170:17–27. https://doi.org/10.1016/j.jbiotec.2013.11.007
11. Spahn PN, Lewis NE (2014) Systems glycobiology for glycoengineering. Curr Opin Biotechnol 30:218–224. https://doi.org/10.1016/j.copbio.2014.08.004
12. Stolfa G, Smonskey MT, Boniface R, Hachmann A-B, Gulde P, Joshi AD, Pierce AP, Jacobia SJ, Campbell A (2018) CHO-omics review: the impact of current and emerging technologies on Chinese hamster ovary based bioproduction. Biotechnol J 13(3):1700227. https://doi.org/10.1002/biot.201700227
13. Sha S, Agarabi C, Brorson K, Lee D-Y, Yoon S (2016) N-glycosylation design and control of therapeutic monoclonal antibodies. Trends Biotechnol 34(10):835–846. https://doi.org/10.1016/j.tibtech.2016.02.013
14. Hossler P (2012) Protein glycosylation control in mammalian cell culture: past precedents and contemporary prospects. Adv Biochem Eng Biotechnol 127:187–219. https://doi.org/10.1007/10_2011_113

15. Zhou J, Gao H, Xie W, Li Y (2020) Bovine serum albumin affects N-glycoforms of murine IgG monoclonal antibody purified from hybridoma supernatants. Appl Microbiol Biotechnol 104(4):1583–1594. https://doi.org/10.1007/s00253-019-10309-z
16. Dordal MS, Wang FF, Goldwasser E (1985) The role of carbohydrate in erythropoietin action*. Endocrinology 116(6):2293–2299. https://doi.org/10.1210/endo-116-6-2293
17. Nose M, Wigzell H (1983) Biological significance of carbohydrate chains on monoclonal antibodies. Proc Natl Acad Sci USA 80(21):6632–6636. https://doi.org/10.1073/pnas.80.21.6632
18. Thacker BE, Sharfstein ST (2018) Metabolic engineering of mammalian cells to produce heparan sulfates. Emerg Top Life Sci 2(3):443–452. https://doi.org/10.1042/ETLS20180007
19. Schiestl M, Li J, Abas A, Vallin A, Millband J, Gao K, Joung J, Pluschkell S, Go T, Kang H-N (2014) The role of the quality assessment in the determination of overall biosimilarity: a simulated case study exercise. Biologicals 42(2):128–132. https://doi.org/10.1016/j.biologicals.2013.11.009
20. Rathore AS, Winkle H (2009) Quality by design for biopharmaceuticals. Nat Biotechnol 27(1):26–34. https://doi.org/10.1038/nbt0109-26
21. Lipsitz YY, Timmins NE, Zandstra PW (2016) Quality cell therapy manufacturing by design. Nat Biotechnol 34(4):393–400. https://doi.org/10.1038/nbt.3525
22. Kirchhoff CF, Wang XM, Conlon HD, Anderson S, Ryan AM, Bose A (2017) Biosimilars: key regulatory considerations and similarity assessment tools. Biotechnol Bioeng 114(12):2696–2705. https://doi.org/10.1002/bit.26438
23. Kesik-Brodacka M (2018) Progress in biopharmaceutical development. Biotechnol Appl Biochem 65(3):306–322. https://doi.org/10.1002/bab.1617
24. Dahodwala H, Sharfstein ST (2017) Biosimilars: imitation games. ACS Med Chem Lett 8(7):690–693. https://doi.org/10.1021/acsmedchemlett.7b00199
25. Fisher AC, Kamga MH, Agarabi C, Brorson K, Lee SL, Yoon S (2019) The current scientific and regulatory landscape in advancing integrated continuous biopharmaceutical manufacturing. Trends Biotechnol 37(3):253–267. https://doi.org/10.1016/j.tibtech.2018.08.008
26. Rathore AS (2009) Roadmap for implementation of quality by design (QbD) for biotechnology products. Trends Biotechnol 27(9):546–553. https://doi.org/10.1016/j.tibtech.2009.06.006
27. Rathore AS (2016) Quality by design (QbD)-based process development for purification of a biotherapeutic. Trends Biotechnol 34(5):358–370. https://doi.org/10.1016/j.tibtech.2016.01.003
28. Butler M, Meneses-Acosta A (2012) Recent advances in technology supporting biopharmaceutical production from mammalian cells. Appl Microbiol Biotechnol 96(4):885–894. https://doi.org/10.1007/s00253-012-4451-z
29. Wang Q, Yin B, Chung CY, Betenbaugh MJ (2017) Glycoengineering of CHO cells to improve product quality. Methods Mol Biol (Clifton, NJ) 1603:25–44. https://doi.org/10.1007/978-1-4939-6972-2_2
30. Mohanty S, Chaudhary BP, Zoetewey D (2020) Structural insight into the mechanism of N-linked glycosylation by oligosaccharyltransferase. Biomol Ther 10(4). https://doi.org/10.3390/biom10040624
31. Stanley P, Taniguchi N, Aebi M (2015) N-Glycans. In: Varki A, Cummings RD et al (eds) Essentials of glycobiology. Cold Spring Harbor (NY), pp 99–111. https://doi.org/10.1101/glycobiology.3e.009
32. Araki K, Nagata K (2011) Protein folding and quality control in the ER. Cold Spring Harb Perspect Biol 3(11):a007526. https://doi.org/10.1101/cshperspect.a007526
33. Wang Q, Stuczynski M, Gao Y, Betenbaugh MJ (2015) Strategies for engineering protein N-glycosylation pathways in mammalian cells. Methods Mol Biol (Clifton, NJ) 1321:287–305. https://doi.org/10.1007/978-1-4939-2760-9_20
34. Huang HH, Hassinen A, Sundaram S, Spiess AN, Kellokumpu S, Stanley P (2015) GnT1IP-L specifically inhibits MGAT1 in the Golgi via its luminal domain. elife 4. https://doi.org/10.7554/eLife.08916

35. Carraway KL, Hull SR (1989) O-glycosylation pathway for mucin-type glycoproteins. BioEssays 10(4):117–121. https://doi.org/10.1002/bies.950100406
36. Hart GW, Housley MP, Slawson C (2007) Cycling of O-linked β-N-acetylglucosamine on nucleocytoplasmic proteins. Nature 446(7139):1017–1022. https://doi.org/10.1038/nature05815
37. Lalonde M-E, Durocher Y (2017) Therapeutic glycoprotein production in mammalian cells. J Biotechnol 251:128–140. https://doi.org/10.1016/j.jbiotec.2017.04.028
38. Tejwani V, Andersen MR, Nam JH, Sharfstein ST (2018) Glycoengineering in CHO cells: advances in systems biology. Biotechnol J 13(3):e1700234. https://doi.org/10.1002/biot.201700234
39. Mimura Y, Katoh T, Saldova R, O'Flaherty R, Izumi T, Mimura-Kimura Y, Utsunomiya T, Mizukami Y, Yamamoto K, Matsumoto T, Rudd PM (2018) Glycosylation engineering of therapeutic IgG antibodies: challenges for the safety, functionality and efficacy. Protein Cell 9(1):47–62. https://doi.org/10.1007/s13238-017-0433-3
40. Pacis E, Yu M, Autsen J, Bayer R, Li F (2011) Effects of cell culture conditions on antibody N-linked glycosylation – what affects high mannose 5 glycoform. Biotechnol Bioeng 108(10):2348–2358. https://doi.org/10.1002/bit.23200
41. Alessandri L, Ouellette D, Acquah A, Rieser M, Leblond D, Saltarelli M, Radziejewski C, Fujimori T, Correia I (2012) Increased serum clearance of oligomannose species present on a human IgG1 molecule. MAbs 4:509–520. https://doi.org/10.4161/mabs.20450
42. Goetze AM, Liu YD, Zhang Z, Shah B, Lee E, Bondarenko PV, Flynn GC (2011) High-mannose glycans on the Fc region of therapeutic IgG antibodies increase serum clearance in humans. Glycobiology 21(7):949–959. https://doi.org/10.1093/glycob/cwr027
43. Yu M, Brown D, Reed C, Chung S, Lutman J, Stefanich E, Wong A, Stephan J-P, Bayer R (2012) Production, characterization, and pharmacokinetic properties of antibodies with N-linked mannose-5 glycans. MAbs 4(4):475–487. https://doi.org/10.4161/mabs.20737
44. Seo N, Polozova A, Zhang M, Yates Z, Cao S, Li H, Kuhns S, Maher G, McBride H, Liu J (2018) Analytical and functional similarity of Amgen biosimilar ABP 215 to bevacizumab. MAbs 10:1–58. https://doi.org/10.1080/19420862.2018.1452580
45. Miyoshi E, Noda K, Yamaguchi Y, Inoue S, Ikeda Y, Wang W, Ko JH, Uozumi N, Li W, Taniguchi N (1999) The alpha1-6-fucosyltransferase gene and its biological significance. Biochim Biophys Acta 1473(1):9–20. https://doi.org/10.1016/s0304-4165(99)00166-x
46. Kamoda S, Nomura C, Kinoshita M, Nishiura S, Ishikawa R, Kakehi K, Kawasaki N, Hayakawa T (2004) Profiling analysis of oligosaccharides in antibody pharmaceuticals by capillary electrophoresis. J Chromatogr A 1050:211–216. https://doi.org/10.1016/s0021-9673(04)01399-8
47. Imai-Nishiya H, Mori K, Inoue M, Wakitani M, Iida S, Shitara K, Satoh M (2007) Double knockdown of alpha1,6-fucosyltransferase (FUT8) and GDP-mannose 4,6-dehydratase (GMD) in antibody-producing cells: a new strategy for generating fully non-fucosylated therapeutic antibodies with enhanced ADCC. BMC Biotechnol 7:84. https://doi.org/10.1186/1472-6750-7-84
48. Shinkawa T, Nakamura K, Yamane N, Shoji-Hosaka E, Kanda Y, Sakurada M, Anazawa H, Satoh M, Yamasaki M, Hanai N, Shitara K (2003) The absence of fucose but not the presence of galactose or bisecting N-acetylglucosamine of human IgG1 complex-type oligosaccharides shows the critical role of enhancing antibody-dependent cellular cytotoxicity. J Biol Chem 278:3466–3473. https://doi.org/10.1074/jbc.M210665200
49. Natsume A, Wakitani M, Yamane-Ohnuki N, Shoji-Hosaka E, Niwa R, Uchida K, Satoh M, Shitara K (2005) Fucose removal from complex-type oligosaccharide enhances the antibody-dependent cellular cytotoxicity of single-gene-encoded antibody comprising a single-chain antibody linked the antibody constant region. J Immunol Methods 306(1):93–103. https://doi.org/10.1016/j.jim.2005.07.025
50. Ferrara C, Grau S, Jäger C, Sondermann P, Brünker P, Waldhauer I, Hennig M, Ruf A, Rufer A, Stihle M, Umana P, Benz J (2011) Unique carbohydrate-carbohydrate interactions are

required for high affinity binding between Fc RIII and antibodies lacking core fucose. Proc Natl Acad Sci USA 108:12669–12674. https://doi.org/10.1073/pnas.1108455108
51. Shields RL, Lai J, Keck R, O'Connell LY, Hong K, Meng YG, Weikert SH, Presta LG (2002) Lack of fucose on human IgG1 N-linked oligosaccharide improves binding to human Fcgamma RIII and antibody-dependent cellular toxicity. J Biol Chem 277(30):26733–26740. https://doi.org/10.1074/jbc.M202069200
52. Reusch D, Tejada ML (2015) Fc glycans of therapeutic antibodies as critical quality attributes. Glycobiology 25(12):1325–1334. https://doi.org/10.1093/glycob/cwv065
53. Visser J, Feuerstein I, Stangler T, Schmiederer T, Fritsch C, Schiestl M (2013) Physicochemical and functional comparability between the proposed biosimilar rituximab GP2013 and originator rituximab. BioDrugs 27. https://doi.org/10.1007/s40259-013-0036-3
54. Hodoniczky J, Zheng Y, James D (2005) Control of recombinant monoclonal antibody effector functions by Fc N-glycan remodeling in vitro. Biotechnol Prog 21:1644–1652. https://doi.org/10.1021/bp050228w
55. Peschke B, Keller CW, Weber P, Quast I, Lünemann JD (2017) Fc-galactosylation of human immunoglobulin gamma isotypes improves C1q binding and enhances complement-dependent cytotoxicity. Front Immunol 8:646–646. https://doi.org/10.3389/fimmu.2017.00646
56. Raju T, Jordan R (2012) Galactosylation variations in marketed therapeutic antibodies. MAbs 4:385–391. https://doi.org/10.4161/mabs.19868
57. Sha S, Yoon S (2019) An investigation of nucleotide sugar dynamics under the galactose supplementation in CHO cell culture. Process Biochem 81:165–174. https://doi.org/10.1016/j.procbio.2019.03.020
58. Arnold J, Wormald M, Sim R, Rudd P, Dwek R (2007) The impact of glycosylation on the biological function and structure of human immunoglobulins. Annu Rev Immunol 25:21–50. https://doi.org/10.1146/annurev.immunol.25.022106.141702
59. Byrne B, Donohoe G, O'Kennedy R (2007) Sialic acids: carbohydrate moieties that influence the biological and physical properties of biopharmaceutical proteins and living cells. Drug Discov Today 12:319–326. https://doi.org/10.1016/j.drudis.2007.02.010
60. Fukuta K, Yokomatsu T, Abe R, Asanagi M, Makino T (2000) Genetic engineering of CHO cells producing human interferon-γ by transfection of sialyltransferases. Glycoconj J 17(12):895–904. https://doi.org/10.1023/a:1010977431061
61. Schjoldager KT, Narimatsu Y, Joshi HJ, Clausen H (2020) Global view of human protein glycosylation pathways and functions. Nat Rev Mol Cell Biol 21(12):729–749. https://doi.org/10.1038/s41580-020-00294-x
62. Costa AR, Rodrigues ME, Henriques M, Oliveira R, Azeredo J (2014) Glycosylation: impact, control and improvement during therapeutic protein production. Crit Rev Biotechnol 34(4):281–299. https://doi.org/10.3109/07388551.2013.793649
63. Goh JB, Ng SK (2018) Impact of host cell line choice on glycan profile. Crit Rev Biotechnol 38(6):851–867. https://doi.org/10.1080/07388551.2017.1416577
64. Yeo JHM, Ho SCL, Mariati M, Koh E, Tay SJ, Woen S, Zhang P, Yang Y (2017) Optimized selection marker and CHO host cell combinations for generating high monoclonal antibody producing cell lines. Biotechnol J 12(12):1700175. https://doi.org/10.1002/biot.201700175
65. Baik JY, Gasimli L, Yang B, Datta P, Zhang F, Glass CA, Esko JD, Linhardt RJ, Sharfstein ST (2012) Metabolic engineering of Chinese hamster ovary cells: towards a bioengineered heparin. Metab Eng 14(2):81–90. https://doi.org/10.1016/j.ymben.2012.01.008
66. Thacker BE, Thorne KJ, Cartwright C, Park J, Glass K, Chea A, Kellman BP, Lewis NE, Wang Z, Di Nardo A,Sharfstein ST, Jeske W, Walenga J, Hogwood J, Gray E, Mulloy B, Esko JD, Glass CA, (2022) Multiplex genome editing of mammalian cells for producing recombinant heparin. Metab Eng 70:155–165 https://doi.org/10.1016/j.ymben.2022.01.002
67. Walsh G (2010) Post-translational modifications of protein biopharmaceuticals. Drug Discov Today 15(17–18):773–780. https://doi.org/10.1016/j.drudis.2010.06.009
68. Yang Z, Wang S, Halim A, Schulz MA, Frodin M, Rahman SH, Vester-Christensen MB, Behrens C, Kristensen C, Vakhrushev SY, Bennett EP, Wandall HH, Clausen H (2015)

Engineered CHO cells for production of diverse, homogeneous glycoproteins. Nat Biotechnol 33(8):842–844. https://doi.org/10.1038/nbt.3280
69. Angata T, Varki A (2002) Chemical diversity in the sialic acids and related alpha-keto acids: an evolutionary perspective. Chem Rev 102(2):439–469
70. Lee EU, Roth J, Paulson JC (1989) Alteration of terminal glycosylation sequences on N-linked oligosaccharides of Chinese hamster ovary cells by expression of beta-galactoside alpha 2,6-sialyltransferase. J Biol Chem 264(23):13848–13855
71. Li Y, Chen X (2012) Sialic acid metabolism and sialyltransferases: natural functions and applications. Appl Microbiol Biotechnol 94(4):887–905. https://doi.org/10.1007/s00253-012-4040-1
72. Takeuchi M, Takasaki S, Miyazaki H, Kato T, Hoshi S, Kochibe N, Kobata A (1988) Comparative study of the asparagine-linked sugar chains of human erythropoietins purified from urine and the culture medium of recombinant Chinese hamster ovary cells. J Biol Chem 263(8):3657–3663. https://doi.org/10.1016/s0021-9258(18)68975-6
73. Ashwell G, Harford J (1982) Carbohydrate-specific receptors of the liver. Annu Rev Biochem 51:531–554. https://doi.org/10.1146/annurev.bi.51.070182.002531
74. Zhang M, Koskie K, Ross JS, Kayser KJ, Caple MV (2010) Enhancing glycoprotein sialylation by targeted gene silencing in mammalian cells. Biotechnol Bioeng 105(6):1094–1105. https://doi.org/10.1002/bit.22633
75. Pearce OM, Laubli H, Verhagen A, Secrest P, Zhang J, Varki NM, Crocker PR, Bui JD, Varki A (2014) Inverse hormesis of cancer growth mediated by narrow ranges of tumor-directed antibodies. Proc Natl Acad Sci USA 111(16):5998–6003. https://doi.org/10.1073/pnas.1209067111
76. Padler-Karavani V, Yu H, Cao H, Chokhawala H, Karp F, Varki N, Chen X, Varki A (2008) Diversity in specificity, abundance, and composition of anti-Neu5Gc antibodies in normal humans: potential implications for disease. Glycobiology 18(10):818–830. https://doi.org/10.1093/glycob/cwn072
77. Hedlund M, Padler-Karavani V, Varki NM, Varki A (2008) Evidence for a human-specific mechanism for diet and antibody-mediated inflammation in carcinoma progression. Proc Natl Acad Sci USA 105(48):18936–18941. https://doi.org/10.1073/pnas.0803943105
78. Noguchi A, Mukuria CJ, Suzuki E, Naiki M (1995) Immunogenicity of N-glycolylneuraminic acid-containing carbohydrate chains of recombinant human erythropoietin expressed in Chinese hamster ovary cells. J Biochem 117(1):59–62. https://doi.org/10.1093/oxfordjournals.jbchem.a124721
79. Yu C, Gao K, Zhu L, Wang W, Wang L, Zhang F, Liu C, Li M, Wormald MR, Rudd PM, Wang J (2016) At least two Fc Neu5Gc residues of monoclonal antibodies are required for binding to anti-Neu5Gc antibody. Sci Rep 6(1):20029. https://doi.org/10.1038/srep20029
80. Fischer S, Mathias S, Stadermann A, Yang S, Schmieder V, Zeh N, Schmidt N, Richter P, Wright S, Zimmermann E, Ley Y, van der Meer J, Hartsch T, Bernloehr C, Otte K, Bradl H, Gamer M, Schulz P. Loss of a newly discovered microRNA in Chinese hamster ovary cells leads to upregulation of NGNA sialylation on monoclonal antibodies. Biotechn Bioeng. https://doi.org/10.1002/bit.28015
81. Kurogochi M, Mori M, Osumi K, Tojino M, Sugawara S, Takashima S, Hirose Y, Tsukimura W, Mizuno M, Amano J, Matsuda A, Tomita M, Takayanagi A, Shoda S, Shirai T (2015) Glycoengineered monoclonal antibodies with homogeneous glycan (M3, G0, G2, and A2) using a chemoenzymatic approach have different affinities for FcgammaRIIIa and variable antibody-dependent cellular cytotoxicity activities. PLoS One 10(7):e0132848. https://doi.org/10.1371/journal.pone.0132848
82. Anthony RM, Nimmerjahn F, Ashline DJ, Reinhold VN, Paulson JC, Ravetch JV (2008) Recapitulation of IVIG anti-inflammatory activity with a recombinant IgG Fc. Science 320(5874):373–376. https://doi.org/10.1126/science.1154315
83. Lin CW, Tsai MH, Li ST, Tsai TI, Chu KC, Liu YC, Lai MY, Wu CY, Tseng YC, Shivatare SS, Wang CH, Chao P, Wang SY, Shih HW, Zeng YF, You TH, Liao JY, Tu YC, Lin YS, Chuang

HY, Chen CL, Tsai CS, Huang CC, Lin NH, Ma C, Wu CY, Wong CH (2015) A common glycan structure on immunoglobulin G for enhancement of effector functions. Proc Natl Acad Sci USA 112(34):10611–10616. https://doi.org/10.1073/pnas.1513456112
84. Kaneko Y, Nimmerjahn F, Ravetch JV (2006) Anti-inflammatory activity of immunoglobulin G resulting from Fc sialylation. Science 313(5787):670–673. https://doi.org/10.1126/science.1129594
85. Raymond C, Robotham A, Spearman M, Butler M, Kelly J, Durocher Y (2015) Production of α2,6-sialylated IgG1 in CHO cells. mAbs:7. https://doi.org/10.1080/19420862.2015.1029215
86. Jeong YT, Choi O, Lim HR, Son YD, Kim HJ, Kim JH (2008) Enhanced sialylation of recombinant erythropoietin in CHO cells by human glycosyltransferase expression. J Microbiol Biotechnol 18(12):1945–1952
87. Varki A, Schauer R (2009) Sialic Acids. In: Varki A, Cummings RD, Esko JD et al (eds) Essentials of glycobiology. Cold Spring Harbor Laboratory Press, Cold Spring Harbor
88. Wong NS, Yap MG, Wang DI (2006) Enhancing recombinant glycoprotein sialylation through CMP-sialic acid transporter over expression in Chinese hamster ovary cells. Biotechnol Bioeng 93(5):1005–1016. https://doi.org/10.1002/bit.20815
89. Gu X, Wang DI (1998) Improvement of interferon-gamma sialylation in Chinese hamster ovary cell culture by feeding of N-acetylmannosamine. Biotechnol Bioeng 58(6):642–648
90. Yin B, Wang Q, Chung CY, Bhattacharya R, Ren X, Tang J, Yarema KJ, Betenbaugh MJ (2017) A novel sugar analog enhances sialic acid production and biotherapeutic sialylation in CHO cells. Biotechnol Bioeng 114(8):1899–1902. https://doi.org/10.1002/bit.26291
91. Jeong YT, Choi O, Son YD, Park SY, Kim JH (2009) Enhanced sialylation of recombinant erythropoietin in genetically engineered Chinese-hamster ovary cells. Biotechnol Appl Biochem 52(Pt 4):283–291. https://doi.org/10.1042/ba20080044
92. Son YD, Jeong YT, Park SY, Kim JH (2011) Enhanced sialylation of recombinant human erythropoietin in Chinese hamster ovary cells by combinatorial engineering of selected genes. Glycobiology 21(8):1019–1028. https://doi.org/10.1093/glycob/cwr034
93. Marx N, Grünwald-Gruber C, Bydlinski N, Dhiman H, Ngoc Nguyen L, Klanert G, Borth N (2018) CRISPR-based targeted epigenetic editing enables gene expression modulation of the silenced beta-galactoside alpha-2,6-sialyltransferase 1 in CHO cells. Biotechnol J 13(10):1700217. https://doi.org/10.1002/biot.201700217
94. Monti E, Preti A, Venerando B, Borsani G (2002) Recent development in mammalian sialidase molecular biology. Neurochem Res 27(7–8):649–663
95. Seyrantepe V, Landry K, Trudel S, Hassan JA, Morales CR, Pshezhetsky AV (2004) Neu4, a novel human lysosomal lumen sialidase, confers normal phenotype to sialidosis and galactosialidosis cells. J Biol Chem 279(35):37021–37029. https://doi.org/10.1074/jbc.M404531200
96. Kakugawa Y, Wada T, Yamaguchi K, Yamanami H, Ouchi K, Sato I, Miyagi T (2002) Up-regulation of plasma membrane-associated ganglioside sialidase (Neu3) in human colon cancer and its involvement in apoptosis suppression. Proc Natl Acad Sci USA 99(16):10718–10723. https://doi.org/10.1073/pnas.152597199
97. Hinek A, Pshezhetsky AV, von Itzstein M, Starcher B (2006) Lysosomal sialidase (neuraminidase-1) is targeted to the cell surface in a multiprotein complex that facilitates elastic fiber assembly. J Biol Chem 281(6):3698–3710. https://doi.org/10.1074/jbc.M508736200
98. Gramer MJ (2014) Product quality considerations for mammalian cell culture process development and manufacturing. Adv Biochem Eng Biotechnol 139:123–166. https://doi.org/10.1007/10_2013_214
99. Ferrari J, Gunson J, Lofgren J, Krummen L, Warner TG (1998) Chinese hamster ovary cells with constitutively expressed sialidase antisense RNA produce recombinant DNase in batch culture with increased sialic acid. Biotechnol Bioeng 60(5):589–595
100. Krapp S, Mimura Y, Jefferis R, Huber R, Sondermann P (2003) Structural analysis of human IgG-Fc glycoforms reveals a correlation between glycosylation and structural integrity. J Mol Biol 325(5):979–989

101. Suzuki E, Niwa R, Saji S, Muta M, Hirose M, Iida S, Shiotsu Y, Satoh M, Shitara K, Kondo M, Toi M (2007) A nonfucosylated anti-HER2 antibody augments antibody-dependent cellular cytotoxicity in breast cancer patients. Clin Cancer Res 13(6):1875–1882. https://doi.org/10.1158/1078-0432.ccr-06-1335
102. Junttila TT, Parsons K, Olsson C, Lu Y, Xin Y, Theriault J, Crocker L, Pabonan O, Baginski T, Meng G, Totpal K, Kelley RF, Sliwkowski MX (2010) Superior in vivo efficacy of afucosylated trastuzumab in the treatment of HER2-amplified breast cancer. Cancer Res 70(11):4481–4489. https://doi.org/10.1158/0008-5472.can-09-3704
103. Becker DJ, Lowe JB (2003) Fucose: biosynthesis and biological function in mammals. Glycobiology 13(7):41r–53r. https://doi.org/10.1093/glycob/cwg054
104. Malphettes L, Freyvert Y, Chang J, Liu PQ, Chan E, Miller JC, Zhou Z, Nguyen T, Tsai C, Snowden AW, Collingwood TN, Gregory PD, Cost GJ (2010) Highly efficient deletion of FUT8 in CHO cell lines using zinc-finger nucleases yields cells that produce completely nonfucosylated antibodies. Biotechnol Bioeng 106(5):774–783. https://doi.org/10.1002/bit.22751
105. Yamane-Ohnuki N, Kinoshita S, Inoue-Urakubo M, Kusunoki M, Iida S, Nakano R, Wakitani M, Niwa R, Sakurada M, Uchida K, Shitara K, Satoh M (2004) Establishment of FUT8 knockout Chinese hamster ovary cells: an ideal host cell line for producing completely defucosylated antibodies with enhanced antibody-dependent cellular cytotoxicity. Biotechnol Bioeng 87(5):614–622. https://doi.org/10.1002/bit.20151
106. Chan KF, Shahreel W, Wan C, Teo G, Hayati N, Tay SJ, Tong WH, Yang Y, Rudd PM, Zhang P, Song Z (2016) Inactivation of GDP-fucose transporter gene (Slc35c1) in CHO cells by ZFNs, TALENs and CRISPR-Cas9 for production of fucose-free antibodies. Biotechnol J 11(3):399–414. https://doi.org/10.1002/biot.201500331
107. Ferrara C, Brunker P, Suter T, Moser S, Puntener U, Umana P (2006) Modulation of therapeutic antibody effector functions by glycosylation engineering: influence of Golgi enzyme localization domain and co-expression of heterologous beta1, 4-N-acetylglucosaminyltransferase III and Golgi alpha-mannosidase II. Biotechnol Bioeng 93(5):851–861. https://doi.org/10.1002/bit.20777
108. Schachter H (1986) Biosynthetic controls that determine the branching and microheterogeneity of protein-bound oligosaccharides. Biochem Cell Biol = Biochimie et biologie cellulaire 64(3):163–181
109. Schuster M, Umana P, Ferrara C, Brunker P, Gerdes C, Waxenecker G, Wiederkum S, Schwager C, Loibner H, Himmler G, Mudde GC (2005) Improved effector functions of a therapeutic monoclonal Lewis Y-specific antibody by glycoform engineering. Cancer Res 65(17):7934–7941. https://doi.org/10.1158/0008-5472.can-04-4212
110. Abdel Rahman AM, Ryczko M, Nakano M, Pawling J, Rodrigues T, Johswich A, Taniguchi N, Dennis JW (2015) Golgi N-glycan branching N-acetylglucosaminyltransferases I, V and VI promote nutrient uptake and metabolism. Glycobiology 25(2):225–240. https://doi.org/10.1093/glycob/cwu105
111. Fukuta K, Abe R, Yokomatsu T, Kono N, Asanagi M, Omae F, Minowa MT, Takeuchi M, Makino T (2000) Remodeling of sugar chain structures of human interferon-gamma. Glycobiology 10(4):421–430
112. Yin B, Gao Y, Chung CY, Yang S, Blake E, Stuczynski MC, Tang J, Kildegaard HF, Andersen MR, Zhang H, Betenbaugh MJ (2015) Glycoengineering of Chinese hamster ovary cells for enhanced erythropoietin N-glycan branching and sialylation. Biotechnol Bioeng 112(11):2343–2351. https://doi.org/10.1002/bit.25650
113. Mellahi K, Brochu D, Gilbert M, Perrier M, Ansorge S, Durocher Y, Henry O (2019) Assessment of fed-batch cultivation strategies for an inducible CHO cell line. J Biotechnol 298:45–56. https://doi.org/10.1016/j.jbiotec.2019.04.005
114. Mellahi K, Cambay F, Brochu D, Gilbert M, Perrier M, Ansorge S, Durocher Y, Henry O (2019) Process development for an inducible rituximab-expressing Chinese hamster ovary cell line. Biotechnol Prog 35(1):e2742. https://doi.org/10.1002/btpr.2742

115. McHugh K, Xu J, Aron K, Borys M, Li ZJ (2020) Effective temperature shift strategy development and scale confirmation for simultaneous optimization of protein productivity and quality in Chinese hamster ovary cells. Biotechnol Prog. https://doi.org/10.1002/btpr.2959
116. Ivarsson M, Villiger TK, Morbidelli M, Soos M (2014) Evaluating the impact of cell culture process parameters on monoclonal antibody N-glycosylation. J Biotechnol 188:88–96. https://doi.org/10.1016/j.jbiotec.2014.08.026
117. Seo J, Kim Y, Cho J, Baek E, Lee G (2013) Effect of culture pH on recombinant antibody production by a new human cell line, F2N78, grown in suspension at 33.0 C and 37.0 C. Appl Microbiol Biotechnol 97. https://doi.org/10.1007/s00253-013-4849-2
118. Jiang R, Chen H, Xu S (2018) pH excursions impact CHO cell culture performance and antibody N-linked glycosylation. Bioprocess Biosyst Eng 41(12):1731–1741. https://doi.org/10.1007/s00449-018-1996-y
119. Zheng C, Zhuang C, Chen Y, Fu Q, Qian H, Wang Y, Qin J, Wu X, Qi N (2018) Improved process robustness, product quality and biological efficacy of an anti-CD52 monoclonal antibody upon pH shift in Chinese hamster ovary cell perfusion culture. Process Biochem 65:123–129. https://doi.org/10.1016/j.procbio.2017.11.013
120. Reinhart D, Damjanovic L, Kaisermayer C, Sommeregger W, Gili A, Gasselhuber B, Castan A, Mayrhofer P, Grunwald-Gruber C, Kunert R (2019) Bioprocessing of recombinant CHO-K1, CHO-DG44, and CHO-S: CHO expression hosts favor either mAb production or biomass synthesis. Biotechnol J 14(3):e1700686. https://doi.org/10.1002/biot.201700686
121. Wang Q, Yang G, Wang T, Yang W, Betenbaugh MJ, Zhang H (2019) Characterization of intact glycopeptides reveals the impact of culture media on site-specific glycosylation of EPO-Fc fusion protein generated by CHO-GS cells. Biotechnol Bioeng 116(9):2303–2315. https://doi.org/10.1002/bit.27009
122. Wang C, Wang J, Chen M, Fan L, Zhao L, Tan WS (2018) Ultra-low carbon dioxide partial pressure improves the galactosylation of a monoclonal antibody produced in Chinese hamster ovary cells in a bioreactor. Biotechnol Lett 40(8):1201–1208. https://doi.org/10.1007/s10529-018-2586-4
123. Zhang L, Castan A, Stevenson J, Chatzissavidou N, Vilaplana F, Chotteau V (2019) Combined effects of glycosylation precursors and lactate on the glycoprofile of IgG produced by CHO cells. J Biotechnol 289:71–79. https://doi.org/10.1016/j.jbiotec.2018.11.004
124. Ghaffari N, Jardon M, Krahn N, Butler M, Kennard M, Turner R, Gopaluni B, Piret J (2019) Effects of cysteine, asparagine or glutamine limitations in CHO cell batch and fed-batch cultures. Biotechnol Prog 36. https://doi.org/10.1002/btpr.2946
125. Baik JY, Dahodwala H, Oduah E, Talman L, Gemmill TR, Gasimli L, Datta P, Yang B, Li G, Zhang F, Li L, Linhardt RJ, Campbell AM, Gorfien SF, Sharfstein ST (2015) Optimization of bioprocess conditions improves production of a CHO cell-derived, bioengineered heparin. Biotechnol J 10(7):1067–1081. https://doi.org/10.1002/biot.201400665
126. Nguyen Dang A, Mun M, Rose CM, Ahyow P, Meier A, Sandoval W, Yuk IH (2019) Interaction of cell culture process parameters for modulating mAb afucosylation. Biotechnol Bioeng 116(4):831–845. https://doi.org/10.1002/bit.26908
127. Shalel Levanon S, Aharonovitz O, Maor-Shoshani A, Abraham G, Kenett D, Aloni Y (2018) An efficient method to control high mannose and core fucose levels in glycosylated antibody production using deoxymannojirimycin. J Biotechnol 276-277:54–62. https://doi.org/10.1016/j.jbiotec.2018.04.006
128. Bruhlmann D, Muhr A, Parker R, Vuillemin T, Bucsella B, Kalman F, Torre S, La Neve F, Lembo A, Haas T, Sauer M, Souquet J, Broly H, Hemberger J, Jordan M (2017) Cell culture media supplemented with raffinose reproducibly enhances high mannose glycan formation. J Biotechnol 252:32–42. https://doi.org/10.1016/j.jbiotec.2017.04.026
129. Ehret J, Zimmermann M, Eichhorn T, Zimmer A (2019) Impact of cell culture media additives on IgG glycosylation produced in Chinese hamster ovary cells. Biotechnol Bioeng 116(4):816–830. https://doi.org/10.1002/bit.26904

130. Yin B, Wang Q, Chung CY, Ren X, Bhattacharya R, Yarema KJ, Betenbaugh MJ (2018) Butyrated ManNAc analog improves protein expression in Chinese hamster ovary cells. Biotechnol Bioeng 115(6):1531–1541. https://doi.org/10.1002/bit.26560
131. Wells E, Song L, Greer M, Luo Y, Kurian V, Ogunnaike B, Robinson AS (2020) Media supplementation for targeted manipulation of monoclonal antibody galactosylation and fucosylation. Biotechnol Bioeng. https://doi.org/10.1002/bit.27496
132. Prabhu A, Gadgil M (2019) Nickel and cobalt affect galactosylation of recombinant IgG expressed in CHO cells. Biometals 32(1):11–19. https://doi.org/10.1007/s10534-018-0152-0
133. Pande S, Rahardjo A, Livingston B, Mujacic M (2015) Monensin, a small molecule ionophore, can be used to increase high mannose levels on monoclonal antibodies generated by Chinese hamster ovary production cell-lines. Biotechnol Bioeng 112(7):1383–1394. https://doi.org/10.1002/bit.25551
134. Grainger RK, James DC (2013) CHO cell line specific prediction and control of recombinant monoclonal antibody N-glycosylation. Biotechnol Bioeng 110(11):2970–2983. https://doi.org/10.1002/bit.24959
135. Wang Q, Chung CY, Yang W, Yang G, Chough S, Chen Y, Yin B, Bhattacharya R, Hu Y, Saeui CT, Yarema KJ, Betenbaugh MJ, Zhang H (2019) Combining butyrated ManNAc with glycoengineered CHO cells improves EPO glycan quality and production. Biotechnol J 14(4):e1800186. https://doi.org/10.1002/biot.201800186
136. Jing Y, Qian YM, Li ZJ (2010) Sialylation enhancement of CTLA4-Ig fusion protein in Chinese hamster ovary cells by dexamethasone. Biotechnol Bioeng 107(3):488–496. https://doi.org/10.1002/bit.22827
137. Batra J, Rathore AS (2016) Glycosylation of monoclonal antibody products: current status and future prospects. Biotechnol Prog 32(5):1091–1102. https://doi.org/10.1002/btpr.2366
138. Bas M, Terrier A, Jacque E, Dehenne A, Pochet-Beghin V, Beghin C, Dezetter AS, Dupont G, Engrand A, Beaufils B, Mondon P, Fournier N, de Romeuf C, Jorieux S, Fontayne A, Mars LT, Monnet C (2019) Fc sialylation prolongs serum half-life of therapeutic antibodies. J Immunol 202(5):1582–1594. https://doi.org/10.4049/jimmunol.1800896
139. Li D, Lou Y, Zhang Y, Liu S, Li J, Tao J (2021) Sialylated immunoglobulin G: a promising diagnostic and therapeutic strategy for autoimmune diseases. Theranostics 11(11):5430–5446. https://doi.org/10.7150/thno.53961

Chapter 7
Even a Worm Will Turn: Immunity Following AAV Vector Administration

Kruti Patel, Arpana Khatri, and Suryanarayan Somanathan

Abstract While several approaches have been considered for in vivo delivery of therapeutic genes, vectors based on a small defective parvovirus, adeno-associated virus (AAV), have proven to be safer in achieving durable therapeutic transgene expression. Nevertheless, the use of vectors based on a virus that infects humans carries the risk of pre-existing host immunity from previous exposure to the virus. Even in the absence of prior infection, administration of large doses of a viral vector can activate host immune responses that can reduce, or prevent, therapeutic gene expression. Hence, careful monitoring of host immune responses before and after vector administration is required when delivering AAV-based gene therapies. Here we review anti-AAV immune responses and the approaches currently being explored to mitigate host immunity and achieve successful therapeutic gene expression.

Keywords Gene therapy · Adeno-associated virus · Innate immunity · Adaptive immunity · Antibodies · Cytotoxic T cells

7.1 Introduction

> Extreme remedies are most appropriate for extreme diseases
>
> Hippocrates, Greek Physician, c.460–357 BCE

Genetic medicines are expected to be the next generation of transformational therapies. In contrast to treatments with small molecules and monoclonal antibodies that address disease symptoms, genetic medicines treat the root cause of the disease. These novel therapies were first touted to be the panacea for fixing or altering disease-causing genes and alleviating patient suffering. Integration of knowledge from the fields of human genetics, molecular and cell biology, virology, and immunology in the 1970s led to early proposals to develop genetic medicines [1]. Nucleic

K. Patel · A. Khatri · S. Somanathan (✉)
Rare Disease Research Unit, Pfizer, Inc., Cambridge, MA, USA

K. Gadamasetti, S. A. Kolodziej (eds.), *Bioprocessing, Bioengineering and Process Chemistry in the Biopharmaceutical Industry*,
https://doi.org/10.1007/978-3-031-62007-2_7

acids encoding a therapeutic gene were proposed to correct, replace, or augment a diseased gene. However, translating these proposals into genetic medicines involved circumventing several obstacles that had evolved over billions of years to resist invading pathogens. These anatomical, biochemical, and immunological barriers exist to prevent cellular entry of viruses and nucleic acids or eliminate cells that do take them up. The field of gene therapy then required a viral or non-viral vector that could traverse these extracellular and intracellular barriers to deliver therapeutic genes into cells. A key breakthrough was the discovery of latent viruses; pathogens that infect host cells yet remain undetected by host immune systems [2]. This gave the impetus to develop viral vectors to deliver therapeutic genes in vivo to cells and more importantly into the nucleus, a feat that is still difficult with non-viral-based gene delivery methods [3].

In vivo viral vector-based gene therapies have now been approved to treat several monogenic disorders [4]. Especially, gene delivery based on adeno-associated viral (AAV) vectors has demonstrably measured stable transgene expression at therapeutic levels for several years [5]. However, initial forays in AAV in vivo gene delivery to treat patients with hemophilia B were met with loss of transgene expression stemming from host immune responses [6]. An understanding of host–vector interactions that triggered host immunity and the therapeutic interventions available to mitigate immune activation later led to successful gene transfer [7]. Besides limiting the success of gene therapy, immune responses pose a serious safety concern that in extreme cases can be fatal [8]. This review summarizes the host immune response activated by AAV gene therapy vectors and the approaches currently being pursued to mitigate immune recognition and activation.

7.1.1 AAV Gene Therapy

Risk of insertional mutagenesis [9] and immunogenicity [8] led to viral vectors based on lentiviruses and adenoviruses being shunned for in vivo gene delivery. Instead, vectors based on adeno-associated viruses (AAV) were selected as the viral vector of choice for in vivo gene transfer. AAV is a small, single-stranded, DNA parvovirus that demonstrated durable transgene expression with a notable absence of immune activation in several preclinical studies [5]. However, the clinical experience with AAV gene therapy was different (Table 7.1) and early clinical trials observed activation of host immune responses that eliminated transgene expression [10]. These findings were unexpected based on previous experience in animal models [11]. Strategies were soon deployed to manage host immune responses and achieve durable transgene expression [10]. Here we review host immune responses following AAV vector administration and discuss the approaches being pursued to reduce immune activation by redesigning the vector as well as using immunomodulatory drugs. This review will mainly focus on the aspects of host immunity that are activated following vector administration. Immune responses that preexist in a

Table 7.1 Immune responses in AAV clinical trials

Disease	Sponsor	Vector	Dose and ROA	Adverse events
DMD	Pfizer [74]	AAV9	2E14 vg/kg; IV	Complement activation, persistent vomiting, acute kidney injury, thrombocytopenia
	Sarepta [75]	AAVRh74	1.3E14 vg/kg; IV	Myositis
	Solid [76]	AAV9	0.5 and 2E14 vg/kg; IV	Reduced platelets and RBCs, complement activation, acute kidney injury, cardiopulmonary insufficiency
Hemophilia A	Biomarin [77]	AAV5	4 to 6E13 vg/kg; IV	Transient transaminitis at 4–8 weeks after gene transfer
	Pfizer [78]	AAV6	3E13 vg/kg; IV	Transient transaminitis at 4–8 weeks after gene transfer
Hemophilia B	St Jude/UCL [7]	AAV8	6E11 vg/kg; IV	Transient transaminitis at 6–10 weeks after gene transfer
	Uniqure [79]	AAV5	2E13 vg/kg; IV	Transient transaminitis at 6–10 weeks after gene transfer
	Pfizer [80]	Spark-100	5E11 vg/kg; IV	Transient transaminitis at 4–8 weeks after gene transfer
	Dimension therapeutics [81]	AAVrh10	1.6 to 5.0E12 vg/kg	Transient transaminitis at 4–8 weeks after gene transfer
FH	REGENXBIO [82]	AAV8	2.5 and 7.5E12 vg/kg, IV	Transaminases 4–6 weeks post-dosing, T-cell response to the AAV capsid
OTCD	Ultragenyx [83]	AAV8	0.2 to 1E13 vg/kg, IV	None
PD	Asklepios [84]	AAV2/8	1.6E12 vg/kg; IV	None
SMA	Novartis [85, 86]	AAV9	1.1 E14 vg/kg	Transient transaminitis between 3 and 7 days after gene transfer, secondary increase after reduction in prednisolone Thrombocytopenia was observed between days 6 and 8 after gene transfer

DMD Duchenne muscular dystrophy, *FH* Familial hypercholesterolemia, *OTCD* Ornithine transcarbamylase deficiency, *PD* Pompe disease, *UCL* University College London, *UPenn* University of Pennsylvania, *UF* University of Florida, *ROA* Route of administration, *IV* Intravenous, *vg* Vector genomes

patient from a wild-type AAV infection and its impact on vector administration are beyond the scope of this review.

7.1.2 Immune Response to AAV Vectors

AAV drugs comprise three components that can potentially activate host immunity; the outer shell or capsid comprising viral proteins, the nucleic acid, and the therapeutically expressed protein (Fig. 7.1). Activation of immunity to these components has been noted in AAV clinical trials and can be classified temporally as those that occur during (innate) or after vector uptake (adaptive). The innate immune system is the first arm that recognizes and responds to pathogens based on common molecular patterns that occur across microorganisms, nucleic acids, and compounds [12]. Adaptive immunity, as the name suggests, is based on selection and training a subset of antibodies or T cells specific to a particular pathogen. Adaptive immune responses are slower and evolve over several days but are poised to respond faster to a challenge with the same or similar pathogen [13].

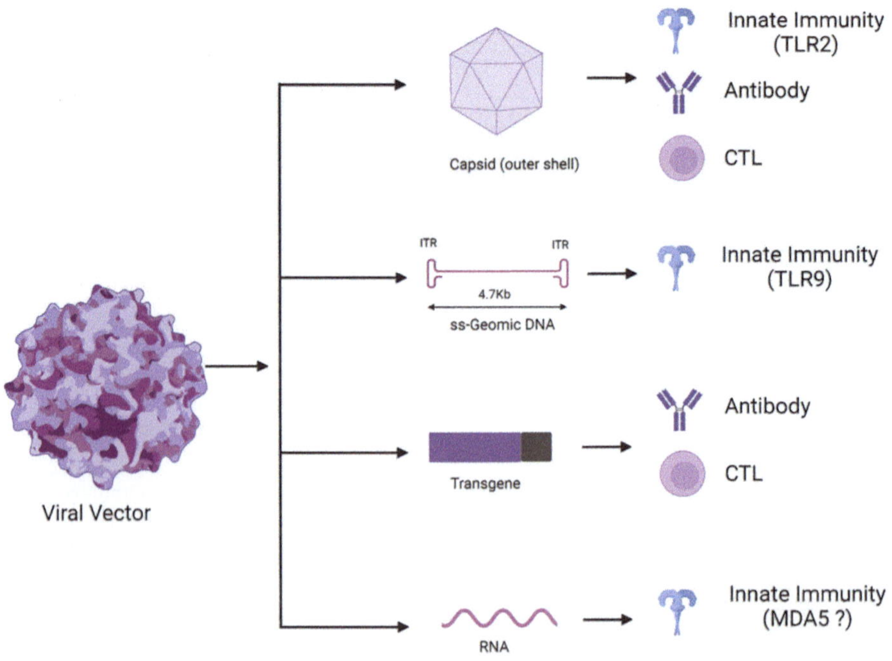

Fig. 7.1 AAV drug components that activate immunity

7.2 Innate Immune Sensing of AAV Vectors

Innate immune recognition leads to slowing down viral replication while also sending "danger" signals. The danger or inflammatory signals further attract other immune effector cells to amplify the antiviral response [14]. The pathogen sensors of innate immunity were first proposed [15] and later proved to involve receptors that recognize invariant molecular patterns shared across pathogens [16]. These receptors collectively called pattern recognition receptors (PRRs) sense pathogen-associated molecular patterns (PAMPs) or danger-associated molecular patterns (DAMPs). While PAMPs are biochemical signatures unique to pathogens, DAMPs are molecules released by dying cells that are not commonly found in circulation [17].

The most well-studied PRRs are the toll-like receptors (TLRs), an evolutionarily conserved set of proteins found in species as diverse as flies and man [16]. TLRs recognize triacylated (TLR1/2) and diacylated (TLR2/6) lipopeptides, double-stranded RNA (TLR3), LPS (TLR4), flagellin (TLR5), single-stranded RNA (TLR7 and 8), and CpG dinucleotides (TLR9) [18]. In general, PAMP/DAMP recognition leads to a cascade of events coordinated to alert other arms of the immune system via secreted factors called cytokines and chemokines [19]. For viral infections, this establishes an early inflammatory response blocking viral replication, inducing cellular destruction, and attracting other effector cells [20]. The antiviral state is maintained until the pathogen is eliminated or suppressed [14].

AAV-based vectors were initially found to be safe and rarely induced cytokines or chemokines when administered to animals [21]. However, later studies demonstrated activation of a potent inflammatory response via multiple engagement [22, 23]. Innate sensing of AAV vectors was demonstrated to involve the detection of molecular patterns in both the capsid and the viral genome via TLR2 and TLR9 recognition, respectively. While the AAV capsid signature that leads to TLR2 activation is currently unknown, CpG dinucleotides in the genome have been implicated in activating the TLR9 receptor pathway [24]. These motifs are suppressed in the mammalian genome but overrepresented in pathogens that increase their sensing by TLR9 [25]. Activation often leads to secretion of Type I interferons; cytokines that block viral replication while amplifying inflammatory signals [26]. Knowledge of TLR9 activation by CpG dinucleotide motifs within AAV vectors has led to efforts to reduce their numbers when designing AAV vectors [27]. The strategy has often been to reduce CpG content only in the transgene and not the regulatory elements. The benefits of also reducing CpG content in the regulatory elements are currently unknown. While CpG reduction has demonstrated some benefits in mouse models using highly immunogenic AAV vectors (AAVRh32.22), it is unclear whether these benefits extend to less immunogenic vectors. Nevertheless, CpG reduction appears to be a strategy to reduce the overall immunogenicity of an AAV vector.

A caveat of CpG depletion is the impact on transgene expression. Reports indicate that CpG depletion of transgene sequences reduces gene expression [28]. CpG reduction in other regulatory elements may further negatively impact transgene expression [29]. An alternate approach has relied on the inclusion of inhibitory sequences within AAV vectors to block TLR9 activation [30]. The inclusion of such sequences as part of a vector genome reduced TLR9 activation in the absence of CpG depletion [31]. This approach bypasses the need to CpG deplete the transgene and may avoid the issues of lower transgene expression when reducing CpGs from the transgene or regulatory elements.

TLR9 activation may also occur from bacterial DNA that gets mispackaged into AAV vectors. As mentioned earlier, bacterial DNA has higher CpG content and its presence in AAV preparations can activate TLR9. The most common approach to produce AAV vectors utilizes three plasmids each encoding a different function. While packaging plasmid DNA into AAV is unintended, vector production is prone to mispackaging plasmid DNA that can be transferred and persist in host tissue after gene delivery [32, 33]. Even if not incorporated into the AAV genome, bacterial DNA contamination can occur due to inefficient removal of these sequences during production. Alternatively, fragments of plasmid or bacterial genomic DNA may bind directly to the surface of the capsids and copurify during vector production.

Construct design and vector production thus need to be optimized to reduce plasmid DNA incorporation and effectively remove impurities. For instance, wild-type AAV genome is ~4.7 kb; however, subgenomic size vectors can be efficiently produced and packaged. A small percentage of these undersized constructs are likely to incorporate plasmid DNA adjacent to the ITRs in a process called "reverse packaging" [34]. Such mispackaged constructs containing plasmid DNA are likely to pose a risk of immune activation. Reverse packaging risk of vector constructs <4.7 kb can be minimized by including inert DNA sequences into AAV constructs to optimize packaging length [35]. However, this requires a careful selection of sequences to ensure inertness. Of concern is the risk of insertional mutagenesis. While AAV vectors integrate at significantly lower frequency [36], the risk of activating a protooncogene close to the site of AAV integration by promoter/enhancer activities within any non-coding sequences included to optimize vector size is a concern that needs to be eliminated [37]. Reducing the amounts of plasmids used in AAV vector manufacturing or CpG depleting these sequences can be alternative approaches to mitigate the risks from packed or contaminating plasmid sequences.

Finally, pharmacological approaches can mitigate innate immune activation. Several modulators of innate immunity are currently used to treat immune disorders [38]. These drugs target common pathways that may be shared by AAV vectors in activating innate immunity and hence be amenable for use with AAV administration. The dose and time course of intervention with these immunomodulatory drugs will depend on the pharmacokinetics and pharmacodynamics of vector after infusion.

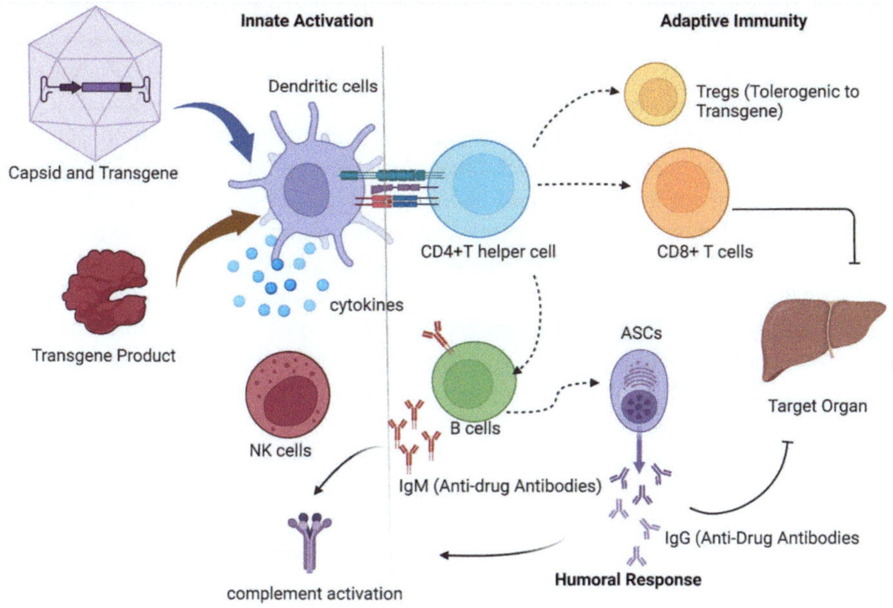

Fig. 7.2 Interplay of innate and adaptive immunity upon AAV recognition

7.3 Adaptive Immune Responses to AAV

While innate immunity relies on the recognition of shared molecular identities among pathogens, the adaptive immune response is customized to recognize and respond to unique non-self peptides (Fig. 7.2). Adaptive immunity results in long-lasting humoral (antibody) and cell-mediated (cytotoxic T cells, CTLs) immunity that is both rapid and more potent.

7.3.1 Humoral Immunity

Humoral immunity is activated when B cell receptors (BCRs), membrane-bound antibodies, bind circulating capsid [39]. BCR engagement and subsequent activation result in switching antibodies from a membrane-bound form to a secreted form [40]. In humans, AAV administration leads to an increase in circulating anticapsid immunoglobulin M (IgM) and high-affinity immunoglobulin G (IgG) antibodies that persist for several weeks [41].

In the course of a natural viral infection, the result of de novo antibody production most likely slows down viral replication. However, AAV is a nonreplicating vector and the impact of these de novo anticapsid antibodies detected after vector administrations is unclear. Nevertheless, patients administered Rituximab, a B cell-depleting monoclonal antibody, along with AAV vectors resulted in higher circulating levels of vector over an extended period compared to control patients [42]. This finding suggests that de novo antibodies may have a role in reducing the amount of circulating vector.

De novo antibodies have also been implicated in complement activation. Muruve et al. were the first to report AAV vectors activated complement; an ancient system of proteins involved in binding and lysing pathogens or pathogen-infected cells [21]. However, for many years after this initial publication, there were no reports of complement activation in AAV preclinical studies or in the clinic. A general assumption was that while the potential always exists for complement activation, this was an uncommon event. This changed with reports of complement activation following high-dose systemic vector administration in AAV clinical trials for Duchenne muscular dystrophy (DMD). Two independent trials that infused high-dose AAV vector into pediatric patients to treat DMD noted complement activation [43]. Patients enrolled in these trials were selected to be NAb negative and therefore should not have activated complement via preexisting anti-AAV antibodies. Emerging data suggests that complement activation may be related to de novo antibodies [44].

Vector antibody immune complexes could also trigger a strong inflammatory response that in some instances could be fatal. Such "antibody dependent enhancement" leading to increased inflammation has been observed for several viral infections [45]. Of note, a NAb-positive patient administered high-dose adenoviral vector as part of a clinical trial for ornithine transcarbamylase deficiency developed a fatal systemic inflammatory response [8]. Posterior analysis of the patient serum before vector infusion demonstrated ADE of adenoviral infection that triggered strong inflammatory responses in APCs [46]. While ADE of AAV transduction has been reported, its role in enhancing in vivo inflammatory responses is unclear. It must be noted that these reports of ADE of viral infectivity have been related to preexisting immunity. The risk of ADE from de novo antibodies is less clear. Since ADE is more strongly associated with IgG and not IgM antibodies, the rapid clearance of vector from circulation before the formation of high-affinity IgGs may reduce ADE risk [47].

Humoral immunity to the transgene product can also occur especially when treating patients with missense or null mutations. However, there is no evidence of antitransgene antibodies in Hemophilia B patients treated with AAV.FIX gene therapy [48, 49]. The risk of developing antitransgene antibodies may have been reduced by excluding patients with a history of developing anti-FIX antibodies following factor replacement therapy [48].

Pharmacological approaches to reduce the risk of de novo antibody generation following AAV vector administration have focused on the use of immunosuppression strategies. A combination of immunosuppressive drugs, rapamycin, and rituximab, was demonstrated to block the emergence of antibodies after AAV administration to patients [50]. Similarly, coadministration of rapamycin nanoparticles also effectively prevented antibody generation after vector administration in preclinical animal models and in the clinic [51]. This approach relies on tolerance induction to the AAV capsids following the uptake of vector and rapamycin nanoparticles by the same APC. Alternatively, coadministration of the CTLA4 mimetic, abatacept, was also demonstrated to reduce humoral immunity to AAV8 capsids in preclinical trials for hemophilia B [52]. Transient depletion of immunoglobulins can also avoid the formation of capsid–antibody complexes until antigen is cleared [53–55]. Finally, complement activation in patients treated with AAV vectors has been managed by administering the anticomplement factor monoclonal antibody Eculizumab [56].

7.3.2 Cell-Mediated Immunity

Cell-mediated immunity (CMI) follows the activation of cytotoxic T cells (CTLs) that recognize and eliminate AAV transduced cells, and T helper cells (Th cells) that promote CTL or B cell responses [57]. These distinctions are not always strict and functional overlap can occur between cell types. CTL killing requires T cell receptor (TCR) recognition of antigen-derived peptides presented on Class I major histocompatibility complex molecules (human HLA or mouse MHC) in transduced cells. NHP studies demonstrated that AAV transduced hepatocytes present capsid antigen on Class I molecules that rendered them targets for CTL killing. In the clinic, hemophilia B patients treated with an AAV2 vector expressing human FIX demonstrated transgene expression loss after activation of capsid-specific CTLs [6]. Similarly, hemophilia B patients administered a self-complimentary AAV8 vector expressing human FIX also demonstrated activation of AAV capsid-specific T cells at 8 weeks that coincided with a drop in FIX levels [7]. These findings differ from a muscle-directed AAV gene therapy for alpha-1 antitrypsin deficiency that demonstrated stable transgene expression in the presence of capsid-specific T cells [58]. It is possible that the routes of vector administration (intravenous vs intramuscular) or capsids (AAV2 vs AAV1) that were employed in these trials differentially impact CTL activation.

CTLs can also be activated following Class I presentation of the encoded therapeutic transgene product. However, the response to transgene has been different and varies among animal models. In mice, the presence of transgene-specific CTL cells was insufficient to eliminate AAV transduced cells. A second inflammatory or "danger" signal via TLR activation was additionally required for efficient CTL homing and elimination of transduced cells [59]. An exception was immunogenic AAV vectors that could bypass the need for additional inflammatory signals [60]. While the exact role of these inflammatory signals is not clear, the resulting cytokines are known to increase the expression of both Class I MHC molecules on hepatocytes [61], and adhesion molecules on endothelial cells [59]. An increase in adhesion molecules is expected to help with CTL extravasation; a process where leukocytes migrate across the endothelial barrier to access inflamed tissue [62]. Transgene CTL responses in larger animals are complicated compared to genetically identical inbred mice. For instance, while several studies have reported stable non-self GFP transgene expression in non-human primates following AAV.GFP vector administration, there is also evidence of GFP CTL responses that resulted in hepatitis and loss of transgene expression in cynomolgus macaques intraportally administered with an AAV7.GFP vector [63].

There is currently limited evidence for the loss of AAV-transduced cells in the clinic following a transgene-specific immune response. CTL responses were observed in three out of six DMD patients administered with an AAV vector expressing mini-dystrophin [64]. Epitope mapping confirmed that CTL responses were directed against exons present in the therapeutic mini-dystrophin that were missing in the patient. In some instances, transgene-specific CTL responses have been observed to polymorphic residues following AAV vector administration to A1AT patients [65]. This finding underscores the risk of gene transfer to patients with polymorphism that differs from the therapeutic transgene sequence and can elicit immunity. More recently, T-cell responses to AAV transduced dorsal root ganglions (DRGs) have been implicated in the loss of these lower motor neurons within a few months of vector administration in NHPs [66]. Initial findings suggest CTL responses to the transgene caused DRG cytotoxicity.

Besides CTLs and Th cells, a specialized type of lymphocyte, the T regulatory cell (Treg), suppresses AAV transgene immune responses [67]. In rodents, hepatic T regulatory cells were involved in maintaining tolerance to the transgene following liver-directed gene transfer [68]. Similarly, muscle biopsies of patients treated with AAV.A1AT gene therapy demonstrated Treg infiltration into muscles [69]. Interestingly, these patients had sustained A1AT expression despite the presence of capsid-specific CD8+ and CD4+ T cells. These findings strongly suggest a role for Tregs in mitigating anti-AAV immunity and helping maintain tolerance to capsid and/or transgene.

Several approaches have been adopted in the clinic before and after AAV vector administration to minimize the impact of developing capsid and/or transgene-specific CTLs (Table 7.2). Reactive corticosteroid administration after an increase in LFTs that occurred 6–7 weeks after vector infusion mitigated the CTL response and rescued transgene expression in AAV FIX clinical trials [7]. Prophylactic corticosteroid administrations starting a day before AAV administration have also been used to suppress CTL activation [70]. A combination of immunomodulatory drugs that suppressed both T and B cells transiently has also proved successful in keeping CTL responses lower in Pompe patients treated with an AAV vector [50]. However, the most effective combination of immunosuppressive drugs to transiently curtail CTL responses and achieve stable transgene expression is still being explored. Of concern are the different arms of the immune system that need to be suppressed and the duration of immunosuppression. For instance, CTL responses have occurred 6–8 weeks after vector administration in the clinic. Thus, one option would be to immunosuppress patients until the risk of activating a CTL response has passed after which immunosuppression can be withdrawn. Whether the window of immunosuppression needs to be longer or shorter has been explored and may vary based on several factors, including the choice of capsid, route of administration, and vector dose. There is also the risk to patients from opportunistic infections that require limiting the duration of immunosuppression [71]. Immunosuppression regimens

Table 7.2 Immunosuppression protocols in AAV clinical trials

Disease	Sponsor	Age	Vector	Dose	ROA	IS protocol	Steroid
LPL	AMT [87]	Adult	Glybera	1E12 vg/kg	IM	12 weeks Cs + MMF	Daily
Hemophilia A	Pfizer [78]	Adult	AAV6	3E13 vg/kg	IV	–	Reactive
	Biomarin [77]		AAV5	6E13 vg/kg			
Hemophilia B	St.Jude/UCL [7]	Adult	AAV8	6E11 vg/kg	IV	–	Reactive
	Pfizer [80]		Spark100	5E11 vg/kg			
	Uniqure [79]		AAV5	2E13 vg/kg			
DMD	Solid [88]	Pediatric	AAV9	0.5 & 2E14 vg/kg	IV		Reactive
	Pfizer [74]		AAV9	2E14 vg/kg			Daily
	Sarepta [75]		Rh74	1.3E14 vg/kg			Reactive
PD	UF [89]	Pediatric	AAV9	4.6E13 vg	IM (diaphragm)	Rx + Rapa	Daily
Canavan	Myrtelle [90]	Adult	AAV9	4.5E13 + 5E13 (IT)	IV + IT		Daily
GM1	NIH [91]	Adult	AAV9	1.5E13 vg/kg	IV	Rx + Rapa	Daily
ALS	Apic Bio [92]	Adult	AAVrh10	4.20E14	IT	Rx + Rapa	Daily
Danon disease	Rocket [93]	Pediatric	RP-A501	(A)1.1E14 vg/kg (P) 5E13 vg/kg	IV	Rx + tac Rx + Rapa	MP tapered at day 10 for pediatric patients

LPLD Lipoprotein lipase deficiency, *DMD* Duchenne muscular dystrophy, *PD* Pompe disease, *ALS* Amyotropic lateral sclerosis, *UF* University of Florida, *IM* Intramuscular, *IV* Intravenous, *IT* Intratracheal, *MP* Methylprednisolone, *MMF* Mycophenolate mofetil, *Rx* Rituximab, *Tac* Tacrolimus, *Rapa* Rapamycin/sirolimus

may have the added benefit of increasing the circulating half-life of the vector or modulating apoptotic or proteasome pathways that increase transgene expression [42, 72]. A caveat of immunomodulation is the effect on Tregs and strategies to dampen host immunity should take into consideration the beneficial effect of Treg cell activation and expansion [73].

7.4 Conclusion

The infusion of trillions of vector particles is expected to alert the immune system that can impact both the safety and efficacy of AAV gene therapy. Furthermore, targeting some diseased organs is likely to exacerbate the pathophysiology associated with AAV gene transfer. Hence, strategies are warranted to reduce the provocation of host immunity by creatively designing the vector. This starts with avoiding molecular signatures that have a propensity to agonize PRRs and alert the immune system. Next, increasing vector potency can lower the dose required to achieve therapeutic transgene expression levels and thereby reduce immune insult. These approaches may still need to be combined with immunosuppression strategies to further attenuate immune responses and achieve durable transgene expression. The problems of host immunity against AAV-transduced tissues are akin to those that once plagued the field of solid organ transplantation. Nevertheless, immunosuppression strategies developed and employed in recent years have reduced the risk of allograft rejection. The adoption of these and other newer approaches will likely bolster the chances of achieving the goal of treating all patients safely with AAV gene therapy.

Acknowledgments The authors thank Dr. Pan Clark and Dr. Seng H. Cheng for their comments on the chapter. Images were created with BioRender.com

Author Disclosure All authors were employees of Pfizer, Inc. at the time of writing and have no conflicts of interest to declare.

References

1. Friedmann T, Roblin R (1972) Gene therapy for human genetic disease? Science 175(4025):949–955
2. Muzyczka N, Berns KI (2015) AAV's golden jubilee. Mol Ther 23(5):807–808
3. Durymanov M, Reineke J (2018) Non-viral delivery of nucleic acids: insight into mechanisms of overcoming intracellular barriers. Front Pharmacol 9:971
4. Mendell JR, Al-Zaidy SA, Rodino-Klapac LR, Goodspeed K, Gray SJ, Kay CN et al (2021) Current clinical applications of in vivo gene therapy with AAVs. Mol Ther 29(2):464–488
5. Muhuri M, Levy DI, Schulz M, McCarty D, Gao G (2022) Durability of transgene expression after rAAV gene therapy. Mol Ther 30(4):1364–1380

6. Manno CS, Pierce GF, Arruda VR, Glader B, Ragni M, Rasko JJ et al (2006) Successful transduction of liver in hemophilia by AAV-factor IX and limitations imposed by the host immune response. Nat Med 12(3):342–347
7. Nathwani AC, Reiss UM, Tuddenham EG, Rosales C, Chowdary P, McIntosh J et al (2014) Long-term safety and efficacy of factor IX gene therapy in hemophilia B. N Engl J Med 371(21):1994–2004
8. Raper SE, Chirmule N, Lee FS, Wivel NA, Bagg A, Gao GP et al (2003) Fatal systemic inflammatory response syndrome in a ornithine transcarbamylase deficient patient following adenoviral gene transfer. Mol Genet Metab 80(1–2):148–158
9. Milone MC, O'Doherty U (2018) Clinical use of lentiviral vectors. Leukemia 32(7):1529–1541
10. Mingozzi F, High KA (2013) Immune responses to AAV vectors: overcoming barriers to successful gene therapy. Blood 122(1):23–36
11. Snyder RO, Miao CH, Patijn GA, Spratt SK, Danos O, Nagy D et al (1997) Persistent and therapeutic concentrations of human factor IX in mice after hepatic gene transfer of recombinant AAV vectors. Nat Genet 16(3):270–276
12. Turvey SE, Broide DH (2010) Innate immunity. J Allergy Clin Immunol 125(2 Suppl 2):S24–S32
13. Chaplin DD (2010) Overview of the immune response. J Allergy Clin Immunol 125(2 Suppl 2):S3–S23
14. Kabelitz D, Medzhitov R (2007) Innate immunity-cross-talk with adaptive immunity through pattern recognition receptors and cytokines. Curr Opin Immunol 19(1):1–3
15. Janeway CA Jr (1989) Approaching the asymptote? Evolution and revolution in immunology. Cold Spring Harb Symp Quant Biol 54(Pt 1):1–13
16. Medzhitov R, Preston-Hurlburt P, Janeway CA Jr (1997) A human homologue of the Drosophila Toll protein signals activation of adaptive immunity. Nature 388(6640):394–397
17. Medzhitov R, Janeway CA Jr (2002) Decoding the patterns of self and nonself by the innate immune system. Science 296(5566):298–300
18. Macedo AB, Novis CL, Bosque A (2019) Targeting cellular and tissue HIV reservoirs with toll-like receptor agonists. Front Immunol 10:2450
19. Medzhitov R (2007) TLR-mediated innate immune recognition. Semin Immunol 19(1):1–2
20. Pasare C, Medzhitov R (2005) Toll-like receptors: linking innate and adaptive immunity. Adv Exp Med Biol 560:11–18
21. Zaiss AK, Liu Q, Bowen GP, Wong NC, Bartlett JS, Muruve DA (2002) Differential activation of innate immune responses by adenovirus and adeno-associated virus vectors. J Virol 76(9):4580–4590
22. Hosel M, Broxtermann M, Janicki H, Esser K, Arzberger S, Hartmann P et al (2012) Toll-like receptor 2-mediated innate immune response in human nonparenchymal liver cells toward adeno-associated viral vectors. Hepatology 55(1):287–297
23. Zhu J, Huang X, Yang Y (2009) The TLR9-MyD88 pathway is critical for adaptive immune responses to adeno-associated virus gene therapy vectors in mice. J Clin Invest 119(8):2388–2398
24. Faust SM, Bell P, Cutler BJ, Ashley SN, Zhu Y, Rabinowitz JE et al (2013) CpG-depleted adeno-associated virus vectors evade immune detection. J Clin Invest 123(7):2994–3001
25. Krieg AM (2002) CpG motifs in bacterial DNA and their immune effects. Annu Rev Immunol 20:709–760
26. Lukhele S, Boukhaled GM, Brooks DG (2019) Type I interferon signaling, regulation and gene stimulation in chronic virus infection. Semin Immunol 43:101277
27. Wright JF (2020) Quantification of CpG motifs in rAAV genomes: avoiding the toll. Mol Ther 28(8):1756–1758
28. Bauer AP, Leikam D, Krinner S, Notka F, Ludwig C, Langst G et al (2010) The impact of intragenic CpG content on gene expression. Nucleic Acids Res 38(12):3891–3908
29. Deaton AM, Bird A (2011) CpG islands and the regulation of transcription. Genes Dev 25(10):1010–1022

30. Trieu A, Roberts TL, Dunn JA, Sweet MJ, Stacey KJ (2006) DNA motifs suppressing TLR9 responses. Crit Rev Immunol 26(6):527–544
31. Chan YK, Wang SK, Chu CJ, Copland DA, Letizia AJ, Costa Verdera H et al (2021) Engineering adeno-associated viral vectors to evade innate immune and inflammatory responses. Sci Transl Med 13(580):eabd3438
32. Chadeuf G, Ciron C, Moullier P, Salvetti A (2005) Evidence for encapsidation of prokaryotic sequences during recombinant adeno-associated virus production and their in vivo persistence after vector delivery. Mol Ther 12(4):744–753
33. Nony P, Chadeuf G, Tessier J, Moullier P, Salvetti A (2003) Evidence for packaging of rep-cap sequences into adeno-associated virus (AAV) type 2 capsids in the absence of inverted terminal repeats: a model for generation of rep-positive AAV particles. J Virol 77(1):776–781
34. Wright JF (2014) Product-related impurities in clinical-grade recombinant AAV vectors: characterization and risk assessment. Biomedicine 2(1):80–97
35. Lee SHS, Chang H, Kim HJ, Choi JS, Kim J, Kim JH et al (2019) Effects of stuffer DNA on the suppression of choroidal neovascularization by a rAAV expressing a mTOR-inhibiting shRNA. Mol Ther Methods Clin Dev 14:171–179
36. Smith RH (2008) Adeno-associated virus integration: virus versus vector. Gene Ther 15(11):817–822
37. Deyle DR, Russell DW (2009) Adeno-associated virus vector integration. Curr Opin Mol Ther 11(4):442–447
38. Hu X, Li J, Fu M, Zhao X, Wang W (2021) The JAK/STAT signaling pathway: from bench to clinic. Signal Transduct Target Ther 6(1):402
39. Treanor B (2012) B-cell receptor: from resting state to activate. Immunology 136(1):21–27
40. Peterson ML (2007) Mechanisms controlling production of membrane and secreted immunoglobulin during B cell development. Immunol Res 37(1):33–46
41. Murphy SL, Li H, Mingozzi F, Sabatino DE, Hui DJ, Edmonson SA et al (2009) Diverse IgG subclass responses to adeno-associated virus infection and vector administration. J Med Virol 81(1):65–74
42. Byrne BJ (2020) NIH virtual workshop on systemic immunogenicity considerations for adeno-associated virus (AAV)-mediated gene therapy (Day 2)
43. Ronzitti G, Gross DA, Mingozzi F (2020) Human immune responses to adeno-associated virus (AAV) vectors. Front Immunol 11:670
44. Fidler B (2020) Safety questions linger as Pfizer preps Duchenne gene therapy for final test. Biopharma Dive
45. Tirado SM, Yoon KJ (2003) Antibody-dependent enhancement of virus infection and disease. Viral Immunol 16(1):69–86
46. Somanathan S, Calcedo R, Wilson JM (2020) Adenovirus-antibody complexes contributed to lethal systemic inflammation in a gene therapy trial. Mol Ther 28(3):784–793
47. Anam V, Sebayang AA, Fahlena H, Knopoff D, Stollenwerk N, Soewono E et al (2022) Modeling dengue immune responses mediated by antibodies: insights on the biological parameters to describe dengue infections. Comput Math Method 2022:8283239
48. Colella P, Ronzitti G, Mingozzi F (2018) Emerging issues in AAV-mediated in vivo gene therapy. Mol Ther Methods Clin Dev 8:87–104
49. Crudele JM, Finn JD, Siner JI, Martin NB, Niemeyer GP, Zhou S et al (2015) AAV liver expression of FIX-Padua prevents and eradicates FIX inhibitor without increasing thrombogenicity in hemophilia B dogs and mice. Blood 125(10):1553–1561
50. Corti M, Elder M, Falk D, Lawson L, Smith B, Nayak S et al (2014) B-cell depletion is protective against anti-AAV capsid immune response: a human subject case study. Mol Ther Methods Clin Dev 1:14033
51. Peter G, Traber SG, Petr O, Ilyinskii P, Leung S, Capela T, Michaud A, Schroeder H, Rizzo TW, Jiang C, Farve D, Johnston L, Samulski RJ, Kishimoto TK (2022) Effect of tolerogenic ImmTOR nanoparticles on the formation of anti-AAV8 antibodies in mice, nonhuman primates, and healthy human volunteers. ASGCT, Washington

52. Frentsch M, Japp AS, Dingeldey M, Matzmohr N, Thiel A, Scheiflinger F et al (2020) Blockade of the costimulatory CD28-B7 family signal axis enables repeated application of AAV8 gene vectors. J Thromb Haemost 18(5):1075–1080
53. Jordan SC, Lorant T, Choi J (2017) IgG endopeptidase in highly sensitized patients undergoing transplantation. N Engl J Med 377(17):1693–1694
54. Blumberg LJ, Humphries JE, Jones SD, Pearce LB, Holgate R, Hearn A et al (2019) Blocking FcRn in humans reduces circulating IgG levels and inhibits IgG immune complex-mediated immune responses. Sci Adv 5(12):eaax9586
55. Leborgne C, Barbon E, Alexander JM, Hanby H, Delignat S, Cohen DM et al (2020) IgG-cleaving endopeptidase enables in vivo gene therapy in the presence of anti-AAV neutralizing antibodies. Nat Med 26(7):1096–1101
56. Keating GM (2013) Eculizumab: a review of its use in atypical haemolytic uraemic syndrome. Drugs 73(18):2053–2066
57. Apasov SG, Sitkovsky MV (2005) T cell-mediated immunity. In: Nijkamp FP, Parnham MJ (eds) Principles of immunopharmacology. Birkhäuser Basel, Basel, pp 19–28
58. Gruntman AM, Flotte TR (2015) Progress with recombinant adeno-associated virus vectors for gene therapy of alpha-1 antitrypsin deficiency. Hum Gene Ther Methods 26(3):77–81
59. Somanathan S, Breous E, Bell P, Wilson JM (2010) AAV vectors avoid inflammatory signals necessary to render transduced hepatocyte targets for destructive T cells. Mol Ther 18(5):977–982
60. Mays LE, Vandenberghe LH, Xiao R, Bell P, Nam HJ, Agbandje-McKenna M et al (2009) Adeno-associated virus capsid structure drives CD4-dependent CD8+ T cell response to vector encoded proteins. J Immunol 182(10):6051–6060
61. Yang Y, Xiang Z, Ertl HC, Wilson JM (1995) Upregulation of class I major histocompatibility complex antigens by interferon gamma is necessary for T-cell-mediated elimination of recombinant adenovirus-infected hepatocytes in vivo. Proc Natl Acad Sci USA 92(16):7257–7261
62. Muller WA (2013) Getting leukocytes to the site of inflammation. Vet Pathol 50(1):7–22
63. Gao G, Wang Q, Calcedo R, Mays L, Bell P, Wang L et al (2009) Adeno-associated virus-mediated gene transfer to nonhuman primate liver can elicit destructive transgene-specific T cell responses. Hum Gene Ther 20(9):930–942
64. Mendell JR, Campbell K, Rodino-Klapac L, Sahenk Z, Shilling C, Lewis S et al (2010) Dystrophin immunity in Duchenne's muscular dystrophy. N Engl J Med 363(15):1429–1437
65. Calcedo R, Somanathan S, Qin Q, Betts MR, Rech AJ, Vonderheide RH et al (2017) Class I-restricted T-cell responses to a polymorphic peptide in a gene therapy clinical trial for alpha-1-antitrypsin deficiency. Proc Natl Acad Sci USA 114(7):1655–1659
66. Hordeaux J, Buza EL, Jeffrey B, Song C, Jahan T, Yuan Y et al (2020) MicroRNA-mediated inhibition of transgene expression reduces dorsal root ganglion toxicity by AAV vectors in primates. Sci Transl Med 12(569):eaba9188
67. Vignali DAA, Collison LW, Workman CJ (2008) How regulatory T cells work. Nat Rev Immunol 8(7):523–532
68. Breous E, Somanathan S, Vandenberghe LH, Wilson JM (2009) Hepatic regulatory T cells and Kupffer cells are crucial mediators of systemic T cell tolerance to antigens targeting murine liver. Hepatology 50(2):612–621
69. Mueller C, Chulay JD, Trapnell BC, Humphries M, Carey B, Sandhaus RA et al (2013) Human Treg responses allow sustained recombinant adeno-associated virus-mediated transgene expression. J Clin Invest 123(12):5310–5318
70. Mendell JR, Al-Zaidy S, Shell R, Arnold WD, Rodino-Klapac LR, Prior TW et al (2017) Single-dose gene-replacement therapy for spinal muscular atrophy. N Engl J Med 377(18):1713–1722
71. Giancane G, Swart JF, Castagnola E, Groll AH, Horneff G, Huppertz HI et al (2020) Opportunistic infections in immunosuppressed patients with juvenile idiopathic arthritis: analysis by the Pharmachild Safety Adjudication Committee. Arthritis Res Ther 22(1):71
72. Hosel M, Huber A, Bohlen S, Lucifora J, Ronzitti G, Puzzo F et al (2017) Autophagy determines efficiency of liver-directed gene therapy with adeno-associated viral vectors. Hepatology 66(1):252–265

73. Furukawa A, Wisel SA, Tang Q (2016) Impact of immune-modulatory drugs on regulatory T cell. Transplantation 100(11):2288–2300
74. Pfizer I (2020) Pfizer's new phase 1b results of gene therapy in ambulatory boys with duchenne muscular dystrophy (DMD) support advancement into pivotal phase 3 study. Pfizer Inc
75. Sarepta therapeutics' SRP-9001 shows sustained functional improvements in multiple studies of patients with duchenne [press release] (2021)
76. Hale C (2019) Solid bio sees yet another clinical hold for its DMD gene therapy
77. Rangarajan S, Walsh L, Lester W, Perry D, Madan B, Laffan M et al (2017) AAV5-factor VIII gene transfer in severe hemophilia A. N Engl J Med 377(26):2519–2530
78. Pfizer and Sangamo announce updated phase 1/2 results showing sustained bleeding control in highest dose cohort through two years following hemophilia A gene therapy (2021) https://www.pfizer.com/news/press-release/press-release-detail/pfizer-and-sangamo-announce-updated-phase-12-results-1 [press release]
79. Miesbach W, Meijer K, Coppens M, Kampmann P, Klamroth R, Schutgens R et al (2018) Gene therapy with adeno-associated virus vector 5-human factor IX in adults with hemophilia B. Blood 131(9):1022–1031
80. George LA, Sullivan SK, Giermasz A, Rasko JEJ, Samelson-Jones BJ, Ducore J et al (2017) Hemophilia B gene therapy with a high-specific-activity factor IX variant. N Engl J Med 377(23):2215–2227
81. Azevedo M (2017) Dimension to end development of DTX101 as gene therapy for hemophilia B. Hemophilia News Today
82. RGX–501 Phase I/II clinical trial interim results and program update [press release] (2019)
83. Ultragenyx announces positive topline cohort 2 results from phase 1/2 clinical study of DTX301 gene therapy in ornithine transcarbamylase (OTC) deficiency and progression to higher dose [press release] (2018)
84. Koeberl DD (2020) Phase I study of gene therapy in late onset Pompe disease: analyses of safety and secondary endpoints. ASGCT
85. Chand D, Mohr F, McMillan H, Tukov FF, Montgomery K, Kleyn A et al (2021) Hepatotoxicity following administration of onasemnogene abeparvovec (AVXS-101) for the treatment of spinal muscular atrophy. J Hepatol 74(3):560–566
86. Friese J, Geitmann S, Holzwarth D, Muller N, Sassen R, Baur U et al (2021) Safety monitoring of gene therapy for spinal muscular atrophy with onasemnogene abeparvovec–a single centre experience. J Neuromuscul Dis 8(2):209–216
87. Ferreira V, Twisk J, Kwikkers K, Aronica E, Brisson D, Methot J et al (2014) Immune responses to intramuscular administration of alipogene tiparvovec (AAV1-LPL(S447X)) in a phase II clinical trial of lipoprotein lipase deficiency gene therapy. Hum Gene Ther 25(3):180–188
88. Bryson S (2021) Gene therapy SGT-001 shows continued benefit in DMD boys. Muscular Dystrophy News Today
89. Elder ME, Nayak S, Collins SW, Lawson LA, Kelley JS, Herzog RW et al (2013) B-cell depletion and immunomodulation before initiation of enzyme replacement therapy blocks the immune response to acid alpha-glucosidase in infantile-onset Pompe disease. J Pediatr 163(3):847–54 e1
90. Myrtelle (2021) rAAV-Olig001-ASPA gene therapy for treatment of children with typical canavan disease (CAN-GT). Available from: https://clinicaltrials.gov/ct2/show/NCT04833907
91. Institute NHGR (2019) A phase 1/2 study of intravenous gene transfer with an AAV9 vector expressing human beta-galactosidase in type I and type II GM1 gangliosidosis. Available from: https://clinicaltrials.gov/ct2/show/NCT03952637?recrs=abd&cond=Gm1+gangliosidosis&draw=2
92. Mueller C (2020) Suppression of SOD1 in SOD1 familial ALS using AAV-RNAI. Available from: https://events-support.com/events/NCATS_Gene_Therapies_November_2020/page/2282
93. Barry Greenberg JR, Epstein S, Battiprolu P, Urey MA, Hong KN, Kahn AM, Waldron A, Ricks D, Yarabe P, Patel K, Schwartz J, Shah G, Trevejo J (2022) Extended results from first-in-human clinical trial of RP-A501 (AAV9:LAMP2B) gene therapy treatment for Danon disease. ASGCT

Chapter 8
COVID-19 Vaccine Manufacturing Processes: Making the Molecules to Solve the Pandemic

Jennifer A. S. Romine, Stephen A. Kolodziej, Tarl Vetter, and Michael P. Dux

Abstract Since the first report of the full genome sequence of the severe acute respiratory syndrome coronavirus 2 (SARS-CoV-2), the causative agent of the coronavirus disease 2019 (COVID-19) pandemic, many universities and pharmaceutical/biotechnology companies have worked to identify potential vaccine candidates to end this pandemic. With respect to the identity of the actual drug substance (DS), approved COVID-19 vaccines fall into one of four general categories: mRNA-, viral vector-, whole inactivated virus-, and protein-based vaccines. The manufacturing processes to deliver these vaccines vary broadly, with production from *Escherichia coli* to Sf9 insect cells, Chinese hamster ovary (CHO) cells, or enzymatic reactions, while purification for these products can be as simple as buffer exchange or require extensive recovery and chromatographic efforts. Here, we explain which processes are used for each vaccine category, discuss their individual challenges, and look at how the portfolio of vaccine candidates can come together to help solve the pandemic.

Keywords Vaccine manufacture · mRNA vaccine · Viral vector · Whole inactivated virus · Protein · Upstream process · Downstream process · Adenovirus · Bioreactor · Cell culture · Chromatography · Viral inactivation · Virus-like particle

J. A. S. Romine (✉) · S. A. Kolodziej
Bioprocess Research and Development, Biotherapeutics Pharmaceutical Sciences, Pfizer Inc., Chesterfield, MO, USA
e-mail: jennifer.romine@pfizer.com; steve.a.kolodziej@pfizer.com

T. Vetter
Novartis Pharmaceuticals Corporation, Durham, NC, USA

BX-CGT Technical R&D, Novartis Pharmaceuticals Corporation, Durham, NC, USA

M. P. Dux
Pfizer Global Supply, Pfizer Inc., Sanford, NC, USA

K. Gadamasetti, S. A. Kolodziej (eds.), *Bioprocessing, Bioengineering and Process Chemistry in the Biopharmaceutical Industry*,
https://doi.org/10.1007/978-3-031-62007-2_8

8.1 Introduction

The coronavirus disease 2019 (COVID-19) pandemic impacted all aspects of human health and society as it spread across the globe after its emergence in December 2019 with a global death toll of over 6.6 million as of November 11, 2022 [1, 2]. The causative agent of COVID-19 has been identified as the severe acute respiratory syndrome coronavirus-2 (SARS-CoV-2) [3], and its full genome was reported in January 2020 [4].

Coronaviruses (CoVs) are enveloped, positive-sense single-stranded RNA viruses that infect mammals and birds. Their virions contain numerous surface protrusions consisting of trimers of the viral spike (S) protein, and it is these protrusions that give these viruses their characteristic crown-like appearance in electron micrographs [3, 5]. CoVs are subdivided into four different genera (α-CoV, β-CoV, γ-CoV, and δ-CoV) [6]. SARS-CoV-2 is a member of the β-CoV genus, which also includes other important human respiratory pathogens like human coronavirus-OC43 (HCoV-OC43) and HCoV-HKU1 (causes of the common cold), as well as MERS-CoV (middle east respiratory syndrome coronavirus) and SARS-CoV (severe acute respiratory syndrome coronavirus) [7]. The S protein plays a critical role in viral entry and consists of two subunits: S1 (enables viral attachment to the host cell) and S2 (promotes membrane fusion) [8]. One key structural feature of the S1 subunit is a receptor binding domain (RBD) with high affinity for angiotensin-converting enzyme 2 (ACE2), the cellular receptor for SARS-CoV and SARS-CoV-2 [5]. Neutralizing antibodies from convalescent SARS-CoV patients were shown to be directed largely against the S protein, and these antibodies also reduced SARS-CoV-2 viral entry, albeit with lower efficiency [8]. Antibodies are produced against other structural proteins present in SARS-CoV-2 (the envelope, membrane, and nucleocapsid proteins), but the response to the S protein is far greater [9]. Thus, the S protein has become the most frequently targeted antigen for both protein and nucleic acid-based vaccine design [9].

Vaccine design and production have coevolved over the past two centuries due to leaps in understanding in the fields of vaccinology, cell biology, and biotechnology [10–14]. Early vaccines were composed of whole pathogens, and commercial-scale manufacture required the use of large animal farms [15]. Advances in growing bacteria and viruses in culture enabled robust and safe manufacture of whole pathogen and viral vector vaccines at commercial scales. The advent of component vaccines, whose design was based on surface proteins and capsular polysaccharides, as well as secreted protein toxins, resulted in vaccines based on a single biomolecule, though use of adjuvants was typically required. Manufacture of component vaccines capitalized on advances in biotechnology for large-scale production of purified proteins and advances in bioanalytical characterization techniques. Further, groundbreaking work in nucleic acid vaccines set the stage for this vaccine modality to demonstrate its potential enabling rapid construct design and production [16].

The race to develop a COVID-19 vaccine began with the identification of SARS-CoV-2 as the causative agent [3] and the subsequent publication of the genome sequence of the virus. The urgency of the need prompted groups to take chances with multiple different modalities, which would ultimately result in the development of highly effective vaccines that would have a major impact on human health [17]. This chapter highlights the public information currently available on the drug substance processes behind each of those modalities and sheds light on the challenges of delivering these vaccines at the scales needed to put an end to the pandemic focusing on vaccines that have achieved full regulatory approval for human use (Table 8.1).

8.2 mRNA Vaccines

8.2.1 Background

As the world rushed to find a COVID-19 vaccine, candidates were developed using a diverse assortment of modalities. One of the most exciting new approaches was based on mRNA. In this approach, mRNA molecules are encased within lipid

Table 8.1 Vaccines against SARS-CoV-2 approved for human use

Vaccine type	Company/Institution	Candidate name	Country where first approved	References
mRNA	BioNTech/Pfizer	Comirnaty, BNT162b2	US	[18, 19]
	Moderna	Spikevax, mRNA-1273	US, Canada	[20, 21]
Viral vector	CanSino Biologics	Convidecia, Ad5-nCoV	China	[22, 23]
	Gamaleya Research Inst	Sputnik V, Gam-Covid-Vac	Russia	[24, 25]
	Univ. of Oxford/ AstraZeneca	Vaxzevria, AZD1222	Brazil, India	[26, 27]
	Johnson and Johnson/ Beth Israel Deaconess Medical Center	Ad26.COV2.S, JNJ-78436735	Canada	[28, 29]
Whole inactivated virus	Sinovac	CoronaVac, PiCoVacc	China	[30, 31]
	Sinopharm/Beijing Institute of Biological Products	BBIBP-CorV	China, Bahrain	[32, 33]
	Bharat Biotech	Covaxin, BBV152	India	[34–36]
	Sinopharm/Wuhan Institute of Biological Products	NA	China	[37, 38]
Protein-based	Novavax/Serum Institute of India	Covovax, NVX-CoV2373	Canada, South Korea	[39–41]
	Medicago	Covifenz, CoVLP	Canada	[42–44]
	Vector Institute, BEKTOP	EpiVacCorona, Aurora-CoV	Turkmenistan	[45, 46]

nanoparticles and delivered to a person's cells. The cell then uses the mRNA to express the encoded protein to stimulate immune recognition. As shown in Table 8.2, the first two mRNA vaccines that were approved for FDA Emergency Use Authorization were from BioNTech/Pfizer and Moderna, both of which targeted the viral S protein.

Regarding the type of mRNA used, these vaccines can be divided into three categories: unmodified, modified, and self-amplifying. BioNTech/Pfizer and another manufacturer, CureVac, both pursued an unmodified mRNA vaccine where all the nucleotide bases are the native structures [47–49]. However, neither company has obtained approval for the unmodified mRNA vaccine: CureVac has begun on a second-generation unmodified mRNA [50], while BioNTech/Pfizer chose to move forward to Phase 3 clinical trials with the modified mRNA instead. For the modified mRNA, prior research has shown that using a modified uridine (N1-methylpseudouridine, Fig. 8.1) provides better effectiveness with lower immunogenicity [51–53]. This strategy was used by both BioNTech/Pfizer and Moderna. The last category, self-amplifying RNA (saRNA) is an exciting strategy for the field because it provides a polymerase in the mRNA coding sequence that allows for amplification of the transcript inside the cell [54]. Thus, lower doses can potentially be delivered. This strategy was employed by Arcturus Therapeutics (in combination with pseudouridine) [47, 55] and was an initial BioNTech/Pfizer candidate [47].

Table 8.2 Details for approved mRNA vaccines against SARS-CoV-2

Vaccine name	Developer	mRNA category	Reference
Comirnaty, BNT162b2	BioNTech/Pfizer	Modified	[18, 19]
Spikevax, mRNA-1273	Moderna	Modified	[20, 21]

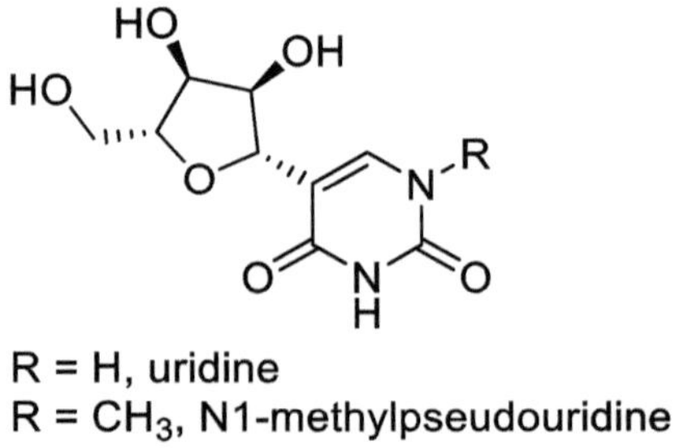

Fig. 8.1 N1-Methylpseudouridine structure

8.2.2 *Production Process*

8.2.2.1 Overview of Steps

The general process for production-scale mRNA synthesis is relatively similar to the small-scale in vitro transcription (IVT) process performed in labs around the world—an enzymatic reaction where nucleotides are synthesized into mRNA from a DNA template, followed by enzymatic steps for capping and/or initial removal of the enzymes and DNA (Fig. 8.2). The biggest difference is in the final purification. While labs are able to perform this step with functionalized magnetic beads or precipitation by lithium chloride, more scalable approaches based on chromatography columns and ultrafiltration/diafiltration (UF/DF) are used for the large volumes processed in commercial scale. Overall, the synthesis step is completed on the order of hours and the purification steps can be completed in 1–2 days, making this process remarkably fast compared to products produced using cell culture.

8.2.2.2 Production

The first production steps where the drug substance (DS) is produced are often referred to as the upstream process (USP). While other processes described in this chapter require cell culture for their USP, the mRNA production process is exclusively driven by enzymatic reactions, often referred to as IVT. These reactions can be performed in a batch or fed batch mode. In the reaction, DNA is the template for the RNA sequence, most often in the form of plasmid or PCR template. The nucleoside triphosphates (NTPs), adenosine, cytidine, guanosine, and uridine (modified or unmodified), and an RNA polymerase (usually T7, T3, or Sp6 RNA polymerase) [56] are added to a buffered reaction. In the fed batch mode, these NTPs can also be fed during the course of the reaction. Pyrophosphatase can also be added to increase yield in the reactions. It does this by cleaving pyrophosphate (a by-product of the addition of each nucleotide to the mRNA transcript) into phosphate [57]. The 5′-cap, described next, can also be added into this reaction for co-transcriptional capping or can be added after the IVT reaction, post-transcriptionally [56]. Once all of the initial raw materials are added to the reactor, the in vitro transcription process

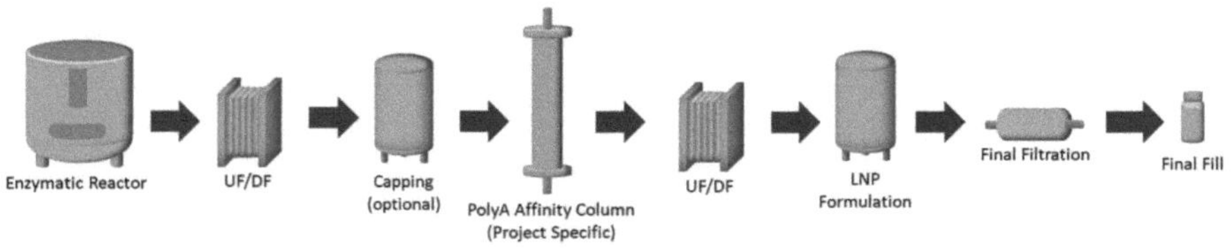

Fig. 8.2 Example process flow diagram for mRNA vaccine production—depending on the project specific contaminants, the included polyA affinity column may not be needed for all mRNA vaccines

proceeds until the nucleotides are consumed or purification steps are started and finishes within hours. IVT reactions have been demonstrated at manufacturing scales over 100 L [48].

As mentioned above, a key aspect of the RNA is the 5′-cap. The 5′-cap can be added co-transcriptionally, using a molecule that is added to the IVT reaction as the first step in building the mRNA chain, or post-transcriptionally, using enzymes to modify the first nucleotide on the 5′ end. For co-transcriptional capping, there are several commercially available options. The two most common are the antireverse cap analog (ARCA) and CleanCap. The ARCA produces a cap 0 structure, while the CleanCap produces a cap 1 structure (Fig. 8.3) [58]. These different structures interact differently with the immune system, with the cap 1 structure less likely to trigger an innate immune response. Post-transcriptional enzymatic capping is commonly performed by the Vaccinia capping enzyme system [59]. This requires Vaccinia virus capping enzyme to achieve a cap 0 structure [60] and 2′-O-methyltransferase if adding the cap 1 structure [61]. The methyl donors required for the process are GTP and S-adenosylmethionine (SAM). While post-transcriptional capping adds an additional unit operation and more raw materials, it can have benefits over co-transcriptional capping for the capping efficiency (percentage of capped mRNA transcripts) and potential for reverse capping (capping in the wrong orientation) [62].

8.2.2.3 Purification

For the isolation process, also known as the downstream process (DSP), biologics requiring cell culture for production often rely on complex purification trains designed to remove host-cell contaminants that impact the safety and efficacy of the product [63]. Because the production of mRNA-based vaccines doesn't require cell

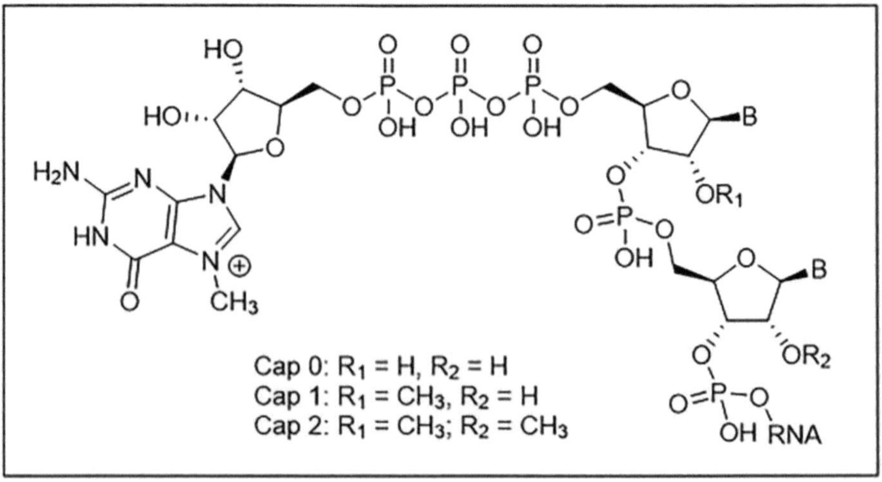

Fig. 8.3 Cap 0–2 structures

culture, the purification processes used are typically much simpler and of shorter duration. Contaminants from the synthesis of mRNA consist of enzymes and unreacted feed materials such as plasmids or partially reacted species that don't have the size or structure of the target product. Thus, a combination of enzymatic, chromatographic, and UF/DF steps is sufficient for clearing these contaminants.

Once the IVT reaction is complete, the reaction can be stopped by the addition of DNase [64]. This enzyme degrades the DNA template in the reaction, halting any further transcription into mRNA and allows the DNA to be broken down into a smaller size. This step can also be used if the purification process is unable to separate RNA from DNA [65].

After degrading the DNA, in the simplest cases, a UF/DF operation has the potential to remove the smaller contaminants, concentrate the mRNA, and replace the buffering species from the reaction with those needed for final formulations. In more complex processes, the UF/DF step would be preceded by orthogonal purification steps, such as reverse phase or ion exchange chromatography [66]. Regardless of the approach, the material is sterile filtered a final time before filling into the final drug substance container. The drug substance is then used for formulation in lipid nanoparticles [48]. However, that process is out of scope for this review.

For situations where the contaminants are not easily separated based on steric interactions, affinity capture media have been developed that selectively bind the polyadenylated (polyA) tail of RNA using hydrogen bonding [67]. The high selectivity of this interaction allows all non-RNA contaminant species or RNA fragments without a polyA tail to flow through during loading, regardless of size, while the mRNA is retained for later elution. This more extensive clearance of contaminants makes polyA affinity a useful platform for mRNA, although some form of UF/DF may be useful to achieve the desired product concentration and buffer replacement for final formulation. Despite being an additional operation, this approach should be universally applicable to other mRNA products and is less burdensome than the multi-column purification trains used for the purification of recombinant proteins.

8.3 Viral Vectors

8.3.1 Background

Viral vector vaccines work by modifying an existing virus to delete essential viral genes that prevent replication of the vector, and instead replace them with DNA encoding for the gene of interest for immune stimulation. The viral vector vaccines that have been approved for SARS-CoV-2 contain DNA encoding for the S protein. These vectors then enter target cells and deliver the genetic material, resulting in expression of the S protein and eliciting an immune response. Thus far, four viral vector vaccines have been approved and can be seen in Table 8.3.

Table 8.3 Details for approved viral vector vaccines against SARS-CoV-2

Vaccine name	Developer	Vector	Cell line	Reference
Ad5-nCoV, Convidecia	CanSino Biologics	Adenovirus type 5	HEK293SF-3F6	[22, 23]
Gam-Covid-Vac Sputnik V	Gamaleya Research Inst	Adenovirus type 5 and 26	Not disclosed	[24, 25]
ChAdOx1 nCoV-19; AZD1222; Vaxzevria	Univ of Oxford/ AstraZeneca	Chimpanzee adenovirus (ChAdOx1)	HEK293	[26, 27]
Ad26.COV2.S	Johnson & Johnson	Adenovirus type 26	PER.C6	[28, 29]

For the approved adenoviral vector vaccines, the vectors chosen are human adenovirus serotype 5 (Ad5), serotype 26 (Ad26), and a chimpanzee adenovirus (specifically, ChAdOx1). Ad5 is the most characterized adenoviral vector [68], making it a common choice for vaccines. However, Ad5 produces a strong antivector immunity after dosing that may inhibit repeat dosing, and it is also limited by high seroprevalence in human populations, which can limit its effectiveness [69]. Both Ad26 and ChAdOx1 (and other simian adenoviruses) are chosen due to their low seroprevalence [70, 71]. With this benefit, the clinical trial from Oxford, AstraZeneca, and Vaccitech has shown promise for a two-dose, prime-boost delivery of the chimpanzee adenoviral vector [72]. Additionally, Gamaleya Research Institution has chosen to use Ad26 to prime and Ad5 to boost in order to deliver two separate serotypes for its vaccine [25].

8.3.2 Production Process

8.3.2.1 Overview of Steps

All of the adenovirus processes start with human cell lines. However, for the HEK293 and PER.C6 cell lines, the USP differs. For HEK293, the adenovirus is produced by infection with the virus where the PER.C6 process can use either infection or transfection with plasmid and cosmid DNA. In either approach, the virus stays contained within the living cell until release during apoptosis. For the manufacturing process, chemical or physical means can be used to cause cell lysis and the release of the viruses into the bulk media for further purification [73]. In general, cell disruption is followed by clarification to remove cell debris, further purification using chromatography, filter-based buffer exchange, and final filtration [73].

8.3.2.2 Production

HEK293 and PER.C6 cells are the most commonly used cell lines for adenovirus production because they both contain two key adenoviral genes, E1A and E1B [74, 75]. This allows for the replication of adenoviral vectors in which those essential genes have been deleted and replaced with the gene of interest [71]. Despite that similarity, the process of introducing the DNA to make the product of interest is different for both cell lines.

In the case of production in HEK293 cells, a fed batch process is typically used to achieve cell densities on the order of 2–4e6 cells/mL, at which point the DNA is introduced by adding purified recombinant viral seed stock to the bioreactor [76]. These viruses infect the cells and begin to make the product of interest. Depending on the multiplicity of infection (MOI; the ratio of virus to cells in the culture), the produced adenovirus from the initial infection may go on to infect additional cells in the culture. For Oxford and AstraZeneca's process, they evaluated both a high MOI (>1) process where cells were infected at a higher density and infected with high amounts of viral seed stock and a low MOI (<1) process with low cell density and low viral seed stock where cells were cultured longer after infection to allow for reinfection inside the bioreactor [76]. In the end, the low MOI process was chosen because it generated comparable volumetric productivity but with lower demand for viral seed stock. While HEK293 cells are capable of being grown to much higher densities (greater than 80e6 cells/mL by perfusion processes) [77], the use of the infection process leads to several challenges such as the necessity for large quantities of viral seed stock and the "cell density effect" wherein the viral particles produced per cell becomes progressively lower with higher cell densities [78, 79].

For the PER.C6 process, cells can be grown to higher densities, up to 100e6 cells/mL [71]. High densities for adenoviral production have been obtained through fed batch and perfusion processes [80]. To introduce the DNA, cells can either be infected similarly to the HEK293 process, as described above, or the DNA can be introduced by transfection. The transfection method uses a cosmid that encodes most of the adenovirus genome and an adapter plasmid that contains the left inverted terminal repeat (ITR) of the adenovirus genome, a packaging signal, the transgene expression cassette (in the place of the E1 gene), and a small portion of residual E1 sequence to support homologous recombination with the cosmid to produce the full-length recombinant genome [71]. Janssen has demonstrated this process up to 1000 L [71].

8.3.2.3 Purification

Virus purification will vary based on the specific viral vector of interest, as each virus, even compared to those with similar size and structure, has unique characteristics that may make it behave differently. Accordingly, even when limited to adenoviruses, purification trains differ significantly.

For a prototypical adenovirus, the purification process begins with the lysis of the host cells using a detergent such as polysorbate 20 (PS20) or Triton X-100 [81]. Benzonase, a highly active endonuclease, is also added to degrade any DNA or RNA in free solution, as shown in Fig. 8.4 [82]. Additional chemicals like sucrose or magnesium chloride may also be added to increase recovery of the virus from the cell debris and improve stability. The bulk harvest is then filtered using a depth filter, often with diatomaceous earth, to remove particulate cell debris [83]. Depending on the pH and conductivity of the clarified material, it can then either be loaded directly onto the capture chromatography media, or buffer exchanged using UF/DF to provide the proper binding conditions for capture. The capture media varies widely, with resins (Poros 50 HQ, Phenyl Sepharose FF, Source 15Q), membranes (Mustang Q, Sartobind Q), and monolithic materials (Cimmultus QA-1) finding use in this application [84]. The eluted material is then concentrated and buffer exchanged into the final formulation buffer using UF/DF.

8.4 Whole Inactivated Virus Vaccines

8.4.1 Background

A more traditional approach to vaccine design is the use of a whole pathogen, and many historical vaccines have been developed this way [12–14]. Whole-virus vaccines fall into one of two classes: live-attenuated and whole inactivated virus (WIV) [12]. Live-attenuated virus vaccines consist of a weakened form of a virus that

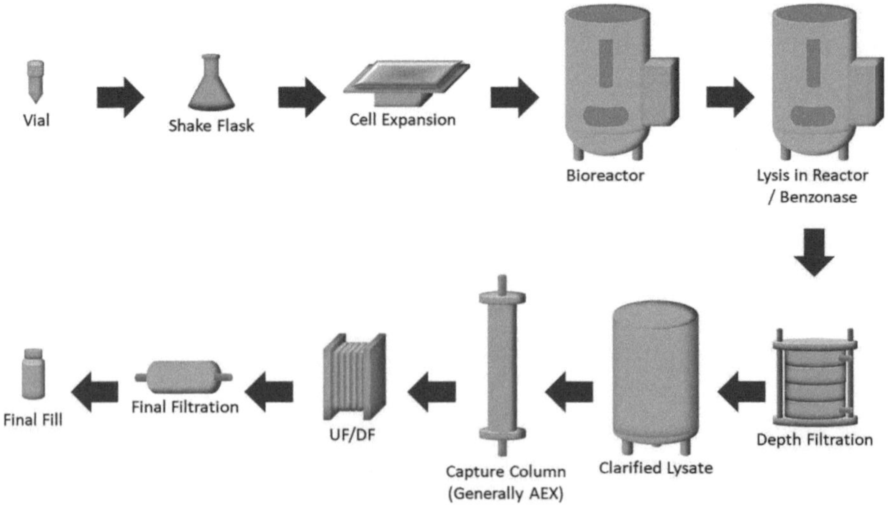

Fig. 8.4 Example process flow diagram for adenovirus vaccine DSP—depending on the specific project, the capture step may be completed using a monolith or membrane

stimulates the immune system without causing severe illness. Live-attenuated virus vaccines can reproduce in the vaccinated subject, effectively increasing the amount of antigen present [10]. WIV vaccines use a form of the virus which cannot reproduce in cells due to chemical or physical inactivation, yet still triggers an immune response [10]. They are generally safe and effective because they cannot induce an infection, but often require larger doses of viral antigen to elicit an immune response [10]. As shown in Table 8.4, four WIV vaccines against SARS-CoV-2 have gained full approval in at least one country, and more than 10 additional WIVs candidates are in clinical development [85]. While no live-attenuated SARS-CoV-2 vaccines have been approved for human use at the time of authoring this manuscript, Codagenix has developed a live-attenuated virus vaccine, CoviLiv, that is given intranasally and has advanced to Phase 2 testing [86]. A second live-attenuated SARS-CoV-2 vaccine is being developed at Speransa Therapeutics and is in a Phase 1 clinical trial [85]. Focusing on the approved vaccines, the manufacturing process for WIV will be described in detail in this review.

8.4.2 Production Process

8.4.2.1 Overview of Steps

The manufacture of WIV vaccines begins with selection of appropriate cell lines for viral propagation of the viral pathogen, specifically SARS-CoV-2 [88]. Once a large population of cells susceptible to the virus is grown, virus is added to the reactor to infect the cells and initiate viral replication. Once a large amount of virus is produced inside the cell, the cells may begin to lyse naturally, or a mechanical force is applied to release the virus. Once cell disruption has occurred, the lysate is collected and clarified to remove cell debris. The whole virus is then inactivated physically or chemically to ensure that no further replication can occur. Additional purification, including chromatography, filter-based buffer exchange, and final filtration can then be performed to yield a final inactivated viral drug substance. A generic process flow for the production of WIVs is shown in Fig. 8.5.

Table 8.4 Details for approved whole inactivated viral vaccines against SARS-CoV-2

Vaccine name (developer)	Cell culture	Virus strain	Inactivation agent	Reference
CoronaVac, PiCoVacc (Sinovac)	Vero cells	CN2	β-propiolactone	[30]
BBIBP-CorV (Sinopharm/Bejing Inst.)	Vero cells	HB02	β-propiolactone	[32]
Covaxin, BBV152 (Bharat Biotech)	Vero cells	NIV-2020-770	β-propiolactone	[34]
(Sinopharm/Wuhan Inst)	Vero cells	WIV-04	β-propiolactone	[87]

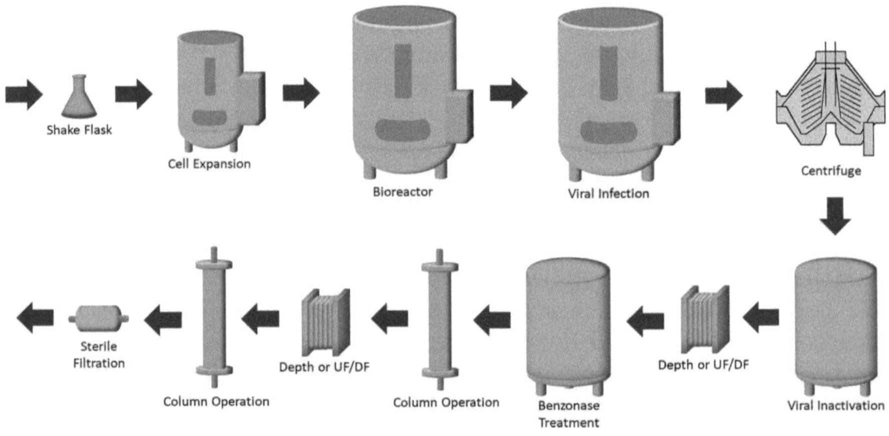

Fig. 8.5 Generic process flow diagram for whole inactivated virus vaccines

8.4.2.2 Production

While whole virus vaccines have been historically produced in eggs, Vero cells have become a common choice for culturing the viruses, with the list of approved COVID-19 vaccines (Table 8.4) all using this approach. The Vero cell line, derived from the kidney of an African green monkey, is a well-characterized, continuous cell line that was found to be nontumorigenic (below a specific passage number) and free of adventitious viruses and proteins [88], and was shown to be particularly susceptible to viral infection due to an impaired antiviral response [89]. Growth of these cells has historically required an adherent cell culture platform. This requires the use of surfaces to support cell growth, typically roller bottles, cell factories, or microcarriers in a bioreactor. Microcarrier processes have been demonstrated up to the 6000 L scale; however, the process still achieves low cell densities (<3e6 cells/mL) due to the surface area limitations of the microcarriers [90]. It has also been shown that this cell line can be adapted to full suspension culture, without the need for cell attachment, and able to achieve cell densities of up to 8e6 cells/mL in batch and perfusion culture processes [91]. This allows for more cells to be produced and in greater volumes [91]. Once the cells are grown to the intended target cell density, the cells are then infected with the SARS-CoV-2 virus [91].

There are some details available regarding the viral replication step for the approved WIV vaccines against SARS-CoV-2. In the case of CoronaVac, Vero cell cultures were inoculated with the CN2 SARS-CoV-2 viral strain (obtained from bronchoalveolar lavage of hospitalized patients) using a MOI of 0.0001–0.01, reaching a peak titer by 3–4 days postinfection [30]. After allowing adequate time for the virus to replicate within the cells, the supernatant was separated by centrifugation and taken into the inactivation step [30]. For the BBIBP-CorV vaccine, the MOI was 0.01–0.3, and the virus titer peaked at 48–72 hours postinfection. The entire bioreactor cell culture process lasted 4–8 days [32]. The starting virus strain was selected by cultivating throat swabs from several patients with COVID-19, and the SARS-CoV-2 virus strain that was the highest yielding was chosen for production [33]. The viral strain used for Covaxin was NIV-2020-770, and virus harvest occurred 36–72 hours postinfection. Because of the presence of the SARS-CoV-2 virus itself, Bharat Biotech operated their manufacturing facilities under Biosafety Level (BSL) 3 requirements [35].

8.4.2.3 Viral Inactivation

For WIVs, a viral inactivation step follows separation of cell debris to ensure that no further replication can occur. Each of the four whole inactivated SARS-CoV-2 vaccines listed in Table 8.4 was chemically inactivated with β-propiolactone. This reagent was first reported by LoGrippo et al. in 1955, and it represented an improvement over the then current standard (formalin and phenol). With the updated approach, inactivation occurs very fast (within 10–15 minutes at 37 °C versus days for the formalin and phenol) with no loss of antigenicity [92]. Modern viral inactivation processes are typically performed at lower temperatures (2–8 °C) for 24–48 hours, which was reported to preserve major neutralizing antigens due to selective inactivation of viral RNA over protein [87].

The structure of β-propiolactone is shown in Fig. 8.6, along with the proposed reaction mechanism that leads to viral inactivation [93]. It is proposed that hard nucleophiles (e.g., basic amines) react with β-propiolactone to yield the acylation products *1*, while soft nucleophiles (e.g., thiols) react to give the alkylation product *2* [93]. RNA and DNA (single and double stranded) were shown to be rapidly alkylated by β-propiolactone [94], as were peptides side chains [93]. While it is impossible to determine the precise chemical modifications leading to viral inactivation with current technology, these studies have illuminated possible adducts with biologically relevant nucleophiles producing potential molecules of interest.

In the case of the WIV vaccine developed by Wuhan Institute with Sinopharm, a chemical inactivation with β-propiolactone was performed twice during the DSP with a depth filtration between the two inactivations [87]. Inactivation of the SARS-CoV-2 virus by physical means has been extensively studied using heat, detergent and UV treatment as well as methanol and paraformaldehyde, although antigenic properties of the inactivated viruses were not evaluated [95].

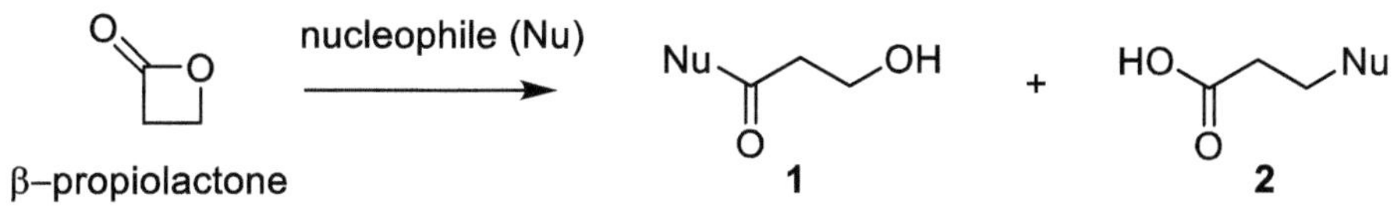

Fig. 8.6 Structure of β-propiolactone and reaction with nucleophiles

8.4.2.4 Purification

The large-scale DSP for viruses can face a range of obstacles different from those of many biotherapeutics. These challenges mostly arise from the size and complexity of the virus, which can affect product purity and recovery. Often the first step includes Benzonase treatment to reduce free DNA and RNA, as was described for viral vector vaccines [82]. Subsequent downstream steps can involve ultra-centrifugation, UF/DF, and different types of chromatography methods to purify the virus. These are chosen based on the specific properties of the virus. For example, the DSP for CoronaVac consisted first of concentration and buffer exchange of the inactivated material using UF/DF in preparation for downstream purification [30]. Following Benzonase treatment, the purification train proceeded to ion exchange chromatography (IEX), followed by a second concentration using UF/DF, and finally size-exclusion chromatography (SEC) [30]. For Covaxin, the DSP included a column chromatography step followed by concentration using tangential flow filtration (TFF) [34], while purification of BBIBP-CoV-2 consisted of buffer exchange of the inactivated virus supernatant by UF/DF followed by endonuclease digestion and a chromatography step [32]. The doubly inactivated WIV vaccine from Sinopharm/Wuhan Institute was purified by gel-chromatography and IEX chromatography followed by sterile filtration [87]. Once the virus is in the purified form, the final step is putting it in a matrix that keeps it stable until it is ready to be formulated into the final drug product.

8.5 Protein-Based Vaccines

8.5.1 Background

Protein-based vaccines contain purified protein antigens lacking DNA or RNA and come in a variety of different presentations. The earliest protein-based vaccines were chemically detoxified bacterial toxins, e.g., diphtheria toxoid (DT) and tetanus toxoid (TT) [14]. Recombinant protein vaccines composed of surface antigens of viruses and bacteria are considered component or subunit vaccines [96]. Some modern examples include Trumenba [97] and VLA15 [98]. Protein antigens can also be presented on nanoparticles or virus-like particles (VLPs), such as human papillomavirus vaccines [99].

While protein vaccines are one of the most established vaccine modalities, they also have the most diversity in approach [100]. This modality provides a unique challenge compared to the other modalities: how do we make the protein in the right conformation with all of the right post-translational modifications? This can be seen by the breadth of different constructs and the cell lines that have been chosen [101]. The constructs range from full S proteins (monomeric and trimeric) to portions of the protein like the receptor binding domain, to virus like particles with S proteins incorporated [101]. Some constructs also have modifications of the sequence to improve protease resistance and antigen stability. For protein production, the cell lines chosen are important for not only getting sufficient yields, but also the post-translational modifications that ensure proper protein folding and contribute to protection from protease cleavage and recognition by neutralizing antibodies [39, 102].

While at least 16 different protein-based vaccines against SARS-CoV-2 have advanced to clinical testing so far [96], only three have been fully approved for human use (Table 8.5). Both NVX-CoV2373 and Covifenz utilize the full-length trimeric form of the S protein with sequence modifications to increase metabolic stability by removing a furin cleavage site and increase conformational stability of the prefusion form [39, 42, 96, 103]. The design of EpiVacCorona is very different in that it consists of three peptides from the S protein sequence conjugated to a carrier protein [45, 46]. The design was intended to prevent the potential for antibody-dependent enhancement (ADE) of infection by avoiding epitopes thought to be responsible for ADE on the SARS-CoV-2 S protein [104]. With many differences in process for these vaccines, each will be discussed in its own subsection.

8.5.2 *Production Processes*

8.5.2.1 NVX-CoV2373 (Novavax)

The protein drug substance for NVX-CoV2373 is the trimeric full-length S protein from SARS-CoV-2 with mutations of the furin cleavage site (682-RRAR-685 to 682-QQAQ-685) and two proline substitutions (K986P and V987P) in a nanoparticle form (27.2 nm) [39]. A high level process flow diagram for NVX-CoV2373 is shown in Fig. 8.7. The double mutant SARS-CoV-2 *S-gene* was cloned into

Table 8.5 Details for approved protein-based vaccines against SARS-CoV-2

Vaccine name (developer)	Construct/Modality	Cell line	Downstream process (DSP)	References
NVX-CoV2373, Covovax, Nuvaxovid (Novavax)	Full-length S protein incorporated into nanoparticles	*Spodoptera frugiperda* (Sf9, insect cells)	TMAE anion exchange column, lentil lectin affinity column, hollow fiber TFF	[39]
CoVLP, Covifenz (Medicago)	Full-length S protein incorporated into VLPs	*Nicotiana benthamiana* (plant cells)	Similar to hemagglutinin based VLPs	[42, 105]
EpiVacCorona, Aurora-CoV (BEKTOP)	Three 21–30 residue synthetic peptides from the S protein covalently bound to MBP-6xHis-N-CoV-2019	*E. coli* (carrier protein)	No details provided	[46]

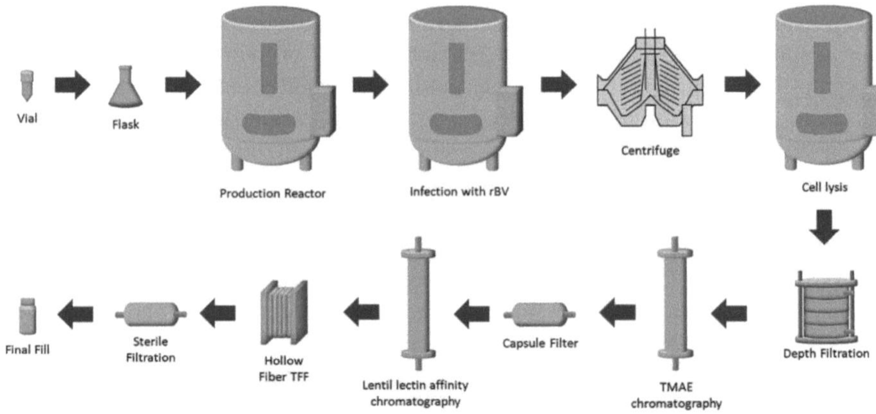

Fig. 8.7 High-level process flow diagram for NVX-CoV2372

recombinant baculovirus (rBV) which was then used to infect *Spodoptera frugiperda* (Sf9) cells that had been expanded to 2–3e6 cells/mL density (MOI $\leq$ 0.1 pfu/cell). Cell culture was maintained at 27 °C for 68–72 hours after infection, and the cells were harvested by centrifugation. Cell pellets were suspended in Tris buffer, and S proteins were extracted into Tris buffer with NP-9 surfactant. The downstream purification process of the S protein included trimethylaminoethyl (TMAE) anion exchange chromatography followed by lentil lectin affinity chromatography. Buffer exchange with hollow fiber TFF yielded the final drug substance.

8.5.2.2 CoVLP (Medicago)

The protein antigen for CoVLP is the trimeric full-length S protein of SARS-CoV-2 with mutations of the furin cleavage site (R667G, R668S, and R670S) and mutations to stabilize the prefusion conformation (K971P and V972P) presented on a VLP surface [42]. VLP-vaccines are manufactured in cell lines capable of producing particles that mimic viral structures and demonstrate increased immunogenicity over purified recombinant protein vaccines. The CoVLP manufacturing process follows Medicago's plant-based platform for VLP production, which is based on agroinfiltration, a method of inducing transient gene expression in plants [105]. The overall process is represented in Fig. 8.8. Briefly, the USP starts with batches of *Nicotiana benthamiana* plants grown for 5 weeks prior to agroinfiltration with an *Agrobacterium tumefaciens* vector coding for the full-length S protein. Harvest occurs 6 days after transfection using Medicago's proprietary extraction method [42].

The purification process begins with removal of cellular debris by centrifugation followed by clarification using chemical and physical methods [105]. Concentration of the clarified supernatant precedes several chromatographic steps including IEX and affinity chromatography. The final purified eluent is concentrated by TFF, formulated and sterile filtered prior to final fill [105].

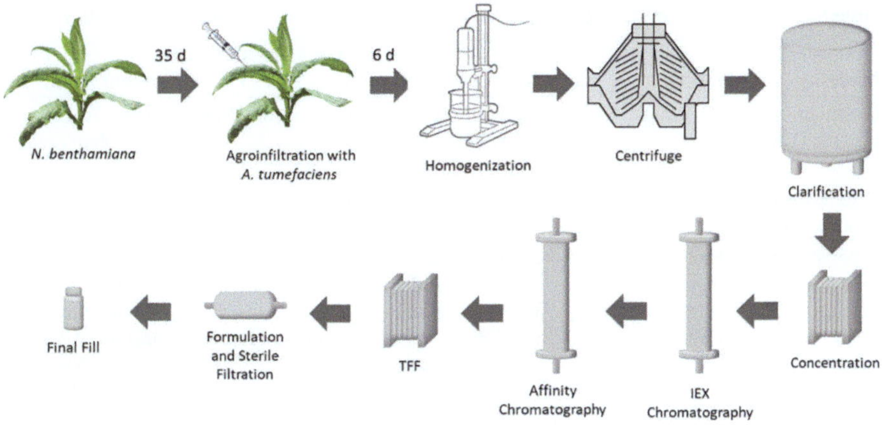

Fig. 8.8 High-level CoVLP process flow diagram

8.5.2.3 EpiVacCorona (Vector Institute)

The EpiVacCorona vaccine comprises three chemically synthesized peptide sequences from the SARS-CoV-2 S protein conjugated to a carrier protein, which was reported to be the N-protein of SARS-CoV-2 [45, 46]. While no details of the synthetic methods utilized to produce the peptides were provided, the reader is directed to several recent reviews on the subject [106, 107]. Likewise, the specific chemistry employed for conjugation of the peptides to the N-protein was not reported for EpiVacCorona. A similar conjugate vaccine against SARS-CoV-2 has been developed at the Finlay Vaccine Institute (Soberana 1) which consists of the SARS-CoV-2 S protein RBD conjugated to TT [108]. The RBD containing an unpaired cysteine was produced in CHO cells and was conjugated to maleimido-activated TT using known conjugation techniques [108]. Capsular polysaccharide-conjugate vaccines utilize similar manufacturing techniques and have been reviewed elsewhere [109–112].

8.6 Comparison of Manufacturing Processes for the Different Modalities

While each of the COVID-19 vaccine modalities has been highlighted individually, Table 8.6 shows a comparison of several manufacturing parameters for the mRNA, viral vectors, WIV, and protein-based vaccines. Manufacturing processes for viral vaccines vary greatly depending on the identity of the antigen drug substance for a given modality.

The USP typically requires cell culture to produce the antigen, whether it is a whole pathogen or a purified component. Both the viral vector and whole pathogen

Table 8.6 Comparison of different manufacturing parameters for viral vaccine modalities

Modality	mRNA	Viral vectors (adenovirus)	Whole inactivated virus (WIV)	Protein-based
Upstream process (USP)	Enzymatic	Cell culture, HEK293/PER.C6	Cell culture, Vero	Varies (insect cells, plant cells, *E. coli*)
Approximate upstream scale	50 to >100 L	~1000 L	Up to 6000 L	Up to 10,000 L
Downstream process (DSP)	Enzymatic digestion, chromatography, UF/DF	Varies with different products: depth filtration, UF/DF, chromatography	Inactivation, depth filtration, UF/DF, chromatography	Depth filtration, UF/DF, chromatography, viral filtration (mammalian culture only)
Duration of manufacturing batch	~Hours-days for enzymatic reaction, several days for purification (3–4 unit ops)	Cell culture scale up requires several weeks, several days for purification (5 unit ops)	Cell culture scale up requires several weeks, several days for purification (7 unit ops)	Cell culture scale up requires several weeks, several days for inactivation and purification
Manufacturing safety considerations	None—enzymes don't pose risk	Viral product, infection process requires viral seed banks	Intermediate is infectious SARS-CoV-2 virus, toxic materials used for inactivation	Low risk: endotoxin (*E. coli*), viruses (mammalian)

vaccines use mammalian cells (HEK293, PER.C6, and Vero) to produce the viruses, while the protein-based vaccines use insect, plant, and microbial cells. With this, the raw materials required for the cell culture are typically inexpensive and abundant such as glucose in contrast to the less abundant and more expensive purified enzymes and NTPs for the mRNA process. One exception to the cell culture process is mRNA vaccine, where the DS is mRNA produced enzymatically from NTPs using a DNA template. This approach affords multiple advantages over traditional cell culture processes, including shorter production times. The scale of the USPs is also notable, with the cell culture reactors requiring much larger volumes to produce large quantities of product as compared to the mRNA process.

The DSPs for the modalities may appear to employ similar unit ops, but each have unique nuances. Most of the DSPs for cell culture processes start with a lysis step since the product is contained within the cells. These are either chemical detergents (viral vectors, NVX-CoV2373) or mechanical (CoVLP). The exception to this is WIVs, since the product is in the supernatant. Following lysis, depth filtration is required to remove cell debris. Several of the processes also employ enzymatic steps for digestion of nucleic acid impurities: Benzonase to remove free DNA and RNA (viral vectors, WIVs) and DNase to remove template DNA (mRNA). After depth filtration, chromatography is typically the next step. However, the WIVs add in an additional inactivation step, which can be either chemical or physical. The chromatography steps can take many different iterations, directed by qualities of the modality: affinity, IEX, and SEC. Multiple chromatography steps may be required to reach the purity needed. Notably, a viral filtration step at the end of the DSP is used only for the mammalian culture for the protein-based vaccines. This step is not performed for the viral vectors, WIVs, or the CoVLP process because the product is a virus or virus-like particle that would be retained by the filter. Overall, a simplified DSP once again benefits the mRNA vaccines since there is no need to remove cellular debris and enzymatic degradation of process-based impurities can reduce the separation to simple chromatography and UF/DF operations. Fewer unit operations also mean shorter processing times overall compared to DSP for a conventional cell culture process with multiple chromatography steps.

Considering safety in the manufacturing process, the production of WIVs necessitates the use of the pathogenic virus which requires special containment facilities to ensure safety during manufacture. Furthermore, inactivation step for WIVs requires the use of toxic chemicals like β-propiolactone and formalin. Vaccines that utilize recombinant proteins as antigens avoid the use of highly infective viruses but risk contamination with adventitious viruses when mammalian cell cultures are required. Viral removal filtration methods have been developed to ensure the safety of biotherapeutics derived from mammalian cell culture, but this adds to the complexity of process development and requires an additional step in the DSP. Enzymatic production of mRNA circumvents exposure to viruses by avoiding the use of cell culture in the drug substance process.

8.7 Conclusions

Overall, many skilled pharmaceutical development groups used the full range of technologies and tools in their portfolios to answer the critical need for a vaccine against SARS-CoV-2. To respond to this challenge, time was the most critical factor for the manufacturing processes. Faster process development times meant more material could be generated quickly for clinical trials to evaluate safety and efficacy and, if successful, more material could be made to dose the population. With this, advantages are given to shorter processes with fewer unit operations. The cell culture processes inherently add more time due to the need for cell expansion and growth. These processes also require more complex harvest streams, adding additional unit operations and more manufacturing time. Additionally, more unit operations mean more development to make sure yield and product quality requirements are met for each operation, while shorter manufacturing processes give the advantage of quicker process validation.

Based on the outlined constraints, mRNA had significant advantages over the other approaches. Despite the necessity for more complex feed materials, the enzymatic USP reduces time over the cell culture processes and fewer contaminants reduce the number of DSP unit operations and shorten the manufacturing run time. With these advantages combined with the high efficacy demonstrated in their clinical studies, mRNA vaccines were able to reach FDA Emergency Use Authorization first. Early availability of the vaccines was a crucial development in amplifying their impact on human health. The lessons learned while searching for a vaccine for SARS-CoV-2 should help make the response to any future pandemics more effective.

Acknowledgements The authors thank Adam Burrell, Anna Chau, Rodney Combs, Nathaniel Leachman, Chris Murphy, Lara Payne, and Joe Russo for their valuable comments on this chapter.

Author Disclosure All authors are employees of Pfizer, Inc. and have no conflicts of interest to declare.

References

1. Dong E, Du H, Gardner L (2020) An interactive web-based dashboard to track COVID-19 in real time. Lancet Infect Dis 20:533–534. https://doi.org/10.1016/S1473-3099(20)30120-1
2. John Hopkins Coronavirus Resource Center (2022) John Hopkins University & Medicine
3. Zhu N, Zhang D, Wang W, Li X, Yang B, Song J, Zhao X, Huang B, Shi W, Lu R, Niu P, Zhan F, Ma X, Wang D, Xu W, Wu G, Gao GF, Tan W (2020) A novel coronavirus from patients with pneumonia in China, 2019. N Engl J Med 382:727–733. https://doi.org/10.1056/NEJMoa2001017
4. Chan JF-W, Kok K-H, Zhu Z, Chu H, To KK-W, Yuan S, Yuen K-Y (2020) Genomic characterization of the 2019 novel human-pathogenic coronavirus isolated from a patient with atypical pneumonia after visiting Wuhan. Emerg Microbes Infect 9:221–236. https://doi.org/10.1080/22221751.2020.1719902
5. Yan R, Zhang Y, Li Y, Xia L, Guo Y, Zhou Q (2020) Structural basis for the recognition of SARS-CoV-2 by full-length human ACE2. Science 367:1444–1448. https://doi.org/10.1126/science.abb2762
6. Vakulenko Y, Deviatkin A, Drexler JF, Lukashev A (2021) Modular evolution of coronavirus genomes. Viruses 13:1270
7. Corman VM, Muth D, Niemeyer D, Drosten C (2018) Chapter 8 – Hosts and sources of endemic human coronaviruses. In: Kielian M, Mettenleiter TC, Roossinck MJ (eds) Advances in virus research. Academic Press, pp 163–188
8. Hoffmann M, Kleine-Weber H, Schroeder S, Krüger N, Herrler T, Erichsen S, Schiergens TS, Herrler G, Wu N-H, Nitsche A, Müller MA, Drosten C, Pöhlmann S (2020) SARS-CoV-2 cell entry depends on ACE2 and TMPRSS2 and is blocked by a clinically proven protease inhibitor. Cell 181:271–280.e8. https://doi.org/10.1016/j.cell.2020.02.052
9. Ahmed SF, Quadeer AA, McKay MR (2020) Preliminary identification of potential vaccine targets for the COVID-19 coronavirus (SARS-CoV-2) based on SARS-CoV immunological studies. Viruses 12:254
10. Burrell CJ, Howard CR, Murphy FA (2017) Chapter 11 – Vaccines and vaccination. In: Burrell CJ, Howard CR, Murphy FA (eds) Fenner and White's medical virology, 5th edn. Academic Press, London, pp 155–167
11. Gomez PL, Robinson JM (2018) Vaccine manufacturing. Plotkin's Vaccin:51–60e1. https://doi.org/10.1016/B978-0-323-35761-6.00005-5
12. Hussein IH, Chams N, Chams S, Sayegh SE, Badran R, Raad M, Gerges-Geagea A, Leone A, Jurjus A (2015) Vaccines through centuries: major cornerstones of global health. Front Public Health 3:1–16
13. Plotkin S (2014) History of vaccination. Proc Natl Acad Sci 111:12283–12287. https://doi.org/10.1073/pnas.1400472111

14. Plotkin SA, Plotkin SL (2011) The development of vaccines: how the past led to the future. Nat Rev Microbiol 9:889–893. https://doi.org/10.1038/nrmicro2668
15. Esparza J, Lederman S, Nitsche A, Damaso CR (2020) Early smallpox vaccine manufacturing in the United States: introduction of the "animal vaccine" in 1870, establishment of "vaccine farms", and the beginnings of the vaccine industry. Vaccine 38:4773–4779. https://doi.org/10.1016/j.vaccine.2020.05.037
16. Li M, Wang Z, Xie C, Xia X (2022) Chapter 7 – Advances in mRNA vaccines. In: Aranda F, Berraondo P, Galluzzi L (eds) International review of cell and molecular biology. Academic Press, pp 295–316
17. Zimmer C, Corum J, Wee S-L, Kristoffersen M (2020) Coronavirus vaccine tracker. The New York Times
18. Vogel AB, Kanevsky I, Che Y, Swanson KA, Muik A, Vormehr M, Kranz LM, Walzer KC, Hein S, Güler A, Loschko J, Maddur MS, Tompkins K, Cole J, Lui BG, Ziegenhals T, Plaschke A, Eisel D, Dany SC, Fesser S, Erbar S, Bates F, Schneider D, Jesionek B, Sänger B, Wallisch A-K, Feuchter Y, Junginger H, Krumm SA, Heinen AP, Adams-Quack P, Schlereth J, Kröner C, Hall-Ursone S, Brasky K, Griffor MC, Han S, Lees JA, Mashalidis EH, Sahasrabudhe PV, Tan CY, Pavliakova D, Singh G, Fontes-Garfias C, Pride M, Scully IL, Ciolino T, Obregon J, Gazi M, Carrion R, Alfson KJ, Kalina WV, Kaushal D, Shi P-Y, Klamp T, Rosenbaum C, Kuhn AN, Türeci Ö, Dormitzer PR, Jansen KU, Sahin U (2020) A prefusion SARS-CoV-2 spike RNA vaccine is highly immunogenic and prevents lung infection in non-human primates. bioRxiv:2020.09.08.280818. https://doi.org/10.1101/2020.09.08.280818
19. Polack FP, Thomas SJ, Kitchin N, Absalon J, Gurtman A, Lockhart S, Perez JL, Pérez Marc G, Moreira ED, Zerbini C, Bailey R, Swanson KA, Roychoudhury S, Koury K, Li P, Kalina WV, Cooper D, Frenck RW, Hammitt LL, Türeci Ö, Nell H, Schaefer A, Ünal S, Tresnan DB, Mather S, Dormitzer PR, Şahin U, Jansen KU, Gruber WC (2020) Safety and efficacy of the BNT162b2 mRNA Covid-19 vaccine. N Engl J Med 383:2603–2615. https://doi.org/10.1056/NEJMoa2034577
20. Corbett KS, Flynn B, Foulds KE, Francica JR, Boyoglu-Barnum S, Werner AP, Flach B, O'Connell S, Bock KW, Minai M, Nagata BM, Andersen H, Martinez DR, Noe AT, Douek N, Donaldson MM, Nji NN, Alvarado GS, Edwards DK, Flebbe DR, Lamb E, Doria-Rose NA, Lin BC, Louder MK, O'Dell S, Schmidt SD, Phung E, Chang LA, Yap C, Todd J-PM, Pessaint L, Van Ry A, Browne S, Greenhouse J, Putman-Taylor T, Strasbaugh A, Campbell T-A, Cook A, Dodson A, Steingrebe K, Shi W, Zhang Y, Abiona OM, Wang L, Pegu A, Yang ES, Leung K, Zhou T, Teng I-T, Widge A, Gordon I, Novik L, Gillespie RA, Loomis RJ, Moliva JI, Stewart-Jones G, Himansu S, Kong W-P, Nason MC, Morabito KM, Ruckwardt TJ, Ledgerwood JE, Gaudinski MR, Kwong PD, Mascola JR, Carfi A, Lewis MG, Baric RS, McDermott A, Moore IN, Sullivan NJ, Roederer M, Seder RA, Graham BS (2020) Evaluation of the mRNA-1273 vaccine against SARS-CoV-2 in nonhuman primates. N Engl J Med 383:1544–1555. https://doi.org/10.1056/NEJMoa2024671
21. Baden LR, El Sahly HM, Essink B, Kotloff K, Frey S, Novak R, Diemert D, Spector SA, Rouphael N, Creech CB, McGettigan J, Khetan S, Segall N, Solis J, Brosz A, Fierro C, Schwartz H, Neuzil K, Corey L, Gilbert P, Janes H, Follmann D, Marovich M, Mascola J, Polakowski L, Ledgerwood J, Graham BS, Bennett H, Pajon R, Knightly C, Leav B, Deng W, Zhou H, Han S, Ivarsson M, Miller J, Zaks T (2020) Efficacy and safety of the mRNA-1273 SARS-CoV-2 vaccine. N Engl J Med 384:403–416. https://doi.org/10.1056/NEJMoa2035389
22. Zhu F-C, Li Y-H, Guan X-H, Hou L-H, Wang W-J, Li J-X, Wu S-P, Wang B-S, Wang Z, Wang L, Jia S-Y, Jiang H-D, Wang L, Jiang T, Hu Y, Gou J-B, Xu S-B, Xu J-J, Wang X-W, Wang W, Chen W (2020) Safety, tolerability, and immunogenicity of a recombinant adenovirus type-5 vectored COVID-19 vaccine: a dose-escalation, open-label, non-randomised, first-in-human trial. Lancet 395:1845–1854. https://doi.org/10.1016/S0140-6736(20)31208-3
23. Halperin SA, Ye L, MacKinnon-Cameron D, Smith B, Cahn PE, Ruiz-Palacios GM, Ikram A, Lanas F, Lourdes Guerrero M, Muñoz Navarro SR, Sued O, Lioznov DA, Dzutseva V, Parveen G, Zhu F, Leppan L, Langley JM, Barreto L, Gou J, Zhu T, Mao H, Gagnon L, Tran S-P,

Khan ST, Becerra Aquino AG, Saldaña Montemayor EE, Rivera Martínez NE, Bohórquez López VC, Simón Campos JA, Pineda Cárdenas FJ, Chen W, Hou L, Zhang Z, Corral G, López E, Teijeiro R, Alzogaray MF, Zaidman C, Lopardo G, Goecke B, Feijooó Seoane RM, Mahmood SF, Khan EA, Akram J, Abbas S, Salahuddin N, Rozhkova E, Zubkova T (2022) Final efficacy analysis, interim safety analysis, and immunogenicity of a single dose of recombinant novel coronavirus vaccine (adenovirus type 5 vector) in adults 18 years and older: an international, multicentre, randomised, double-blinded, placebo-controlled phase 3 trial. Lancet 399:237–248. https://doi.org/10.1016/S0140-6736(21)02753-7

24. Logunov DY, Dolzhikova IV, Zubkova OV, Tukhvatulin AI, Shcheblyakov DV, Dzharullaeva AS, Grousova DM, Erokhova AS, Kovyrshina AV, Botikov AG, Izhaeva FM, Popova O, Ozharovskaya TA, Esmagambetov IB, Favorskaya IA, Zrelkin DI, Voronina DV, Shcherbinin DN, Semikhin AS, Simakova YV, Tokarskaya EA, Lubenets NL, Egorova DA, Shmarov MM, Nikitenko NA, Morozova LF, Smolyarchuk EA, Kryukov EV, Babira VF, Borisevich SV, Naroditsky BS, Gintsburg AL (2020) Safety and immunogenicity of an rAd26 and rAd5 vector-based heterologous prime-boost COVID-19 vaccine in two formulations: two open, non-randomised phase 1/2 studies from Russia. Lancet 396:887–897. https://doi.org/10.1016/S0140-6736(20)31866-3
25. Logunov DY, Dolzhikova IV, Shcheblyakov DV, Tukhvatulin AI, Zubkova OV, Dzharullaeva AS, Kovyrshina AV, Lubenets NL, Grousova DM, Erokhova AS, Botikov AG, Izhaeva FM, Popova O, Ozharovskaya TA, Esmagambetov IB, Favorskaya IA, Zrelkin DI, Voronina DV, Shcherbinin DN, Semikhin AS, Simakova YV, Tokarskaya EA, Egorova DA, Shmarov MM, Nikitenko NA, Gushchin VA, Smolyarchuk EA, Zyryanov SK, Borisevich SV, Naroditsky BS, Gintsburg AL (2021) Safety and efficacy of an rAd26 and rAd5 vector-based heterologous prime-boost COVID-19 vaccine: an interim analysis of a randomised controlled phase 3 trial in Russia. Lancet 397:671–681. https://doi.org/10.1016/S0140-6736(21)00234-8
26. van Doremalen N, Lambe T, Spencer A, Belij-Rammerstorfer S, Purushotham JN, Port JR, Avanzato VA, Bushmaker T, Flaxman A, Ulaszewska M, Feldmann F, Allen ER, Sharpe H, Schulz J, Holbrook M, Okumura A, Meade-White K, Pérez-Pérez L, Edwards NJ, Wright D, Bissett C, Gilbride C, Williamson BN, Rosenke R, Long D, Ishwarbhai A, Kailath R, Rose L, Morris S, Powers C, Lovaglio J, Hanley PW, Scott D, Saturday G, de Wit E, Gilbert SC, Munster VJ (2020) ChAdOx1 nCoV-19 vaccine prevents SARS-CoV-2 pneumonia in rhesus macaques. Nature 586:578–582. https://doi.org/10.1038/s41586-020-2608-y
27. Voysey M, Clemens SAC, Madhi SA, Weckx LY, Folegatti PM, Aley PK, Angus B, Baillie VL, Barnabas SL, Bhorat QE, Bibi S, Briner C, Cicconi P, Collins AM, Colin-Jones R, Cutland CL, Darton TC, Dheda K, Duncan CJA, Emary KRW, Ewer KJ, Fairlie L, Faust SN, Feng S, Ferreira DM, Finn A, Goodman AL, Green CM, Green CA, Heath PT, Hill C, Hill H, Hirsch I, Hodgson SHC, Izu A, Jackson S, Jenkin D, Joe CCD, Kerridge S, Koen A, Kwatra G, Lazarus R, Lawrie AM, Lelliott A, Libri V, Lillie PJ, Mallory R, Mendes AVA, Milan EP, Minassian AM, McGregor A, Morrison H, Mujadidi YF, Nana A, O'Reilly PJ, Padayachee SD, Pittella A, Plested E, Pollock KM, Ramasamy MN, Rhead S, Schwarzbold AV, Singh N, Smith A, Song R, Snape MD, Sprinz E, Sutherland RK, Tarrant R, Thomson EC, Török ME, Toshner M, Turner DPJ, Vekemans J, Villafana TL, Watson MEE, Williams CJ, Douglas AD, Hill AVS, Lambe T, Gilbert SC, Pollard AJ, Aban M, Abayomi F, Abeyskera K, Aboagye J, Adam M, Adams K, Adamson J, Adelaja YA, Adewetan G, Adlou S, Ahmed K, Akhalwaya Y, Akhalwaya S, Alcock A, Ali A, Allen ER, Allen L, Almeida TCDSC, Alves MPS, Amorim F, Andritsou F, Anslow R, Appleby M, Arbe-Barnes EH, Ariaans MP, Arns B, Arruda L, Azi P, Azi L, Babbage G, Bailey C, Baker KF, Baker M, Baker N, Baker P, Baldwin L, Baleanu I, Bandeira D, Bara A, Barbosa MAS, Barker D, Barlow GD, Barnes E, Barr AS, Barrett JR, Barrett J, Bates L, Batten A, Beadon K, Beales E, Beckley R, Belij-Rammerstorfer S, Bell J, Bellamy D, Bellei N, Belton S, Berg A, Bermejo L, Berrie E, Berry L, Berzenyi D, Beveridge A, Bewley KR, Bexhell H, Bhikha S, Bhorat AE, Bhorat ZE, Bijker E, Birch G, Birch S, Bird A, Bird O, Bisnauthsing K, Bittaye M, Blackstone K, Blackwell L, Bletchly H, Blundell CL, Blundell SR, Bodalia P, Boettger BC, Bolam E, Boland E, Bormans D, Borthwick N, Bowring

F, Boyd A, Bradley P, Brenner T, Brown P, Brown C, Brown-O'Sullivan C, Bruce S, Brunt E, Buchan R, Budd W, Bulbulia YA, Bull M, Burbage J, Burhan H, Burn A, Buttigieg KR, Byard N, Cabera Puig I, Calderon G, Calvert A, Camara S, Cao M, Cappuccini F, Cardoso JR, Carr M, Carroll MW, Carson-Stevens A, Carvalho YM, Carvalho JAM, Casey HR, Cashen P, Castro T, Castro LC, Cathie K, Cavey A, Cerbino-Neto J, Chadwick J, Chapman D, Charlton S, Chelysheva I, Chester O, Chita S, Cho J-S, Cifuentes L, Clark E, Clark M, Clarke A, Clutterbuck EA, Collins SLK, Conlon CP, Connarty S, Coombes N, Cooper C, Cooper R, Cornelissen L, Corrah T, Cosgrove C, Cox T, Crocker WEM, Crosbie S, Cullen L, Cullen D, Cunha DRMF, Cunningham C, Cuthbertson FC, Da Guarda SNF, da Silva LP, Damratoski BE, Danos Z, Dantas MTDC, Darroch P, Datoo MS, Datta C, Davids M, Davies SL, Davies H, Davis E, Davis J, Davis J, De Nobrega MMD, De Oliveira Kalid LM, Dearlove D, Demissie T, Desai A, Di Marco S, Di Maso C, Dinelli MIS, Dinesh T, Docksey C, Dold C, Dong T, Donnellan FR, Dos Santos T, dos Santos TG, Dos Santos EP, Douglas N, Downing C, Drake J, Drake-Brockman R, Driver K, Drury R, Dunachie SJ, Durham BS, Dutra L, Easom NJW, van Eck S, Edwards M, Edwards NJ, El Muhanna OM, Elias SC, Elmore M, English M, Esmail A, Essack YM, Farmer E, Farooq M, Farrar M, Farrugia L, Faulkner B, Fedosyuk S, Felle S, Feng S, Ferreira Da Silva C, Field S, Fisher R, Flaxman A, Fletcher J, Fofie H, Fok H, Ford KJ, Fowler J, Fraiman PHA, Francis E, Franco MM, Frater J, Freire MSM, Fry SH, Fudge S, Furze J, Fuskova M, Galian-Rubio P, Galiza E, Garlant H, Gavrila M, Geddes A, Gibbons KA, Gilbride C, Gill H, Glynn S, Godwin K, Gokani K, Goldoni UC, Goncalves M, Gonzalez IGS, Goodwin J, Goondiwala A, Gordon-Quayle K, Gorini G, Grab J, Gracie L, Greenland M, Greenwood N, Greffrath J, Groenewald MM, Grossi L, Gupta G, Hackett M, Hallis B, Hamaluba M, Hamilton E, Hamlyn J, Hammersley D, Hanrath AT, Hanumunthadu B, Harris SA, Harris C, Harris T, Harrison TD, Harrison D, Hart TC, Hartnell B, Hassan S, Haughney J, Hawkins S, Hay J, Head I, Henry J, Hermosin Herrera M, Hettle DB, Hill J, Hodges G, Horne E, Hou MM, Houlihan C, Howe E, Howell N, Humphreys J, Humphries HE, Hurley K, Huson C, Hyder-Wright A, Hyams C, Ikram S, Ishwarbhai A, Ivan M, Iveson P, Iyer V, Jackson F, De Jager J, Jaumdally S, Jeffers H, Jesudason N, Jones B, Jones K, Jones E, Jones C, Jorge MR, Jose A, Joshi A, Júnior EAMS, Kadziola J, Kailath R, Kana F, Karampatsas K, Kasanyinga M, Keen J, Kelly EJ, Kelly DM, Kelly D, Kelly S, Kerr D, Kfouri RÁ, Khan L, Khozoee B, Kidd S, Killen A, Kinch J, Kinch P, King LDW, King TB, Kingham L, Klenerman P, Knapper F, Knight JC, Knott D, Koleva S, Lang M, Lang G, Larkworthy CW, Larwood JPJ, Law R, Lazarus EM, Leach A, Lees EA, Lemm N-M, Lessa A, Leung S, Li Y, Lias AM, Liatsikos K, Linder A, Lipworth S, Liu S, Liu X, Lloyd A, Lloyd S, Loew L, Lopez Ramon R, Lora L, Lowthorpe V, Luz K, MacDonald JC, MacGregor G, Madhavan M, Mainwaring DO, Makambwa E, Makinson R, Malahleha M, Malamatsho R, Mallett G, Mansatta K, Maoko T, Mapetla K, Marchevsky NG, Marinou S, Marlow E, Marques GN, Marriott P, Marshall RP, Marshall JL, Martins FJ, Masenya M, Masilela M, Masters SK, Mathew M, Matlebjane H, Matshidiso K, Mazur O, Mazzella A, McCaughan H, McEwan J, McGlashan J, McInroy L, McIntyre Z, McLenaghan D, McRobert N, McSwiggan S, Megson C, Mehdipour S, Meijs W, Mendonça RNÁ, Mentzer AJ, Mirtorabi N, Mitton C, Mnyakeni S, Moghaddas F, Molapo K, Moloi M, Moore M, Moraes-Pinto MI, Moran M, Morey E, Morgans R, Morris S, Morris S, Morris HC, Morselli F, Morshead G, Morter R, Mottal L, Moultrie A, Moya N, Mpelembue M, Msomi S, Mugodi Y, Mukhopadhyay E, Muller J, Munro A, Munro C, Murphy S, Mweu P, Myasaki CH, Naik G, Naker K, Nastouli E, Nazir A, Ndlovu B, Neffa F, Njenga C, Noal H, Noé A, Novaes G, Nugent FL, Nunes G, O'Brien K, O'Connor D, Odam M, Oelofse S, Oguti B, Olchawski V, Oldfield NJ, Oliveira MG, Oliveira C, Oosthuizen A, O'Reilly P, Osborne P, Owen DRJ, Owen L, Owens D, Owino N, Pacurar M, Paiva BVB, Palhares EMF, Palmer S, Parkinson S, Parracho HMRT, Parsons K, Patel D, Patel B, Patel F, Patel K, Patrick-Smith M, Payne RO, Peng Y, Penn EJ, Pennington A, Peralta Alvarez MP, Perring J, Perry N, Perumal R, Petkar S, Philip T, Phillips DJ, Phillips J, Phohu MK, Pickup L, Pieterse S, Piper J, Pipini D, Plank M, Du Plessis J, Pollard S, Pooley J, Pooran A, Poulton I, Powers C, Presa FB, Price DA, Price V, Primeira M, Proud PC,

Provstgaard-Morys S, Pueschel S, Pulido D, Quaid S, Rabara R, Radford A, Radia K, Rajapaska D, Rajeswaran T, Ramos ASF, Ramos Lopez F, Rampling T, Rand J, Ratcliffe H, Rawlinson T, Rea D, Rees B, Reiné J, Resuello-Dauti M, Reyes Pabon E, Ribiero CM, Ricamara M, Richter A, Ritchie N, Ritchie AJ, Robbins AJ, Roberts H, Robinson RE, Robinson H, Rocchetti TT, Rocha BP, Roche S, Rollier C, Rose L, Ross Russell AL, Rossouw L, Royal S, Rudiansyah I, Ruiz S, Saich S, Sala C, Sale J, Salman AM, Salvador N, Salvador S, Sampaio M, Samson AD, Sanchez-Gonzalez A, Sanders H, Sanders K, Santos E, Santos Guerra MFS, Satti I, Saunders JE, Saunders C, Sayed A, Schim van der Loeff I, Schmid AB, Schofield E, Screaton G, Seddiqi S, Segireddy RR, Senger R, Serrano S, Shah R, Shaik I, Sharpe HE, Sharrocks K, Shaw R, Shea A, Shepherd A, Shepherd JG, Shiham F, Sidhom E, Silk SE, da Silva Moraes AC, Silva-Junior G, Silva-Reyes L, Silveira AD, Silveira MBV, Sinha J, Skelly DT, Smith DC, Smith N, Smith HE, Smith DJ, Smith CC, Soares A, Soares T, Solórzano C, Sorio GL, Sorley K, Sosa-Rodriguez T, Souza CMCDL, Souza BSDF, Souza AR, Spencer AJ, Spina F, Spoors L, Stafford L, Stamford I, Starinskij I, Stein R, Steven J, Stockdale L, Stockwell LV, Strickland LH, Stuart AC, Sturdy A, Sutton N, Szigeti A, Tahiri-Alaoui A, Tanner R, Taoushanis C, Tarr AW, Taylor K, Taylor U, Taylor IJ, Taylor J, te Water Naude R, Themistocleous Y, Themistocleous A, Thomas M, Thomas K, Thomas TM, Thombrayil A, Thompson F, Thompson A, Thompson K, Thompson A, Thomson J, Thornton-Jones V, Tighe PJ, Tinoco LA, Tiongson G, Tladinyane B, Tomasicchio M, Tomic A, Tonks S, Towner J, Tran N, Tree J, Trillana G, Trinham C, Trivett R, Truby A, Tsheko BL, Turabi A, Turner R, Turner C, Ulaszewska M, Underwood BR, Varughese R, Verbart D, Verheul M, Vichos I, Vieira T, Waddington CS, Walker L, Wallis E, Wand M, Warbick D, Wardell T, Warimwe G, Warren SC, Watkins B, Watson E, Webb S, Webb-Bridges A, Webster A, Welch J, Wells J, West A, White C, White R, Williams P, Williams RL, Winslow R, Woodyer M, Worth AT, Wright D, Wroblewska M, Yao A, Zimmer R, Zizi D, Zuidewind P (2021) Safety and efficacy of the ChAdOx1 nCoV-19 vaccine (AZD1222) against SARS-CoV-2: an interim analysis of four randomised controlled trials in Brazil, South Africa, and the UK. Lancet 397:99–111. https://doi.org/10.1016/S0140-6736(20)32661-1

28. Mercado NB, Zahn R, Wegmann F, Loos C, Chandrashekar A, Yu J, Liu J, Peter L, McMahan K, Tostanoski LH, He X, Martinez DR, Rutten L, Bos R, van Manen D, Vellinga J, Custers J, Langedijk JP, Kwaks T, Bakkers MJG, Zuijdgeest D, Rosendahl Huber SK, Atyeo C, Fischinger S, Burke JS, Feldman J, Hauser BM, Caradonna TM, Bondzie EA, Dagotto G, Gebre MS, Hoffman E, Jacob-Dolan C, Kirilova M, Li Z, Lin Z, Mahrokhian SH, Maxfield LF, Nampanya F, Nityanandam R, Nkolola JP, Patel S, Ventura JD, Verrington K, Wan H, Pessaint L, Van Ry A, Blade K, Strasbaugh A, Cabus M, Brown R, Cook A, Zouantchangadou S, Teow E, Andersen H, Lewis MG, Cai Y, Chen B, Schmidt AG, Reeves RK, Baric RS, Lauffenburger DA, Alter G, Stoffels P, Mammen M, Van Hoof J, Schuitemaker H, Barouch DH (2020) Single-shot Ad26 vaccine protects against SARS-CoV-2 in rhesus macaques. Nature 586:583–588. https://doi.org/10.1038/s41586-020-2607-z
29. Sadoff J, Gars ML, Shukarev G, Heerwegh D, Truyers C, de Groot AM, Stoop J, Tete S, Van Damme W, Leroux-Roels I, Berghmans P-J, Kimmel M, Van Damme P, de Hoon J, Smith W, Stephenson KE, Barouch DH, De Rosa SC, Cohen KW, McElrath MJ, Cormier E, Scheper G, Hendriks J, Struyf F, Douoguih M, Van Hoof J, Schuitemaker H (2020) Safety and immunogenicity of the Ad26.COV2.S COVID-19 vaccine candidate: interim results of a phase 1/2a, double-blind, randomized, placebo-controlled trial. medRxiv:2020.09.23.20199604. https://doi.org/10.1101/2020.09.23.20199604
30. Gao Q, Bao L, Mao H, Wang L, Xu K, Yang M, Li Y, Zhu L, Wang N, Lv Z, Gao H, Ge X, Kan B, Hu Y, Liu J, Cai F, Jiang D, Yin Y, Qin C, Li J, Gong X, Lou X, Shi W, Wu D, Zhang H, Zhu L, Deng W, Li Y, Lu J, Li C, Wang X, Yin W, Zhang Y, Qin C (2020) Development of an inactivated vaccine candidate for SARS-CoV-2. Science 369:77–81. https://doi.org/10.1126/science.abc1932
31. Zhang Y, Zeng G, Pan H, Li C, Hu Y, Chu K, Han W, Chen Z, Tang R, Yin W, Chen X, Hu Y, Liu X, Jiang C, Li J, Yang M, Song Y, Wang X, Gao Q, Zhu F (2021) Safety, tolerability, and

immunogenicity of an inactivated SARS-CoV-2 vaccine in healthy adults aged 18–59 years: a randomised, double-blind, placebo-controlled, phase 1/2 clinical trial. Lancet Infect Dis 21:181–192. https://doi.org/10.1016/S1473-3099(20)30843-4
32. Wang H, Zhang Y, Huang B, Deng W, Quan Y, Wang W, Xu W, Zhao Y, Li N, Zhang J, Liang H, Bao L, Xu Y, Ding L, Zhou W, Gao H, Liu J, Niu P, Zhao L, Zhen W, Fu H, Yu S, Zhang Z, Xu G, Li C, Lou Z, Xu M, Qin C, Wu G, Gao GF, Tan W, Yang X (2020) Development of an inactivated vaccine candidate, BBIBP-CorV, with potent protection against SARS-CoV-2. Cell 182:713–721.e9. https://doi.org/10.1016/j.cell.2020.06.008
33. Xia S, Zhang Y, Wang Y, Wang H, Yang Y, Gao GF, Tan W, Wu G, Xu M, Lou Z, Huang W, Xu W, Huang B, Wang H, Wang W, Zhang W, Li N, Xie Z, Ding L, You W, Zhao Y, Yang X, Liu Y, Wang Q, Huang L, Yang Y, Xu G, Luo B, Wang W, Liu P, Guo W, Yang X (2021) Safety and immunogenicity of an inactivated SARS-CoV-2 vaccine, BBIBP-CorV: a randomised, double-blind, placebo-controlled, phase 1/2 trial. Lancet Infect Dis 21:39–51. https://doi.org/10.1016/S1473-3099(20)30831-8
34. Yadav PD, Ella R, Kumar S, Patil DR, Mohandas S, Shete AM, Vadrevu KM, Bhati G, Sapkal G, Kaushal H, Patil S, Jain R, Deshpande G, Gupta N, Agarwal K, Gokhale M, Mathapati B, Metkari S, Mote C, Nyayanit D, Patil DY, Sai Prasad BS, Suryawanshi A, Kadam M, Kumar A, Daigude S, Gopale S, Majumdar T, Mali D, Sarkale P, Baradkar S, Gawande P, Joshi Y, Fulari S, Dighe H, Sharma S, Gunjikar R, Kumar A, Kalele K, Srinivas VK, Gangakhedkar RR, Ella KM, Abraham P, Panda S, Bhargava B (2021) Immunogenicity and protective efficacy of inactivated SARS-CoV-2 vaccine candidate, BBV152 in rhesus macaques. Nat Commun 12:1386. https://doi.org/10.1038/s41467-021-21639-w
35. Ella R, Vadrevu KM, Jogdand H, Prasad S, Reddy S, Sarangi V, Ganneru B, Sapkal G, Yadav P, Abraham P, Panda S, Gupta N, Reddy P, Verma S, Kumar Rai S, Singh C, Redkar SV, Gillurkar CS, Kushwaha JS, Mohapatra S, Rao V, Guleria R, Ella K, Bhargava B (2021) Safety and immunogenicity of an inactivated SARS-CoV-2 vaccine, BBV152: a double-blind, randomised, phase 1 trial. Lancet Infect Dis 21:637–646. https://doi.org/10.1016/S1473-3099(20)30942-7
36. Ella R, Reddy S, Blackwelder W, Potdar V, Yadav P, Sarangi V, Aileni VK, Kanungo S, Rai S, Reddy P, Verma S, Singh C, Redkar S, Mohapatra S, Pandey A, Ranganadin P, Gumashta R, Multani M, Mohammad S, Bhatt P, Kumari L, Sapkal G, Gupta N, Abraham P, Panda S, Prasad S, Bhargava B, Ella K, Vadrevu KM, Aggarwal P, Aglawe V, Ali A, Anand N, Awad N, Bafna V, Balasubramaniyam G, Bandkar A, Basha P, Bharge V, Bhate A, Bhate S, Bhavani V, Bhosale R, Chalapathy DV, Chaubal C, Chaudhary D, Chavan A, Desai P, Dhodi D, Dutta S, Garg R, Garg K, George M, Goyal P, Guleria R, Gupta S, Jain M, Jain MK, Jindal S, Kalra M, Kant S, Khosla P, Kulkarni P, Kumar P, Kumar Y, Majumdar A, Meshram P, Mishra V, Mohanty S, Nair J, Pandey S, Panigrahi SK, Patil B, Patil V, Rahate P, Raj V, Ramanand S, Rami K, Ramraj B, Rane S, Rao EV, Rao N, Raphael R, Reddy G, Redkar V, Redkar S, Sachdeva A, Saha J, Sahoo J, Sampath P, Savith A, Shah M, Shanmugam L, Sharma R, Sharma P, Sharma D, Singh A, Singh J, Singh P, Sivaprakasam S, Subramaniam S, Sudheer D, Tandon S, Tariq M, Tripathi V, Vable M, Verma R, Waghmare S (2021) Efficacy, safety, and lot-to-lot immunogenicity of an inactivated SARS-CoV-2 vaccine (BBV152): interim results of a randomised, double-blind, controlled, phase 3 trial. Lancet 398:2173–2184. https://doi.org/10.1016/S0140-6736(21)02000-6
37. Xia S, Duan K, Zhang Y, Zhao D, Zhang H, Xie Z, Li X, Peng C, Zhang Y, Zhang W, Yang Y, Chen W, Gao X, You W, Wang X, Wang Z, Shi Z, Wang Y, Yang X, Zhang L, Huang L, Wang Q, Lu J, Yang Y, Guo J, Zhou W, Wan X, Wu C, Wang W, Huang S, Du J, Meng Z, Pan A, Yuan Z, Shen S, Guo W, Yang X (2020) Effect of an inactivated vaccine against SARS-CoV-2 on safety and immunogenicity outcomes: interim analysis of 2 randomized clinical trials. JAMA 324:951–960. https://doi.org/10.1001/jama.2020.15543
38. Al Kaabi N, Zhang Y, Xia S, Yang Y, Al Qahtani MM, Abdulrazzaq N, Al Nusair M, Hassany M, Jawad JS, Abdalla J, Hussein SE, Al Mazrouei SK, Al Karam M, Li X, Yang X, Wang W, Lai B, Chen W, Huang S, Wang Q, Yang T, Liu Y, Ma R, Hussain ZM, Khan T, Saifuddin

Fasihuddin M, You W, Xie Z, Zhao Y, Jiang Z, Zhao G, Zhang Y, Mahmoud S, ElTantawy I, Xiao P, Koshy A, Zaher WA, Wang H, Duan K, Pan A, Yang X (2021) Effect of 2 inactivated SARS-CoV-2 vaccines on symptomatic COVID-19 infection in adults: a randomized clinical trial. JAMA 326:35–45. https://doi.org/10.1001/jama.2021.8565

39. Tian J-H, Patel N, Haupt R, Zhou H, Weston S, Hammond H, Logue J, Portnoff AD, Norton J, Guebre-Xabier M, Zhou B, Jacobson K, Maciejewski S, Khatoon R, Wisniewska M, Moffitt W, Kluepfel-Stahl S, Ekechukwu B, Papin J, Boddapati S, Jason Wong C, Piedra PA, Frieman MB, Massare MJ, Fries L, Bengtsson KL, Stertman L, Ellingsworth L, Glenn G, Smith G (2021) SARS-CoV-2 spike glycoprotein vaccine candidate NVX-CoV2373 immunogenicity in baboons and protection in mice. Nat Commun 12:372. https://doi.org/10.1038/s41467-020-20653-8
40. Keech C, Albert G, Cho I, Robertson A, Reed P, Neal S, Plested JS, Zhu M, Cloney-Clark S, Zhou H, Smith G, Patel N, Frieman MB, Haupt RE, Logue J, McGrath M, Weston S, Piedra PA, Desai C, Callahan K, Lewis M, Price-Abbott P, Formica N, Shinde V, Fries L, Lickliter JD, Griffin P, Wilkinson B, Glenn GM (2020) Phase 1–2 trial of a SARS-CoV-2 recombinant spike protein nanoparticle vaccine. N Engl J Med 383:2320–2332. https://doi.org/10.1056/NEJMoa2026920
41. Dunkle LM, Kotloff KL, Gay CL, Áñez G, Adelglass JM, Barrat Hernández AQ, Harper WL, Duncanson DM, McArthur MA, Florescu DF, McClelland RS, Garcia-Fragoso V, Riesenberg RA, Musante DB, Fried DL, Safirstein BE, McKenzie M, Jeanfreau RJ, Kingsley JK, Henderson JA, Lane DC, Ruíz-Palacios GM, Corey L, Neuzil KM, Coombs RW, Greninger AL, Hutter J, Ake JA, Smith K, Woo W, Cho I, Glenn GM, Dubovsky F (2021) Efficacy and safety of NVX-CoV2373 in adults in the United States and Mexico. N Engl J Med 386:531–543. https://doi.org/10.1056/NEJMoa2116185
42. Ward BJ, Gobeil P, Séguin A, Atkins J, Boulay I, Charbonneau P-Y, Couture M, D'Aoust M-A, Dhaliwall J, Finkle C, Hager K, Mahmood A, Makarkov A, Cheng MP, Pillet S, Schimke P, St-Martin S, Trépanier S, Landry N (2021) Phase 1 randomized trial of a plant-derived virus-like particle vaccine for COVID-19. Nat Med 27:1071–1078. https://doi.org/10.1038/s41591-021-01370-1
43. Hager KJ, Pérez Marc G, Gobeil P, Diaz RS, Heizer G, Llapur C, Makarkov AI, Vasconcellos E, Pillet S, Riera F, Saxena P, Geller Wolff P, Bhutada K, Wallace G, Aazami H, Jones CE, Polack FP, Ferrara L, Atkins J, Boulay I, Dhaliwall J, Charland N, Couture MMJ, Jiang-Wright J, Landry N, Lapointe S, Lorin A, Mahmood A, Moulton LH, Pahmer E, Parent J, Séguin A, Tran L, Breuer T, Ceregido M-A, Koutsoukos M, Roman F, Namba J, D'Aoust M-A, Trepanier S, Kimura Y, Ward BJ (2022) Efficacy and safety of a recombinant plant-based adjuvanted COVID-19 vaccine. N Engl J Med 386:2084–2096. https://doi.org/10.1056/NEJMoa2201300
44. Pillet S, Arunachalam PS, Andreani G, Golden N, Fontenot J, Aye PP, Röltgen K, Lehmicke G, Gobeil P, Dubé C, Trépanier S, Charland N, D'Aoust M-A, Russell-Lodrigue K, Monjure C, Blair RV, Boyd SD, Bohm RP, Rappaport J, Villinger F, Landry N, Pulendran B, Ward BJ (2022) Safety, immunogenicity, and protection provided by unadjuvanted and adjuvanted formulations of a recombinant plant-derived virus-like particle vaccine candidate for COVID-19 in nonhuman primates. Cell Mol Immunol 19:222–233. https://doi.org/10.1038/s41423-021-00809-2
45. Ryzhikov AB, Ryzhikov EA, Bogryantseva MP, Danilenko ED, Imatdinov IR, Nechaeva EA, Pyankov OV, Pyankova OG, Susloparov IM, Taranov OS, Gudymo AS, Danilchenko NV, Sleptsova ES, Bodnev SA, Onkhonova GS, Petrov VN, Moiseeva AA, Torzhkova PY, Pyankov SA, Tregubchak TV, Antonets DV, Gavrilova EV, Maksyutov RA (2021) Immunogenicity and protectivity of the peptide vaccine against SARS-CoV-2. Ann Rus Acad Med Sci 76:5–19. https://doi.org/10.15690/vramn1528
46. Ryzhikov AB, Ryzhikov EA, Bogryantseva MP, Usova SV, Danilenko ED, Nechaeva EA, Pyankov OV, Pyankova OG, Gudymo AS, Bodnev SA, Onkhonova GS, Sleptsova ES, Kuzubov VI, Ryndyuk NN, Ginko ZI, Petrov VN, Moiseeva AA, Torzhkova PY, Pyankov

SA, Tregubchak TV, Antonec DV, Gavrilova EV, Maksyutov RA (2021) A single blind, placebo-controlled randomized study of the safety, reactogenicity and immunogenicity of the "EpiVacCorona" vaccine for the prevention of COVID-19, in volunteers aged 18–60 years (phase I–II). Rus J Infect Immun 11:283–296. https://doi.org/10.15789/2220-7619-asb-1699

47. Pious N, Ingole SD (2020) Race for COVID-19 vaccine. Trends Biomater Artif Organs 34:62–65
48. Thorn CR, Sharma D, Combs R, Bhujbal S, Romine J, Zheng X, Sunasara K, Badkar A (2022) The journey of a lifetime—development of Pfizer's COVID-19 vaccine. Curr Opin Biotechnol:102803. https://doi.org/10.1016/j.copbio.2022.102803
49. Kremsner PG, Ahuad Guerrero RA, Arana-Arri E, Aroca Martinez GJ, Bonten M, Chandler R, Corral G, De Block EJL, Ecker L, Gabor JJ, Garcia Lopez CA, Gonzales L, Granados González MA, Gorini N, Grobusch MP, Hrabar AD, Junker H, Kimura A, Lanata CF, Lehmann C, Leroux-Roels I, Mann P, Martinez-Reséndez MF, Ochoa TJ, Poy CA, Reyes Fentanes MJ, Rivera Mejia LM, Ruiz Herrera VV, Sáez-Llorens X, Schönborn-Kellenberger O, Schunk M, Sierra Garcia A, Vergara I, Verstraeten T, Vico M, Oostvogels L, Lovesio L, Diez F, Grazziani F, Ganaha MC, Zalatnik VJ, Dittrich RJ, Espínola L, Lambert S, Longhi A, Vecchio C, Mastruzzo M, Fernandez A, Borchowiek S, Potito R, Ahuad Guerrero RA, Guardiani FM, Castella S, Foccoli M, Pedernera A, Braida A, Durigan V, Martella C, Bobat A, Boggia BE, Nemi SA, Tartaglione JG, Piedimonte FC, De Bie J, Reynales Londoño H, Rodríguez Ordoñez PA, García Cruz JM, Bautista Toloza L, Ladino González MC, Zambrano Ochoa AP, Prieto Pradera I, Torres Hernandez D, Mazo Elorza DP, Collazos Lennis MF, Vanegas Dominguez B, Solano Mosquera LM, Fendel R, Fleischmann WA, Koehne E, Kreidenweiss A, Köhler C, Esen M, Horn C, Eberts S, Kroidl A, Huber K, Thiel V, Mazara Rosario S, Reyes G, Rivera L, Donastorg Y, Lantigua F, Torres Almanzar D, Candelario R, Peña Mendez L, Rosario Gomez N, Portolés-Pérez A, Ascaso del Río A, Laredo Velasco L, Bustinduy Odriozola MJ, Larrea Arranz I, Martínez Alcorta LI, Durán Laviña MI, Imaz-Ayo N, Meijide S, García-de-Vicuña A, Santorcuato A, Gallego M, Aguirre-García GM, Olmos Vega J, González Limón P, Vázquez Villar A, Chávez Barón J, Arredondo Saldaña F, Luján Palacios JD, Camacho Choza LJ, Vázquez Saldaña EG, Ortega Dominguez SJ, Vega Orozco KS, Torres Quiroz IA, Martinez Avendaño A, Herrera Sanchez J, Guzman E, Castro Castrezana L, Ruiz Palacios y Santos GM, de Winter RFJ, de Jonge HK, Schnyder JL, Boersma W, Hessels L, Djamin R, van der Sar S, DeAntonio R, Peña M, Rebollon G, Rojas M, Escobar J, Hammerschlag Icaza B, Wong TDY, Barrera Perigault P, Ruiz S, Chan M, Arias Hoo DJ, Gil AI, Celis CR, Balmaceda MP, Flores O, Ochoa M, Peña B, de la Flor C, Webb CM, Cornejo E, Sanes F, Mayorga V, Valdiviezo G, Ramírez Lamas SP, Grandez Castillo GA, Lama JR, Matta Aguirre ME, Arancibia Luna LA, Carbajal Paulet Ó, Zambrano Ortiz J, Camara A, Guzman Quintanilla F, Diaz-Parra C, Morales-Oliva J, Cornejo RE, Ricalde SA, Vidal J, Rios Nogales L, Cheatham-Seitz D, Gregoraci G, Brecx A, Walz L, Vahrenhorst D, Seibel T, Quintini G (2022) Efficacy and safety of the CVnCoV SARS-CoV-2 mRNA vaccine candidate in ten countries in Europe and Latin America (HERALD): a randomised, observer-blinded, placebo-controlled, phase 2b/3 trial. Lancet Infect Dis 22:329–340. https://doi.org/10.1016/S1473-3099(21)00677-0
50. Gebre MS, Rauch S, Roth N, Yu J, Chandrashekar A, Mercado NB, He X, Liu J, McMahan K, Martinot A, Martinez DR, Giffin V, Hope D, Patel S, Sellers D, Sanborn O, Barrett J, Liu X, Cole AC, Pessaint L, Valentin D, Flinchbaugh Z, Yalley-Ogunro J, Muench J, Brown R, Cook A, Teow E, Andersen H, Lewis MG, Boon ACM, Baric RS, Mueller SO, Petsch B, Barouch DH (2021) Optimization of non-coding regions for a non-modified mRNA COVID-19 vaccine. Nature. https://doi.org/10.1038/s41586-021-04231-6
51. Andries O, Mc Cafferty S, De Smedt SC, Weiss R, Sanders NN, Kitada T (2015) N(1)-methylpseudouridine-incorporated mRNA outperforms pseudouridine-incorporated mRNA by providing enhanced protein expression and reduced immunogenicity in mammalian cell lines and mice. J Control Release 217:337–344. https://doi.org/10.1016/j.jconrel.2015.08.051

52. Svitkin YV, Cheng YM, Chakraborty T, Presnyak V, John M, Sonenberg N (2017) N1-methyl-pseudouridine in mRNA enhances translation through eIF2α-dependent and independent mechanisms by increasing ribosome density. Nucleic Acids Res 45:6023–6036. https://doi.org/10.1093/nar/gkx135
53. Anderson BR, Muramatsu H, Jha BK, Silverman RH, Weissman D, Karikó K (2011) Nucleoside modifications in RNA limit activation of 2′-5′-oligoadenylate synthetase and increase resistance to cleavage by RNase L. Nucleic Acids Res 39:9329–9338. https://doi.org/10.1093/nar/gkr586
54. Bloom K, van den Berg F, Arbuthnot P (2021) Self-amplifying RNA vaccines for infectious diseases. Gene Ther 28:117–129. https://doi.org/10.1038/s41434-020-00204-y
55. de Alwis R, Gan ES, Chen S, Leong YS, Tan HC, Zhang SL, Yau C, Low JGH, Kalimuddin S, Matsuda D, Allen EC, Hartman P, Park K-JJ, Alayyoubi M, Bhaskaran H, Dukanovic A, Bao Y, Clemente B, Vega J, Roberts S, Gonzalez JA, Sablad M, Yelin R, Taylor W, Tachikawa K, Parker S, Karmali P, Davis J, Sullivan BM, Sullivan SM, Hughes SG, Chivukula P, Ooi EE (2021) A single dose of self-transcribing and replicating RNA-based SARS-CoV-2 vaccine produces protective adaptive immunity in mice. Mol Ther 29:1970–1983. https://doi.org/10.1016/j.ymthe.2021.04.001
56. Pardi N, Muramatsu H, Weissman D, Karikó K (2013) In vitro transcription of long RNA containing modified nucleosides. In: Rabinovich PM (ed) Synthetic messenger RNA and cell metabolism modulation: methods and protocols. Humana Press, Totowa, pp 29–42
57. Beckert B, Masquida B (2011) Synthesis of RNA by in vitro transcription. In: Nielsen H (ed) RNA: methods and protocols. Humana Press, Totowa, pp 29–41
58. Vaidyanathan S, Azizian KT, Haque AKMA, Henderson JM, Hendel A, Shore S, Antony JS, Hogrefe RI, Kormann MSD, Porteus MH, McCaffrey AP (2018) Uridine depletion and chemical modification increase Cas9 mRNA activity and reduce immunogenicity without HPLC purification. Mol Ther Nucleic Acids 12:530–542. https://doi.org/10.1016/j.omtn.2018.06.010
59. Martin SA, Paoletti E, Moss B (1975) Purification of mRNA guanylyltransferase and mRNA (guanine-7-) methyltransferase from vaccinia virions. J Biol Chem 250:9322–9329. https://doi.org/10.1016/S0021-9258(19)40646-7
60. Shuman S (1990) Catalytic activity of vaccinia mRNA capping enzyme subunits coexpressed in Escherichia coli. J Biol Chem 265:11960–11966. https://doi.org/10.1016/S0021-9258(19)38494-7
61. Bélanger F, Stepinski J, Darzynkiewicz E, Pelletier J (2010) Characterization of hMTr1, a human Cap1 2′-O-ribose methyltransferase. J Biol Chem 285:33037–33044. https://doi.org/10.1074/jbc.M110.155283
62. Pourseif MM, Masoudi-Sobhanzadeh Y, Azari E, Parvizpour S, Barar J, Ansari R, Omidi Y (2022) Self-amplifying mRNA vaccines: mode of action, design, development and optimization. Drug Discov Today 27:103341. https://doi.org/10.1016/j.drudis.2022.103341
63. Goey CH, Alhuthali S, Kontoravdi C (2018) Host cell protein removal from biopharmaceutical preparations: towards the implementation of quality by design. Biotechnol Adv 36:1223–1237. https://doi.org/10.1016/j.biotechadv.2018.03.021
64. Liu MA (2019) A comparison of plasmid DNA and mRNA as vaccine technologies. Vaccine 7:37
65. Henderson JM, Ujita A, Hill E, Yousif-Rosales S, Smith C, Ko N, McReynolds T, Cabral CR, Escamilla-Powers JR, Houston ME (2021) Cap 1 messenger RNA synthesis with co-transcriptional CleanCap® analog by in vitro transcription. Curr Protoc 1:e39. https://doi.org/10.1002/cpz1.39
66. Rosa SS, Prazeres DMF, Azevedo AM, Marques MPC (2021) mRNA vaccines manufacturing: challenges and bottlenecks. Vaccine 39:2190–2200. https://doi.org/10.1016/j.vaccine.2021.03.038

67. Rio DC, Ares M, Hannon GJ, Nilsen TW (2010) Enrichment of poly(A)+ mRNA using immobilized oligo (dT). Cold Spring Harb Protoc 2010:pdb.prot5454. https://doi.org/10.1101/pdb.prot5454
68. Shen CF, Jacob D, Zhu T, Bernier A, Shao Z, Yu X, Patel M, Lanthier S, Kamen A (2016) Optimization and scale-up of cell culture and purification processes for production of an adenovirus-vectored tuberculosis vaccine candidate. Vaccine 34:3381–3387. https://doi.org/10.1016/j.vaccine.2016.04.090
69. Abbink P, Lemckert AAC, Ewald BA, Lynch DM, Denholtz M, Smits S, Holterman L, Damen I, Vogels R, Thorner AR, O'Brien KL, Carville A, Mansfield KG, Goudsmit J, Havenga MJE, Barouch DH (2007) Comparative seroprevalence and immunogenicity of six rare serotype recombinant adenovirus vaccine vectors from subgroups B and D. J Virol 81:4654–4663. https://doi.org/10.1128/JVI.02696-06
70. Afolabi MO, Tiono AB, Adetifa UJ, Yaro JB, Drammeh A, Nébié I, Bliss C, Hodgson SH, Anagnostou NA, Sanou GS, Jagne YJ, Ouedraogo O, Tamara C, Ouedraogo N, Ouedraogo M, Njie-Jobe J, Diarra A, Duncan CJA, Cortese R, Nicosia A, Roberts R, Viebig NK, Leroy O, Lawrie AM, Flanagan KL, Kampman B, Bejon P, Imoukhuede EB, Ewer KJ, Hill AVS, Bojang K, Sirima SB (2016) Safety and immunogenicity of ChAd63 and MVA ME-TRAP in West African children and infants. Mol Ther 24:1470–1477. https://doi.org/10.1038/mt.2016.83
71. Custers J, Kim D, Leyssen M, Gurwith M, Tomaka F, Robertson J, Heijnen E, Condit R, Shukarev G, Heerwegh D, van Heesbeen R, Schuitemaker H, Douoguih M, Evans E, Smith ER, Chen RT (2021) Vaccines based on replication incompetent Ad26 viral vectors: standardized template with key considerations for a risk/benefit assessment. Vaccine 39:3081–3101. https://doi.org/10.1016/j.vaccine.2020.09.018
72. Folegatti PM, Ewer KJ, Aley PK, Angus B, Becker S, Belij-Rammerstorfer S, Bellamy D, Bibi S, Bittaye M, Clutterbuck EA, Dold C, Faust SN, Finn A, Flaxman AL, Hallis B, Heath P, Jenkin D, Lazarus R, Makinson R, Minassian AM, Pollock KM, Ramasamy M, Robinson H, Snape M, Tarrant R, Voysey M, Green C, Douglas AD, Hill AVS, Lambe T, Gilbert SC, Pollard AJ, Aboagye J, Adams K, Ali A, Allen E, Allison JL, Anslow R, Arbe-Barnes EH, Babbage G, Baillie K, Baker M, Baker N, Baker P, Baleanu I, Ballaminut J, Barnes E, Barrett J, Bates L, Batten A, Beadon K, Beckley R, Berrie E, Berry L, Beveridge A, Bewley KR, Bijker EM, Bingham T, Blackwell L, Blundell CL, Bolam E, Boland E, Borthwick N, Bower T, Boyd A, Brenner T, Bright PD, Brown-O'Sullivan C, Brunt E, Burbage J, Burge S, Buttigieg KR, Byard N, Cabera Puig I, Calvert A, Camara S, Cao M, Cappuccini F, Carr M, Carroll MW, Carter V, Cathie K, Challis RJ, Charlton S, Chelysheva I, Cho J-S, Cicconi P, Cifuentes L, Clark H, Clark E, Cole T, Colin-Jones R, Conlon CP, Cook A, Coombes NS, Cooper R, Cosgrove CA, Coy K, Crocker WEM, Cunningham CJ, Damratoski BE, Dando L, Datoo MS, Davies H, De Graaf H, Demissie T, Di Maso C, Dietrich I, Dong T, Donnellan FR, Douglas N, Downing C, Drake J, Drake-Brockman R, Drury RE, Dunachie SJ, Edwards NJ, Edwards FDL, Edwards CJ, Elias SC, Elmore MJ, Emary KRW, English MR, Fagerbrink S, Felle S, Feng S, Field S, Fixmer C, Fletcher C, Ford KJ, Fowler J, Fox P, Francis E, Frater J, Furze J, Fuskova M, Galiza E, Gbesemete D, Gilbride C, Godwin K, Gorini G, Goulston L, Grabau C, Gracie L, Gray Z, Guthrie LB, Hackett M, Halwe S, Hamilton E, Hamlyn J, Hanumunthadu B, Harding I, Harris SA, Harris A, Harrison D, Harrison C, Hart TC, Haskell L, Hawkins S, Head I, Henry JA, Hill J, Hodgson SHC, Hou MM, Howe E, Howell N, Hutlin C, Ikram S, Isitt C, Iveson P, Jackson S, Jackson F, James SW, Jenkins M, Jones E, Jones K, Jones CE, Jones B, Kailath R, Karampatsas K, Keen J, Kelly S, Kelly D, Kerr D, Kerridge S, Khan L, Khan U, Killen A, Kinch J, King TB, King L, King J, Kingham-Page L, Klenerman P, Knapper F, Knight JC, Knott D, Koleva S, Kupke A, Larkworthy CW, Larwood JPJ, Laskey A, Lawrie AM, Lee A, Ngan Lee KY, Lees EA, Legge H, Lelliott A, Lemm N-M, Lias AM, Linder A, Lipworth S, Liu X, Liu S, Lopez Ramon R, Lwin M, Mabesa F, Madhavan M, Mallett G, Mansatta K, Marcal I, Marinou S, Marlow E, Marshall JL, Martin J, McEwan J, McInroy L, Meddaugh G, Mentzer AJ, Mirtorabi N, Moore M, Moran E, Morey E, Morgan

V, Morris SJ, Morrison H, Morshead G, Morter R, Mujadidi YF, Muller J, Munera-Huertas T, Munro C, Munro A, Murphy S, Munster VJ, Mweu P, Noé A, Nugent FL, Nuthall E, O'Brien K, O'Connor D, Oguti B, Oliver JL, Oliveira C, O'Reilly PJ, Osborn M, Osborne P, Owen C, Owens D, Owino N, Pacurar M, Parker K, Parracho H, Patrick-Smith M, Payne V, Pearce J, Peng Y, Peralta Alvarez MP, Perring J, Pfafferott K, Pipini D, Plested E, Pluess-Hall H, Pollock K, Poulton I, Presland L, Provstgaard-Morys S, Pulido D, Radia K, Ramos Lopez F, Rand J, Ratcliffe H, Rawlinson T, Rhead S, Riddell A, Ritchie AJ, Roberts H, Robson J, Roche S, Rohde C, Rollier CS, Romani R, Rudiansyah I, Saich S, Sajjad S, Salvador S, Sanchez Riera L, Sanders H, Sanders K, Sapaun S, Sayce C, Schofield E, Screaton G, Selby B, Semple C, Sharpe HR, Shaik I, Shea A, Shelton H, Silk S, Silva-Reyes L, Skelly DT, Smee H, Smith CC, Smith DJ, Song R, Spencer AJ, Stafford E, Steele A, Stefanova E, Stockdale L, Szigeti A, Tahiri-Alaoui A, Tait M, Talbot H, Tanner R, Taylor IJ, Taylor V, Te Water Naude R, Thakur N, Themistocleous Y, Themistocleous A, Thomas M, Thomas TM, Thompson A, Thomson-Hill S, Tomlins J, Tonks S, Towner J, Tran N, Tree JA, Truby A, Turkentine K, Turner C, Turner N, Turner S, Tuthill T, Ulaszewska M, Varughese R, Van Doremalen N, Veighey K, Verheul MK, Vichos I, Vitale E, Walker L, Watson MEE, Welham B, Wheat J, White C, White R, Worth AT, Wright D, Wright S, Yao XL, Yau Y (2020) Safety and immunogenicity of the ChAdOx1 nCoV-19 vaccine against SARS-CoV-2: a preliminary report of a phase 1/2, single-blind, randomised controlled trial. Lancet 396:467–478. https://doi.org/10.1016/S0140-6736(20)31604-4

73. Shabram P, Vellekamp G, Scandella C (2002) Chapter 7 – Purification of adenovirus. In: Curiel DT, Douglas JT (eds) Adenoviral vectors for gene therapy. Academic Press, San Diego, pp 167–203
74. Fallaux FJ, Bout A, van der Velde I, van den Wollenberg DJM, Hehir KM, Keegan J, Auger C, Cramer SJ, van Ormondt H, van der Eb AJ, Valerio D, Hoeben RC (1998) New helper cells and matched early region 1-deleted adenovirus vectors prevent generation of replication-competent adenoviruses. Hum Gene Ther 9:1909–1917. https://doi.org/10.1089/hum.1998.9.13-1909
75. Graham FL, Smiley J, Russell WC, Nairn R (1977) Characteristics of a human cell line transformed by DNA from human adenovirus type 5. J Gen Virol 36:59–72. https://doi.org/10.1099/0022-1317-36-1-59
76. Joe CCD, Jiang J, Linke T, Li Y, Fedosyuk S, Gupta G, Berg A, Segireddy RR, Mainwaring D, Joshi A, Cashen P, Rees B, Chopra N, Nestola P, Humphreys J, Davies S, Smith N, Bruce S, Verbart D, Bormans D, Knevelman C, Woodyer M, Davies L, Cooper L, Kapanidou M, Bleckwenn N, Pappas D, Lambe T, Smith DC, Green CM, Venkat R, Ritchie AJ, Gilbert SC, Turner R, Douglas AD (2022) Manufacturing a chimpanzee adenovirus-vectored SARS-CoV-2 vaccine to meet global needs. Biotechnol Bioeng 119:48–58. https://doi.org/10.1002/bit.27945
77. Schwarz H, Zhang Y, Zhan C, Malm M, Field R, Turner R, Sellick C, Varley P, Rockberg J, Chotteau V (2020) Small-scale bioreactor supports high density HEK293 cell perfusion culture for the production of recombinant Erythropoietin. J Biotechnol 309:44–52. https://doi.org/10.1016/j.jbiotec.2019.12.017
78. Kamen A, Henry O (2004) Development and optimization of an adenovirus production process. J Gene Med 6:S184–S192. https://doi.org/10.1002/jgm.503
79. Vellinga J, Smith JP, Lipiec A, Majhen D, Lemckert A, van Ooij M, Ives P, Yallop C, Custers J, Havenga M (2014) Challenges in manufacturing adenoviral vectors for global vaccine product deployment. Hum Gene Ther 25:318–327. https://doi.org/10.1089/hum.2014.007
80. Luitgens A, Lewis JA (2018) Method for the production of adenoviral vaccines. In: Office USP (ed) Janssen Vaccines and Prevention B.V., United States
81. Moleirinho GM, Rosa S, Carrondo JTM, Silva JSR, Hagner-McWhirter Å, Ahlén G, Lundgren M, Alves MP, Peixoto C (2018) Clinical-grade oncolytic adenovirus purification using polysorbate 20 as an alternative for cell lysis. Curr Gene Ther 18:366–374. https://doi.org/10.2174/1566523218666181109141257

82. Klyushnichenko V, Bernier A, Kamen A, Harmsen E (2001) Improved high-performance liquid chromatographic method in the analysis of adenovirus particles. J Chromatogr B Biomed Sci Appl 755:27–36. https://doi.org/10.1016/S0378-4347(00)00597-1
83. Riske F, Berard N, Albee K, Pan P, Henderson M, Adams K, Godwin S, Spear S (2013) Development of a platform process for adenovirus purification that removes human set and nucleolin and provides high purity vector for gene delivery. Biotechnol Bioeng 110:848–856. https://doi.org/10.1002/bit.24742
84. Altaras NE, Aunins JG, Evans RK, Kamen A, Konz JO, Wolf JJ (2005) Production and formulation of adenovirus vectors. In: Gene therapy and gene delivery systems. Springer Berlin Heidelberg, Berlin, Heidelberg, pp 193–260
85. Zimmer C, Corum J, Wee S, Kristoffersen M (2022) Coronavirus Vaccine Tracker. The New York Times. https://www.nytimes.com/interactive/2020/science/coronavirusvaccine-tracker.html
86. Wang Y, Yang C, Song Y, Coleman JR, Stawowczyk M, Tafrova J, Tasker S, Boltz D, Baker R, Garcia L, Seale O, Kushnir A, Wimmer E, Mueller S (2021) Scalable live-attenuated SARS-CoV-2 vaccine candidate demonstrates preclinical safety and efficacy. Proc Natl Acad Sci 118:e2102775118. https://doi.org/10.1073/pnas.2102775118
87. Wang Z-J, Zhang H-J, Lu J, Xu K-W, Peng C, Guo J, Gao X-X, Wan X, Wang W-H, Shan C, Zhang S-C, Wu J, Yang A-N, Zhu Y, Xiao A, Zhang L, Fu L, Si H-R, Cai Q, Yang X-L, You L, Zhou Y-P, Liu J, Pang D-Q, Jin W-P, Zhang X-Y, Meng S-L, Sun Y-X, Desselberger U, Wang J-Z, Li X-G, Duan K, Li C-G, Xu M, Shi Z-L, Yuan Z-M, Yang X-M, Shen S (2020) Low toxicity and high immunogenicity of an inactivated vaccine candidate against COVID-19 in different animal models. Emerg Microbes Infect 9:2606–2618. https://doi.org/10.1080/22221751.2020.1852059
88. Barrett PN, Mundt W, Kistner O, Howard MK (2009) Vero cell platform in vaccine production: moving towards cell culture-based viral vaccines. Expert Rev Vaccines 8:607–618. https://doi.org/10.1586/erv.09.19
89. Sène M-A, Xia Y, Kamen AA (2022) Overview of recent advances in Vero cells genomic characterization and engineering for high-throughput vaccine manufacturing. Clin Transl Discov 2:e40. https://doi.org/10.1002/ctd2.40
90. Kiesslich S, Kamen AA (2020) Vero cell upstream bioprocess development for the production of viral vectors and vaccines. Biotechnol Adv 44:107608. https://doi.org/10.1016/j.biotechadv.2020.107608
91. Kiesslich S, Kim GN, Shen CF, Kang CY, Kamen AA (2021) Bioreactor production of rVSV-based vectors in Vero cell suspension cultures. Biotechnol Bioeng 118:2649–2659. https://doi.org/10.1002/bit.27785
92. LoGrippo GA, Hartman FW (1955) Antigenicity of β-propiolactone-inactivated virus vaccines. J Immunol 75:123–128
93. Uittenbogaard JP, Zomer B, Hoogerhout P, Metz B (2011) Reactions of β-propiolactone with nucleobase analogues, nucleosides, and peptides: implications for the inactivation of viruses. J Biol Chem 286:36198–36214. https://doi.org/10.1074/jbc.M111.279232
94. Hemminki K (1981) Reactions of β-propiolactone, β-butyrolactone and γ-butyrolactone with nucleic acids. Chem Biol Interact 34:323–331. https://doi.org/10.1016/0009-2797(81)90104-6
95. Patterson EI, Prince T, Anderson ER, Casas-Sanchez A, Smith SL, Cansado-Utrilla C, Solomon T, Griffiths MJ, Acosta-Serrano Á, Turtle L, Hughes GL (2020) Methods of inactivation of SARS-CoV-2 for downstream biological assays. J Infect Dis 222:1462–1467. https://doi.org/10.1093/infdis/jiaa507
96. Pollet J, Chen W-H, Strych U (2021) Recombinant protein vaccines, a proven approach against coronavirus pandemics. Adv Drug Deliv Rev 170:71–82. https://doi.org/10.1016/j.addr.2021.01.001
97. Sunasara K, Cundy J, Srinivasan S, Evans B, Sun W, Cook S, Bortell E, Farley J, Griffin D, Bailey Piatchek M, Arch-Douglas K (2018) Bivalent rLP2086 (Trumenba®): development of

a well-characterized vaccine through commercialization. Vaccine 36:3180–3189. https://doi.org/10.1016/j.vaccine.2017.03.100
98. Comstedt P, Schüler W, Meinke A, Lundberg U (2017) The novel Lyme borreliosis vaccine VLA15 shows broad protection against Borrelia species expressing six different OspA serotypes. PLoS One 12:e0184357. https://doi.org/10.1371/journal.pone.0184357
99. Phillips A, Patel C, Pillsbury A, Brotherton J, Macartney K (2018) Safety of human papillomavirus vaccines: an updated review. Drug Saf 41:329–346. https://doi.org/10.1007/s40264-017-0625-z
100. Gräslund S, Nordlund P, Weigelt J, Hallberg BM, Bray J, Gileadi O, Knapp S, Oppermann U, Arrowsmith C, Hui R, Ming J, dhe-Paganon S, Park H-W, Savchenko A, Yee A, Edwards A, Vincentelli R, Cambillau C, Kim R, Kim S-H, Rao Z, Shi Y, Terwilliger TC, Kim C-Y, Hung L-W, Waldo GS, Peleg Y, Albeck S, Unger T, Dym O, Prilusky J, Sussman JL, Stevens RC, Lesley SA, Wilson IA, Joachimiak A, Collart F, Dementieva I, Donnelly MI, Eschenfeldt WH, Kim Y, Stols L, Wu R, Zhou M, Burley SK, Emtage JS, Sauder JM, Thompson D, Bain K, Luz J, Gheyi T, Zhang F, Atwell S, Almo SC, Bonanno JB, Fiser A, Swaminathan S, Studier FW, Chance MR, Sali A, Acton TB, Xiao R, Zhao L, Ma LC, Hunt JF, Tong L, Cunningham K, Inouye M, Anderson S, Janjua H, Shastry R, Ho CK, Wang D, Wang H, Jiang M, Montelione GT, Stuart DI, Owens RJ, Daenke S, Schütz A, Heinemann U, Yokoyama S, Büssow K, Gunsalus KC, Structural Genomics C, Architecture et Fonction des Macromolécules B, Berkeley Structural Genomics C, China Structural Genomics C, Integrated Center for S, Function I, Israel Structural Proteomics C, Joint Center for Structural G, Midwest Center for Structural G, New York Structural Genomi XRCfSG, Northeast Structural Genomics C, Oxford Protein Production F, Protein Sample Production Facility MDCfMM, Initiative RSGP and Complexes S (2008) Protein production and purification. Nat Methods 5:135–146. https://doi.org/10.1038/nmeth.f.202
101. Tripathi NK, Shrivastava A (2019) Recent developments in bioprocessing of recombinant proteins: expression hosts and process development. Front Bioeng Biotechnol 7. https://doi.org/10.3389/fbioe.2019.00420
102. Walls AC, Park Y-J, Tortorici MA, Wall A, McGuire AT, Veesler D (2020) Structure, function, and antigenicity of the SARS-CoV-2 spike glycoprotein. Cell 181:281–292.e6. https://doi.org/10.1016/j.cell.2020.02.058
103. Wrapp D, Wang N, Corbett KS, Goldsmith JA, Hsieh C-L, Abiona O, Graham BS, McLellan JS (2020) Cryo-EM structure of the 2019-nCoV spike in the prefusion conformation. Science (New York, NY) 367:1260–1263. https://doi.org/10.1126/science.abb2507
104. Wang Q, Zhang L, Kuwahara K, Li L, Liu Z, Li T, Zhu H, Liu J, Xu Y, Xie J, Morioka H, Sakaguchi N, Qin C, Liu G (2016) Immunodominant SARS coronavirus epitopes in humans elicited both enhancing and neutralizing effects on infection in non-human primates. ACS Infect Dis 2:361–376. https://doi.org/10.1021/acsinfecdis.6b00006
105. D'Aoust M-A, Couture MM-J, Charland N, Trépanier S, Landry N, Ors F, Vézina L-P (2010) The production of hemagglutinin-based virus-like particles in plants: a rapid, efficient and safe response to pandemic influenza. Plant Biotechnol J 8:607–619. https://doi.org/10.1111/j.1467-7652.2009.00496.x
106. Amblard M, Fehrentz J-A, Martinez J, Subra G (2006) Methods and protocols of modern solid phase peptide synthesis. Mol Biotechnol 33:239–254. https://doi.org/10.1385/MB:33:3:239
107. Palomo JM (2014) Solid-phase peptide synthesis: an overview focused on the preparation of biologically relevant peptides. RSC Adv 4:32658–32672. https://doi.org/10.1039/C4RA02458C
108. Valdes-Balbin Y, Santana-Mederos D, Quintero L, Fernández S, Rodriguez L, Sanchez Ramirez B, Perez-Nicado R, Acosta C, Méndez Y, Ricardo MG, Hernandez T, Bergado G, Pi F, Valdes A, Carmenate T, Ramirez U, Oliva R, Soubal J-P, Garrido R, Cardoso F, Landys M, Gonzalez H, Farinas M, Enriquez J, Noa E, Suarez A, Fang C, Espinosa LA, Ramos Y, González LJ, Climent Y, Rojas G, Relova-Hernández E, Cabrera Infante Y, Losada SL, Boggiano T, Ojito E, León K, Chiodo F, Paquet F, Chen G-W, Rivera DG, Garcia-Rivera D,

Verez Bencomo V (2021) SARS-CoV-2 RBD-tetanus toxoid conjugate vaccine induces a strong neutralizing immunity in preclinical studies. ACS Chem Biol 16:1223–1233. https://doi.org/10.1021/acschembio.1c00272

109. van der Put RMF, Kim TH, Guerreiro C, Thouron F, Hoogerhout P, Sansonetti PJ, Westdijk J, Stork M, Phalipon A, Mulard LA (2016) A synthetic carbohydrate conjugate vaccine candidate against shigellosis: improved bioconjugation and impact of alum on immunogenicity. Bioconjug Chem 27:883–892. https://doi.org/10.1021/acs.bioconjchem.5b00617
110. Dalal J, Rana R, Harale K, Hanif S, Kumar N, Singh D, Chhikara MK (2019) Development and pre-clinical evaluation of a synthetic oligosaccharide-protein conjugate vaccine against Neisseria meningitidis serogroup C. Vaccine 37:5297–5306. https://doi.org/10.1016/j.vaccine.2019.07.053
111. Prasad AK, Kim J-H, Gu J (2018) Design and development of glycoconjugate vaccines. In: Carbohydrate-based vaccines: from concept to clinic. American Chemical Society, pp 75–100
112. Prasad AK, Murphy K, Kolodziej SA (2018) Manufacturing considerations for glycoconjugate vaccines. In: Carbohydrate-based vaccines: from concept to clinic. American Chemical Society, pp 101–122

Part V
Special Topics: CAR-T and CRISPR Technologies and Applications

Chapter 9
CAR-T Bioprocessing

Adebola Adeniran, Salina Handy, and Abdulrahman Baki

Abstract Over the past 30 years, immunotherapy has grown from bench experiments to a viable clinical option in the battle against cancer with nearly 1000 currently registered clinical trials Ivica NA, Young CM. Tracking the CAR-T Revolution: Analysis of Clinical Trials of CAR-T and TCR-T Therapies for the Treatment of Cancer (1997-2020). Healthcare (Basel) 19;9(8):1062.

In chimeric antigen receptor T-cell (CAR-T) immunotherapy, a patient's own immune cells are engineered to express a CAR which recognizes cancer cells, expanded ex vivo and then introduced back into the patient, where the engineered cells will mount an immune response against the targeted cancer cells. Immunotherapy manufacturing requires careful consideration of the collection of immune cells from the patient, introduction of receptor into the cells, method of cell expansion, formulation for cryopreservation, and the unique supply chain that must connect clinics, hospitals, and manufacturers to supply individual patients. In this chapter, we provide a brief overview of the current clinical landscape, discuss current practices and key challenges at each manufacturing step using clinical examples where available, and highlight emerging technologies that may accelerate the progress of CAR-T therapies.

Keywords CAR-T · T-cell · Immunotherapy · Bioprocessing · Apheresis · Vector · Activation · Transduction · Expansion · Automation · Formulation ·

A. Adeniran (✉)
Cell Therapy Process Development, The Center for Breakthrough Medicine, King of Prussia, PA, USA
e-mail: badeniran@cfbm.com

S. Handy
Manufacturing Science and Technology, Lonza, Houston, TX, USA

A. Baki
Cell and Gene Therapy, Medicinal Science and Technology, GlaxoSmithKline, Stevenage, UK

K. Gadamasetti, S. A. Kolodziej (eds.), *Bioprocessing, Bioengineering and Process Chemistry in the Biopharmaceutical Industry*,
https://doi.org/10.1007/978-3-031-62007-2_9

Cryopreservation · Supply chain · Yescarta · Kymriah · Tecartus · Breyanzi · Abecma · Carvykti

9.1 Introduction

9.1.1 Generation and Function

Chimeric antigen receptor T-cells, also known as CAR-T cells, represent an exciting, growing class of anticancer immunotherapies. In CAR-T immunotherapy, a patient's own immune cells are engineered to express a chimeric antigen receptor (CAR) which recognizes cancer cells, expanded ex vivo and then introduced back into the patient, where the engineered cells will mount an immune response against the targeted cancer cells. CAR-Ts have proven themselves to be a promising therapy, with ongoing remissions reported to last as long as a decade in a small subset of leukemia patients [1]. However, it should be noted that success with CAR-T cells is generally limited to hematological malignancies and longer survival rates are observed in patients with lower disease burdens [2].

CARs consist of an ectodomain and intracellular signaling stimulatory domains [3]. The ectodomain contains an antigen-binding domain, which is typically a single-chain fragment variant (scFv) from an antibody, a signal peptide to ensure proper trafficking to the cell membrane, and a hinge region between the signal peptide and recognition domains [4]. The transmembrane domain, a hydrophobic alpha helix, connects the ectodomain and the intracellular signaling domain. The intracellular signaling domain is responsible for initiating the intracellular signaling cascade to produce an anticancer immune response (Fig. 9.1).

To date, CAR-T therapies such as Yescarta, Kymriah, Tecartus, Breyanzi, Abecma, and Carvykti have been approved for a variety of leukemias and lymphomas across different patient populations (Table 9.1). In this chapter, we review the key steps of CAR-T manufacturing, highlighting many available technologies, challenges to the field, and provide relevant examples from clinical trial manufacturing where available.

9.2 Transgene and Vector Bioprocessing

9.2.1 Introduction

The intrinsic ability of viruses to incorporate genetic material into a cell for the purpose of altering cell function for a therapeutic benefit has made viral vectors an extremely useful tool in gene therapy, being considered key raw material required for CAR-T cell manufacturing in the United States [5]. As the amount of success in

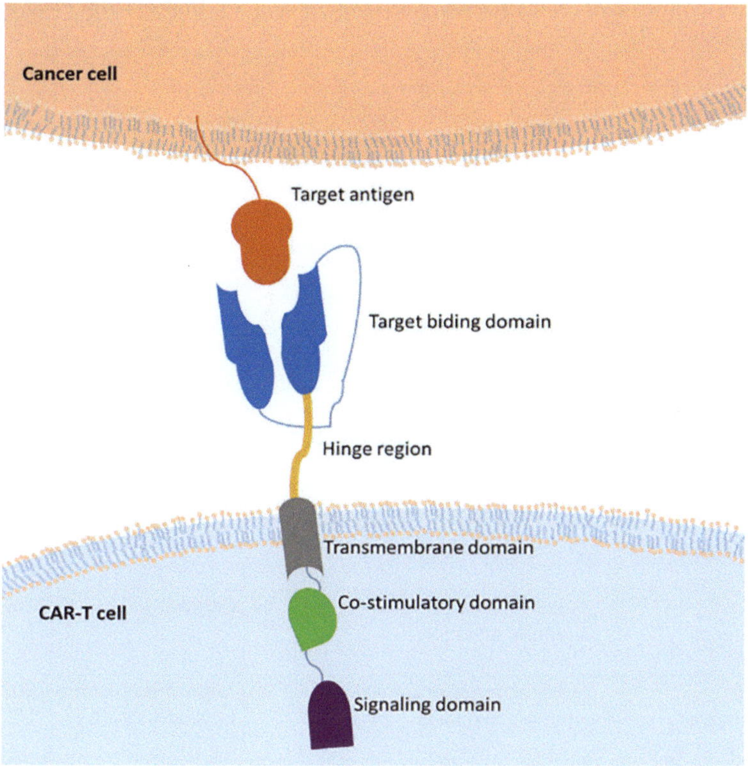

Fig. 9.1 CAR-T cell. Depiction of a CAR-T cell interacting with the target antigen on a cancer cell. The CAR is composed of a target binding domain, a hinge region to connect the target binding domain to the transmembrane domain, the co-stimulatory domain, and the signaling domain

clinical trials and potentially approved CAR-T drug products continue to increase and rapidly advance to mainstream manufacturing, large-scale production and purification of high-titer viral vectors are becoming key [6] to meeting demand and potential global distribution.

Notably three viral vector technologies, adenovirus, adeno-associated viruses (AAV), and lentiviruses, have demonstrated much success in both clinical and pre-clinical studies [7] attracting global attention and focus into the development of viral vector technology. In particular, lentiviral vectors (LVVs), derived from lentiviruses, have become increasingly popular especially for CAR-T drug product development, as the first lentiviral vector-based CAR-T immunotherapy, Kymriah®(Novartis), was FDA approved in 2017. Due to their increased safety profiles, broad tropism, and ability to integrate transgenes of interest into host cell genomes, LVVs have been the major vector of choice for treatments which require introduction of genetic material into host cells. For the purpose of this section and the following overview, the bioprocessing of LVVs will be reviewed as well as key challenges, current methods, and emerging technologies.

Table 9.1 Approved CAR T therapies to date (Wang et al. [20]; United States Food and Drug Administration 2021)

CAR T therapy	Indications	Company	First approval year
Yescarta (Axicabtagene Ciloleucel)	Adult relapsed or refractory large B-cell Lymphoma	Kite Pharma	2017 (USFDA)
Kymriah (tisagenlecleucel)	Adult relapsed or refractory large B-cell Lymphoma Pediatric and young adult relapsed or refractory acute lymphoblastic leukemia (ALL)	Novartis	2017 (USFDA)
Tecartus (brexucabtagene autoleucel)	Adult relapsed or refractory mantle cell lymphoma (MCL) Adult relapsed or refractory B-cell precursor acute lymphoblastic leukemia (ALL)	Kite Pharma	2020 (USFDA)
Breyanzi (Lisocabtagene maraleuce)	Adult relapsed or refractory large B-cell lymphoma	Juno Therapeutics	2021 (USFDA)
Abecma (idecabtagene vicleucel)	Adult relapsed or refractory multiple myeloma	Celgene	2021 (USFDA)
Carvykti (ciltacabtagene autoleucel; ciltacel)	Adult relapsed or refractory multiple myeloma	Janssen Biotech, Legend Biotech	2022 (USFDA)

Abbreviation: *USFDA* United States Food and Drug Administration

9.2.2 Lentiviral Vector Design

Lentiviral vectors (LVVs) are derived from the human immunodeficiency virus type 1 (HIV-1) and have conventionally been produced using a multi-plasmid DNA construct system where separate plasmid constructs are carry essential genomic information required for viral vector production. With time, as our viral vector knowledge and experience have expanded, the LVV construct design has undergone multiple rounds of modifications and enhancements with a particular focus in increasing safety. As a result, three generations of LVVs have been developed over time each with an increasing safety profile, and in addition, genetic modifications suggested to increase gene expression, transduction efficiency, and vector stability during bioprocessing steps. For clinical applications, the third-generation construct design, thought to be the modern packaging system [8], is the vector design of choice and most widely used, due to its increased safety profile which avoids the potential for replicative competent virus to be generated.

To provide some context, the third-generation construct design, modified from earlier generations of LVV construct designs, has undergone the complete removal of regulatory genes, such as the *tat* gene, from packaging plasmid constructs, while the *rev* gene, originally within the packaging plasmid, was placed on a separate fourth plasmid construct of its own. In addition, deletions were introduced into the 3'LTR region of the transfer plasmid to reduce the risk of generating replication competent LVVs (RCL), which thus created the third-generation self-inactivating lentiviral vector (SIN) [7]. Reports of use of earlier generations of LVVs indicated their cause of the development of lymphomas in primate models due to the generation of replicative competent virus [9] which had raised a safety concern around the use of these LVVs. Generation of RCLs can occur as a result of recombination events that reconstitute a pathogenic virus [10] which can lead to adverse events. However, by removing unessential genes and splitting and modifying viral genetic sequences within the construct design used to produce LVVs, the development of the third-generation SIN LVVs has greatly reduced the potential risk of generating RCL rendering it an unlikely event. Nonetheless, to ensure safety to patients, the Food and Drug Administration (FDA), the European Medicines Agency, and most regulatory agencies have required that all cell products transduced with retroviral vectors be tested for the presence of RCL prior to infusion of the cell drug product into patients for treatment. In addition, the FDA suggests that every 3 months patients be tested for the presence of recombinant virus during the first year of having received the gene therapy [3, 9].

The ability of retroviral-based vectors, such as LVVs, to stably integrate genetic information into the host genome is a key advantage with the potential to increase cell drug product efficacy in patients by allowing long-term sustained transgene expression and has been utilized as a powerful technique allowing for gene expression manipulation and engineering capabilities [10]. However, the integration potential of LVVs as retroviral-based vectors has raised additional safety concerns around their use, and their ability to cause insertional mutagenesis. The gammaretroviral vector, a retroviral-based vector, has demonstrated an oncogenic and transformative potential in transduced cells as a result of insertional mutagenesis. Several accounts of patients experiencing adverse events leading to the development of leukemia have been reported [11] with the use of gammaretroviral vectors. Insertional mutagenesis is a phenomenon that occurs when the viral vector integrates genetic material into an area of the host genome that is essential for cellular function causing a genetic mutation or disruption of cell function which has a potential to become cancerous.

The process of integration for most retroviruses is not random, and each class of retrovirus has its characteristics preferential site of integration. In comparison to other retroviruses, lentiviruses have been shown to prefer integration of genetic material within transcriptional units, in addition they also have the unique ability to translocate across the nuclear pore of intact nuclear envelopes which may increase their integration and oncogenic potential [3]. To address this concern, it is imperative to design efficient and safe viral vectors by achieving tissue or cell-type specificity [7].

A potential alternative, which some groups have investigated to address some of the concerns around the use of integrating LVVs, is the development of non-integrating lentiviral vectors (NILVs). Some reports have demonstrated that integration into the host cell genome may not be required for transgene expression in target cells both in vitro and in vivo [11]. NILVs have been developed by inducing genetic mutations into the genes responsible for the integration of viral DNA into the host genome. These modifications result in a vector episomal DNA species that is non-integrative and remains within the nucleus as extrachromosomal molecules. Though there have been reports of many applications for this technology, there has also been low levels of expression associated with NILVs in comparison to their integrating LVVs counterparts. For other clinical applications that require high expression of the transgene further investigation into techniques increasing transgene expression will need to be explored [11], but results using these types of vectors prove promising and there may be therapeutic advantage to look forward to as progression continues.

Some of the notable advantages of using LVs for clinical applications are their ability to transduce both dividing and non-dividing cells in comparison to other viral vectors of the Retroviridae family, such as gammaretroviral vectors, which are only able to transduce actively dividing cells. This characteristic provides a major benefit allowing for a variety of cell types such as neurons, hematopoietic stem cells, and immune system cells, like T-cells, to be transduced and receive therapeutic genetic material [8]. LVVs also exhibit reduced immunogenicity, ideal for use in patients, reducing the likelihood of adverse events and severe immune reactions to occur. LVVs also have the ability to express multiple genes from a single vector which could be beneficial for certain diseases that require high-level expression of multiple genes [7].

An added benefit has been the ability to pseudotype the viral envelope, which has broadened the range of tropism, where the HIV-1 envelope can be exchanged for an envelope from another virus. The more commonly used and widely considered gold standard envelope protein is the vesicular stomatitis virus Indiana glycoprotein envelope (VSV-G) [8]. The VSG-G glycoprotein provides not only a broad tropism for a variety of cell types but increased vector stability which is extremely important for bioprocessing steps, though the VSV-G envelope has displayed cytotoxic effects in host cells when not tightly regulated or when overexpressed which can limit its application to the transient transfection platforms [8]. Other viral envelope proteins have been investigated that exhibit decreased toxicity, comparable tropism and stability, for instance the glycoproteins, Cocal-G, PIRY, Chandipura, and VSV-New Jersey which all derive from the same vesiculovirus family as VSV and the RD114 viral envelope deriving from the feline endogenous retrovirus [8] could all be potential alternatives to circumvent potential toxicity due to the viral envelope.

As we continue to expand our knowledge and acquire more information about these efficient genetic delivery systems, it is important to keep in mind, for the purpose of CAR-T drug product manufacturing, that optimization of the viral vectors must also be conducive to facilitating a scalable production process that will enable robustness, reproducibility, consistency, and cost-effectiveness in producing LVVs.

9.2.3 Cell Culture Technology for the Production of LVVs

Each cell essentially is a miniature LVV production factory, producing a volumetric product yield of LVVs. In theory the more dense the culture is, the higher the productivity or titer of LVV produced. It is important to point out that there are other types of appropriate cells lines available for the production of viruses. The type of cell line used will be dependent on the type of virus being produced, hence the type of cell culturing will then be dependent on the type of cell line used. Specifically, LVVs typically are produced in an appropriate producer cell line or derivative, such as the human embryonic kidney 293 T (HEK293T) cell line. Traditionally, this cell line is adherent-dependent requiring that culturing take place in a single monolayer vessel that provides a surface for attachment and growth. For small-scale production, T-flask vessels can provide the appropriate surface area required allowing for cell attachment and growth, they can come treated or non-treated depending on culture needs. For large-scale production of viral vectors, this vessel type proves unsuitable as the surface area provided does not support the high cell density growth required for high titer viral vector production. Multiple surface cultivation systems, like roller bottles, multi-layer flasks, and cell factories, can alternatively be used to increase the total surface area needed to support high cell density growth but can still be considered low cell density culture systems as they have been reported to result in moderate increases of viral vector titers [12]. Instead, these systems would rather need to be "scaled-out" opposed to "scaled-up" where the number of culture units are increased in parallel requiring large enough facility footprint and personnel to accommodate and operate the numerous culture units which can prove to be both labor and cost intensive.

Alternatively, other types of cell culture systems such as the hollow fiber bioreactor technology, fixed-bed bioreactors, and microcarriers can provide larger surface areas in comparison to the aforementioned systems, supporting high cell density growth and product yield than what has been seen to be possible in flasks and roller bottles. Hollow fiber bioreactors are a closed system that utilizes a hollow fiber bioreactor cartridge that provides a porous support for cells to attach and grow, the cartridge consist of two separate spaces: the extra-capillary and intracapillary space. The cells are cultivated in the extra-capillary space and receive nutrients and metabolites via an exchange from the medium transported through the intracapillary space [12]. Fixed-bed bioreactors or packed-bed bioreactors provide a fixed-bed of microfiber carriers or disks for cell growth where cells receive nutrient and metabolites through a perfusion mode. Microcarriers are small sphere-like beads which provide high surface area for cells to attach and grow but interestingly enough can also be grown in a three-dimensional suspension system, such as stirred tank bioreactors, as opposed to the conventional two-dimensional planar culture systems [12]. Based on clinical trial size, these cell culture technologies can prove sufficient providing a solution to generating increased LVV titers; however, as demand may increase these approaches with the exception of microcarriers still require a scale-out approach, again requiring large enough incubator space and facility footprint.

Recently there has been a shift away from anchorage-dependent adherent cell culture technology to the development of suspension cell cultures to produce LVVs as this has been shown within a limited workspace to be much more amendable to scaling-up. Suspension cell cultures are derived from the adaptation of adherent HEK293T cells to suspension culture conditions and can grow directly in shake flasks or bioreactors as they do not require a surface for attachment allowing for high cell density growth to occur. In addition, these cells do not require serum-containing media. Reports have described that suspension adapted cells in serum-free containing medium are able to retain their cell virus packaging abilities with nearly a five-fold improvement in comparison to original adherent 293 T cells and proved suitable for cultivation in a bioreactor setting yielding comparable LV titers to shake flask induced titers [6]. The serum that has been traditionally used is a derived animal component which is complex and unpredictable in nature and the composition of the serum used in culture media can lead to batch variability resulting in inconsistencies within the process [6] as well as posing a risk of introducing animal components and contaminants into the final clinical grade vector stocks [12] which can complicate further downstream processing steps. The ability to remove the use of the highly variable animal derivative serum in culture media has been highly favorable easing the GMP manufacturing process and satisfying safety compliance regulations.

9.2.4 Mode of Production for LVVs

Predominately the transient transfection method is used for the production of LVVs where plasmid DNA constructs carrying the viral genes are introduced into the appropriate cell line of choice. This is mostly attributed to time as this method can quickly produce the LVVs. Transfection can be carried out using a variety of transfection reagents dependent on the needs of the process. Once such method which has been heavily used is the static transfection process called the calcium phosphate ($CaPO_4$) co-precipitation method. Cell cultures are required to remain static as precipitates formed via $CaPO_4$ co-precipitation settle on the cell monolayer for a period of time to allow for endocytosis, uptake of the DNA, to occur [12]. A media exchange typically occurs 24 h post-transfection (hpt), to reduce cytotoxic effects induced by $CaPO_4$, and multiple harvests are carried out to increase the final titer produced. Though considered a cost-effective transfection reagent, $CaPO_4$ is cytotoxic to cells requiring serum or albumin in culture media to reduce cytotoxic effects, which as mentioned can pose a safety and compliance issue.

Other methods of transfection utilize reagents such as polyethyleneimine (PEI) or lipofectamine for the production of LVVs. However, for large-scale production, lipofectamine proves costlier than PEI. PEI is a linear polyethylenimine high-charge cationic polymer that will bind to the anionic plasmid DNA creating a positively charged transfection complex that will enter into cells by endocytosis. PEI can increase consistency across batches, is non-toxic to cells thus eliminating the need for serum, and simpler to use as there is an unnecessary requirement for a tight regulation of transfection conditions as opposed to the $CaPO_4$ method which requires pH to be regulated at a certain value to increase transfection efficiency. Though it is important to keep in mind that many other factors can also affect the efficiency of transfection, theoretically all vector produced would be of similar quality, but due to per cell scale factors such as plasmid ratio, reagent: DNA ratio, incubation and mixing time, and host cell viability the vector produced is typically heterologous and further optimization would be required [8] for scale-up purposes to increase transfection efficiency.

The development of a packaging cell line which constitutively expresses the vector components and viral envelope has the potential to streamline the production process by replacing the need for costly GMP grade plasmid DNA and eliminating the need for DNA preparation steps reducing the potential for source contamination. However, cytotoxicity effects have been reported with constitutive expression of some of the viral vector components. To mitigate those toxic effects, a potential alternative is the tetracycline-inducible system to control expression or use of packaging cell lines that can express a number of less toxic envelope proteins [7, 8]. Stable cell lines in comparison to transient transfection can reduce cost, increase reproducibility, and enable scalability ensuring manufacturability [7, 8].

Recently developed, a suspension adapted producer cell line derived from 293 T has been stably transfected with a single bacteria artificial chromosome (BAC) DNA construct that expresses all of the lentiviral vector components. This stable cell line technology utilizes an inducible system where the addition of doxycycline to culture media initiates transcription and production of LVVs yielding high titers comparable to what has been achieved with transient transfection. This cell line technology is suggested to be readily scalable in single-use stirred-tank bioreactors maintaining genetic and functional stability [13].

For the purpose of producing high titer clinical grade LVVs at scales to meet commercial demand, the use of suspension cell cultures has been the go-to choice circumventing much of the challenges associated with the limited scalability that previously mentioned cell culture technologies have not yet been able to overcome.

9.2.5 Upstream Bioprocessing

As the industry moves toward the use of suspension cell cultures for large-scale production of LVVs, the typical vessels of choice to support the growth of cell cultures are bioreactors. Bioreactors provide an operator-controlled cell culture environment ideal for production processes that require larger and more consistent batches. Cell cultures and viral production take place during what is called the upstream bioprocessing portion of the process. The purpose of the upstream bioprocessing steps is to generate a sufficient amount of LVV bulk for the downstream processing portion, typically clarification and concentration of the virus to standards that meet compliant safety, purity, and efficacy regulations.

The upstream process begins at vial thaw and continues on through a series of cell expansion steps to generate a sufficient amount of cell inoculum to inoculate the production vessel (N), the vessel the LVVs will be produced within (Fig. 9.2). Cell expansion can be carried out in shake flask vessels that are incubated in incubators that support cell aeration and temperature requirements for growth. Alternatively, in an effort to increase biomanufacturing capacity other alternatives such as the disposable wave bioreactor can prove promising in providing ease of operation by eliminating the need to manipulate and manage multiple shake flasks, no cross-contamination, while still ensuring good mixing and oxygen transfer without cell damage [14]. In simple wave bioreactors allow for cell cultures to be inoculated into a disposable sterile plastic bag that sits on a rocking platform that allows for efficient mixing and increased oxygen transfer due to the available high surface area conditions.

Once the appropriate cell culture inoculum is generated, cells are then inoculated into the production vessels where LVV production will occur. Bioreactors are intended to provide suitable environments to maintain culture homogeneity while ensuring the production of a product with desired quality attributes. Vector production is highly dependent on the producer cell and its` health during culturing. Decline in cell health can have an effect on productivity and production of LVVs thus resulting in low, if any, vector titers. When scaling up a bioprocess using bioreactors it is important to keep in mind that optimization of key bioengineering principles is critical to maintaining a desirable culture environment for cells to thrive. Some of these parameters are the power input per volume ratio (P/V), impeller tip speed, and the volumetric mass-transfer coefficient ($k_{L}a$) as these all can have an effect on culture conditions. In brief, the P/V is the rate of impeller energy transferred into the cell culture; this parameter can be indicative of shear stress to the cells and can influence mixing and oxygen transfer capabilities. Impeller tip speed also influences mixing time and if not optimized can lead to concentration gradients within the culture, and $k_{L}a$ is the aeration capability of the bioreactor system, the ability to transfer oxygen to the cell culture. In practice, it is recommended to use bioreactors of similar design or geometry as this will facilitate the scaling up process.

Once culture conditions or cell expansion are optimized, then mode of production is the next step in the upstream LVV bioprocessing. Whether the mode of production is performed via the transient transfection method or the use of a packaging

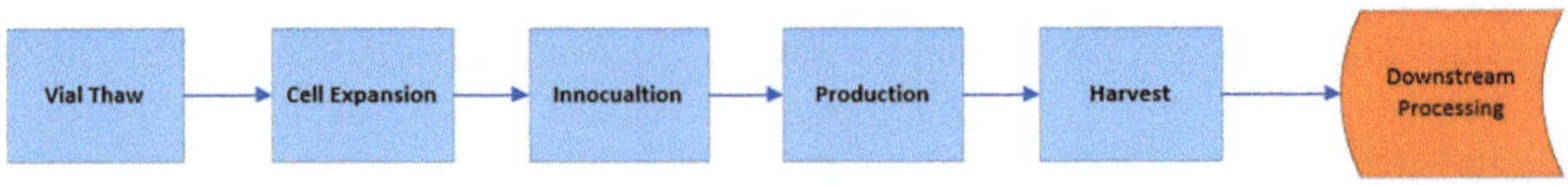

Fig. 9.2 Flow diagram of upstream bioprocessing steps for lentiviral vector production

or stable cell line is dependent on the process and dosage requirements. There are both pros and cons to the using either method. In one aspect transient transfection is an attractive method due to its shorter process development times, though for large-scale production this method is associated with higher running costs due to the costly cGMP-grade plasmid DNA. On the other hand, stable producer cell lines are associated with longer process development timelines and higher upfront costs with lower flexibility, but their use has the potential to generate more consistent harvest titers and higher quality material while reducing running costs due to the elimination of the plasmid DNA requirement [15].

A decisional tool, such as one developed by Comisel et al., could be used to determine the most cost-effective cell culture technology for the manufacturing of LVVs. The group describes a cost of goods (COG) breakdown and rankings of all the currently available cGMP-grade LVV manufacturing cell culture technologies across a wide range of LVV products, clinical and commercial demand, and harvest titer. Their analysis has suggested that single-use bioreactors (SUBs) are predicted to be the most cost-effective cell culture technology with cost benefits increasing with increased demand, mostly attributed to the broad scalability of SUBs with the largest scale at 2000 L making SUBs superior for the purpose of commercialization and large-scale production compared to other cell culture technologies [15].

Typically, the mode of production for fed batch cultures span a timeframe of 48 h post-transfection (hpt) or induction (hpi), with viral vector production peaking at 48 hpt and declining from 72 h on, limiting the timeframe of when harvesting of the LVVs can occur [8]. However, another area of interest in the production of LVVs is process intensification. Process intensification or perfusion is another process that can be used for the production of virus. The perfusion process extends over a period of time without compromising cell health and LVV stability. It has enabled a scalable production process for the production of LVVs while increasing total vector titer. Advantages of perfusion cultures are the ability to remove cell waste products as they develop while maintaining nutrient and metabolite levels [8] required to support high cell density growth over extended periods of time. In addition, by retaining cells within the bioreactor this mode can facilitate the harvest step and subsequent downstream processing. Some groups have demonstrated that LVVs are able to be produced in 3 L bioreactors operated under perfusion mode using novel approaches that are scalable and are able to achieve increased viral titers of 30-fold compared to a fed batch process [16]. Process intensification involves the use of a cell retention device that retains cells inside the bioreactor while allowing for continuous media exchange. As fresh media is continuously added, spent media containing the LVV is removed and collected at the same rate.

The upstream process for the production of virus can vary dependent on the type of cell line used, virus produced, and demand. Once the intended product, the virus, is produced, next begins the downstream processing portion.

9.2.6 Downstream Bioprocessing

The timepoint at which the upstream process ends is the timepoint when downstream processing steps commence. Once produced, LVVs within the cell culture must then be harvested and processed through a series of downstream processing steps to concentrate and generate a final bulk vector that is of high purity and titer. As high titers are important to meet the demand of doses required to treat patients, purity of the product is also critically important to prevent potential inflammatory reactions both in vitro and in vivo [3] as a result of common process and product related impurities such as host cell proteins, residual DNA, transfection reagents, media components, viral aggregates, or free vector components [8].

The harvest step begins by using a process that will aim to remove large host cell proteins and cellular debris clarifying the vector (Fig. 9.3). Two known methods can be employed for the harvest clarification step; microfiltration and/or centrifugation. Centrifugation can act as a pre-filtration step to help prevent filter fouling during subsequent filtration steps, but not all processes involve a centrifugation step and is dependent on the process. Cells and debris are separated from LVV based on the mass properties of these components. Typically, as cells tend to weigh more than LVV, at certain centrifugal speeds the cells and larger debris will sediment toward the bottom, while the LVV will remain in the supernatant of the culture broth. Whether a centrifuge step is carried out or not the harvest material will still be passed through a membrane or depth filter for clarification of the virus. Filter choice can impact the efficiency of the clarification step, [8] therefore careful considerations should be applied in selecting a filter uniquely appropriate for the process at hand as not all filter sizes and properties are compatible with all virus types. Furthermore, seamless execution of the clarification step can be dependent on optimizing the upstream production. Take for instance culture viability at the time of harvest, a lower cell viability will complicate the clarification step, due to the increased amounts of cellular debris as a result of cell fragmentation during cell death. This can potentially lead to filter clogging, increased pressures, and ultimately pre-mature filter fouling. These types of considerations should be kept in mind not only for efficient processing of the vector but also to drive down costs of purchasing additional filters. With proper optimization and choice of filters for the process, this can be avoided.

Once the vector has been clarified, the next step is to treat the harvest material with a nuclease enzyme to digest residual DNA/RNA contaminants. Once treated, the viral vector bulk can then be further purified through a series of sequential chromatography, or in some cases nonchromatographic steps that can either bind or

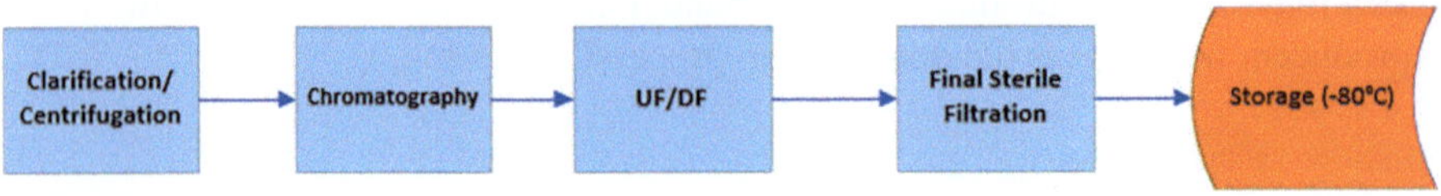

Fig. 9.3 Flow diagram of downstream bioprocessing steps for lentiviral vector production

elute the vector, depending on the conductivity of the buffer solution used, separating the vector from impurities. Chromatography, in the case of ion exchange (IEX) chromatography, separates the vector from impurities based on the charge characteristics of the components, while size exclusion chromatography (SEC) separates based on size. Again, the choice of chromatography methods will be dependent on the chosen process and virus produced. For LVV purification multiple types of chromatography have been applied resulting in varying recoveries [8]. Once purified the vector is concentrated and exchanged into a final fill formulated buffer of choice. This step can be carried out using tangential flow filtration in an ultrafiltration/diafiltration (UF/DF) mode that enables buffer exchange while removing low molecular weight impurities [8]. The final step is to sterile filter the final vector product and freeze for long-term storage until quality testing is carried for these products to be dispositioned for use in CAR-T manufacturing. The next few sections will discuss how the viral vectors are used to produce the therapeutic product once produced.

9.3 Cell Product Bioprocessing

9.3.1 End-to-End Systems

Generally speaking, the steps in cell product processing can be performed on different process unit operations, discussed in the following sections. However, there is a growing interest in end-to-end production that requires minimal manipulation and reduces the number of instruments needed per production. To this end, the CliniMACS Prodigy from Miltenyi Biotec and the Cocoon from Lonza are two available closed end-to-end systems that can accommodate selection, activation, transduction, expansion, and harvest of the CAR-T product. The Prodigy has been used for clinical trials at University College London [17] and is a popular option in product development departments. The Prodigy can accommodate the isolation of up to 3E9 CD3+ cells. While not yet as widely adopted, the Cocoon can accommodate real-time monitoring of cell culture characteristics like pH and dissolved oxygen, which is a key advantage over the monitoring available with the Prodigy system.

While it is possible to complete a fully closed CAR-T manufacturing process, manufacturers need to ensure that all inputs, such as media, buffers and apheresis, are available in weldable bags to maintain a functionally closed system. If necessary, open manipulations and transfer of material to a weldable bag can be performed in a BSC (biological safety cabinet), and then welded on to the closed system. Weld compatibility of different materials should be taken into consideration when designing a closed process. For example, closed reagents are available from Miltenyi Biotec to ensure compatibility with the CliniMACS Prodigy system.

These end-to-end integrated systems will be discussed among the instruments available for the different process units, but their integration of different process units is worth highlighting to the reader (Table 9.2).

9.3.2 *Apheresis Collection and Preservation*

To begin CAR-T manufacturing, a patient's T cells must first be isolated using leukapheresis, which separates leukocytes from a patient's blood and returns the remaining blood components to the patient.

Leukapheresis is typically performed with machines such as the Spectra Optia, COBE Spectra (Terumo), or Amicus (Fresnius-Kabi) by highly trained nursing staff at hospitals and clinical collection centers. Geothe University, one of the first German centers to be authorized for pediatric CAR-T treatment, used the Spectra Optia for apheresis collection for Kymriah production [18].

In an apheresis machine such as the Spectra Optia, blood is first mixed with an anticoagulate like citrate before being separated by continuous-flow centrifugation. The blood is separated into three layers: a red blood cell layer on the bottom, a buffy

Table 9.2 Example technologies for CAR-T bioprocessing

Cell processing step	Example commercial technologies
Apheresis collection	Spectra Optia (Terumo) COBE Spectra (Terumo) Amicus (Fresenius-Kabi)
Lymphocyte purification	BCT Elutra (Terumo) X-Lab (Corning)
Leukapheresis thawing	Plasmatherm (Barkey) Varitherm (Barkey) ZipThaw (FreMon) VIAThaw (Cytiva)
T-cell isolation	Dynabead (ThermoFisher) CD4, CD8 immunomagnetic reagents (Miltenyi Biotec) RoboSep-C (Stemcell) Sepax C-Pro (Cytiva) WOLF cell sorter (NanoCellect) MACSQuant Tyto (Miltenyi Biotec)
T-cell activation	Dynabead (ThermoFisher) TransACT (Miltenyi Biotec)
T-cell transduction	LentiBOOST (Sirion Biotech) Vectofusin-1 (Miltenyi Biotec)
T-cell transfection	4D Nucleofector (Lonza)
Cytokines for expansion	GMP ProDots (Biotechne) Lyophilized in vials (Miltenyi Biotec, CellGenix)
Closed processing systems	CliniMACS Prodigy (Miltenyi Biotec) Quantum (Terumo) Cocoon (Lonza) G-Rex (Wilson Wolf) Xuri Wave (Cytiva)
1 L+ scale cell expansion	Hyperforma (ThermoFisher) Biostat RM (Satorius)
Scale up systems	G-Rex (Wilson Wolf) Ambr (Sartorius)

coat layer in the middle, and a plasma layer on top. The quality of layer separation is determined by the packing factor, a numerical relationship between the inlet flow speed and the centrifuge speed. The buffy coat contains the lymphocytes needed as input for CAR-T manufacturing, as well as other contaminating cell types such as monocytes and granulocytes. The nontarget layers, and replacement fluids such as plasma or saline, are returned to the patient to prevent electrolyte imbalances and hypotension [19]. Plasma is often added to the isolated cells before further processing to promote cell healthy and viability.

Elutriation, the separation of cell types based on density and size, can be used if further purification of the lymphocytes is desired before downstream processing. The method is particularly good for separating lymphocytes from monocytes [20], which can be beneficial as monocytes can decrease final product quality [21, 22]. Terumo's BCT Elutra is a fully closed device available for performing elutriation. The X-Lab from thermogenesis is available for a sedimentation-based approach to cell separation.

To allow flexibility in the supply chain and maintain health of the collected cells, the apheresis is typically cryopreserved at the clinical collection site or a manufacturing facility. Both cryopreservation methods and supply chain challenges are further discussed in detail later in this chapter. Cryopreservation has been demonstrated to eliminate myeloid-derived suppressor cells which have been shown to inhibit T-cell proliferation in CAR-T manufacturing [23]. While cryopreservation is common, use of fresh apheresis material as an input material may help decrease overall manufacturing time if logistical barriers to accommodate the required "just-in-time" supply chain can be overcome. Kite pharmaceuticals has applied to the EMA (European Medicines Agency) requesting use of fresh material to start manufacturing [24].

Apheresis collections can be highly variable from patient to patient and this variability can complicate manufacturing success. Factors contributing to variability can include patient age, clinical indication, prior treatments, tolerance for anticoagulants, method of venous access, and apheresis center practices. Collecting apheresis before treatment with frontline chemotherapy or radiation regimens has been suggested to combat challenges due to clinical indication and prior treatments, though it remains to be seen if this would result in higher quality CAR-T drug products [20].

9.3.3 T-Cell Thawing and Isolation

Cryopreserved apheresis must be thawed before T-cell isolation can begin.

Though cryopreserved leukapheresis can be thawed in a water bath, dry thawing reduces the likelihood of contamination and should be used in GMP facilities. The Barkey Plasmatherm, used in by Novartis for Kymriah production [25], passes warmed water through two cushions and the cryopreserved material is placed between the two cushions until thawed. The ZipThaw from Fremon Scientific and the VIA Thaw and Smart-Max, both from Cytiva, are alternative dry thawing devices. The ZipThaw differentiates itself for its light weight, the VIA Thaw for its dry conduction technology, and the Smart-Max for its pneumatic mixing technology.

The T-cells must be separated from contaminating cells such as monocytes and granulocytes in the apheresis as contaminating material can inhibit activation and proliferation [20]. Additionally, there is emerging evidence that suggests defined ratios of CD4 to CD8 cells yields uniform potency of CAR-T products [26]. Breyanzi is dosed as a 1:1 ratio of CD4 to CD8 cells [27]. However, Yescarta, Kymriah, Tecartus, Abecma, and Carvykti do not specify a CD4:CD8 ratio for dosing [28–32]. T-cells can be specifically isolated from the thawed apheresis using immunomagnetic separation, fluorescence-activated cell sorting (FACS), or acoustics.

With immunomagnetic technology, positive and/or negative selection can be used to isolate T-cells using beads. The beads can simultaneously display an antibody or ligand specific to the desired cell type and a magnetic particle, or the beads themselves can be magnetic. In positive immunomagnetic separation, the beads enable the desired population to be cross-linked to magnetic particle. In negative immunomagnetic separation, the undesired cell types are targeted. Separation of the bead-bound population can occur over a magnetic column or can be column-free.

ThermoFisher's Dynabead technology is a column-free option for cell separation. With the Dynabead technology, the bead itself is magnetic, rather than being linked to magnetic particle. Anti-CD3 and anti-CD4 Dynabeads are commercially available, as well as the DynaMag CTS magnet, which can accommodate 50–330 mL for static separations, and >10 L for continuous flow separations.

Miltenyi Biotec offers anti-CD4 and anti-CD8 immunomagnetic reagents, which are compatible for use in the CliniMACS Prodigy. The reagent vial is spiked onto the Prodigy tubing set and filtered before making contact with the cells. The cells are filtered on a pre-separation column to reduce any clumped material from entering the magnetic column. The selected cells are then transferred to another bag where they can be readily dispensed to a chamber to begin cell culture.

Another closed system for cell separation is the RoboSep-C from Stemcell Technologies. The RoboSep-C can process up to 2x10E7 cells but unlike the CliniMACS Prodigy, the RoboSep-C is column-free. The cells of interest are isolated in a bag with PVC tubing that can be welded on to other downstream equipment. A unique feature of the RoboSep-C is its cell washing cartridge, which washes and concentrates cells without pelleting or centrifugation [33].

Closed cell separation is also possible on Sepax C-Pro from Cytiva Life Sciences. The Sepax C-Pro is multifunctional like the CliniMACS Prodigy, but uses a separate, closed kit for each of its isolation, transduction, harvest, and formulation functions. One kit is available for immunomagnetic separation and can accommodate up to 880 mL of input volume.

Fluorescence-activated cell sorting (FACS) can also be used to separate cell types. For this flow cytometry-based separate method to be clinically feasible, the flow cytometer must be in a closed, aseptic system. Sorting cytometers such as the WOLF Cell sorter or the cartridge-based MACSQuant Tyto can be used to aseptically separate cells.

Finally, buoyancy-activated cell separation (BACS) uses functionalized microbubbles to bind cells of interests, and after a gentle agitation, the microbubbles and bound cells float to top of the solution. Cesca Therapeutics and Akadeum Life Sciences offer BACS kits. However, these kits are currently only available at the mL scale and would need further scale-up development to accommodate the volumes required for clinical manufacturing.

9.3.4 Activation

Activation of the selected T-cells is necessary for cells to be able to expand and accept the vector encoding the CAR-T receptor. The TCR and CD28 receptors are often targeted for cellular activation as this mimics the manner by which antigen-presenting cells naturally activate T-cells [34], and activation of the TCR alone by anti-CD3 antibodies does not induce full activation [35]. There are a variety of activation technologies available that utilize components such as matrices, beads, and cells.

Matrix- and bead-based activation reagents are advantageous for commercial manufacturing as they allow for a consistent activation reagent to be used across multiple manufacturing batches.

Novartis reported the use of Dynabeads for Kymriah manufacturing. Available from ThermoFisher, Dynabeads are magnetic beads covalently linked to anti-CD3, anti-CD28 and, if desired, anti-CD137 antibodies. Dynabeads can be removed in a magnetic field after activation is complete [36]. These beads can also be used for cell isolation, streamlining the isolation and activation steps and reducing the number of reagents needed.

TransACT technology from Miltenyi Biotec is a polymeric matrix conjugated to CD3 and CD28 agonists compatible with Miltenyi's closed system Prodigy unit. Following activation, TransACT is removed from the culture by exchanging culture medium.

Expamer technology has been introduced to temporally fine-tune T-cell activation by utilizing the reversible Twin-Strep tag interaction with Strep-Tactin. Anti-CD3 and anti-CD28 antibody fragments are functionalized with Twin-Strep tag, which binds a soluble Strep-Tactin multimer. This allows for muti-valent binding by one Strep-Tactin multimer. Adding D-biotin to the cell culture disrupts the interaction between the Twin-Strep tag and the Strep-Tactin multimer, terminating the activation signal at the user's discretion [35].

9.3.5 Cell Transduction and Transfection

The purified, activated T-cells must be transduced with gene to express the CAR by a viral vector, or, less commonly, transfected directly with the gene. Vector production is discussed previously in this chapter. If packaged in a weld compatible bag, the viral vector may simply be thawed and connected to a closed system for transduction. To date, all approved CAR-T therapies use viral vectors for transduction.

Chemical transduction enhancers can be used to reduce the amount of viral vector needed by improving vector–cell interactions. Reducing the amount of vector needed will increase the number of patients that can be treated with a single vector batch, but the cost of the transduction enhancer must also be considered to understand if the use benefits patients economically. LentiBOOST is a commercially available poloxamer-based transduction enhancer from Sirion Biotech. Protransduzin is a peptide-based transduction enhancer that has been reported to enhance transduction by 24% [37]. Vectofusin-1 is another peptide-based transduction enhancer compatible with primary human T-cells.

An alternative to transduction via a viral vector to the culture is electroporation of the DNA encoding the CAR. Though electroporation exposes the T-cells to a harsher environment and negatively impacts cell viability as compared to viral vector transduction, it is a more efficient means of introducing genetic material into the cells. Electroporation does not promote genomic integration of the CAR [38], which should be carefully taken into consideration depending on application of the CAR-T therapy as long-term persistence of CAR-T cells seems to positively influence clinical outcomes [39].

The CliniMACS Prodigy has an electroporator module to expand the closed system capabilities of the Prodigy. Similarly, the 4D Nucelofector can be integrated with the Cocoon unit from Lonza for closed transfection.

9.3.6 Cell Expansion

Two key considerations in the expansion of CAR-T products are the media and instrumentation. Regarding media, cytokines are used in ex vivo culture to promote expansion and a preserve memory-enriched phenotype in CAR-T clinical trials. Cytokines frequently used include IL-2, IL-7, IL-15, and IL-21 [40]. Cytokines are typically available in a lyophilized form in vials, which can complicate addition of cytokines in a closed manner. Biotechne offers GMP ProDots, lyophilized cytokines produced in weldable bags and convenient for addition to closed systems. The number of cytokines added to media must be fine-tuned, as an excess of cytokines can inhibit proliferation and induce apoptosis [40]. The type of cytokine used must also be carefully considered, as certain cytokines may lead to differentiated phenotypes that may negatively impact CAR-T clinical performance [39].

Human serum is another common component in CAR-T expansion media as it promotes expansion and viability in vivo [41]. However, using human serum introduces considerable variability into the manufacturing process as each serum lot may either be from a single donor or pooled from multiple donors, and lot-to-lot comparability should be established. Higher concentrations of serum are not necessary for successful production of all CAR-T products. Novartis reported a reduction in serum concentration in clinical batches of Kymriah resulted in a process with equivalent growth kinetics, product quality, viable cell number and transduction efficiency [42]. During clinical trial manufacturing optimization, Kite successfully removed serum from cell culture media to minimize risk of viral contamination of Yescarta [43].

Ideally, media components would be added in a closed manner in the manufacturing facility. As this is not always possible, adding components to media may require additional manipulation in a BSC, increasing opportunities for contamination. One consideration is to prepare media, ensure sterility, and package into containers compatible with closed systems ahead of manufacturing; however, this would be difficult with media that has a short shelf life.

The instrumentation used for cell expansion can be closed, open, static, shaking, automated, or any combination thereof. In 2019, Roddie et al. reported that 43% of clinical trials use rocking bioreactors, 35% use static culture bags, and 22% use T-flasks. Open systems are more prone to contamination than fully closed systems and can have limited scalability due to the labor needed. The CliniMACS Prodigy closed system has been reported to lower overall manufacturing costs of CAR-T therapies [44]. Another option for closed expansion is the Terumo Quantum, which uses a hollow fiber bioreactor.

One option for static cultures is the G-Rex system from Wilson Wolf. There are 24- and 6-well G-Rex plates that can be used in development, and larger flasks appropriate for clinical manufacturing, which can accommodate 5 L of media and expansion up to a final count of 10–20 billion cells. The larger flasks can be connected to a fluid management system like the GatherEx from Wilson Wolf for closed liquid handling. Gas exchange in the G-Rex occurs at the bottom of the flask where the cells settle. Convection throughout the media allows for nutrients to constantly reach the cell layer as cells continue to expand in the G-Rex systems [45].

T-flasks and cell factories are other options for static culture. Cell factories have the advantage of increased surface area by utilizing vertical space to stack culture plates on top of each other. The Nunc cell factory system from ThermoScientific is available in stacks of 1–52 layers, and a closed version of the cell factory is available in up to 10 layers. There are liquid handlers, shakers, and incubators specific to the cell factory system to ease handing of the stacked plates [46].

The Xuri Wave System by Cytvia Life Sciences is a common option for rocking cultures. Rocking cultures induce mixing and oxygen transfer. The Xuri is a closed and automated system for cell expansion with single-use bags that have a working volume from 300 mL to 25 L. Other rocking culture vessels appropriate for T-cell culture include the HyPerforma from ThermoScientific [47] and the Biostat RM from Sartorius [48].

The duration of cell culture should be fine-tuned to the clinical dose. However, maintaining CAR-T cells too long in ex vivo culture may drive the cells toward an exhausted phenotype, as a shorter culturing time has been shown to be associated with positive clinical outcomes [39]. During manufacturing optimization, Kite shortened the expansion time of Yescarta to 6–8 days and still met clinical dose of 2E6 CAR-T cells/kg of body weight [43].

9.3.7 In-Process and Release Testing

The lack of approved and well-defined standardized testing is a continuing challenge for the growing field of CAR-T manufacturing despite regulatory requirements to test for safety, purity, and potency [17]. As many advancements in CAR-T development occur at academic centers [49] which do not often interface with regulatory agencies [50], it is difficult to centralize the numerous testing protocols and approaches for clinical manufacturing. Post-manufacturing quality control is a particular area of concern as any delays in release testing will translate to delays for CAR-T administration to patients. While validated assays are not typically required during product development, the relatively quick approval that can happen with CAR-T therapies (i.e., Kymriah received marketing approval in the United States after just 106 patients were treated) [50] necessitates a robust testing strategy be identified as quickly as possible.

Over the course of manufacturing, it is critical to quantify and/or qualify cell identity and functionality. Flow cytometry is a standard analytical method in CAR-T manufacturing due to its high specificity and ability to identify many cell types simultaneously. However, flow cytometry is notoriously complex and can be subjective. The choice of flow cytometer should be guided by the number of dyes to be detected in the process, and the lasers and detectors that are appropriate for the number and color of dyes to be used. To reduce variation and improve standardization, flow cytometry cartridge-based systems that use machine learning to automatically identify populations are in development by companies like Accellix. A simple to use and automated flow cytometry assay may be one way to reduce costs and the number of highly trained full-time employees (FTEs) required for manufacturing.

Weil et al. propose applying diagnostic pathology assay standards to cell therapy quality control testing and have suggested detailed lists of acceptability criteria for flow cytometry and hematology blood analyzers including guidelines around instrument maintenance, assay controls and sample running, and analysis guidelines [50].

Proving cellular potency is a key requirement for release of CAR-T cells and remains a challenge not encountered in more traditional biopharmaceuticals. Potency assay development can be a particularly challenging part of CAR-T development due to variability of autologous starting materials, limited sample availability, lack of reference standards, distinguishing cytotoxic contributions of multiple cell subtypes, and discrepancies between in vivo and in vitro behavior [51, 52]. There is a regulatory challenge as well as potency measurements are uniquely designed for each product. In the United States, the FDA does not provide recommendations regarding specific assay types or acceptance criteria and notes that potency assays may evolve as more knowledge is gained about the product during development. Additionally, the FDA does suggest orthogonal assay development, as a singular assay alone may not fully demonstrate potency [51]. While this allows for considerable flexibility for the manufacturer, the lack of defined standards can be challenging. It may also be worthwhile to measure a wide range of product attributes as early as possible, and such exploratory studies can help identify which attributes correlate with cellular potency.

The key assumption of the potency assay is that the killing ability of the CAR-T cell product measured in vitro is an appropriate proxy of the in vivo killing capabilities. This assumption is fundamentally complicated by the biological variability of the patient population.

Cytotoxicity is measured by exposing CAR-T cells, known as effector cells, to the target cancer cell, and using target cell viability to determine the killing ability of the effector cells [52]. The effector and target cells can be mixed at different effector to target ratios (E:T). Assays to measure CAR-T cytotoxicity include chromium release, luciferase-mediated bioluminescence imaging, flow cytometry, cytokine measurement, and impedance [53].

In the chromium release assay, the target cells are labeled with radioactive chromium. As the target cells are killed by the CAR-T cells, chromium is released, and the amount of target cells killed is proportional to measured radioactivity. With the luciferase assay, no radioactive materials are required. Target cells are transduced with a luciferase reporter gene and only viable cells produce luciferase. As effector cells kill target cells, bioluminescence decreases since luciferase can no longer be produced [53]. Flow cytometry-based cytotoxicity assays uses a live/dead staining reagent, such as 7-AAD, to measure dead target cells. As suggested previously in this section, flow cytometry is not necessarily a preferred method due to the level of technical expertise required and subjectivity that can influence data analysis.

Cytokine measurement assays can be used as an indirect measure of proxy potency. One class of cytokine-based functionality assays measure cytokines, such as interferon gamma, released by activated effector cells. Granzyme B and perforin detecting assays can also be used to measure functionality. A noteworthy recent development suggests that interferon gamma does not necessarily correlate with better killing ability of effector cells in liquid tumors [54].

One advantage of the impedance-based assays for cytotoxicity is that the target cells do not need to be labeled. However, this assay class measures impedance due to target cells, thereby requiring the target cell population to be either naturally adherent or made to be adherent. In this assay, the presence of microelectrodes on the bottom of a microtiter well allows for the measurement of impedance due to the target cells. The effector cells are in suspension. As effector cells kill target cells, impedance changes due to the decrease in target cell population [53].

Specialized instruments for the impedance assay are the xCELLigence, Incucyte, and Maestro ZHT platforms. These instruments can be of particular use to pursue to simultaneous development of multiple potency assays over the course of drug development as suggested by the FDA [51]. In addition to providing impedance measurements, these instruments can also be used for cytokine secretion assays. The supernatant from the co-culture can be isolated and evaluated for the presence of cytokines. The manufacturer should ensure any instrument and software used in the evaluation of potency is 21 CFR compliant. To date, the Maestro ZHT and xCELLigence platforms meet 21 CFR compliance.

A critical aspect of potency assays to consider is how representative the in vitro assay is of the in vivo environment. In vivo, CAR-T cells must overcome the tumor microenvironment, which can be characterized by the presence of immuno-suppressive factors and hypoxia. As the CAR-T field builds upon the success of treating hematological cancers and expands toward treatment of solid tumors, it is important to consider the 3D environment in which the CAR-T needs to demonstrate functionality. Interestingly, a weakened cytotoxic response is observed from CAR-T cells applied to a spheroid of target cells as opposed to a monolayer of target cells at the same E:T ratios [52].

Discussion around types of qualification assays needed for the production and release of GMP lentiviral vectors has been debatable. Qualification assays should demonstrate titer, potency, identity, purity, and safety of the vectors. In developing types of qualification assays, it is important to understand that these assays need to demonstrate that the vector is safely carrying out its intended role for which it is developed for, while exhibiting high efficacy. Characteristics for release can vary, and sometimes more than one type of assay, dual testing, may be required. White et al. summarizes a approaching regulatory considerations for the use of lentiviral vectors in respect to the types of characteristics that should be assessed and how to develop proof-of-concept studies to demonstrate vector attributes which satisfy regulatory requirements.

As described earlier in this chapter, these vectors are derived from pathogenic viruses which pose the risk of RCR/RCL generating during recombination events. Testing for replication-competent lentiviruses and retroviruses safeguards against any undesired recombination events postinfusion that may result in new replication-competent viruses. The utility of this test is debatable, as to date, no CAR-T products have reported a positive RCR/RCL result [55]. Previous recommendations from the FDA suggest testing material, including vector supernatant and cell products transduced with vector, from multiple stages of product manufacture as RCR may develop at any stage during manufacturing [56]. This can vary depending on the mode of production, whether via transient transfection or via stably transfected producer cell lines, type of vector envelopes used and if the effector CAR-T cells will be administered ex vivo or in vivo so it is important to remain current on regulatory recommendations.

Assays to assess for RCR may consist of the vector supernatant being co-cultured with a permissive cell line to allow for amplification of potential RCR. The amplified material can then be assessed for the presence of RCR by utilizing appropriate indicator cell assays such as PERT which measure the activity of reverse transcriptase, PCR detecting viral sequences, or the commercially available p24 ELISA-based assay. It is critical to optimize these assays and to develop appropriate standards and controls to increase assay specificity, sensitivity, and reproducibility. FDA has recommended two methods: (1) serologic detection of RCR-specific antibodies; and (2) analysis of patient peripheral blood mononuclear cells by PCR and RCR-specific DNA sequences. Again the choice of the assay will depend on the type of vector [56].

Assays may assess the properties and biological effects of the promoter sequence, transgene, and gene product delivered for the intended clinical application [57]. For CAR-T cells transformed by viral vectors that integrate the CAR-T in the cell genome, vector copy number (VCN) should be measured as too high of VCN carries an increased risk of oncogenesis. The FDA recommends a VCN <5 copies per cell [13]. Digital droplet PCR and real-time quantitative PCR are both feasible options for VCN measurement [58, 59].

CAR-T products must be tested for sterility prior to patient infusion. The standard 14-day sterility test is not feasible when CAR-T products should be administered as quickly as possible to patients. Therefore, a conditional release may be made after a 3- or 4-day sterility test showing no contamination with the caveat that the full 14-day sterility test will be completed [55]. Undoubtedly, patient safety must remain the highest concern in the production and administration of CAR-T therapies. However, streamlining the testing requirements may get medicines to patients faster and at a lower cost.

9.3.8 T-Cell Cryopreservation

Once the cell expansion step achieves the target number of nucleated cells within the mixture, the final formulation of T cells starts with two cell washing cycles followed by a buffer exchange step. The T-cell mixture is first centrifuged within the same closed chamber used for cell expansion, then the supernatant cell expansion buffer is aspirated out of the chamber and replaced with a formulation buffer made of an extracellular protective agent such as human serum albumin (HSA) and saline solution suitable for intravenous (IV) infusion. The formulation buffer helps retain the cells in a balanced solution in terms of osmolality and pH to suit IV infusion; however, it provides limited protection from ice crystal formation during the freezing stage required for the proposed storage conditions for biologics in liquid nitrogen.

Successful cryopreservation limits crystal formation during the freezing stage while keeping the osmolality and pH values within the acceptable limits of IV infusion, thus maintaining CAR-T cell phenotype and function while retaining viability with 50–90% recovery post-thaw [55, 60]. Therefore, cryopreservation methods developed for hematopoietic cells are mostly adopted for CAR-T cell manufacturing as well even though some recent studies have pointed out to the potential adverse effects of cryopreservation on the vital characteristics of CAR-T cells [25]. These methods generally involve resuspension of cells in isotonic buffers (e.g., saline for IV infusion) containing 10% dimethyl sulfoxide (DMSO) before transferring to controlled rate freezer prior to storing in vapor phase liquid nitrogen (VPLN) [61]. Lower concentrations of DMSO can be also used once mixed with extracellular protective agents (e.g., hydroxyethyl starch (HES) solutions, HSA, serum, and plasma), and they can be available commercially as preformulated Cryostor at final DMSO ratios of 2%, 5%, and 10%. Recent studies have also pointed out to the potential of using pentaisomaltose to help reduce DMSO ratios down to 1% in the final formulation for effective cryopreservation of CAR-T cells [62]. Considering the adverse effects of DMSO on cell viability, it is clinically regulated to not exceed 1 gr/kg as per European Society of Bone Marrow Transplantation (EBMT) and the American Association of Blood Banks (AABB) [63].

The freezing rate is also another important factor in controlling the formation of ice crystals, so controlled rate freezers are used to tune the temperature profile during freezing to manipulate the physical state of water molecules. The fine manipulation of the physical state of water molecules along with the use of cryopreserving agents can significantly limit the formation of ice crystals in the final cell formulations during freezing stage thus maintaining cell viability and stability for extended periods of storage. Previous methods suggest freezing CAR-T cell products passively in −80 °C deep freezers, but the more recent approaches adopted for GMP manufacturing involve using controlled rate freezers (CRF) to achieve a temperature reduction rate at 1 °C per minute to better limit the formation of intracellular ice crystals [64]. Cells can be later transferred to VPLN for storage at temperatures less than −150 °C once sufficiently hydrated to withstand the sudden reduction in temperatures [17, 65]. The main advantage of using CRFs from a quality perspective is the consistency of freeze with the versatility of programming and data traceability for individual products, and new mechanical CRFs (e.g., asymptotes) can overcome the high costs and safety concerns associated with the heavy use of LN2 to run traditional CRFs in closed spaces [17].

Validating the cryopreservation step is a crucial quality requirement for large-scale manufacturing as suboptimal cryopreservation can lead to reduced numbers of viable cells with increased levels impaired cell phenotype and functions [66]. Therefore, stability testing is routinely performed on individual products to ensure preserved viability, expression of CAR transgene, and functional capability (cytotoxicity or cytokine release data) over time according to established assays. To maintain traceability, cryopreserved products are retrieved from VPLN and transported in temperature logged and validated LN dry containers by contracted couriers under principles of good distribution practice (GDP) prior to infusion in patients in hospitals [1, 67].

9.3.9 Formulation Considerations for the Drug Product

To prepare the final formulation mixture, the dose to be infused—which correlates to the number of $CD3^+$ transduced T cells as a fraction of the total number of nucleated cells in a given volume of the formulation buffer—is calculated. Once the final dose is calculated, the formulator then decides on the type and number of final drug product container closure systems (i.e., bags) suitable for the chosen dose. The selection of the suitable bags takes into consideration different attributes of the final drug product according to the Quality Target Product Profile (QTPP) as referred to in the pharmaceutical development quality guidelines Q8(R2). This dose system can be updated over the progress of the clinical trials according to feedbacks from clinicians as they gain more understanding of the cancer treatment regimen. Since T cells can be considered as living drug modalities, the formulation procedure can be a bit different from that of the other non-living drug modalities and must aim to maintain cell metabolism and functionalities instead of physicochemical interactions with other excipients [68].

Dosing strategy should take into consideration the pharmacokinetic profiles of T cells where they can peak in 1–2 weeks after administration and have been shown to persist in some patients for up to 10 years [1]. This in turn reflects therapeutic efficacy with a potential high drug exposure on the short term and a sustained functionality on the long term. Therefore, current dosing practices aim to design clinical trials with patients split in multiple groups of three patients to accommodate a regimen of incremental dose increases to account for any potential toxicity at each incremental increase [69]. To account for the assumption that dose toxicity would only be modelled with simple scaling and monitoring, the first group would start with a dose considered safe (i.e., low concentration of TNC/mL), and the following two groups would receive a dose with predetermined incremental increases and monitored for toxicity. This has been shown in a recent Phase I study of anti-CD19 CAR-T cells where patients with relapsed/refractory B-cell non-Hodgkin's lymphoma received escalating doses of 25×10^6 ($n = 3$), 50×10^6, and 100×10^6 viable CAR-positive T-cells [70].

Next generations of dosing strategies tended to enhance the potency of the DP by adopting new formulation platforms which enhance T-cell stemness, thus significantly reducing the dose level while maximizing the therapeutic effects of the final DP [39]. These strategies also aimed to minimize the in vitro cell expansion time to significantly reduce the total vein-to-vein time and reflect on a reduced cost of the final DP [71, 72]. On the other hand, these next-generation approaches also opened the door toward future T-cell immunotherapeutics with improved activity against solid tumors [73].

Considering the variabilities between different dosing strategies, the total number of cells required to meet the target doses of the high concentration group may well exceed one order of magnitude that of the low concentration at a given clinical study [74]. Therefore, the total volumes required to achieve target doses may be split into several containers (bags or vials) prior to administering to patients. This shows that formulation options can be quite limited to meet the QTTP requirements, and design of these dosage forms can consider only few options such as primary container type, fill volume, and cell concentration per dose to cover a target range of 10^6–10^8 CAR/TCR T-cells/mL. Table 9.3 shows different clinical CAR-T cell therapies in different FDA approval stages with the different doses, primary containers, and fill volumes.

Table 9.3 Comparison of the primary containers used to fill the allocated doses of marketed T-cell drug products

T cell drug product (Proprietary name)	Dose (adults)	Primary container	Fill volume per container	Number of containers per dose
Brexucabtagene autoleucel (Tecartus™)	2×10^6 per kg body weight, max. 2×10^8	Cryogenic infusion bag	~68 mL	1
Axicabtagene ciloleucel (Yescarta®)	2×10^6 per kg body weight, max. 2×10^8	Cryogenic infusion bag	~68 mL	1
Tisagenlecleucel (Kymriah®)	$0.6–6.0 \times 10^8$	Cryogenic infusion bag	10–50 mL	1–3
Lisocabtagene maraleucel (Breyanzi®)	$0.5–1.1 \times 10^8$[a]	Cryogenic vials	4.6 mL	1–4[b]
Ciltacabtagene autoleucel (CARVYKTI ®)	$0.5–1.0\times 10^6$ viable T-cells per kg body weight up to 1×10^8	Cryogenic infusion bag	30 or 70 mL	1

[a] CD4 and CD8 components at 1:1 ratio
[b] For each CD8 and CD4 component

9.3.10 Scale-Up Considerations: Toward Automated Formulation

Qualification of the successful manufacturing process of T-cell drug products takes into consideration different aspects of the procedures and components involved such as raw materials, ancillary components, equipment, facilities, sampling plans for quality control and analytical assays, standard operating procedures, and trained personnel to consider it as a reproducible and robust manufacturing platform [75]. The quality and quantity of the harvested patient's peripheral blood mononuclear cells (PBMCs) will be highly amenable to the disease progression and the immune state of their body following any potential chemotherapeutics. Lymphocytopenia (low count of lymphocytes in blood) can be induced in some cancer patients following chemotherapies and can in turn have negative effects on T-cell proliferation rate, phenotype, and transduction efficiency which can eventually affect the dose to be manufactured. Therefore, manufacturing strategies should take into account the feasibility of translating leukapheresis material from healthy donors used in formulation development to that of the cancer patients with compromised numbers of PBMCs [69].

Considering the complex and lengthy workflows involved in the routine manufacturing of autologous T-cells, developing a strategy to reduce the total costs of goods for the effective delivery of vein-to-vein CAR-T cells becomes a necessity. Since T-cell therapy is highly personalized and specific to each therapeutic area, sourcing patient cells and ensuring their validated transport to GMP manufacturing facilities from and back to patient introduces a lot of challenges with regard to process automation [76]. The current vein-to-vein procedures involve aseptic filling of

T-cell formulations into different containers at room temperature within strict time schedules considering the presence of cryopreserving agents. This time constrains drive the chemistry, manufacturing, and controls (CMS) strategy to adopt fully or semi-automated formulation and filling steps where they can be integrated with upstream cell processing steps to reduce contamination and time constrains [77]. One approach to address these needs is to have multiple validated centers to source patient cells and transport them to central GMP manufacturing facilities to process into appropriate dosed containers of drug products [78]. Considering these challenges, big pharma, and biotech companies are building partnerships with universities to facilitate this process and drive development costs down as well [79].

Adopting closed end-to-end workflows for sterile production within ISO 14644-1 Class 7 cleanrooms (FDA Class 10,000 equivalent) is far more advantageous than open process workflows that require intensive manual steps under ISO 14644-1 Class 5 cleanrooms (FDA Class 100 equivalent). A two-phase process was demonstrated by the National Institute of Health Clinical Center where a modular system involving two manufacturing phases was adopted to provide TCR-T cells to Phase I/II clinical trials. The first phase involved closed lymphocyte enrichment in culture flasks before transferring enriched cells into culture bags for transfection by day 7. The second phase then moved the cells from culture bags into a closed culture apparatus to induce up to 1890-fold expansion by day 14 [80]. Fully closed and automated end-to-end process involving an advanced apparatus has shown high yields of CAR-T cells after just 10 days of expansion [81]. This in turn reflects the amenability of both the semi and the fully automated methods to generate sufficient numbers of T-cells for different therapeutic regimens.

The formulation process starts with the bulk drug substance obtained from the expansion step where the total number of nucleated cells (TNC) is used to calculate the buffer volumes required to perform cell wash/concentration and resuspension/dilution steps required to mix with DMSO and achieve the target TNC concentration in the final drug product. In the manual formulation process, the formulator makes the necessary resuspension/dilution steps to obtain the target TNC concentration before filling into final container closure systems (e.g., bags, vials) [5]. On the other hand, the closed systems with automated fill-finish process calculates the cell concentration and dilution steps out of the TNC concentration value of the drug substance to meet the target TNC concentration of the drug product [82]. The drug product volume remaining minus the volume required for quality control is later split into different bags at the validated fill volumes of the bags [68]. Bag material is chosen to maintain a flexible and robust structure with minimum leachable chemicals while enabling gaseous exchange of O2 and CO2 through permeation, and current preference seems to favor bags made of ethylene-vinyl acetate (EVA) over those made from poly vinyl chloride (PVC) for this purpose [83, 84]. These bags are later over-wrapped and placed into metal cassettes to ensure uniform thickness during freezing for a uniform heat transfer upon freezing and thawing procedures considering the excellent heat conductivity of the metal cassettes [85].

9.4 Supply Chain and Logistic Model

The overview, provided earlier in this chapter, describes the key process steps required for the development of CAR-T drug products. Though these types of drug products have demonstrated both clinical and therapeutic success harnessing much attention, there can be a series of challenges that need to be addressed to bring CAR-T cell drug products to market. It has been projected that demand will continue to grow, with forecasted patient numbers reaching 40,000 by the year of 2031 [86]. Current supply chain models for CAR-T cell drug products, for instance in the UK, only provide treatment to a finite number of patients in the magnitude of hundreds per region/country annually. As patient numbers increase, it is imperative to begin looking ahead and develop an efficient strategy to put in place a supply chain and logistic model that will meet the growing demand. Once the CAR-T cell drug product is formulated, it is then assessed to ensure efficacy, product quality, safety, and potency by Quality Control (QC) and Quality Assurance (QA). Depending on the manufacturing facility, this process can either take place at the site of manufacturing or outsourced to a QC/QA third-party. Once successfully dispositioned, the drug product is transported back to the clinical hospital where the cell drug product can be infused back into the correct patient for treatment. The processes that need to occur for this to happen need to be optimized, controlled, and regulated in a manner that is right first time [86].

For autologous CAR-T cell drug products, patient scheduling is key to ensuring timely leukapheresis collection, manufacturing, product release and proper administration of the cell drug product for treatment. It is critical that the correct CAR-T cell drug product be administered to the correct patient, thus highlighting the importance and need for systems that allow for proper sample tracking, documentation, and patient identification. Recent development of software and digital tools, such as cloud-based platforms and application programming interfaces, provides an efficient low-risk sample tracking and record keeping platform allowing disparate systems and information provided by key stakeholders to be seamlessly shared and connected while keeping patient identity and data private [86]. Further optimization and investigation are still required in developing automated systems that can ensure chain of custody and identity is maintained throughout the entire CAR-T manufacturing process from leukapheresis all the way through to infusion of the cell drug product into the correct patient. Shipping logistics need to be continuously evaluated including the design of cryo-shippers, temperature control mechanisms, and couriers to provide shipment service to ensure integrity of the product and timely delivery [42].

As the manufacturing process for CAR-T drug products is scaled up, important factors influencing the CAR-T cell drug product supply chain emerges. The logistic model design for the manufacturing and distribution of cell products is one paramount factor to consider in determining the outlook of future CAR-T cell therapeutics. A simulation study of the centralized versus decentralized supply chain in the UK has shown limited impact of both models on the total cost of treatment at scale;

however, it concluded that decentralized supply chains (i.e., distributed point of care manufacturing) may prove more efficient for larger geographical regions with some constrained transport facilities [87]. Other factors to consider for scaling up the production of CAR-T cells include allogeneic approaches where abundant cells are retrieved from healthy donors and processed on a larger scale in a centralized manufacturing facility. However, due to the biological and repeatability concerns, this has yet to be evaluated as a viable solution for the ease of access potential [88]. Designing highly personalized CAR-T cell therapeutics holds the most aspiring potential of this technology, and scaling up this potential without using the allogeneic approaches, which would arguably limit this potential, remains a challenging topic in the field [89]. Considering the potential risks associated with such new personalized therapeutic modalities, regulators often require intensive post-marketing evaluation of these therapeutics in patients for extended period of times compared to traditional non-biologic therapeutics [90]. This in turn put some further pressure on the logistics of such cellular-based therapies, and often requires novel models of access to be considered [91]. Though the advancement in this field is quite substantial with therapies being developed from academic to industrial application within short periods of time, the scalability and affordability of these novel therapeutics are the driving factors in determining their outlook as key tools in treating critical tumor conditions [92].

9.5 Considerations for the CAR-T Field

As CAR-T therapies continue to progress in clinical trials, there are key challenges that inhibit widespread adoption (summarized in Table 9.4). One key challenge is the cost of autologous CAR-T therapies. The current cost of approved CAR-T therapies ranges from $373,000 to 475,000 [93–96]. One approach to lowering the cost of CAR-T therapies may be moving from autologous to allogeneic therapies. While allogeneic therapies have their own specific challenges, including needing to establish a master bank that will not induce graft vs host disease, moving toward an "off the shelf" adoptive cellular therapy may reduce costs [97] as a new drug product batch is not needed for every individual patient, greatly reducing the number of processing steps, logistical considerations, release testing required, and wait time for the patient. Second, moving toward closed, automated production may reduce costs of cellular therapies. An 8-day expansion process on the automated and closed CliniMACS Prodigy is reported to cost $25,000, including labor and reagents [44]. Open systems that require frequent intervention by a highly trained labor force are not only more costly than closed systems [98] but also more difficult to scale. It is estimated that producing 10,000 cell therapy doses per year by open systems requires 1700 full time employees across manufacturing, quality assurance, quality control, and logistics [97]. Finally, it has been modeled that viral vector manufacturing is the costliest part of CAR-T therapy [98]. Any progress in reducing the cost of viral vector, whether that be through improving vector potency or economizing

Table 9.4 Opportunities for further innovation

Challenge	Potential solutions
High operation costs	Adopt innovative curriculum and interdisciplinary learning approaches to train highly skilled workforce to meet the highly technical nature of the development and manufacturing process [103] Allogeneic, "off the shelf" therapies reduce the number of batches per patient [97]
Long vein-to-vein time	Decentralized manufacturing [97, 102] Adopt manufacturing platforms to reduce cell expansion time by increasing T-cell stemness [71, 72]
Treating solid tumors	CAR NK cells, CAR macrophages, gene editing [53] Polyclonal T-cells with enhanced stemness [73]
Improve efficiency, optimizing workflows	Automated filling and closed systems [98]
Utilizing allogeneic or iPSC-derived CAR-T cells	Gene editing (CRISPR, TALENs) to overcome immune rejection [99, 100].
Potential immunotoxicity	Develop highly specific antigens [104] Develop CAR "switch-offs" including apoptosis inducers and CAR anti-bodies [105, 106]
Cold supply chains across international borders	Innovative supply chain considerations [86]

vector production, should have a positive impact in increasing affordability of CAR-T products. Proposed strategies for optimizing production workflows are outlined in Table 9.5.

Use of molecular gene editing techniques like CRISPR or TALENs is an area of growing interest in CAR-T manufacturing, particularly in the development of allogeneic therapies. Targeted gene editing allows for the precise removal of genes that elicit an immunosuppressive response in the host. An allogeneic anti-CD19 CAR-T generated with CRISPR from Caribou Biosciences is currently in a Phase I clinical trial for the treatment of relapsed/refractory B-cell non-Hodgkin lymphoma (NCT04637763) [99, 100].

Lentiviral transduction and TALENs have been combined to develop a therapy for pediatric refractory relapsed B-cell acute lymphoblastic leukemia. The lentivirus was used to introduce the CAR, and TALENs were used to edit the native TCR and CD52 gene loci to generate an allogeneic product that achieved remission ahead of stem cell transplantation in 2 infants [101].

Treatment of solid tumors remains a key challenge for CAR-T therapies. Variations in CAR therapies, including CAR natural killer cells and CAR macrophages, are under current clinical evaluation in solid tumor indications. Genetic engineering presents a potential solution regarding solid tumors. Genetic engineering may allow for the fine-tuning of cells to overcome the hypoxic and immunosuppressive microenvironment of solid tumors [53].

Table 9.5 Strategies for optimizing production workflows and quality attributes that may be affected

Production workflow	Strategies	Quality attributes potentially affected
Apheresis collection	Collecting apheresis before treatment with frontline chemotherapy or radiation regimens	Purity, viability
Cell selection and expansion	Dry thaw cryopreserved apheresis Reduction in human-derived media components Implement end-to-end, closed systems from cell selection to final formulation	Purity, sterility, viability
Vector preparation and transduction	Media optimization Closed automated systems Quality control assays Plasmid design Filtration and purification	Viability Sterility, viability, potency Potency, efficacy High titer, transduction efficiency High recovery, stability, potency,
Formulation	Selection of formulation buffers to increase T-cell stemness Optimization of the cryopreservant type and concentration Container closure system integrity Container filling approaches (automated versus manual)	Potency, cell count, transduction efficiency Cell count, viability, potency Cell count, viability, sterility Cell count, viability, potency, transduction efficiency, sterility
Infusion	Cold chain temperature excursions Thawing rate and type Hold time and infusion rate (in-use stability)	Viability, potency, transduction efficiency Cell count, viability, potency Viability, potency, cell count (dose level)

The model of manufacturing for CAR-T therapies requires different demands than a traditional biopharmaceutical therapy. Because CAR-T manufacturing begins after a patient has fallen ill, the time from apheresis to drug product administration ("vein to vein" time) is critical to patient success. To this end, decentralized manufacturing may present a manufacturing model that is more beneficial to patients than centralized manufacturing. In decentralized manufacturing, a central hub can manage multiple, smaller manufacturing facilities located closer to the patient and reduce the footprint of a complicated supply chain associated with centralized manufacturing [97]. Manufacturing at the point of care is another model for consideration. A collaboration between Lonza and CellPoint is exploring point of care manufacturing with Lonza's Cocoon system for cell expansion, reducing vein to vein time to 5–7 days [102]. As the field continues to grow, decentralized manufacturing may be key to expanding access to patients in new geographical areas and delivering medicines as quickly as possible.

As the CAR-T and immunotherapy field continues to make exciting strides and new clinical offerings, manufacturing advancements are also growing to meet demands for scalability, price, and patient access. In this chapter, we have discussed current manufacturing processes for CAR-T therapies and key challenges and concerns for this growing field. CAR-T cells hold great promise as a curative therapy, and as the knowledge of the therapy and field grows, research scientists, manufacturers, and clinical partners can continue to push the bounds of immunotherapy to offer more options for patients.

Acknowledgments The authors thank Dr. Alexandra McGregor for helpful discussions around potency assays.

References

1. Melenhorst JJ et al (2022) Decade-long leukaemia remissions with persistence of CD4+ CAR T cells. Nature 602:503. https://doi.org/10.1038/s41586-021-04390-6
2. Park JH et al (2018) Long-term follow-up of CD19 CAR therapy in acute lymphoblastic leukemia. N Engl J Med 378(5):449–459. https://doi.org/10.1056/nejmoa1709919
3. June CH, O'connor RS, Kawalekar OU, Ghassemi S, Milone MC (2018) CAR T cell immunotherapy for human cancer. Science 359:1361
4. Zhang C, Liu J, Zhong JF, Zhang X (2017) Engineering CAR-T cells. Biomark Res 5(1) BioMed Central Ltd. https://doi.org/10.1186/s40364-017-0102-y
5. Levine BL, Miskin J, Wonnacott K, Keir C (2017) Global manufacturing of CAR T cell therapy. Mol Ther Methods Clin Dev 4:92. https://doi.org/10.1016/J.OMTM.2016.12.006
6. Tang QL, Gu LX, Xu Y, Liao XH, Zhou Y, Zhang TC (2021) Establishing functional lentiviral vector production in a stirred bioreactor for CAR-T cell therapy. Bioengineered 12(1):2095–2105. https://doi.org/10.1080/21655979.2021.1931644
7. Bulcha JT, Wang Y, Ma H, Tai PWL, Gao G (2021) Viral vector platforms within the gene therapy landscape. Signal Transduct Target Ther 6(1):53. https://doi.org/10.1038/s41392-021-00487-6
8. Perry C, Rayat ACME (2021) Lentiviral vector bioprocessing. Viruses 13(2):268. https://doi.org/10.3390/v13020268
9. Skrdlant LM, Armstrong RJ, Keidaisch BM, Lorente MF, DiGiusto DL (2018) Detection of replication competent lentivirus using a qPCR assay for VSV-G. Mol Ther Methods Clin Dev 8:1–7. https://doi.org/10.1016/j.omtm.2017.09.001
10. Schlimgen R et al (2016) Risks associated with lentiviral vector exposures and prevention strategies. J Occup Environ Med 58(12):1159–1166. https://doi.org/10.1097/JOM.0000000000000879
11. Luis A (2020) The old and the new: prospects for non-integrating lentiviral vector technology. Viruses 12(10):1103. MDPI AG. https://doi.org/10.3390/v12101103
12. McCarron A, Donnelley M, McIntyre C, Parsons D (2016) Challenges of up-scaling lentivirus production and processing. J Biotech 240. Elsevier B.V:23–30. https://doi.org/10.1016/j.jbiotec.2016.10.016
13. Chen A, Velickovic Z, Rasko J (2020) Vector copy number quality control testing for CAR T-cells: critical validation parameters. Cytotherapy 22(5):S142. https://doi.org/10.1016/j.jcyt.2020.03.283
14. Rodrigues ME, Costa AR, Henriques M, Azeredo J, Oliveira R (2012) Wave characterization for mammalian cell culture: residence time distribution. New Biotechnol 29(3):402–408. https://doi.org/10.1016/j.nbt.2011.10.006

15. Comisel RM, Kara B, Fiesser FH, Farid SS (2021) Lentiviral vector bioprocess economics for cell and gene therapy commercialization. Biochem Eng J 167:107868. https://doi.org/10.1016/j.bej.2020.107868
16. Cattaneo, M, et al. (2020). Continuous perfusion with a stable producer HEK293 cell line for scaling up lentiviral vector production. Mol Ther 28(4):50
17. Roddie C, O'Reilly M, Pinto JDA, Vispute K, Lowdell M (2019) Manufacturing chimeric antigen receptor T cells: issues and challenges. Cytotherapy 21(3):327–340. https://doi.org/10.1016/j.jcyt.2018.11.009
18. Jarisch A et al (2020) Unstimulated apheresis for chimeric antigen receptor manufacturing in pediatric/adolescent acute lymphoblastic leukemia patients. J Clin Apher 35(5):398–405. https://doi.org/10.1002/jca.21812
19. Terumo (2019) Principles of apheresis on the spectra optia. https://www.terumobct.com/eLearning/306612715A-Principles-Of-Apheresis-CD/application/application.html?tracking=cd&gated=false
20. Fesnak AD (2020) The challenge of variability in chimeric antigen receptor T cell manufacturing. Regen Eng Transl Med 6(3):322–329. https://doi.org/10.1007/s40883-019-00124-3
21. Wang X et al (2021) Depletion of high-content CD14+ cells from apheresis products is critical for successful transduction and expansion of CAR T cells during large-scale cGMP manufacturing. Mol Ther Methods Clin Dev 22(September):377–387. https://doi.org/10.1016/j.omtm.2021.06.014
22. Stroncek DF et al (2017) Elutriated lymphocytes for manufacturing chimeric antigen receptor T cells. J Transl Med 15(1):4–11. https://doi.org/10.1186/s12967-017-1160-5
23. Davis MM et al (2017) Predictors of manufacturing (MFG) success for chimeric antigen receptor (CAR) T cells in Non-Hodgkin Lymphoma (NHL). Cytotherapy 19(5):S118–S119. https://doi.org/10.1016/j.jcyt.2017.02.190
24. Byrne J (2021) Kite: new data offers transparency into CAR T cell therapy manufacturing in real world setting. BioPharma-Reporter. https://www.biopharma-reporter.com/Article/2021/04/07/Kite-New-data-offers-transparency-into-CAR-T-cell-therapy-manufacturing-in-real-world-setting. Accessed 26 Jan 2022
25. Tyagarajan S, Schmitt D, Acker C, Rutjens E (2019) Autologous cryopreserved leukapheresis cellular material for chimeric antigen receptor–T cell manufacture. Cytotherapy 21(12):1198–1205. https://doi.org/10.1016/j.jcyt.2019.10.005
26. Sommermeyer D et al (2016) Chimeric antigen receptor-modified T cells derived from defined CD8+ and. Leukemia 30(2):492–500. https://doi.org/10.1038/leu.2015.247.Chimeric
27. BMS (2022) Breyanzi package insert, June.
28. Kite Pharma (2002) Yescarta package insert, April
29. Novartis Pharmaceuticals (2022) Kymriah package insert, May
30. Kite Pharma (2021) Tecartus package insert
31. Celgene (2021) Abecma package insert
32. Jannsen Biotech (2022) Carvykti package insert, Feb
33. Stemcell (2020) Seamless integration into your changing flawlessly integrate cell isolation into your evolving workflow
34. Trickett A, Kwan YL (2003) T cell stimulation and expansion using anti-CD3/CD28 beads. J Immunol Methods 275(1–2):251–255. https://doi.org/10.1016/S0022-1759(03)00010-3
35. Poltorak MP et al (2020) Expamers: a new technology to control T cell activation. Sci Rep 10(1):17832. https://doi.org/10.1038/s41598-020-74595-8
36. Thermo Fisher Scientific Inc (2017) First FDA-approved cell therapy for leukemia utilizes thermo fisher scientific's CTS Dynabeads technology. https://thermofisher.mediaroom.com/2017-08-30-First-FDA-Approved-Cell-Therapy-for-Leukemia-Utilizes-Thermo-Fisher-Scientifics-CTS-Dynabeads-Technology#:~:text=The. Novartis chimeric antigen receptor, therapy systems (CTS) portfolio
37. Castro A et al (2019) Protransduzin™ – a novel transduction enhancer to accelerate en gros CAR-T production. Cytotherapy 21(5):e14. https://doi.org/10.1016/j.jcyt.2019.04.044

38. Zhang Z, Qiu S, Zhang X, Chen W (2018) Optimized DNA electroporation for primary human T cell engineering. BMC Biotechnol 18(1):1–9. https://doi.org/10.1186/s12896-018-0419-0
39. Gattinoni L, Klebanoff CA, Restifo NP (2012) Paths to stemness: building the ultimate antitumour T cell. Nat Rev Cancer 12(10):671–684. https://doi.org/10.1038/nrc3322
40. Jin J, Cheng J, Huang M, Luo H, Zhou J (2020) Fueling chimeric antigen receptor T cells with cytokines. Am J Cancer Res 10(12):4038–4055
41. Alnabhan R, Gaballa A, Mörk L-M, Mattsson J, Uhlin M, Magalhaes I (2018) Media evaluation for production and expansion of anti-CD19 chimeric antigen receptor T cells. Cytotherapy 20(7):941–951. https://doi.org/10.1016/j.jcyt.2018.04.007
42. Tyagarajan S, Spencer T, Smith J (2020) Optimizing CAR-T cell manufacturing processes during pivotal clinical trials. Mol Ther Methods Clin Dev 16. Cell Press:136–144. https://doi.org/10.1016/j.omtm.2019.11.018
43. Roberts ZJ, Better M, Bot A, Roberts MR, Ribas A (2018) Axicabtagene ciloleucel, a first-in-class CAR T cell therapy for aggressive NHL. Leuk Lymphoma 59(8) Taylor and Francis Ltd:1785–1796. https://doi.org/10.1080/10428194.2017.1387905
44. Zhang W, Jordan KR, Schulte B, Purev E (2018) Characterization of clinical grade CD19 chimeric antigen receptor T cells produced using automated CliniMACS Prodigy system. Drug Des Devel Ther 12:3343–3356. https://doi.org/10.2147/DDDT.S175113
45. Ludwig J, Hirschel M (2020) Methods and process optimization for large-scale CAR T expansion using the G-Rex cell culture platform. In: Chimeric antigen receptor T cells: development and production, vol 2086. Humana Press, New York, pp 165–177
46. ThermoFisher (2017) Nunc cell factory systems for consistent cell culture performance 2 the Thermo Scientific Nunc Cell Factory system difference
47. ThermoFisher (2019) CTS optmizer T cell expansion SFM, no phenol red, displays T cell manufacturing workflow flexibility
48. Hupfeld J, Heinrichs K, Kraft M, Tappe A (2020) Comparison of primary T cells cultured in static conditions versus rocking motion bioreactors. Cytotherapy 22(5):S147. https://doi.org/10.1016/j.jcyt.2020.03.298
49. Hanley PJ, Lowdell MARK (2019) Advancing cellular therapies towards standard of care: a focus on testing of cellular therapy products. Cytotherapy 21(3):275–277. https://doi.org/10.1016/j.jcyt.2019.02.003
50. Weil B, Hanley PJ, Lowdell M (2019) Proposal for the International Society for Cell & Gene Therapy position statement on assays for the quality control and potency assessment of adoptive cellular immunotherapies. Cytotherapy 21(3):367–375. https://doi.org/10.1016/j.jcyt.2019.02.001
51. US FDA (2011) Potency tests for cellular and gene therapy products, vol 27. US FDA, pp 568–577
52. Weems O, Sullivan D, Millard D (2021) CART cell potency assessment with 3D cancer spheroid models key
53. Kiesgen S, Messinger JC, Chintala NK, Tano Z, Adusumill PS (2021) Comparative analysis of assays to measure CAR T cell– mediated cytotoxicity. Nat Protoc 16(3):1331–1342. https://doi.org/10.1038/s41596-020-00467-0
54. Larson RC et al (2022) CAR T cell killing requires the IFNγR pathway in solid but not liquid tumours. Nature 604:563–570. https://doi.org/10.1038/s41586-022-04585-5
55. Li Y, Huo Y, Yu L, Wang J (2019) Quality control and nonclinical research on CAR-T cell products: general principles and key issues. Engineering 5(1):122–131. https://doi.org/10.1016/j.eng.2018.12.003
56. Fda and Cber (2020) Testing of retroviral vector-based human gene therapy products for replication competent retrovirus during product manufacture and patient follow-up; Guidance for Industry [Online]. Available: https://www.fda.gov/vaccines-blood-biologics/guidance-compliance-

57. White M, Whittaker R, Gándara C, Stoll EA (2017) A guide to approaching regulatory considerations for lentiviral-mediated gene therapies. Hum Gene Ther Methods 28(4) Mary Ann Liebert Inc:163–176. https://doi.org/10.1089/hgtb.2017.096
58. Hu Y, Huang J (2020) The chimeric antigen receptor detection toolkit. Front Immunol 11(August):1–16. https://doi.org/10.3389/fimmu.2020.01770
59. Haderbache R et al (2021) Droplet digital PCR allows vector copy number assessment and monitoring of experimental CAR T cells in murine xenograft models or approved CD19 CAR T cell-treated patients. J Transl Med 19(1):1–13. https://doi.org/10.1186/s12967-021-02925-z
60. Lee SY et al (2018) Preclinical optimization of a CD20-specific chimeric antigen receptor vector and culture conditions. J Immunother 41(1):19. https://doi.org/10.1097/CJI.0000000000000199
61. Gee AP (2018) GMP CAR-T cell production. Best Pract Res Clin Haematol 31(2):126–134. https://doi.org/10.1016/J.BEHA.2018.01.002
62. Haastrup EK, Munthe-Fog L, Ballesteros OR, Fischer-Nielsen A, Svalgaard JD (2021) DMSO (Me2SO) concentrations of 1–2% in combination with pentaisomaltose are effective for cryopreservation of T cells. Transfus Apher Sci 60(4):103138. https://doi.org/10.1016/J.TRANSCI.2021.103138
63. Windrum P, Morris TCM, Drake MB, Niederwieser D, Ruutu T (2005) Variation in dimethyl sulfoxide use in stem cell transplantation: a survey of EBMT centres. Bone Marrow Transplant 36(7):601–603. https://doi.org/10.1038/sj.bmt.1705100
64. Morgenstern DA et al (2016) Post-thaw viability of cryopreserved peripheral blood stem cells (PBSC) does not guarantee functional activity: important implications for quality assurance of stem cell transplant programmes. Br J Haematol 174(6):942–951. https://doi.org/10.1111/BJH.14160
65. Anagnostakis I et al (2014) Successful short-term cryopreservation of volume-reduced cord blood units in a cryogenic mechanical freezer: effects on cell recovery, viability, and clonogenic potential. Transfusion (Paris) 54(1):211–223. https://doi.org/10.1111/TRF.12239
66. Germann A, Oh YJ, Schmidt T, Schön U, Zimmermann H, von Briesen H (2013) Temperature fluctuations during deep temperature cryopreservation reduce PBMC recovery, viability and T-cell function. Cryobiology 67(2):193–200. https://doi.org/10.1016/J.CRYOBIOL.2013.06.012
67. Dyson PG, Tocchetti R, Hiwase S (2014) Validation of an alternative transport system for cryopreserved haemopoietic progenitor cells. Cytotherapy 16(4):S56. https://doi.org/10.1016/J.JCYT.2014.01.204
68. van der Walle CF, Godbert S, Saito G, Azhari Z (2021) Formulation considerations for autologous T cell drug products. Pharmaceutics 13(8):1317. https://doi.org/10.3390/pharmaceutics13081317
69. Dasyam N, George P, Weinkove R (2020) Chimeric antigen receptor T-cell therapies: optimising the dose. Br J Clin Pharmacol 86(9):1678–1689. https://doi.org/10.1111/BCP.14281
70. Yan ZX et al (2019) Clinical efficacy and tumor microenvironment influence in a dose-escalation study of anti-CD19 chimeric antigen receptor T cells in refractory B-cell non-Hodgkin's lymphoma. Clin Cancer Res 25(23):6995–7003. https://doi.org/10.1158/1078-0432.CCR-19-0101/74891/AM/CLINICAL-EFFICACY-AND-TUMOR-MICROENVIRONMENT
71. Bu D et al (2021) Identification and development of PHE885: a novel and highly potent fully human anti-BCMA CAR-T manufactured with a novel T-charge TM platform for the treatment of multiple myeloma. Blood 138(Supplement 1):2770–2770. https://doi.org/10.1182/BLOOD-2021-148390
72. Engels B et al (2021) Preservation of T-cell stemness with a novel expansionless CAR-T manufacturing process, which reduces manufacturing time to less than two days, drives enhanced CAR-T cell efficacy. Blood 138(Supplement 1):2848–2848. https://doi.org/10.1182/BLOOD-2021-146246

73. Kishton RJ, Vodnala SK, Vizcardo R, Restifo NP (2022) Next generation immunotherapy: enhancing stemness of polyclonal T cells to improve anti-tumor activity. Curr Opin Immunol 74:39–45. https://doi.org/10.1016/J.COI.2021.10.001
74. Xia Y et al (2018) Treatment of metastatic non-small cell lung cancer with NY-ESO-1 specific TCR engineered-T cells in a phase I clinical trial: a case report. Oncol Lett 16(6):6998–7007. https://doi.org/10.3892/OL.2018.9534/HTML
75. Campbell A et al (2015) Concise review: process development considerations for cell therapy. Stem Cells Transl Med 4(10):1155–1163. https://doi.org/10.5966/SCTM.2014-0294
76. Kaiser AD et al (2015) Towards a commercial process for the manufacture of genetically modified T cells for therapy. Cancer Gene Ther 22(2):72–78. https://doi.org/10.1038/cgt.2014.78
77. Heathman TR, Nienow AW, McCall MJ, Coopman K, Kara B, Hewitt CJ (2015) The translation of cell-based therapies: clinical landscape and manufacturing challenges. Regen Med 10(1):49–64. https://doi.org/10.2217/RME.14.73/ASSET/IMAGES/LARGE/FIGURE6.JPEG
78. Hourd P et al (2014) Qualification of academic facilities for small-scale automated manufacture of autologous cell-based products. Regen Med 9(6):799–815. https://doi.org/10.2217/RME.14.47
79. Wang X, Rivière I (2016) Clinical manufacturing of CAR T cells: foundation of a promising therapy. Mol Ther Oncolytics 3:16015. https://doi.org/10.1038/MTO.2016.15
80. Jin J et al (2018) Enhanced clinical-scale manufacturing of TCR transduced T-cells using closed culture system modules. J Transl Med 16(1):1–13. https://doi.org/10.1186/S12967-018-1384-Z/FIGURES/7
81. Vedvyas Y et al (2019) Manufacturing and preclinical validation of CAR T cells targeting ICAM-1 for advanced thyroid cancer therapy. Sci Rep 9(1):1–15. https://doi.org/10.1038/s41598-019-46938-7
82. Marín Morales JM et al (2019) Automated clinical grade expansion of regulatory T cells in a fully closed system. Front Immunol 10:38. https://doi.org/10.3389/FIMMU.2019.00038/BIBTEX
83. Simmchen J, Ventura R, Segura J (2012) Progress in the removal of di-[2-Ethylhexyl]-phthalate as plasticizer in blood bags. Transfus Med Rev 26(1):27–37. https://doi.org/10.1016/J.TMRV.2011.06.001
84. Wang K, Deng Q (2019) The thermal and mechanical properties of poly(ethylene-co-vinyl acetate) random copolymers (PEVA) and its covalently crosslinked analogues (cPEVA). Polymers 11(6):1055. https://doi.org/10.3390/POLYM11061055
85. Meneghel J, Kilbride P, Morris GJ (2020) Cryopreservation as a key element in the successful delivery of cell-based therapies—a review. Front Med 7:824. https://doi.org/10.3389/FMED.2020.592242/BIBTEX
86. Papathanasiou MM, Stamatis C, Lakelin M, Farid S, Titchener-Hooker N, Shah N (2020) Autologous CAR T-cell therapies supply chain: challenges and opportunities? Cancer Gene Ther 27(10–11):799–809. https://doi.org/10.1038/s41417-019-0157-z
87. Lam C, Meinert E, Yang A, Cui Z (2021) Comparison between centralized and decentralized supply chains of autologous chimeric antigen receptor T-cell therapies: a UK case study based on discrete event simulation. Cytotherapy 23(5):433–451. https://doi.org/10.1016/J.JCYT.2020.08.007
88. Graham C, Jozwik A, Pepper A, Benjamin R (2018) Allogeneic CAR-T cells: more than ease of access? Cells 7(10):155. https://doi.org/10.3390/cells7100155
89. Dai X, Mei Y, Cai D, Han W (2019) Standardizing CAR-T therapy: getting it scaled up. Biotechnol Adv 37(1):239–245. https://doi.org/10.1016/J.BIOTECHADV.2018.12.002
90. Elsallab M, Levine BL, Wayne AS, Abou-El-Enein M (2020) CAR T-cell product performance in haematological malignancies before and after marketing authorisation. Lancet Oncol 21(2) Elsevier:e104–e116. https://doi.org/10.1016/S1470-2045(19)30729-6

91. Karakostas P, Panoskaltsis N, Mantalaris A, Georgiadis MC (2020) Optimization of CAR T-cell therapies supply chains. Comput Chem Eng 139:106913. https://doi.org/10.1016/J.COMPCHEMENG.2020.106913
92. Schultz L, Mackall C (2019) Driving CAR T cell translation forward. Sci Transl Med 11(481):2127. https://doi.org/10.1126/SCITRANSLMED.AAW2127/ASSET/3A6BC78D-0FDC-4AC6-89C9-C33D4A74498D/ASSETS/GRAPHIC/AAW2127-F1.JPEG
93. Novartis Pharmaceuticals (2020) Coding and billing guide for KYMRIAH ® (tisagenlecleucel), pp 14–15
94. Pharma K (2017) Yescarta Hospital billing process, p 80218
95. Pharma K (2009) Tecartus Hospital billing, pp 3817–3817
96. Roy M (2021) Bristol Myers, bluebird bio set multiple myeloma therapy list price at $419,500. https://www.reuters.com/business/healthcare-pharmaceuticals/bluebird-bio-sets-list-price-multiple-myeloma-therapy-419500-2021-03-29/. Accessed 8 Feb 2022
97. Abraham E, Trainor N, Ostrout N, Hewitt MM (2021) Cell therapy products are revolutionizing multiple therapeutic areas; to maintain this pace of innovation, manufacturing solutions must adapt. Curr Opin Biomed Eng 20:100340. https://doi.org/10.1016/j.cobme.2021.100340
98. Jenkins MJ, Farid SS (2018) Cost-effective bioprocess design for the manufacture of allogeneic CAR-T cell therapies using a decisional tool with multi-attribute decision-making analysis. Biochem Eng J 137:192–204. https://doi.org/10.1016/j.bej.2018.05.014
99. ClinicalTrials.gov (2020) CRISPR-edited allogeneic anti-CD19 CAR-T cell therapy for relapsed/refractory B Cell non-Hodgkin lymphoma. https://clinicaltrials.gov/ct2/show/NCT04637763#wrapper
100. Caribou Biosciences (2021) Caribou biosciences announces dosing of first patient in phase 1 clinical trial evaluating CB-010, a CRISPR-edited allogeneic anti-CD19 CAR-T cell therapy. In: Patients with relapsed or refractory B cell non-Hodgkin lymphoma
101. Qasim W et al (2017) Molecular remission of infant B-ALL after infusion of universal TALEN gene-edited CAR T cells. Sci Transl Med 9(374):1–9. https://doi.org/10.1126/scitranslmed.aaj2013
102. Clark M, Mateus M, Lempidaki S (2021) CellPoint and Lonza enter strategic collaboration to deliver CAR-T cells to patients at point-of-care
103. Sterner RM et al (2020) A graduate-level interdisciplinary curriculum in CAR-T cell therapy. Mayo Clin Proc Innov Qual Outcomes 4(2):203–210. https://doi.org/10.1016/J.MAYOCPIQO.2019.12.006
104. Sterner RC, Sterner RM (2021) CAR-T cell therapy: current limitations and potential strategies. Blood Cancer J 11(4):1–11. https://doi.org/10.1038/s41408-021-00459-7
105. di Stasi A et al (2011) Inducible apoptosis as a safety switch for adoptive cell therapy. N Engl J Med 365(18):1673. https://doi.org/10.1056/NEJMOA1106152
106. Jones BS, Lamb LS, Goldman F, di Stasi A (2014) Improving the safety of cell therapy products by suicide gene transfer. Front Pharmacol 5:254. https://doi.org/10.3389/FPHAR.2014.00254

Chapter 10
CRISPR Technology and Its Application in Therapeutics

Rigel Kishton, Montse Morell, Kevin Holden, Meenakshi Prabhune, Rebecca Roberts, Bobby Moon, and Rebecca L. Nugent

Abstract Clustered Regularly Interspaced Short Palindromic Repeats (CRISPR) technology is one of the most significant scientific breakthroughs of this century, revolutionizing therapeutics. A component of bacterial immune systems, CRISPR has been repurposed to precisely edit DNA sequences. The accuracy, efficiency, and simplicity of CRISPR technology have provided exciting new avenues to modify DNA and RNA for the treatment of human diseases. This includes both gene therapy, in which genes are edited in vivo in patients, and gene-edited cell therapy, in which cells are edited ex vivo and transfused into patients. This chapter is a comprehensive guide to CRISPR gene editing and its applications in biotherapeutics. It begins with an introduction to CRISPR-Cas9 technology, examining its discovery and adaptation, mode of action, advantages over previous tools, derivatives of the original CRISPR-Cas9 technology, and gene editing methods. This is followed by a review of the applications of CRISPR technology in biotherapeutics, including both gene therapies and gene-edited cell therapies for the treatment of three disease categories: genetic disorders, infectious diseases, and cancer. Relevant preclinical research and/or clinical trials are discussed for each disease category. The final

R. Kishton
Moonwalk Biosciences, South San Francisco, CA, USA
e-mail: rigel.kishton@moonwalk.bio

M. Morell
EditCo Bio, Redwood City, CA, USA
e-mail: montse.morell@editco.bio

K. Holden · R. Roberts · B. Moon
Synthego Corporation, Redwood City, CA, USA

M. Prabhune
The Scientist, Lab X Media Group, Midland , ON, Canada
e-mail: mprabhune@the-scientist.com

R. L. Nugent (✉)
Tessera Therapeutics, Sommerville, MA, USA
e-mail: rnugent@tesseratx.com

K. Gadamasetti, S. A. Kolodziej (eds.), *Bioprocessing, Bioengineering and Process Chemistry in the Biopharmaceutical Industry*,
https://doi.org/10.1007/978-3-031-62007-2_10

section delves into the future of CRISPR therapeutics, including nascent applications, the industrialization and projected growth of the industry, safety and ethical considerations, and its transformative potential for human health.

Keywords CRISPR · Gene editing · Genome engineering · Base editing · Primeb editing · Guide RNA · Cell therapy · Gene therapy

10.1 Introduction to CRISPR Technology

Clustered Regularly Interspaced Short Palindromic Repeats (CRISPR) gene editing is undoubtedly one of the most important scientific breakthroughs of the last century, allowing scientists to precisely modify DNA and RNA. This technology has driven the creation of a new generation of enhanced biologics for the treatment of diseases, as well as more relevant models of these diseases for research and therapeutic development purposes. To understand its revolutionary applications in bioengineering and therapeutics, we will first explore the basics of CRISPR technology, how it was discovered and adapted as a tool for gene editing, its mode of action, advantages over previous gene editors, and the different types of CRISPR systems currently available.

10.1.1 What Is CRISPR?

CRISPR is a key component of bacterial adaptive immune systems that has been repurposed as a powerful tool for gene editing [1]. Composed of a *C*RISPR-*as*sociated (Cas) nuclease and a customizable single-guide RNA sequence (sgRNA), the nuclease is directed by the sgRNA to a complementary target sequence and creates double-stranded breaks in the genomic DNA. These DNA breaks are repaired using inherent DNA repair mechanisms, resulting in premature stop codons or missense mutations, or creating conditions for inserting novel DNA payloads to precise genomic locations. The first CRISPR system adapted for gene editing used a Cas9 nuclease isolated from the bacteria *Streptococcus pyogenes* (SpCas9) [1]; this remains the most well-described and commonly used system.

10.1.2 Discovery and Adaptation of CRISPR

CRISPR was first discovered in the early 1990s by Dr. Francisco Mojica, a molecular microbiologist at Spain's University of Alicante. Dr. Mojica identified arrays of regularly spaced repeats in the DNA of a halophilic archaea [2], and later in other species of archaea and bacteria. After identifying a DNA sequence in the CRISPR

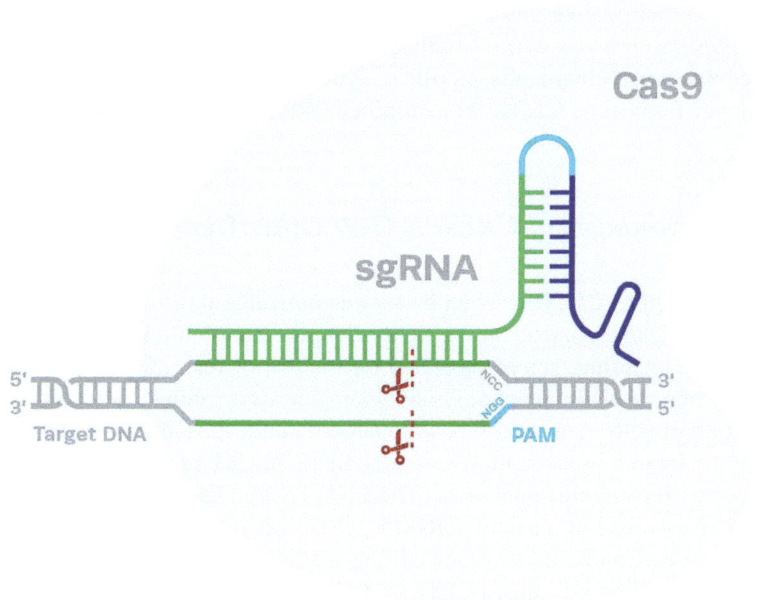

Fig. 10.1 The CRISPR-Cas9 systemThe CRISPR-Cas9 system comprises a guide RNA (gRNA) and Cas9 nuclease, which together form a ribonucleoprotein (RNP) complex (Fig. 10.1). The presence of a specific protospacer adjacent motif (PAM) in the genomic DNA is required for the Cas9 to bind to the target sequence. The Cas9 nuclease then makes a double-strand break in the DNA (denoted by the scissors). Endogenous repair mechanisms triggered by the double-strand break may result in gene knockout via frameshift mutation or knock-in of a desired sequence if a DNA template is present

array of *Escherichia. coli* identical to DNA from a virus that commonly infects these bacteria, he hypothesized that CRISPR loci were a component of the bacterial immune system [3]. This theory was confirmed in 2007 by Dr. Philippe Horvath by integrating a viral resistance sequence from the invading phage into its CRISPR loci [4] (Fig. 10.1).

In 2012, Dr. Emmanuelle Charpentier of Berlin's Max Planck Unit for the Science of Pathogens and Dr. Jennifer Doudna of the University of California, Berkeley, published a landmark paper in the journal *Science*, in which they described the molecular mechanism of the CRISPR-Cas9 system in the pathogenic bacteria *Streptococcus pyogenes* [1]. They explained that activity of the Cas9 nuclease is directed by a dual-RNA structure of a CRISPR RNA (crRNA) and a trans-activating crRNA (tracrRNA) which base-pair together through partial complementary sequences. By engineering the dual-RNA structure as a single RNA chimera, Doudna and Charpentier demonstrated that spCas9 can be reprogrammed and directed to specific sites [1]. They postulated that CRISPR-Cas9 could be readily adapted as a tool for gene editing, and their work resulted in the pair winning the 2020 Nobel Prize for Chemistry. Shortly after, Dr. Feng Zhang made a key contribution by demonstrating that CRISPR-Cas9 could be used to edit DNA in immortalized mammalian cell lines [5]. The translation of CRISPR-Cas9 editing from cell lines to primary cells initially proved challenging; however, a 2015 paper described the use of chemically modified, synthetic sgRNAs to efficiently edit primary human T cells and CD34+ hematopoietic stem and progenitor cells [6].

10.1.3 Advantages of CRISPR Over Other Gene Editors

The development of CRISPR as an easily customizable tool for gene editing was, quite simply, a game changer in the field of engineering biology. In the early days of genome engineering, scientists relied on the use of restriction enzymes, error-prone polymerase chain reactions, and mutagenesis to induce alterations in the genetic code of cells. Several tools were subsequently developed for more precise gene editing, including meganucleases, zinc finger nucleases (ZFNs), and transcription activator-like effector nucleases (TALENs) [7–9]. Due to the simplicity, precision, and relatively low cost of CRISPR, these previous methods are used less commonly [10]. Compared to other editing methods, which are laborious to customize or have a limited range of targets, CRISPR only requires the presence of a specific sequence (called PAM, specific protospacer adjacent motif) near the desired target site [11]. Each new target site simply requires a new sgRNA while maintaining the nuclease constant. Multiple edits within the same cell or organism are possible with relative ease just by introducing two sgRNAs [12, 13].

10.1.4 Mechanism Behind CRISPR Gene Editing

The Cas nuclease searches the genome for a sequence known as the protospacer adjacent motif (PAM), typically found 3–4 nucleotides downstream of a potential target site [1]. PAM requirements differ widely among CRISPR systems in terms of their sequence and stringency. SpCas9, the most widely used Cas9 nuclease, recognizes a 3 bp 5′-NGG-3′ PAM sequence, where N can be any nucleotide. If the PAM sequence is found, Cas9 will search the region upstream for a sequence that is complementary to its bound guide RNA. If both PAM and target sequence are present, Cas9 will create a double-stranded break (DSB) in the DNA [1].

Once CRISPR-Cas9 has generated a DSB in the DNA of a cell, editing can take place by exploiting cellular DNA repair pathways–predominantly the nonhomologous end joining (NHEJ) and homology-directed repair (HDR) mechanisms [14]. NHEJ is an error-prone DNA repair process that frequently results in either insertion or deletion mutations, together referred to as indels. This process can be exploited to render genes nonfunctional, otherwise known as a gene knockout [15]. Conversely, HDR of a DSB can result in the insertion of new genetic material, known as gene knock-in, if a suitable DNA donor template is provided along with the other editing components [16] as seen in Fig. 10.2.

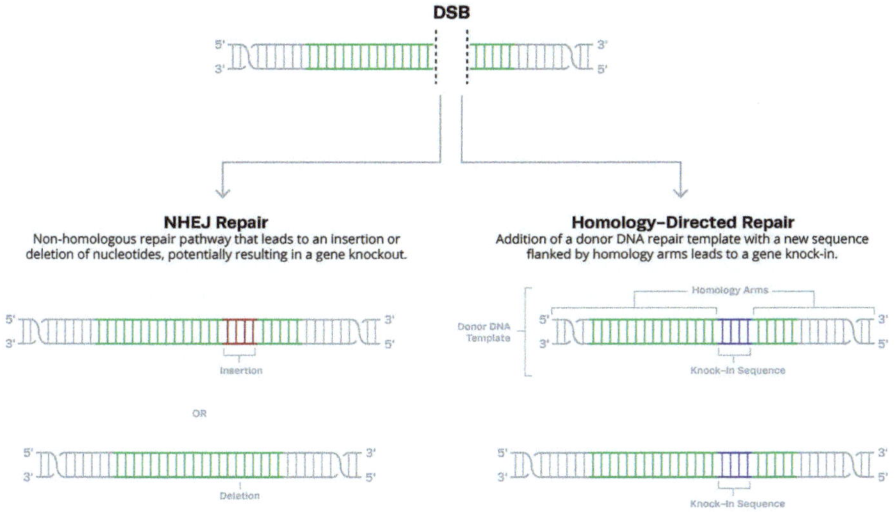

Fig. 10.2 CRISPR repair mechanismsThe two most common repair mechanisms facilitating CRISPR-Cas9 editing are nonhomologous end joining (NHEJ) and homology-directed repair (HDR). NHEJ results in either insertions or deletions (indels) of nucleotides to repair the DSB. HDR repairs the DSB by incorporating the sequence encoded in the DNA donor template into the target sequence

10.2 CRISPR Gene Editing: Techniques and Methods

Using CRISPR, several different techniques and methods to manipulate the genome have been developed, including the deletion and insertion of genetic material, or the upregulation or downregulation of gene expression.

10.2.1 Creating Gene Knockouts

Gene knockout (KO) is one of the most valuable tools in the functional genomics toolkit because it allows researchers to effectively "delete" a particular region of the genome, such that a functional protein is no longer produced. CRISPR-Cas9 is by far the simplest and most cost-effective way to generate gene knockouts in many cell types. If Cas9 creates DSBs in DNA, the cell will repair the break using NHEJ, which is prone to errors. This frequently results in indels that alter the reading frame, preventing a functional protein from being produced [6]. Because there are two copies of each gene, gene knockout experiments have three possible outcomes. The first is known as a heterozygous knockout, sometimes referred to as a monoallelic knockout (+/−); only one allele is successfully edited, leaving the other intact. The second is called a homozygous knockout, otherwise known as a complete or biallelic knockout (−/−); both alleles are edited and no protein can be produced.

The third and final outcome is that both alleles are edited to include frameshift mutations but each allele has different mutations—this is also referred to as a biallelic knockout.

The likelihood of generating a functional knockout increases when two or more guides target different regions of the same gene. Cas9 will create multiple DSBs and can therefore create large-scale deletions or multiple indels. This strategy is known as multiguide knockout and increases the chances of generating complete, functional gene knockout [17]. Additionally, CRISPR can be used to generate double or multiple gene knockouts through the introduction of multiple sgRNA sequences to the cell. Gene knockouts can be controlled spatially and temporally; constitutive knockouts delete the target gene in an entire organism, whereas conditional knockouts can delete the target gene in specific tissues or at a certain time point, for example, a specific developmental stage [18]. Cells and animal models generated using CRISPR are now used for research in labs around the world for studying gene function and disease modeling.

10.2.2 Creating Gene Knock-Ins

A gene knock-in (KI) involves the insertion of new genetic material into the genome of a cell or organism. Unlike knockouts, which rely on the cell's NHEJ-based repair mechanism to create indels, knock-ins take advantage of the other main DNA repair mechanism, HDR. A DNA knock-in template is designed to include homology regions on either side of the cut genomic DNA sequence, allowing for the targeted insertion of the desired genetic material [19]. Like knockouts, knock-ins can be either biallelic (heterozygous) or monoallelic (homozygous).

One of the most impactful therapeutic applications of KIs in therapy is in correcting pathogenic mutations for the treatment of genetic diseases. However, they are significantly more challenging to achieve than knockouts for two key reasons: first, HDR-based repair is less frequent than NHEJ in many cell types [20]; and second, providing a donor template for repair adds to the challenges of delivering editing reagents into cells. Due to the low knock-in editing efficiencies, a variety of techniques have been developed to increase the likelihood of success, including cell cycle synchronization [20] or use of chemical compounds to favor HDR over NHEJ.

10.2.3 CRISPR Activation and CRISPR Inhibition

In addition to inserting and deleting DNA sequences, CRISPR has been used to upregulate or downregulate gene expression using a catalytically dead Cas9 (dCas9) [21]. Because dCas9 cannot induce DSBs, it is used in these systems to specifically bind to DNA. CRISPR activation (CRISPRa) uses dCas9 fused to transcriptional effectors such as VP64 or p65, directing them to promoter or enhancer regions of

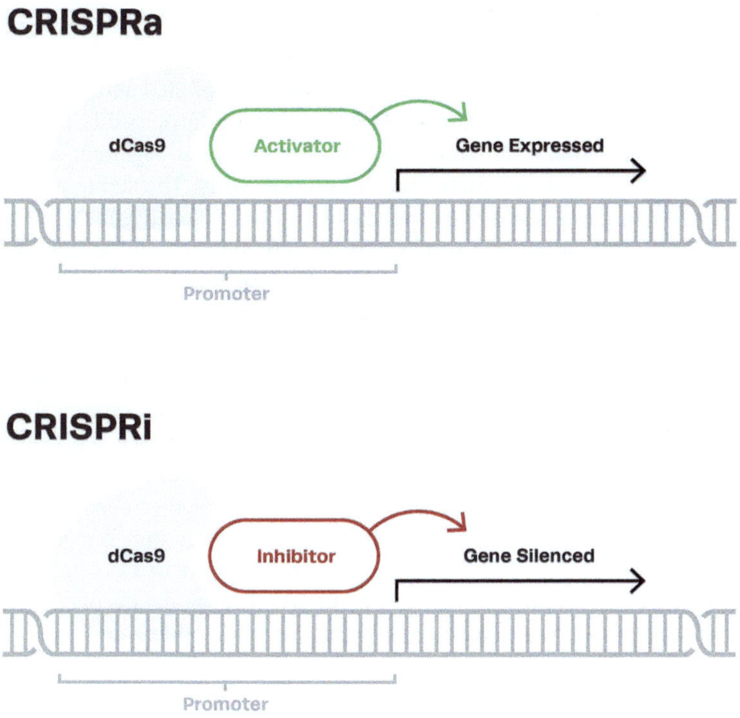

Fig. 10.3 Mechanisms of CRISPR interference and activationCRISPRi and CRISPRa methods use catalytically dead Cas9 (dCas9) to alter the gene expression of targets by directing a transcriptional regulator protein near the gene transcriptional start to (**a**) inhibit or (**b**) activate its expression

DNA upstream of the target gene, resulting in increased transcription, and therefore translation, of the gene. In contrast, CRISPR interference (CRISPRi) can either simply use dCas9 to bind to DNA to interfere with transcription or fuse dCas9 to the Kruppel-associated box (KRAB) repressor domain, directing it to the promoter region to reduce transcription [22]. The latter method provides more consistent and reliable gene silencing than dCas9 alone. Like knockouts and knock-ins, these systems, as seen in Fig. 10.3, can be used to elucidate gene function by modulating their expression and examining the effects on the phenotype of cells and organisms.

10.2.4 CRISPR Screens

CRISPR screens involve large-scale genetic manipulations in cells to examine the impact of gene modifications on cell phenotypes and attributes, therefore determining the function of various genes [23]. CRISPR screens can be loss-of-function, whereby genes are knocked out or downregulated by CRISPR or CRISPRi, or gain-of-function, whereby genes are upregulated using CRISPRa. Loss-of-function screens are useful tools in functional genomics, particularly as they allow researchers to identify and validate targets for drug discovery.

CRISPR screening can be conducted in two formats, pooled and arrayed. Pooled screens commonly involve packaging sgRNA-containing plasmids into lentiviral particles (one per vector) and then combining each virus together equally. This pooled lentiviral library is then transduced into host cells. The stable expression of guide (along with Cas9) facilitates the knockouts of targeted genes. Because the edits occur across all targets in a single tube of cells, it is difficult to link the phenotype of each individual cell with the underlying genetic perturbation. Pooled screens are thus only compatible with binary assays that physically separate edited cells exhibiting a phenotype of interest from those that do not. The enrichment or depletion of particular sgRNAs following selection relays information about the involvement of these genes in a particular phenotype and indicates targets that can be investigated further. Arrayed screens involve targeting each gene separately across wells of a multi-well plate (Fig. 10.4). This format is a newer technology that is more versatile in both methodology and analysis than pooled screens. Library delivery may be accomplished through transfection or transduction. Because gene targets are separated across wells, phenotypes do not need to be selected for and sequencing, and data deconvolution is not required to associate phenotypes with genotypes. Arrayed screens are compatible with both binary and multiparametric assays.

10.3 CRISPR Evolution and Derivative Technologies

Since the discovery of the original CRISPR-Cas9 system, researchers have identified many new CRISPR systems, expanding the range of components that can be delivered for therapeutic purposes. Known CRISPR systems are currently divided into two classes, six types, and 33 subtypes [24]. Class 1 is made up of multi-subunit effector proteins, further divided into types I (subtypes A-G), III (subtypes A-F), and IV (subtypes A-C). Class 2 comprises single effector proteins and is divided into types II (subtypes A-C, which includes the famous spCas9), V (subtypes A-I, K), and VI (subtypes A-D) [24]. These systems differ widely in size, complexity, targeting, and cleaving ability. This includes Cas9 nucleases from other species that are smaller than SpCas9, as well as other nucleases that recognize different PAM sequences, target RNA rather than DNA, target single-stranded DNA rather than double-stranded, or work at different temperatures. Several naturally occurring class 2 Cas nucleases are now offered by commercial vendors like Synthego for creating engineered cells. *Staphylococcus aureus* Cas9 (SaCas9) is now a popular nuclease for therapeutic editing because its corresponding DNA sequence is 1 kb smaller than SpCas9 and therefore easier to deliver in vivo [26]. Cas12a is also increasing in popularity because it creates staggered or "sticky-end" DSBs, making

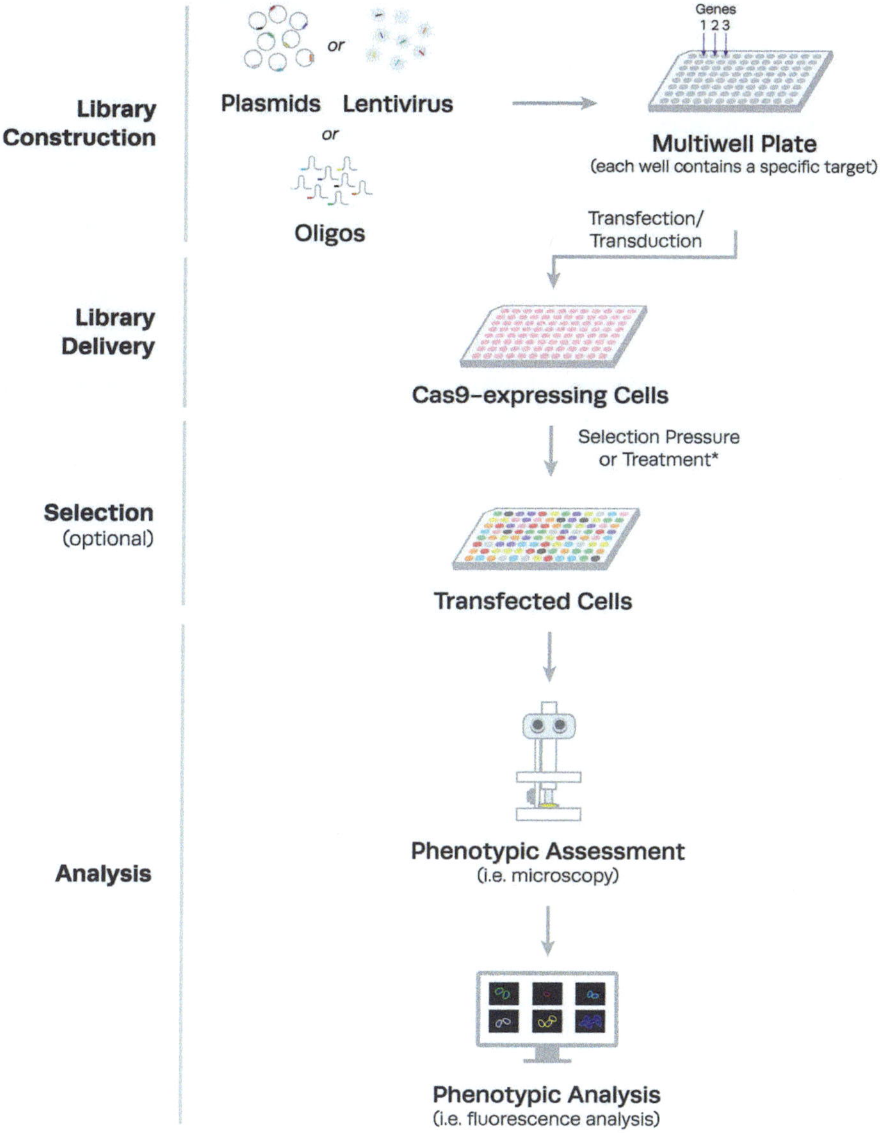

Fig. 10.4 Example of a CRISPR arrayed screening workflowIn a knockout CRISPR arrayed screening, a gRNA targeting each gene is introduced to each well of a multi-well plate. After editing the cells, applying treatment (such as a drug) enables the identification of which genes (when rendered inoperative) that cause the cells to have heightened or diminished sensitivity to the drug. This process can be used to identify potential drug targets

it more suitable for knock-in experiments relying on HDR for the repair [27], as compared to SpCas9, which creates blunt-end DSBs. Cas12a can also be used for multiplexed editing because it can process multiple guides simultaneously [28].

Additionally, many new Cas9 nucleases have been engineered for specific purposes: high-fidelity Cas9 variants, which have increased specificity [29]; Cas9 nickases, which cut only a single strand of the DNA double helix [43]; Cas9 nucleases with different PAM recognition sites or specificities, which have an expanded range of targets [30]; and catalytically dead Cas9, which does not cut DNA but retains its recognition and binding capacity [31]. The original SpCas9 system remains the most widely used, largely because it is the most well-studied and therefore has a more predictable safety profile for therapeutic use. However, several naturally occurring and engineered nucleases are also approaching clinical use. For example, the EDIT-301 ex vivo gene-edited cell therapy (NCT04853576) developed by Editas Medicine aims to treat sickle cell disease (SCD) using Cas12a (see Sec. 10.7.1).

10.3.1 Derivative Technologies

A myriad of technologies has been derived from CRISPR that perform genomic modifications without generating DSBs. They deploy variants of Cas9 that will nick (nCas9) or bind to (dCas9) the DNA. One notable example is base editing, which generates a base substitution. Base editors comprise Cas9 variants (nCas9 or dCas9) fused to enzymes that either convert adenosine to inosine which ultimately results in an A-T to G-C substitution (adenine base editors or ABEs) [25] or cytidine to uracil, generating a C-G to T-A substitution (cytosine base editors or CBEs) [26] Since their development, multiple iterations of these base editors have been created to improve the activity as well as minimize off-target effects.

Prime editing is another derivative of CRISPR, using nCas9 fused to an engineered reverse transcriptase [27]. This system requires the use of a specific type of guide RNA known as a prime editing guide RNA (pegRNA), which contains both a guide sequence and a template encoding for the desired substitution(s). After nCas9 creates a single-stranded break, it is repaired via reverse-transcription of the provided template encoded in the pegRNA.

Another novel technology called Programmable Addition via Site-specific Targeting Elements (PASTE) has been developed to allow for very large insertions of DNA [29]. PASTE editing works via a dual system of prime editors and serine integrase proteins. In a first step, the prime editor knocks in the integrase landing sequence in the desired target location. Next, the integrase performs targeted DNA insertion of the desired sequence in the previously generated landing pad.

An additional nascent application of CRISPR is in epigenome editing (eGE), which uses dCas9 fused to epigenetic enzymes or their catalytic domains, such as DNA methyltransferases (DNMT) or ten-eleven translocation dioxygenases (TETs) to regulate the expression of genes without inducing DSBs [22].

Another level of modifying protein expression without DSBs is targeting messenger RNA (mRNA). Many species of bacteria possess CRISPR systems that are capable of targeting RNA, such as those found in Class 2, type VI CRISPR-Cas13 systems [32]. Unfortunately, the first reported RNA targeting enzyme (Cas13) presents collateral activity; after cleaving the target transcript, it will degrade other transcripts present in the cell nonspecifically. This has led to the search for other RNA-targeting enzymes that can efficiently edit RNA without collateral activity, such as Cas7–11 [33].

10.4 CRISPR Component Format and Delivery Methods

For editing to occur via a nuclease, a guide RNA and knock-in template (for knock-ins) must be delivered into cells. These components can be delivered as a number of biological constructs such as DNA (plasmid), RNA, or protein. The delivery of the components into the cell can be through viral, chemical, or mechanical methods. The choice of biological construct and delivery method is dependent on the final application and therapeutic conditions, such as delivery for in vivo vs. ex vivo therapies.

10.4.1 Gene Editing Components: Cas Nucleases and Cargo Formats

Introduction of the Cas nuclease and sgRNA to cells can take three possible forms: (i) DNA plasmid(s), (ii) mRNA transcripts for translation, or (iii) a ribonucleoprotein (RNP) complex [32]. DNA plasmids were commonly used in the early years of genetic engineering; however, several factors have hamstrung their therapeutic value. These include variable editing efficiencies, cytotoxicity issues, potential random incorporation into the host genome, and difficulties choosing appropriate promoters for nuclease protein expression. The use of DNA plasmids results in delayed editing owing to the time required for the transcription and translation of the editing components, and the long-term expression of Cas9 can result in off-target editing. RNPs have rapidly become a very popular form of CRISPR cargo for therapeutic purposes for several key reasons. RNP cargos result in rapid, precise editing because the components are immediately active, and reduce the risk of toxicity and off-target editing events. Moreover, RNPs are amenable to delivery using almost all of the currently available vehicles. Therapeutic gene knock-in adds another layer of complexity to delivery. In this case, an additional editing component of a DNA donor template for HDR must be provided along with the Cas nuclease and sgRNA.

10.4.2 Delivery Methods

There are three main categories of CRISPR delivery vehicles: viral, chemical, and mechanical. The final application and therapeutic conditions will influence the cargo format as well as the choice of vehicle for delivery. Compared to ex vivo therapies, in which cells are edited outside the body, in vivo therapies present a much more challenging delivery scenario with minimal room for errors owing to patient safety requirements [33]. Moreover, in vivo gene therapies typically require editing components to be targeted to specific organs or tissues, providing an additional delivery challenge.

Viral vectors work on the basis of infecting target cells or tissue to introduce the editing components [34]. These vectors include lentiviruses (LVs; *Retroviridae*), adenoviruses (AdVs; *Adenoviridae*), and adeno-associated viruses (AAVs; *Parvoviridae*). These viral vectors act as chassis for the editing components to enter the cell and contain elements encoding for the nuclease and guide sequences, referred to as cargo. LVs and AdVs are large and can therefore carry considerable cargo; however, they have several undesirable effects, including integration into the host genome, adverse immune responses, and off-target effects. In contrast, AAVs are nonpathogenic and nonintegrating viruses in humans, causing very mild host immune responses. These safety advantages make them one of the most common delivery vehicles for in vivo therapeutic use. However, they cannot package DNA above 4.7 kb, limiting their application in knock-ins. Strategies to overcome this challenge include using smaller Cas9 variants such as SaCas9 to minimize the overall cargo load and splitting the CRISPR components between two vectors. Engineered virus-like particles (eVLPs) have also recently been described as a highly efficient delivery vehicle for CRISPR RNP cargo [35].

Chemical delivery vehicles for CRISPR editing components include liposomes, lipid nanoparticles (LNPs), lipo- and poly-plexes, gold and other inorganic nanoparticles, and cell-penetrating peptides [36]. Nanoparticles such as LNPs are increasing in popularity and have been adapted for use in other non-CRISPR therapeutics, such as the delivery of mRNA vaccines. Mechanical vehicles for CRISPR delivery include microinjection into cells, electroporation, nucleofection, and microfluidics. However, these are not compatible with in vivo therapies and therefore are exclusively used for ex vivo gene-edited cell therapies and in vitro experiments.

10.5 CRISPR Therapeutics: Gene Therapies and Gene-Edited Cell Therapies

The development of CRISPR-based therapeutics has ushered in a new era of personalized medicine, including both gene therapy and gene-edited cell therapy. In the decade since CRISPR was first adapted as a gene editor, the field has made astonishing progress, with the first ex vivo CRISPR gene-edited cell therapy beginning clinical trials in 2016, and the first in vivo gene therapy beginning trials in 2019.

10.5.1 What Are Cell and Gene Therapies?

Cell and gene therapies are a broad category of medical therapeutics based on the transfer of either whole cells (cell therapy) or genetic material (gene therapy) to patients to treat disease. Gene therapies are performed in vivo and involve either introducing new genetic material to patients or disrupting genes that bear pathogenic mutations [37]. There are currently over 5000 known diseases caused by mutations in a single gene [38]. These monogenic disorders can theoretically be treated by either correcting the pathogenic mutation, known as gene correction therapy, or delivering a functional copy of part or all of the gene to alleviate the condition, known as gene replacement therapy [39].

Cell therapies engineered using CRISPR or other technologies are known as gene-edited cell therapies. Gene-edited cell therapies include ex-vivo biologics that can be either autologous (patient-derived) or allogeneic (derived from a healthy donor) [40]. They aim to replace diseased cells, compensate for a lack of a particular cell type, or generate a desired functionality through engineered physiological activity. Many different types of gene-edited cell therapies have been developed, and they can be categorized by the types of cells used, for example, hematopoietic stem and progenitor cells (HSPCs), or primary immune cells like T cells. Gene-edited cell therapies are becoming increasingly common for the treatment of many different medical conditions, including cancer, infectious diseases, and genetic diseases.

10.5.2 Role of CRISPR in Cell and Gene Therapy

The simplicity, accuracy, and wide range of potential targets of the CRISPR system make it ideal as a therapeutic agent for many different diseases. CRISPR's adoption in the field of cell and gene therapy has been rapid. The first clinical trial of an ex vivo CRISPR gene-edited cell therapy began in 2016, only 4 years after the original CRISPR paper was published, followed by another in 2018. The first trial of a CRISPR-based in vivo gene therapy began in 2019. The popularity and rapid clinical translation of CRISPR gene therapies and gene-edited cell therapies have only increased, with new trials beginning regularly. While currently no CRISPR therapies have been approved by the Federal Drug Administration (FDA), it is likely that many cell and gene therapies will be CRISPR-based in the near future.

10.6 CRISPR in Cancer Therapy

Cancer comprises a diverse, heterogeneous group of diseases, making it difficult to model and treat [41]. Cancers are driven by mutations in DNA, and various types of cancer are often characterized by commonly occurring signatures of mutations in oncogenes and tumor suppressor genes [41]. Heterogeneity in the drivers and features of tumors is observed between tumors arising from various tissue types and is also seen within tumors of similar tissue origin. Further, there can be considerable heterogeneity within a single tumor mass due in part to the variety of cell types that are contained within the complex tumor microenvironment [42]. The genetic complexity of cancer influences not only how treatable the disease is but also how it spreads, the potential for drug resistance, and the likelihood of recurrence. Metastatic cancer, which has spread to multiple tissues and organs, presents challenges because each tumor may exhibit a heterogeneous mutational profile, potentially resulting in differential biological features and responses to treatment [42, 43]. Additionally, cancer cells have significant capacity to either avoid detection by or otherwise suppress the immune system [44].

CRISPR technology has quickly accelerated cancer research, both in terms of disease modeling and therapeutic strategies [45]. In therapeutic settings, CRISPR gene editing enables scientists to create better, safer, and more affordable gene-edited cell therapies. These therapies offer the ability to treat cancer in a highly specific manner and are highly desirable over traditional treatments like chemotherapy and radiotherapy, which are toxic to healthy cells.

Cellular immunotherapies that have been broadly used in the clinic as cancer treatments include the infusion of expanded tumor-infiltrating lymphocytes, T cell receptor (TCR) engineered T cells, and chimeric antigen receptor (CAR) engineered T cells [46]. These are often referred to as adoptive cell therapies, and each of these therapies generally involves harvesting T cells from patients and promoting or imbuing antitumor specificity. Following ex vivo expansion to a large number of cells, the T cell products are infused into patients [47]. Cellular immunotherapies for the treatment of hematological cancers have generally shown more clinical efficacy to date than those used to treat solid tumors. CRISPR-mediated gene editing in T cell therapeutic products is now being explored as an avenue to generate cell therapies that are safer and more efficacious than previous iterations.

10.6.1 Tumor-Infiltrating Lymphocyte Therapy

Tumor-infiltrating lymphocytes (TIL) are T cells that are localized within a tumor, with a fraction of TIL exhibiting the capacity to recognize tumor antigens and kill tumor cells. The degree of TIL infiltration in tumors can have positive prognostic value in several types of cancer, and TIL can be expanded ex vivo and reinfused in patients as an autologous cancer therapy. The infusion of expanded TIL in lymphodepleted patients in conjunction with interleukin-2 (IL-2) support can be clinically effective, with high rates of response reported in patients with metastatic melanoma. Despite the efficacy of TIL in certain settings, the expansion and efficacy of TIL can be limited in cancers that evade T cell surveillance and recognition [48]. Additionally, the antitumor activity of TIL may not be sufficient to mediate durable clinical responses in many patients. Consequently, CRISPR gene editing of TIL has been explored as a pathway to enhance T cell expansion and function, such as through the knockout of T cell intrinsic genes that inhibit antitumor function [49, 50].

10.6.2 TCR Engineered T Cell Therapy

T cell receptor (TCR) engineered T cell therapy uses T cells that are modified to express TCRs conferring the capacity for the specific recognition and destruction of tumor cells [51]. TCRs mediate T cell recognition of tumor cells through the major histocompatibility complex (MHC)/human leukocyte antigen (HLA) complex [52] presentation of antigenic peptides. The presented peptides can originate from either intracellular or membrane-bound proteins. TCR engineered T cells are produced by the engineered expression of exogenous TCRs with a known capacity to recognize tumor antigens in therapeutic cell products. The exogenous TCR has historically been introduced into T cells through viral transduction, but recent work has demonstrated that CRISPR-mediated knock in of a therapeutic TCR is a viable strategy to produce clinical products. As seen with TIL therapy, TCR engineered T cells can mediate clinical responses in select patients, but widespread efficacy is lacking. This is due to a number of factors, including tumor downregulation of HLA expression and mispairing of engineered TCR protein components with endogenous TCR proteins. Mispairing of the transgenic TCR with the endogenous TCR is a particularly concerning issue, as it can result in inefficient antigen recognition or even autoimmunity and graft-versus-host disease [52].

CRISPR-Cas9 editing can be used to knock out the endogenous TCR as well as knock-in the transgenic TCR, eliminating mispairing and toxicity issues [47]. Recent research used CRISPR-Cas9 to knock out the endogenous TRAC (T cell Receptor Alpha Chain) locus in primary T cells and knock in a TCR with specificity to epitopes of Wilm's tumor antigen 1, which is overexpressed in many types of cancer [53]. The resulting TCR cell therapy displayed potent activity against both hematological malignancies and solid tumors in mouse models, including acute myeloid leukemia, acute lymphoblastic leukemia, and glioblastoma, with no observable off-target editing [53].

10.6.3 Chimeric Antigen Receptor T Cell Therapy

Chimeric antigen receptor T (CAR-T) cell therapy involves engineering T cells to express chimeric receptors on their surface to mediate recognition and killing of cancer cells [54]. Unlike TCRs, which are naturally occurring and recognize MHC/HLA, chimeric antigen receptors are engineered proteins that recognize specific antigens expressed on the surface of cancer cells.

Similar to other cell immunotherapies, CAR-T is often inhibited by the cancer cells overexpressing immune checkpoint inhibitors, such as programmed cell death protein 1 ligand (PD-L1) on the cell surface. PD-L1 binds to the PD-1 receptor expressed on the surface of T cells, inhibiting immune cells [55]. Overcoming the PD-L1 inhibition pathway provides a promising avenue for next-generation cell therapies via a CRISPR knockout of PD-1 [55]. CRISPR can also be used for targeted knock-in of the CAR, which is then continuously expressed on the cell surface. Results of a Phase I trial of mesothelin-specific CAR-T cells with *PD-1* and *TCR* gene knockouts have established the safety and feasibility of this method in human subjects [56].

Allogeneic CAR-T cells, although advantageous to manufacture relative to autologous cell products, are often unable to persist in hosts. This is due to recognition and rejection by the host immune system or the presence of cytotoxic drugs. CRISPR has been employed in attempts to circumvent this issue by engineering universal allogeneic CAR-T therapy; by knocking out key genes involved in recognition by the host immune system, the cells can theoretically be used for any patient. Unlike previous technologies, CRISPR-Cas9 can be used to perform these complex edits; currently, several clinical trials are underway, in which multi-knockout allogeneic CAR-T cells are being tested in human patients with various types of cancer [57, 58]. By generating CAR-T cells from iPSCs rather than from healthy donors, large stocks of "off-the-shelf" CAR-T cells are ready to treat patients with minimal wait times.

10.6.4 In Vivo CRISPR Editing of Solid Tumors

In recent years, CRISPR has also been adapted for in vivo editing of cancers by targeting cancer-specific mutations [59] or otherwise disrupting oncogenes. CRISPR can be used to disrupt the expression of key proteins that are overexpressed in cancer cells, such as Nuclear Factor Erythroid 2-Related Factor 2 (NRF2) [60]. NRF2 regulates hundreds of other genes, and its overexpression in cancer cells can result in resistance to chemotherapy. However, CRISPR-based knockout of the NRF2 protein can restore chemosensitivity of the cancer cells. This strategy is being investigated in preclinical studies by researchers at the ChristianaCare Gene Editing Institute [60]. Other recent research in tumor editing includes improving delivery systems to efficiently target tumors in vivo. For example, one key study from Tel

Aviv University developed targeted lipid nanoparticles (LNPs) for the delivery of CRISPR-Cas9 components [61] to target the KO of the *PLK1* gene, resulting in high (~80%) editing efficiencies, potent inhibition of tumor growth, and improved survival rates [61].

10.7 CRISPR Treatments for Genetic Disease

There are currently over 7000 known monogenic disorders that are ideal targets for CRISPR-mediated gene therapy [38]. CRISPR therapies for monogenic diseases can take several different forms based on the type of mutation(s) which cause the disease. For example, if a disease is caused by a gain-of-function mutation in a particular gene, CRISPR knockout gene therapy can be used to delete the gene or the specific regions which bear the pathogenic mutations. Alternatively, if the disorder is caused by loss-of-function mutations in a gene that cause it to produce an abnormal protein or no protein at all, CRISPR knock-in can be used to correct the mutations, which is also referred to as gene correction therapy. Disorders in which a particular gene is overexpressed or underexpressed can also be therapeutically targeted using CRISPRi and CRISPRa, respectively. Many monogenic disorders are caused by single nucleotide polymorphisms (SNPs), making them possible to correct using base- and prime-editing techniques which do not induce DSBs. Depending on the target organ or tissue, these methods can be performed in vivo as gene therapies, or ex vivo as gene-edited cell therapies. In this section, we cover these different therapeutic options with examples from proof-of-concept and preclinical studies.

10.7.1 CRISPR Gene Correction for Monogenic Diseases

SCD is perhaps one of the best-known monogenic disorders and was the first genetic disease for which researchers attempted to generate a CRISPR cure. SCD is caused by a SNP mutation in the ß-globin gene (*HBB*), which normally produces adult hemoglobin (hemoglobin A) and oxygenates the blood [62]. The traditional treatments are blood transfusions or a bone marrow transplant from a healthy donor. CRISPR has offered several promising avenues for treating this debilitating condition via autologous gene-edited HPSC therapy.

The first and currently most popular method is sometimes referred to as gene switching or gene substitution. Fetal hemoglobin (hemoglobin F) is produced until ~6 months of age in humans, at which time it is switched off by activation of the *BCL11A* transcription factor and replaced by hemoglobin A. Hemoglobin F is an adequate substitute for hemoglobin A [63]. Therefore, by performing CRISPR-Cas9 knockout of *BCL11A*, hemoglobin F production can be restored. This approach has been championed by several biotechnology companies, including CRISPR Therapeutics and Vertex Pharmaceuticals, who jointly ran the first clinical trial based on this approach using their CTX-001 therapy [64]. Alternative approaches are being developed by Editas Medicine [65], using the Cas12a nuclease, and Beam Therapeutics, using base editing [66].

Therapies for SCD based on upregulating hemoglobin F can also be used to treat a related condition called beta-thalassemia. While they have similar symptoms and limited treatment options, beta-thalassemia is caused by multiple mutations in *HBB* rather than just one SNP. This makes it more difficult to cure with gene therapy because repairing many mutations is more challenging than correcting a single one. The ability to treat both diseases with a single therapy is why many companies have focused on the therapeutic upregulation of hemoglobin F. The CTX-001 therapy developed by CRISPR Therapeutics and Vertex Pharmaceuticals has also been tested on beta-thalassemia patients with similar levels of success [67].

Many genetic disorders require the correction of one or more pathogenic mutations in a particular gene. X-linked agammaglobulinemia (XLA) is an immunodeficiency disorder caused by a variety of mutations in the Bruton's tyrosine kinase (*BTK*) gene [68]. Researchers at the University of California, Los Angeles created a CRISPR gene-edited cell therapy approach for XLA, correcting the majority of pathogenic mutations in the BTK gene in peripheral blood stem cells (PBSCs) [68]. The corrected PBSCs could therefore be used as an autologous cell therapy for XLA, with preclinical research being continued for this strategy.

10.7.2 CRISPR Knockout of Repeat Hyperexpansions

Some genetic diseases are caused by multiple repeats of certain nucleotides that form hyperexpansions and interfere with the normal production of protein from the gene. Friedreich's ataxia (FA) is an autosomal recessive disorder caused by mutations in the *FXN* gene which encodes frataxin, a mitochondrial protein responsible for aspects of energy production and iron metabolism [67]. The *FXN* gene bears a GAA trinucleotide repeat in its first intron. Healthy individuals have five to 30 repeats. FA patients have hyperexpansion of between 44 to 1700 repeats of this trinucleotide sequence [69]. The number of GAA repeats is correlated with the severity of FA, as well as the age of onset.

CRISPR offers the potential to cure FA by excising the pathogenic trinucleotide hyperexpansion; researchers from the University of California, San Diego School of Medicine developed a CRISPR method to delete large sections of the expansion in patient-derived hematopoietic stem and progenitor cells (HSPCs). The treatment successfully increased the production of frataxin protein in the HSPCs, which then differentiated normally with no noticeable off-target editing or cellular toxicity [69]. Additional therapies are being developed by CRISPR Therapeutics and Capsida Biotherapeutics for FA [70]. As part of the collaboration, these companies are also creating a similar treatment for amyotrophic lateral sclerosis (ALS), a large number of cases of which are caused by a GGGGCC hexanucleotide repeat within the *C9orf72* gene [71].

10.7.3 CRISPR In Vivo Gene Therapies

Leber's congenital amaurosis (LCA) is a retinal disease caused by loss-of-function mutations in multiple genes affecting the rod and cone cells of the retina [72]. The most common causes are mutations in the *RPE65* and *CEP290* genes, making these genes prime targets for gene therapy. An early attempt to generate a cure for LCA was developed by Spark Therapeutics in a therapy known as Luxterna, which simply provided a functional copy of *RPE65* cDNA in an AAV vector via sub-retinal injection [73]. However, this therapy is not a one-time cure, and the patient group suitable for treatment with Luxterna is limited.

Editas Medicine more recently developed a CRISPR-based gene therapy known as EDIT-101, an RNP cargo packaged in an AAV5 delivery vehicle and administered via sub-retinal injection [74]. The BRILLIANCE clinical trial of EDIT-101 was the world's first-ever trial of an in vivo CRISPR gene therapy, using CRISPR-Cas9 to edit the *CEP290* gene, restoring healthy expression of the CEP290 protein and photoreceptor function. Interestingly, rather than correct the IVS26 mutation in *CEP290*, EDIT-101 uses two guide RNAs to create a DSB on either side of the mutation and delete it [75]. After repair, transcription and RNA splicing are restored, and the protein can be translated normally. Preliminary results of the ongoing Phase I/II trial of EDIT-101 indicate no serious toxicity or adverse events of the therapy and evidence for clinical benefit [76]. Base editing is also being explored as an alternative therapy for LCA by researchers at the University of California, Irvine (UCI), with a proof-of-concept study demonstrating the correction of the *RPE65* mutations in vivo in a mouse model of the disease [77].

10.8 CRISPR in Infectious Disease Treatment

While cancer and genetic disease may appear the most obvious targets for CRISPR therapies, there is also significant potential for this technology in the treatment of infectious diseases. Pathogenic microorganisms, including viruses, bacteria, fungi, and parasites, can all be targeted using CRISPR, and there has been significant progress in this field in recent years.

10.8.1 Circumventing Antibiotic Resistance in Bacteria

CRISPR's role in protecting bacteria from invading bacteriophages is now being used to turn the tables, targeting bacteria themselves. This can involve killing the target pathogenic bacteria or deleting or repressing the transcription of key genes in the bacterium, such as antibiotic resistance genes. A key strategy for the targeted killing of pathogenic bacteria is to engineer bacteriophages using CRISPR systems [78]. The key benefit of these therapies is that they eliminate the need for antibiotic treatments and can be used to target multi-drug resistant "superbugs" [78]. This approach has been pioneered by Locus Biosciences in their clinical trial of LBP-EC01, which uses CRISPR-Cas3 enhanced bacteriophages (termed crPhage™) for the therapeutic removal of *Escherichia coli* strains that cause urinary tract infections [79].

10.8.2 CRISPR-Edited Cell Therapies for the Treatment of Viral Pathogens

CRISPR technology offers the possibility of targeting a variety of viral pathogens [80], including DNA viruses, such as human papillomavirus (HPV) and Epstein-Barr virus, as well as RNA viruses, such as SARS-CoV-2. The E6 and E7 genes of HPV have been the subject of many recent CRISPR studies due to their roles in causing cervical cancer, making them key targets for CRISPR-based knockout [81]. In preclinical research, CRISPR-Cas9 was used to knock out the E7 gene in a mouse model bearing HPV16 tumors; treatment was effective at clearing the tumors from the animals without causing liver or spleen toxicity [82].

Another promising option for the treatment of infectious diseases is CRISPR-edited B cell therapy. B cells produce antibodies and are an essential component of the adaptive immune system [83]. Vaccines are designed to prime the adaptive immune system by encouraging B cells to recognize new pathogens and produce antibodies for them [84]. CRISPR-edited B cells have been suggested as an alternative to vaccines because they can be used to produce large quantities of neutralizing antibodies for pathogens via targeted gene knock-in [83, 84]. This strategy has been adopted to treat the respiratory syncytial virus, HIV, Epstein-Barr virus, and influenza viruses [85].

10.8.3 In Vivo CRISPR Therapy to Excise Integrated Viruses from the Human Genome

HIV is a retrovirus that infects human CD4 helper T cells, in which it reverse-transcribes its RNA genome and then inserts itself into the DNA genome of the host cells [86]. Due to its integration into the human genome and its ability to hide in dormant, noncirculating helper T cells for long periods of time, even antiretroviral drugs can never completely remove HIV from the body [87]. CRISPR now offers a potential cure for HIV, cutting it out of the human genome with minimal risk of off-target editing using a multi-guide approach successful for excising large DNA sequences [20]. Excision Bio is currently testing their in vivo EBT-101 therapy for HIV in a Phase I/II trial [88] and is also developing similar therapies for other viruses. This includes EBT-103 therapy for JC virus, which can cause progressive multifocal leukoencephalopathy, EBT-104 for herpes simplex virus, and EBT-107 for hepatitis B virus [89].

10.9 The Future of CRISPR Therapeutics

Given the current trends around FDA approvals, the prediction is that in the near future, approximately one-third of medical treatments will be cell and gene therapies. Given the advantages of CRISPR over previous gene editing technologies, it is likely that many of these cell and gene therapies will be CRISPR-based. In the next section, the critical factors that will determine the future of CRISPR technology in medical therapeutics will be explained.

10.9.1 Industrialization of CRISPR Editing: Increasing Safety and Lowering Costs

Many of the CRISPR cell and gene therapies currently in trials show promise. However, the size of patient groups in clinical studies is generally limited. If a trial is successful and the therapy in question gains FDA approval, there are often significant obstacles in scaling up these therapeutics for widespread use in larger patient populations [90]. This includes accessing viral vectors for delivery, the necessary Good Manufacturing Practice (GMP)-grade reagents, and adequate production facilities. Significant costs are associated with cell and gene therapies, making these medicines difficult for many patients to access. Due to the long wait times and high costs, cell and gene therapy developers often decide whether to build out GMP facilities for viral vector production. Commercial entities like Synthego, Editas and Agilent manufacture GMP sgRNA. Moreover, the inherent variability of gene-edited cell therapies and issues associated with in vivo gene editing means that great care must be taken in order to ensure these medicines are safe and appropriately regulated [91].

10.9.2 Safety Considerations

An emerging concern is the safety of CRISPR therapies. Significant efforts have been made by the scientific community to increase the safety of CRISPR therapies, including engineered Cas9 and other nuclease variants with increased accuracy. In late March, the FDA recently released updated guidelines on the topic of *Human Gene Therapy Products Incorporating Human Genome Editing* [92]. These guidelines include the identification of off-target editing activity, including the type, frequency, and location of all off-target editing events, and assessment of genomic integrity including chromosomal rearrangements, large insertions or deletions, integration of exogenous DNA, and potential oncogenicity or insertional mutagenesis. For gene-edited cell therapies, advances in single-cell DNA sequencing enabled more accurate and reliable characterization of these therapies. In-depth analysis of off-target editing and the assessment of large-scale chromosomal rearrangements that can result from inducing DSBs in DNA are also critical for safe CRISPR therapies [93].

The coming years are likely to see an increase in the use of CRISPR derivatives to treat a wider variety of diseases. This includes RNA editing systems for the treatment of RNA-mediated diseases and removal of RNA viruses, and CRISPR systems that can modify DNA without inducing DSBs, such as base, prime, PASTE, and epigenetic editing. The potentially increased safety profiles of these systems may generate more public support for these advanced biologics they are used to create.

10.9.3 Regulatory Issues and Developing New Clinical Frameworks for CRISPR Therapies

As the field of CRISPR gene editing continues to grow, an increasing number of therapies are approaching clinical trials. Despite this progress, a key stumbling block remains in the form of the outdated clinical development pipeline; the current framework for developing a therapeutic product was simply not designed to accommodate the novelty and complexity of next-generation CRISPR cell and gene therapies. Many CRISPR scientists have stated that an entirely new clinical framework should be developed by the FDA with these therapies in mind.

The FDA has recently voiced its support for CRISPR-based therapies, announcing several new draft guidance and initiatives for their improved development. This includes the bespoke Gene Therapy Consortium, which aims to provide clearer information on basic and clinical research, manufacturing, and regulatory requirements in order to streamline the development of small-batch therapies [94]. As part of this initiative, the FDA also suggested that financial incentives may be provided to encourage companies to generate gene therapies that may not otherwise be financially viable, such as diseases with very small patient populations. Other FDA initiatives include a gene therapy pilot program, which would provide real-time feedback to sponsors throughout development, and the *IN*itial *T*argeted *E*ngagement for *R*egulatory *A*dvice on *C*BER produc*T*s (INTERACT) program, which allows for informal meetings between the FDA's Center for Biologics Evaluation and Research (CBER) and trial sponsors in the early stages of development [95]. The FDA's increased commitment to ensuring the streamlined development and success of CRISPR-based therapies suggests that this field will only continue to grow in the coming years, likely with more positive outcomes for patients.

References

1. Martin J et al (2012) A programmable dual-RNA–guided DNA endonuclease in adaptive bacterial immunity. Science 337(6096):816–821
2. Mojica FJM, Juez G, Rodriguez-Valera F (1993) Transcription at different salinities of Haloferax mediterranei sequences adjacent to partially modified PstI sites. Mol Microbiol 9(3):613–621
3. Mojica FJM, Díez-Villaseñor C, García-Martínez J, Soria E (2005) Intervening sequences of regularly spaced prokaryotic repeats derive from foreign genetic elements. J Mol Evol 60(2):174–182
4. Barrangou R et al (2007) CRISPR provides acquired resistance against viruses in prokaryotes. Science 315(5819):1709–1712
5. Ran FA et al (2013) Genome engineering using the CRISPR-Cas9 system. Nat Protoc 8(11):2281–2308
6. Hendel A et al (2015) Chemically modified guide RNAs enhance CRISPR-Cas genome editing in human primary cells. Nat Biotechnol 33(9):985–989
7. Urnov FD, Genome Editing BC (2018) (Before CRISPR): lasting lessons from the 'Old Testament'. CRISPR J 1(1):34–46
8. Joung JK, Sander JD (2013) TALENs: a widely applicable technology for targeted genome editing. Nat Rev Mol Cell Biol 14(1):49–55
9. Carroll D (2011) Genome engineering with zinc-finger nucleases. Genetics 188(4):773–782
10. Uddin F, Rudin CM, Sen T (2020) CRISPR gene therapy: applications, limitations, and implications for the future. Front Oncol 10:1387
11. Hsu PD, Lander ES, Zhang F (2014) Development and applications of CRISPR-Cas9 for genome engineering. Cell 157(6):1262–1278
12. Jao L-E, Wente SR, Chen W (2013) Efficient multiplex biallelic zebrafish genome editing using a CRISPR nuclease system. Proc Natl Acad Sci USA 110(34):13904–13909
13. Yang H et al (2013) One-step generation of mice carrying reporter and conditional alleles by CRISPR/Cas-mediated genome engineering. Cell 154(6):1370–1379
14. Mali P, Esvelt KM, Church GM (2013) Cas9 as a versatile tool for engineering biology. Nat Methods 10(10):957–963
15. Song B, Yang S, Hwang G-H, Yu J, Bae S (2021) Analysis of NHEJ-based DNA repair after CRISPR-mediated DNA cleavage. Int J Mol Sci 22(12):6397
16. Lin S, Staahl BT, Alla RK, Doudna JA (2014) Enhanced homology-directed human genome engineering by controlled timing of CRISPR/Cas9 delivery. elife 3:e04766
17. Synthego Corporation (2020) Synthego's multi-guide design strategy for gene knockout kits and arrayed CRISPR libraries. https://www.synthego.com/help/synthego-multi-guide-design
18. Ma Y et al (2014) Generating rats with conditional alleles using CRISPR/Cas9. Cell Res 24(1):122–125
19. Wang B et al (2015) Highly efficient CRISPR/HDR-mediated knock-in for mouse embryonic stem cells and zygotes. BioTechniques 59(4):201–202, 204, 206–208
20. Yang H et al (2020) Methods favoring homology-directed repair choice in response to CRISPR/Cas9 induced-double strand breaks. Int J Mol Sci 21(18):6461
21. Synthego Corporation (2022) CRISPRa and CRISPRi. https://www.synthego.com/guide/crispr-methods/crispri-crispra
22. Lo A, Qi L (2017) Genetic and epigenetic control of gene expression by CRISPR-Cas systems. F1000Res 6:F1000 Faculty Rev-747
23. Shalem O et al (2014) Genome-scale CRISPR-Cas9 knockout screening in human cells. Science 343(6166):84–87
24. Makarova KS et al (2020) Evolutionary classification of CRISPR–Cas systems: a burst of class 2 and derived variants. Nat Rev Microbiol 18(2):67–83
25. Yan WX et al (2019) Functionally diverse type V CRISPR-Cas systems. Science 363(6422):88–91

26. Ran FA et al (2015) In vivo genome editing using Staphylococcus aureus Cas9. Nature 520(7546):186–191
27. Zetsche B et al (2015) Cpf1 is a single RNA-guided endonuclease of a class 2 CRISPR-Cas system. Cell 163(3):759–771
28. McCarty NS, Graham AE, Studená L, Ledesma-Amaro R (2020) Multiplexed CRISPR technologies for gene editing and transcriptional regulation. Nat Commun 11(1):1281
29. Kleinstiver BP et al (2016) High-fidelity CRISPR-Cas9 nucleases with no detectable genome-wide off-target effects. Nature 529(7587):490–495
30. Kleinstiver BP et al (2015) Broadening the targeting range of Staphylococcus aureus CRISPR-Cas9 by modifying PAM recognition. Nat Biotechnol 33(12):1293–1298
31. Qi LS et al (2013) Repurposing CRISPR as an RNA-guided platform for sequence-specific control of gene expression. Cell 152(5):1173–1183
32. Taha EA, Lee J, Hotta A (2022) Delivery of CRISPR-Cas tools for in vivo genome editing therapy: trends and challenges. J Control Release 342:345–361
33. Behr M, Zhou J, Xu B, Zhang H (2021) In vivo delivery of CRISPR-Cas9 therapeutics: progress and challenges. Acta Pharm Sin B 11(8):2150–2171
34. Xu CL, Ruan MZC, Mahajan VB, Tsang SH (2019) Viral delivery systems for CRISPR. Viruses 11(1):28
35. Banskota S, Raguram A, Suh S et al (2022) Engineered virus-like particles for efficient in vivo delivery of therapeutic proteins. Cell 185(2):250–265.e16
36. Lino CA, Harper JC, Carney JP, Timlin JA (2018) Delivering CRISPR: a review of the challenges and approaches. Drug Deliv 25(1):1234–1257
37. Landhuis E (2021) The definition of gene therapy has changed. Innov Gene Ther. https://doi.org/10.1038/d41586-021-02736-8
38. Prakash V, Moore M, Yáñez-Muñoz RJ (2016) Current progress in therapeutic gene editing for monogenic diseases. Mol Ther 24(3):465–474
39. Cring MR, Sheffield VC (2022) Gene therapy and gene correction: targets, progress, and challenges for treating human diseases. Gene Ther 29(1–2):3–12
40. Maeder ML, Gersbach CA (2016) Genome-editing technologies for gene and cell therapy. Mol Ther 24(3):430–446
41. Chakraborty S, Rahman T (2012) The difficulties in cancer treatment. Ecancermedicalscience 6:ed16
42. Zhang A, Miao K, Sun H, Deng C-X (2022) Tumor heterogeneity reshapes the tumor, so microenvironment to influence drug resistance. Int J Biol Sci 18(7):3019–3033
43. Nguyen B et al (2022) Genomic characterization of metastatic patterns from prospective clinical sequencing of 25,000 patients. Cell 185(3):563–575.e11
44. Whiteside TL (2006) Immune suppression in cancer: effects on immune cells, mechanisms and future therapeutic intervention. Semin Cancer Biol 16(1):3–15
45. Yang Y, Xu J, Ge S, Lai L (2021) CRISPR/Cas: advances, limitations, and applications for precision cancer research. Front Med (Lausanne) 8:649896
46. Berraondo P et al (2017) Cellular immunotherapies for cancer. Onco Targets Ther 6:e1306619
47. Ghaffari S, Khalili N, Rezaei N (2021) CRISPR/Cas9 revitalizes adoptive T-cell therapy for cancer immunotherapy. J Exp Clin Cancer Res 40(1):269
48. Linette GP, Carreno BM (2019) Tumor-infiltrating lymphocytes in the checkpoint inhibitor era. Curr Hematol Malig Rep 14(4):286–291
49. Palmer DC et al (2015) Cish actively silences TCR signaling in CD8+ T cells to maintain tumor tolerance. J Exp Med 212(12):2095–2113
50. Palmer DC et al (2022) Internal checkpoint regulates T cell neoantigen reactivity and susceptibility to PD1 blockade. Med (N Y):S2666-6340(22)00318-X
51. Tsimberidou A-M et al (2021) T-cell receptor-based therapy: an innovative therapeutic approach for solid tumors. J Hematol Oncol 14(1):102
52. Singh NK et al (2017) Emerging concepts in TCR specificity: rationalizing and (maybe) predicting outcomes. J Immunol 199(7):2203–2213

53. Ruggiero E et al (2022) CRISPR-based gene disruption and integration of high-avidity, WT1-specific T cell receptors improve antitumor T cell function. Sci Transl Med 14(631):eabg8027
54. Jackson HJ, Rafiq S, Brentjens RJ (2016) Driving CAR T-cells forward. Nat Rev Clin Oncol 13(6):370–383
55. Hu W et al (2019) CRISPR/Cas9-mediated PD-1 disruption enhances human mesothelin-targeted CAR T cell effector functions. Cancer Immunol Immunother 68(3):365–377
56. Wang Z et al (2021) Phase I study of CAR-T cells with PD-1 and TCR disruption in mesothelin-positive solid tumors. Cell Mol Immunol 18(9):2188–2198
57. ClinicalTrials.gov (2021) TT52CAR19 therapy for B-cell Acute Lymphoblastic Leukaemia (B-ALL) (PBLTT52CAR19). https://clinicaltrials.gov/ct2/show/NCT04557436
58. ClinicalTrials.gov (2022) Programmed allogeneic CRISPR-edited T cells engineered to express anti-CD19 chimeric antigen receptor (PACE CART19) in patients with relapsed or refractory CD19+ leukemia and lymphoma. https://clinicaltrials.gov/ct2/show/NCT05037669
59. Kwon T et al (2022) Precision targeting tumor cells using cancer-specific InDel mutations with CRISPR-Cas9. Proc Natl Acad Sci USA 119(9):e2103532119
60. Banas K et al (2022) Exon skipping induced by CRISPR-directed gene editing regulates the response to chemotherapy in non-small cell lung carcinoma cells. Gene Ther 29(6):357–367
61. Rosenblum D et al (2020) CRISPR-Cas9 genome editing using targeted lipid nanoparticles for cancer therapy. Sci Adv 6(47):eabc9450
62. Wilson M, Forsyth P, Whiteside J (2010) Haemoglobinopathy and sickle cell disease. Contin Educ Anaesth Crit Care Pain 10:24–28
63. Paikari A, Sheehan VA (2018) Fetal haemoglobin induction in sickle cell disease. Br J Haematol 180(2):189–200
64. ClinicalTrials.gov (2022) A safety and efficacy study evaluating CTX001 in subjects with severe sickle cell disease. https://clinicaltrials.gov/ct2/show/NCT03745287
65. ClinicalTrials.gov (2022) EDIT-301 for autologous HSCT in subjects with severe sickle cell disease. https://clinicaltrials.gov/ct2/show/NCT04853576
66. ClinicalTrials.gov (2022) BEACON: a study evaluating the safety and efficacy of BEAM-101 in patients with severe sickle cell disease (BEACON)
67. ClinicalTrials.gov (2022) A safety and efficacy study evaluating CTX001 in subjects with transfusion-dependent β-Thalassemia. https://clinicaltrials.gov/ct2/show/NCT03655678
68. Gray DH et al (2021) Optimizing integration and expression of transgenic Bruton's tyrosine kinase for CRISPR-Cas9-mediated gene editing of X-linked agammaglobulinemia. CRISPR J 4(2):191–206
69. Rocca CJ et al (2020) CRISPR-Cas9 gene editing of hematopoietic stem cells from patients with Friedreich's ataxia. Mol Ther Methods Clin Dev 17:1026–1036
70. CRISPR Therapeutics (2021) CRISPR therapeutics and capsida biotherapeutics announce strategic collaboration to develop gene-edited therapies for amyotrophic lateral sclerosis and Friedreich's ataxia. http://www.crisprtx.com/about-us/press-releases-and-presentations/crispr-therapeutics-and-capsida-biotherapeutics-announce-strategic-collaboration-to-develop-gene-edited-therapies-for-amyotrophic-lateral-sclerosis-and-friedreichs-ataxia
71. Fratta P et al (2012) C9orf72 hexanucleotide repeat associated with amyotrophic lateral sclerosis and frontotemporal dementia forms RNA G-quadruplexes. Sci Rep 2:1016
72. den Hollander AI, Roepman R, Koenekoop RK, Cremers FPM (2008) Leber congenital amaurosis: genes, proteins and disease mechanisms. Prog Retin Eye Res 27(4):391–419
73. Prado DA, Acosta-Acero M, Maldonado RS (2020) Gene therapy beyond luxturna: a new horizon of the treatment for inherited retinal disease. Curr Opin Ophthalmol 31(3):147–154
74. ClinicalTrials.gov (2022) Single ascending dose study in participants with LCA10. https://clinicaltrials.gov/ct2/show/NCT03872479
75. Maeder ML et al (2019) Development of a gene-editing approach to restore vision loss in Leber congenital amaurosis type 10. Nat Med 25(2):229–233
76. Editas Medicine (2021) Editas Medicine announces positive initial clinical data from ongoing phase 1/2 BRILLIANCE Clinical Trial of EDIT-101 for LCA10

77. Suh S et al (2021) Restoration of visual function in adult mice with an inherited retinal disease via adenine base editing. Nat Biomed Eng 5(2):169–178
78. Barnard AML, Fairhead HIM (2021) A commentary on the development of engineered phage as therapeutics. Drug Discov Today 26(9):2095–2098
79. ClinicalTrials.gov (2022) Safety, tolerability, and PK of LBP-EC01 in patients with lower urinary tract colonization caused by E. Coli. https://clinicaltrials.gov/ct2/show/NCT04191148
80. de Buhr H, Lebbink RJ (2018) Harnessing CRISPR to combat human viral infections. Curr Opin Immunol 54:123–129
81. Zhen S, Li X (2017) Oncogenic Human Papillomavirus: application of CRISPR/Cas9 therapeutic strategies for cervical cancer. Cell Physiol Biochem 44(6):2455–2466
82. Jubair L, Lam AK, Fallaha S, McMillan NAJ (2021) CRISPR/Cas9-loaded stealth liposomes effectively cleared established HPV16-driven tumours in syngeneic mice. PLoS One 16(1):e0223288
83. Johnson MJ, Laoharawee K, Lahr WS, Webber BR, Moriarity BS (2018) Engineering of primary human B cells with CRISPR/Cas9 targeted nuclease. Sci Rep 8(1):12144
84. Ueda N et al (2021) Immunotherapy perspectives in the new era of B-cell editing. Blood Adv 5(6):1770–1779
85. Moffett HF et al (2019) B cells engineered to express pathogen-specific antibodies protect against infection. Sci Immunol 4(35):eaax0644
86. Vajpayee M, Negi N, Kurapati S (2013) The enduring tale of T cells in HIV immunopathogenesis. Indian J Med Res 138(5):682–699
87. Dampier W, Nonnemacher MR, Sullivan NT, Jacobson JM, Wigdahl B (2014) HIV excision utilizing CRISPR/Cas9 technology: attacking the proviral quasispecies in reservoirs to achieve a cure. MOJ Immunol 1(4):00022
88. ClinicalTrials.gov (2022) Study of EBT-101 in aviremic HIV-1 infected adults on stable ART
89. Excision Biotherapeutics (2022) Technology: therapeutic pipeline. https://www.excision.bio/technology
90. Shupe J, Zhang A, Odenwelder DC, Dobrowsky T (2022) Gene therapy: challenges in cell culture scale-up. Curr Opin Biotechnol 75:102721
91. Silverman LI et al (2019) Identifying and managing sources of variability in cell therapy manufacturing and clinical trials. Regen Eng Transl Med 5:354–361
92. U.S. Food and Drug Administration. Human gene therapy products incorporating human genome editing. https://www.fda.gov/regulatory-information/search-fda-guidance-documents/human-gene-therapy-products-incorporating-human-genome-editing
93. Aquino-Jarquin G (2021) Current advances in overcoming obstacles of CRISPR/Cas9 off-target genome editing. Mol Genet Metab 134(1–2):77–86
94. U.S. Food and Drug Administration (2021) FDA, NIH, and 15 private organizations join forces to increase effective gene therapies for rare diseases. https://www.fda.gov/news-events/press-announcements/fda-nih-and-15-private-organizations-join-forces-increase-effective-gene-therapies-rare-diseases
95. U.S. Food and Drug Administration (2020) INTERACT Meetings (Initial Targeted Engagement for Regulatory Advice on CBER Products). https://www.fda.gov/vaccines-blood-biologics/industry-biologics/interact-meetings#:~:text=INitial Targeted Engagement for Regulatory Advice on CBER producTs (INTERACT, Research (CBER) at FDA

Part VI
Fusion Proteins, Antibody Drug Conjugates and Process Chemistry

Chapter 11
Fusion Proteins: Current Status and Future Perspectives

Stefan R. Schmidt

Abstract Fusion proteins, consisting of the joined peptide chains of two or more proteins that are naturally not connected, are popular and highly successful recombinant protein drugs with a wide range of different functionalities and therapeutic applications. This chapter summarizes the main features of this protein class and explains the general design principles. Typical building blocks comprise the constant domain of immunoglobulins, linkers with specific properties, and a second protein module that carries some extra functionality such as an enzyme for instance. In some cases, the inherent ability to form higher order complexes is utilized to generate oligomers with novel characteristics. As the artificially combined protein subunits not necessarily share the same physicochemical characteristics, manufacturing becomes challenging. Often even the orientation of the fusion partners has an effect on activity and yield. Therefore, some efforts were undertaken to optimize the constructs through protein engineering. These unnatural proteins are unknown to the human body and immune system, although their components are derived from existing proteins. This can lead to immunogenic reactions particularly through novel epitopes at the junction positions between linkers and fusion partners. The different therapeutic concepts rely on some specific features such as half-life extension by either enlarging the diameter of the proteins to protect them from rapid kidney clearance or through recycling by Fc gamma receptor (FcRn). Both approaches have successfully been used to generate so-called biobetters with enhanced functions. The other important category of fusion proteins contains targeting abilities often conferred trough antibody derivatives. Some proteins from that subtype are immunocytokines that deliver cytokines to specific cell types or organs to exert the corresponding pharmaceutical effect. Overall fusion proteins represent a highly successful but heterogeneous class of recombinant therapeutic.

S. R. Schmidt (✉)
evitria AG, Zurich, Switzerland
e-mail: ssc@evitria.com

K. Gadamasetti, S. A. Kolodziej (eds.), *Bioprocessing, Bioengineering and Process Chemistry in the Biopharmaceutical Industry*,
https://doi.org/10.1007/978-3-031-62007-2_11

Keywords Fusion protein · Half-life extension · Glycosylation · Fc-fusion · Aggregation · Albumin · Stability · Cytokine · Protein engineering · Orientation · Biobetters

11.1 Introduction

11.1.1 Definition

This chapter focuses on fusion proteins engineered from two or more genes that code for separate proteins joined by genetic techniques. The result is a single polypeptide sharing functional properties of both parent proteins. These recombinant proteins are combinations of unrelated domains that do not occur in nature. Therapeutic fusion proteins belong to the class of next generation biologicals as they are designed exclusively through human creativity and are not evolved from a natural process. Very often they represent so-called biobetters, molecules with improved functionalities such as, for instance, extended half-life.

Although this definition would also cover multi-epitope recombinant vaccines, they are excluded from the content of this chapter as well as naturally occurring fusion proteins resulting from chromosomal rearrangements that can be observed in many cancer cells or fusion tags utilized for protein affinity purification. Bispecific functionalities in antibodies have also become very popular in the last decade but are out of scope for this chapter as they typically are not forming a continuous polypeptide, but rather come together via disulfide bridges.

11.1.2 Categories

The simplest classification of these new proteins might be based on the function of their integrated domains. Typically, one fragment plays a role in molecular recognition or binding, while the other fragment adds a specific function, such as half-life extension or stability, cytotoxicity, or new targeting or delivery pathways [1]. Most fusion proteins fulfill three main goals, extend the half-life, and introduce a targeting function or a specific bioactivity. At least two of these three elements coexist in the fusion protein. Interestingly, a natural molecule, the immunoglobulin G also known as antibody, incorporates all three elements in one molecule. Antibody derivatives, fragments, or domains are widely used as building blocks of fusion proteins, constituting a huge share of the protein portfolio discussed here. High affinity and selectivity for specific epitopes, which are key functions of antibodies, have been generated from many nonantibody constructs that can be used as a single module or by combining two units with different specificities [2]. Fusion proteins are very attractive due to several advantages. The combination of the two functions in a single molecule facilitates the production and delivery of drugs. Two molecules linked together automatically have the same biodistribution profile instead of two separate molecules with completely different profiles. Interestingly new functions not found in natural or free proteins can be generated. This includes changes in half-life or targeting specificity [3]. There are also economic opportunities by creating new intellectual properties for novel and unnatural protein combinations. Therapeutic benefits due to reduced side effects or longer dosing intervals together with enhanced activity are strong motifs for fusion protein production. However, in addition to these important benefits, there are also many issues. In some cases, the combination of unrelated proteins can be difficult because the fusion partners have incompatible properties. For instance, misfolding or aggregation can occur for one domain while the same conditions might be perfect for the other domain. Although some fusion protein modules are elements of other well-established molecules such as antibodies, platform processes may not apply because other elements might replace the necessary domain. This can prohibit a successful formulation because of conflicting features. In addition, it could be complicated to control and fine-tune the relative amounts of each component, making it difficult to administer in the optimal efficacy and safety window. Perhaps the most important issue is the high potential for immunogenicity due to the formation of new epitopes at the junctions between the fusion partners, even when using only human protein domains. A summary can be seen in Table 11.1.

Table 11.1 Advantages and challenges of fusion proteins

Advantages	Challenges
Two functionalities in one molecule simplify manufacture and drug delivery Truly identical bio-distribution profile for both entities Designing of novel „unnatural "functionalities with clear therapeutic benefits (e.g., half-life and targeting) Potential of life cycle extension of products with expiring patents	Combination of two different proteins → different properties (not always compatible) pH preferences, hydrophobicity, glycosylation-isoforms, cellular localization Sensitivity to virus inactivation method at low pH Typically lower expression level than antibodies No platform process (neither USP nor DSP) Rarely affinity chromatography applicable (exception: Fc and HSA-Fusions) Tendency to aggregate → avoiding and removing aggregates Difficult fine tuning of functions (e.g., affinity) Immunogenicity risk

11.2 Current Status

11.2.1 Approved Products

In 1998, the first fusion protein, Etanercept, was approved. It belongs to the class of (fragment crystallizable) Fc fusions and receptor traps, consisting of the extracellular domain of tumor necrosis factor-alpha (TNF-α) receptor connected to a sequence encoding the Fc portion and hinge region of an immunoglobulin G1 (IgG1) heavy chain. This drug has been marketed under the name Enbrel® and is the best-selling fusion protein to date. Since this first regulatory approval, multiple other fusion proteins successfully passed clinical trials and became available for patients.

So far 22 19 fusion proteins received approval by the U.S. Food and Drug Administration (FDA). The most successful year was 2014, with four new fusion proteins getting commercial licenses. Twelve proteins contain an Fc-domain that is not only a typical building block for receptor traps but also serves as module to extend the plasma half-life to generate biobetters. An alternative approach is the fusion to human serum albumin (HSA) that binds as well to the neonatal Fc gamma receptor (FcRn) thus enabling recycling and improved pharmacokinetics. Another category with a few other examples is immunotoxins. Here a bacterial toxin is fused to a targeting moiety that allows to selectively kill malignant cells bearing the corresponding target. A summary of the approved fusion protein is displayed in Table 11.2.

11.2.2 Market

The therapeutic success of the first approved fusion protein, Enbrel®, triggered the interest of many scientists. For the first time, a non-naturally existing protein designed from unrelated building blocks was used for the treatment of severe diseases. Some fusion proteins are real blockbuster drugs with multibillion USD sales per year. The four most successful therapeutic fusion proteins are Eylea®, Trulicity®, Enbrel®, and Orencia® with annual revenues of more than 9.2, 6.5, 4.5, and 3.3 billion USD in 2021, respectively, as described in Fig. 11.1.

Due to their unique functionality and the difficulty to copy their mode of action, fusion proteins are interesting for investors and thus became the target of mergers and acquisitions. Over the last two decades, multiple transactions took place that included single molecules, platform technologies, and whole pipelines at different development levels as displayed in Table 11.3. Only a minority contained approved fusion proteins, indicating that most companies were taken over for expectations of future success for the molecules. Interestingly, 2021 has been the year with the highest number of deals in that context.

Table 11.2 Overview on currently approved fusion proteins

Brand	Approval	Protein 1 N-Term	Protein 2 C-Term	Indication	Host	Mode of action
Enbrel®	11/98	TNFαR	IgG1 Fc	Arthritis	CHO	Receptor trap
Ontak®	02/99	Diphtheria toxin	IL2	Lymphoma	*Escherichia coli*	Immunotoxin
Amevive®	01/03	LFA-3	IgG1 Fc	Psoriasis	CHO	
Orencia®	02/06	CTLA-4	IgG1 Fc	Arthritis	CHO	
Arcalyst®	02/08	IL1R & AP	IgG1 Fc	Cryopyrin syndrome	CHO	Receptor trap
Nplate®	08/08	IgG1 Fc	Peptide	ITP	*E. coli*	Peptibody
Elonva®	02/10	FSH	CTP	Fertility	CHO	Biobetter
Nulojix®	06/11	CTLA-4	IgG1 Fc	Transplant rejection	CHO	
Eylea®/Zaltrap®	11/11	VEGF R1/2	IgG1 Fc	Macular degen./Cancer	CHO	Receptor trap
Alprolix®	03/14	FIX	IgG1 Fc	Hemophilia B	HEK	Biobetter
Tanzeum®	04/14	GLP-1	HSA	Diabetes	*Saccharomyces cerevisiae*	Biobetter
Eloctate®	06/14	FVIII	IgG1 Fc	Hemophilia A	HEK	Biobetter
Trulicity®	09/14	GLP-1	IgG4 Fc	Diabetes	CHO	Biobetter
Strensiq®	10/15	Alkaline phosphatase	IgG1 Fc	Hypophosphatasia	CHO	
Idelvion®	03/16	FIX	HSA	Hemophilia B	CHO	Biobetter
Lumoxiti®	09/18	aCD22 scFv	Pseudomonas toxin	Hairy cell leukemia	*E. coli*	Immunotoxin
Elzonris®	12/18	Diphtheria toxin	IL3	BPDCN	*E. coli*	Immunotoxin
Reblozyl®	08/19	Activin receptor IIB	IgG1 Fc	Beta-thalassemia	CHO	Receptor trap
Altuviiio®	02/23	IgG1-Fc	FVIII-VWF-XTEN	Hemophilia A	HEK293	Biobetter
Ngenla®	06/23	CTP	hGH	Growth hormone deficiency	CHO	Biobetter
Ryzneuta®	09/23	G-CSF	IgG2 Fc	Neutropenia	CHO	Biobetter
Anktiva®	04/24	IL15/IL15R	IgG1 Fc	Bladder cancer	CHO	

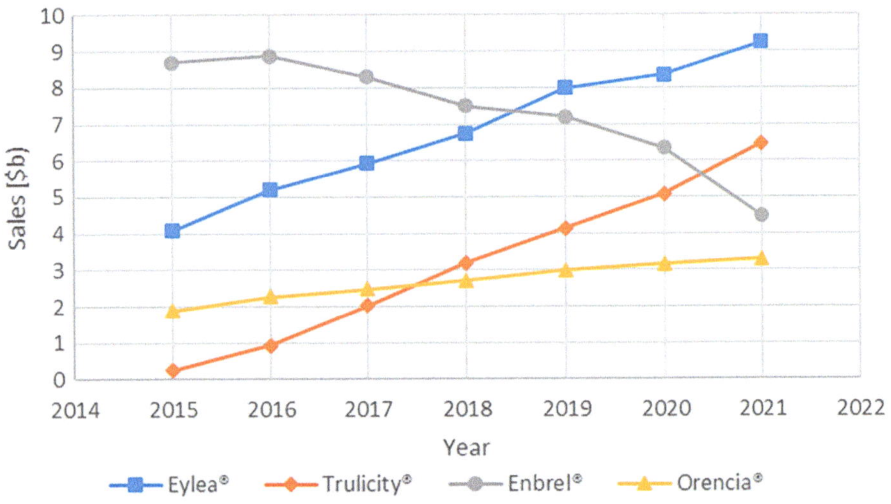

Fig. 11.1 Global sales of top selling fusion proteins

The huge revenues generated by some fusion proteins are obviously attractive for biosimilar developers. For instance, Enbrel® as one of the blockbusters has lost patent protection in Europe and other parts of the world in 2015. As a consequence, a wave of biosimilar variants appeared in those markets, causing a significant drop of sales for the originator drug since then. Since its peak sales in 2015 with close to $b 9, its sales dropped significantly. However, in the US, Amgen's Enbrel® patent will expire only in 2029.

11.3 Design

11.3.1 Building Blocks

The design of fusion proteins relies on the combination of several modules. One source of these building blocks are antibodies or their derivates such as the constant Fc domain [4] or the variable domains often used as single chain (scFv). Also all kinds of protein scaffolds containing some loops that are accessible for mutagenesis to generate novel affinity binders have frequently been used [5]. Fully synthetic therapeutic fusion proteins based on de novo design are still rather uncommon. Interestingly, synthetic fusion proteins generated in vitro allow the insertion of chemical modification not possible with the cellular translational apparatus. This could comprise non-natural amino acids, chemical conjugation of nonprotein molecules as, for instance, in the case of antibody drug conjugates (ADC) [6]. Chemical synthesis is only advantageous if it really allows cost efficient manufacturing of non-natural moieties. The topic of this chapter, however, focuses on fusion proteins

Table 11.3 Mergers and acquisitions involving companies with fusion protein technology

Buyer	Target	Date	Value [$M]	Technology	Fusion protein	Indication	Phase (at acquis.)
Boehringer Ingelheim	Trutino Biosciences	14.06.2022	n.a.	On demand cytokine	Cytokines	Oncology	Preclinical
Sanofi	Amunix	21.12.2021	1000	Masked T-cell engager, XTEN	AMX-818	Oncology	1
MSD	Acceleron	30.09.2021	11,500	Receptor traps	Reblozyl® Sotatercept	TGF-β blocker	Approved3
Sanofi	Kadmon	08.09.2021	1900	aPD-L1/IL15 immunocytokine	KD033	Solid tumors	1
Pfizer	Trillium Therapeutics	23.08.2021	2260	SIRPα-Fc fusion protein	TTI-621TTI-622	Hematological malignancies	1b/2
Amgen	Teneobio	27.07.2021	900	T-cell engagers	TNB-585	Prostate cancer	1
Takeda	Maverick Therapeutics	09.03.2021	525	T-cell engagers	MVC-101	Solid tumors	1b/2
MSD	Pandion Therapeutics	25.02.2021	1850	IL-2 mutein	PT-101	Colitis	1a/2b
AstraZeneca	Alexion	12.12.2020	39,000	multiple	Strensiq®	Rare diseases, immunology	Various
MSD	OncoImmune	23.11.2020	425	CD24-Fc	MK-7110	COVID-19	3
Menarini	Stemline Therapeutics	05.05.2020	677	Immunotoxin	Elzonris®	Blastic plasmacytoid dendritic cell neoplasm	Approved
Sotio	Cytune	30.08.2018	n.a.	IL-2/IL-15Rβγ agonist	SO-C101	Oncology	Preclinical
Sanofi	Bioverativ	22.01.2018	11,000	monomeric Fc-fusions	Eloctate® Alprolix®	Hemophilia	Approved
Celgene	Delinia	26.01.2017	300	IL-2 mutein Fc fusion protein	DEL106	Autoimmune	Preclinical

(continued)

Table 11.3 (continued)

Buyer	Target	Date	Value [$M]	Technology	Fusion protein	Indication	Phase (at acquis.)
Eleven Biotherapeutics	Viventia	21.09.2016	14	Immunotoxin	Vicinium Proxinium	Bladder cancer Squamous cell carcinoma	3 2
Celgene	EngMab	03.10.2013	600	T-cell engagers		Oncology	Preclinical
Opko Health	Prolor	24.04.2013	480	Half-life extension	hGH-CTP	Growth hormone deficiency	3
Amgen	Micromet	26.01.2012	1160	T-cell engagers	Blincyto®	Acute lymphoblastic leukemia	2
Alexion	Enobia	07.02.2012	610	Bone targeting	ENB-0040	Hypophosphatasia	2
Alexion	Taligen	31.01.2011	111	C3d targeting	TT30	Autoimmune	Preclinical
Insys	Neopharm	29.10.2010	135	Immunotoxins	NK-408	Oncology	3
Bristol-Myers Squibb	Zymogenetics	07.09.2010	885	Fc-Fusion	Atacicept	Autoimmune	2
Biomarin	Zystor	07.09.2010	22	GILT	ZC-701	Pompe's disease	1
Emergent	Trubion	18.08.2010	97	SMIP and Scorpion	SBI-087 TRU-016	Autoimmune oncology	2
Teva	CoGenesys	12.08.2010	400	Albumin fusion	Neugranin Cardeva	Febrile neutropenia Congestive heart failure	2 2
Pfizer	CovX	22.01.2008	n.a.	CovX body (peptide-mAb fusion)	CVX-045 CVX-060 CVX-096	Oncology Oncology Type 2 diabetes	1 1 1
Topotarget	Apoxis	18.12.2007	19	Mega Ligand	APO010	Oncology	1
Pfizer	Biorexis	11.04.2007	n.a.	Transferrin fusion	BRX-0585	Type 2 diabetes	Preclinical
Biogen Idec	Syntonix	01.02.2007	40	Monomeric Fc fusion	FIX-Fc	Hemophilia B	IND
Amgen	Avidia	29.09.2006	290	Avimer (modular binding domains)	IL-6 inhibitor	Inflammation, autoimmune	1
Medigene	Avidex	31.08.2006	62	TCR technology	EsoDex	Oncology	Preclinical

FIX factor IX, *hGH* human growth hormone, *GILT* glycosylation independent lysosomal targeting, *IL* interleukin, *SMIP* small modular immunopharmaceutical, *TCR* T-cell receptor, *TGF* transforming growth factor) generated by recombinant biological processes. Recently a new trend to improve biological functions such as higher selectivity, stability, half-life, and/or lower toxicity/immunogenicity can be observed. Most of these so-called biobetters belong to the class of fusion proteins [7]. Building blocks that contribute to the extension of plasma half-life are discussed in a later paragraph.

11.3.2 Linkers

Connecting two nonrelated protein domains is usually done by inserting a linker between both partners to generate a continuous uninterrupted poly-peptide chain. Besides bridging the gap between the two modules, these spacers can not only contribute to improve expression by enabling independent folding of the two domains but can also add specific functions as summarized in Fig. 11.2.

For instance, flexibility between the different domains could be required. A high degree of flexibility can be obtained by multiple glycine-serine $(G_4S)_n$ repeats, whereas higher rigidity is observed for helical linkers containing $(EA_3K)_n$ sequences. By alternating between both elements, the degree of flexibility and distance can be fine-tuned. Combining five of these modules in different ratios resulted in a variability of helical content from 100% for $(EA_3K)_n$ only motifs to 0% for pure $(G_4S)_n$ structures [8]. The same set of rigid (R) $(EA_3K)_n$ and flexible (F) $(G_4S)_n$ linker was applied to evaluate the impact of the linker on the enzymatic activity of alkaline phosphatase (PhoA) when fused to green fluorescent protein (GFP). PhoA can catalyze hydrolysis or exert the reverse activity on the substrate inosine. The rate of phosphatase and phosphotransferase varies under the influence of the linker flexibility. The best ratio between both activities could be obtained by a FRRRR linker with just one flexible module [9].

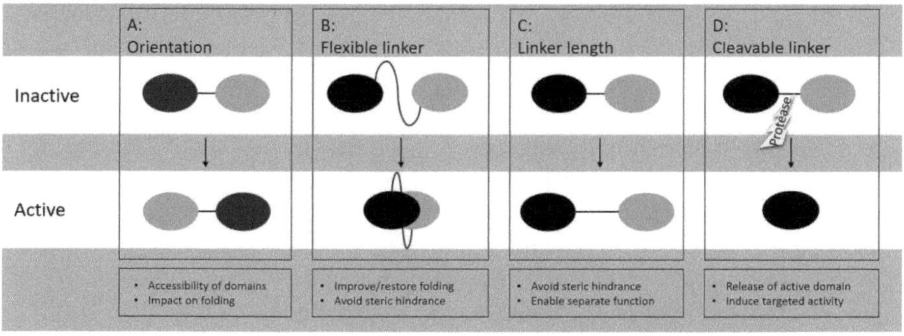

Fig. 11.2 Linker variants connecting fusion protein partners. The orientation of linkers has an impact on folding by allowing access of the domains, which is even more pronounced by flexible linkers that help avoiding steric hindrance. The same is true for longer linkers that furthermore enable the execution of separate functions. Cleavable linkers allow the activation of domains in proximity to certain proteases as, for instance, in tumors

In some cases, it can be beneficial to enable cleaving the linker to release both subunits. One version is dithiocyclopeptide, connecting transferrin and granulocyte colony-stimulating factor (G-CSF). As this linker contains as well a thrombin cleavage site, G-CSF is released in a two-step mode in the liver. This linker can be applied for other fusion partners as well [10]. A variation in that theme is the use of a different dithiocyclopeptide including a target sequence for secretion signal proteases. This means during secretion the peptide is cleaved leaving the fusion protein just joined through the disulfide bridge [11]. The furin recognition motifs is a quite useful sequence to release anticancer domains in proximity to malignant cells with high extracellular concentration of this protease. This furin sensitive spacer was successfully tested with an immunotoxin [12]. Another motif that was applied in the context of tumor targeting is the recognition sequence of matrix metalloproteinase-2 (MMP2). In combination with a fibrin clot binding peptide interferon-y could thus be delivered and released to tumors [13]. The protease-rich tumor microenvironment can act as activator for masked antibodies which cover their binding sites by a peptide that can be released after cleavage and thus allow binding of the antibody to its target. This prodrug concept is particularly useful for antibodies against epitopes that are not fully cancer specific and therefore would be limited due to unwanted toxicity on normal tissue [14]. Further examples can be found with fusion proteins acting in the blood coagulation cascade that anyhow contain protease activities. The long half-life of FXIII-B subunit can be transferred to FIX by fusing both molecules together. Using a short linker sequence cleavable by activated FX (FXa), FIX can be released in the presence of trace amounts of FXa to exert its coagulating function when needed and remains in the inactive long half-life variant until then [15]. The release of a bioactive compound from a carrier or a half-life extension module such as HSA through a cleavable linker was exemplified with IFN-HSA fusions. Versions with a cleavable linker generally showed a lower concentration distribution over time but a significantly higher bioactivity [11]. Besides extracellular proteases, intracellular ones such as furin have been exploited to split fusion proteins like immunotoxins at the linker region after their internalization to allow their endosomal escape [12]. As unwanted protease degradation can destroy fusion proteins, it is preferable to screen for cryptic protease sites already during the design phase. Besides the elimination of protease recognition motifs, these sequences can be masked instead to make them inaccessible for the enzyme. A typical building block supporting that approach is the utilization of glycosylation sites. In the case of a peptibody containing the GLP-1 peptide, clipping at the transition to the hinge region of the Fc fusion partner was observed. Interestingly the degree of clipping varied in different cell lines. The underlying reason for that was not only the expression level of host cell proteases but rather the degree of N-glcosylation in that region. In addition, highly glycosylated versions had a longer serum half-life [16].

A further parameter to take into account is the linker length. This can be significant to spatially separate subunits of a fusion protein but might also contribute to the expression yield. For instance, an onconase albumin fusion protein exhibited higher activity and expression with longer linkers [17]. Positive effects of long linkers on bioactivity were also reported for Fynomer Fc fusions [18]. The impact of linker length and type of linkers on biological functions was systematically evaluated on the example human serum albumin (HSA) fused to interferon a2b (IFN-a2b). Independent of their structural features (flexible: $(G_4S)_n$, rigid: $(PA)_n$, helical: $(AEA_3K)_n$) linkers with at least five amino acids inserted between both fusion partners were sufficient to prevent aggregation, heterogeneity, and instability of the fusion protein and increased purification yield twofold. Interestingly rigid linkers showed the best resistance to hydrolysis, whereas helical linkers improved antiviral activity [19]. Often the linker length contributes to multiple effects. For instance, a complex fusion protein consisting of the BH3 domain of BH3 interacting-domain death agonist (BID) fused to the N-terminus of Tumor Necrosis Factor Related Apoptosis Inducing Ligand (TRAIL) via multiple protease recognition sites showed better bioactivity if a cysteine containing linker was added. However, the shorter version without that cysteine containing module was easier to express [20]. TRAIL was also part of an experiment as bifunctional protein connected to VAS by either a rigid, a flexible, or a helical linker. Surprisingly, the rigid linker allowed the detection of activity already in inclusion bodies, indicating that this construct contains some properly folded structures. Consequently, this variant resulted in best bioactivity after refolding [21]. The impact of linker design on enzymatic efficiency was also demonstrated in a glucanase-xylanase fusion. Here the bifunctional catalytic activity was enhanced to more than 300% when using a $(G_4S)_2$ linker. Interestingly helical linkers improved thermal stability [22].

An early study based on three-dimensional (3D) structures evaluated the amino acid content of naturally occurring spacers between different protein domains. Frequently Ser, Thr, Gly, Ala were constituents of linkers. The native sequences were often composed of short peptides that on the one hand had enough conformational stability, while on the other hand offered enough flexibility. Incorporation of polar residues such as Ser or Thr supports solubility in water by enabling hydrogen bonding. The average length of linkers of that study was 6.5 amino acids [23]. Too long linkers could elicit an immune response to the novel epitope. A threshold for that seems to be around 15 amino acids. An investigation with an even larger sample group classified linkers into helical and nonhelical types. It was concluded that the primary function of natural linkers is the spatial separation of domains to prevent unfavorable interactions during folding. This structural rigidity will isolate the attached protein modules from each other [24].

Besides the initial purpose of joining protein domains, linkers can also add a wide range of functionalities to the fusion proteins. They can, for instance, improve folding, stability, expression by maintaining sufficient distance between domains to minimize interference and to allow proper folding. But also pharmacokinetics, bioactivity, and targeting tissues or certain cell types can be positively influenced [25]. Functional parameters like bioavailability were assessed with the examples of either human growth hormone (hGH) or granulocyte colony stimulating factor (GCSF) fused to transferrin (Tf). Linker length affected the affinity of both fusion partners to their respective receptors which also contributed to pharmacokinetic differences as only the Tf receptor mediates recycling that contributes to extended half-life [26].

Recently the application of short peptides as switches and sensors was systematically evaluated. Based on well-defined motifs, a molecular toolbox was generated that contains linkers to mediate conformational coupling or linkers to act as tethers facilitating domain movements. However, none of these functions has yet been inserted into therapeutic fusion proteins [27].

Besides the many advantages linker sequences can introduce into fusion proteins, some unexpected post-translational modifications were observed with the linkers. The Ser position in the standard linker $(G_4S)_n$ was spontaneously O-xylosylated to 30% when the construct was expressed in HEK293 cells. Changing the host cells to Chinese hamster ovaries (CHO) reduced the level ten-fold. The basis for that event is the presence of a cryptic xylosyltransferase motif (GSG) in the $(G_4S)_n$ sequence. Replacing the Ser residue by Gly completely abolished this aberrant glycosylation [28]. The same effect was detected in an antibody fusion protein containing either $(G_4S)_3$ or $(G_4S)_4$ linkers. Interestingly the intensity of xylosylation is directly correlated to the accessibility of the sequence to xylosyltransferase and not the number of (G_4S) motifs. Inserting a Pro either before or after the Ser eliminated the glycosylation [29]. Replacement of Gly by Pro, however, can lead to another variation if Pro is followed by Ala, as this creates a recognition site for prolyl hydroxylase and introduces hydroxyproline [30]. The unwanted effect is further complicated by the heterogeneity of the additional glycan structures ranging up to penta-saccharides attached to the Ser. Rarely even phosphorylation of xylose could be found [31]. Ser in (G_4S) motifs can be directly phosphorylated as well. As in the case of xylosylation the level of modification was higher in HEK cells than in CHO cells [32]. Undesired O-glycosylation was also detected in Fc-fusion proteins containing the hinge region of IgG1. The insertion of the bulky glycan structure at this position has a negative impact on aggregation levels as it disturbs the formation of disulfide-bridges below the hinge region. The problem was solved by eliminating potential O-glycosylation sites via mutagenesis, hereby improving protein stability and homogeneity [33].

11.3.3 Oligomerization

Oligomerization or the generation of multimers is one of the advantages that are introduced by the de novo design of fusion proteins. A graphical overview can be found in Fig. 11.3.

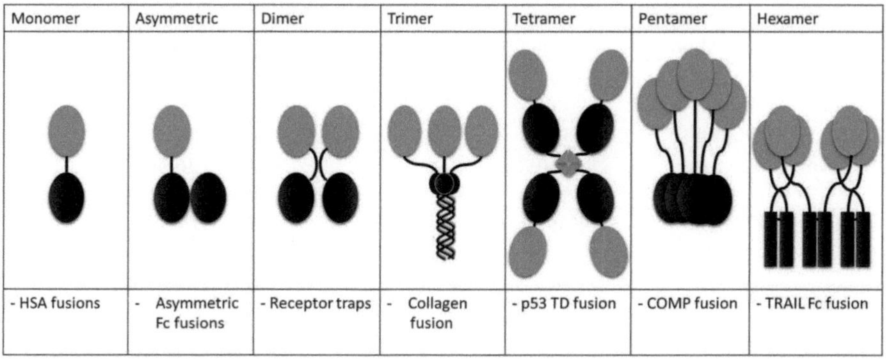

Fig. 11.3 Examples of fusion protein oligomers. One important element of fusion proteins is the possibility to generate oligomers containing more than one copy of the fusion partners to increase avidity, activity or trigger responses by receptor crosslinking. (HSA human serum albumin, TD tetramerization domain, COMP cartilage oligomeric matrix protein, TRAIL tumor necrosis factor–related apoptosis-inducing ligand)

11.3.3.1 Monomer

A downside to that approach is the potential steric hindrance of the fusion partners that might prevent proper binding to the corresponding receptor or exert stress on the hinge region of the Fc domain. This is particularly true for naturally occurring monomeric proteins as, for instance, the large blood coagulation factors [34]. Interestingly, monomeric Fc-fusion proteins have been used to enable pulmonary uptake of proteins [35]. Furthermore, monomeric Fc-fusions of EPO or IFN exhibited significantly better bioactivity when compared to their homodimer format. On the other hand, heterodimers as fusion proteins can be advantageous when they resemble naturally occurring heterodimers. For instance, proteins like FSH consisting of two subunits can be displayed either as tandem homodimer or heterodimer. Another protein class benefitting from an asymmetric conformation are cytokines. Here therapeutic effects such as biodistribution and tumor localization of single cytokines per fusion protein are improved. Overall there are many applications that favor heterodimers [36]. Heterodimerization has become an important prerequisite for the generation of bispecific antibodies. Therefore, a great body of knowledge can be found there which is applicable for Fc fusion proteins as well [37]. Table 11.4 summarizes these findings.

11.3.3.2 Dimer

The simplest version of fusion proteins consists of two connected subunits. As soon as a Fc domain is utilized, multimerization comes into play. The normal case is a homodimer pairing two identical polypeptide chains.

Table 11.4 Overview on point mutations to enable heterodimerization of Fc domains

Company	Technology	Mutations HC1	Mutations HC2
Genentech	Knobs-into-holes	T366W	T366S, L368A, Y407V
Genmab	DuoBody	F405L	K409R
Zymeworks	Azymetric	T350V, L351Y, F405A, Y407V	T350V, T366L, K392L, T394W
Amgen	Charge Pair	K392D, K409D	E356K, D399K
Rinat-Pfizer	Charge Pair	D221E, P228E, L368E	D221R, P228R, K409R
Xencor	HA-TF	S364H, F405A	Y349T, T394F

11.3.3.3 Trimer

Trimeric formats can be generated by utilizing naturally occurring proteins based on trimers such as collagen. Some years ago, the combination of scFV with the collagen-like peptide scaffold $(GPP)_{10}$ was evaluated. Fusions at the N-terminus of the peptide resulted in stable homotrimers. Due to the increased avidity, the trimer bound 20 and 1000-fold stronger than the bivalent or monovalent versions of the same scFv, respectively. When expressed in NSO cells, the murine prolyl 4-hydroxylase converted approximately 61% of the 10 available prolines into hydroxyproline, thus giving a thermo-stability comparable to the naturally occurring collagen. The so-called Collabody is currently evaluated as a CD3-specific T-cell engager for therapeutic applications [38].

11.3.3.4 Tetramer

The next oligomeric level are tetramers. A naturally available building block is the self-assembling tetramerization domain from p53 (p53TD) that can surpass the naturally occurring affinity threshold of antibodies through higher avidity. In multiple examples, modules such as dAbs, scFv, Fabs, and extracellular protein domains were fused to p53TD either at its N- or C-terminus. Due to its self-assembling capabilities active tetramers could be successfully expressed, secreted, and purified from a single poly-peptide chain. As the p53 domain is deeply embedded in the overall structure, no detrimental immunogenic effects are expected. Involving a dimerization unit such as the Fc-domain, even an octamer could be obtained. Octameric versions of Humira were approximately 50-fold more potent than the original divalent Humira. Due to the higher potency the so-called Quads could either be used at lower doses than the original antibody or allow a switch from intravenous to subcutaneous application [39].

Tetrameric antibodies have been described since many years. One recent example is the evaluation of tetrameric formats to fight COVID infections. Here the variable domain of heavy chain is either duplicated at the N-terminus or fused to the C-terminus of the Fc-domain, while maintaining the light chain. This antibody with four binding domains neutralizes the virus with enhanced potency so limiting its ability to achieve resistance though escape mutations [40].

11.3.3.5 Pentamer

Some receptor ligand interaction happens at low affinity and therefore requires multiple ligands to bind simultaneously. For instance, CD200 binds to its corresponding receptor at approximately 1 μM. When generating a pentamer by fusing CD200 with an eleven amino acid linker to cartilage oligomeric matrix protein (COMP), the receptor binding was highly improved [41].

11.3.3.6 Hexamer

Larger multimers have also been described in the form of tandem repeats of Fc domains to increase the antibody-dependent cellular cytotoxicity (ADCC) and complement-dependent cytotoxicity (CDC) [42]. Even larger hexameric complexes have been designed by adding small 18-mer peptides to the C-terminus. These tailpieces connect six Fc-dimers to form a large multi-subunit molecule with the ability to crosslink low affinity FcγR binding. The hexameric format exhibits stronger avidity with 12 potential binding sites or fusion partners [43].

Looking at other fusion partners beyond Fc-domains, a dimerization can be enforced by leucine zippers, naturally occurring peptide sequences. For instance, a noncovalently bound heterodimeric Fab fragment, a so-called "Zipbody" was designed by fusing leucine zipper pairs LZA and LZB or c-Jun and c-Fos to the C-terminus of heavy and light chain thus enhancing correct pairing and resulting in a properly folded active Fab [44]. A similar example was described earlier by creating bispecific homodimers based on two different designed ankyrin repeat proteins (DARPins) and a leucine zipper linker [45].

More complicated trivalent molecules based on tumor necrosis factor-related apoptosis-inducing ligand (TRAIL) utilized leucine zippers forming trimers. One case incorporated the yeast GCN4-pII leucine zipper in fusion to the N-terminus of TRAIL. The Ile substitutions in the "a" and "d' positions stabilizes the trimer formation. Replacing the yeast protein GCN4-pII by the human ATF7-pII greatly reduced the immunogenicity while maintaining high antitumor potency [46]. To enable an even more oligomeric molecule, the building blocks human IgG4 Fc-domain, the TNF receptor-associated factor 2 (TRAF2) coiled coil domain, and the extracellular receptor binding domain (RBD) of human OX40 ligand (OX40L) were fused together to obtain a 300 kDa large hexameric fusion protein [47]. A functional OX40L multimer can also be obtained by using similar building blocks, but using trimerizing isoleucine zipper domain GCN4 instead of TRAF2 and IgG1 Fc instead of IgG4 Fc. The final hexameric molecule consists of three Fc-induced disulfide linked dimers that forming two connected trimers [48]. A more elegant and less complicated solution to generating trimeric fusion proteins was described in the hexavalent receptor agonist (HERA) Technology. These hexameric fusion proteins comprise three receptor binding domains in a single chain arrangement, connected to an Fc-silenced human IgG1 not requiring FcγR-mediated crosslinking for tumor killing activity [49]. Other approaches with the class of homo-multimeric proteins focused on finding the most active version of CD40L as adjuvant for DNA vaccines. Three different self-assembling versions were compared. The starting point was a leucine zipper-based trimer. The next variant contained two trimers connected through fusion to the N-terminal part of adiponectin (Acrp30). The third molecule was a tetra-trimer containing 12 individual CD40L arranged through fusion to surfactant protein D. Their immunostimulatory activity increased in direct proportion to the valency of the trimers [50]. Naturally occurring trimeric proteins such as the noncollagenous domain (NC1) of human collagen XVIII NC1 can be utilized as well in fusion protein applications. NC1 is composed of a trimerization domain (TD), a hinge region, and the endostatin domain. Proteolytic cleavage on the hinge region releases the endostatin domains to establish a fully functional TRAIL-TD trimer [51].

11.3.3.7 Octamer

So far, the highest described order of fusion proteins are octamers. Here the naturally occurring affinity threshold can be surpassed by higher avidity. The base element in that context is the self-assembling tetramerization domain from p53.

11.3.4 Orientation

On the first glance, the orientation of both fusion partners seems to be irrelevant as long as a single polypeptide chain can be obtained. However, as soon as a free N- or C- terminus is required for activity this parameter needs to be considered in the design. An overview with current examples for an orientation preference can be seen in Table 11.5.

Sometimes simply following the natural configuration is giving the best results. For instance, this is true for all receptor traps consisting of the extracellular domain of membrane bound receptors fused at their C-terminus to an IgG Fc domain. Peptibodies are another class of Fc fusions but with opposite orientation. Here small peptides or concatemers are fused to the C-terminus of Fc domains. One underlying reason for this configuration is the protection from peptidases that attack free N-termini. Furthermore, for so far unknown reasons, a C-terminally fused NGF peptibody was 1500 times more potent than its N-terminal counterpart [52].

A wide range of other examples can be found in the literature about human serum albumin (HSA) fusions. Interestingly in few cases N-terminal fusion of HSA was preferable. Although the real reason for that has rarely been properly identified, but in the here mentioned examples the N-terminal fusion typically resulted in better

Table 11.5 Examples of fusion proteins with a clear preference of the orientation of their fusion partners

N-Term	C-Term	Effect (better)	Reason
TNF-R	Fc	Functionality	Natural configuration
Fc	FGF21	Activity	Unknown
Fc	EPO	Half-life and activity	Endosomal stability
X	ELP	Activity, expression	Mis-folding in N-Term fusion
scFv	ANG	Functionality	Free C-Term required
ANG	END	Potency	Unknown
IFN-a2b	HSA	Reduced aggregation	Improved S-S formation
TNF-RI	HSA	Activity	Improved S-S formation
HSA	IL28B	Activity and expression	Unknown
HSA	BNP_2	Activity and expression	Lower degradation
HM3	HSA	Half-life, stability, activity	Unknown
hLF	HSA	Stability, low proteolysis	Steric hindrance
HSA	TMP_2	Expression	Unknown
$SS28_2$	HSA	Activity and expression	Lower degradation

expression levels and higher activity. For instance, only IL28B fused to HSA at its N-terminus was active and exhibited better stability than its C-terminal counterpart [53].

On the contrary for IFNα2b, lower heterogeneity and instability were observed for this orientation but could be significantly enhanced by changing to a C-terminal fusion [54]. The underlying reason for that effect seems to be the improved ability to form di-sulfide bridges. The same cause was identified for the better TNF-α neutralizing activity of a receptor trap consisting of TNF-RI fused to HSA on its C-terminus [55].

Sometimes the benefit of one version is different in the other version. For instance, evaluating the ideal orientation for a HSA lactoferrin fusion, it was found out that placing HSA at the N-terminus of the construct results in sixfold increased half-life, but for the cost of reduced thermal and proteolytic stability [56]. Peptides typically suffer from a short half-life that can be extended by HSA-fusion. On the one hand, there are examples such as the integrin antagonistic peptide HM-3, where the best result in terms of higher stability and activity was achieved with connecting HSA to the C-terminus of the peptide [57]. On the other hand, cases such as the human brain natriuretic peptide (BNP) obtained best biological potency when two copies are fused to the C-terminus of HSA [58]. Similarly TPO mimetic peptide (TMP) in tandem could significantly better expressed having HSA at its N-terminus [59]. Interestingly another example utilizing tandem peptides showed lower degradation and higher activity for the variant with somastatin at the N-terminus [60].

The orientation of bifunctional molecules not relying on the classical building blocks Fc domain or albumin is even more difficult to predict. When connecting angiostatin (ANG) and endostatin (ENG) to inhibit vascularization a clear preference of the arrangement is not obvious. However, in biological assays the

ANG-ENG version was significantly more potent than its ENG-ANG counterpart [61]. Another special case is the fusion to elastin like peptides (ELP). When testing the position of ELP in fusion to four different proteins, it was found out that the ELP at the C-terminus resulted in higher expression and better activity than the N-terminal fusion of ELP. The underlying reason for that effect is most probably an increase in misfolded protein when the translation starts with ELP [62]. The class of immunotoxins consisting of a targeting ligand such as scFv and a toxin was also evaluated for the optimal orientation of both elements. Chimeric toxins containing the scFv at the C-terminus of restrictocin were shown to be more active than those having a ligand at the N-terminus of the toxin, most probably due to folding differences [12].

The multitude of examples here clearly demonstrates the fact that it is very difficult to predict the ideal orientation upfront, but rather requires empirical results testing both variations. Typical parameters to differentiate between the options are expression levels, activity, stability, and improved pharmacokinetics.

11.3.5 Protein Engineering

The paragraph above highlighted design details on linker sequences. But protein engineering can also help in solving other issues such as efficacy, production yield that impact the therapeutic effect of fusion proteins.

11.3.5.1 Modifications of Fc as Building Block

When designing Fc-fusion proteins, it is worthwhile to have a closer look on the inbuilt features of the constant region. The Fc domain has multiple functions such as defining half-life through recycling via the FcRn receptor, triggering effects such as ADCC or CDC via FcγR, and enabling oligomerization from asymmetrical monomer and homo- or heteromeric dimer. Over the last decade, the underlying amino acid interactions were revealed. In some cases, when bivalency is not possible due to steric hindrance of too large fusion partners or not desirable to avoid receptor crosslinking, the availability of a truly monomeric Fc would be beneficial. Unfortunately, most attempts were unsuccessful and resulted in instable molecules or suffered from decreased half-life. The first stable Fc monomer was obtained by introducing two additional N-glycosylation sites at the CH3-CH3 interface with the double replacement Ser364Asn and Phe405Asn [63]. The contemporary approaches on Fc optimizations to address receptor binding have been summarized in a comprehensive review [64]. Details can be found in Table 11.6.

Fc receptor functions have often been used to optimize therapeutic effects [65]. For instance, an improved version of Enbrel® had higher ADCC and CDC activity than the original fusion protein that could extend its use to other chronic disease such as ulcerative colitis or Morbus Crohn. Additionally, higher affinity to the target TNF-α was observed. Taken together, these modifications can lead to a better

Table 11.6 Point mutations in the Fc domain modulating FcyR-mediated functions

Company	ADCC	ADCP	CDC	IgG	Mutation
Genentech	+	+		1	S298A, E333A, K334A
Xencor	+	+		1	S239D, I332E
Xencor	+	+	–	1	S239D, A330L, I332E
Applied Molecular Evolution	+	+		1	P247I, A339D/Q
Applied Molecular Evolution	+	+		1	D280H, K290S +/– S298D/V
Macrogenics	+	+		1	F243L, R292P, Y300L +/– P396L or V305I
Xencor		+		1	G236A, S239D, I332E
Abgenix/Genentech	=		+	1	K326A, E333A
Abgenix/Genentech	0		+	1	K326W, E333S
GlaxoSmithKline/Tolerx	–	–		1	N297A
Ortho Biotech	–	–		1	L234A, L235A
Protein Design labs	–	–		2	V234A, G237A
Welcome Labs				4	L235A, G237A, E318A
Merck	–	–	–	2	H268Q, V309L, A330S, A331S
Bristol-Myers Squibb	–	–		1	C220S, C226S, C229S,
Seattle Genetics	–	–	–	1	C226S, C229S, E233P, L234V, L235A
Medimmune	–	–	–	1	L234F, L235E, P331S

efficacy and a wider applications [66]. A different approach to improve Enbrel® against its primary indication rheumatoid arthritis was taken with a hybrid Fc domain that consists of a immunological silent IgG4 domain connected to the highly flexible hinge region of IgD, hiding the junction site inside the structure to avoid unwanted immunogenicity. The molecule showed a 1.5-fold higher neutralizing activity than Enbrel® [67]. The three-dimensional structure of the Fc domain is governed by S-S bridges [68].

11.3.5.2 Modifications of Human Serum Albumin as Building Block

Protein engineering has been applied to HSA fusion proteins as well. An obvious target is the free Cys in HSA that causes aggregation and reduces stability. Cys34 was replaced by Ser by site directed mutagenesis in a IFNα2b-HSA fusion protein. After incubation at 60 °C or prolonged agitation still 90% of the fusion protein was in a monomeric state [69]. Interestingly additional free Cys residues could be useful for chemical conjugation of drugs. Therefore, two more thiol groups were introduced at position 93 and 294 resulting in the replacements Lys93Cys and Glu294Cys [70].

As the primary reason for HSA fusion is the extension of half-life, several attempts were undertaken to optimize its binding to FcRn which is the main contributor. Typically, albumin has a serum half-life of 3 weeks. The first described site directed exchange of Lys573Pro at the C-terminus in domain DIIIb improved affinity to FcRn by a factor of 12, leading to a 1.5-fold prolonged half-life [71]. As the binding to FcRn happens at the C-terminus of HSA, it is obvious that fusions to the N-terminus of HSA will have less impact on recycling through FcRn and are therefore preferable.

Combining yeast display and co-crystallization of FcRn with HSA variants, a double-mutant Glu505Gly and Val547Ala was identified that exhibited a ten-fold higher affinity to FcRn leading to a 1.3-fold improved half-life in cynomolgus monkeys [72].

Although the DIII domain is the main contributor for FcRn binding, it was found out that two loops in DI at the N-terminus modulate the interaction [73]. It will be worthwhile to utilize that region as well to optimize FcRn affinity at different pH values together with further mutations in DIII to fine tune new designer HSA molecules as fusion partners.

Another group selected different amino acid residues to not only improve half-life but also enabling transcellular delivery. The new albumin variant QMP contains three substitutions: Glu505Gln, Thr527Met and the known mutation Lys573Pro. Compared to the wild-type, QMP was translocated twofold more efficient over mucosal barriers at intranasal application. Fusing the triple mutant HSA to coagulation factor VII, the half-life improved 3.6-fold [74]. Table 11.7 summarizes the mutagenesis approaches based on albumin.

11.3.5.3 Modification of Other Functions

Serum half-life is typically influenced either by the diameter of the molecule that prevents kidney filtration, charge density that would cause repulsion or the ability of recycling through FcRn or transferrin receptors. On the example of immunocytokines (fusions between antibodies and cytokines), it was demonstrated that the faster than expected clearance from the bloodstream can be influenced by altering the affinity to other FcRn either by changing the Fc isotype or mutagenesis of the FcRn binding site on the Fc domain [75]. In a second study, the half-life was further improved by optimizing the junction between IL-2 and the antibody to prevent intracellular degradation during the FcRn-mediated recycling [76].

Stability issues cannot only be attributed to aggregation but can also be the result of degradation. Peptidbodies with N-terminal fusion of short amino acid sequences

Table 11.7 Point mutations in human serum albumin to optimize certain functionalities

Additional functions	Half-life extension	Translocation
Cys34Ser Improved stability	Lys573Pro 1.5-fold longer half-life	Glu505Gln, Thr527Met, Lys573Pro Twofold better translocation 3.6-fold longer half-life
Lys93Cys, Glu294Cys Additional conjugation sites	Glu505Gly, Val547Ala 1.3-fold longer half-life	

can suffer from enzymatic proteolysis. This can be avoided if a glutamate residue is introduced at the first position of the polypeptide which forms pyroglutamate after cyclization, protecting against protease attacks. Studying the functionality of this peptidbody, an unexpected O-glycosylation site was removed by exchanging Thr to Ala, thus significantly increasing potency [77]. Similarly, the negative impact of proteolysis and aggregation on an FGF21-Fc fusion protein could be successfully abolished through the replacements Pro171Gly and Leu98Ar, respectively [78].

The next level of engineering targets three-dimensional alterations. One example in that context is the introduction of four point-mutations into the EPO domain within an Fc-EPO fusion protein to improve multiple features. Overall, the amino acid exchanges were His32Gly, Cys33Pro, Trp88Cys and Pro90Ala that resulted in a disulfide bridge rearrangement from Cys29-Cys33 to Cys29-Cys88. The modified fusion protein was secreted as monomer with better pharmacokinetic and stability after the removal of N-linked glycosylation [79].

11.3.6 Immunogenicity

Some unwanted effects like glycosylation at novel positions introduced through linker sequences, nonhuman glycosylation or degradation at cryptic protease recognition motifs have been described in the paragraphs above. However, another even more deleterious biologic consequence of a novel, non-natural protein design is the eliciting immunogenicity, particularly at novel epitopes created at the junction between both fusion partners, even if only proteins of human origin are being used. Furthermore, fusion proteins might be taken up or processed differently than their native counterparts which could lead to an increased immune response. Therefore, it is beneficial to apply rational design of fusion proteins to minimize immunogenicity of the new construct. The current approach to identify amino acid positions for targeted mutagenesis is in silico analysis to predict T- and B-cell epitopes [80].

Particularly Fc-fusion proteins could show increased immunogenicity as they bind via Fcγ receptors (FcγR) to antigen presenting cells (APC) thus targeting antigens to these cells. If that results in enhanced uptake to APCs, or if the fusion protein stimulates these cells, then a stronger immune response could be triggered. Current deimmunization strategies based on prediction of particularly immunogenic epitopes focus on modifying epitope sequences to disrupt HLA binding to eliminate T-cell recognition [81].

The effect of the Fc domain in fusion proteins has also been studied as means to modulate immunogenicity. Primarily CD4 T-cells and B-cells are involved with an important participation of APCs. In case of crosslinking activating FcγR on APCs the antigen presentation to T-cells increases, strengthening the immune response. The other case of signaling through inhibitory FcγR can also occur and will down-regulate immune responses. Interestingly the nature of the fusion partner can influence the affinity of the corresponding Fc- domain to the FcγR [82]. The hypothesis of lowered immunogenicity of Fc-fusion proteins by binding to inhibitory Fc receptors on B lymphocytes, the FcγRIIb was experimentally proven with the injection of DNA coding for a Exendin-4 Fc-fusion protein that did not result in the generation of neutralizing antibodies, which was the case for Exendin-4 alone [83]. Just recently a Fc receptor trap fusion protein consisting of fully human sequences only was risk assessed for immunogenicity. Although fully human sequences reduce the risk of antidrug antibodies (ADA), the effects of ADA might be higher if they occur, as they target the corresponding component on cells with all potential negative consequences [84].

Ideally B- and T-cell epitopes are removed, as both contribute to immunogenicity. It is easier to remove B-cell epitopes, as they are not restricted to major histocompatibility complex (MHC). In contrast to that T-cell epitopes rely on the high polymorphism of MHC [85].

11.4 Manufacturing

On the first glance, the production of fusion proteins cannot be distinguished from that of regular therapeutic proteins. Manufacturing typically comprises the steps of upstream processing including all aspects of cell culture, downstream processing covering steps to capture and purify the protein of interest from impurities and finally formulation steps that convert the protein into a storable and administrable form. One of the many advantages of fusion proteins is the uninterrupted manufacturing process of a single but multifunctional molecule. However, fusion proteins suffer from a lot of issues that can complicate the production. The combination of two different proteins with different properties might not always be compatible with respect to their pH preferences, hydrophobicity, native cellular localization, and glycol-isoforms. Furthermore, they are sometimes sensitive to low pH which then does not allow conventional virus inactivation. Fusion proteins are often larger than well-studied molecules like antibodies and therefore might have lower expression levels. Figure 11.4 gives a comprehensive overview on current manufacturing practices.

11.4.1 Upstream

Due to their size and complexity, most modern fusion proteins are expressed in eukaryotic cells. This allows the proper formation of post-translational modifications such as glycosylation or disulfide bridges. The major host cell is derived from Chinese hamster ovaries (CHO) which is frequently used for Fc-fusions. Yeast cells are applied primarily for albumin fusions. Prokaryotic expression is favored either for relatively small proteins with a molecular mass below 30 kDa or proteins that form inclusion bodies to enable the production of toxic proteins like immunotoxins.

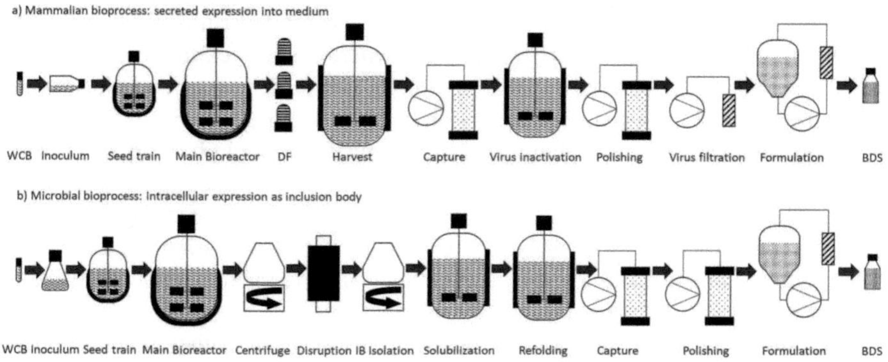

Fig. 11.4 Comparison of mammalian and microbial bioprocesses for the manufacturing of fusion proteins. (**a**) The upstream part of cultivation and expanding the cells takes typically several weeks in a mammalian bioprocess. Here the protein of interest is secreted to the medium; therefore, cells are separated from the supernatant by either depth filtration or centrifugation. Protein from the cell free harvest is then captured on the first column and potential virus impurities are inactivated by typical low pH exposure. The protein is further purified in several polishing steps to remove impurities before any remaining viruses are removed by virus filtration. In the last step, the pure protein is formulated by the addition of excipients and filled as bulk drug substance. (**b**) Microbial cultivation processes are usually finished in days. Here the protein of interest often accumulates in the form of inclusion bodies in the host cell. Therefore, these cells are collected by centrifugation, disrupted, and the inclusion bodies are washed and solubilized by denaturing agents. Then the unfolded protein has to be refolded before the regular purification process is started. With the exception of the absence of virus removal steps, the downstream procedure is quite similar. (WCB working cell bank, DF depth filtration, BDS bulk drug substance, IB inclusion body)

In principle, the upstream process contains three elements: the DNA construct coding for the gene of interest plus regulating sequences, the host cell, and the cultivation conditions. All three elements contribute to the expression level, or titer that can be achieved during upstream processing. Prokaryotic expression can directly work by introducing a plasmid that codes for the fusion protein, accompanied by at least a promoter sequence, a replication origin, and typically a selection marker such as an antibiotic resistance gene into the microbial cell. However, in mammalian cells the expression construct is integrated into the cell genome. Often this integration happens at multiple sites simultaneously increasing the expression level. To generate the producer cell line, a selection process is frontloaded that identifies the best cell clone based on expression yield, growth characteristics, and genomic stability.

Temperature influences cell growth and product titer. It was observed that in the case of Etanercept growth at 30 °C resulted in a threefold higher titer than at 37 °C. However, increasing the viable cell density at 30 °C decreased the productivity. So, essentially both effects have to be systematically assessed to find optimal expression conditions [86]. A more detailed analysis of cultivation parameters pH, temperature (T), and dissolved oxygen (DO) has been undertaken for EPO-Fc expressed in CHO. Lowering T and pH reduce cell growth and metabolism but increase the viable cell integral and consequently enhance the specific productivity while reducing volumetric productivity, leading to extended cultivation times. The solution to this dilemma is a biphasic process where initially culture conditions are established that support rapid expansion of cells followed by a second phase when switching to parameters that reduce growth but improve productivity [87]. The disadvantage of incomplete sialylation at low temperatures could be compensated by simultaneously reducing the pH [88]. A further benefit of a biphasic cultivation strategy is the reduction in EPO-Fc aggregates from 75% at standard conditions at 37 °C and pH 7.05 to less than 1% at 30 °C and pH 6.75. This synergistic effect even led to reduced glucose and glutamine consumption while simultaneously lowering lactate and ammonium generation [89].

The other significant eukaryotic expression host is yeast. Either *Saccharomyces* or *Pichia pastoris* cells are used to manufacture fusion proteins containing albumin or transferrin [90]. For instance, a IL2-HSA fusion could be produced in P. pastoris using glucose as carbon source. Besides the feeding strategy optimization, it was found out that lower fermentation temperature reduced the presence of proteases and a neutral pH of 7.0 was best for maximizing expression [91]. Surprisingly this fusion protein did only achieve 50% of the expression levels compared to HSA alone. Therefore, the co-expression of five different secretion helper proteins, Er1, Kar2, Pdi1, Sec1, and Sly1, was evaluated. All them could restore the initially observed expression level of approximately 1 g/L [92]. Expression yield in P. pastoris can also be influenced negatively by the structure of the fusion protein to be expressed. For instance, a triple copy of somatostatin fused to HSA resulted in significantly lower expression than its tandem copy, indication that gene multiplication not necessarily is a suitable approach to improve fusion protein manufacturing [93].

11.4.2 Downstream

The downstream process (DSP) starts with the material containing the product of interest. Depending on the expression system this can either be extracellular or intracellular material.

Microbial expression utilizing *Escherichia coli* typically results in inclusion bodies (IB) that contain denatured aggregated proteins. IBs can represent 10–50% of the total cell protein containing up to 95% of a single protein species [94]. This means during harvest the cells are collected and the later mechanically or chemically disrupted to set the IBs free. Typical impurities of that step are endotoxins that are present in huge amounts as the cells wall is destroyed. The IBs are then washed and solubilized by chaotropic substances such as guanidinium hydrochloride or urea under reducing conditions to break all disulfide bridges. After this denaturation, the so-called refolding process is initiated that essentially means gradual removal of the denaturing agents while slowly establishing oxidizing conditions to re-form disulfide bridges, thus achieving the proper natural secondary and tertiary structure [95]. Alternatively, recombinant proteins can be secreted to the periplasmatic space of *E. coli*, which avoids refolding, however, with lower expression yields.

Still some process and product related impurities are present that have to be removed by typical downstream unit operations such as chromatography and filtration.

In the case of extracellular production, the protein of interest is secreted to the cell culture medium. Here the harvest process means the removal of cells either by centrifugation or depth filtration or a combination of both to obtain a cell free fluid (CFF) for further processing. Hosts for secretion are typically eukaryotic cells such as yeast or mammalian animal cells. Secretion is certainly preferred because in modern serum free, chemically defined media only few host cell proteins are present together with the secreted product [96].

Fusion partners sometimes help in establishing platform purification processes. In the case of elastin like peptides (ELP), the temperature-dependent aggregation allows simple capturing by sedimentation as an initial enrichment [97]. The most frequently used fusion partner IgG Fc can usually be treated just like regular antibodies. However, there are reports, where protein A resins exhibit a lower dynamic binding capacity for Fc-fusion proteins primarily due to steric hindrance caused by bulky fusion partners [98]. The rest of the capture process is comparable to an antibody process with binding at neutral pH, and elution at acidic pH below 3.7 that also inactivates potential viruses. Unfortunately, sometimes aggregation during elution at low pH might occur that is triggered either by high local protein concentrations or the unnatural charges induced by low pH. Here the addition of a chaotropic agent during elution can help maintaining monomeric species [99].

Other antibody derivatives like Fab fragments or scFv do not bind to a protein A ligand. They require specific affinity matrices, for instance, protein L which is *kappa* light chain specific [100, 101] or a synthetic ligand such as FabSorbent® [102]. Recently a procedure to isolate Fab fragments with a series of protein G and L chromatography was described, where some product-related impurities like fragmented light chains were co-purified on protein L resin [103].

Albumin from plasma used to be isolated by dye-affinity chromatography. Nowadays advanced highly selective resins like Albupure™[1] based on a synthetic triazine or Captureselect™ with single domain antibody fragments are available to capture recombinant albumin fusion proteins.

All other potentially more unorthodox fusion proteins without any of the well understood fusion partners cannot be enriched from crude supernatants by affinity chromatography but have to be collected by a more unspecific method with high capacity but limited selectivity, ion exchange chromatography. Here the protein of interest binds to immobilized charges of the stationary phase at a pH which is either lower or higher than its isoelectric point to expose sufficient surface charges. Elution is achieved by either changing the pH to modulate the charges of the protein or by adding salt ions that disrupt the ionic binding. Besides bind and elute mode, proteins

[1] https://www.astreabioseparations.com/resources/albumin-fusion-proteins/albupure-pc-3151

can also be purified by a flow through method, that allows the protein of interest to pass, while impurities are bound to the resin.

In the case of disulfide containing proteins scrambling during purification might occur [104] that requires careful monitoring. Obviously wrongly formed disulfide bonds have a negative impact on potency as described for Etanercept [105].

A high purity of the fusion protein is then achieved by combining different purification principles that selectively remove all kinds of impurities, while enriching the protein of interest. Although it might seem preferable to execute a two-step process from an economical perspective, a three-step approach could result in improved purity [106]. So always cost must be balanced to a reasonable level. Finally, the protein just has to be concentrated and mixed with suitable excipients to be stabilized for its administration. Particularly the stabilization of fusion proteins can be quite challenging as the two different fusion partners might have different preferences for additives.

11.4.3 Glycosylation

Besides the proper formation of disulfide bridges that determine the three-dimensional structure, glycosylation is the other important post-translational modification of proteins. Glycosylation improves the solubility of proteins by adding highly hydrophilic glycans. These large carbohydrate structures can also enhance the stability by either covering protease sensitive regions or prohibiting aggregation through hydrophobic patches. Another biological function is their recognition by specific receptors that enable tissue targeting or the recruitment of immune cells.

Glycosylated proteins contain carbohydrates either N-linked to the nitrogen in asparagine side chains or O-linked to the oxygen in serine or threonine. Interestingly N-linked glycans require a recognition motif Asn-x-Ser/Thr, while O-linked glycosylation depends on the secondary structure and the accessibility of Ser or Thr. The often observed heterogeneity of glycans is a result of the expression level of the required enzymes and the availability of their substrates in the ER and Golgi [107]. The glycosylation can be controlled on several levels, starting at the protein sequence level, the choice of host cells, the cultivation process parameters, and finally the ability of the downstream process to select certain isoforms as described in Fig. 11.5.

The CH2 part of the Fc-domain contains a canonical position for N-glycosylation at Asparagine 297 (N297). N-glycans have important structural functions. They stabilize the CH2 domain of IgGs and the lack of glycans decreases the thermal stability of mAbs and make them more susceptible to unfolding. Furthermore, deglycosylated mAbs are more prone to aggregation. Depending on the size of the attached glycans, the structure of the CH2 domain can turn into either an open or closed conformation which influences the binding properties to Fc receptors. These receptors transfer the antibody-dependent cell-mediated cytotoxicity (ADCC) or complement-dependent cytotoxicity (CDC). The absence of fucose greatly enhances ADCC, whereas the presence of terminal galactose improves the CDC activity.

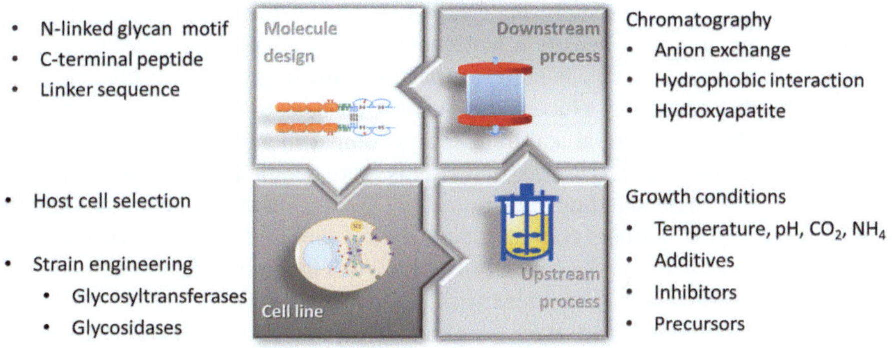

Fig. 11.5 Controlling glycosylation at different process steps. Glycosylation should be designed into the protein from the start. The next important choice is the host cell or potential strain engineering approaches to improve homogeneity and intensity of glycosylation. During the upstream process, the glycan pattern can be influenced by the growth conditions or the presence of certain substances in the culture medium. During downstream processing, the ideal glycoisoform must be separated from wrong or incomplete forms to obtain a product with a glycosylation as homogeneous as possible

Capping of the terminal glycans with sialic acid positively influences the pharmacokinetics, which is caused by prohibiting the clearance of desialylated proteins through receptors the recognize mannose or galactose [108]. As glycosylation has these wide-ranging effects, different approaches were undertaken to specifically address this functionality by either engineering the producer cells though eliminating fucosyltransferase or adding glycosyltransferases or optimizing the cultivation conditions that support the enhancement of one or the other glycol-isoform. For instance, the content of sialic acid could be increased through the addition of dexamethasone into the culture medium of CHO cells, expressing a CTLA4-Ig fusion protein [109]. A similar effect was achieved by adding hydrocortisone to cell cultures of Fc-fusion protein which also improved the titer [110]. In one example, the production level of a Fc-fusion protein was improved through a hyperosmotic culture medium, which unfortunately reduced the sialic acid content. This was due to differential down-regulation of a series of genes, which could be restored through the addition of betaine [111]. A more extensive recent study identified oxidative stress resulting from gas transfer limitations in large bioreactors as source for low or highly variable sialylation of a Fc-fusion protein. On a metabolic level, the hexosamine pathway became limited and thus the key building blocks for sialylation were not sufficiently available [112]. The obvious counteraction to the lack of educts is their supplementation. Simply the addition of galactose enhanced the sialylation of an Fc-fusion protein [113]. Beside the supply of precursors also extracellular factors such as degradation via hydrolysis of α 2–3 sialic acids through sialidase activity released from destroyed cells during cultivation can contribute to low levels of sialic acid as seen during the cultivation of an Fc -fusion protein [114]. Modern tools nowadays allow the monitoring of the complete N-glycosylation pathway on a molecular level. A recent study on CTLA4-Fc fusion protein revealed that, for instance, at maximal cell density a bottleneck arises between endoplasmic reticulum (ER) and the cis-Golgi, leading to increased mannose 8 and 9 glycans, that could be partly reduced by glucose deprivation. At later timepoints in the culture, the transition between medial and trans-Golgi compartment becomes rate limiting, resulting in decrease of sialylated species. This more detailed information will help in better process understanding and control [115]. The situation gets more complicated if both fusion partners contain glycosylation sites and even vary between N- and O-linked species as in the case of EPO-Fc. With that molecule the influence of media composition on type and intensity of glycosylation was analyzed. Both fusion protein partners maintained their original N-glycan characteristics. However, depending on the media used for culture, either bi- and tri-antennary glycans were generated, while the other medium delivered tetra-antennary glycans. With regard to O-linked glycans, three species were preferred [116]. As O-glycans do not require the presence of transferase recognition motifs, they can in principle occur at any serine or threonine position. This unpredictability has recently been described on the example of a CTLA4-Fc fusion protein that surprisingly contained a number of unexpected new O-glycans. It could be attributed to the linker sequence derived from the hinge region of IgG1. This specific post-translational modification resulted in a complex formation of the hinge sequence and a glycan transferase. As the undesired O-glycans hindered proper inter-chain disulfide bridge formation, O-glycan-rich species tended to aggregation. Therefore, all potential O-glycosylation site were eliminated by point mutations [33].

All examples above discuss experiments in CHO cells. But using a different host such as *Pichia Pastoris* can have a dramatic effect on glycan pattern.

11.4.4 Aggregation

When investigating the causes of aggregation, it can be distinguished between extrinsic and intrinsic factors. For instance, temperature changes during freeze-thawing can induce high molecular weight species. The same is true for physical stress like stirring, filtration, or high flow rates. Finally, solution factors such as pH, ionic strength, organic solvents or metal ions can lead to aggregation. Intrinsic factors are molecule derived and can be dependent on the presence of residues sensitive to clipping, oxidation, deamidation, or isomerization. Furthermore, hydrophobic patches or charges residues can trigger aggregation as well as the presence of unpaired cysteines or scrambled disulfides [117].

As mentioned before, fusion proteins can be prone to aggregation due to domain instability as in the case of HSA-hGH fusion. Modulating HSA stability with octanoic acid, it was revealed that colloidal instability but not conformational stability of HSA contributes to aggregation of that molecule at low pH [118]. Orencia®, a receptor trap consisting of several domains with different conformational stabilities tends to aggregation due to partial unfolding of the CTLA-4 domain at pH 6 [119].

Aggregation can also be caused by disulfide rearrangements resulting in intermolecular aggregation as demonstrated with a Fc-fusion protein. Here the presence of N-linked glycosylation can reduce the tendency to form aggregates [120]. Proteins containing a free thiol can generate also disulfide crosslinking leading to massive aggregation even under relatively inert conditions such as −30 °C. Here during storage at elevated concentrations, salting out effects can occur that intensify aggregate formation [121]. More detailed understanding of aggregation can be obtained by hydrogen/deuterium exchange that identifies surfaces with a tendency to aggregate [122].

Very often sensitivity to low pH has been observed for Fc fusion proteins. Adding chaotropic substances such as urea during elution helps to prevent high molecular weight species generation [123]. The so-called on-column disaggregation can reduce aggregates by approximately 50% [124]. Interestingly aggregation can be beneficial, for instance, for the oral delivery. Here it prevents the degradation during the stomach and intestine passage [125].

11.4.5 Analytics

Although many of the standard analytical methods for antibodies can still be applied, the level of complexity of fusion proteins often requires a more in-depth assessment. Particularly Fc-fusion proteins with a conserved N-linked glycosylation site represent a difficult sample if the conjugated partner contains additional glycosylation sites. Therefore, site-specific glycan analysis is a key element of quality control.

The typical assays to characterize any therapeutic protein comprised the assessment of identity, purity, and potency as displayed in Fig. 11.6. Usually, platform methods are established for the first two assay, whereas potency determination often requires highly specific bioassays. Identity determination nowadays relies primarily on mass spectrometric methods. The content of variants is quantified with size exclusion chromatography (SEC) or capillary electrophoresis (CE-SDS) with respect to the content of high molecular weight species, monomers, or fragments or capillary isoelectric focusing (cIEF) for charge variations that represent glycosylation isoforms. In case of high heterogeneity due to glycosylation, an enzymatic pre-treatment for deglycosylation is done to establish the correct mass of the naked polypeptide. Furthermore, large proteins can be digested by proteases into smaller peptides to obtain either a peptide mass fingerprint or to identify modifications that prohibit a proper digest. Previously amino acid sequencing was applied to confirm the correct N-terminus, but nowadays this method is replaced by liquid-chromatography tandem mass spectrometry (LC-MS/MS). The advantage of LC-MS/MS is the simultaneous detection of individual modifications such as methionine or tryptophan oxidation, asparagine deamidation, aspartic isomerization, and lysine glycation. In principle, fusion proteins can be seen as multispecific molecules; therefore, a big part of the analytics developed for bispecific antibodies

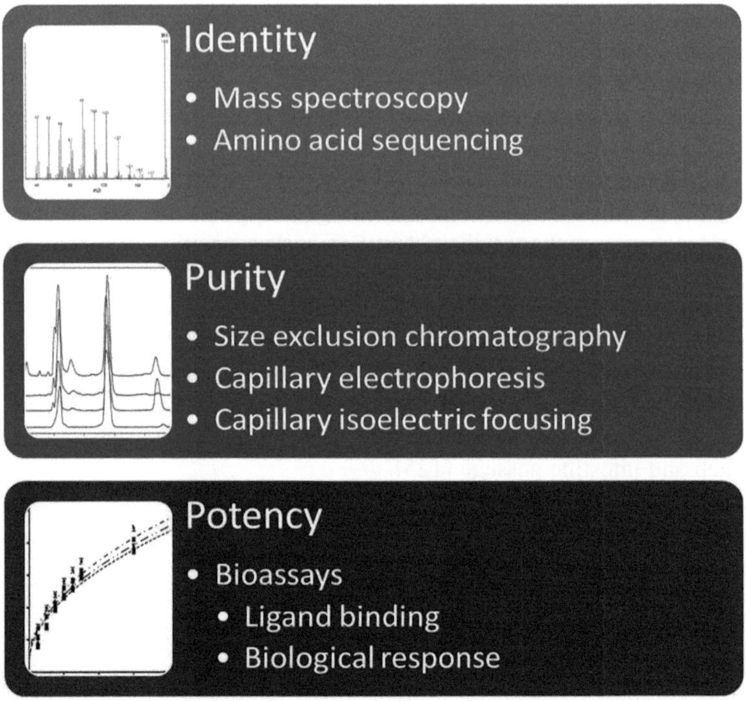

Fig. 11.6 Overview on typical analytical methods for fusion protein characterization

can be applied as well. For instance, it can be relevant to determine the thermal stability by differential scanning calorimetry (DSC) and differential scanning fluorimetry (DSF) that indicates correct pairing of polypeptide chains [126].

Particularly receptor traps require extensive analytics for S-S scrambling, incomplete folding, or aggregates. Another class of difficult molecules are enzyme fusions that are sensitive to aberrant glycoisoforms and coagulation factors requiring carboxylation. For all these cases, specific detailed analytical methods must be established [127].

11.5 Therapeutic Concepts

11.5.1 Half-Life Extension

Therapeutic proteins have gained a wide acceptance in the last decades, but still some of the most attractive biological substances in the class of hormones, cytokines, coagulation, and growth factors are small and suffer from a quick clearance from the bloodstream. In the past, this small therapeutic window was compensated for with frequent dosing, making treatment difficult for patients and causing

Table 11.8 Half-life of approved fusion proteins

Brand	[kDa]	T½ range	Average t½ [h]
Enbrel®	150	3–6 d	120
Ontak®	58	70–80 min	1.2
Amevive®	92	11 d	264
Orencia®	92	8–25 d	400
Arcalyst®	251	9 d	216
Nplate®	60	1–34 d	408
Elonva®	47	59–82 h	70
Nulojix®	90	8–10 d	220
Eylea®/Zaltrap®	115	5–6 d	130
Alprolix®	98	3.5 d	86
Tanzeum®	93	5–8 d	120
Eloctate®	220	10–23 h	16
Trulicity®	63	5 d	120
Strensiq®	161	1.7–2.8	55
Idelvion®	125	4.3 d	104
Lumoxiti®	63	0.8–1.8 h	1.4
Elzonris®	58	0.7–1 h	1
Reblozyl®	76	13 d	312

dose-limiting concentration peaks. Some of the currently approved fusion proteins have significantly increased half-life as described in Table 11.8. The pharmacokinetic of drugs is influenced by target-mediated drug disposition, antidrug antibodies, nonspecific catabolism, or kidney excretion [128].

Interestingly, receptor binding aids in recycling, keeping proteins in circulation longer and protecting them from lysosomal degradation. This is actually a concept utilized for some fusion protein types. Globular proteins smaller than 70 kDa are removed very rapidly by glomerular filtration in the kidneys. In addition to the 60 Å pore diameter of the glomerulus, another important parameter that regulates the excretion of macromolecules is the negative charge on the cell surface that repels anionic macromolecules. Multiple strategies to extend the plasma half-life have now been feasible with fusion protein as summarized in Fig. 11.7.

11.5.1.1 Half-Life Extension by Size and Recycling

Albumin Fusions

Human serum albumin (HSA), the most abundant serum protein with a size of 66 kDa and a diameter of 90 Å, exceeds the renal clearance threshold. However, its exceptionally long half-life of 19 days is more influenced by recycling through the neonatal Fc-receptor (FcRn). Its primary function during infancy is the capture protective maternal antibodies from milk by absorption through the intestinal

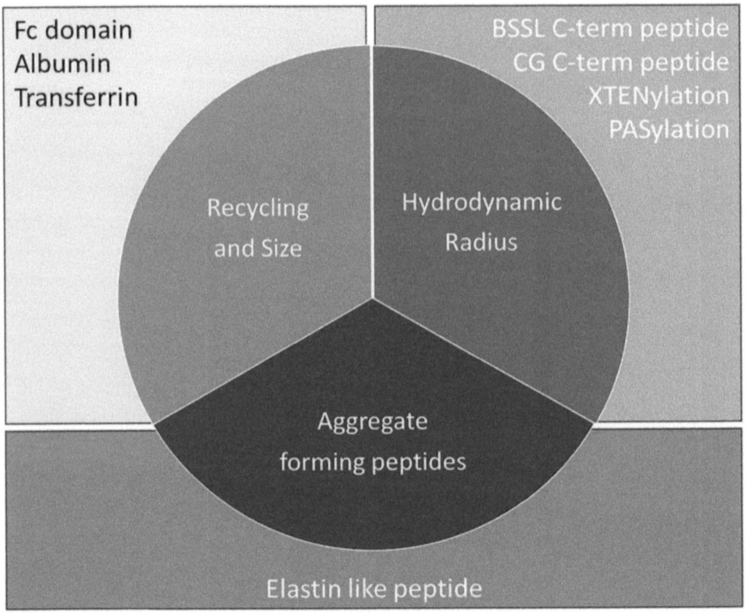

Fig. 11.7 Fusion protein-based strategies for half-life extension (BSSL Bile salt stimulated-lipase, CG chorionic gonadotropin)

epithelium. In adults, FcRn is downregulated on intestinal cells but remains present on vascular cells. The mechanism of recycling relies on several conserved histidines that can change their charge distribution at different pH. HSA, or IgG internalized by pinocytosis bind strongly to the FcRn under acidic conditions in the endosome thus being protected from degradation and remain bound to the FcRn until they reach the cell surface again, where they are released at neutral pH.

Furthermore, the phenomenon of albumin accumulation in tumors, together with the intrinsic ability to enhance the half-life of fusion or attachment proteins, could be exploited for therapeutic approaches [129]. Another advantage is the ability of HSA to stabilize proteins despite their own structural instability. For instance, the fusion of HSA to granulocyte colony-stimulating factor (G-CSF) eliminated the aggregation tendency of G-CSF [118]. As HSA can be attached either to the N-or C- terminus, it gives design flexibility even for fusion partners requiring an accessible terminus. Several HSA fusion proteins are now commercially available drugs as can be seen in Table 11.8. Designing adaptations of HSA fusions often involve the modifications of linkers. For instance, the blood clotting factor FIX not only benefitted from an increased plasma half-life but also from 10 to 30-fold higher clotting activity when inserting a cleavable linker. This linker was split by proteases simultaneously to FIX activation, thus triggering release from HSA when blood clotting was needed [130].

Fc Fusions

Probably the most successful class of fusion proteins are those containing an Fc domain. The fragment crystallizable (Fc) region of an antibody consists of the second and third constant domains (CH2, CH3) of the heavy chain and forms a dimer through two disulfide bridges in the n-terminus of the hinge region. The first ever approved fusion protein, Enbrel (R), is a Fc-fusion protein. More specifically it belongs to the family of receptor traps that combine the extracellular domains of receptors with the constant domain of antibodies.

The therapeutic success of Fc fusion proteins is primarily based on the well-understood concept of combining the high specific affinity of the extracellular receptor domain and receptor-mediated antibody recycling through the Fc portion. Although the addition of a 50 kDa units helps preventing rapid renal filtration, the contribution of FcRn-mediated recycling has a much more profound effect. Nevertheless, it was tried to correlate the contribution of FcRn-mediated recycling to half-life with the affinity of antibodies and Fc fusion proteins for FcRn. As expected, low affinity correlates to short half-life and vice versa. This was impressively demonstrated by optimizing Avastin® with a threefold longer half-life through 11-fold higher affinity. The same effect was proven by modifying Erbitux® [131].

Great efforts have been undertaken to identify critical residues in the Fc domain that could be modified to increase FcRn binding and consequently extend the elimination half-life. Table 11.9 summarizes the results [132].

But having a good FcRn affinity alone does not always help, as it could be shown that some Fc-fusion proteins displayed shorter half-life than expected. The reduced value could be attributed to sterical or conformational effects of the molecules fused to the N-terminus [133]. For dimeric molecules, Fc fusion proteins are preferable because the interaction between the antibody heavy chain and the disulfide backbone of the hinge region automatically forces the formation of homodimers. Sometimes the dimerization does not work due to exceptionally bulky fusion partners. Then a monomeric fusion might be advisable, where just one Fc-heavy chain is combined with its fusion partner, while another Fc-only domain binds in an asymmetric way. Monomeric Fc-fusions might also be beneficial in cases where

Table 11.9 Mutations in the Fc domain of IgG1 that influence the plasma half-life

Company	Half-life	Mutation
Medimmune	+	M252Y, S254T, T256E
Protein Design Labs	+	T250Q, M428L
Genentech	+	N434A
Sally Ward	+	H433K, N434Y
Derry Roopenian	+	T307A, E380A, N434A
Xencor	+	M428L, N434S
Derry Roopenian	–	I253A
Eli Lilly	–	P257I, N434H or D376V, N434H

pulmonary delivery is desired. There have been extensive studies on monomeric erythropoietin (EPO) Fc fusion that exerted a twofold higher affinity to the FcRn and a 30% better half-life than the dimeric variant [134].

Conformational stability is often pH dependent. In the case of Orencia®, which consists of the Fc portion of IgG1 and the soluble portion of the T-cell receptor CTLA-4, destabilization of the CH2 domain can lead to aggregation [119].

Therapeutic peptides suffer from a fast renal clearance due to their small size and could, therefore, benefit from fusion to a larger protein such as the Fc domain. Their typical half-lives are in the minutes range, so any improvement will be helpful. One example from that group is Nplate®, a molecule approved for the treatment of chronic idiopathic (immune) thrombocytopenic purpura with a half-life of 3.5 days in the Fc fusion version. Interestingly the peptide is fused to the C-terminus of Fc. Cytokines are another therapeutic class of small proteins where Fc-fusion could contribute to a prolonged circulation. Although there are numerous examples in clinical studies, there is currently no approved Cytokine fusion available.

Transferrin Fusions

Transferrin (Tf) is one of the ten most abundant serum proteins and delivering iron from intestine and liver into proliferating cells. As fusion partner, it certainly prevents rapid elimination by renal filtration due to its size of 79 kDa. But interestingly it can also be recycled by receptors as part of its normal function. The observed half-live of nonglycosylated Tf is 14–17 days. In contrast to FcRn recycling, transferrin receptor (TfR) recycling is pH independent. The acidic pH of endosomes separates iron from Tf, but not Tf from receptors [135]. A transferrin fusion protein targeting a cell surface receptor first binds to the fusion partner receptor in the presence of abundant transferrin. After that, secondary binding between Tf and its receptor can happen. Secondary binding can also occur after endocytosis in the early endosome. When the fusion protein is attached to the TfR, it stays bound to the TfR even under acidic conditions and is recycled back to the cell surface [26]. Transferrin-mediated trafficking through endosomes can also support the activation of protein pro-drugs through proteolysis. This was exemplified with proinsulin–transferrin fusion that was transformed into active insulin-Tf when incubated with hepatocytes [136]. In addition to its positive effect on the half-life of fusion proteins, transferrin has another important property. Although iron is essential for nerve function, it must be balanced to avoid dangerous side effects. Homeostasis is regulated by iron transporters such as TfR at the blood–brain barrier (BBB) between the blood and the cerebrospinal. Hence, transferrin can serve as vehicle for transport across the BBB. An early example is nerve growth factor (NGF) fused to transferrin [137]. However, recent progress focused on the oral bioavailability of transferrin fusions. Here TfR acts as mediator of transcytosis. Proteins such as G-CSF [138] pro-insulin [139] or FIX [140] have been fused to Tf and applied orally to rodents to detect bioactivity in the blood stream of test animals. All three examples could demonstrate successfully uptake and functionality through oral passage. Interestingly the protection from acidic stomach fluids was primarily achieved through aggregation of the fusion proteins [125]. However, transcytosis was clearly Tf dependent as the co-administration of free Tf abolished the effect. A crucial element of all Tf-versions is the linker between Tf and the payload. In the case of FIX fusion, the traditional flexible $(G_4S)_2$ proved to have best expression and oral availability [140].

11.5.1.2 Half-Life Extension by Increasing the Hydrodynamic Radius

Repetitive Peptide Fusions

As discussed earlier, the duration of circulation in the bloodstream is partly depending on the size of the respective molecule. Small proteins with a hydrodynamic radius below 60 Å are quickly removed by renal filtration. Therefore, concepts for fusion proteins have been developed using this phenomenon. Most are based on the addition of peptides (repeats) attached to one or both ends. Ideally the peptides are uncharged to prevent aggregation, hydrophilic, and lack an ordered structure to generate a cloud like shape with a large radius. The dimension of the cloud is only dependent on the length of the peptide chain. This phenomenon exists in nature and is utilized by parasites to increase the bioavailability of virulence factors. This observation prompted research to develop artificial sequences that could be even more effective. Initial attempts with poly-glycine repeats unfortunately resulted in limited solubility. To balance that, serine was introduced to improve the hydrophilicity, a fact that is well known from shorter G_4S peptide linkers. First studies involved up to 40 copies of the G_4S sequence attached to a Fab fragment. Interestingly the half-life of that fusion construct was threefold longer than with the Fab alone. While the mass increased only by 9% due to the additional peptide, the apparent size determined by size exclusion chromatography was expanded to 120%. Unfortunately, larger polymers suffered from aggregation and the difficulty to generate stable genetic constructs [141]. Stimulated by this observation it was evaluated if the insertion of proline with its cis/trans isomerization ability could prevent aggregation. Furthermore, glycine was replaced by alanine. Randomizing the sequence of the three amino acids further prevents the formation of stable secondary structure. This technology, called PASylation®, increases the hydrodynamic radius significantly when fused to other proteins. PAS sequences are designed to be devoid of T-cell epitopes and protease recognition motifs. The physicochemical properties of PAS sequences have been extensively studied. This artificial biopolymer represents a strongly hydrophilic and structurally disordered polypeptide with expanded hydrodynamic volume that closely resembles polyethylene glycol and offers attractive features for the pharmacokinetic modulation of therapeutic proteins [142].

Alternatively, random peptide polymers can be engineered with a larger subset of amino acids including alanine, glycine, glutamate, proline, serine, and threonine, but excluding larger, hydrophobic, positively charged or sulfur containing amino acids. These building blocks were systematically assembled to reveal suitable sequences that are nonrepetitive and unstructured. The selected candidate polypeptide called XTEN® contains 864 amino acids and increased the half-life of Exenatide in mice by a factor of 71 [143]. In contrast to PAS, XTEN® contains multiple negative charged residues that could lead to repulsion at the glomeruli in kidney adding another factor that could positively influence half-life. Unfortunately, XTEN® effects are not always predictable, as, for instance, glucagon-XTEN exhibited only 15% of the normal potency [144].

In a recent study, both polypeptides, PAS and XTEN®, were compared. Pharmacokinetic analysis demonstrated a clear linear relationship between serum half-life and hydrodynamic radius with both polymers. As both fusion partners behaved quite similar, no influence of charge on any evaluated parameter could be proven [145].

Bile salt stimulated-lipase (BSSL), an enzyme that did not reach its primary endpoint in a phase 3 study, contains 17 repetitive segments of 11 residues (PVPPTGDSGAP) that are connected to form a continuous domain at the C-terminus. The peptide is proline rich and O-glycosylated. As part of an digestive enzyme in fresh breastmilk, it displays improved stability and secretion while being protected from proteolysis. The 11 residues can be multiplied and the size increase results in an extended half-life and improved solubility. When fusing 17 repeats to an Affibody, SEC detected a sixfold larger molecular mass. Pharmacokinetic profiles can be fine-tuned by varying the number of repeats. Renal clearance was slowed more than 100-fold when using 51 repeats.[2]

Glycosylated Peptides

Branched glycosylation forms a cloud like structure around proteins. The artificial introduction of glycosyltransferase recognition motifs into the original peptide sequence can lead to hyperglycosylation as in the case of Aranasep® that had a threefold longer half-life than the wild-type erythropoietin [146]. Instead of changing the sequence of the target molecule, peptides with multiple glycosylation sites can be fused at either the N- or C-terminus. The natural carboxyl terminal domain (CTP) of chorionic gonadotropin (CG) contains 4 O-linked glycosylation sites [147]. Interestingly, the naked peptide has a molecular mass of 2.9 kDa, while the glycosylated version reaches 8 kDa. The versatility of CTP was demonstrated with human growth hormone hGH in multiple fusion constructs. The most successful variant with CTP at both termini was fourfold more potent than the hGH alone and showed a more than ten-fold prolonged half-life (fares 2010). The first product containing CTP, Elonva®, a follicle stimulant composed of the α subunit of human follicle-stimulating hormone (FSH) was approved in 2010 and allows a once weekly injection instead of daily administration.

Besides the clinically proven CTP peptide, there have been numerous other examples described, utilizing increased glycosylation for half-life extension. For

[2] Patent: US2021/0284690A1.

instance, the hydroxyproline Hyp-rich repetitive peptide (HypRP) tag that directs the formation of proline hydroxylation and subsequent Hyp-*O*-glycosylation in plant cells has been fused to hGH to obtain a sixfold improved plasma circulation [148]. Another approach was described for polymers containing the N-linked glycosylation motif Asn-Xaa-Ser/Thr. The so-called Genetic Polymers™ were tested in as 155 triplet repeats fused to G-CSF, resulting in fourfold extended half-life. However, neither HypRP nor Genetic Polymers made clinical progress in the last years.[3]

11.5.1.3 Aggregate Forming Peptides

Repeated sequences of other naturally occurring peptides can be found in gelatin and elastin. Elastin-like peptides (ELPs) were initially applied as a downstream processing aid due to their unique ability to form aggregates at high temperatures and convert to soluble monomers at low temperatures. This phenomenon is called a reversible thermal phase transition, and the critical temperature of the phase transition varies with the length of the polymer. The longer the polymer, the higher the transition temperature [149]. ELP fusion proteins form a depot in the body where the active molecule slowly dissociates which represents a totally different mode of action for half-life extension. Currently, the most advanced candidate drug Pemziviptadil, called Vasomera™ or PB1046, the fusion between the Vasoactive Intestinal Peptide (VIP) and ELP, is in clinical phase 2 to be evaluated as treatment against pulmonary arterial hypertension, allowing once weekly subcutaneous dosing.

Peptide sequences used for half-life extension are collected in Table 11.10.

Table 11.10 Peptide fusion-based half-life extension strategies

Mode of action	Name	Amino acid composition	Company
Soluble peptide polymer	PASylation	P, A, S	xl-protein
	Elvera	$(PVPPTGDSGAP)_n$	Swedish Orphan Biovitrium
	XTENylation	P, A, S, E, T, G	Amunix (Sanofi)
Glycosylation	C-terminal peptide	$[SSSKAPPPSLPSPSRLPGPSDTPILPQ]_n$	Prolor (Opko)
	Genetic polymers	G, N, Q & A, S, T, D, E	Aequus Biopharma
Aggregate forming	Elastin like peptide	$[VPGX^aG]_n$	Phasebio

X^a: any amino acid except P

[3] Patent: US2016/0296632A1.

11.5.2 Targeting Functions

Fusion proteins have been used extensively to target their partners to specific organs or certain cell types. Particularly for cancer therapeutic it is of utmost relevance to only address malignant cells to avoid unwanted adverse effects. In other cases, like enzyme replacement therapy it is important not to lose too much of the dose to irrelevant organs. Targeted drug delivery is a very valuable approach, particularly for large therapeutic molecules, to keep them away from areas in the body where they are quickly degraded or because systemic distribution might never reach a sufficiently efficacious dose. The fusion protein approaches for organ targeting are exemplified in Fig. 11.8.

To get biologics into the brain, the molecules must cross the blood–brain barrier (BBB). Current strategies comprise primarily receptor-mediated transcytosis (RMT) via the receptors for insulin (IR) and transferrin (TfR). The most advanced procedure is to shuttle payloads across the BBB with TfR and IR directed antibodies. So far neurotrophic factors, lysosomal enzymes. and antibodies have been fused to the molecular Trojan horse (MTH) antibodies [150].

Although liver as the main metabolic organ should be of high interest, not many successful examples on targeted drug delivery exist. A recent example utilizes transferrin as targeting moiety to get proinsulin as fusion protein to hepatocytes. Overall this fusion protein showed three distinct effects: TfR-mediated binding and uptake of the prodrug on the cell surface, liver-specific, TfR-mediated conversion of the prodrug into its active form, and the bifunctional binding of the active fusion protein to both Tf and INS receptors in the liver to achieve prolonged retention [151].

The lung can be reached through normal respiration and therefore the focus of targeting was on pulmonary delivery of biologics. The research focused on utilizing the neonatal Fc receptor (FcRn) in lung epithelia. Most of the Fc fusion proteins were monomeric or asymmetric molecules with the fusion partner just on one of the Fc heavy chains. For instance, an asymmetric erythropoietin (EPO) Fc fusion exhibited twofold better affinity to the FcRn and a 30% improved half-life than the standard dimeric variant. Compared to the original EPO, monomeric EPO-Fc achieved

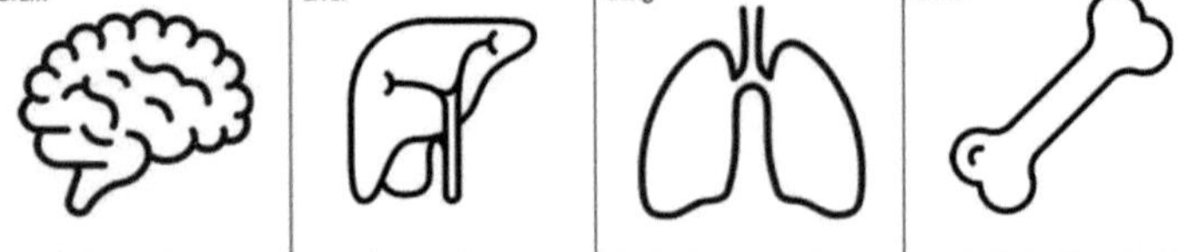

	Brain	Liver	Lung	Bone
Approach	• Receptor mediated transcytosis	• Receptor mediated transcytosis	• Immunoglobulin mediated transcytosis	• Acidic peptide targeting
Target	• Transferrin receptor • Insulin receptor	• Transferrin receptor	• Neonatal Fc receptor • polymeric immuno-globulin receptor	• Hydroxyapatite
Examples	• Neurotrophic factors • Lysosomal enzymes	• Proinsulin	• Ertyhropoietin	• Tissue-nonspecific alkaline phosphatase

Fig. 11.8 Fusion protein-based strategies for organ targeting

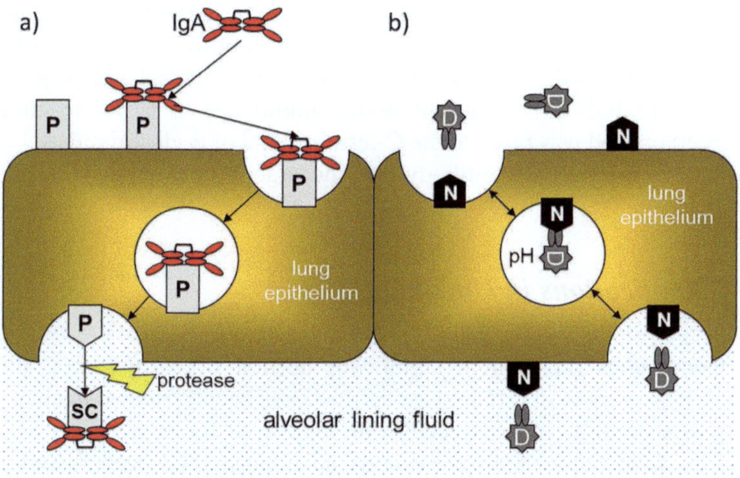

Fig. 11.9 Lung epithelium targeting and translocation of fusion proteins. (**a**) In cases where pulmonary delivery is impossible, transport to the lung can be facilitated through the polymeric immunoglobulin receptor (P). It transfers polymeric antibodies like IgA from the basal to the apical side. This passage is essentially irreversible, because the secretory component (SC) to which IgA is attached to is cleaved from the receptor. (**b**) The neonatal (N) Fc Receptor binds antibodies, Fc-fusion proteins (D) or albumin containing formulations. This transcytosis can transfer proteins bi-directionally and relies on pH dependent binding to N

a much higher systemic concentration [134]. Besides the FcRn also the polymeric immunoglobulin receptor (pIgR) can be the entry point into lung epithelium. pIgR internalizes and transfers polymeric antibodies (IgA or IgM) from the basolateral to the apical surface of the cell where the antibodies are released to the lumen. As the secretory component (SC) to which the polymeric Igs are attached to, is clipped from the receptor, the transfer is essentially irreversible [152]. A comparison of both pathways is described in Fig. 11.9.

A large constituent of the vertebrate body are bones. There are a number of diseases that affect bones; therefore, targeted therapies for bones are highly relevant. One of the most recently approved Fc fusion proteins, Strensiq®, represents a bone targeted tissue-nonspecific alkaline phosphatase (TNALP) to treat hypophosphatasia (HPP). This molecule consists of a truncated TNALP without its hydrophobic C-terminus connected to an Fc-part and a deca-aspartate sequence (D10) to target mineralizing tissue [153].

Several enzyme replacement therapies (ERT) must reach an intracellular location to treat lysosomal storage diseases. This is typically achieved by high mannose glycosylation as a recognition motif for translocation. Instead of modifying the glycosylation, a plant expression system can be used that specifically mannosylates four of the five potential N-linked glycosylation positions in the b-glucocerebrosidase. The short C-terminal targeting and retention signal encoding the sequence DLLVDTM direct the protein to the vacuole. Glucocerebrosidase with terminal mannose glycans from plants was approved by the FDA under the trade name

Elelyso® 2012 [154]. A similar effect could be achieved with the glycosylation-independent lysosomal targeting (GILT) approach. It utilizes the ability of insulin-like growth factor 2 (IGF-2), to bind to the bifunctional, IGF-2 cation-independent Man6-P receptor. IGF-2 was truncated to the amino acids 8–67 to prevent binding to other receptors and was fused to the C-terminal end of β-glucuronidase [155] and a-glucosidase [156]. Recently antibody-enzyme targeting was applied for ERT [157].

11.5.3 Applications in Oncology

Cancer therapeutics represent a significant portion of current drug discovery research. A huge variety of different modalities have been evaluated in clinical trials and every year multiple new molecules are approved for human use. Antibodies dominate the field due to their convincing advantages. However, fusion proteins allow new modes of action that are difficult to achieve by other means. This paragraph summarizes the various approached to eliminate malignant cells by fusion proteins.

11.5.3.1 Fc Domain Receptor-Mediated Toxicity

Antibodies already represent a "magic bullet" by on the one hand selectively addressing a molecule through the variable domain and on the other hand triggering antibody-dependent cellular cytotoxicity (ADCC) or complement dependent cytolysis (CDC) through the Fc domain [158].

ADCC depletes cells through binding of the antibody to the respective antigen on the target cell. Then the CH3 domain of the antibody interacts with natural killer (NK) cells via their CD16 Fc receptor (FcγRIIIa). The cross linking of the CD16 receptors ultimately leads to degranulation at a so-called lytic synapse releasing of granzyme and perforin to destroy the target cells.

Antibody-dependent cell-mediated phagocytosis (ADCP) is primarily dependent on FcγRIIa present on macrophages that attack cells marked with antibodies whose Fc domain can bind that receptor.

CDC is caused by the interaction of the CH2 domain of antibodies with the C1q complement factor, consisting of six heterotrimeric subunits. That activates a proteolytic cascade, triggering the formation of a membrane-attack complex (MAC) that generates a transmembrane channel to lyse the cells.

Marking the cell surface of the target cell with C3b as opsonin induces cell killing as well. By the interaction of C3b with its corresponding receptor (C3bR) on macrophages or NK cells phagocytosis is initiated. This process is called complement-dependent cell-mediated cytotoxicity (CDCC). Additionally, the chemoattractant C5a, facilitating an inflammatory response, supports cell killing. An overview of these activities is displayed in Fig. 11.10. Some fusion proteins have these functions incorporated in their molecular design.

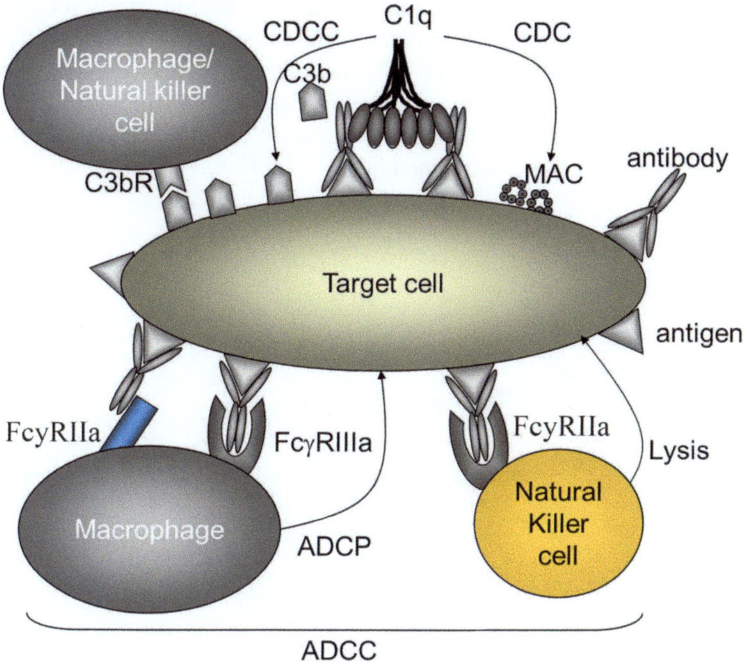

Fig. 11.10 Fc-domain mediated cytotoxicity. C1q complement factor interacts with the CH2 constant region of an antibody bound to its respective antigen on the surface of a target cell. This induces the formation of a membrane-attack complex (MAC) that lyses the cell through complement-dependent cytotoxicity (CDC). Alternatively, C3b that is generated during this cascade facilitates phagocytosis and cytolysis by labelling the cell as opsonin and interacting with the C3b receptor (C3bR) on a macrophage or natural killer cell. This reaction is called complement-dependent cell-mediated cytotoxicity (CDCC). Antibody-dependent cellular cytotoxicity (ADCC) is induced through the FcγRIIIa on natural killer cells or antibody-dependent cell mediated phagocytosis (ADCP) by macrophages that additionally contain FcγRIIb

11.5.3.2 Toxins

Interestingly the second ever approved fusion protein for cancer application was an immunotoxin. Since the market introduction of Ontak® in 1999 it took until 2018 to get the next two immunotoxins Lumoxiti® and Elzonris® successfully through the regulatory process. The toxins derived either from bacteria or plants always exert their catalytic function inside the cell. Native toxins bind to receptors to be internalized. To exert their function often they are split by proteases and finally they have to escape from endosomes to avoid degradation in lysosomes. Instead of the natural targeting domain a much higher specificity to certain cell types can be achieved by replacing this domain by antibodies or their fragments that recognize typical markers of malignant cells. In general, immunotoxins are mainly used for the treatment of hematological malignancies due to the accessibility of malignant cells and the immunocompromised state of patients [159]. As natural toxins are highly immunogenic and would be eliminated fast, site directed mutagenesis is required to reduce the number of immune response triggering epitopes. The most frequently used bacterial toxins are Diphtheria toxin (DT) released by gram-positive aerobic *Corynebacterium diphtheria* as prototype for ADP-ribosylating toxin, or Exotoxin A of the aerobic gram-negative bacterium *Pseudomonas aeruginosa* (PE). The function of both toxins is explained in Fig. 11.11.

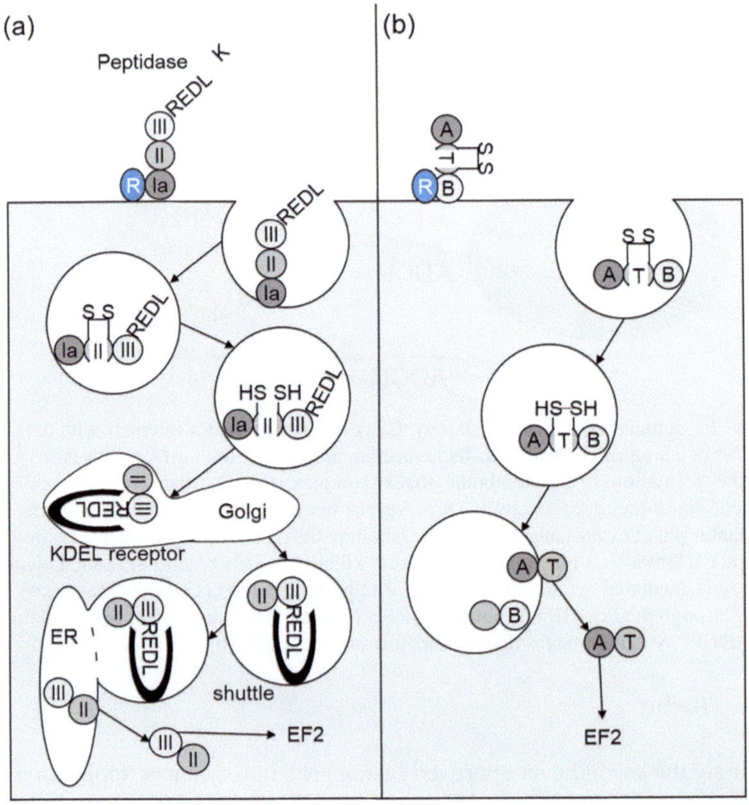

Fig. 11.11 Mechanism of bacterial toxins utilized in immunotoxin fusion proteins. (**a**) The C-terminal lysine (K) of pseudomonas exotoxin is cleaved off by a serum peptidase before internalization through coated pits. Reduction in its disulfide bridge and furin cleavage separates binding domain (Ia) from the catalytic domain (III) that is still bound to the translocation domain (II). In the Golgi apparatus, the exposed REDL-sequence is captured by the KDEL-receptor and the toxin is shuttled to the endoplasmatic reticulum (ER). There the translocation domain mediates the escape to the cytosol where it inactivates EF2. (**b**) After binding to its receptor (R) and cleavage, diphtheria toxin is internalized. The reduction in the remaining disulfide bond separates the binding domain (B) from the catalytic (A) and translocation (T) domain. A pH-induced conformational change enables the transfer of A and T to the cytosol where A blocks the elongation factor 2 (EF2)

11.5.3.3 Immunocytokines

Besides using nonhuman proteins to eliminate tumors by direct action, malignant cells can also be destroyed by attracting and stimulating immune cells. To activate the immune response, certain cytokines can be administered. Unfortunately, systemic administration of cytokines can lead to unwanted side effects, such as fever, inflammation or other flu-like symptoms [160]. Therefore, a more targeted approach is desirable for patient safety. Furthermore, cytokine targeting can also improve the duration of the cytokine response as the small cytokines are relatively rapidly removed from the body. Overall, immunocytokines can enrich the cytokine in the tumor tissue, decrease systemic distribution and, thus, reduce the side effects. Typically targeting is achieved by fusing the cytokines to full length antibodies, their Fab- or scFv fragments or other affinity binders. The most frequently used cytokine is interleukin 2 (IL-2), a signaling molecule that regulates the activity of lymphocytes. IL-2 triggers the differentiation of T-cells and can induce cell killing through NK-cells and cytotoxic T-cells. As the high potency of wild-type IL-2 often overshadows the affinity of the fused antibody, tumor targeting could be limited. Therefore, IL-2 muteins with reduced binding are becoming the cytokine of choice. For instance, recently an IL-2 with increased CD122 binding and decreased CD25 binding fused to a tumor specific antibody was designed [161]. A monomeric CEA-targeted IL-2 variant cytokine used a similar approach, weakening CD25 binding to avoid T-reg activation while improving binding to the CEA antigen [162]. Currently there are more than ten different IL-2 based immunocytokines in clinical trials, addressing antigens such as GD2, CD20, EDB, Tenascin C, CEA, FAP, EpCAM, or D7 [163]. Although the obvious design would place the antibody at the N-terminus of the fusion protein, while the cytokine is found at the C-terminal end, many other variants have been developed in the last two decades utilizing both ends or a combination between fusion or the heavy or light chain [164]. Overall three effects of immunocytokines have to be orchestrated to obtain a long-lasting cure; first, immune cells have to infiltrate the tumor attracted by chemokine and antibody-mediated targeting; second, these immune cells must be activated, ideally in a co-stimulatory way to generate a memory effect; third, tumor-mediated immune tolerance must be eliminated. All that can be achieved by a careful design of immune modulating fusion proteins [165].

11.5.3.4 Human Enzymes

A totally different concept of cancer therapy relies on the killing of aberrant cells through enzymes. Enzymes are not saturated through binding such as antibodies; therefore, lower doses can be applied, making them attractive therapeutics. Their activity can be directed either to external or internal targets.

Internal targets for enzymes can be all biological functions which either induce cell death through apoptosis or that are essential for survival. Many of the toxins described before address the protein translation, blocking either transcription factors or the ribosome. Ribonucleic acid (RNA) is the key component for the translation process involved either as messenger RNA (mRNA) or as amino acid carrying transfer RNA (tRNA). This highly sensitive molecule can be easily damaged by RNA degrading enzymes, the RNases. So in principle targeted RNases, the so-called ImmunoRNases can act just like toxins [166]. However, a limiting factor to the efficacy of RNAses is the presence of RNase inhibitor (RI) in the cytosol that protects RNA from enzymatic destruction. Therefore, besides human pancreatic RNase, angiogenin (ANG) or amphibian RNases, as, for instance, Onconase® (ONC), that is insensitive to RI, have been utilized. Alternatively, mutations can be introduced that abolish the inhibition by RI or multimerization of RNases can prevent RI blockade simply by steric hindrance. A further hurdle to overcome is the difficulty of design as RNases either need a free N- or C-terminus for activity. ONC must be preferably fused to its C-terminus, while ANG can be connected to its N-terminus as the opposite end contains the active center. Efficacy can be further improved by designing a fusion protein based on a diabody that could bear two ANG moieties [167]. This type of construct can help to counteract the depletion of the target antigen on the cell surface. It could be demonstrated that a traditional $(G_4S)_3$ linker combining an anti-CD22 scFv with ANG increased the activity, as it probably improved folding when expressed and secreted from mammalian cells [168]. In general, it seems that the folding apparatus of mammalian cells can better cope with the complexity of RNase fusion proteins as microbial expression systems. But nevertheless, production cost and production scale still represent a limitation to the clinical application of ImmunoRNases. Additionally, as RNA is an intracellular target, internalizing surface markers must be selected for the targeting fusion partner to be efficiently translocated across the cell membrane. Interestingly ONC does not require a targeting moiety, as it binds to tumor cells through a still unidentified receptor which is the starting point in the routing of ONC to the cytosol via the Golgi apparatus. Therefore, ONC can also be used as fusion partner to affinity binders against noninternalizing to increase the local concentration of ONC on cancer cells. The proper pathway into the cytosol is another important prerequisite for the cytotoxic activity as the default transfer to endosome results in much lower activity. To avoid potential immunogenicity of amphibian RNases, human RNases are being used more frequently nowadays.

Optimizing cell killing potency through increasing the catalytic activity or resisting inhibition while reducing collateral toxicity are the key parameters for therapeutic success. Although ImmunoRNases are a research topic for more than 20 years, none of the proposed molecules has reached clinical trials yet [169].

Cellular proliferation is regulated through signal transduction cascades relying on phosphorylation. Aberrant phosphorylation often causes uncontrolled cell expansion; therefore, an obvious countermeasure to trigger self-destruction or to halt proliferation would be to introduce kinases. Ideally a kinase is chosen that can trigger death signals, as, for instance, death-associated protein kinases (DAPK). Within cancer cells DAPK is often silenced; therefore, the restoration of its activity could initiate cell death. Pairing a constitutively highly active DAPK mutant at its C-terminus with a scFv targeting CD22 a so-called immunokinase could be created.

When internalized in malignant B-cells, this fusion enzyme successfully caused apoptosis and autophagy. To avoid any negative effect on the HEK293T cells that were used for the production, a strong secretion signal was used that immediately pushed the fusion protein to the supernatant of the cell culture. The immunokinase showed efficacy on malignant B-cells isolated from CLL patients [170].

Another natural factor that is used by immune cells to specifically destroy cancer cells is Granzyme B (GrB) which is released from cytoplasmic granules of natural killer (NK) cells and cytotoxic T lymphocytes (CTLs). It enters cells with the help of co-secreted pore forming proteins, the so-called perforins. The dependency on perforin to reach the cytosol is probably the major hurdle to overcome, as new ways to facilitate endosomal escape must be established. This could be achieved by incorporating the translocation domain of bacterial toxins. Granzyme B is a protease that activates multiple caspase family members by proteolytic cleavage, thus inducing apoptosis. Unfortunately, GrB contains several positively charged amino acids that could lead to unspecific accumulation on any negatively charged cell surface. Furthermore, GrB activity could be inhibited by a protease inhibitor (PI-9) that can be produced by cancer cells to evade destruction. Therefore, mutant variants of GrB resistant to PI-9 inhibition and without the basic amino acids were developed and fused to anticancer scFv molecules, creating immunoproteases or cytolytic fusion proteins (CFP). A further modification that allows cancer cell specific activation and thus reducing unspecific side effects is the replacement of the N-terminal Gly-Glu dipeptide by a sequence recognized by proteases highly expressed in cancer cells [171].

Alternatively, to using GrB as activator of caspases, these proteases themselves can be used directly to initiate apoptosis. Several attempts have been made with Caspase 3 or 6, and the most advanced molecule is a fusion of an anti-HER2 scFv connected to a constitutively active caspase 3 via the translocation domain of a bacterial toxin [172].

11.5.3.5 Apoptosis Induction

Instead of utilizing surface markers to identify cancer cells and then translocate apoptosis inducing agents like proteases or RNases, the natural function of death receptors (DR) that belong to the tumor necrosis factor (TNF) receptor superfamily can be utilized. The DR transduce an extracellular signal triggered by binding a corresponding ligand through their cytoplasmic death domain (DD) into the cytosol. Suitable ligands are TNF, apoptosis stimulating fragment ligand (FasL), and the tumor necrosis factor–related apoptosis-inducing ligand (TRAIL) which all induce receptor oligomerization. Initially these ligands were utilized in their natural conformation, but due to short half-life and high unspecific activity most clinical programs were abandoned. In the meantime, fusion proteins that, on the one hand, target the ligands to cancer cells while enabling multimerization of receptors have become the molecules of choice. The first progress was to design trimeric single chain ligands that are connected through short linkers in a linear fashion which also simplified their manufacturing. Then either a scFv or full IgG was fused to enable a targeting function. In the case of TNF, it was fused to an anti-L19 antibody and tested in clinical trials in melanoma [173]. Molecules harboring FasL showed promising results in preclinical studies when fused to scFv and targeting CD40, CD20, CD7, or TAG72, but did not yet reach clinical studies. FasL fusion proteins require the ligand positioned at the N-terminus. Most data exist for TRAIL fusion proteins that exert their activity by binding to DR4 and DR5. Again the typical fusion partners are scFvs, addressing, for instance, EGFR, ErbB2, CD40, CD27, or other cancer-specific antigens. A hexameric single chain TRAIL variant (ABBV-621) resulting from a fusion with the Fc domain of IgG1 with improved clustering of TRAIL receptors is currently in clinical phase 1 trials against solid tumors and hematologic malignancies [174].

As in all cases the apoptotic activity is induced through receptor clustering, the crosslinking can be further enhanced by using full antibodies instead of scFv. This can either happen through binding on the same cell (autocrine activation) or through neighboring cells (paracrine activation) Sometimes even a bystander effect could be observed, killing cells that are not presenting the corresponding antigen. This observation could on the one hand be beneficial for heterogeneous tumors, but might be a safety concern on the other hand. A further safety control can be establishing by so-called prodrugs that shield the active ligand through receptor fragments included in the molecule but being able to be released through the cleavage by tumor associated protease. Additional modifications can be introduced through modifying affinity and selectivity to the respective receptors through mutations in the ligands [175].

As alternative to the extrinsic pathway also an intrinsic pathway can be utilized to trigger apoptosis. The two molecules Bak (Bcl-2 homologous antagonist/killer) and Bax (Bcl-2–associated X protein) destroy mitochondria by forming pores, thus destroying cells. Therefore, it was attempted to introduce Bak or Bax into malignant cells. One example is the fusion of Bax to endostatin (ES) that targets endothelia cells and supports translocation into the cytoplasm. The combination of ES and Bax is more powerful in destructing cancer cells in vivo and in vitro compared to ES alone [176]. Both extrinsic and intrinsic apoptosis pathways are interlinked, and the protein BH3-interacting domain death agonist (BID) is positioned at the interface between them. A molecule consisting of an anti-HER2 scFv connected by a minimal endosomal translocation sequence derived from *Pseudomonas* containing a furin cleavage site and truncated BID resulted in efficient cell killing of HER2 positive cells [177]. Figure 11.12 shows how the different pathways can interact to induce apoptosis.

11.6 Summary

Currently there are 22 original fusion proteins approved and additionally numerous biosimilar variants have reached the market. Most frequently, the Fc-domain of IgG has been utilized as the fusion partner of choice. Nevertheless, there are in principle

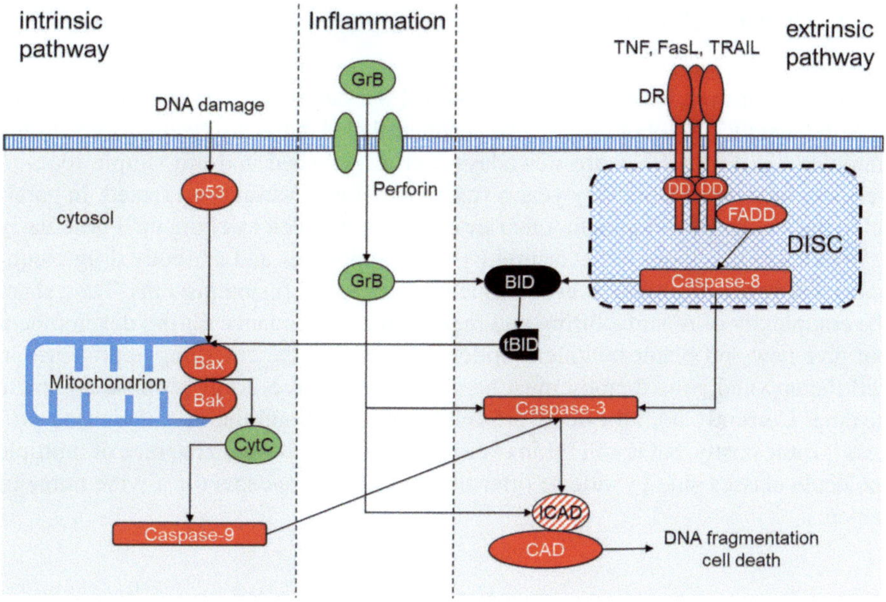

Fig. 11.12 Apoptosis induction as utilized by some fusion proteins. Binding of death receptor (DR) ligands such as tumor necrosis factor (TNF), TRAIL, or FasL, trigger the generation of a death-inducing signaling complex (DISC) at their death domains (DD) that interact with FAS-associated death-domain (FADD) and activate caspase-8. This protease cleaves BID to form tBID which connects the extrinsic with the intrinsic pathway. tBID or the p53-induced oligomerization of Bax and Bak causes cytochrome C (CytC) release which triggers ultimately the activation of caspase-9. This protease links the intrinsic with the extrinsic pathway by activating caspase-3 that degrades ICAD to release CAD which in turn causes cell death by DNA fragmentation. Several important molecules such as BID, caspase-3, and ICAD can also be activated by granzyme B (GrB) that can enter the cell during inflammation through perforin pores

endless choices to design fusion proteins from natural building blocks. Some potential limitations of a widespread use are the still not completely understood mechanisms of aggregation and the risk of immunogenicity due to unnatural combinations or novel epitopes. Manufacturing of fusion proteins is benefitting from the huge body of knowledge generated from the production of antibodies but still has its unique challenges such as low expression levels. Fusion proteins are often designed to solve some shortcomings of their unfused building blocks. For instance, many fusion proteins address the issue of short half-life or low bioavailability due to lack of targeting. Particularly in the context of oncology there is a wide range of therapeutic concepts trying to eliminate malignant cells by different means. Despite the various approaches, currently immunotoxins, representing the "magic bullet" idea, are the only examples that have reached approval.

11.7 Future Perspectives

Fusion proteins represent a highly successful but heterogeneous class of recombinant therapeutics. A lot of progress has been achieved in the last twenty years. Some modules like Fc or HSA are nowadays well established and are simple tools to generate biobetters. But every year a vast range of new variants is created. In parallel to the positive development, other modalities have been catching up. Particularly interesting is the boom of bi- or multispecific antibodies and antibody drug conjugates that partly resemble or even refine the concept of fusion proteins. They share the complexity of manufacturing and require much attendance during development but give new and better treatment options. In parallel the emerging approaches in cell therapy and gene therapy must be considered. These drugs promise complete healing. Currently not all effects are fully understood and the manufacturing process is quite costly, but it will be an exciting experience to watch the race of multiple molecule classes side by side to offer new treatment concepts for a wide range of patients.

References

1. Schmidt SR (2009) Fusion-proteins as biopharmaceuticals – applications and challenges. Curr Opin Drug Discov Dev 12(2):284–295
2. Gebauer M, Skerra A (2015) Alternative protein scaffolds as novel biotherapeutics. In: Rosenberg A, Demeule B (eds) Biobetters—protein engineering to approach the curative. 1. Springer, New York
3. Strohl WR (2015) Fusion proteins for half-life extension of biologics as a strategy to make biobetters. BioDrugs 29(4):215–239
4. Im SJ, Yang SI, Yang SH, Choi DH, Choi SY, Kim HS et al (2011) Natural form of noncytolytic flexible human Fc as a long-acting carrier of agonistic ligand, erythropoietin. PLoS One 6(9):1–13
5. Skerra A, Schmidt SR (2015) Pharmaceutical 'Engineered protein scaffolds': have they lived up to expectations? Pharm Bioprocess 3(6):383–385
6. Yu K, Liu C, Kim BG, Lee DY (2015) Synthetic fusion protein design and applications. Biotechnol Adv 33(1):155–164
7. Torres-Obreque KM, Meneguetti GP, Muso-Cachumba JJ, Feitosa VA, JHPM S, SPM V et al (2022) Building better biobetters: from fundamentals to industrial application. Drug Discov Today 27(1):65–81
8. Li G, Huang Z, Zhang C, Dong BJ, Guo RH, Yue HW et al (2016) Construction of a linker library with widely controllable flexibility for fusion protein design. Appl Microbiol Biotechnol 100:9733
9. Huang Z, Li G, Zhang C, Xing XH (2016) A study on the effects of linker flexibility on acid phosphatase PhoC-GFP fusion protein using a novel linker library. Enzym Microb Technol 83:1
10. Chen X, Bai Y, Zaro JL, Shen WC (2011) Design of an in vivo cleavable disulfide linker in recombinant fusion proteins. BioTechniques 49(1):513–518
11. Zhao HL, Xue C, Du JL, Ren M, Xia S, Liu ZM (2012) Balancing the pharmacokinetics and pharmacodynamics of interferon-α2b and human serum albumin fusion protein by proteolytic or reductive cleavage increases its in vivo therapeutic efficacy. Mol Pharm 9(3):664–670

12. Goyal A, Batra JK (2000) Inclusion of a furin-sensitive spacer enhances the cytotoxicity of ribotoxin restrictocin containing recombinant single-chain immunotoxins. Biochem J 345:247–254
13. Ando M, Fujimoto M, Takahashi Y, Nishikawa M, Hamana A, Takakura Y (2017) Targeted delivery of interferon gamma using a recombinant fusion protein of a fibrin clot–binding peptide with interferon gamma for cancer gene therapy. J Pharm Sci 106(3):892–897
14. Autio KA, Boni V, Humphrey RW, Naing A (2020) Probody therapeutics: an emerging class of therapies designed to enhance on-target effects with reduced off-tumor toxicity for use in immuno-oncology. Clin Cancer Res 26(5):984–989
15. Quellec SL, Enjolras N, Perot E, Girard J, Negrier C, Dargaud Y (2018) Fusion of factor IX to factor XIII-B sub-unit improves the pharmacokinetic profile of factor IX. Thromb Haemost 118(12):2053–2063
16. Dorai H, Nemeth JF, Cammaart E, Wang Y, Tang QM, Magill A et al (2009) Development of mammalian production cell lines expressing CNTO736, a glucagon like peptide-1-MIMETIBODY: factors that influence productivity and product quality. Biotechnol Bioeng 103(1):162–176
17. Yang GG, Xu XY, Ding Y, Cui QQ, Wang Z, Zhang QY et al (2015) Linker length affects expression and bioactivity of the onconase fusion protein in Pichia pastoris. Genet Mol Res 14(4):19360–19370
18. Silacci M, Baenziger-Tobler N, Lembke W, Zha W, Batey S, Bertschinger J et al (2014) Linker length matters: Fynomer – Fc Fusion with an optimized linker displaying picomolar IL-17A inhibition potency. J Biol Chem 289(20):14392–14398
19. Zhao HL, Yao XQ, Xue C, Wang Y, Xiong XH, Liu ZM (2008) Increasing the homogeneity, stability and activity of human serum albumin and interferon-alpha2b fusion protein by linker engineering. Protein Expr Purif 61(1):73–77
20. Figiel M, Bonarek P, Górecki A, Pawlak SD, Zerek B, Chęcińska B et al (2018) Improved cytotoxicity of novel TRAIL variants produced as recombinant fusion proteins. Protein Eng Des Sel 31(2):37–46
21. Fan J, Huang L, Sun J, Qiu Y, Zhou J, Shen Y (2015) Strategy for linker selection to enhance refolding and bioactivity of VAS-TRAIL fusion protein based on inclusion body conformation and activity. J Biotechnol 209:16
22. Lu P, Feng MG (2008) Bifunctional enhancement of a b-glucanase-xylanase fusion enzyme by optimization of peptide linkers. Appl Microbiol Biotechnol 79(4):579–587
23. Argos P (1990) An investigation of oligopeptides linking domains in protein tertiary structures and possible candidates for general gene fusion. J Mol Biol 211(4):943–958
24. George R, Heringa J (2002) An analysis of protein domain linkers: their classification and role in protein folding. Protein Eng 15(11):871–879
25. Chen X, Zaro JL, Shen W-C (2013) Fusion protein linkers: property, design and functionality. Adv Drug Deliv Rev 65(10):1357–1369
26. Chen X, Lee HF, Zaro JL, Shen WC (2011) Effects of receptor binding on plasma half-life of bifunctional transferrin fusion proteins. Mol Pharm 8(2):457–465
27. Gräwe A, Stein V (2021) Linker engineering in the context of synthetic protein switches and sensors. Trends Biotechnol 39:731
28. Spencer D, Novarra S, Zhu L, Mugabe S, Thisted T, Baca M et al (2013) O-xylosylation in a recombinant protein is directed at a common motif on glycine-serine linkers. J Pharm Sci 102(11):3920–3924
29. Wen D, Foley SF, Hronowski XL, Gu S, Meier W (2013) Discovery and investigation of O-xylosylation in engineered proteins containing a (GGGGS)n linker. Anal Chem 85(9):4805–4812
30. Spahr C, Gunasekaran K, Walker KW, Shi SD (2017) High-resolution mass spectrometry confirms the presence of a hydroxyproline (Hyp) post- translational modification in the GGGGP linker of an Fc-fusion protein post-translational modi fi cation in the GGGGP linker of an Fc-fusion protein. MAbs 9(5):812–819

31. Spahr C, Shi SDH, Lu HS (2014) O-glycosylation of glycine-serine linkers in recombinant Fc-fusion proteins: attachment of glycosaminoglycans and other intermediates with phosphorylation at the xylose sugar subunit. MAbs 6(4):904–914
32. Tyshchuk O, Völger HR, Ferrara C, Bulau P, Koll H, Mølhøj M (2017) Detection of a phosphorylated glycine-serine linker in an IgG-based fusion protein. MAbs 9(1):94–103
33. Song Y, Qian Y, Huang Z, Khattak SF, Li ZJ (2020) Computational insights into O-glycosylation in a CTLA4 Fc-fusion protein linker and its impact on protein quality attributes. Comput Struct Biotechnol J 18:3925–3935
34. Peters RT, Toby G, Lu Q, Liu T, Kulman JD, Low SC et al (2013) Biochemical and functional characterization of a recombinant monomeric factor VIII-Fc fusion protein. J Thromb Haemost 11(1):132–141
35. Dumont JA, Low SC, Peters RT, Bitonti AJ (2006) Monomeric Fc fusions: Impact on pharmacokinetic and biological activity of protein therapeutics. BioDrugs 20(3):151–160
36. Ha J-H, Kim J-E, Kim Y-S (2016) Immunoglobulin Fc heterodimer platform technology: from design to applications in therapeutic antibodies and proteins. Front Immunol 7(October):1–16
37. Spiess C, Zhai Q, Carter PJ (2015) Alternative molecular formats and therapeutic applications for bispecific antibodies. Mol Immunol 67(2):95–106
38. Fan C-Y, Huang C-C, Chiu W-C, Lai C-C, Liou G-G, Li H-C et al (2008) Production of multivalent protein binders using a self-trimerizing collagen-like peptide scaffold. FASEB J 22(11):3795–3804
39. Miller A, Carr S, Rabbitts T, Ali H (2020) Multimeric antibodies with increased valency surpassing functional affinity and potency thresholds using novel formats. MAbs 12(1):e1752529
40. Miersch S, Li Z, Saberianfar R, Ustav M, Case JB, Blazer L et al (2021) Tetravalent SARS-CoV-2 neutralizing antibodies show enhanced potency and resistance to escape mutations. J Mol Biol 433(167177):1–21
41. Voulgaraki D, Mitnacht-Kraus R, Letarte M, Foster-Cuevas M, Brown MH, Barclay AN (2005) Multivalent recombinant proteins for probing functions of leucocyte surface proteins such as the CD200 receptor. Immunology 115(3):337–346
42. Nagashima H, Kaneko K, Yamanoi A, Motoi S, Konakahara S, Kohroki J et al (2011) TNF receptor II fusion protein with tandemly repeated Fc domains. J Biochem 149(3):337–346
43. Mekhaiel DNA, Czajkowsky DM, Andersen JT, Shi J, El-faham M, Doenhoff M et al (2011) Polymeric human Fc-fusion proteins with modified effector functions. Sci Rep 1:1–11
44. Ojima-Kato T, Fukui K, Yamamoto H, Hashimura D, Miyake S, Hirakawa Y et al (2016) "Zipbody" leucine zipper-fused Fab in E. coli in vitro and in vivo expression systems. Protein Eng Des Sel 29(4):149–157
45. Boersma YL, Chao G, Steiner D, Wittrup KD, Plučkthun A. (2011) Bispecific designed ankyrin repeat proteins (DARPins) targeting epidermal growth factor receptor inhibit A431 cell proliferation and receptor recycling. J Biol Chem 286(48):41273–41285
46. Rozanov D, Spellman P, Savinov A, Strongin AY (2015) A humanized leucine zipper-TRAIL hybrid induces apoptosis of tumors both in vitro and in vivo. PLoS One 10(4):1–13
47. Oberst MD, Auge C, Morris C, Kentner S, Mulgrew K, McGlinchey K et al (2018) Potent immune modulation by medi6383, an engineered human ox40 ligand igg4p fc fusion protein. Mol Cancer Ther 17(5):1024–1038
48. Morris NP, Peters C, Montler R, Hu HM, Curti BD, Urba WJ et al (2007) Development and characterization of recombinant human Fc:OX40L fusion protein linked via a coiled-coil trimerization domain. Mol Immunol 44(12):3112–3121
49. Merz C, Sykora J, Marschall V, Richards DM, Heinonen K, Müller MR et al (2018) The hexavalent CD40 agonist HERA-CD40L induces T-cell-mediated antitumor immune response through activation of antigen-presenting cells. J Immunother 41(9):385–398
50. Stone GW, Barzee S, Snarsky V, Kee K, Spina C, Yu XF et al (2006) Multimeric soluble CD40L and GITR L as adjuvants for HIV DNA vaccines. J Virol 80(4):in press:1762
51. Pan LQ, Xie ZM, Tang XJ, Wu M, Wang FR, Naranmandura H et al (2013) Engineering and refolding of a novel trimeric fusion protein TRAIL-collagen XVIII NC1. Appl Microbiol Biotechnol 97(16):7253–7264

52. Shimamoto G, Gegg C, Boone T, Quéva C (2012) A flexible alternative format to antibodies. MAbs 4(5):586–591
53. Zhao J, Si Y, Cheng M, Yang Y, Niu Y, Li X et al (2013) Albumin fusion of interleukin-28B: production and characterization of its biological activities and protein stability. PLoS One 8(5):e64301
54. Zhao HL, Xue C, Wang Y, Li XY, Xiong XH, Yao XQ et al (2007) Circumventing the heterogeneity and instability of human serum albumin-interferon-alpha2b fusion protein by altering its orientation. J Biotechnol 131(3):245–252
55. Zhan J, Chen Y, Yuan HY, Li H, Lu H (2012) Fusion of HSA influences TNF-α neutralizing activity of shTNFRs. Biotechnol Lett 34:417–423
56. Ueda K, Shimizu M, Ohashi A, Murata D, Suzuki T, Kobayashi N et al (2020) Albumin fusion at the N-terminus or C-terminus of human lactoferrin leads to improved pharmacokinetics and anti-proliferative effects on cancer cell lines. Eur J Pharm Sci 155(September):105551
57. Li T, Zhang HZ, Ge GF, Yue ZR, Wang RY, Zhang Q et al (2021) Albumin fusion at the N-terminus or C-terminus of HM-3 leads to improved pharmacokinetics and bioactivities. Biomedicine 9(9):1–16
58. Ding Y, Peng Y, Deng L, Wu Y, Fu Q, Jin J (2014) The effects of fusion structure on the expression and bioactivity of human brain natriuretic peptide (BNP) albumin fusion proteins. Curr Pharm Biotechnol 15(9):856–863
59. Wang M, Zhi D, Xian J, Ru Y, Wei S, Wang N et al (2016) Functional expression of human serum albumin-tandem thrombopoietin mimetic peptide fusion protein as a novel thrombopoietin analog in Pichia pastoris. Biotechnol Lett 38(5):779–785
60. Ding Y, Fan J, Li W, Peng Y, Yang R, Deng L et al (2014 Jun) The effect of albumin fusion structure on the production and bioactivity of the somatostatin-28 fusion protein in Pichia pastoris. J Ind Microbiol Biotechnol 41(6):997–1006
61. Paek S-Y (2010) The orientation-dependent expression of angiostatin-endostatin hybrid proteins and their characterization for the synergistic effects of antiangiogenesis. J Microbiol Biotechnol 20(10):1430–1435
62. Christensen T, Amiram M, Dagher S, Trabbic-Carlson K, Shamji MF, Setton LA et al (2009) Fusion order controls expression level and activity of elastin-like polypeptide fusion proteins. Protein Sci 18(7):1377–1387
63. Ishino T, Wang M, Mosyak L, Tam A, Duan W, Svenson K et al (2013) Engineering a monomeric Fc domain modality by N-glycosylation for the half-life extension of biotherapeutics. J Biol Chem 288(23):16529–16537
64. Strohl WR (2009) Optimization of Fc-mediated effector functions of monoclonal antibodies. Curr Opin Biotechnol 20(6):685–691
65. Levin D, Golding B, Strome SE, Sauna ZE (2015) Fc fusion as a platform technology : potential for modulating immunogenicity. Trends Biotechnol 33(1):27–34
66. Shen Y, Li G, Gu C, Chen B, Chen A, Li H et al (2017) T0001, a variant of TNFR2-Fc fusion protein, exhibits improved Fc effector functions through increased binding to membrane-bound TNFα. PLoS One 12(5):1–15
67. Lee JH, Cho JH, Yeo J, Lee SH, Yang SH, Sung YC et al (2013) The pharmacology study of a new recombinant TNF receptor-hyFc fusion protein. Biologicals 41(2):77–83
68. Yang C, Gao X, Gong R (2018) Engineering of Fc fragments with optimized physicochemical properties implying improvement of clinical potentials for Fc-based therapeutics. Front Immunol 8(January):1860
69. Zhao HL, Xue C, Wang Y, Sun B, Yao XQ, Liu ZM (2009) Elimination of the free sulfhydryl group in the human serum albumin (HSA) moiety of human interferon-α2b and HSA fusion protein increases its stability against mechanical and thermal stresses. Eur J Pharm Biopharm 72(2):405–411
70. Schelde KK, Nicholls K, Dagnæs-Hansen F, Bunting K, Rawsthorne H, Andersen B et al (2019) A new class of recombinant human albumin with multiple surface thiols exhibits stable conjugation and enhanced FcRn binding and blood circulation. J Biol Chem 294(10):3735–3743

71. Andersen JT, Dalhus B, Viuff D, Ravn BT, Gunnarsen KS, Plumridge A et al (2014) Extending serum half-life of albumin by engineering neonatal Fc receptor (FcRn) binding. J Biol Chem 289(19):13492–13502
72. Schmidt MM, Townson SA, Andreucci AJ, King BM, Schirmer EB, Murillo AJ et al (2013) Crystal structure of an HSA/FcRn complex reveals recycling by competitive mimicry of HSA ligands at a pH-dependent hydrophobic interface. Structure 21(11):1966–1978
73. Sand KMK, Bern M, Nilsen J, Dalhus B, Gunnarsen KS, Cameron J et al (2014) Interaction with both domain I and III of albumin is required for optimal pH-dependent binding to the neonatal Fc receptor (FcRn). J Biol Chem 289(50):34583–34594
74. Bern M, Nilsen J, Ferrarese M, Sand KMK, Gjølberg TT, Lode HE et al (2020) An engineered human albumin enhances half-life and transmucosal delivery when fused to protein-based biologics. Sci Transl Med 12(565):1–14
75. Gillies SD, Lan Y, Lo K, Super M, Wesolowski J (1999) Improving the efficacy of antibody-interleukin 2 fusion proteins by reducing their interaction with Fc receptors. Cancer Res 59:2159–2166
76. Gillies SD, Lo K, Burger C, Lan Y, Dahl T, Wong W (2002) Improved circulating half-life and efficacy of an antibody – interleukin 2 immunocytokine based on reduced intracellular proteolysis. Clin Cancer Res 8(January):210–216
77. Zhong X, Kieras E, Sousa E, D'Antona A, Baber JC, He T et al (2013) Pyroglutamate and O-linked glycan determine functional production of anti-IL17A and anti-IL22 peptide-antibody bispecific genetic fusions. J Biol Chem 288:1409–1419
78. Hecht R, Li Y-S, Sun J, Belouski E, Hall M, Hager T et al (2012) Rationale-based engineering of a potent long-acting FGF21 analog for the treatment of type 2 diabetes. PLoS One 7(11):e49345
79. Way JC, Lauder S, Brunkhorst B, Kong SM, Qi A, Webster G et al (2005) Improvement of Fc-erythropoietin structure and pharmacokinetics by modification at a disulfide bond. Protein Eng Des Sel 18(3):111–118
80. De Groot AS, Moise L (2007) Prediction of immunogenicity for therapeutic proteins: state of the art. Curr Opin Drug Discov Devel 10(3):332–340
81. Jawa V, Cousens L, De Groot A (2013) Immunogenicity of therapeutic fusion proteins: contributory factors and clinical experience. In: Schmidt SR (ed) Fusion protein technologies for biopharmaceuticals: applications and challenges, 1st edn. Wiley, pp 75–90
82. Levin D, Golding B, Strome SE, Sauna ZE (2014) Fc fusion as a platform technology: potential for modulating immunogenicity. Trends Biotechnol 1–8:27
83. Liang Y, Qiu H, Glinka Y, Lazarus AH, Ni H, Prudhomme GJ et al (2011) Immunity against a therapeutic xenoprotein/Fc construct delivered by gene transfer is reduced through binding to the inhibitory receptor Fc g RIIb. J Gene Med 13(May):470–477
84. Sperinde G, Montgomery D, Mytych DT (2020) Clinical Immunogenicity Risk Assessment for a Fusion Protein. AAPS J 22(3):64
85. Onda M (2009) Reducing the immunogenicity of protein therapeutics. Curr Drug Targets 10(2):131–139
86. Fan L, Zhao L, Ye Z, Sun Y, Kou T, Zhou Y et al (2010 Sep) Effect of culture temperature on TNFR-Fc productivity in recombinant glutamine synthetase-chinese hamster ovary cells. Biotechnol Lett 32(9):1239–1244
87. Trummer E, Fauland K, Seidinger S, Schriebl K, Lattenmayer C, Kunert R et al (2006) Process parameter shifting: part I. Effect of DOT, pH, and temperature on the performance of Epo-Fc expressing CHO cells cultivated in controlled batch bioreactors. Biotechnol Bioeng 94(6):1033–1044
88. Trummer E, Fauland K, Seidinger S, Schriebl K, Lattenmayer C, Kunert R et al (2006) Process parameter shifting: part ii. biphasic cultivation—a tool for enhancing the volumetric productivity of batch processes using Epo-Fc expressing CHO cells. Biotechnol Bioeng 94(6):1045–1052
89. Kaisermayer C, Reinhart D, Gili A, Chang M, Aberg PM, Castan A et al (2016) Biphasic cultivation strategy to avoid Epo-Fc aggregation and optimize protein expression. J Biotechnol 227:3–9

90. Sheffield WP, McCurdy T, Bhakta V (2005) Fusion to albumin as a means to slow the clearance of small therapeutic proteins using the Pichia pastoris expression system. Methods Mol Biol 308:145–154
91. Guan B, Chen F, Lei J, Li Y, Duan Z, Zhu R et al (2013) Constitutive expression of a rhIL-2-HSA fusion protein in pichia pastoris using glucose as carbon source. Appl Biochem Biotechnol 171(7):1792–1804
92. Guan B, Chen F, Su S, Duan Z, Chen Y, Huazhong L et al (2016) Effects of co-overexpression of secretion helper factors on the secretion of a HSA fusion protein (IL2-HSA) in pichia pastoris. Yeast 33(11):587–600
93. Ding Y, Fan J, Li W, Yang R, Peng Y, Deng L et al (2013) The effect of albumin fusion patterns on the production and bioactivity of the somatostatin-14 fusion protein in Pichia pastoris. Appl Biochem Biotechnol 170:1637–1648
94. Baneyx F, Mujacic M (2004) Recombinant protein folding and misfolding in Escherichia coli. Nat Biotechnol 22(11):1399–1408
95. Fahnert B, Lilie H, Neubauer P (2004) Inclusion bodies: formation and utilisation. Adv Biochem Eng Biotechnol 89:93–142
96. Li Y (2017) Effective strategies for host cell protein clearance in downstream processing of monoclonal antibodies and Fc-fusion proteins. Protein Expr Purif 134:96–103
97. Hassouneh W, Christensen T, Chilkoti A (2010) Elastin-like polypeptides as a purification tag for recombinant proteins. Curr Protoc Protein Sci (Suppl 61):1–16
98. Ghose S, Hubbard B, Cramer SM (2007) Binding capacity differences for antibodies and Fc-fusion proteins on protein A chromatographic materials. Biotechnol Bioeng 96(4):768–779
99. Xu X, Didio DM, Leister KJ, Ghose S (2009) Disaggregation of High-molecular weight species during downstream processing to recover functional monomer. Biotechnol Prog 26(3):717–726
100. Rodrigo G, Gruvegård M, Van Alstine JM (2015) Antibody fragments and their purification by protein L affinity chromatography. Antibodies 4(3):259–277
101. Nilson BHK, Solomon A, Bjorck L, Akerstrom B (1992) Protein L from Peptostreptococcus magnus binds to the κ light chain variable domain. J Biol Chem 267(4):2234–2239
102. Pigeon C, Jackson M, Baines D (2009) Application of fabsorbenttm F1P HF, a synthetic ligand adsorbent for capture and purification of a single-domain antibody fragment expressed in Escherichia coli. Bioprocess Int 7(7):90–91
103. Schimek C, Kubek M, Scheich D, Fink M, Brocard C, Striedner G et al (2021) Three-dimensional chromatography for purification and characterization of antibody fragments and related impurities from Escherichia coli crude extracts. J Chromatogr A 1638:461702
104. Hapuarachchi S, Fodor S, Apostol I, Huang G (2011) Use of capillary electrophoresis-sodium dodecyl sulfate to monitor disulfide scrambled forms of an Fc fusion protein during purification process. Anal Biochem 414(2):187–195
105. Lamanna WC, Mayer RE, Rupprechter A, Fuchs M, Higel F, Fritsch C et al (2017) The structure-function relationship of disulfide bonds in etanercept. Sci Rep 7(1):1–8
106. Herzer S, Bhangale A, Barker G, Chowdhary I, Conover M, O'Mara BW et al (2015) Development and scale-up of the recovery and purification of a domain antibody Fc fusion protein-comparison of a two and three-step approach. Biotechnol Bioeng 112(7):1417–1428
107. Schmidt SR (2017) Controlling glycosylation in fusion protein manufacturing to generate potent biobetters. Bioprocess Int 15(8):12–21
108. Higel F, Seidl A, Sörgel F, Friess W (2016) N-glycosylation heterogeneity and the influence on structure, function and pharmacokinetics of monoclonal antibodies and Fc fusion proteins. Eur J Pharm Biopharm 100:94–100
109. Jing Y, Qian Y, Li ZJ (2010) Sialylation enhancement of CTLA4-Ig fusion protein in Chinese hamster ovary cells by dexamethasone. Biotechnol Bioeng 107(3):488–496
110. Rouiller Y, Périlleux A, Marsaut M, Stettler M, Vesin MN, Broly H (2012) Effect of hydrocortisone on the production and glycosylation of an Fc-fusion protein in CHO cell cultures. Biotechnol Prog 28:803–813

111. Lee JH, Jeong YR, Kim YG, Lee GM (2017) Understanding of decreased sialylation of Fc-fusion protein in hyperosmotic recombinant Chinese hamster ovary cell culture: N-glycosylation gene expression and N-linked glycan antennary profile. Biotechnol Bioeng 114(8):1721–1732
112. Lewis AM, Croughan WD, Aranibar N, Lee AG, Warrack B, Abu-Absi NR et al (2016) Understanding and controlling sialylation in a CHO Fc-Fusion process. PLoS One 11(6):1–24
113. Liu J, Wang J, Fan L, Chen X, Hu D, Deng X et al (2015) Galactose supplementation enhance sialylation of recombinant Fc-fusion protein in CHO cell: an insight into the role of galactosylation in sialylation. World J Microbiol Biotechnol 31(7):1147–1156
114. Chen X, Liu X, Xiao Z, Liu J, Zhao L, Tan WS et al (2019) Insights into the loss of protein sialylation in an fc-fusion protein-producing CHO cell bioprocess. Appl Microbiol Biotechnol 103(12):4753–4765
115. Bora de Oliveira K, Spencer D, Barton C, Agarwal N (2017) Site-specific monitoring of N-Glycosylation profiles of a CTLA4-Fc-fusion protein from the secretory pathway to the extracellular environment. Biotechnol Bioeng 114(7):1550–1560
116. Wang Q, Yang G, Wang T, Yang W, Betenbaugh MJ, Zhang H (2019) Characterization of intact glycopeptides reveals the impact of culture media on site-specific glycosylation of EPO-Fc fusion protein generated by CHO-GS cells. Biotechnol Bioeng 116(9):2303–2315
117. Schmidt SR (2017) Production challenges for complex biologics: fusion proteins. Am Pharm Rev (February):1–4
118. Cordes AA, Platt CW, Carpenter JF, Randolph TW (2012) Selective domain stabilization as a strategy to reduce fusion protein aggregation. J Pharm Sci 101(4):1400–1409
119. Fast JL, Cordes A, a, Carpenter JF, Randolph TW. (2009) Physical instability of a therapeutic Fc fusion protein: domain contributions to conformational and colloidal stability. Biochemistry 48(49):11724–11736
120. Strand J, Huang C, Xu JIN (2013) Characterization of Fc-fusion protein aggregates derived from extracellular domain disulfide bond rearrangements. J Pharm Sci 102(2):441–453
121. Wang T, Fodor S, Hapuarachchi S, Jiang XG, Chen K, Apostol I et al (2013) Analysis and characterization of aggregation of a therapeutic Fc-fusion protein. J Pharm Biomed Anal 72:59–64
122. Huang RY-C, Iacob RE, Krystek SR, Jin M, Wei H, Tao L et al (2016) Characterization of aggregation propensity of a human Fc-fusion protein therapeutic by hydrogen/deuterium exchange mass spectrometry. J Am Soc Mass Spectrom 28:795
123. Shukla AA, Gupta P, Han X (2007) Protein aggregation kinetics during Protein A chromatography. Case study for an Fc fusion protein. J Chromatogr A 1171(1–2):22–28
124. Xu X, Didio DM, Leister KJ, Ghose S (2010) Disaggregation of high-molecular weight species during downstream processing to recover functional monomer. Biotechnol Prog 26(3):717–726
125. Bobst CE, Wang S, Shen W-C, Kaltashov IA (2012) Mass spectrometry study of a transferrin-based protein drug reveals the key role of protein aggregation for successful oral delivery. Proc Natl Acad Sci USA 109(34):13544–13548
126. Amaral M, Hölper S, Lange C, Jung J, Sjuts H, Weil S et al (2020) Engineered technologies and bioanalysis of multispecific antibody formats. J Appl Bioanal 6(1):26–51
127. Duivelshof BL, Murisier A, Camperi J, Fekete S, Beck A, Guillarme D et al (2021) Therapeutic Fc-fusion proteins: current analytical strategies. J Sep Sci 44(1):35–62
128. Wu B, Sun YN (2014) Pharmacokinetics of peptide-Fc fusion proteins. J Pharm Sci 103(1):53–64
129. Wunder A, Stehle G, Sinn H, Schrenk HH, Hoff-Biederbeck D, Bader F et al (1997) Enhanced albumin uptake by rat tumors. Int J Oncol 11(3):497–507
130. Schulte S (2009) Half-life extension through albumin fusion technologies. Thromb Res 124(Suppl):6–8
131. Zalevsky J, Chamberlain AK, Horton HM, Karki S, Leung IWL, Sproule TJ et al (2010) Enhanced antibody half-life improves in vivo activity. Nat Biotechnol 28(2):157–159

132. Presta LG (2008) Molecular engineering and design of therapeutic antibodies. Curr Opin Immunol 20(4):460–470
133. Suzuki T, Ishii-watabe A, Tada M, Kanayasu-toyoda T, Kawanishi T, Yamaguchi T et al (2010) Importance of neonatal FcR in regulating the serum half-life of therapeutic proteins containing the Fc domain of human IgG1: a comparative study of the affinity of monoclonal antibodies and Fc-fusion proteins to human neonatal FcR. J Immunol 184:1968
134. Bitonti AJ, Dumont JA, Low SC, Peters RT, Kropp KE, Palombella VJ et al (2004) Pulmonary delivery of an erythropoietin Fc fusion protein in non-human primates through an immunoglobulin transport pathway. Proc Natl Acad Sci USA 101(26):9763–9768
135. Mayle KM, Le AM, Kamei DT (2012) The intracellular trafficking pathway of transferrin. Biochim Biophys Acta, Gen Subj 1820(3):264–281
136. Wang Y, Chen YS, Zaro JL, Shen WC (2011) Receptor-mediated activation of a proinsulin-transferrin fusion protein in hepatoma cells. J Control Release 155:386–392
137. Park E, Starzyk RM, McGrath JP, Lee T, George J, Schutz AJ et al (1998) Production and characterization of fusion proteins containing transferrin and nerve growth factor. J Drug Target 6(1):53–64
138. Bai Y, Shen W-C (2006) Improving the oral efficacy of recombinant granulocyte colony-stimulating factor and transferrin fusion protein by spacer optimization. Pharm Res 23(9):2116–2121
139. Chen YS, Zaro JL, Zhang D, Huang N, Simon A, Shen WC (2018) Characterization and oral delivery of proinsulin-transferrin fusion protein expressed using expresstec. Int J Mol Sci 19(2):1–13
140. Xie C, Wang Z, Su Y, Wang J, Shen WC (2020) Discovery of an orally effective factor IX-transferrin fusion protein for hemophilia B. Int J Mol Sci 21(1)
141. Schlapschy M, Theobald I, Mack H, Schottelius M, Wester H-J, Skerra A (2007) Fusion of a recombinant antibody fragment with a homo-amino-acid polymer: effects on biophysical properties and prolonged plasma half-life. Protein Eng Des Sel 20(6):273–284
142. Breibeck J, Skerra A (2017) The polypeptide biophysics of proline / alanine-rich sequences (PAS): recombinant biopolymers with PEG-like properties. Biopolymers (July):1–12
143. Schellenberger V, Wang C, Geething NC, Spink BJ, Campbell A, To W et al (2009) A recombinant polypeptide extends the in vivo half-life of peptides and proteins in a tunable manner. Nat Biotechnol 27(12):1186–1190
144. Geething NC, To W, Spink BJ, Scholle MD, Wang CW, Yin Y et al (2010) Gcg-XTEN: An improved glucagon capable of preventing hypoglycemia without increasing baseline blood glucose. PLoS One 5(4):e10175
145. Brandl F, Merten H, Zimmermann M, Béhé M, Zangemeister-Wittke U, Plückthun A (2019) Influence of size and charge of unstructured polypeptides on pharmacokinetics and biodistribution of targeted fusion proteins. J Control Release 307(May):379–392
146. Egrie JC, Browne JK (2001) Development and characterization of novel erythropoiesis stimulating protein (NESP). Br J Cancer 84(s1):3–10
147. Fares F, Suganuma N, Nishimori K, LaPolt P, Hsueh A, Boime I (1992) Design of a long-acting follitropin agonist by fusing the C-terminal sequence of the chorionic gonadotropin beta subunit to the follitropin beta subunit. Proc Natl Acad Sci USA 89(10):4304–4308
148. Xu J, Okada S, Tan L, Goodrum KJ, Kopchick JJ, Kieliszewski MJ (2010) Human growth hormone expressed in tobacco cells as an arabinogalactan-protein fusion glycoprotein has a prolonged serum life. Transgenic Res 19(5):849–867
149. Betre H, Liu W, Zalutsky MR, Chilkoti A, Kraus VB, Setton LA (2006) A thermally responsive biopolymer for intra-articular drug delivery. J Control Release 115(2):175–182
150. Boado RJ (2022) IgG fusion proteins for brain delivery of biologics via blood–brain barrier receptor-mediated transport of the transporting. Pharmaceutics 14(7):14071476
151. Liu Y, Wang HY, Zhou L, Su Y, Shen WC (2018) Biodistribution, activation, and retention of proinsulin-transferrin fusion protein in the liver: mechanism of liver-targeting as an insulin prodrug. J Control Release 275:186–191

152. Kim K-J, Malik AB (2003) Protein transport across the lung epithelial barrier. Am J Phys Lung Cell Mol Phys 284(2):L247–L259
153. Hofmann C, Seefried L, Jakob F (2016) Asfotase alfa: enzyme replacement for the treatment of bone disease in hypophosphatasia. Drugs Today 52(5):271–285
154. Zimran A, Wajnrajch M, Hernandez B, Pastores GM (2018) Taliglucerase alfa: Safety and efficacy across 6 clinical studies in adults and children with Gaucher disease. Orphanet J Rare Dis 13(1):1–10
155. LeBowitz JH, Grubb JH, Maga JA, Schmiel DH, Vogler C, Sly WS (2004) Glycosylation-independent targeting enhances enzyme delivery to lysosomes and decreases storage in mucopolysaccharidosis type VII mice. Proc Natl Acad Sci USA 101(9):3083–3088
156. Maga JA, Zhou J, Kambampati R, Peng S, Wang X, Bohnsack RN et al (2013) Glycosylation-independent lysosomal targeting of acid a-glucosidase enhances muscle glycogen clearance in pompe mice. J Biol Chem 288(3):1428–1438
157. Zhou Z, Austin GL, Shaffer R, Armstrong DD, Gentry MS (2019) Antibody-mediated enzyme therapeutics and applications in glycogen storage diseases. Trends Mol Med 25(12):1094–1109
158. Correale P, Cusi MG, Tagliaferri P (2011) Immunomodulatory properties of anticancer monoclonal antibodies: Is the "magic bullet" still a reliable paradigm? Immunotherapy 3(1):1–4
159. Shapira A, Benhar I (2010) Toxin-based therapeutic approaches. Toxins (Basel) 2(11):2519–2583
160. Baldo BA (2014) Side effects of cytokines approved for therapy. Drug Saf 37(11):921–943
161. Sun Z, Ren Z, Yang K, Liu Z, Cao S, Deng S et al (2019) A next-generation tumor-targeting IL-2 preferentially promotes tumor-infiltrating CD8+ T-cell response and effective tumor control. Nat Commun 10(1):3874
162. Klein C, Waldhauer I, Nicolini VG, Freimoser-Grundschober A, Nayak T, Vugts DJ et al (2017) Cergutuzumab amunaleukin (CEA-IL2v), a CEA-targeted IL-2 variant-based immunocytokine for combination cancer immunotherapy: overcoming limitations of aldesleukin and conventional IL-2-based immunocytokines. Onco Targets Ther 6(3):e1277306
163. Murer P, Neri D (2019) Antibody-cytokine fusion proteins: a novel class of biopharmaceuticals for the therapy of cancer and of chronic inflammation. New Biotechnol 52(April):42–53
164. Hutmacher C, Neri D (2019) Antibody-cytokine fusion proteins: biopharmaceuticals with immunomodulatory properties for cancer therapy. Adv Drug Deliv Rev 141:67–91
165. Lechner MG, Russell SM, Bass RS, Epstein AL (2011) Chemokines, costimulatory molecules and fusion proteins for the immunotherapy of solid tumors. Immunotherapy 3(11):1317–1340
166. Schirrmann T, Krauss J, Arndt MAE, Rybak SM, Dübel S (2008) Targeted therapeutic RNases (ImmunoRNases). Expert Opin Biol Ther 9(1):79–95
167. Arndt MAE, Vu BK, Newton DL, Rybak SM (2005) A dimeric angiogenin immunofusion protein mediates selective toxicity toward CD221 tumor cells. J Immunother 28(3):245–251
168. Krauss J, Arndt MAE, Vu BK, Newton DL, Rybak S (2005) Targeting malignant B-cell lymphoma with a humanized anti-CD22 scFv-angiogenin immunoenzyme. Br J Haematol 128(5):602–609
169. Jordaan S, Akinrinmade O, Nachreiner T, Cremer C, Naran K, Chetty S et al (2018) Updates in the development of ImmunoRNases for the selective killing of tumor cells. Biomedicine 6(1):28
170. Lilienthal N, Lohmann G, Crispatzu G, Vasyutina E, Zittrich S, Mayer P et al (2016) A novel recombinant anti-CD22 immunokinase delivers proapoptotic activity of death-associated protein kinase (DAPK) and mediates cytotoxicity in neoplastic B cells. Mol Cancer Ther 15(5):971–984
171. Hlongwane P, Mungra N, Madheswaran S, Akinrinmade OA, Chetty S, Barth S (2018) Human granzyme B based targeted cytolytic fusion proteins. Biomedicine 6(72):1–11
172. Zhang DX, Zhao PT, Xia L, Liu LL, Liang J, Zhai HH et al (2010) Potent inhibition of human gastric cancer by HER2-directed induction of apoptosis with anti-HER2 antibody and caspase-3 fusion protein. Gut 59(3):292–299

173. Papadia F, Basso V, Patuzzo R, Maurichi A, Di Florio A, Zardi L et al (2013) Isolated limb perfusion with the tumor-targeting human monoclonal antibody-cytokine fusion protein L19-TNF plus melphalan and mild hyperthermia in patients with locally advanced extremity melanoma. J Surg Oncol 107(2):173–179
174. Diaz Arguello OA, Haisma HJ (2021) Apoptosis-inducing tnf superfamily ligands for cancer therapy. Cancers (Basel) 13(7):1543–1565
175. Wajant H, Gerspach J, Pfizenmaier K (2013) Engineering death receptor ligands for cancer therapy. Cancer Lett 332(2):163–174
176. Chura-Chambi RM, Bellini MH, Jacysyn JF, Andrade LN, Medina LP, Prieto-Da-Silva ARB et al (2014) Improving the therapeutic potential of endostatin by fusing it with the BAX BH3 death domain. Cell Death Dis 5(8):1–10
177. Wang F, Ren J, Qiu XC, Wang LF, Zhu Q, Zhang YQ et al (2010) Selective cytotoxicity to HER2-positive tumor cells by a recombinant e23sFv-TD-tBID protein containing a furin cleavage sequence. Clin Cancer Res 16(8):2284–2294

Chapter 12
Development of Antibody-Drug Conjugates

David Y. Jackson

Abstract Antibody drug conjugates (ADCs) are a rapidly growing class of targeted cancer drugs in which a highly toxic small molecule (payload) is conjugated to a tumor-selective antibody. Over the past two decades, a total of 11 ADCs have been approved by the FDA in the United States; including: gemtuzumab ozogamicin (Mylotarg™), brentuximab vedotin (Adcetris™), ado-trastuzumab emtansine (Kadcyla™), inotuzumab ozogamicin (Besponsa™), polatuzumab vedotin (Polivy™), enfortumab vedotin (Padcev™), trastuzumab deruxtecan (Enhertu™), sacituzumab govitecan (Trodelvy™), belantamab mafodotin (Blenrep™), loncastuximab tesirine-lpyl (Zynlonta™), and tisotumab vedotin-tftv (Tivdak™). The path to commercial success for these ADCs has been challenging however, and new ADC approvals were rare prior to 2017 when only three ADCs had been approved by the FDA. Then in 2019 three more ADCs were approved, followed by two approvals in 2020 and two more in 2021. Dozens of ADCs are now in late-stage clinical trials and new approvals are expected to remain consistent for the near future. Following closely behind are over a hundred new ADCs in early clinical or preclinical development. This chapter will summarize the history of currently approved ADCs with emphasis on the challenges that were overcome during development and new technology that likely contributed to their success. The safety and efficacy of each ADC will be discussed from a critical but honest perspective based on personal experience. My intention in writing this chapter is to encourage readers to educate themselves about the real benefits and risks of ADC therapeutics so that informed decisions can be made by cancer patients in collaboration with their doctors.

Keywords Antibody-drug conjugate · Toxic payload · FDA approved · Cancer drug · oncology · Targeted therapy

---

D. Y. Jackson (✉)
DYJ Pharma, San Mateo, CA, USA

K. Gadamasetti, S. A. Kolodziej (eds.), *Bioprocessing, Bioengineering and Process Chemistry in the Biopharmaceutical Industry*,
https://doi.org/10.1007/978-3-031-62007-2_12

12.1 Introduction

Antibody-drug conjugates (ADCs) are a rapidly growing class of targeted therapeutic agents for the treatment of cancer [1, 2]. Cancer is the second most prevalent cause of death in the US and many other countries; only heart disease kills more people. Most conventional drugs for the treatment of cancer are highly potent small molecules that kill cells or inhibit cell growth and often have poor therapeutic windows due to their lack of selectivity for tumor cells [3]. In the last three decades however, new monoclonal antibody (mAb) therapeutics that bind to specific antigens on tumor cells have demonstrated improved tumor selectivity and safety [4]. These mAbs, however, frequently lack sufficient potency for use as single-agent drugs and are often used in combination with conventional chemotherapeutic agents to improve patient treatment outcomes [5]. Although multidrug treatment regimens have been moderately successful in treating some forms of cancer, the increased costs associated with conventional chemotherapy, combined with complicated dosing schedules and potential toxicity due to adverse drug interactions, demonstrate a need for safer more effective cancer drugs.

Antibody-drug conjugates combine the target specificity of a monoclonal antibody with the potency of a small molecule drug (payload) by connecting them into a single ADC molecule that retains the properties of both [6]. The improved selectivity and potency of ADCs lead to superior safety and efficacy resulting in broader therapeutic windows compared to conventional chemotherapeutic drugs (Fig. 12.1). In fact, ADCs have been described as the "*smart bombs*" of cancer therapy because they deliver highly toxic payloads directly to tumor cells while minimizing "collateral damage" to the surrounding noncancerous or normal tissues [7]. The concept is simple, but in reality, tumor cells are not easy to target because they are derived from normal cells and express similar proteins on their cell membrane. Moreover, cancer cells are usually very good at camouflage by mimicking their surroundings; making them difficult targets to hit, even for antibodies.

The key to developing an effective ADC therapeutic is to find the optimal combination of components; target, antibody, linker, and payload for a specific cancer

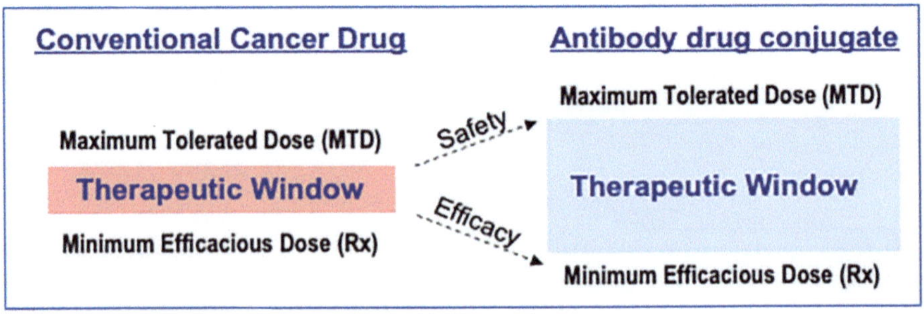

Fig. 12.1 ADCs improve the safety (higher MTD) & efficacy (lower Rx) to broaden the therapeutic window of conventional cancer drugs

type. This chapter will discuss these components and their desired properties in order to give the reader a better understanding of the benefits and limitations of the currently approved ADC therapeutics. Emerging new technologies such as site-specific conjugation methods designed to improve the safety and efficacy of future ADCs will also be discussed. ADCs do not yet offer a cure for cancer but they do offer an alternative to conventional chemotherapeutic drugs and should continue to be investigated as targeted therapeutic agents. Cancer may never disappear completely from the human condition, but each small step toward finding a cure might eventually save the life of someone you care about.

12.2 ADC History

Most antibody-drug conjugates have experienced a rugged path to commercial success and have overcome numerous challenges along the journey. For example, the first ADC to reach the market (Mylotarg) was approved in 2000 but was later discontinued due to lack of efficacy [8]. Nearly 10 years elapsed before the second ADC, Adcetris was approved in 2011; followed by a third ADC, Kadcyla, which was approved in 2013. With only three approvals in the first decade of ADC development, their approval rate was lower than that for small molecules and only a fraction of the approval rate for most antibody therapeutics [9]. Fortunately, this trend was temporary and the rate of ADC approvals has increased dramatically in recent years with seven new ADCs approved for cancer between 2017 and 2022 [10].

Given the rapid increase in approvals, one might conclude that the lessons learned from early ADCs may have facilitated the development pathways of those that followed. Closer analysis however, indicates that the increase in ADC approvals likely reflects the growing number of ADCs entering clinical trials, rather than new breakthroughs in ADC technology. During the past decade (between 2011 and 2021), the number of clinical trials has steadily increased by nearly tenfold to over 300 trials in 2021 (Fig. 12.2a), yet the number of new ADC approvals during the same time period only doubled (Fig. 12.2b). Although the number of new clinical trials continues to grow each year, the discrepancy between the number of phase I/II and phase III trials indicates that most ADCs fail to progress beyond the early stages of clinical development.

The reasons for these clinical failures are difficult to assess and seldom reported, even if they've been determined by the companies that sponsored them. Based on the currently available information, however, most ADC researchers would probably agree that systemic exposures to toxic payloads and off-target toxicity are primary contributors to ADC failures in the clinic [11]. Methods for reducing ADC toxicity would therefore be expected to improve the overall safety of ADCs and will likely be a major focus for future research and development efforts.

In general, there are four major components (target, antibody, linker, and drug payload) that impact the commercial success or failure of an antibody-drug

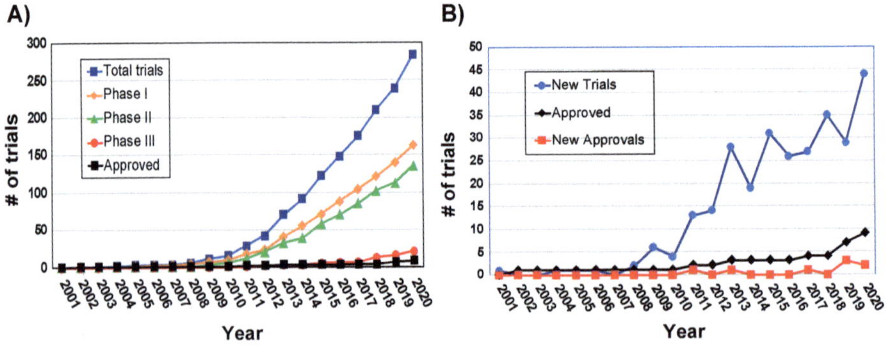

Fig. 12.2 (**a**) Number of ADC clinical trials by phase of development since 2001. (**b**) Number of new clinical trials and approvals since 2001. (Source: clinical trials.gov)

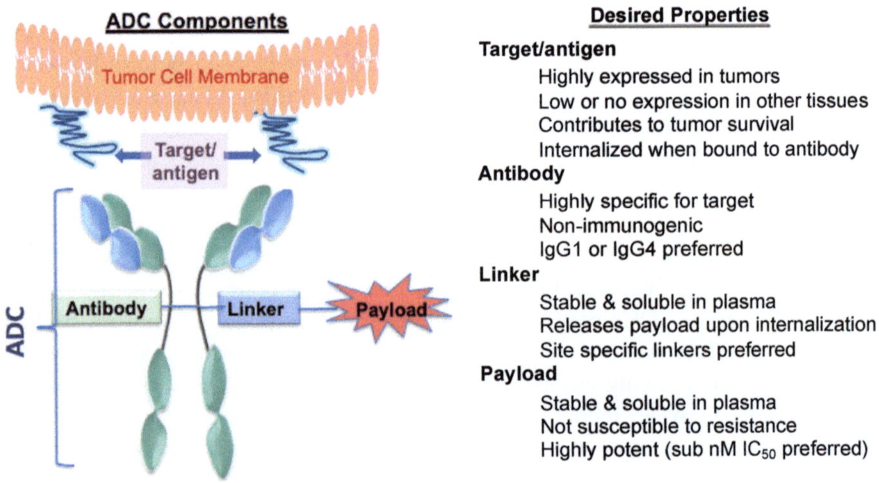

Fig. 12.3 Components and desired properties of an antibody-drug conjugate (ADC)

conjugate (Fig. 12.3). Pharmaceutical companies focus significant efforts on selecting components with optimal properties in order to improve the effectiveness of ADCs. Ironically the component that likely has the greatest impact on the success or failure of an ADC resides in the tumor and is not part of the ADC [12]. The selectivity of an ADC for killing tumor cells over normal adjacent tissues is almost solely dependent upon the expression profile of the antigen target, so target selection is an essential part of the ADC early development process.

12.3 Target Selection

ADC targets are often described as "tumor antigens," or they are said to be "overexpressed" on tumors, but these terms are misleading for a number of reasons. First, there are currently over a hundred different antigens being targeted by ADCs in clinical development, yet very few are differentially expressed on tumors at higher levels than on normal tissues [13]. A majority of the data used to evaluate potential tumor-associated antigens are derived from mRNA expression analysis, which is known to be highly variable, irreproducible, and not suitable for accurate quantification of proteins. Researchers frequently use complimentary methods such as IHC to confirm the expression of a selected antigen on tumor cells, but IHC also has significant limitations for quantifying protein expression and the results are dependent on the samples being tested.

Problems with cross-reactivity are common with IHC and other antibody-based detection methods (ELISAs, FACs ...etc.), and heterogeneous protein expression in tumors creates additional uncertainties when trying to analyze expression data. Furthermore, the quantitation of specific antigens via protein isolation, purification, and analysis is not practical in a clinical setting. In general, it is very difficult to accurately measure protein expression in tumors because proteins and mRNA often continue to degrade in tumor samples, even after they are removed from the patient, Ideally, the expression levels of a specific antigen in tumors would be compared with expression levels in normal tissue, and higher target expression in tumors than in surrounding normal tissues would be considered desirable because it would enable relatively more drug to be delivered to the tumors by the ADC. Unfortunately, it is also very difficult to obtain truly normal tissue samples to use as controls, since most people are reluctant to give away parts of their healthy organs. As a result, the availability of normal tissue samples in the US relies on organs donated by people who suffer untimely deaths.

The prevalence of target expression, or the number of tumors that express a specific target antigen also varies widely and is dependent on the type of cancer being treated. To further complicate matters, target expression is often not uniform throughout the tumor and is seldom restricted to specific organs or systems [14]. Consequently, the therapeutic windows of most ADCs are quite narrow, and selecting an optimal target is likely the most important step toward developing an effective ADC.

12.4 Antibody Selection

In addition to the target, the antibody component of an ADC can also significantly impact its commercial success (Fig. 12.3). Factors such as affinity for the target, the specific binding epitope, the internalization rate, and the antibody subtype can all affect the success of an ADC. Processes used for antibody production, the species origin of the parent antibody, and methods used for humanization also influence the performance of the ADC [15]. Even the cell culture media used for antibody production can affect the activity of an ADC. For example, different glycosylation patterns have been observed in antibodies produced in different media that resulted in significant changes in immunogenicity [16]. Large manufacturing batches worth millions of dollars have been ruined because of subtle changes in the cell culture media and stories of process scientists losing their jobs over such incidents are common. Most antigen targets are not patentable because they are not human inventions so researchers have begun patenting specific epitopes on the targets in order to claim ownership of specific binding sites. Antibodies that bind to different epitopes on a target can have profoundly different activities and, in some cases, different epitopes can provide freedom to operate (FTO) for targets being investigated by competitor companies [17]. New epitopes are therefore often treated like new targets from a legal and regulatory perspective.

Occasionally the antibody component used in an ADC possesses antitumor activity without any payloads attached. Two of the currently approved ADCs, Kadcyla and Enhertu, fall into this category and utilize a previously approved cancer drug, trastuzumab, as their antibody component. Trastuzumab, which binds to growth-stimulating receptor (ERbB2 *aka* HER2) expressed on breast and other solid tumors, blocks binding to its ligand and inhibits tumor growth. Most other ADCs contain antibodies that lack inherent anti-tumor activity, and the antibody only serves as a delivery vehicle for the cytotoxic payload. As a result, some researchers believe that the antibody component of an ADC primarily serves as a slow-release mechanism to help maintain steady-state concentrations and improve the overall pharmacological profile of the payloads. This perspective is well supported since clinical data have shown that systemic concentration of the payloads remains relatively constant in patients dosed with ADCs [18]. In contrast, there is very little clinical evidence to support claims that ADCs can selectively deliver their payloads to tumors while sparing normal tissues.

Many other factors need to be considered when selecting an appropriate antibody to use for constructing an ADC therapeutic. For example, minor variations in antibody subtypes can significantly impact the conjugation process, especially when it involves the reduction of disulfide bonds. Different IgG subtypes vary in the number of interchain disulfides and afford different numbers of available cysteines upon reduction. Usually, antibodies of the IgG_1 subtype are preferred for cysteine conjugation; due to the relatively mild conditions required for disulfide reduction necessary to yield an optimal number of eight cysteines for conjugation. Species cross-reactivity is another important property to consider since the efficacy of the ADC in tumor models is more predictive of the efficacy expected in humans if the antibody also binds the analogous mouse antigen with equal affinity [19].

12.5 Selection of Linkers and Payloads

The remaining two components that significantly impact the success of an ADC are the linker and the drug or payload (Fig. 12.3). Finding the right combination of linker and payload has proven to be very difficult because the potency of an ADC often does not correlate with the potency of the unconjugated payload [20]. In addition, the conjugation of a small molecule payload to an antibody, which is often 100× larger in size, limits the amount of payload that can be administered to patients. Conventional small molecule drugs can often reach plasma concentrations of 20 μM or more, even when relatively small doses (10–100 mg) are administered. In order for an antibody to reach a plasma concentration of 20 μM (1.5 mg/mL), a dose of at least 20 grams would be required. Treatments of this size would not be safe or cost-effective so the payload must be extremely potent (IC_{50} < 10 nM) to enable more practical doses of ADC in the 50–500 mg range.

The payload potency alone is not sufficient to ensure ADC activity, however, since there are numerous other factors that contribute to the activity of an ADC. These include the payload's mechanism of action (MoA), the location of its target within the cell (cytoplasm or nucleus), the membrane permeability, and the linker used for conjugation. The conjugation of a highly potent drug payload to an antibody can result in an ADC with poor activity if the payload never reaches its target or falls off of the ADC prematurely. Conversely, highly potent payloads with low membrane permeability will have poor activity in their unconjugated state but can yield extremely potent ADCs because the permeability of the payload is determined by the antibody.

A classic example of the lack of correlation between the potency of an ADC with its unconjugated payload is shown in Fig. 12.4. Two structurally and mechanistically similar tubulin inhibitor payloads (MMAE and MMAF) differ in potency by roughly 200-fold in their unconjugated state (avg. IC_{50} = 0.4 and 100 nM, respectively), but when conjugated to trastuzumab, the relative potencies of the resulting ADCs are reversed and the MMAF ADC is actually more potent than the analogous MMAE ADC [21]. The potency of the conjugated MMAF increased by orders of

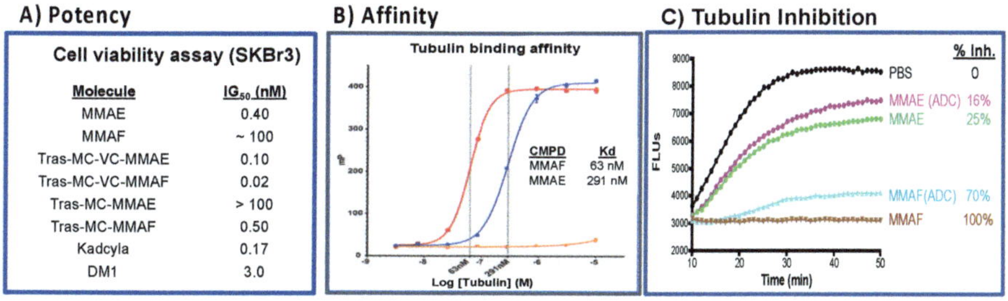

Fig. 12.4 (**a**) Cytotoxicity of unconjugated MMAE & MMAF payloads and trastuzumab ADCs against HER2 positive SkBr3 cells. (**b**) Tubulin binding affinity of MMAE vs MMAF. (**c**) Tubulin inhibition assay showing the superior potency of MMAF over MMAE

magnitude relative to the unconjugated drug (Fig. 12.4a), while the MMAE ADC had roughly equal potency with the unconjugated payload. Both of the MMAE ADCs are less potent than analogous MMAF ADCs regardless of the linkers used for conjugation. ADC potency does correlate with the tubulin binding affinity of the payloads however, and with their potency in tubulin polymerization assays where MMAF is the clear winner (Fig. 12.4b, c).

The dramatic increase in the potency of the MMAF ADCs relative to unconjugated MMAF (Fig. 12.4a) is consistent with the fact that unconjugated MMAF has extremely poor membrane permeability. Conjugation to trastuzumab gives both payloads equal access to their tubulin target because the payloads are internalized with the antibody at similar rates. MMAF is actually a more potent tubulin inhibitor than MMAE (Fig. 12.4b), which is consistent with the superior potency of the MMAF ADCs. Moreover, the poor permeability of MMAF causes it to accumulate to lethal concentrations inside the cell, while MMAE is free to diffuse out into nearby cells. The superior permeability of MMAE is believed to contribute to the enhanced in vivo activity of MMAE ADCs against tumors with low antigen expression. This property has been named "Bystander effect" and some researchers believe that the increased permeability of MMAE makes it more effective against certain tumors because it is able to diffuse into nearby cancer cells.

In contrast, MMAF remains in the intercellular space and is cleared more rapidly than MMAE. Some ADC researchers believe the "bystander effect" is also responsible for the systemic off-target toxicity observed in patients treated with ADCs [22]. This hypothesis is well founded since the free payloads are no longer selective for antigen-expressing cells after they are released from the ADC.

12.6 ADC Technology

The benefits of combining small molecule potency with the specificity of antibodies are often accompanied by new challenges during the development process. ADCs can be considered either as turbo-charged antibody therapeutics with improved potency or as tumor targeting chemotherapeutics enhanced by antibodies. The dual complexicity of ADCs raises numerous questions regarding CMC and regulatory requirements such as whether to file an NDA or BLA [23]. Some researchers believe that the antibody component of an ADC simply serves as a slow-release delivery system for the payload and therefore ADCs should be classified as small molecule drugs. Others believe that the payloads just enhance the activity of an existing antibody therapeutic; like Kadcyla & Enhertu do for trastuzumab. Regardless of which side of the fence you stand on, ADCs are currently considered to be biologics by the FDA and therefore follow a similar development path as antibody therapeutics [24]. In reality, the ADC development process is more rigorous than that for antibodies (or small molecules) because additional safety considerations are necessary due to the cytotoxic payloads that are released. Some researchers say that "developing an

ADC is similar to developing a new drug and a new biologic simultaneously and therefore requires twice the amount of work."

Another major factor that further complicates ADC development is heterogeneity. Eight of the 11 approved ADCs are heterogeneous mixtures of distinct molecules that differ in the number of payloads per antibody and their location [10]. There can be thousands of different molecular entities in a single ADC drug (Fig. 12.5). Kadcyla (*aka* T-DM1), an ADC approved for breast cancer in 2013, is a prime example of a heterogeneous ADC [25]. Kadcyla is composed of a payload conjugated to an antibody via lysine side chains to afford a complex mixture of different molecules that contain between 0 and 10 payloads per antibody and an average drugs/antibody ratio (DAR) of 3.7. Since the payloads are dispersed over multiple conjugation sites, the ADC is composed of a cocktail of thousands of different molecules, each with its own unique properties. Like Kadcyla, most currently approved ADCs are heterogeneous mixtures that range from dozens to thousands of different molecules. ADC heterogeneity arises from conventional conjugation methods that use lysine or cysteine reactive linkers that are not site-specific. Since a typical antibody has up to 12 cysteines and more than 70 lysines available for conjugation and a typical ADC contains only three to four drugs per antibody, most conventional ADCs differ in both DAR and conjugation sites (Fig. 12.5).

Over the past decade, substantial effort has been focused on reducing ADC heterogeneity resulting in a plethora of new site-specific conjugation methods [26]. Significant progress has been made toward producing homogeneous ADCs, but only two of the currently approved ADCs (Enhertu & Polivy) can be considered homogeneous [27, 28]. The homogeneity of Enhertu was not achieved through recombinant antibody engineering or other site-specific methods, but through new linker technology that enabled the conjugation of eight payloads per antibody without precipitation or aggregation of the ADC.

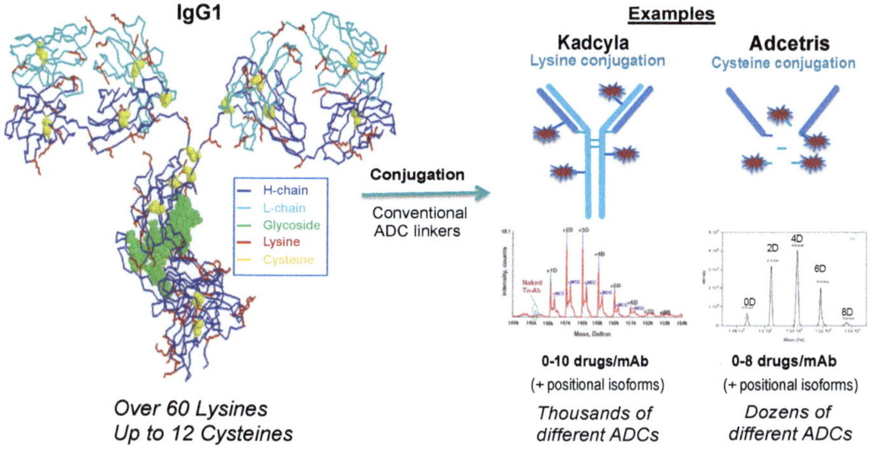

Fig. 12.5 Heterogeneity of ADCs with payloads conjugated to lysine or cysteine

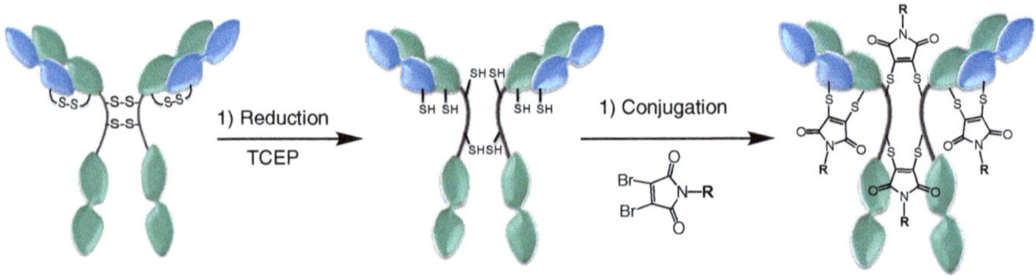

Fig. 12.6 ADCs containing dibromomaleimide linkers maintain covalent bonds between the heavy and light chains and yield homogeneous ADCs with 4 drugs/mAb

Other linker-based approaches for synthesizing homogeneous ADCs include the use of dibromomaleimide linkers that crosslink interchain cysteines and yield homogeneous ADCs with 4 drugs per antibody (Fig. 12.6) [29]. This innovative approach does not require recombinant engineering and maintains covalent bonds between the antibody heavy and light chains for improved stability in vivo. Similar linker-based approaches for generating homogeneous ADCs have been reported, but have not yet progressed to clinical development [30].

12.7 ADC Clinical Development

Despite the difficulties that many companies have faced during ADC development, 11 ADCs have now been approved for cancer (Table 12.1), and at least 20 more are in late-stage clinical trials (Table 12.2). A vast majority of these ADCs carry tubulin inhibitor payloads, which suggests that tubulin may be an ideal target for ADC payloads. Alternatively, the frequent use of tubulin inhibitor payloads might be a reflection of the risk-averse nature of pharmaceutical companies that chose to follow a similar path as their predecessors. Indeed, the popularity of MMAE as an ADC payload probably arises from the fact that it was used in Adcetris, the third approved ADC and arguably the most efficacious ADC to date.

When considering the unprecedented clinical activity involving ADCs, it is important to remember that more ADCs fail in the clinic than those that succeed. Dozens of ADC clinical trials have been terminated, discontinued, or withdrawn. The large number of ADC failures illuminates the significant challenges that the ADC field has experienced over the last 20 years [31]. A partial list of ADC failures is shown in Table 12.3 which suggests that there is ample room for improving the therapeutic properties of ADCs.

Eight of the 11 ADCs currently approved for cancer share one or more of the four major ADC components discussed previously (see Fig. 12.3). For example, Mylotarg and Besponsa have identical linkers and payloads but target different antigens (CD33 & CD22) [32, 33]. Kadcyla and Enhertu share the same target (HER2) and the same antibody (trastuzumab) but have different linkers and payloads; while four

Table 12.1 ADCs currently approved for cancer indications

Drug	ADC	Company	Year	Indications	Target	Payload	MOA
Mylotarg	Gemtuzumab ozogamicin	Pfizer/Wyeth	2000	AML	CD33	Calicheamicin	DNA alkylator
Adcetris	Brentuximab vedotin	SEAGEN	2011	Hodgkin's lymphoma	CD30	MMAE	Tubulin inh.
Kadcyla	Trastuzumab emtansine	Genentech	2013	Breast cancer	HER2	DMI	Tubulin inh.
Besponsa	Inotuzumab ozogamicin	Pfizer/Wyeth	2017	ALL	CD22	Calicheamicin	DNA alkylator
Enhertu	Trastuzumab deruxtecan	Daiichi/AZ	2019	Breast cancer	HER2	Dxd	Topo 1 inh.
Padcev	Enfortumab vedotin	Astellas/SEAGEN	2019	Urothelial cancer	Nectin-4	MMAE	Tubulin inh.
Polivy	Polatuzumab vedotin	Genentech/Roche	2019	Large Bcl	CD79b	MMAE	Tubulin inh.
Trodelvy	Sacituzumab govitecan	Immunomedics	2020	TNBC	TROP-2	SN38	Topo 1 inh.
Blenrep	Belantamab mafodotin	GSK	2020	Multiple myeloma	BCMA	MMAF	Tubulin inh.
Tivdak	Tisotumab vedotin	SEAGEN	2021	Cervical cancer	TF	MMAE	Tubulin inh.
Zynlonta	Loncastuximab tesirine	ADC Therapeutics	2021	B-cell lymphoma	CD19	PBD	DNA crosslinker

Table 12.2 ADCs in late stage development in 2021

ADC name	Sponsor	Indications	Status	Target	Linker (type)	Payload	MOA
Depatuxizumab Mafodotin	Abbvle	Glioblastoma	Phase III	EGFR	MC	MMAF	Tubulin inh.
Mirvetuximab Soravtansine	ImmunoGen	Ovarian	Phase III	FolR alpha	SPDP (disulfide)	DM4	Tubulin inh.
Vadastuximab Talirine	Seattle Genetics	AML	Phase III	CD33A	VCP (cleavable)	PBD	DNA alkylator
Trastuzumab duocarmazine	Synthon	MBC (breast)	Phase III	Her2	VCP (cleavable)	Duocarmycin	DNA alkylator
Rovalpituzumab tesirine	AbbVie	SCLC	Phase III	DLL3	VCP (cleavable)	PBD	DNA crosslinker
Telisotuzumab vedotin	Abbvie	Solid tumors	Phase III	c-Met	VCP (cleavable)	MMAE	Tubulin inh.
Camidanlumab tesirine	ADC Therapeutics	AML	Phase II	CD25	VCP (cleavable)	PBD	DNA crosslinker
Loncastuximab tesirine	ADC Therapeutics	DLBCL	Phase II	CD19	VCP (cleavable)	PBD	DNA crosslinker
Indatuximab ravtansine	Biotest	MM	Phase II	CD138	SPDP (disulfide)	DM4	Tubulin inh.
BMs-986148	BMS	Solid tumors	Phase II	Mesothelin	VCP (cleavable)	Tubulysin	Tubulin Inh.
Glembatumumab vedotin	Celldex	Melanoma	Phase II	gpNMB	VCP (cleavable)	MMAE	Tubulin inh.
Patritumab Deruxtecan	Daiichi Sankyo	Breast cancer	Phase II	Her3	GGFG (cleavable)	DXd	Topo I inh.
Pinatuzumab vedotn	Genentech/Roche	DLBCL, NHL	Phase II	CD22	VCP (cleavable)	MMAE	Tubulin inh.
Enapotamab vedotin	GenMab	Solid tumors	Phase II	AxI	VCP (cleavable)	MMAE	Tubulin inh.
Tetosumab vedotin	GenMab	Ovarian/solid	Phase II	TF	VCP (cleavable)	MMAE	Tubulin inh.
Labetuzumab govitecan	Immunomedics	CRC	Phase II	CEACAM5	PEG (cleavable)	Camptothecin	Topo 1 inh.
OBI-999	OBI Pharma	Solid tumors	Phase II	Globo H	MCVC (thiobridge)	MMAE	Tubulin inh.
PSMA-ADC	Progenics	Prostate	Phase II	PSMA	VCP (cleavable)	MMAE	Tubulin inh.
Disitamab vedotin	Remegen/Mabplex	Solid tumors	Phase II	Her2	VCP (cleavable)	MMAE	Tubulin
STRO-002	Sutro	Ovarian	Phase II	FoI-Ra	p-AMF/D8CO	Hemiasterlin	Tubulin inh.
CX-2029	Cytomx/Abbvie	Solid tumor	Phase II	CD71	VCP (cleavable)	MMAE	Tubulin

Source: Clinical trials.gov

Table 12.3 Examples of ADCs discontinued or terminated due to inadequate safety

ADC Name	Sponsor	Indications	Status	Target	Linker	Payload	MOA
Coltuximab ravtansine	Immunogen	ALL, NHL	Phase II	CD19	SPDP	DM4	Tubulin inh.
MLN2704	Takeda	Prostate	Phase II	PSMA	SPP	DM1	Tubulin inh.
MDX-1203, BMS936561	BMS	RCC	Phase I	CD70	MC-VC	Duocarmycin	DNA binder
Cantuzumab ravtansine	Immunogen	Gastric	Phase II	CANAG/MUC1	SPDP	DM4	Tubulin inh.
Anetumab ravtansine	Bayer/NCI	Solid tumors	Phase II	Mesothelin	SPDP (cleavable)	DM4	Tubulin inh.
Denintuzumab mafodotin	SEAGEN	DLBCL	Phase II	CD19A	MC (stable)	MMAF	Tubulin inh.
Glembatumumab vedotin	Celldex	BC, melanoma	Phase II	gpNMB	VCP (cleavable)	MMAE	Tubulin inh.
Pinatuzumab vedotin	Genentech	DLBCL, NHL	Phase II	CD22	VCP (cleavable)	MMAE	Tubulin inh.
Rovalpituzumab tesirine	Abbvie	Small cell lung	Phase II	DLL-3	VCP (cleavable)	PBD	DNA crosslinker
Vadastuximab talirine	SEAGEN	AML	Phase II	CD33A	VCP (cleavable)	PBD	DNA alkylator
ABBV-838	Abbvie	MM	Phase I	CD319	VCP (cleavable)	MMAE	Tubulin inh.
AGS67E	Astellas	AML	Phase I	CD37	VCP (cleavable)	MMAE	Tubulin inh.
BAY 1129980	Bayer	Solid tumors	Phase I	LYPD3	VC (cleavable)	MMAE	Tubulin inh._
BAY 1187982	Bayer	Gastric	Phase I	FGFR2	NHS-caproyl	MMAW	Tubulin inh.
CDX-014	Celldex	RCC	Phase I	TIM-1	VCP (cleavable)	MMAE	Tubulin inh._
HKT288	Novartis	Ovarian	Phase I	Cadherin-6	SPDP (disulfide)	DM4	Tubulin inh.
IMGN289	ImmunoGen	NSCLC	Phase I	EGFR	MCC (stable)	DMI	Tubulin inh._
LOP628	Novartis	Solid tumors	Phase I	CD117	MCC (stable)	DMI	Tubulin inh._
SGN-CD123A	SEAGEN	AML	Phase I	CD123A	VCP (cleavable)	PBD	DNA alkylator
SGN-CD19B	SEAGEN	ALL, NHL	Phase I	CD19B	VCP (cleavable)	MMAE	Tubulin inh.
SGN-CD48A	SEAGEN	Mult. Myeloma	Phase I	CD48A	glucuronidase	MMAE	Tubulin inh._
SGN-CD70A (SGN-75)	SEAGEN	NHL, RCC	Phase I	CD70	MC (stable)	MMAF	Tubulin inh._

ADCs: Adcetris, Polivy, Padcev, and Tivdak share identical linkers and payloads but have different antibodies directed against different targets [34–36].

Only one ADC property is shared by all 11 approved ADCs; they are all designed to release toxic payloads in tumors. The following section will discuss the pros and cons of the currently approved ADCs, the synthetic processes that were used to make them, and the technology that contributed to their approval. The remaining sections of this chapter will focus on the evolution of ADCs starting with a review of those currently approved. New ADCs currently in late-stage development and predictions for future ADC therapeutics will also be discussed. The objective of this chapter is to provide information about the risks and benefits that are associated with ADCs for the treatment of cancer in addition to summarizing their development history.

12.8 ADCs Approved for Cancer

12.8.1 Mylotarg

Gemtuzumab ozogamicin (Mylotarg) was the first immunoconjugate to receive FDA approval to treat patients with acute myeloid leukemia (AML) [32]. It was originally developed by Wyeth in the 1990s as an immunoconjugate, before the term "ADC" was coined, and was later approved in the year 2000. Mylotarg is an ADC in which a humanized anti-CD33 mAb is covalently linked to N-acetyl gamma calicheamicin through an acid-labile linker (Fig. 12.7). The presence of the hydrazone function in the linker allows the rapid release of calicheamicin from its conjugated state under the acidic conditions found in a lysosome [8].

The target of gemtuzumab ozogamicin, CD33, is a 67 Kd adhesion molecule expressed on both normal and malignant cells that bind 2,3-linked sialic acid residues. Upon binding to its CD33 target, Mylotarg is rapidly internalized, and the payload is released in the lysosome. Once released the calicheamicin payload induces double-strand DNA breaks, subsequently inducing cell cycle arrest and apoptotic cell death. The safety of Mylotarg was questioned very early in development due to 8 fatalities out of 131 patients in the dosing arm of the pivotal clinical trial. Although there is a 6% chance that a patient will not survive the treatment,

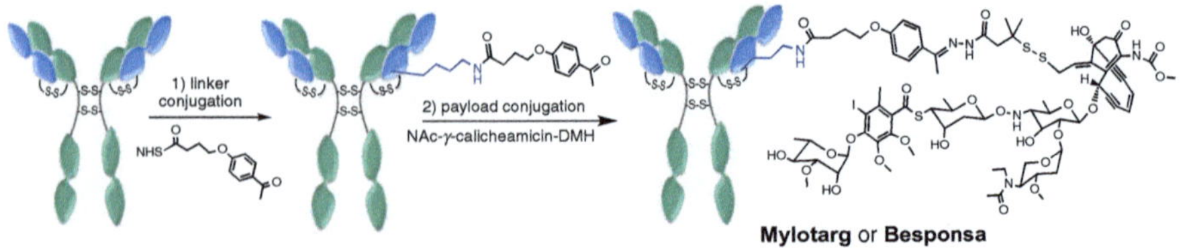

Fig. 12.7 Synthesis of gemtuzumab ozogamicin (*Mylotarg*) and inotuzumab ozogamicin (*Besponsa*), which contain identical calicheamicin payloads

Mylotarg was granted accelerated approval from the FDA based on phase 1 data. Considering the high fatality rate and cost of treatment, combined with marginal efficacy (there was no statistically significant difference between the treatment arm and the control group in overall survival), it is not surprising that Mylotarg was removed from the market in 2010.

Numerous other problems contributed to the temporary demise of Mylotarg. For instance, the original ADC contained an average of only two to three payloads per antibody conjugated stochastically to over 70 lysines on the antibody surface. The low drug/antibody ratio (DAR) resulted in a highly heterogeneous mixture of ADC molecules with up to 30% unconjugated antibody [37]. The presence of an unconjugated antibody is problematic because it can function as a competitive inhibitor of the ADC by occupying target binding sites. These factors together with the inherent linker instability of Mylotarg likely contributed to its discontinuation. Mylotarg was eventually reapproved in 2017 after several more clinical trials demonstrated that the benefits outweigh the risk [38]. Most ADCs that have followed the approval of Mylotarg have shown modest, though somewhat gradual improvement, and the benefits they offer to cancer patients are becoming more apparent.

12.8.2 Adcetris

Nearly a decade after the initial approval of Mylotarg, a second ADC, brentuximab vedotin (*aka Adcetris*), was approved for Hodgkin's lymphoma in 2011 [39]. Adcetris is composed of an anti-CD30 (TNFRSF8) antibody conjugated to a potent anti-mitotic agent, monomethyl auristatin E (MMAE) via a protease cleavable dipeptide linker [40]. Adcetris differs from Mylotarg in many ways that have led to its continued commercial success. Perhaps the most important difference is that Adcetris contains an average of four drugs/antibody conjugated to interchain cysteines after partial reduction of the antibody (Fig. 12.8a) [41]. Since there are only eight interchain cysteines available for conjugation compared to 70 or more lysines used for conjugation on Mylotarg, the heterogeneity of Adcetris is significantly reduced and the fraction of unconjugated antibody is less than 5%.

The use of a tubulin inhibitor payload in Adcetris may provide other less tangible benefits such as bystander killing of nearby tumor cells [42]. Another component that was improved in Adcetris is the linker, which is significantly more stable than Mylotarg's linker. As a result, Adcetris is less prone to premature release of the payloads. In addition, the linker used in Adcetris contains a p-aminobenzyl functional group that self-destructs upon cleavage of the aryl amide bond to release the unmodified payload (Fig. 12.8b) [43]. In contrast to Adcetris, the payload on Mylotarg is released by acid hydrolysis of the hydrazone bond, and later in the nucleus the payload is activated by attack of the glutathione on the disulfide.

Lastly, the expression profile for the Adcetris target, CD30, is more restricted to specific cell types while CD33 is more broadly expressed. When the improved

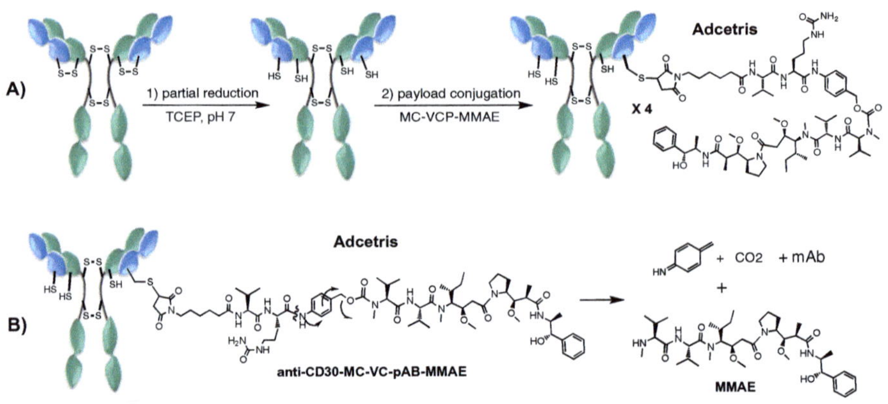

Fig. 12.8 (**a**) Synthesis of Adcetris, Polivy, Padcev and Tivdak via partial reduction of disulfides followed by conjugation with MMAE. (**b**) Cathepsin-mediated cleavage of the val-cit linker followed by elimination of a self-immolative p-aminobenzyl group to release unmodified MMAE

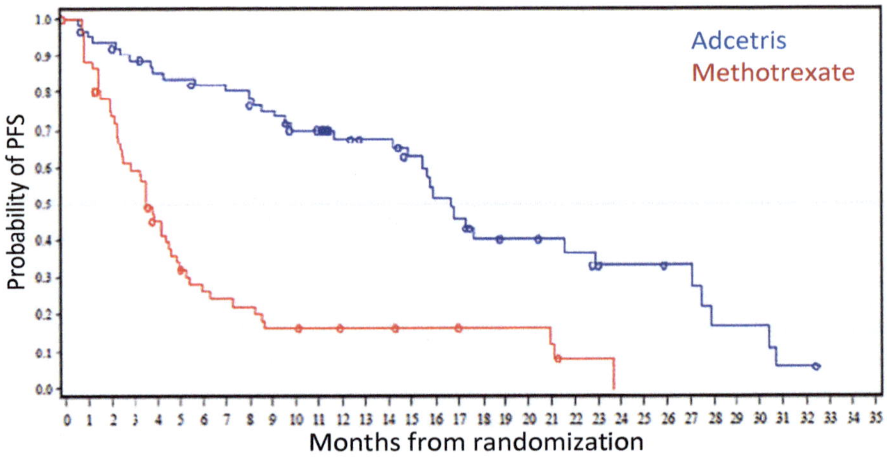

Fig. 12.9 Kaplan Meier Curve of progression-free survival in patients treated with Adcetris versus standard of care, Methotrexate. (Source: Adcetris prescribing information)

properties of Adcetris are combined, it demonstrates a real benefit to patients without the severe toxicity observed with Mylotarg. Adcetris now serves as a benchmark ADC for many others in the clinic and three additional approved ADCs share the same MMAE payload. Adcetris continues to be one of the most efficacious ADCs approved thus far (Fig. 12.9).

12.8.3 Kadcyla

Trastuzumab emtansine (*aka* Kadcyla) was the third ADC to reach the market and was approved in 2013 [44]. Kadcyla was the first ADC to be directed against solid tumor types and it differs from earlier ADCs in a number of other ways. Like Adcetris, Kadcyla carries a tubulin inhibitor payload (DM1), but the payload is conjugated to lysines (instead of cysteines) to afford a highly heterogeneous ADC with an average DAR of 3.7 drugs/antibody (Fig. 12.10a) [45]. Kadcyla's linker also differs from most other ADCs in that it is noncleavable and does not release its payload until the entire ADC is degraded to peptide fragments [46]. Kadcyla's target antigen, human epidermal growth factor receptor 2 (HER2), is expressed in breast tumors but is also found in normal breast tissue and numerous other solid tissues and organs. Consequently, the broad expression profile of its target results in a rather narrow therapeutic window for most patients [47].

Kadcyla's developer Roche (formerly Genentech) had a clever solution to this problem that eventually spawned a whole new approach to personalized medicine. In order to increase the likelihood of clinical success, cancer patients were prescreened for HER2 expression and only patients with favorable expression profiles were admitted into the trial. Roughly a third of breast cancer patients expressed Her2 at levels considered to be adequate for effective treatment, which left the majority of patients looking for other treatment options. This prescreening approach was first used during the development of Kadcyla's parent antibody component known as Herceptin (approved for breast cancer in 1998) and has contributed significantly to the commercial success of both products [48].

To avoid competition between two of their own products, Roche marketed Kadcyla to extend the product lifetime of Herceptin and timed its release to coincide with the expiration of Herceptin's patent. Herceptin is now a generic drug known as trastuzumab and Kadcyla has replaced it as one of Roche's top selling drugs for

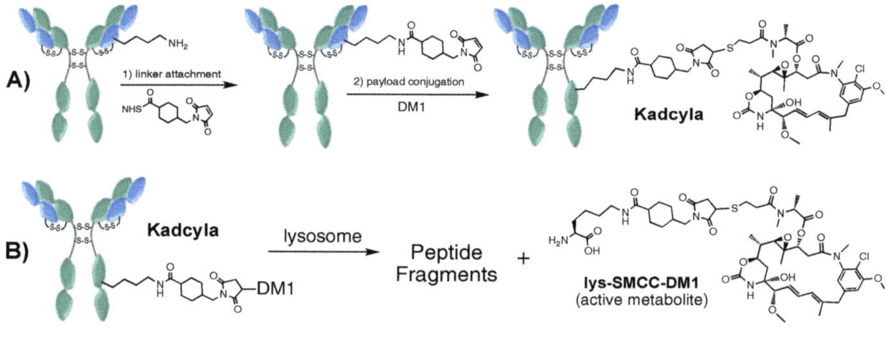

Fig. 12.10 (**a**) Synthesis of Kadcyla via a two-step conjugation process. (**b**) Lysosomal degradation of Kadcyla releases the DM1 payload still connected to lysine

cancer. Nonetheless, trastuzumab remains one of the most successful antibody therapeutics to date. Based on the clinical results for both Herceptin and Kadcyla however, it is widely accepted that neither drug would have been approved without prescreening patients for high HER2 expression. Ironically, the success of HER2 as a cancer target can be attributed primarily to patient selection rather than target selection. This "personalized medicine" is unpopular with some doctors but has become a common practice in oncology and will likely continue to play a part in the future success of many other ADC therapeutics [49].

Another significant difference between Kadcyla and earlier ADCs involves the linker. Kadcyla contains a relatively stable noncleavable (SMCC) linker that minimizes the release of the DM1 payload. Interestingly, Kadcyla's SMCC linker was originally intended to serve as a negative control for comparison with less stable disulfide linkers so researchers were surprised when Kadcyla demonstrated efficacy in xenograft tumor models. Since Kadcyla's linker does not actually release its payload, researchers identified a metabolite that contains the DM1 payload and the SMCC linker still connected to the lysine to which it was originally conjugated (Fig. 12.10b) [50]. This led to the conclusion that activity was dependent upon the lysosomal degradation of Kadcyla into peptide fragments. Whether this metabolite is solely responsible for Kadcyla's antitumor activity is unknown and other potential explanations such as the potential contribution of the intact ADC to Kadcyla's activity were not investigated [51].

The presence of lysine in the active metabolite increases its hydrophilicity relative to the unmodified payload and reduces its permeability. The reduced permeability diminishes the potency of the metabolite and enables rapid clearance after its released from the cell. It also reduces potential bystander effects on surrounding cancer cells that are thought to be beneficial in some cases. Studies conducted after Kadcyla's approval have shown that Kadcyla's linker is actually less stable than originally designed [52]. Researchers found that the maleimide group linking DM1 to the MCC linker can undergo thiol exchange with serum albumin, which results in premature release of the DM1 payload and contributes to systemic off-target toxicity. Since most ADCs currently in clinical development contain maleimide groups in the linkers, they are likely to possess similar linker instability. The fact that most ADC researchers continue to use maleimide linkers, suggests that they are not aware of these stability issues, or they believe that maleimide instability is necessary for optimal performance.

Overall Kadcyla's efficacy was determined to be statistically significant and patients with breast cancer gained about 3 months survival over the standard of care. Whether or not Kadcyla offers improved efficacy over Herceptin (trastuzumab) is debatable since the two have never been compared head-to-head in a clinical trial. Based on Kaplan-Meyer plots for both drugs, Kadcyla may afford a slightly improved response over trastuzumab (Fig. 12.11).

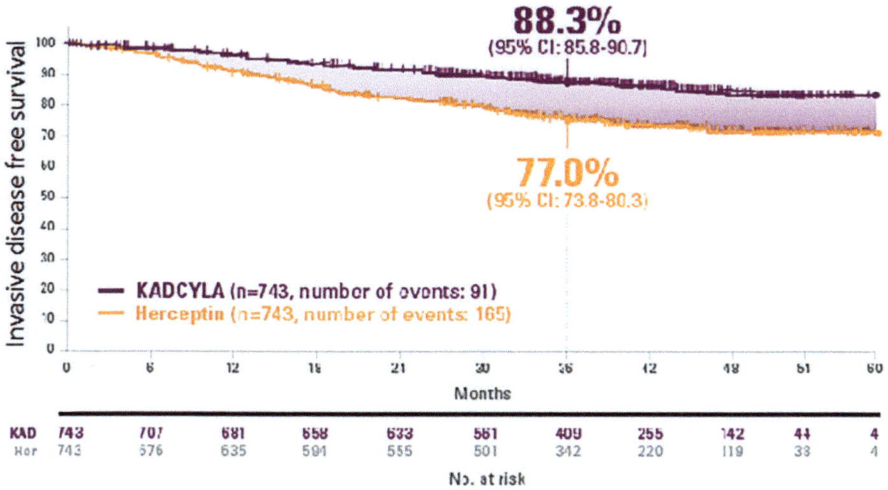

Fig. 12.11 Kaplan Meyer plot showing survival of patients treated with Kadcyla or Herceptin [53]. (Source: KADCYLA Prescribing Information. Genentech, Inc. 2022)

12.8.4 Besponsa

Four years after the approval of Kadcyla, Inotuzumab ozogamicin (Besponsa) gained FDA approval in 2017 and was the fourth ADC approved for cancer indications. Besponsa is developed by Pfizer for the treatment of adults with relapsed or refractory B-cell precursor acute lymphoblastic leukemia. The ADC is composed of an anti-CD22 monoclonal antibody conjugated to the payload N-acetyl-gamma-calicheamicin via a bi-functional 4-(4-acetylphenoxy) butanoic acid linker. The linker and payload components of Besponsa are identical to those used in Mylotarg and are conjugated via lysine residues on the antibody (Fig. 12.7). Besponsa however demonstrates better homogeneity and stability compared to Mylotarg due to improvements in the conjugation process that enabled higher DARs and reduced the amount of unconjugated antibody. Nonclinical data suggest that the anticancer activity of inotuzumab ozogamicin is due to the binding of the ADC to CD22-expressing tumor cells, internalization of the ADC, and intracellular release of the calicheamicin payload after hydrolytic cleavage of the linker in the lysosome. Release and activation of calicheamicin induce double-strand DNA breaks and eventually lead to cell cycle arrest and apoptosis [54].

.According to the prescribing information provided by the drug's manufacturer, Pfizer, Besponsa increases overall survival from 6.2 to 7.7 months (Fig. 12.12). However, the product insert states, "*The analysis of overall survival (OS) did not meet a prespecified boundary for statistical significance of P=0.0104.*" In Laymen's terms, they are not sure if Besponsa actually works, because the treatment group was too small to support their claims statistically.

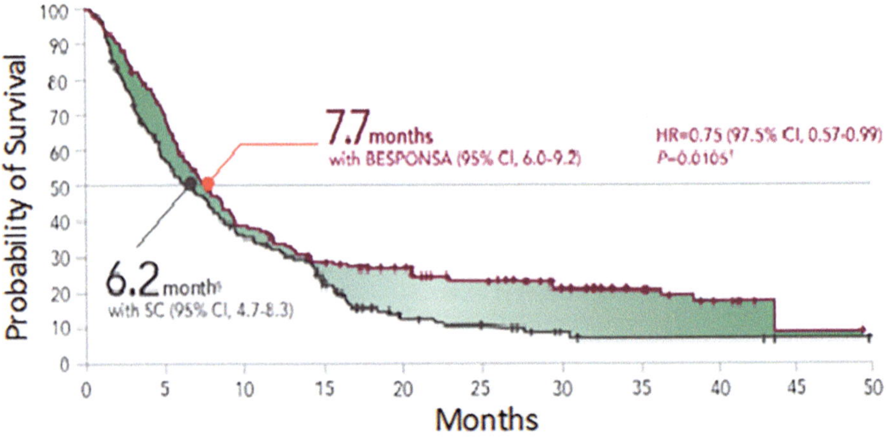

Fig. 12.12 Kaplan-Meyer plot for survival of Besponsa treated patients. (Source: BESPONSA Prescribing Information. New York, NY: Pfizer Inc.)

Since a typical treatment course can cost tens of thousands of dollars or more, patients are again faced with a difficult decision.

12.8.5 Polivy

Polatuzumab vedotin (aka Polivy) is an ADC developed by Roche and SEAGEN for diffuse large B-cell lymphoma. It was the fifth ADC approved in the US and it contains an identical linker and payload to Adcetris, Padcev, and Tivdak (see Fig. 12.8) [55]. The FDA granted accelerated approval for Polivy in 2019 based on a single clinical trial. Because the results were based on only 40 patients, additional clinical trials are required to confirm any clinical benefit. Polivy's target is CD79b, a component of the B-cell antigen receptor that is highly expressed in B-cell lymphoma. The ADC contains a CD79b-specific antibody with an average of 3.5 MMAE molecules per antibody linked via a cathepsin cleavable valine-citrulline inker. Like Adcetris, Polivy's linker is designed to release the antimitotic tubulin inhibitor, MMAE.

The conjugation process for Polivy (and most other ADCs conjugated through cysteine residues) utilizes a thiol-reactive maleimide linker. The reaction of the maleimide linker with cysteine proceeds rapidly with high yields. Consequently, maleimide linkers are used in a majority of ADCs currently in clinical development [56]. Ironically, maleimide-based linkers are known to be inherently unstable, which has likely contributed to the inadequate safety of many ADCs [57]. In fact, most ADCs currently in development (including seven of the approved ADCs) contain unstable maleimide-derived linkers, which are prone to thiol exchange or elimination reactions in plasma [58]. Thiol exchange often results in transfer of the

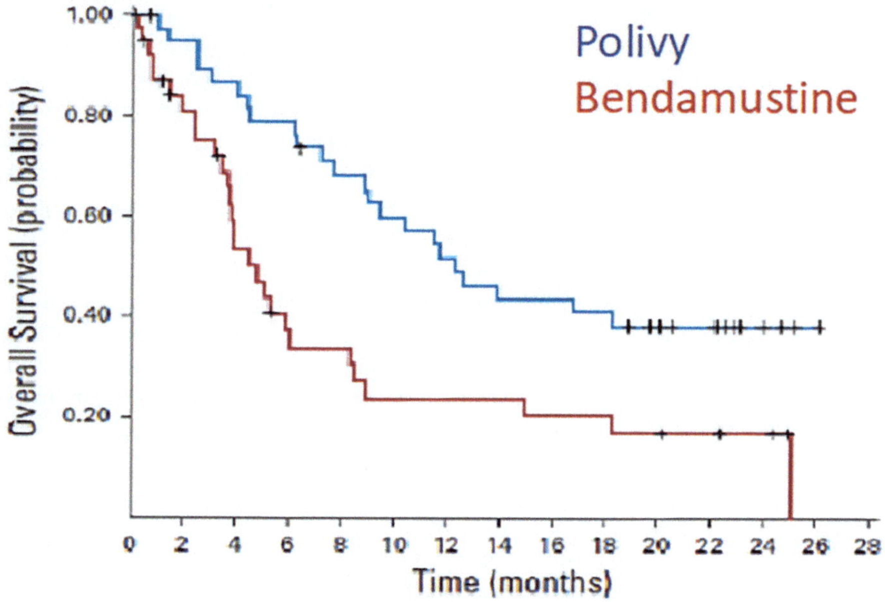

Fig. 12.13 Overall survival of patients with DLBCL treated with Polivy vs chemotherapy

payloads to an unpaired cysteine in serum albumin, the most abundant protein in the blood. Once transferred, the payload is no longer selective for tumor antigens resulting in systemic exposure and off-target toxicity. Although stable maleimide alternatives are available for cysteine conjugation, they haven't been utilized thus far [59].

.Unlike most other ADCs, the recommended number of treatments for Polivy is five cycles and off-target toxicity is a major issue. The extreme toxicity of Polivy is reflected by a 7% mortality rate, but the overall survival probability was considered sufficient to justify its approval in 2019 (Fig. 12.13). It is currently unclear whether Polivy will demonstrate meaningful efficacy in future clinical trials, yet annual gross sales are expected to reach one billion by 2023 [35].

12.8.6 Padcev

Enfortumab vedotin (PADCEV) is a nectin-4-directed antibody and microtubule inhibitor conjugate approved in 2019 for the treatment of adult patients with locally advanced or metastatic urothelial cancer and bladder cancer [36]. Like Adcetris, Polivy, and Tivdak, Padcev carries a tubulin inhibitor payload (MMAE) conjugated to an antibody via a protease cleavable dipeptide linker (see Fig. 12.8). Padcev was developed by Astellas in a partnership with SeaGen [60].

Padcev's target is nectin-4, a cell adhesion molecule found in a wide variety of tissues. Nectin-4, also known as polio virus receptor-like 4 (PVRL4), is thought to contribute to tumor proliferation and metastasis. The broad expression of the Nectin-4 on epithelial tissues is consistent with the toxicity profile, which includes peripheral neuropathy, ocular disorders, and skin reactions. In fact, there were serious adverse reactions in 46% of patients treated with Padcev. The most common reactions were urinary tract infection (6%), cellulitis (5%), febrile neutropenia (4%), diarrhea (4%), sepsis (3%), acute kidney injury (3%), dyspnea (3%), and rash (3%). Fatal adverse reactions occurred in 3.2% of patients, including acute respiratory failure, aspiration pneumonia, cardiac disorder, and sepsis (each 0.8%) [61].

Padcev was the sixth ADC approved by the FDA and the third ADC to use MMAE as a payload. Consequently, two thirds of the approved ADCs carried tubulin inhibitor payloads. The success of these ADCs either reflects the risk-averse business practices of most pharmaceutical companies, or it suggests that tubulin inhibitors may have unique properties that are optimal for use as ADC payloads. The fact that almost 75% of the ADCs currently in the clinic also contain tubulin inhibitor payloads supports the notion that tubulin is a superior target for ADC payloads.

Padcev was granted accelerated approval in 2019 based on an overall response rate of 44% and a median duration of 7.6 months (Table 12.4). This might seem impressive to most patients but closer inspection reveals that the results are based on the data from 55 patients. Like most ADC clinical trials, the lack of a placebo control group makes it very difficult to determine the true benefit of treatment. A Kaplan Meier plot indicates a median survival benefit of almost four months compared to chemotherapy, but after 20 months both treatment groups had an equal number of survivors (Fig. 12.14) [36].

12.8.7 *Enhertu*

Trastuzumab Deruxtecan (*aka Enhertu*) is an ADC that was approved for metastatic Her2-positive breast cancer in 2020. Enhertu targets the same antigen (Her2) as Kadcyla and even uses the same antibody trastuzumab. Trastuzumab (*aka Herceptin*) was approved in 1998 for Her2-positive metastatic breast cancer and later for gastric cancer. It was one of the first antibodies to be approved for solid tumors and is

Table 12.4 Efficacy results for Padcev from EV201 trial

Endpoint	PADCEV $n = 125$
Confirmed ORR (95% CI)	44% (35.1, 53.2)
Complete response rate (CR)	12%
Partial response rate (PR)	32%
Median[a] duration of response, months (95% CI)	7.6 (6.3, NE)

Source: Prescribing information
NE not estimable
[a]Based on patients ($N = 55$) with a response by BICR

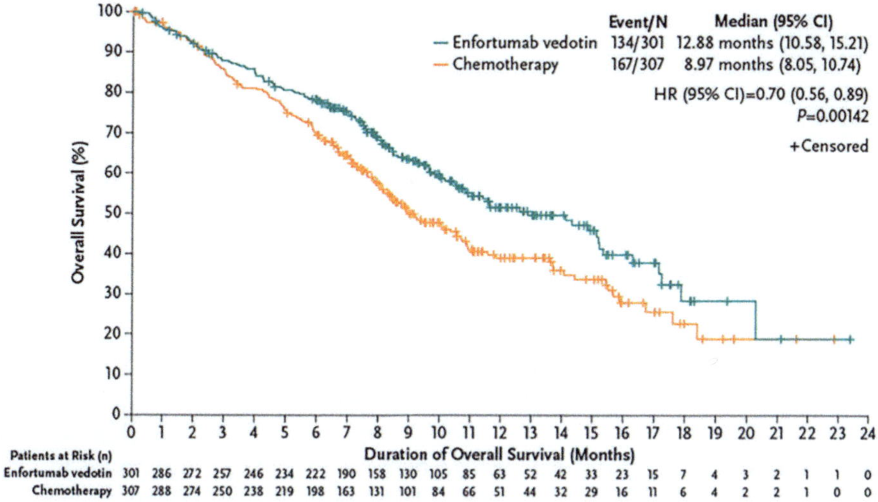

Fig. 12.14 Kaplan Meier plot showing % overall survival of patients treated with Padcev. (Source: *N Engl J Med*. 2021 March 25; 384(12): 1125–1135)

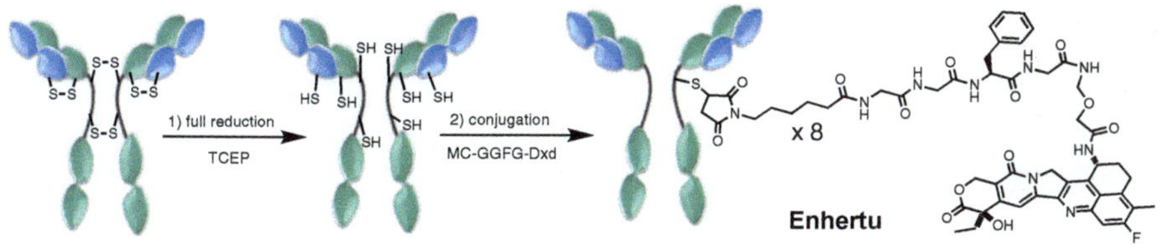

Fig. 12.15 Synthesis of Enhertu via reduction of interchain disulfides followed by conjugation with a tetrapeptide linker to yield a homogeneous ADC with 8 drugs/mAb

thought to inhibit the dimerization and signaling of Her2 (ERBB2) and its other ERBB family receptors. When Roche (formerly Genentech) realized that the commercial lifetime of Herceptin could be prolonged by attaching a cytotoxic payload to yield a more potent drug, they developed Kadcyla, an ADC that was discussed previously (Fig. 12.10). Daiichi-Sankyo timed its release of Enhertu perfectly to coincide with the expiration of Herceptin's patent in the US [62]. This gave them freedom to operate with trastuzumab but since Kadcyla, the first trastuzumab ADC, is patent protected until 2028 Daiichi needed to prove the superiority of Enhertu over Kadcyla. They also needed to find an alternative payload that could outperform Kadcyla's tubulin inhibitor payload, DM1.

The synthesis of Enhertu is shown in Fig. 12.15. One of the properties that distinguishes Enhertu from Kadcyla (and other ADCs) is the linker. Unlike Kadcyla, Enhertu utilizes a protease-cleavable tetrapeptide linker that releases its payload in the lysosome. The use of the GGFG linker enabled Daiichi to circumvent SEAGEN's patents on cathepsin-cleavable dipeptides. More importantly, Enhertu's peptide

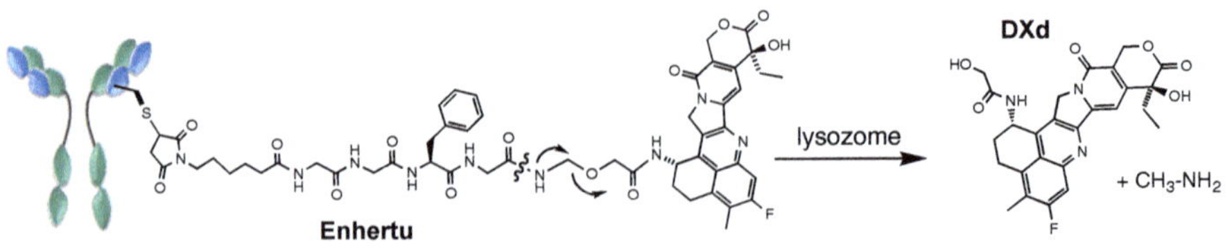

Fig. 12.16 Cleavage of Enhertu GGFG peptide linker to release the DXd payload

linker improves the overall hydrophilicity of the ADC and enables the conjugation of eight payloads per antibody. Since all eight cysteines are conjugated to a payload, Enhertu was the first approved homogeneous ADC [63].

Enhertu was also the first ADC to carry a topoisomerase 1 inhibitor as its payload. The payload (known as DXd) is a close derivative of a well-known topoisomerase 1 inhibitor, Exatecan, that had been investigated previously in clinical trials for a variety of solid tumor indications. Daiichi discovered the DXd payload while testing ADCs designed to release exatecan-like molecules. The challenge faced by Daiichi's chemists was how to release an unmodified payload from the GGFG tetrapeptide without leaving behind remnants of the linker. Attachment of the exatecan payload directly to the C-terminus of GGFG afforded a highly stable amide linkage but the payload was a poor substrate for lysosomal proteases [64]. To facilitate proteolytic cleavage of the GGFG linker, researchers added a 4-aminobutyric acid spacer that resulted in high levels of aggregate when more than 3 drugs per antibody were conjugated. Substitution of the middle methylene group in 4-aminobutyrate with a single oxygen atom resulted in efficient cleavage of the tetrapeptide, followed by spontaneous loss of methylamine, and release of the highly potent topoisomerase 1 inhibitor payload, known as DXd (Fig. 12.16) [65]. Enhertu has outperformed Kadcyla in breast cancer patients with low Her2 expression and has shown promise in treating gastric cancers. It also has a superior safety profile to most other ADCs [66].

12.8.8 Trodelvy

Sacituzumab govitecan or Trodelvy is an ADC developed for the treatment of metastatic triple-negative breast cancer. It was originally developed by Immunomedics and is now marketed by Gilead who acquired them in 2020. Trodelvy has the structure shown in Fig. 12.17 and targets Trop-2 (tumor-associated calcium signal transducer 2), a protein expressed in a variety of different tissues and on numerous solid tumors. Sacituzumab is the humanized antibody component of Trodelvy that is specific for Trop-2. The payload, SN-38, is a derivative of camptothecin, a potent topoisomerase 1 inhibitor [67] and it is the active metabolite of Irinotecan, a chemo-therapeutic drug that was approved in 1996 for the treatment of colon cancer and small cell lung cancer [28].

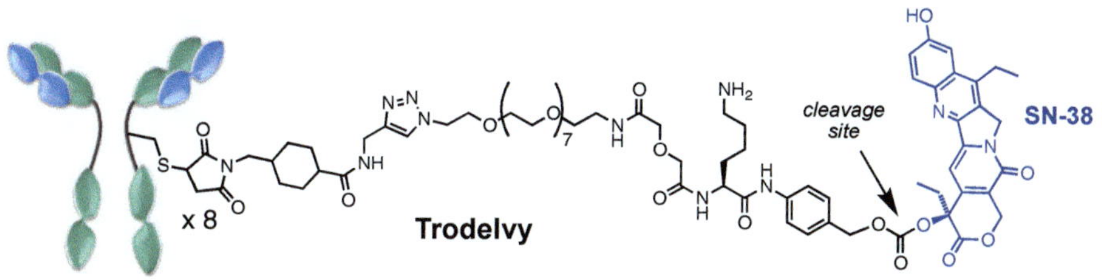

Fig. 12.17 Structure of Trodelvy showing cleavage of the carbonate ester and release of the SN-38 payload, a potent topoisomerase1 inhibitor (shown in blue)

Like Enhertu, Trodelvy contains eight payloads per antibody and is nearly homogeneous. The most distinguishing feature of Trodelvy compared to other approved ADCs is its unusually complex linker. It is unclear why Trodelvy's linker contains so many different functional groups when the cleavage mechanism for releasing the payload is the hydrolysis of the carbonate ester directly attached to the payload (Fig. 12.17). One can only speculate about the linker selection criteria that were used, but a possible explanation for the complex linker design is that it provided Immunomedics with the freedom to operate (FTO). Indeed, the chances of finding a similar design in competitor patent claims would be very small. Linker stability was probably not considered to be important since carbonate esters of this type are susceptible to serum esterases that cause premature release of the SN38 payload. This instability is reflected in the pharmacological properties of Trodelvy, which has a relatively short half-life in plasma of about 15 hours. Some researchers believe that Trodelvy's activity arises primarily from a bystander effect rather than conventional internalization and lysosomal degradation [68, 69].

The efficacy of Trodelvy was determined in the ASCENT trial of 529 patients with metastatic triple-negative breast cancer. Based on the results in which patients treated with Trodelvy survived an average of 5 months longer than those on standard chemotherapy, Trodelvy received accelerated approval in 2020 (Fig. 12.18). Later in 2021, Trodelvy was approved for metastatic urothelial cancer based on a single-arm trial of 112 patients. In both trials, the median overall survival was nearly double that of patients who received chemotherapy.

12.8.9 Blenrep

Belantamab mafodotin (aka BLENREP) is a recently approved ADC [70] and the first that uses MMAF (monomethyl-auristatin Phe) as a payload. MMAF is a closely related analog to MMAE, the payload found in Adcetris, Padcev, Polivy, and Tivdak. Both MMAF and MMAE are anti-mitotic tubulin polymerization inhibitors that disrupt microtubule formation during mitosis. The two payloads share a common tetrapeptide core structure and differ only at the C-terminus where MMAF contains a phenylalanine moiety and MMAE contains a pseudoephedrine group (Fig. 12.19a).

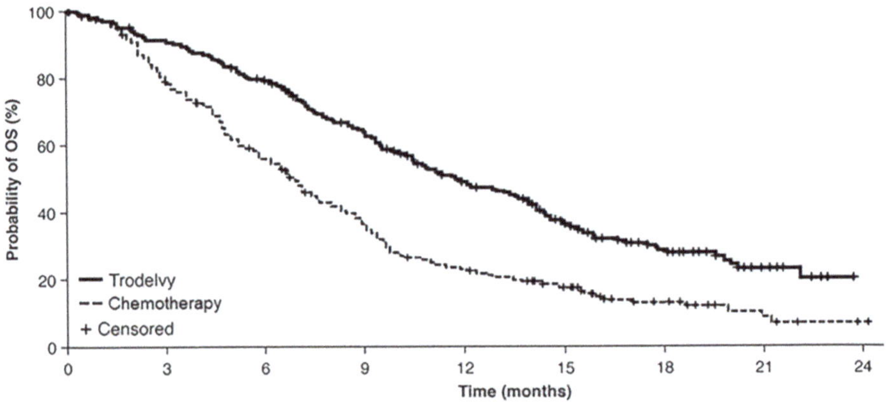

Fig. 12.18 Kaplan-Meier Plot of overall survival in ASCENT trial. (Source: Prescribing information, June 2022)

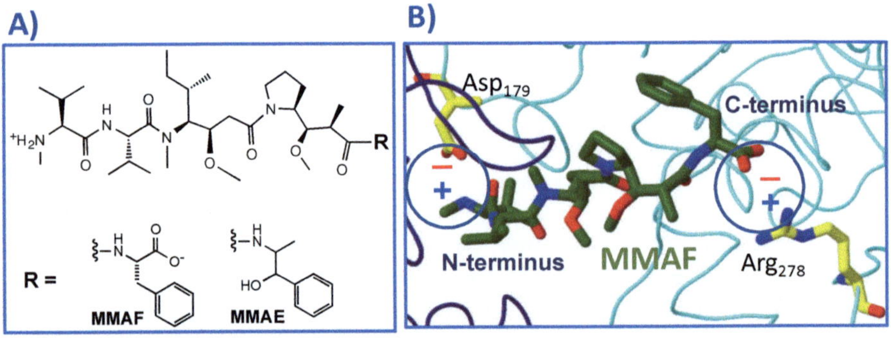

Fig. 12.19 (**a**) Structures of auristatins MMAE and MMAF. (**b**) X-ray structure of MMAF bound to tubulin showing favorable electrostatic interactions at both N and C terminus

The small difference in structure impacts tubulin binding affinity due to the interaction of the negative charge on MMAF with Arg278 on tubulin (Fig. 12.19b) [71]. The absence of a negative charge on the C-terminus of MMAE results in reduced tubulin binding affinity and reduced potency in tubulin polymerization assays.

The negative charge on MMAF also limits membrane permeability and results in a 200-fold decrease in cytotoxicity relative to MMAE in cell proliferation assays. Ironically, the property that increases tubulin binding and antimitotic activity of MMAF also decreases permeability and cytotoxicity. When the payloads are conjugated to antibodies, however the differences in permeability are insignificant and the more potent tubulin inhibitor (MMAF) yields a more potent ADC. Once released from the ADC, MMAF accumulates inside cells to reach high local concentrations while MMAE is able to diffuse out and enter nearby cells. The superior cell permeability of MMAE enables a “bystander effect” in that the payload is able to kill neighboring cells in the tumor microenvironment. Unfortunately, this bystander effect also leads to off-target toxicity and contributes to ADC toxicity.

Another interesting feature of Blenrep is the relatively stable "non cleavable" MC linker. The MC linker is commercially available and can be attached to the payload in a single step which simplifies manufacturing (Fig. 12.20). Metabolic studies support the "non cleavable" linker description and demonstrate that free MMAF is not released from the ADC [72]. In fact, ADC metabolites containing MMAF are not released until the antibody backbone is digested by lysosomal proteases and releases Cys-MC-MMAF. Researchers originally assumed that the cysteine residue was derived from the original antibody conjugation site. Subsequent studies with radiolabeled MMAF demonstrated that the cysteine residue was actually derived from serum albumin based on sequence analysis of cysteine peptide fragments attached to the MMAF payloads [57]. Serum albumin is the most abundant protein in plasma and happens to contain a single unpaired cysteine.

Additional experiments have confirmed that incubation of ADCs containing maleimide linkers in human plasma results in thiol exchange of the maleimide linker and transfer of the toxic payload to serum albumin. Once they are transferred to albumin, the payloads are no longer tumor selective and are distributed systemically to other tissues. The reversible nature of maleimide linkers makes them prone to exchange with other thiol-containing molecules found in serum such as glutathione and cysteine. Researchers have known about maleimide instability since before the second ADC was approved, yet ADC companies continue to use them for conjugation. Even Kadcyla, which is conjugated via lysine, contains a maleimide group connecting the linker to the payload instead of the antibody. The widespread use of maleimide linkers in ADC likely contributes to the toxicity of most ADCs [50].

The first paragraph of the prescribing information for BLENREP, Belantamab mafodotin, reads, "BLENREP is a B-cell maturation antigen (BCMA) directed antibody and microtubule inhibitor (MMAF) conjugate indicated for the treatment of adult patients with relapsed or refractory multiple myeloma who have received at least 4 prior therapies including an anti-CD38 monoclonal antibody, a proteasome inhibitor, and an immunomodulatory agent." The second paragraph is a statement that Blenrep was approved under an accelerated approval based on its response rate. The next 16 pages are dedicated to describing a variety of potential side effects and adverse reactions ranging from ocular disorders to respiratory tract infections. The multiple side-effects might be considered acceptable by some patients and their

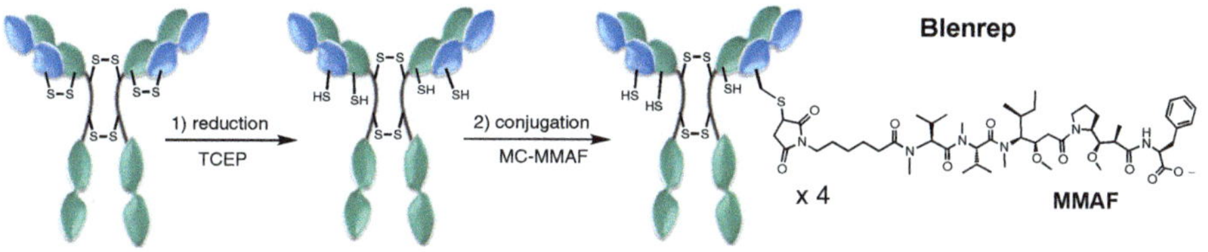

Fig. 12.20 Synthesis of Blenrep via partial reduction and conjugation to yield a heterogeneous ADC with four payloads per antibody

physicians if BLENREP demonstrated significant efficacy. The accelerated approval of Blenrep was based on a clinical trial of 97 patients with relapsed or refractory multiple myeloma (Table 12.5) [70]. The overall response rate of 31% with only one complete response was considered sufficient benefit to justify approval. Like most clinical trials for cancer, the lack of a meaningful control group makes it difficult to assess the real benefit of treatment.

12.8.10 *Zynlonta*

Zynlonta (aka loncastuximab tesirine-lpyl) was developed by ADC therapeutics and approved in April of 2021. It is a CD19-specific ADC for the treatment of patients with relapsed or refractory large B-cell lymphoma. Like most ADCs, Zynlonta was granted accelerated approval by the FDA based on a relatively small data set of 145 patients. In a clinical trial known as LOTIS-2, Zynlonta demonstrated a 48% overall response in DLBCL patients with complete responses observed in 24% of patients. These results sound impressive, however, the median response duration is only 10 months which means patients are likely to be back in the hospital in less than a year. Zynlonta is also safer than most other ADCs with a mortality rate of only 2%. As more data become available for ADCs with PBD payloads, the benefits of using Zynlonta will become clearer.

Zynlonta differs from other ADCs in that it is the first ADC that contains a PBD (pyrrolobenzodiazepine) dimer (*aka* SG3249) as its payload (Fig. 12.21). The PBD payload is an extremely toxic DNA alkylating agent that binds to the DNA minor groove and forms DNA interstrand crosslinks leading to cell death. The released PBD dimer (SG3199) has cytotoxic activity in the picomolar range against numerous tumor cell types and is the most potent cytotoxic payload used in an approved ADC. The payloads are conjugated to antibody cysteines using a protease-cleavable valine-alanine linker similar to that used for MMAE conjugation in Adcetris and

Table 12.5 Efficacy of Blenrep in DREAMM-2 clinical trial

	BLENREP $N = 97$
Overall response rate (ORR), *n* (%) (97.5% CI)	30 (31%) (21%, 43%)
Stringent complete response (sCR), *n* (%)	2 (2%)
Complete response (CR), *n* (%)	1 (1%)
Very good partial response (VGPR), *n* (%)	15 (15%)
Partial response (PR), *n* (%)	12 (12%)
Median duration of response in months[a] (range)	NR [NR to NR]

Source: Prescribing information
[a]*NR* not reached

other ADCs. A distinguishing feature of the linker used in Zynlonta is the insertion of a polyethylene glycol (PEG) spacer that extends the PBD dimer away from the antibody. The PBD payloads are very hydrophobic and have poor solubility in aqueous solution, which limits the number of payloads that can be conjugated per antibody. ADCs conjugated with PBD payloads are prone to aggregation and often precipitate during conjugation. The hydrophilic PEG spacer was designed to reduce aggregate formation and enables the conjugation of up to two PBD payloads per antibody.

12.8.11 *Tivdak*

Tivdak, (aka tisotumab vedotin-tftv) is a tissue factor (TF) directed antibody-drug conjugate for the treatment of metastatic cervical cancer. It is composed of a human anti-TF antibody conjugated to the tubulin inhibitor, MMAE. The payload is identical to that used in Adcetris, Polivy, and Padcev and is conjugated to the antibody via interchain cysteines to yield a heterogeneous mixture of ADC isoforms with an average of 4 drugs per antibody (see Fig. 12.8). Tivdak is the third ADC developed by Seagen and was approved in late 2021 making it the newest FDA-approved ADC. Approval was based on safety and efficacy data from the innovaTV 204 clinical trial with 101 patients. The results are summarized in Table 12.6.

The results indicate that Tivdak demonstrated a modest 24% overall response rate and 7% complete responses. As with other MMAE or MMAF ADCs, Tivdak can cause severe ocular toxicity leading to vision loss and corneal ulceration. Adverse ocular reactions occurred in 60% of patients treated with Tivdak, but most adverse reactions were temporary and resolved when Tivdak administration was discontinued. Other adverse side effects include peripheral neuropathy (42%), hemorrhage (62%), and pneumonitis (1.3%). Fatal adverse reactions occurred in 4% of patients and adverse reactions leading to dose interruption occurred in 47% of patients. Again, the lack of a meaningful control group for comparison raises questions about the benefits of treatment.

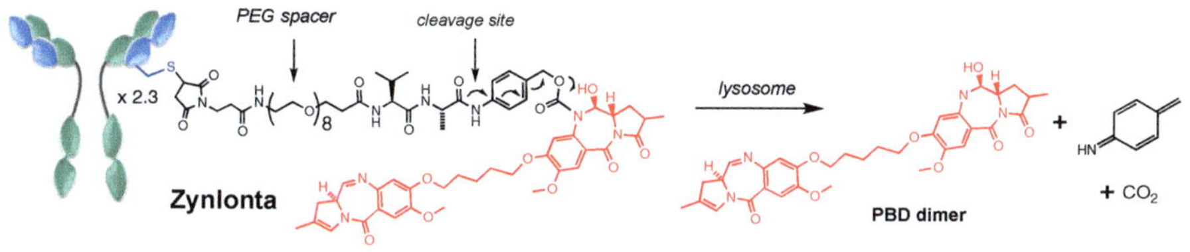

Fig. 12.21 Structure of Zynlonta and release of the PBD dimer upon lysosomal protease cleavage

Table 12.6 Efficacy results from innovaTV 204 clinical trial

Endpoint	$N = 101$
Confirmed ORR **(95% CI)**	24% (15.9, 33.3)
Complete response rate	7%
Partial response rate	17%
Duration of response	
Median duration of response, months[a] (95% CI)	8.3 (4.2, NR)

Source: Prescribing information, 2021
CI confidence interval, *NR* not reached
[a]Based on patients ($n = 24$) with a response by IRC

12.9 Concluding Remarks

The first paragraph in the clinical experience section of the prescribing information for Tivdak is a disclaimer that reads; "*Because clinical trials are conducted under widely varying conditions, adverse reaction rates observed in the clinical trials of a drug cannot be directly compared to rates in the clinical trials of another drug and may not reflect the rates observed in practice.*" This statement could be applied to most cancer drugs and highlights one of the major challenges facing the oncology field. How can we be certain that a drug is efficacious without proper negative control groups? Most doctors will respond to this question by saying that it would be unethical to give cancer patients a placebo or deny them the best possible treatment options. In reality, there are numerous patients who choose not to undergo treatment and could therefore serve as part of a negative control group. Untreated patients could be monitored for tumor progression, survival probability, and other relevant endpoints that would provide a benchmark for comparison with patients who received treatment. The problem is that patients who decline treatment are not likely to show up for appointments even if they were available. In addition, pharmaceutical companies would rather avoid the additional cost of monitoring untreated patients in clinical trials; perhaps because of the potential for unfavorable results.

The unusually high incidence of ADC side effects and the relatively poor response rates offered by most ADCs, combined with the high cost of treatment, can make it difficult to understand why so many patients (and their physicians) would opt to undergo treatment with an ADC. I have asked this question to cancer patients frequently and their answers usually convey a belief that there are not any better options available. Most patients place unquestionable trust in their doctors and seldom question their recommendations for treatment. I suspect that many patients make their decision to undergo treatment based on hope, rather than scientific data. For these patients, clinical trial data for cancer indications are meaningless because the trials involve too many variables or they are too complex for patients to fully understand. A subset of patients decide not to be treated and usually never return to the doctor's office.

Most people know someone who has survived cancer and perhaps know someone who did it without treatment. I personally know several people, including my father, who were diagnosed with cancer and chose to let the disease run its course. These patients were told by their doctors that their survival was dependent upon extensive surgery, followed by radiation and/or chemotherapy. In my father's case, he was diagnosed with stage 3 bladder cancer when he was 70 years old. His doctor advised him to have his bladder removed and undergo chemotherapy. He declined both options and lived 14 more years in relative comfort until he died of a heart attack at 84.

For cancer patients who choose to undergo chemotherapy, ADCs are a welcome alternative to conventional chemotherapy. There are still many challenges to overcome but the steady progress of the last two decades is a promising sign that their benefit will someday outweigh the risk of treatment. The ADCs of the future will carry payloads that do not need to be released and will therefore have better safety. ADCs with highly stable linkers have recently appeared which suggests payload release might not be required for some ADCs. Future ADCs will be capable of delivering multiple payloads to improve their therapeutic window and overcome resistance mechanisms. One undisputable truth about the current status of targeted therapeutics is that ADCs will continue to improve and continue to bring hope to cancer patients.

References

1. Lambert JM, Morris CQ (2017) Antibody-drug conjugates (ADCs) for personalized treatment of solid tumors: a review. Adv Ther 34(5):1015–1035. https://doi.org/10.1007/s12325-017-0519-6
2. Polakis P (2016) Antibody drug conjugates for cancer therapy. Pharmacol Rev 68(1):3–19. https://doi.org/10.1124/pr.114.009373
3. Chari RVJ (2016) Expanding the reach of antibody-drug conjugates. ACS Med Chem Lett 7(11):974–976. https://doi.org/10.1021/acsmedchemlett.6b00312
4. Carter PJ, Lazar GA (2018) Next generation antibody drugs: pursuit of the 'high-hanging fruit'. Nat Rev Drug Discov 17(3):197–223. https://doi.org/10.1038/nrd.2017.227
5. Goli N, Bolla PK, Talla V (2018) Antibody-drug conjugates (ADCs): potent biopharmaceuticals to target solid and hematological cancers – an overview. J Drug Delivery Sci Technol 48:106–117. https://doi.org/10.1016/j.jddst.2018.08.022
6. Sievers EL, Senter PD (2013) Antibody-drug conjugates in cancer therapy. Annu Rev Med 64:15–29. https://doi.org/10.1146/annurev-med-050311-201823
7. Gerber H-P, Sibener LV, Lee LJ, Gee M (2019) Intracellular targets as source for cleaner targets for the treatment of solid tumors. Biochem Pharmacol (Amsterdam, Neth) 168:275–284. https://doi.org/10.1016/j.bcp.2019.07.015
8. Ricart AD (2011) Antibody-drug conjugates of calicheamicin derivative: gemtuzumab ozogamicin and inotuzumab ozogamicin. Clin Cancer Res 17(20):6417–6427. https://doi.org/10.1158/1078-0432.ccr-11-0486
9. Khongorzul P, Ling CJ, Ullah Khan F, Ullah Ihsan A, Zhang J (2020) Antibody-drug conjugates: a comprehensive review. Mol Cancer Res 18(1):3–19. https://doi.org/10.1158/1541-7786.mcr-19-0582

10. Jackson DY (2016) Processes for constructing homogeneous antibody drug conjugates. Org Process Res Dev 20(5):852–866. https://doi.org/10.1021/acs.oprd.6b00067
11. Donaghy H (2016) Effects of antibody, drug and linker on the preclinical and clinical toxicities of antibody-drug conjugates. MAbs 8 (4):659-671:659. https://doi.org/10.1080/19420862.2016.1156829
12. Criscitiello C, Morganti S, Curigliano G (2021) Antibody-drug conjugates in solid tumors: a look into novel targets. J Hematol Oncol 14(1):20. https://doi.org/10.1186/s13045-021-01035-z
13. Sharma S, Li Z, Bussing D, Shah DK (2020) Evaluation of quantitative relationship between target expression and antibody-drug conjugate exposure inside cancer cells. Drug Metab Dispos 48(5):368–377. https://doi.org/10.1124/dmd.119.089276
14. Williams M, Spreafico A, Vashisht K, Hinrichs MJ (2020) Patient selection strategies to maximize therapeutic index of antibody-drug conjugates: prior approaches and future directions. Mol Cancer Ther 19(9):1770–1783. https://doi.org/10.1158/1535-7163.mct-19-0993
15. Singh SK, Luisi DL, Pak RH (2015) Antibody-drug conjugates: design, formulation and physicochemical stability. Pharm Res 32(11):3541–3571. https://doi.org/10.1007/s11095-015-1704-4
16. Hock MB, Thudium KE, Carrasco-Triguero M, Schwabe NF (2015) Immunogenicity of antibody drug conjugates: bioanalytical methods and monitoring strategy for a novel therapeutic modality. AAPS J 17(1):35–43. https://doi.org/10.1208/s12248-014-9684-6
17. Wang H, Wang Y, Xiao Z, Li W, Dimitrov DS, Chen W (2019) Human domain antibodies to conserved epitopes on HER2 potently inhibit growth of HER2-overexpressing human breast cancer cells in vitro. Antibodies 8(1):25. https://doi.org/10.3390/antib8010025
18. Li C, Zhang C, Li Z, Samineni D, Lu D, Wang B, Chen S-C, Zhang R, Agarwal P, Fine BM, Girish S (2020) Clinical pharmacology of vc-MMAE antibody-drug conjugates in cancer patients: learning from eight first-in-human Phase 1 studies. MAbs 12(1):1699768/1699761. https://doi.org/10.1080/19420862.2019.1699768
19. Qu Z, Lyu J, Liu Y, Wang X, Lin Z, Yang Y, Zhang D, Geng X, Li B (2020) Tissue cross-reactivity studies of CPGJ701 in humans, cynomolgus monkeys and Sprague-Dawley rats and correlation analysis with in vivo toxicity. Ann Transl Med 8 (6):325. doi:https://doi.org/10.21037/atm.2020.02.106
20. Doronina SO, Mendelsohn BA, Bovee TD, Cerveny CG, Alley SC, Meyer DL, Oflazoglu E, Toki BE, Sanderson RJ, Zabinski RF, Wahl AF, Senter PD (2006) Enhanced activity of monomethylauristatin F through monoclonal antibody delivery: effects of linker technology on efficacy and toxicity. Bioconjug Chem 17(1):114–124. https://doi.org/10.1021/bc0502917
21. Doronina SO, Bovee TD, Meyer DW, Miyamoto JB, Anderson ME, Morris-Tilden CA, Senter PD (2008) Novel peptide linkers for highly potent antibody-auristatin conjugate. Bioconjug Chem 19(10):1960–1963. https://doi.org/10.1021/bc800289a
22. Khera E, Thurber GM (2018) Pharmacokinetic and immunological considerations for expanding the therapeutic window of next-generation antibody-drug conjugates. BioDrugs 32(5):465–480. https://doi.org/10.1007/s40259-018-0302-5
23. Ricci SM, Angelo De Claro R, Simpson NE (2017) Path to market approval: regulatory perspective of ADC nonclinical safety assessments. Wiley, p 267
24. Kaplon H, Reichert JM (2019) Antibodies to watch in 2019. MAbs 11(2):219–238. https://doi.org/10.1080/19420862.2018.1556465
25. Girish S, Lewis Phillips GD, Jacobson FS, Junutula JR, Guardino E (2015) Antibody-drug conjugates: design and development of trastuzumab emtansine (T-DM1). Wiley, p 213
26. Zhou Q (2017) Site-specific antibody conjugation for ADC and beyond. Biomedicine 5(4):64. https://doi.org/10.3390/biomedicines5040064
27. Conilh L, Fournet G, Fourmaux E, Murcia A, Matera E-L, Joseph B, Dumontet C, Viricel W (2021) Exatecan antibody drug conjugates based on a hydrophilic polysarcosine drug-linker platform. Pharmaceuticals (Basel) 14(3). https://doi.org/10.3390/ph14030247

28. Seligson JM, Patron AM, Berger MJ, Harvey RD, Seligson ND (2020) Sacituzumab govitecan-hziy: an antibody-drug conjugate for the treatment of refractory, metastatic, triple-negative breast cancer. Ann Pharmacother 55:921. https://doi.org/10.1177/1060028020966548
29. Behrens CR, Ha EH, Chinn LL, Bowers S, Probst G, Fitch-Bruhns M, Monteon J, Valdiosera A, Bermudez A, Liao-Chan S, Wong T, Melnick J, Theunissen J-W, Flory MR, Houser D, Venstrom K, Levashova Z, Sauer P, Migone T-S, van der Horst EH, Halcomb RL, Jackson DY (2015) Antibody-drug conjugates (ADCs) derived from interchain cysteine cross-linking demonstrate improved homogeneity and other pharmacological properties over conventional heterogeneous ADCs. Mol Pharm 12(11):3986–3998. https://doi.org/10.1021/acs.molpharmaceut.5b00432
30. Forte N, Chudasama V, Baker JR (2018) Homogeneous antibody-drug conjugates via site-selective disulfide bridging. Drug Discov Today Technol 30:11–20. https://doi.org/10.1016/j.ddtec.2018.09.004
31. Abdollahpour-Alitappeh M, Lotfinia M, Gharibi T, Mardaneh J, Farhadihosseinabadi B, Larki P, Faghfourian B, Sepehr KS, Abbaszadeh-Goudarzi K, Abbaszadeh-Goudarzi G, Johari B, Zali MR, Bagheri N (2019) Antibody-drug conjugates (ADCs) for cancer therapy: Strategies, challenges, and successes. J Cell Physiol 234(5):5628–5642. https://doi.org/10.1002/jcp.27419
32. Norsworthy KJ, Ko C-W, Lee JE, Liu J, John CS, Przepiorka D, Farrell AT, Pazdur R (2018) FDA approval summary: Mylotarg for treatment of patients with relapsed or refractory CD33-positive acute myeloid leukemia. Oncologist 23(9):1103–1108. https://doi.org/10.1634/theoncologist.2017-0604
33. Al-Salama ZT (2018) Inotuzumab ozogamicin: a review in relapsed/refractory B-cell acute lymphoblastic leukaemia. Target Oncol 13(4):525–532. https://doi.org/10.1007/s11523-018-0584-z
34. Gravanis I, Tzogani K, Van Hennik P, De Graeff P, Schmitt P, Mueller-Berghaus J, Salmonson T, Gisselbrecht C, Laane E, Bergmann L, Pignati F (2016) The European medicines agency review of Brentuximab vedotin (Adcetris) for the treatment of adult patients with relapsed or refractory CD30+ Hodgkin lymphoma or systemic anaplastic large cell lymphoma: summary of the scientific assessment of the committee for medicinal products for human use. Oncologist 21(1):102–109. https://doi.org/10.1634/theoncologist.2015-0276
35. Yip V, Lee MV, Saad OM, Ma S, Khojasteh SC, Shen B-Q (2021) Preclinical characterization of the distribution, catabolism, and elimination of a Polatuzumab Vedotin-Piiq (POLIVY®) antibody-drug conjugate in Sprague Dawley rats. J Clin Med 10(6). https://doi.org/10.3390/jcm10061323
36. Halford Z, Anderson MK, Clark MD (2020) Enfortumab vedotin-ejfv: a first-in-class anti-nectin-4 antibody-drug conjugate for the management of urothelial carcinoma. Ann Pharmacother 55:772. https://doi.org/10.1177/1060028020960402
37. Peipp M, Gramatzki M (2014) Calicheamicin conjugates: gemtuzumab ozogamicin (Mylotarg), inotuzumab ozogamicin. Wiley, p 1545
38. Jen EY, Ko C-W, Lee JE, Del Valle PL, Aydanian A, Jewell C, Norsworthy KJ, Przepiorka D, Nie L, Liu J, Sheth CM, Shapiro M, Farrell AT, Pazdur R (2018) FDA approval: gemtuzumab ozogamicin for the treatment of adults with newly diagnosed CD33-positive acute myeloid leukemia. Clin Cancer Res 24(14):3242–3246. https://doi.org/10.1158/1078-0432.ccr-17-3179
39. Minich SS (2012) Brentuximab vedotin: a new age in the treatment of Hodgkin lymphoma and anaplastic large cell lymphoma. Ann Pharmacother 46:3. https://doi.org/10.1345/aph.1q680
40. Senter PD, Sievers EL (2012) The discovery and development of brentuximab vedotin for use in relapsed Hodgkin lymphoma and systemic anaplastic large cell lymphoma. Nat Biotechnol 30(7):631–637. https://doi.org/10.1038/nbt.2289
41. Hamblett KJ, Senter PD, Chace DF, Sun MMC, Lenox J, Cerveny CG, Kissler KM, Bernhardt SX, Kopcha AK, Zabinski RF, Meyer DL, Francisco JA (2004) Effects of drug loading on the antitumor activity of a monoclonal antibody drug conjugate. Clin Cancer Res 10(20):7063–7070. https://doi.org/10.1158/1078-0432.ccr-04-0789

42. Brown MP, Staudacher AH (2014) Could bystander killing contribute significantly to the antitumor activity of brentuximab vedotin given with standard first-line chemotherapy for Hodgkin lymphoma? Immunotherapy 6(4):371–375. https://doi.org/10.2217/imt.14.13
43. Francisco JA, Cerveny CG, Meyer DL, Mixan BJ, Klussman K, Chace DF, Rejniak SX, Gordon KA, DeBlanc R, Toki BE, Law C-L, Doronina SO, Siegall CB, Senter PD, Wahl AF (2003) cAC10-vcMMAE, an anti-CD30-monomethyl auristatin E conjugate with potent and selective antitumor activity. Blood 102(4):1458–1465. https://doi.org/10.1182/blood-2003-01-0039
44. Ballantyne A, Dhillon S (2013) Trastuzumab emtansine: first global approval. Drugs 73(7):755–765. https://doi.org/10.1007/s40265-013-0050-2
45. Lambert JM, Chari RVJ (2014) Ado-trastuzumab Emtansine (T-DM1): an antibody-drug conjugate (ADC) for HER2-positive breast cancer. J Med Chem 57(16):6949–6964. https://doi.org/10.1021/jm500766w
46. Wakankar AA, Feeney MB, Rivera J, Chen Y, Kim M, Sharma VK, Wang YJ (2010) Physicochemical stability of the antibody-drug conjugate trastuzumab-DM1: changes due to modification and conjugation processes. Bioconjug Chem 21(9):1588–1595. https://doi.org/10.1021/bc900434c
47. Burris HA III (2011) Trastuzumab emtansine: a novel antibody-drug conjugate for HER2-positive breast cancer. Expert Opin Biol Ther 11(6):807–819. https://doi.org/10.1517/14712598.2011.580273
48. Burris HA (2012) Trastuzumab emtansine (T-DM1): hitching a ride on a therapeutic antibody. Am Soc Clin Oncol Educ Book:159–161. https://doi.org/10.14694/EdBook_AM.2012.32.109
49. Gupta M, LoRusso PM, Wang B, Yi J-H, Burris HA, Beeram M, Modi S, Chu Y-W, Agresta S, Klencke B, Joshi A, Girish S (2012) Clinical implications of pathophysiological and demographic covariates on the population pharmacokinetics of trastuzumab emtansine, a HER2-targeted antibody-drug conjugate, in patients with HER2-positive metastatic breast cancer. J Clin Pharmacol 52(5):691–703. https://doi.org/10.1177/0091270011403742
50. Erickson HK, Lambert JM (2012) ADME of antibody-maytansinoid conjugates. AAPS J 14(4):799–805. https://doi.org/10.1208/s12248-012-9386-x
51. Shen B-Q, Bumbaca D, Saad O, Yue Q, Pastuskovas CV, Khojasteh SC, Tibbitts J, Kaur S, Wang B, Chu Y-W, LoRusso PM, Girish S (2012) Catabolic fate and pharmacokinetic characterization of trastuzumab emtansine (T-DM1): an emphasis on preclinical and clinical catabolism. Curr Drug Metab 13(7):901–910. https://doi.org/10.2174/138920012802138598
52. Dere R, Yi J-H, Lei C, Saad OM, Huang C, Li Y, Baudys J, Kaur S (2013) PK assays for antibody-drug conjugates: case study with ado-trastuzumab emtansine. Bioanalysis 5(9):1025–1040. https://doi.org/10.4155/bio.13.72
53. Barginear MF, John V, Budman DR (2012) Trastuzumab-DM1: a clinical update of the novel antibody-drug conjugate for HER2-overexpressing breast cancer. Mol Med 18(11):1473–1479. https://doi.org/10.2119/molmed.2012.00302
54. Dushin RG (2019) Calicheamicins as antibody-drug conjugate (ADC) payloads. RSC Drug Discov Ser 71:259–278
55. Deeks ED (2019) Polatuzumab vedotin: first global approval. Drugs 79(13):1467–1475. https://doi.org/10.1007/s40265-019-01175-0
56. Stefano JE, Busch M, Hou L, Park A, Gianolio DA (2013) Micro- and mid-scale maleimide-based conjugation of cytotoxic drugs to antibody hinge region thiols for tumor targeting. Methods Mol Biol 1045:145–171. https://doi.org/10.1007/978-1-62703-541-5_9
57. Alley SC, Benjamin DR, Jeffrey SC, Okeley NM, Meyer DL, Sanderson RJ, Senter PD (2008) Contribution of linker stability to the activities of anticancer immunoconjugates. Bioconjug Chem 19(3):759–765. https://doi.org/10.1021/bc7004329
58. Alley SC, Benjamin D, Law C-L (2009) Antibody-drug conjugate therapy. Wiley, p 821
59. Toda N, Asano S, Barbas CF III (2013) Rapid, stable, chemoselective labeling of thiols with Julia-Kocienski-like reagents: a serum-stable alternative to maleimide-based protein conjugation. Angew Chem Int Ed 52(48):12592–12596. https://doi.org/10.1002/anie.201306241

60. McGregor BA, Sonpavde G (2019) Enfortumab Vedotin, a fully human monoclonal antibody against Nectin 4 conjugated to monomethyl auristatin E for metastatic urothelial Carcinoma. Expert Opin Investig Drugs 28(10):821–826. https://doi.org/10.1080/13543784.2019.1667332
61. Hanna KS (2020) Clinical overview of enfortumab vedotin in the management of locally advanced or metastatic urothelial carcinoma. Drugs 80(1):1–7. https://doi.org/10.1007/s40265-019-01241-7
62. Yver A, Agatsuma T, Soria JC (2020) The art of innovation: clinical development of trastuzumab deruxtecan and redefining how antibody-drug conjugates target HER2-positive cancers. Ann Oncol 31(3):430–434. https://doi.org/10.1016/j.annonc.2019.11.019
63. Nagai Y, Oitate M, Shiozawa H, Ando O (2019) Comprehensive preclinical pharmacokinetic evaluations of trastuzumab deruxtecan (DS-8201a), a HER2-targeting antibody-drug conjugate, in cynomolgus monkeys. Xenobiotica 49(9):1086–1096. https://doi.org/10.1080/00498254.2018.1531158
64. Hashimoto Y, Koyama K, Kamai Y, Hirotani K, Ogitani Y, Zembutsu A, Abe M, Kaneda Y, Maeda N, Shiose Y, Iguchi T, Ishizaka T, Karibe T, Hayakawa I, Morita K, Nakada T, Nomura T, Wakita K, Kagari T, Abe Y, Murakami M, Ueno S, Agatsuma T (2019) A novel HER3-targeting antibody-drug conjugate, U3-1402, exhibits potent therapeutic efficacy through the delivery of cytotoxic payload by efficient internalization. Clin Cancer Res 25(23):7151–7161. https://doi.org/10.1158/1078-0432.ccr-19-1745
65. Takegawa N, Nonagase Y, Yonesaka K, Sakai K, Maenishi O, Ogitani Y, Tamura T, Nishio K, Nakagawa K, Tsurutani J (2017) DS-8201a, a new HER2-targeting antibody-drug conjugate incorporating a novel DNA topoisomerase I inhibitor, overcomes HER2-positive gastric cancer T-DM1 resistance. Int J Cancer 141(8):1682–1689. https://doi.org/10.1002/ijc.30870
66. Kotani D, Shitara K (2021) Trastuzumab deruxtecan for the treatment of patients with HER2-positive gastric cancer. Ther Adv Med Oncol 13:1758835920986518. https://doi.org/10.1177/1758835920986518
67. Kalinsky K, Diamond JR, Vahdat LT, Tolaney SM, Juric D, O'Shaughnessy J, Moroose RL, Mayer IA, Abramson VG, Goldenberg DM, Sharkey RM, Maliakal P, Hong Q, Goswami T, Wegener WA, Bardia A (2020) Sacituzumab govitecan in previously treated hormone receptor-positive/HER2-negative metastatic breast cancer: final results from a phase I/II, single-arm, basket trial. Ann Oncol 31(12):1709–1718. https://doi.org/10.1016/j.annonc.2020.09.004
68. Faltas B, Goldenberg DM, Ocean AJ, Govindan SV, Wilhelm F, Sharkey RM, Hajdenberg J, Hodes G, Nanus DM, Tagawa ST (2016) Sacituzumab govitecan, a novel antibody--drug conjugate, in patients with metastatic platinum-resistant urothelial carcinoma. Clin Genitourin Cancer 14(1):e75–e79. https://doi.org/10.1016/j.clgc.2015.10.002
69. Bardia A, Mayer IA, Diamond JR, Moroose RL, Isakoff SJ, Starodub AN, Shah NC, O'Shaughnessy J, Kalinsky K, Guarino M, Abramson V, Juric D, Tolaney SM, Berlin J, Messersmith WA, Ocean AJ, Wegener WA, Maliakal P, Sharkey RM, Govindan SV, Goldenberg DM, Vahdat LT (2017) Efficacy and safety of anti-trop-2 antibody drug conjugate sacituzumab govitecan (IMMU-132) in heavily pretreated patients with metastatic triple-negative breast cancer. J Clin Oncol 35(19):2141–2150. https://doi.org/10.1200/jco.2016.70.8297
70. Markham A (2020) Belantamab mafodotin: first approval. Drugs 80(15):1607–1613. https://doi.org/10.1007/s40265-020-01404-x
71. Waight AB, Bargsten K, Doronina S, Steinmetz MO, Sussman D, Prota AE (2016) Structural basis of microtubule destabilization by potent auristatin anti-mitotics. PLoS One 11(8):e0160890. https://doi.org/10.1371/journal.pone.0160890
72. Alley SC, Zhang X, Okeley NM, Anderson M, Law C-L, Senter PD, Benjamin DR (2009) The pharmacologic basis for antibody-auristatin conjugate activity. J Pharmacol Exp Ther 330(3):932–938. https://doi.org/10.1124/jpet.109.155549

Chapter 13
Mylotarg: The Journey to FDA Reapproval and Broad International Approval

Leo Letendre, Durgesh Nadkarni, and Frank Kotch

Abstract Mylotarg (gemtuzumab ozogamicin) was the first cytotoxic antibody-drug conjugate approved by the FDA. After its approval, it was withdrawn from the US market in 2010 although it continued to be marketed in Japan. As new clinical data were amassed, an effort to have it reapproved by the FDA and approved in other major markets was initiated. The efforts to bring the Chemistry, Manufacturing, and Controls portion of the regulatory filings up to contemporary standards while still maintaining the original product profile provide an interesting journey through the advances in chemistry, analysis, and regulatory science since the late twentieth century.

Keywords Mylotarg · Calicheamicin · Gemtuzumab ozogamicin · Registration · Amino acid substitution · Antibody-drug conjugate

13.1 Introduction

Mylotarg® (gemtuzumab ozogamicin) is an antibody-drug conjugate (ADC) [1, 2] used to treat CD-33 positive acute myeloid leukemia (AML) in adults and pediatric patients [3]. Gemtuzumab ozogamicin consists of a recombinant humanized

Dedication: This chapter is dedicated to the memory of Tok Han who skillfully shepherded both the reintroduction of Mylotarg and the introduction of Besponsa through technology transfer, manufacturing, and regulatory inspections driven by his empathy toward the patients who would benefit from these medicines. His loss at a young age was felt by all who worked with him.

L. Letendre (✉)
Leo Letendre Consulting, LLC, Oakdale, CT, USA

D. Nadkarni · F. Kotch
Bioprocess Research & Development, Biotherapeutics Pharmaceutical Sciences, Pfizer Inc., Chesterfield, MO, USA

K. Gadamasetti, S. A. Kolodziej (eds.), *Bioprocessing, Bioengineering and Process Chemistry in the Biopharmaceutical Industry*,
https://doi.org/10.1007/978-3-031-62007-2_13

anti-CD33 IgG4 antibody (hP67.6) covalently linked to the cytotoxic agent N-acetyl calicheamicin [4]. Mylotarg originally received accelerated approval from the US FDA in 2000 and later received Japanese approval in 2005. With the original accelerated approval from the FDA came a requirement for the completion of a confirmatory clinical trial. In 2010 Pfizer voluntarily withdrew Mylotarg from the US market when the confirmatory clinical trial failed to demonstrate its clinical benefit [5]. The product was not withdrawn from the Japanese market at the request of the Japanese authorities.

During this initial period of marketing as well as after the US withdrawal, independent researchers continued to evaluate Mylotarg for the treatment of AML. These studies focused on the use of Mylotarg with other anticancer agents as well as variations in the dosing regimens. The resulting phase 3 ALFA-0701 trial, which became the registrational trial, concluded that a "fractionated" Mylotarg dosing regimen in conjunction with daunorubicin and cytarabine provided an overall benefit [6, 7]. The fractionated dosing regimen consists of splitting the typical original dose of 9 mg/m^3 into three smaller doses over 1 week followed by lower doses during continuation therapy. This dosing regimen led to a decrease in side effects while maintaining efficacy [6]. In September 2017, the FDA reapproved Mylotarg for the US market after a positive recommendation by the Oncologic Drugs Advisory Committee (ODAC). Subsequently, Mylotarg has been approved in most major markets.

With the decision to pursue reapproval, an assessment of the state of the original Chemistry, Manufacturing and Controls (CMC) section of the original application was conducted. Mylotarg was the first antibody-drug conjugate to be approved by the FDA [1]. The application, therefore, was breaking new ground for both Wyeth/Pfizer (the applicant) and the agency. One note of the difference between today's antibody-drug conjugate filings and Mylotarg's is that the original application was filed as a New Drug Application (NDA) while current applications are in the form of a Biological License Application (BLA). While CDER was responsible for the review of both the NDA and the BLA, the additional requirements for approval of a BLA were followed. During the intervening years much had changed in the regulatory expectations for new drugs and biologics as analytical methods, in particular, and medical knowledge have progressed. The first question to be addressed after the decision to seek reapproval was whether to go forward with the existing CMC development and characterization data or to attempt to bring it up to contemporary standards. (Drug substance and drug product historical data, all 17 or so years' worth, was always assumed to be included.) It was in this context that the assessment of our knowledge about Mylotarg and the gaps between what was acceptable for the initial filing and that for the refiling was conducted. In the end, the decision was made to supplement the information within the original application with increased knowledge of the manufacturing process in order to contemporize and make it consistent with BLA expectations.

13.2 Gemtuzumab Ozogamicin

Gemtuzumab ozogamicin (the Mylotarg drug substance) consists of a recombinant humanized IgG4 antibody linked to the modified cytotoxic natural product γ-calicheamicin using an acid-labile linker [4].· The structure of gemtuzumab ozogamicin is shown in Fig. 13.1. γ-Calicheamicin is present in the form of its N-acetyl derivative, which provides some additional stability to that portion of the molecule. The antibody is linked to the N-acetyl calicheamicin through a linker, which attaches to the antibody through several of its lysines and to the N-acetyl calicheamicin through a disulfide moiety. The hydrazide portion of the linker provides acid lability allowing the antibody drug conjugate to release the calicheamicin payload in the acidic environment of the lysosome once it is internalized [8].

As Mylotarg was developed during the early days of seeking modified antibody drugs, it was designed using the simplest chemistry available for connecting the payload to the antibody, amide linkage through the lysine side-chain amines. Since the gemtuzumab antibody contains 44 lysines, a kinetically controlled distribution of the linker attachment is expected. Indeed, the attachment of the linker payload is predominated by four lysine sites [9]. However, the distribution of linker payload is still more complicated than a simple explanation of kinetic control would describe. With an average drug-to-antibody ratio of approximately 2.5, the expectation would be that typically two or three calicheamicin moieties would be attached to each antibody with a normal distribution of species with higher and lower loading. What has been observed, however, is a bimodal distribution; approximately half of the antibodies bear twice the average loading while the other half are unconjugated or have very low loading [9]. An explanation for this result still awaits convincing experimental results.

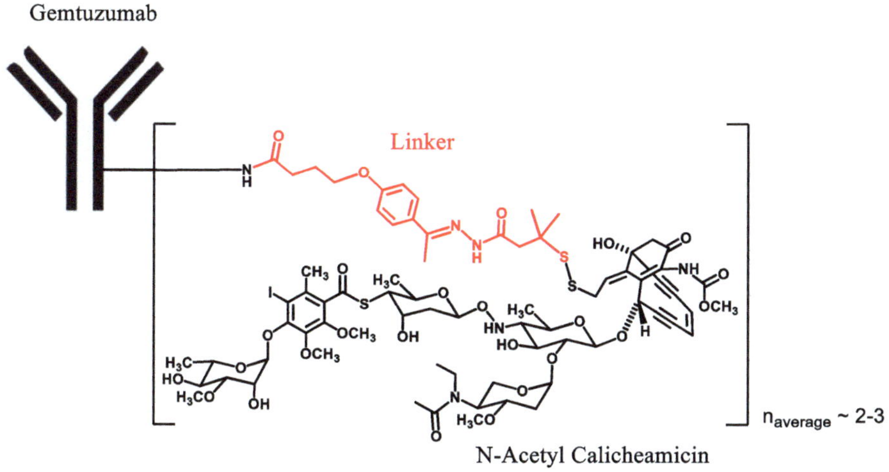

Fig. 13.1 Gemtuzumab ozogamicin (Mylotarg)

13.3 Gemtuzumab Antibody

The gemtuzumab antibody (also referred to as hP67.6) is a humanized murine anti-CD33 IgG4 monoclonal antibody. The antibody sequence comprises a murine complementarity-determining region (CDR) grafted onto a human IgG4 kappa framework [4, 10] to avoid undesired immune response. Gemtuzumab does not have either antibody-dependent cell-mediated cytotoxicity (ADCC) or complement dependent cytotoxicity activity. Additionally, the antibody's effector function was shown to not be part of the mode of action.

13.4 Calicheamicin

γ-Calicheamicin (also known as calicheamicin γ_1^{I}) is an enediyne-containing antibiotic produced by the fermentation of *Micromonospora echinospora* ssp. *calichensis*, which was isolated from a sample of caliche soil from Kerrville, Texas [11, 12]. γ-Calicheamicin is one of several enediyne antibiotics produced by this organism (Fig. 13.2). Other calicheamicins include δ-calicheamicin and β-calicheamicin, both active compounds, and the inactive ε-calicheamicin. The first isolated calicheamicin compounds were brominated on the aromatic ring, including calicheamicin γ_1^{Br}. During optimization of the fermentation conditions, it was found that supplementing the media with sodium iodide resulted in a significant increase in yield and afforded the iodinated derivative, calicheamicin γ_1^{I}, as the major product and core structure for linker-payload development [12].

γ-Calicheamicin is attached to the amine of lysine residues on the gemtuzumab antibody as an N-acetylated derivative through an acid-labile linker. The linker was designed to be hydrolyzed in the lysosome [8] (low pH) once the antibody-drug conjugate has been internalized, releasing the toxin from the antibody as N-acetyl calicheamicin DMH [4] (Fig. 13.3). Mediated by the aromatic containing sugar portion of the molecule ("Glycan" in Fig. 13.3), calicheamicin binds to specific sites within the minor groove of DNA [13, 14]. Cleavage of the disulfide bond leads to an intramolecular hetero-Michael addition followed by a Bergman cyclization [15] to

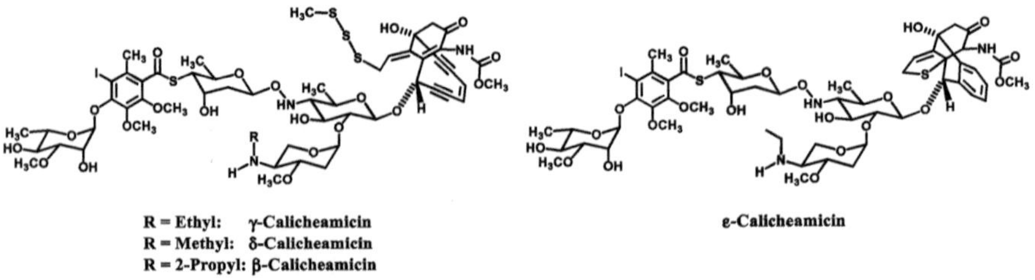

Fig. 13.2 Major calicheamicins produced by *M. echinospora* ssp. calichensis fermentation

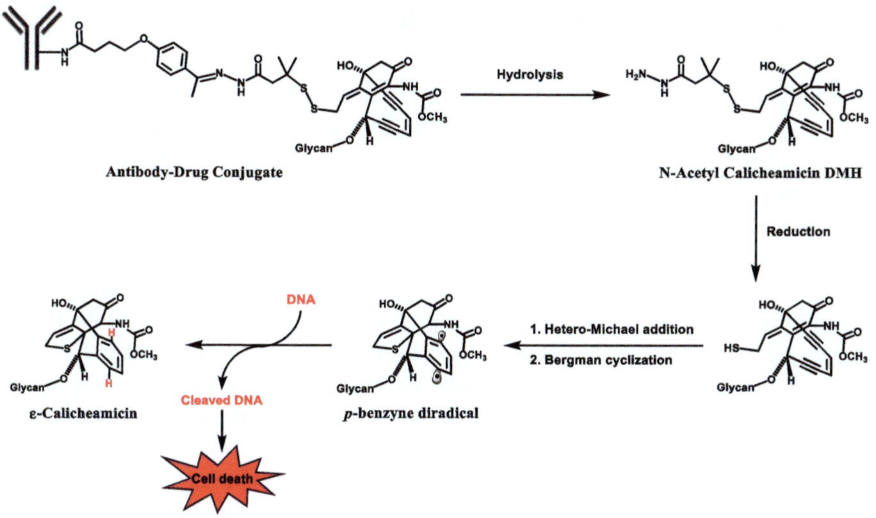

Fig. 13.3 Mode of action of calicheamicin-based ADCs

give a *p*-benzyne diradical. When bound to DNA, the *p*-benzyne diradical causes double-strand cleavage ultimately leading to cell death [13] (Fig. 13.3).

13.5 Assessing the Regulatory and Commercial Needs

An analysis of the state of knowledge about the production of Mylotarg identified a number of areas where the knowledge obtained for the initial approval and current expectations was different. Additionally, even though Mylotarg was withdrawn from the US market, it has remained in the Japanese market since its approval in 2005. Therefore, unlike developing a drug de novo, any redefinition of the processes used to manufacture Mylotarg based upon enhanced understanding would have to maintain comparability with the product produced over 17 years of commercial manufacture.

The primary commercial need was identified as a capacity issue in the production of the linker payload, activated calicheamicin derivative. Two factors were important. First, the yield of the linker payload from a contract manufacturing organization had decreased over the history of the production to an unsatisfactory level. The second factor complicating this issue was that the imminent introduction of another ADC, Besponsa® (inotuzumab ozogamicin), which contained the same linker payload, would significantly increase demand for activated calicheamicin derivative [16]. Clearly, an improved activated calicheamicin derivative supply chain was necessary.

13.6 Production of Gemtuzumab Antibody

The gemtuzumab antibody was originally developed by Celltech Ltd. [10]. The antibody was produced from mammalian myeloma NS0 cell lines and then purified. In planning to re-register Mylotarg, it was anticipated that the production of the gemtuzumab antibody would not require extensive work but that the characterization and release methods would need to be brought to contemporary standards. In the process of scrutinizing the characterization data, a discrepancy in a protein digest indicated that the antibody had contained an amino acid substitution (AAS) at multiple sites. Further analysis indicated that the AAS was present in the initially registered product at low levels and that the level had increased significantly at multiple sites in the contemporary antibody. After determining that the issue was not one of clonal purity, efforts were undertaken to determine both the cause of the substitution and the effects upon efficacy.

Comparison of retained antibody samples with contemporary samples indicated that the amino acid substitution had always been present but that its level had significantly increased and had done so in a stepwise manner [17]. The substitution occurred on both the light and heavy chains including the CDR. Eventually, the root cause of the amino acid substitution was traced to a change in the supplier of a complex raw material used in the gemtuzumab cell culture. Given the nature of this raw material, it was not possible to return to using the original raw material source. This discovery led to an extensive re-investigation of the effects of various components used in the production of gemtuzumab as well as the process parameters controlling the cell culture. Eventually, a control strategy was established to produce the antibody in a robust manner with the amino acid substitution level comparable to that obtained prior to the change in the raw material supplier.

The discovery of amino acid substitution in gemtuzumab was not without its regulatory impact and required additional quality data to be submitted to address concerns raised by regulatory authorities. The concerns raised related to the AAS levels of the materials used in the clinical studies, the control strategy, the potential impact of the AAS on functional activity on Mylotarg, and the overall comparability throughout development. The clinical basis for returning Mylotarg to the world market was largely dependent upon the results of numerous studies run by independent investigators which, in aggregate, took a number of years to conduct. Over this time period, most of the Mylotarg used in these studies did not contain elevated amino acid substitution. Thus, the European Medicines Agency, for example, found that the amino acid substitution led to multiple major deficiencies [17] including joint deficiencies within the Quality and Clinical sections of the application to confirm the comparability of the clinical material with the proposed commercial material.

In order to address these deficiencies, the development team pursued a number of additional studies that further confirmed that the amino acid substitutions did not affect the efficacy of Mylotarg. The team demonstrated that no new sites of conjugation to calicheamicin were introduced with the amino acid substitution and the antibody glycosylation was found to be comparable before and after the amino acid substitution increased. Biophysical and biological activity was also addressed. The level of cytotoxicity before and after increased substitution was demonstrated to be comparable and, supporting this conclusion, binding data were also found to be comparable for both CD33 and FcRn. Taken together, this provided a strong data package for submission to regulatory agencies, which supported the conclusion that, while the amino acid substitution affected the composition of Mylotarg, conjugates with variable amino acid substitution were functionally equivalent and the clinical results were acceptable.

The regulatory response to the amino acid substitution and its correction was varied. With Pfizer submitting the data discussed above, several agencies next sought to understand how quickly a change in the antibody manufacturing conditions could be implemented. As part of the review, Pfizer indicated that process improvements were already in development, but that it would take longer than the regulatory review process to produce antibody under the restorative conditions. Based upon the overall comparability data, the proposed control strategy to affect remediation, and the realities of commercial antibody production, agencies approved the application with the understanding that Pfizer would subsequently introduce the restored antibody. Post approval, an improved gemtuzumab process which resulted in reduced amino acid substitution was implemented and approved. As part of this submission, Pfizer communicated its intent to continue to monitor the level of amino acid substitution as part of the control strategy.

13.7 Production of the Calicheamicin Linker Payload

A number of forces were in play when the decision was taken to revisit the production of the linker payload, activated calicheamicin derivative. With the desire to return Mylotarg to the US market, and expansion to the European Union and Canada as well as the rest of the world, regulatory success would increase the demand significantly. Additionally, success in the registration of Besponsa would even further increase the demand. As mentioned earlier, the original preparation was conducted by a third party. Pfizer determined that the third party would not be able to meet the anticipated combined demand and that the preparation of activated calicheamicin derivative needed to be revisited. Lastly, Pfizer had built a facility in anticipation of bringing the manufacturing in-house and any new preparation would be run in this new facility. The new facility was constructed with a series of large isolators to protect workers from exposure to the highly toxic material [18] which has an occupational exposure limit (OEL) in the tens of ng/m^3. While the isolators were relatively large, any new equipment would need to be compatible with the floor and bench space available.

Activated calicheamicin derivative has been manufactured by three different production routes. The first and last, (Processes 1 and 3 respectively) have been used in the preparation of gemtuzumab ozogamicin. Process 2 was practiced during the development of Besponsa and was being considered for use with the re-launched Mylotarg in addition to the commercial production of Besponsa. The comparison of these routes follows.

13.7.1 Isolation of γ-Calicheamicin

The preparation of the activated calicheamicin derivate begins with fermentation of *M. echinospora* ssp. *calichensis*. A mixture of calicheamicin-related compounds is found in the fermentation media and is isolated by a multi-step process. The fermentation and isolation could not be changed due to the limited availability of alternate equipment in the dedicated Pfizer facility. Also, the existing process was quite efficient and alleviated any push to do so. Thus, no changes were instituted. γ-Calicheamicin was delivered from the fermentation as an oil consisting of the various calicheamicin-related compounds and fermentation materials largely consisting of antifoam [19].

The isolation of γ-calicheamicin from the fermentation isolate differed slightly from Process 1 to Process 2 (Fig. 13.4). The desired calicheamicin was obtained largely free of the fermentation-related byproducts by precipitation from a large volume of di-isopropyl ether, a solvent we wished to remove for safety reasons. Process 1 followed the precipitation with a normal phase chromatography step to further purify the fermentation product. While this chromatography step may have removed some calicheamicin-related products, it did not remove the major ones (Fig. 13.2). Process 2 did not include the chromatography and thus indicated that

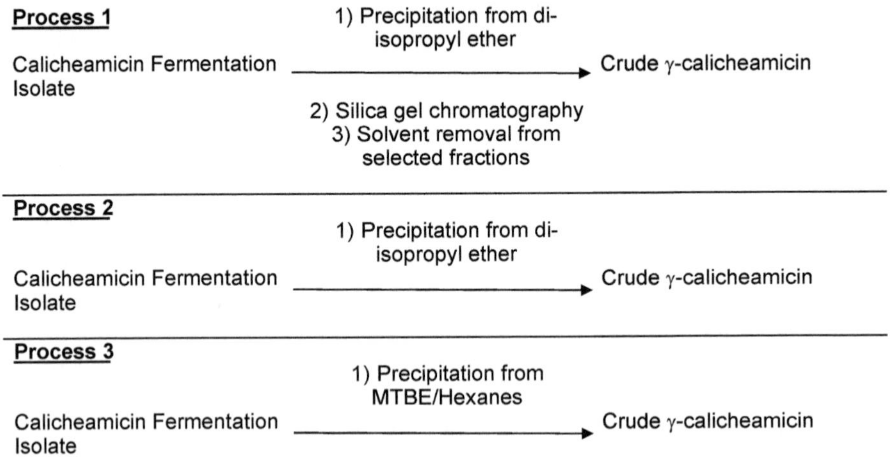

Fig. 13.4 Isolation of γ-calicheamicin

downstream processes could remove the calicheamicin-related impurities. With the major goal of removing di-isopropyl ether in mind, alternative precipitation solvents were investigated. A mixture of methyl *tert*-butyl ether (MTBE) and hexanes was found to give equivalent or better recovery of γ-calicheamicin when compared to the original di-isopropyl ether and was therefore utilized in Process 3.

13.7.2 Formation of N-Acetyl Calicheamicin

The first chemical step in the preparation of activated calicheamicin derivative is the acetylation of the ethyl amino sugar (Fig. 13.5). The challenge in adding an acetyl group to the secondary amine is to prevent over-acetylation by reaction with the five alcohols and the alkoxyamine. During structure elucidation of the calicheamicins, Lee et al. demonstrated that conducting the acetylation in methanol allowed for highly selective acetylation on nitrogen [20]. Processes 1 and 2 essentially used the method of Lee by adding acetic anhydride to a methanol solution containing triethylamine. The methanol effectively competes with the hydroxy groups on the sugars; however, running the reaction in methanol required a large excess of both acetic anhydride and triethylamine. On a production scale, this led to issues controlling significant exotherms during triethylamine addition and the aqueous quench due to heat generated by methanol-water mixing. Both triethylamine addition and water quenching had to be done very slowly to maintain the temperature and minimize the degradation of the product. The Process 3 development team recognized the need to reduce reagent amounts, prevent over-acetylation of γ-calicheamicin, and reduce or avoid the exotherms. Acetonitrile was chosen as the reaction solvent and provided multiple benefits. Acetylation in acetonitrile with only a few percent of methanol allowed significantly lower amounts of acetic anhydride and triethylamine, affording N-acetyl calicheamicin with only trace formation of over-acetylated products. Also, due to the reduced reagent amounts, triethylamine addition could be accomplished almost instantaneously without increasing the reaction temperature. Lastly, the mixing of acetonitrile and water is endothermic, so the water quench could be done rapidly and resulted in a decrease in temperature, eliminating any concerns of product degradation during an exotherm.

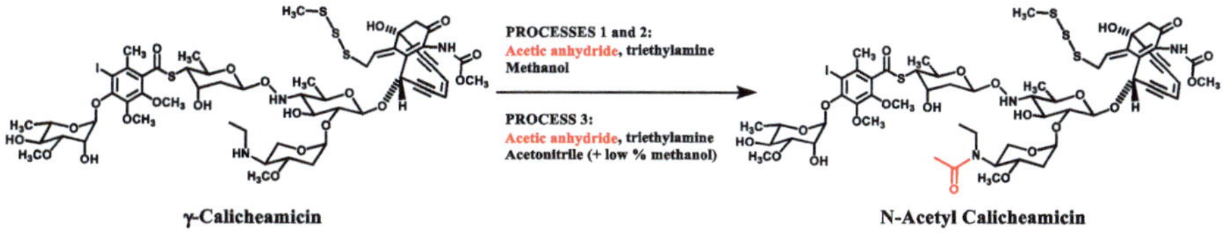

Fig. 13.5 Preparation of N-acetyl calicheamicin

13.7.3 Isolation of N-Acetyl Calicheamicin

The isolation of N-acetyl calicheamicin from both Process 1 and Process 2 was resource intensive (Fig. 13.6). In Process 1, the reaction mixture was reduced in volume prior to normal phase chromatography. Product-containing fractions were evaporated to dryness and the product was then precipitated from ethyl acetate/heptane. Process 2 relied upon an aqueous workup, extraction with dichloromethane, and evaporation of solvent followed by purification using reverse-phase chromatography with aqueous acetonitrile. Product-containing fractions were extracted with dichloromethane followed by concentration and precipitation with ethyl acetate/heptane. Process 3 sought to reduce the number of manipulations, maximize yield, and improve manufacturability. Having conducted the acetylation reaction in mainly acetonitrile and quenched with water, all that was needed to prepare the reaction mixture for purification by preparative reverse-phase chromatography was to target the appropriate acetonitrile-water ratio for column loading. So, the aqueous quench directly generated the chromatography loading solution. Fraction selection was controlled by an in-process assay purity specification.

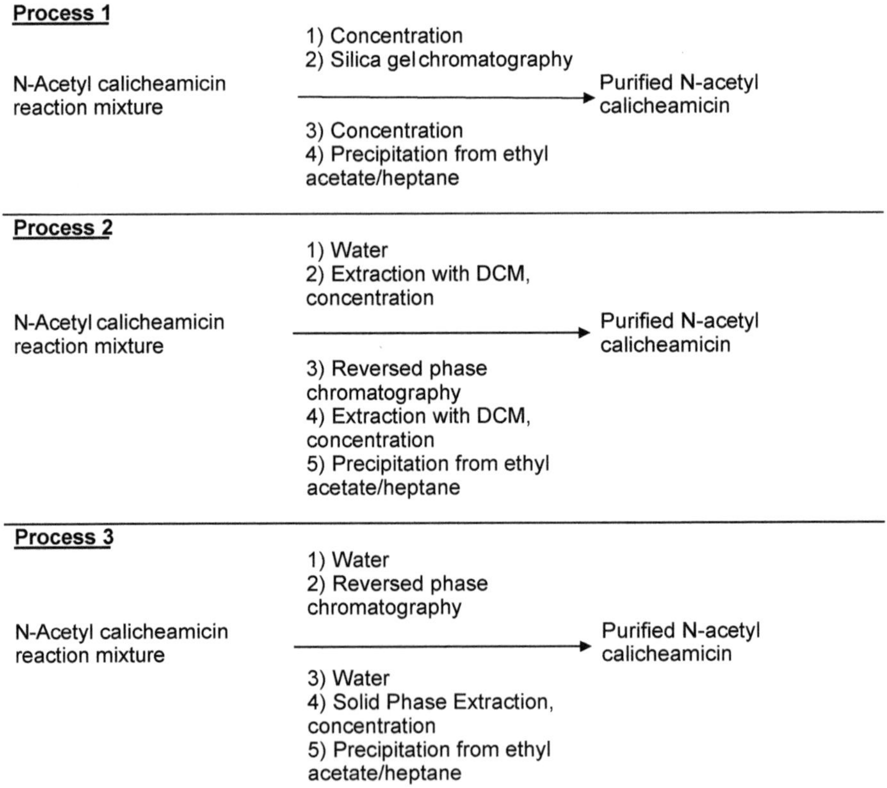

Fig. 13.6 Isolation of N-acetyl calicheamicin

In place of the Process 2 extractive workup of purified fractions, an improved method was sought which reduced unit operations and solvent usage and ideally eliminated dichloromethane. It was found that near quantitative recovery could be achieved by solid-phase extraction. Using the same chromatography stationary phase, the fractions could be loaded onto the column by appropriately choosing the proportion of water in the load solution. Fractions were pooled, diluted to achieve a higher water content, and loaded onto the column. The bound product was washed with water (containing a small percent acetonitrile to prevent column dewetting) and eluted with acetonitrile providing the product in essentially dry solvent. The acetonitrile was evaporated from the high-concentration eluate. Finally, a tractable solid was obtained by precipitation of the product from ethyl acetate/heptane and drying under vacuum in a manner similar to Processes 1 and 2.

The combined effects of the process changes introduced in Process 3 compared to Process 2 were a significant reduction in unit operations, fewer days to complete manufacture, and an approximately 30% increase in average yield.

13.7.4 Production of Activated Calicheamicin Derivative

The conversion of N-acetyl calicheamicin to activated calicheamicin derivative requires the addition of the activated linker. This conversion is the point where Processes 1, 2, and 3 differ most significantly, with each iteration decreasing the number of steps used to install the activated linker. Reducing the amount of chemistry performed on the calicheamicin-containing molecule preserves the highly valuable natural product and was likely to improve overall yield.

Process 1

Process 1 utilized three steps to add the activated linker: two steps to assemble the cleavable linker on the calicheamicin core and one to activate the carboxylic acid for reaction with the antibody (Fig. 13.7, clockwise path). This approach first reacts half of the linker, 3-methyl-3-mercaptobutanoic acid hydrazide (often referred to as DMH in publications [4], red in Fig. 13.7), with the methyl trisulfide of N-acetyl calicheamicin. This addition was a relatively low-yielding step, which was finally improved in Process 3. The addition was conducted in acetonitrile in the presence of triethylamine followed by concentration and silica gel chromatography. The selected fractions were concentrated, dissolved in methanol, and precipitated with water to afford solid N-acetyl calicheamicin DMH.

The second portion of the linker, 4-(4-acetylphenoxy)-butanoic acid (often referred to as AcBut in publications [4], blue in Fig. 13.7), was then appended to N-acetyl calicheamicin DMH under acidic conditions in methanol. After removal of the solvent and silica gel chromatography to afford N-acetyl calicheamicin DMH-AcBut, the linker payload needed only to be activated for subsequent addition to gemtuzumab. The acid moiety was converted to the N-hydroxysuccinimide (NHS) ester by treatment with NHS (green in Fig. 13.7) in acetonitrile and

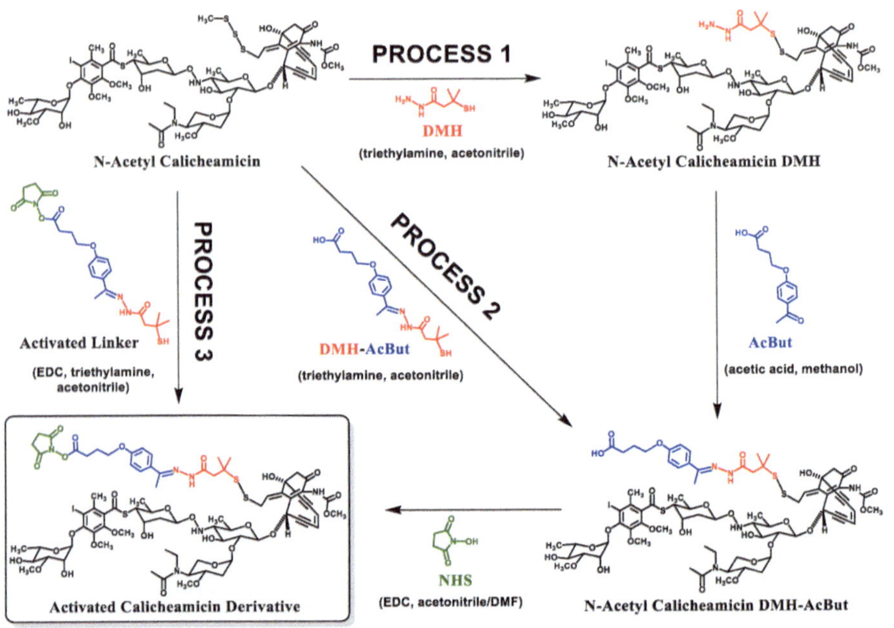

Fig. 13.7 Addition of the activated linker by Processes 1, 2, and 3

N,N-dimethylformamide (DMF) in the presence of 1-ethyl-3-(3-dimethylaminopropyl)carbodiimide (EDC). Subsequent silica chromatography and precipitation from ethyl acetate/hexane led to the desired activated calicheamicin derivative.

As can be readily surmised from the Process 1 description above, this synthesis is lacking since it performs several steps on the valuable calicheamicin core, involves multiple chromatography steps, and several steps involve solvent swaps. All of these limitations were to various degrees addressed in Processes 2 and 3.

Process 2

Process 2 (Fig. 13.7, diagonal path) was initially implemented in order to supplement the supply of activated calicheamicin derivative to support the clinical development of inotuzumab ozogamicin with the intention of subsequently adding the preparation to the gemtuzumab ozogamicin supply chain. The primary improvement included in Process 2 was the assembly of the linker prior to attachment to N-acetyl calicheamicin. Using conditions similar to Process 1, the linker (DMH-AcBut in Fig. 13.7) was added followed by a precipitative workup to provide N-acetyl calicheamicin DMH-AcBut. Conversion to the NHS ester remained the same as in Process 1. However, the isolation of activated calicheamicin derivative was modified to first purify the material by reverse phase chromatography, extraction with dichloromethane, and purification by silica gel chromatography and concentration. As was done in Process 1, the material was precipitated from ethyl acetate/hexane. While this process was more convergent, it still suffered from the

NHS activation step involving the penultimate calicheamicin intermediate and the need for two chromatographic steps, a less than optimal purification process for activated calicheamicin derivative. Process 3, developed some 20 years after Process 1, aimed to address these issues with contemporary knowledge and equipment.

Process 3

Process 3 (Fig. 13.7, left direct path) became the manufacturing process for the US relaunch of Mylotarg and the launch of Besponsa. The primary improvement was the attachment of the activated linker to N-acetyl calicheamicin in one step. Along the way, novel chemistry was discovered that boosted the crude yield which, coupled with an optimized purification procedure, provided a significant increase in overall yield. Preparation of the activated linker is not discussed here other than to say that it proceeds smoothly with the major improvements arising from replacing acid chlorides with less reactive analogs for the formation of the hydrazide and the NHS ester [21].

Attachment of the activated linker to N-acetyl calicheamicin was originally investigated with the relatively crude activated linker. The yields were significantly higher than what had been achieved at the time for the reaction of DMH-AcBut with N-acetyl calicheamicin. However, as the preparation of the activated linker improved with concomitant increase in its purity, the yield of activated calicheamicin derivative decreased. Contemplating upon what could cause this, the development group posited that there was a remaining coupling agent (carbodiimide) in the activated linker from the NHS ester-forming step. The hypothesis was tested and proven to be correct by including EDC in the coupling reaction mixture leading to an improved yield of activated calicheamicin derivative.

A mechanistic understanding of this observation was aided by the work of Myers et al. [22]. The reaction of the sulfur on the activated linker with the trisulfide of N-acetyl calicheamicin is a complicated mix of equilibria and dead-end reactions. As shown in Fig. 13.8, the linker sulfur could, in theory, attack any of the three sulfurs of the trisulfide giving four different products of which only one is desired. Work by Evans and Saville [23] indicated that the two reactions which displace a leaving group from the central sulfur are slow which is consistent with our observation of relatively low levels of ε-calicheamicin, the ultimate product of cleaving the sulfur–sulfur bond between the calicheamicin moiety and methyl disulfide. The formation of ε-calicheamicin results from the reaction of the allylic sulfur via a Michael addition into the enone followed by Bergmann cyclization [22] (Fig. 13.8, lower left). The second product of attack on the central sulfur is the linker trisulfide (Fig. 13.8, lower right). Of the two remaining displacements, one provides the desired disulfide affording activated calicheamicin derivative (Fig. 13.8, upper right) while the other produces a dead-end disulfide (Fig. 13.8, upper left). As shown in Fig. 13.9, a carbodiimide (blue, EDC for this case) can convert the dead-end N-acetyl calicheamicin disulfide to an activated disulfide, allowing another molecule of activated linker to attack the allylic sulfur to produce the desired product [21].

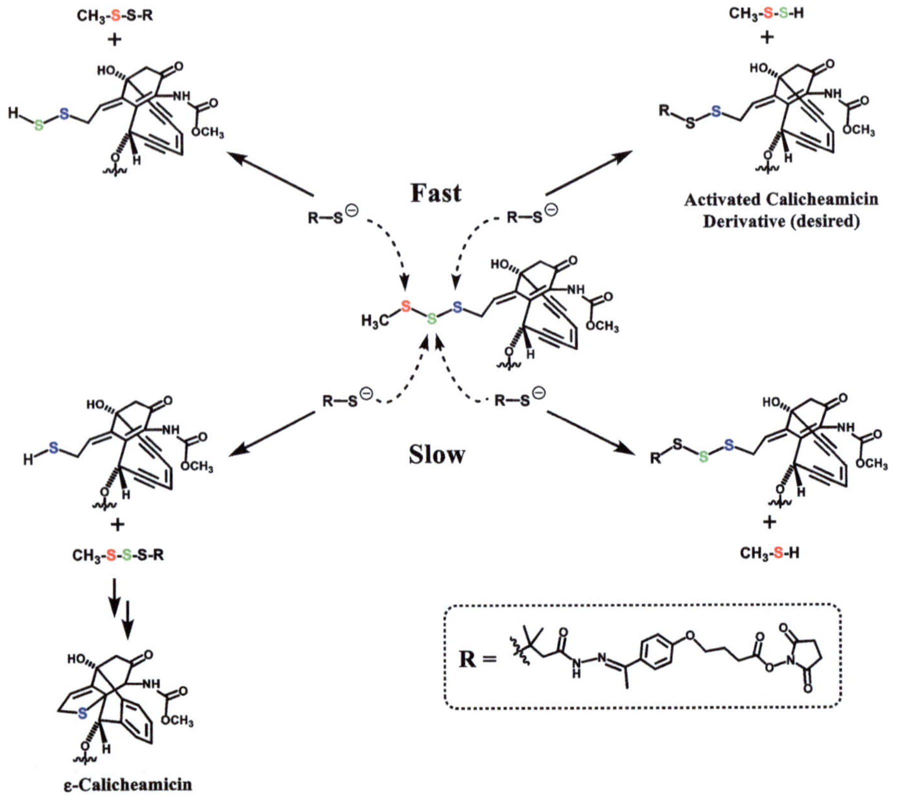

Fig. 13.8 Thiolate attack on the methyl trisulfide. (Adapted from Myers et al. [22])

With improved reaction conditions in hand as well as a mechanistic understanding of the chemistry available, the final efforts focused on optimizing the purification and isolation of the product.

The activated calicheamicin derivate contains both acid-sensitive (hydrazide) and base-sensitive (NHS ester) groups. Thus, any chromatographic purification of the molecule must take this into account. Also included in γ-calicheamicin are a number of ionizable groups which, depending upon the pH of any mobile phase used in reverse phase chromatography, could affect the ability to separate the desired product from impurities. An extensive study of the stability of activated calicheamicin derivative in aqueous acetonitrile identified the optimal pH which allowed for efficient purification of the product. However, the elution conditions could not fully prevent degradation at ambient temperature and therefore fractions were immediately sampled as they were collected and frozen during in-process analysis. Fractions that met the purity requirements were combined and additional water was added to decrease the solvent strength. The solution was then concentrated using solid phase extraction on a reverse phase column similar to the isolation of N-acetyl calicheamicin. Once the material was loaded, the column was washed with water and then

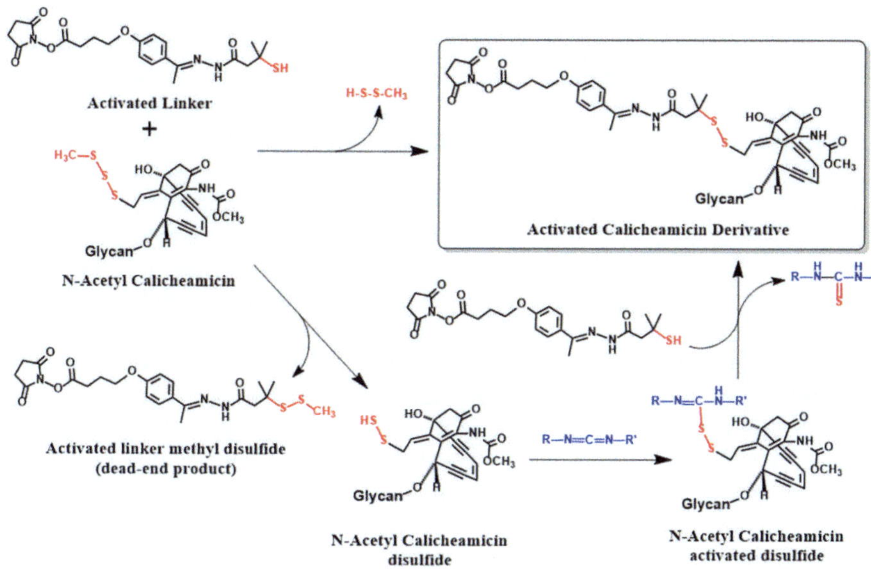

Fig. 13.9 Role of EDC in the addition of activated linker

eluted with acetonitrile. The early portion of the product peak eluted in acetonitrile containing a small amount of water, which could degrade the product if allowed to stand. Therefore, the first few percent of solvent was isolated separately and immediately reduced in volume relying on the acetonitrile/water azeotrope to dry the material. After all of the solvent was removed, the material was precipitated from ethyl acetate/hexane to provide a tractable solid after drying under vacuum.

Manufacture of activated calicheamicin derivative by Process 3 has many improvements compared to Process 2. Changes made in Process 3 have greatly improved manufacturability, reduced the number of unit operations by almost 50% including the removal of multiple time-intensive chromatography steps, removed chlorinated solvents from the process, and doubled the average yield. Compared to Process 1, the Process 3 yield improvement approached an order of magnitude and, to date, the yield and quality of the product have been consistent demonstrating that the process is well controlled.

13.7.5 Regulatory Challenges Involving Activated Calicheamicin Derivative

Two major challenges were presented by the activated calicheamicin derivative. One was chemical and the other was based upon the application of the ICH guidelines. The former presented itself as we attempted to fully understand the fate of the various impurities encountered during the preparation. The latter presented itself in our efforts to reach uniformity in the designation of the regulatory starting material.

The study of the purge and fate of impurities in the preparation of activated calicheamicin derivative began, of course, with the crude calicheamicin fermentation product. Using mainly LC-MS, most of the previously known analogs were identified as well as about a dozen new related compounds. As reported previously, the major analogs were β-calicheamicin and δ-calicheamicin, which differ from γ-calicheamicin by the alkyl group on the nitrogen of the amino sugar, and ε-calicheamicin which is derived from γ-calicheamicin having undergone the Bergmann cyclization [22]. Following these various impurities through the preparation, we were gratified to learn that essentially all impurities were removed during the purification of N-acetyl calicheamicin and those that remained were removed during the purification of activated calicheamicin derivative.

Despite our detailed understanding of the purge of impurities, HPLC analysis of activated calicheamicin derivative always showed a small peak eluting just before the desired product (Fig. 13.10, left) even though the chromatographic fractionation during purification excluded this peak. Believing that the peak must be from some related compound, an LC-MS analysis across both the minor and major peaks provided mass spectral data that were identical. An additional HPLC experiment provided an explanation for the phenomena. The major peak was isolated and reinjected immediately followed by several more injections in sequence thereafter. The resulting chromatograms demonstrated that the major peak was being converted to the minor peak over time. Understanding that the major peak did not completely convert to the minor and that the minor peak always represented the same proportion of the major, we concluded that the two peaks represented isomers in equilibrium.

The source of the isomerization is not known with certainty. The activated linker alone demonstrated similar behavior thus pointing to the isomers originating from the linker portion of activated calicheamicin derivative. The acyl hydrazone core of the activated linker was identified as the likely isomerization center (Fig. 13.10).

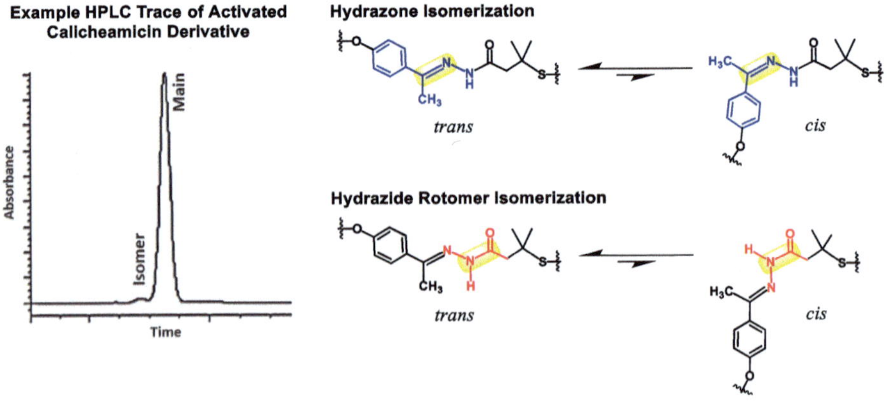

Fig. 13.10 Activated calicheamicin derivative isomers

Interconverting isomerization of acyl hydrazone compounds has been observed by HPLC [24], NMR [25, 26], and IR spectroscopy [27]. The HPLC observation with activated calicheamicin derivative could be explained by dynamic isomerization around the hydrazone bond, the hydrazide bond, or a combination of the two (Fig. 13.10, right). Luckily, narrowing the mechanistic explanation down to one was not required because it had been demonstrated that the two peaks were from the same molecule albeit from different, interconverting forms. Since the hydrazone is hydrolyzed to release the toxin in vivo, the released toxin would be the same regardless of the isomer's origin.

The purge and fate experiments also included the fate of the impurities in the activated linker. In a manner similar to that of N-acetyl calicheamicin, it was demonstrated that none of the impurities in the activated linker contributed to impurities in the activated calicheamicin derivative. The ICH Q11 guideline [28] speaks to several points relevant to the selection of starting materials for the preparation of activated calicheamicin derivative. Among others, these include:

- The activated calicheamicin derivative is a semisynthetic material. In selecting a starting material Q11 states: "*However, if it can be demonstrated that one of the isolated intermediates in the synthetic process complies with the principles outlined above for the selection of starting materials for synthetic drug substances, that isolated intermediate can be proposed as the starting material.*"
- The inclusion of manufacturing steps in Section 3.2.S.2.2 of an NDA or BLA is guided by: "*Manufacturing steps that impact the impurity profile of the drug substance should normally be included in the manufacturing process described in Section 3.2.S.2.2 of the application.*"

A regulatory strategy was thus developed proposing starting materials for activated calicheamicin derivative. The strategy included:

- that the preparation of the semi-synthetic activated calicheamicin derivative fulfilled the requirement to be treated as a standard synthetic compound
- that the impurities in both N-acetyl calicheamicin and the activated linker did not contribute to the impurities found in the activated calicheamicin derivative
- that activated calicheamicin derivative was not a drug substance but an intermediate in the production of antibody-drug conjugates.

Pfizer therefore approached the various regulatory agencies proposing N-acetyl calicheamicin and the activated linker as starting materials. The initial responses from the agencies were varied. Some accepted the proposal. Some indicated that the activated linker and crude γ-calicheamicin fermentation product would be acceptable. And some agencies on final review changed their requirements, differing from Pfizer in their interpretations of the ICH guidelines on starting materials. During the review of the final submission, a number of agencies indicated that the calicheamicin starting material should be earlier in the synthesis than N-acetyl calicheamicin. The argument for requiring an earlier starting material, reminiscent of the propinquity arguments of draft versions of ICH Q11, was cited by some agencies. In the end, the regulatory starting material ranged from the bacterial cell banks in a few markets to N-acetyl calicheamicin in most markets. The activated linker was accepted everywhere as a starting material.

13.8 Production of Gemtuzumab Ozogamicin Drug Substance

Conjugation of the activated calicheamicin derivative to the gemtuzumab antibody followed by purification constitutes the drug substance process. While the drug substance process for this ADC, the first on the market and the only one for over a decade, may seem unsophisticated by contemporary standards, it has provided consistent drug substance for over two decades.

One interesting aspect of gemtuzumab ozogamicin which engendered a number of questions from knowledgeable colleagues as well as others in the industry is whether we would attempt to adjust the drug loading profile to align more closely with contemporary ADCs. Mylotarg contains about 50% unconjugated or "low conjugate" antibody [9, 29]. This was not realized until a number of years after its initial launch when improved analytical methods were able to identify this fraction. Pfizer did not and could not easily change the composition of Mylotarg because the original composition was the basis of the clinical trial experience as well as the US commercial experience and the product remained on the market in Japan. All work to further our understanding of the manufacturing process must ultimately keep the unconjugated fraction unchanged.

The manufacturing process involves the addition of activated calicheamicin derivative in an alcoholic solvent to an aqueous solution of the gemtuzumab antibody, a surfactant, and a buffering system. The process development used for the original approval of Mylotarg explored the process space using one factor at a time (OFAT) experiments. While OFAT justification for some process parameters is acceptable if shown to be unaffected by other process parameters, it is not surprising that many conjugation processes are best described using multivariate analyses.

While the number of ingredients used in the conjugation reaction may seem few, there are multiple parameters associated with them. Complicating the development of this conjugation chemistry, the addition of the activated calicheamicin derivative to the lysine amines is analogous to the Shotten-Baumann reaction, which is known to be both rapid and very exothermic. In the present case, the NHS ester, a less reactive acid derivative, replaces the acid chloride of the Shotten-Baumann reaction but is still quite reactive and the kinetics can be nearly as fast. Therefore, a number of additional reaction parameters such as antibody to activated calicheamicin derivative ratio, antibody concentration, and reaction temperature, which are related to the conjugation kinetics and drug substance quality attributes, needed to be included in the re-examination of the process parameters.

Several parameters were studied for their effect on the level of aggregate and the average drug loading (drug to antibody ratio, DAR) with target values set based upon the capability of the subsequent purification to bring the material within the drug substance specification.

The parameters were investigated in a series of multivariable experiments (design of experiments or DOEs), which demonstrated that several parameters either did not affect the outcome or their effect was not correlated with changes to some other parameter. Given that it had been demonstrated that the kinetics of the addition of the NHS ester of activated calicheamicin derivate was fast, the decision was made that mixing could be set such that it would be sufficient for all subsequent experimental conditions and thus not included in any DOE analysis. In order to determine an acceptable blend time and translate it to the scaled-down model used in the DOE, the blend times and concentration distributions within the commercial scale and lab scale reactors were modeled both experimentally and with in silico fluid dynamics calculations. The modeling exercise first determined that which could be achieved in the commercial setting then the results were scaled down to the lab scale and held constant for the OFAT and DOE experiments. The remaining process parameters were tested in a multivariable environment, which demonstrated some interaction between several of the factors. These interactions would not have been discovered if the parameter space had been investigated as they were originally with OFAT experiments. The goal of these experiments was to determine a set of parameters that would keep the drug substance attributes, including the amount of unconjugated antibody and drug-to-antibody ratio, within the historical ranges dating back to the initial introduction of Mylotarg to the market. Thus, the operating conditions were redefined while keeping specifications unchanged. The practical effect of this deeper knowledge was to adjust the manufacturing instructions to meet the better understood requirements.

Purification of the drug substance reaction mixture to produce bulk drug substance was performed by size exclusion chromatography using the final excipient solution as the eluting buffer. Since the quality of the crude drug substance after the conjugation reaction was well within the historical range, the chromatographic purification process was not modified.

13.9 Production of Mylotarg Drug Product

The Mylotarg drug product is a lyophilized powder, which is reconstituted with water for injection. Since the bulk drug substance contains all of the excipients in the drug product, no changes to the drug product were necessary.

13.10 Conclusions

The story of the reintroduction of Mylotarg into the US market is one of looking back at what was developed and what data were available at the time to support the registration. This analysis provided Pfizer the opportunity to apply the many scientific advances, which had occurred since the original filing. These advances touched the science that is used in characterizing both processes and products and allowed Pfizer to provide a control strategy containing a deeper understanding of the production and control of Mylotarg than was available originally.

Re-developing/investigating a drug manufacturing process with a strictly defined set of drug product attributes and, in this case, with a 17-year history, is different than defining a process as a drug goes into a Phase 1 study and refining it during clinical development. The Mylotarg drug substance is an antibody-drug conjugate; however, it was the drug substance intermediates, gemtuzumab and activated calicheamicin derivative, that benefited from manufacturing optimization. The intermediates could be more easily modified to address their shortcomings while maintaining the final drug substance quality.

Just as Pfizer had no experience returning a nearly 20-year-old drug to market, neither did the regulatory agencies. They were receptive to the possibility of a more advanced intermediate as a regulatory starting material than originally filed for activated calicheamicin derivative. While the FDA and EMA did ultimately require visibility of the intermediate's manufacturing process back to the bacterial cell bank, still other global health authorities accepted N-acetyl calicheamicin as a regulatory starting material.

The multiple functions that worked on all aspects of the re-registration of Mylotarg ultimately succeeded and it was returned to the US market on September 1, 2017, over 17 years after it was originally approved and almost 7 years after it was voluntarily withdrawn. Mylotarg continued the regulatory success with approvals throughout the world including Europe where it had been rejected in prior years. Mylotarg's re-introduction in the US and introduction in other major markets offered a new alternative treatment to AML patients globally to manage this devastating disease.

Acknowledgements With a project that has spanned two decades, it is impossible to acknowledge all those who have contributed to bringing Mylotarg to patients. Similarly, the number of colleagues who have contributed to just the effort to return Mylotarg to the market is also quite large. We acknowledge the following colleagues while realizing that the credit goes well beyond this list.

Pfizer Global Medicine Team who shepherded the vast amount of clinical data that allowed the rest of the team to pursue their chemical/biochemical specialties as their contributions.

Contributors to the calicheamicin process development portion of the project include: Nataliya Bazhina, Eric Bortell, Lawrence Chen, Joe Collins, Robert Dugger, Brad Evans, Stephen Freese, Xi Hu, Julius Lagliva, Mark Maloney, Jim Mo, Vimal Patel, Amar Prashad, Wesley Swanson, April Xu, Chunchun Zhang, Sen Zhang.

Contributors to the antibody portion of the project include: Daniel Boisvert, William Daniels, Mary Denton, Chris Gallo, Heyi Li, Joshua Ochocki, Mary Switzer.

Contributors to the drug substance portion of the project include: Eric Bortell, Brooke Czapkowski, Brad Evans, Qingping Jiang.

Contributors to the drug product portion of the project include: Bakul Bhatnagar, Serguei Tchessalov, Nathalie Hayduk, who led the joint Pharmaceutical Sciences and Pfizer Global Supply development teams.

Contributors to the calicheamicin, drug substance, and drug product manufacturing include: Pfizer Global Supply Pearl River—Technology team, QA team, QC team, Tok Han, Michael Hripcsak, David Merkooloff, Sonal Shah, Paramanathan Guruparan.

Contributors to the antibody manufacturing include: Pfizer Global Supply Andover Team.

Contributors to the CMC regulatory filings include: Jana Cook, Caroline Kinross, Jacqueline Macaulay, Jacklyn Moxham, Pfizer Global CMC team.

References

1. Beck A, Goetsch L, Dumontet C, Corvaïa N (2017) Strategies and challenges for the next generation of antibody–drug conjugates. Nat Rev Drug Discov 16:315–337
2. Tong JTW, Harris PWR, Brimble MA, Kavianinia I (2021) An insight into FDA approved antibody-drug conjugates for cancer therapy. *Molecules* 26:5847
3. Mylotarg (2017) Highlights of prescribing information. FDA. [Online] 1 Sept 2017. Available at: https://www.accessdata.fda.gov/drugsatfda_docs/label/2017/761060lbl.pdf. Accessed Oct 2022
4. Hamann PR et al (2002) Gemtuzumab ozogamicin, a potent and selective anti-CD33 antibody-calicheamicin conjugate for treatment of acute myeloid leukemia. Bioconjug Chem 13:47–58
5. Petersdorf SH et al (2013) A phase 3 study of gemtuzumab ozogamicin during induction and postconsolidation therapy in younger patients with acute myeloid leukemia. Blood 121:4854–4860
6. Castaigne S et al (2012) Effect of gemtuzumab ozogamicin on survival of adult patients with de-novo acute myeloid leukaemia (ALFA-0701): a randomised, open-label, phase 3 study. Lancet 379:1508–1516
7. Rowe JM, Löwenberg B (2013) Gemtuzumab ozogamicin in acute myeloid leukemia: a remarkable saga about an active drug. Blood 121:4838–4841
8. Press OW et al (1996) Comparative metabolism and retention of iodine-125, yttrium-90, and indium-111 radioimmunoconjugates by cancer cells. Cancer Res 56:2123–2129
9. Hamann P (2011) The genesis of the antibody conjugate gemtuzumab ozogamicin (Mylotarg®) for acute myeloid leukemia. In: Accounts in drug discovery: case studies in medicinal chemistry. RSC, pp 103–119
10. Baker TS et al (1994) Humanization of an anti-mucin antibody for breast and ovarian cancer therapy. Adv Exp Med Biol 353:61–82
11. Ellestad GA (2011) Structural and conformational features relevant to the anti-tumor activity of calicheamicin γ_1^I. Chirality 23:660–671
12. Lee MD, Ellestad GA, Borders DB (1991) Calicheamicins: discovery, structure, chemistry, and interaction with DNA. Acc Chem Res 24:235–243
13. Walker S, Landovitz R, Ding WD, Ellestad GA, Kahne D (1992) Cleavage behavior of calicheamicin $\gamma 1$ and calicheamicin T. Proc Natl Acad Sci U S A 89:4608–4612
14. Kumar RA, Ikemoto N, Patel DJ (1997) Solution structure of the calicheamicin γ_1^I-DNA complex. J Mol Biol 265:187–201
15. Bergman RG (1973) Reactive 1,4-dehydroaromatics. Acc Chem Res 6:25–31
16. Hu X, Bortell E, Kotch FW, Xu A, Arve B, Freese S (2017) Development of commercial-ready processes for antibody drug conjugates. Org Process Res Dev 21:601–610

17. European Medicines Agency (2018) Mylotarg: EPAR public assessment report. [Online] 22 Feb 2018. Available at: https://www.ema.europa.eu/en/documents/assessment-report/mylotarg-epar-public-assessment-report_en.pdf. Accessed Feb 2023
18. Hensgen MI, Stump B (2013) Safe handling of cytotoxic compounds in a biopharmaceutical environment. Methods Mol Biol 1045:133–143
19. European Medicines Agency (2017) Besponsa: EPAR public assessment report. [Online] 21 Apr 2017. Available at: https://www.ema.europa.eu/en/documents/assessment-report/besponsa-epar-public-assessment-report_en.pdf. Accessed Feb 2023
20. Lee MD, Dumme TS, Chang CC, Siegal MM, Morton GO, Ellestad GA, McGahre WJ, Borders DB (1992) Calicheamicins, a novel family of antitumor antibiotics. 4. Structure of calicheamicins β_1^{Br}, γ_1^{Br}, α_2^{I}, α_3^{I}, β_1^{I}, γ_1^{I} and δ_1^{I}. J Am Chem Soc 114:985–997
21. Dugger RW, Letendre LJ, Patel VB, Prashad AS, Zhang C (2015) Intermediates and methods for synthesizing calicheamicin derivatives. WO 2015/063680 A1, 7 May 2015
22. Meyers AG, Cohen SB, Kwon BM (1994) A study of the reaction of calicheamicin gamma-1 with glutathione in the presence of double stranded DNA. J Am Chem Soc 116:1255–1271
23. Evans MB, Saville B (1962) Nucleophilic displacements by thioanions on trisulfides. Proc Chem Soc 1962:18–19
24. Wu J, Huang C, Wang J, Hu D, Jin L, Yang S, Song B (2013) Separation, interconversion, and insecticidal activity of the cis- and trans-isomers of novel hydrazone derivatives. J Sep Sci 36:602–608
25. Zelenin KN, Oleinik SV, Alekseev VV, Potekhin AA (2001) Structure of cyanoacetylhydrazones of aldehydes and ketones. Russ J Gen Chem 71:1116–1120
26. Lopes AB, Miguez E, Kümmerle AE, Rumjanek VM, Fraga CAM, Barreiro EJ (2013) Characterization of amide bond conformers for a novel heterocyclic template of N-acylhydrazone derivatives. Molecules 18:11683–11704
27. Hadži D, Jan J (1967) The NH stretching band and the molecular configuration of acyl hydrazones and semicarbazones. Spectrochim Acta A 23A:571–577
28. ICH Expert Working Group (2012) ICH Q11 Development and manufacture of drug substances. Current Step 4 version, 1 May 2012. https://ich.org. Accessed Oct 2022
29. Beck A, D'Atri V, Ehkirch A, Fekete S, Hernandez-Alba O, Gahoual R, Leize-Wagner E, François Y, Guillarme D, Cianférani S (2019) Cutting-edge multi-level analytical and structural characterization of antibody-drug conjugates: present and future. Expert Rev Proteomics 16:337–362

Part VII
Biopharmaceutical Informatics and Analytics

Chapter 14
Biopharmaceutical Informatics: A Strategic Vision for Discovering Developable Biotherapeutic Drug Candidates

Joschka Bauer, Sebastian Kube, Pankaj Gupta, and Sandeep Kumar

Abstract Biotherapeutics are rapidly emerging as a successful class of pharmaceuticals, even though significant challenges to their discovery and development remain. In this book chapter, we establish a conceptual framework for potential computational interventions at every stage of biologic drug discovery and early development. We call this framework biopharmaceutical informatics. This chapter provides a comprehensive overview of our strategic vision. This vision calls for closer collaboration between drug discovery and development functions of biopharmaceutical industry by integrating the considerations of developability during early stages of drug discovery. Computational tools already available to enable biopharmaceutical informatics are reviewed in this work. While our focus is on monoclonal antibody-based biologics, the concepts discussed in this work are also applicable to novel formats such as multispecific biologics.

Keywords Biotherapeutics · Drug design · Drug development · Developability · Function · Computation · In silico · Bioinformatics

J. Bauer · S. Kube
Pharmaceutical Development Biologicals, Boehringer Ingelheim Pharma GmbH & Co. KG, Biberach/Riss, Germany

P. Gupta
In Silico Team, Biotherapeutics Discovery, Boehringer Ingelheim Pharmaceutical Inc., Ridgefield, CT, USA

S. Kumar (✉)
Molecule Design and Modeling, Computational Science, Moderna Therapeutics, Cambridge, MA, USA
e-mail: Sandeep.Kumar@modernatx.com

K. Gadamasetti, S. A. Kolodziej (eds.), *Bioprocessing, Bioengineering and Process Chemistry in the Biopharmaceutical Industry*,
https://doi.org/10.1007/978-3-031-62007-2_14

14.1 Introduction

14.1.1 Growing Demand for Improved Developability of Biotherapeutics

Biotherapeutics, especially, the monoclonal antibodies (mAb) have now been industrialized. This success is reflected in the fact that the load in the development pipelines of the biopharmaceutical industry is ever increasing with more than 800 antibody therapeutics in the early-stage and 115 in the late-stage phases with an average R&D cost to launch per project of about two to four billion US dollars [1]. While we celebrate rise of monoclonal antibodies as the fastest growing class of biotherapeutics (or biologics as they are often referred as), one should, nevertheless, recall that the chance for a monoclonal antibody to be approved for marketing is about 5–10% [2]. Improved molecular understanding of the new drug and fit to the engineering and development platforms at preclinical R&D stages might increase this chance of success [3] and therefore contribute towards even greater success of biologics. This desire to improve the rate of translation of biologic drug candidates into drug products available in the market is often summarized under the term "developability," which describes the feasibility of a drug candidate to successfully progress from discovery to clinical drug product development, and can be traced as early as the first decade of the 2000s [4, 5]. The developability concept is based on experience gained from previous Chemistry, Manufacturing, and Control (CMC) development programs [6], along with the realization that several of the CMC issues encountered during the development stages could have been mitigated via minor primary sequence modifications during discovery.

14.1.2 Developability Issues Arising from Unfavorable Biophysical Properties

In comparison to small molecule drug products, biologics are more complex and heterogenous in their three-dimensional structures and physicochemical properties, which are highly dependent on their respective manufacturing process [7]. A subset of those properties is relevant for drug efficacy and safety and therefore needs to be monitored and controlled [8]. They are referred to as Critical Quality Attributes (CQAs) [9]. As manufacturing processes for biologics are generally more complex and have a greater influence on the biophysical properties of the active pharmaceutical ingredient (API), the regulatory bar for drug manufacturers to demonstrate their ability to make biologic drug products via well controlled, safe, and reproducible processes is much higher in comparison to that for the small molecule drug products [10].

In an ideal biologic drug product development program, an excellent mAb candidate with optimal biophysical properties and a perfect fit to standard development platforms would be reproducibly expressed and purified with a high yield and product quality resulting in low production costs. Subsequently, the active pharmaceutical ingredient would be formulated in liquid form with high chemical, conformational, colloidal, and physical stability during shipping and storage over its shelf life and would be easily administrable at the concentrations required for effective dosing [11]. However, the real circumstances of the biopharmaceutical industry deviate substantially from this idealized state, since the majority of lead candidates experience development issues that are related to their unfavorable biophysical properties, but the specific reasons behind a drug failure in the development phase remain often poorly documented and therefore not fully understood [4]. Increasing the complexity, next-generation drugs of multi-specific mAb derivatives tend to have even greater CMC hurdles [12, 13]. Additionally, the quality target product profile defining the key design requirements of the final drug product is not fully defined in the early development stages [14]. The therapeutic dose and regimen for clinical studies are not yet defined and need to be established in preclinical experiments with the same material that is used in the clinical studies [15]. The anticipated administration route might change during the development as well as the primary packaging and/or the inclusion of a medical device [16]. All these design choices bring different and additional requirements on different physicochemical properties of the biologics and the formulation that can also modify the platform manufacturing processes. Examples are solution viscosity, resistance to shear stress, interaction on air: water and interfaces, compatibility to a liquid or lyophilized formulation, compatible excipients, and active pharmaceutical ingredient (API) concentration. Therefore, the biopharmaceutical industry aspires to meet the increasing demand for modern biotherapeutics by improving their developability via rational design.

Table 14.1 summarizes the most commonly occurring issues in formulation development of a liquid formulation for parenteral application.

14.1.3 Biopharmaceutical Informatics: An Integrated Approach to Discovery and Development of Biotherapeutics

As described above, biotherapeutic drug candidates often face several developability issues and therefore warrant greater scrutiny from the regulatory agencies. Remarkably, however, each of the development risks described above can be traced back to the intrinsic physicochemical properties of the drug candidate encoded in its primary sequence and three-dimensional structure [6], both of which are specifically targeted by the promising approach of biopharmaceutical informatics. Kumar et al. previously defined this emerging field as any computational effort geared to advance efficient and cost-effective translation of biologic drug candidates into drug products [17]. Figure 14.1 describes the motivation for biopharmaceutical informatics. This interdisciplinary area explicitly covers both in silico tools alone and in combination with experimental studies such as developability assessment.

Table 14.1 Major challenges and opportunities for application of computation at various stages of antibody-based biotherapeutic drug discovery and development

Stage	Major challenge(s)	Opportunities for computational applications
In vitro production of immunogen(s)	1. Immunogen proteins may have unknown structure with varying degree of sequence homology with proteins of known structures 2. Protein insolubility and aggregation leading to low material yields	1. Protein structure prediction 2. Sequence/structure-based optimization for improved conformational stability and solubility can help improve quantity as well as quality of material available for immunization
Antibody generation	1. Antibody generation by immunizing animals is costly, time consuming, yields variable results, and may require humanization 2. Humanized mice do not capture fully human immune repertoire 3. Phage and yeast display technologies can rapidly identify binders but their developability may need further optimization	1. Development of computational techniques analogous to virtual screening of human antibody repertoires can lead to rapid identification of binders. It can also potentially alleviate the need to produce immunogens in large quantities 2. Computational design of phage and yeast display libraries for improved developability 3. Computational redesign of antibodies with good developability to binding different specificities 4. In silico generation of antigen-specific/antigen-agnostic antibody libraries via artificial intelligence and machine learning
Hit selection and lead identification (LI)	1. Sequencing of the hits 2. Epitope mapping of the hits to assure desirable therapeutic effect in absence of structural models for Ag:Ab complex 3. Experimental testing of several hundreds of hits for function and developability can be time and resource consuming	1. Development of appropriate Sanger and NGS sequencing pipelines 2. Computational prediction of epitopes and paratopes 3. In silico assessments of hits for developability and manufacturability can guide selection of developable hits and identify lead candidate(s) with good developability
Lead optimization (LO)	Identified lead candidates may require humanization (if from a nonhuman source such as wild type mouse), affinity optimization, and removal of physicochemical liabilities for improved developability	Structure-based modeling of the lead candidates can guide their humanization, affinity maturation, and identification of potential sequence/structural motifs that may drive their physicochemical degradation. Availability of this information can help guide protein engineering strategies for lead optimization (LO)
Early stage developability assessment	1. Molecular stability and fitness of the drug candidate nominated by drug discovery to platform processes used in drug development 2. Dynamic development targets	1. Multiscale molecular and mathematical simulations to predict fit to platform and understand molecular response to stresses faced during manufacturing, storage, and shipping 2. Development of predictive algorithms for identifying the appropriate bioprocess conditions and formulation ingredients for drug product development 3. Decision-support systems

(continued)

Table 14.1 (continued)

Stage	Major challenge(s)	Opportunities for computational applications
Process and drug product development	1. Up-scaling 2. Comparability 3. Customization of processes for manufacturing (bioprocess), formulation development, and analytical characterization	Machine learning and mechanistic modelling techniques to model bioprocesses conditions and yields, and development of digital twins
Clinical development	1. Trial design and criteria for patient recruitment 2. Safety, efficacy, and pharmacology	1. Biostatics and genomics technologies 2. Immunogenicity predictions and their clinical relevance 3. Safety and toxicology data capture and analyses 4. Modeling and simulation of pharmacology of drug candidates
All stages	Numerous challenges related to function and physicochemical stability of the biotherapeutic drug candidates	1. Data capture into databases that can be easily analyzed by following FAIR (Findable, accessible, interoperable, and recyclable) principles 2. Data analyses and creation of digital tools to predict experimental outcomes 3. Biopharmaceutical informatics

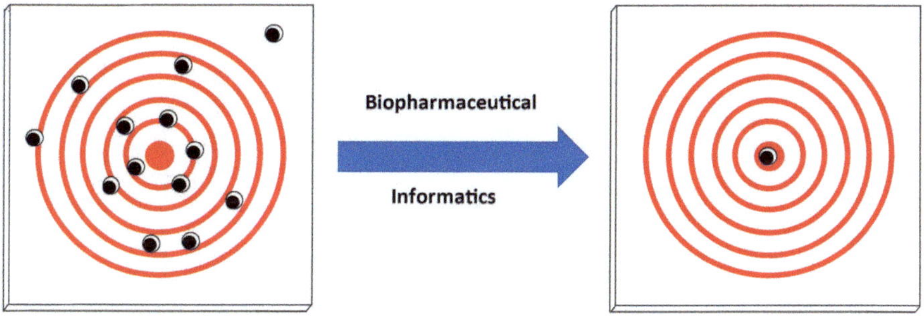

Fig. 14.1 Motivation for biopharmaceutical informatics

The correlation of "macroscopic" experimentally determined attributes of a biologic with its "microscopic" sequence-structure aspects computed in silico is the major unsolved problem in biopharmaceutical informatics. Depending on the amount of data available, several statistical as well as machine learning approaches can be applied to obtain mathematical models capable of predicting solution behaviors of monoclonal antibodies solely from their sequence-structure information [18–26]. Below, we stitch together a diverse set of concepts in bioinformatics, molecular modelling, and simulations along with the experiments to demonstrate practical feasibility of our vision of biopharmaceutical informatics and Table 14.2 lists several opportunities to apply computation at every stage of biopharmaceutical drug discovery and development. However, the field has not matured uniformly in

Table 14.2 Most common developability challenges in biologics

Category	Mechanism	Liability	Initiator	Effect	References
Chemical instability	Oxidation	Trp, Met (in the complementarity determining regions (CDRs))	Light-induced, free radicals, metal-catalysis	Reduced conformational stability, reduced melting temperature, increased levels of fragments, aggregates, and particles, immunogenicity, reduced biological activity, and coloration	[27]
	Deamidation	Asn, Gln (in the CDRs)	Basic pH	A decreased pI, altered hydrophobic and anionic properties, antibody self-association, aggregation, structural changes, loss of function	[28–33]
	Isomerization	Asp	Acidic pH	Altered conformational flexibility, biological activity, and immunogenicity	[29, 34–39]
	Glycation	Lys	Reducing monosaccharides	Increased charge heterogeneity	[40–42]
Conformational instability	Changes in the secondary, tertiary, and quaternary structural features of the native fold	Chemical modifications, destabilized domain folds	Air-water interface, freeze-thaw, shear, agitation, temperature, and light stresses	Aggregation, adsorption, increased viscosity, decreased solubility, and loss of function	[43–50]
Colloidal stability	Reversible self-association of natively folded mAbs	Nonuniform distributions of hydrophobic and charged regions on molecular surface		Aggregation, low solubility, high viscosity, liquid-liquid and liquid-solid phase separation, off-target binding, fast antibody clearance	[51–54]
	Non-native aggregation	Chemical modifications, destabilized domain folds			[49, 55–59]

all the stages described in this table. In the next section, we discuss the opportunities that are beginning to mature.

14.2 In Silico Assessments of Biologics in Research and Development

14.2.1 Antibody Generation

Purified antigens can be used to generate antibodies against them either by immunizing animals (typically laboratory mice, humanized/transgenic mice or other animals like chicken, rabbit, or cows), using hybridoma techniques, or screening of natural and/or synthetic antibody libraries via display technologies such as phage or yeast. Promising hits are selected and validated via antigen binding assays. Currently available methods for antibody generation are almost entirely experimental in nature. Depending on the methods used to generate antibodies against a given antigen, it commonly takes several months before an initial set of antibody-based binders becomes available for further investigations and for lead identification. However, computational technologies that have been originally developed for small molecule drug discovery can be also transferred to antibody-based drug discovery. Once fully developed and deployed, these in silico methods will open another set of means to generate antibody binders against a target antigen.

A potential computational tool originates from the strategy of pharmacophore-based screening of small molecules. Transferring the idea to biologics, the molecular surface features of a desired antigen epitope must be known to screen a database of antibody structures, and to identify potential hits based on molecular shape and electrostatic complementarity. The hits can be prioritized by modelling antigen: antibody complexes via protein: protein docking. However, protein: protein docking scoring methods and selection of the "correct" poses remain a topic of debate, although considerable progress is being made [60]. Further, first antibody binding mutational databases have been established that will help to improve computational affinity predictions [61]. Subsequently, the most promising candidates can be selected for experimental confirmation of binding, and computational optimization of the affinity and developability of the antibody. A crucial part of this innovative approach is the database of antibody structures and molecular models required for the pharmacophore search. Importantly, the database does not require the structures of full-length antibodies, but structures and homology-based molecular models of the variable regions only. As of now, Protein Data Bank (PDB) contains thousands of high-resolution crystal structures of Fragment variable (Fv) as well as fragment antigen-binding (Fab) regions. In addition, several homology-based antibody modelling methods have been developed in past few years and, although some issues remain, the field has matured enough to provide highly accurate antibody models capable of supporting pharmacophore like searches. The second potential approach focuses on the design of computational libraries for phage or yeast display experiments. The availability and growth of large and heterogeneous databases of antibody sequences and structures provide an ideal starting point for the design of computational libraries [62–67].

From the perspective of experimental approaches, fully synthetic human antibody libraries comprising Fabs that have been selected for biophysical characteristics favorable to development have already been created [68]. Thereby, a special focus was placed on the selection of molecules with increased chemical, conformational, and colloidal stability [68]. The idea of optimized antibody libraries for the generation of developable antibodies could be hybridized with de novo computational databases of an extremely large number of diverse combinations of humanoid light and heavy chains [69]. Mutations targeted at specific sequence positions (e.g., CDRs) in the antibody sequences could further expand the library either to make it recognize different antigens, or to optimize its binding affinity toward a given antigen. Recently, a generative adversarial network was successfully applied to create a diverse library of novel antibodies that mimic somatically hypermutated human repertoire response [70]. This in silico approach further unraveled the residue diversity throughout the variable region [70] that might be helpful for further computational tools such as CDR redesign. This approach uses a highly developable antibody framework that is altered in the original CDRs, i.e., paratope to recognize a novel antigen. In the last years, significant advances were made in the design of not only thermodynamically stable but also biologically functional antibodies [71].

Remarkably, first computational methods offer the possibility to design humanized antibody variable regions against targeted antigen epitopes de novo [72, 73]. Alternatively, it was shown that the design can also start with a structural model of an antigen: antibody (Ag: Ab) complex generated using molecular docking of the structures of the Ag and Ab [73]. In the next step, the affinity of the antigen toward the antibody could be either varied by randomly introducing sequence variations [73] or selectively re-designed via structure-based approaches. Subsequently, interfacial residues in the Ab and Ag structures which contribute significantly toward instability of the Ag: Ab complex could be identified via computational alanine (Ala) scanning. In the next step, the identified residue positions could be scanned for mutations that can increase/decrease the stability of Ag: Ab complex and improve or lower affinity of the Ab toward the Ag [74], as per project requirements. Another attractive alternative for the rational antibody design refers to hotspot grafting with CDR loop swapping, which only requires information about the interactions with the antigen [75].

14.2.2 Hit Selection and Lead Identification

After production of antigen-binding antibodies by immunized animals, hybridoma cells, or phage and yeast display techniques, the variable regions of the antibodies are sequenced, and the binders are validated. The wide variety of hits must then be prioritized, and the most promising lead candidates identified (LI, lead identification). Consequently, extensive resources are required to experimentally test each hit and confirm antigen binding.

Several bioinformatic techniques can support the prioritization and selection of hits for in vitro confirmation of antigen binding as well as lead identification. A commonly used strategy is to cluster the hits into bins of high, medium, and low binding affinity, based on the initial estimates, analyze each bin for the diversity of heavy and light chain germlines followed by diversity of the CDRs, and select multiple but few representatives from each germline pair in each bin for experimental testing. Alternatively, one could directly bin the hits based on the germline pairings and CDR diversity and select multiple but few of them based on their estimated antigen binding. In addition to the antigen binding, the aspect of developability can already be considered early at this point of hit selection by employing computational tools. In a simple application, one can score heavy (HC) and light chain (LC) sequences of hits based on presence of potential chemical degradation motifs, aggregation prone regions (APRs), and T-cell immune epitopes present or overlapping with the CDRs of the heavy and light chains. Such scoring schemes can be further optimized by adding different weights based on which CDRs contain these motifs and whether they are present in the at the beginning/end or in the middle of the CDRs. In a more structure-based approach, three-dimensional homology models of all or a subset of hits can be analyzed regarding physicochemical descriptors such as pI, charge, dipole moment, and solvent exposed hydrophobic and ionic patches [22, 24]. In subsequent studies, one or few of the best hits are experimentally tested thoroughly for biological function, cross-reactivity across species, nonspecific binding, and pharmacology indicators such as serum stability. This process culminates into identification of one or more lead candidates.

14.2.3 Lead Humanization and Optimization

Lead optimization (LO) is conducted once one or more lead candidates have been identified and revalidated for function. During LO, the Fv regions may need to be humanized, if necessary, corrected for post-translational modification (PTM) sites, optimized for affinity, and in the best case, for developability (Fig. 14.2). Below it is described, how the LO process of therapeutic antibodies can be supported in every aspect by computational biophysics [76].

Humanization aims for an optimal amino acid sequence in the Fv by converting as many nonhuman residues to human germline residues to decrease the likelihood of immunogenicity and antidrug antibodies (ADAs) which can impact drug efficacy and safety [77, 78]. If necessary, mouse residues essential for binding are identified by back mutations to retain Ag binding similar to the mouse/human chimeric candidate [79]. A better understanding of the structure-function relationship helps to identify fitting templates and back-mutations critical to preserve CDR conformations [80]. Consequently, computational protein design methods have been

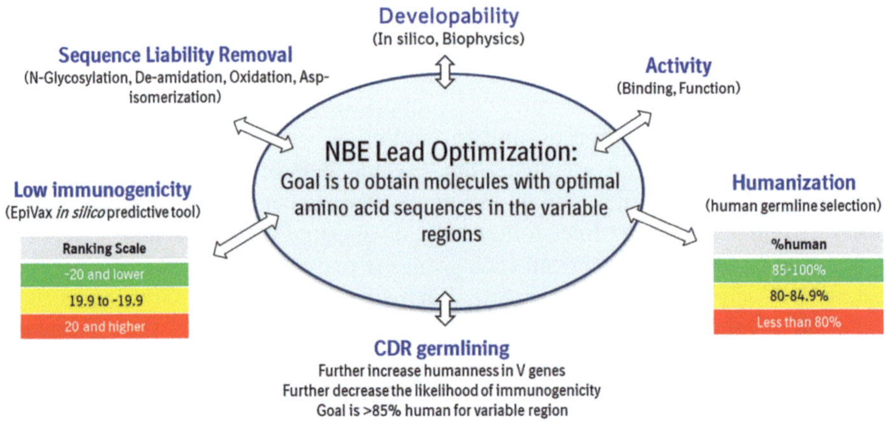

Fig. 14.2 Process of lead optimization (LO)

successfully applied to efficiently increase the humanness of antibodies while maintaining their structural stability [81]. In line with that, a retrospective analysis of a humanization campaign suggested that hotspots in both, the FW regions and Vernier Zones, can affect the antigen binding and thermodynamic stability [82]. State-of-the-art software such as MOE from Chemical Computing Group [83] can facilitate CDR grafting via identification of appropriate human FWs and back-mutations by considering large databases of human sequences [63, 84]. Furthermore, the rating and optimization of humanness can be done via in silico calculations [19, 84–89]. Besides that, statistical interference approaches have been created to characterize the statistical distribution of human Fv sequences [90]. Furthermore, bioinformatic studies have shed light on subtle structural differences between the lambda (VL) and kappa (VK) isotypes that need to be considered during (re-)engineering [91]. Structure-guided approaches can help improve the biophysical properties of a therapeutic mAb by switching from a problematic lambda framework (FW) region to a more stable kappa FW [92].

The humanized sequence(s) are then proceeded with liability engineering campaigns. The cumulative recommendations based on pre-formulation assessment, forced degradation studies analyzing PTMs and stability, and in silico assessments can be considered along with CDR germline residues in the engineering design plan. At this time, the power of phage display or other screening technologies can be used to screen a large panel of variants (typically 10^3–10^9 *Escherichia coli* expressed Fabs). A panel of final lead optimized variants (~50–100) might be formatted as immunoglobulin Gs (IgGs), expressed, and purified at small scales, and characterized by binding and pre-formulation assessments. Complementary, in silico assessments help to investigate intrinsic differences between the candidates. Various predictive in silico tools are available that help to monitor and guide the redesign of the candidate's individual weak points that mediate chemical, conformational, colloidal, and physical instability (see Sect. 14.3). In addition, comparison of the molecular characteristics of the lead candidate against marketed antibodies allows us to estimate the medicine-likeness, which can be further improved by modification of relevant properties such as the extent and magnitude of surface hydrophobicity and charged patches [24].

The resulting recommendations of single or combinations of distinct amino acid exchanges help to design a highly individual engineering strategy of humanization and optimization that is specifically tailored to the mAb candidate. Based on the criteria of the research target profile (RTP), the best LO candidates (~3–6) are then selected for large-scale production to supply material for the next phase. These processes are mostly applied for the optimization of conventional mAbs but can also be extended for multi-specifics along with the additional engineering required to optimize the second and/or third Fv or scFv domains. In addition, identifying an optimal multi-specific format that combines the individually optimized variable domains is required to finalize the LO candidate(s).

14.2.4 Formatting of Conventional and Next-Generation Antibodies

After optimizing the Fv portions, the engineering of conventional mAbs continues with the formatting of the Fvs in the desired antibody format. In this step, the Fv is combined with the Fc of a desired IgG isotype. In this phase, engineering of the Fc might be required to adapt the receptor-mediated functions of the mAb such as ADCC, ADCP, CDC, and endosomal recycling [93]. Depending upon the therapeutic concept, the design of next-generation biotherapeutics as bi- and multi-specific antibodies might require another intermediate formatting step to assess physicochemical compatibility of individual specificities with one another, assuring that the multi-specific modalities have desirable developability properties. Individual case studies already reported how structure-based engineering can support antibody formatting. One case study showed that after converting from scFv to IgG, the affinity of a TGFβ1 (Transforming growth factor β1) binder could be successfully restored by structure-guided reengineering of elbow region [94]. Analogous computational-guided approaches have the power to support formatting of more complex next-generation antibodies such as multi-specific biotherapeutics.

In the last step of the discovery process, a few of the top performing lead variants are assessed in pre-formulation studies prior to transfer to development for cell line generation and early developability assessments to test the fit of the finale molecule(s) to the platforms of upstream, downstream and formulation development [4]. The research phase is completed by the selection of the final candidate for the start of development.

14.2.5 In Silico Assessments in Development

The preliminary stages of drug substance and drug product development are known to be resource intense. Consequently, the full development program comprising of cell line development, upstream and downstream manufacturing process, and formulation development can in most cases only be conducted for the final lead candidate. However, at the time of selection of the final lead candidate, experimental data is only sparsely available due to limitations on quantity as well as quality of material available. At the same time, the sequence of the final lead candidate gets locked at the start of development and not even single point mutations are allowed. This decision inherently puts product development in a disadvantaged situation since real-time data of the candidates' stability are typically not available at the start of development but are essential to meet the regulatory requirements for shelf-life, CQAs, and product heterogeneity. Therefore, there is a particularly strong demand for an early, fast, and reliable prediction of various stability aspects that can be covered by hybrid approaches of in vitro and in silico techniques.

14.3 In Silico Tools for Developability Assessments

Over the last decade, the scientific community has developed a diverse set of computational tools that can be applied in R&D to assess many different developability aspects of biotherapeutics. In discovery, the result obtained from in silico developability assessments can be considered during the selection of hits as well as identification/optimization and engineering of the lead molecule to produce more easily developable antibody drug candidates for drug product development. In the development phase, the in silico tools can help to estimate the final fit of the candidate to standardized platforms, identify potential developability issues, guide adjustments from the platforms that might be required, and finally interpret the complex results of experimental studies [95]. The following section provides an overview of in silico tools that computationally characterize therapeutic candidates and predict important developability properties.

14.3.1 Structure Prediction

Some risk factors such as chemical modification sites can be even identified at the level of the amino acid sequence, but many other developability factors such as biophysical properties depend on the three-dimensional structure of the biotherapeutic. Therefore, most biopharmaceutical companies conduct experimental studies as X-ray crystallography or nuclear magnetic resonance spectroscopy to solve the molecular structures of the lead candidates alone or in complex with the respective target. In case no experimental structures are available at the required time in the project, homology modelling is often used to predict the three-dimensional structure of biologic drug candidates using their amino acid sequence. The modelling abilities have been significantly improved over the last decade by advances in computational power, modelling techniques, and databases of sequences (NGS, next-generation sequencing) and structures [96–101]. Since the modelling of the variable regions of the antibodies can be performed automatically in a high-throughput manner, it is particularly useful toward discovery stages to analyze large sets of candidates [102–105]. Homology modelling is especially suited for the prediction of mAb structures since the framework regions are highly conserved [106]. Despite technological advances in template-based, fragment-based, and template-free modelling, the greatest challenge remains the prediction of diverse CDR canonical classes. Particularly, modelling of the conformations of HCDR3 loops with significantly varying lengths is challenging due to the sequence variability and structure flexibility [103, 107–111]. Currently, the most accurate loop models seem to be achieved via hybrid strategies that combine the benefits of knowledge- and physics-based approaches [107, 112]. For a further improved understanding of the structural characteristics of the dynamic structure of a mAb in solution, additional molecular dynamics (MD) techniques that can capture antibody fluctuations can be applied [69, 113, 114]. Currently, modelling the full-length structures for IgG mAbs, and, for next-generation multi-specific antibodies is challenging because the PDB contains only a handful of such crystal structures.

DeepMind's AlphaFold demonstrated the great potential of deep learning for protein structure prediction [100, 101]. Furthermore, the prediction of flexible systems such as interfaces is still complicated due to the complex balancing of polar and nonpolar interactions as well as solvation effects [97]. At the same time, ML techniques have a great potential to revolutionize the field of template-free prediction of protein structures [97]. Particular attention was attracted by a protein-specific fragment library that has been recently generated via deep neural networks [100, 101]. The analysis identified patterns in protein sequence and co-evolutionary couplings, which have been converted in contact maps [100]. Novel algorithms have further been able to engineer de novo high-order assemblies with therapeutic potential using bioinformatics, even so the design from scratch remains a massive undertaking [98, 115, 116].

14.3.2 Biophysical Properties

The developability of drugs is mostly specified by its biophysical properties. Lipinski's "rule-of-five" represented a substantial leap for the discovery and development campaigns of small molecules since it related calculable physicochemical properties with drug characteristics as solubility and permeability [117]. The complexity of biological entities hampered the definition of similar guidelines for NBEs (New Biological Entities) for decades, but a biophysical screening of clinical stage antibodies guided the empirical definition of analogous boundaries [22, 24]. Subsequently, the heterogeneous biophysical measures have been successfully correlated with sequence features, indicating significant relationships between the sequence of Fv domains and physicochemical properties that define the developability of antibodies [21, 23]. For instance, general correlates have been established between mAb polyreactivity and the presence of basic amino acids in the CDRs as well as Gln in HCDR2 and 3 [23]. Furthermore, mAb aggregation and self-association have been correlated with CDR length and content of aromatic residues [23]. However, since the particular relationship between specific amino acids and other physicochemical properties of the mAb such as conformational stability strongly depends on the structural microenvironment, key residues and their modifications must be individually evaluated under consideration of the structural context [118]. Therefore, another study considered Fv models to establish developability guidelines and develop a Therapeutic Antibody Profiler (TAP) [24]. This tool assesses the developability of candidates via five easily calculable metrics for the total length of CDRs, positively and negatively charged patches as well as hydrophobic patches in the CDRs, and the asymmetry in the surface charges of heavy and light chains [24]. Similar guidelines remain to be developed for multi-specific antibodies that often display suboptimal physical properties impeding the development into therapeutics [119].

14.3.3 Hydrophobicity

The pharmaceutical industry widely recognizes the relevance of hydrophobicity of mAbs for its developability into therapeutics, for which reason Hydrophobic Interaction Chromatography (HIC) is often conducted to compare the apparent hydrophobicity of mAb candidates to assess downstream risks [6, 11, 22]. Even though an accurate prediction of HIC retention times is challenging since it is influenced by the solvent (e.g., salt gradients) and other protein characteristics (e.g., charge) [21], HIC retention times could successfully be correlated with sequence and structure features by using diverse methods as Quantitative Structure Property Relationship (QSPR) modeling or machine learning [120–122]. To rectify homology modeling errors and capture the dynamics of the rather flexible formation and breakage of hydrophobic patches, it is recommended to consider conformational sampling or MD simulations [121–125].

14.3.4 Solution- and Colloidal-State Properties

The solution- and colloidal-state properties are amongst the most important aspects to be considered during discovery and development of a Novel Biologic Entity (NBE) drug candidate, but also amongst the hardest to predict due to multiple influencing factors as the individual patterning of hydrophobic and charged residues. However, recent advances in understanding the complex principles of protein solubility enabled the development of predictive tools. First computational tools such as SOLpro and PROSO II were trained on tens of thousands of proteins and successfully demonstrated their ability to predict solubility upon expression with an accuracy of ~75% [126, 127]. Later, web-based tools (e.g., Protein-Sol) have been developed that offer an easy access to the prediction of protein solubility from sequence [128]. Complementary to sequence-based approaches, CamSol calculates a residue-specific intrinsic solubility profile under consideration of structural influences, and offers to optimize the solubility of a candidate by screening for suitable mutations [129]. This feature makes it particularly interesting for the design of antibody libraries and the selection of lead candidates [130]. CamSol was recently extended to predict the aggregation potential of partially unfolded proteins in temperature ramps of molecular dynamics (MD) simulations [131]. For early development activities, such a tool enables us to assess the candidate's solubility relative to previous molecules without the need for material and laborious experimental activities [132]. An alternative software called SODA estimates changes in the protein solubility via the propensity of the sequence to aggregate (via PASTA) [133] and disorder (via ESpritz) [134], and takes further properties as hydrophobicity and secondary structure (via FELLS) [135] into account [136]. Several computational approaches have already successfully guided the rational design of mAbs with increased solution- and colloidal-state properties [137–142].

A currently evolving field focuses on the prediction of mAb specificity, which refers not only to nonspecific binding but also self-interaction. Recently, a method has been described that predicts the overall specificity of antibodies [26]. Individual and combined sets of chemical rules have been defined that recommend limits of certain amino acids exposed to the surface of the variable region.

While mechanistic tools are most valuable for the screening and minimization of APRs in the hits and lead candidates during discovery [137–139, 143–149], the kinetic predictors are helpful to estimate the rate of aggregation which is key during the development of liquid formulations that must meet the regulatory requirements for the shelf life of the drug product [150–153]. The kinetic models can be trained by ML on large data sets of combined experimental and sequence/structure information [150, 152]. Desirable predictors could thereby screen different formulations to identify the optimal composition (pH, salt and excipients) for minimal kinetics.

14.3.5 Isoelectric Point (pI)

The isoelectric point (pI) is an important physicochemical property for mAbs, and it has been shown to correlate with specific developability aspects as thermostability, viscosity and resistance to HMW formation at low pH [4, 154]. Typically, IgG1s with weakly basic isoelectric points between 8 and 8.5 and Fv isoelectric points between 7.5 and 9 typically display the best combinations of strong repulsive self-interactions and weak nonspecific interactions [154]. Calculation of the pI on the basis of the mAb sequences is offered by multiple software tools as MassLynx [], Vector NTI [155], and EMBOSS [156], and the calculated pIs have been shown to be within a range of 15% compared to experimentally determined ones [157]. Tools that predict the pI on basis of the protein structure can provide a more accurate result, since the underlying residue p*K*a values are calculated by taking into account the residual microenvironments, e.g., shifts in p*K*a values in case the side chain forms a salt-bridge. A case study confirmed that structure-based pI calculations (via PROPKA) [158] corresponded to the measured values at least by ±1 pI unit, whereby the remaining discrepancies result from protein interactions with salt ions that leave room for improvements [11, 159].

14.3.6 Viscosity and Diffusion Interaction Parameter ($\mathbf{k}_D$)

Predictors of the concentration-dependent viscosity of formulations need to consider the pairwise and higher order self-association of antibodies [19]. Consequently, the main driving forces behind viscosity, i.e., electrostatics and hydrophobicity [20] have been linked to predictable characteristics of the Fv sequence and structure [159, 160]. While the viscosity of mAb formulations increases with hydrophobicity, charge dipole distribution, and aggregation propensity, it decreases with net charge [159, 160]. Considering these principles, the in silico tool SCM (spatial charge map) offers to detect highly viscous antibodies on the basis of the Fv structure [161]. The high-throughput approach makes it especially attractive for the lead identification and optimization performed at discovery. An alternative mathematical model considers again the hydrophobicity and charges of antibody regions to predict concentration-dependent viscosity curves [162]. Charge-related in silico descriptors (e.g., pI, net charge, charge on Fv, zeta potential) could be used to calculate the diffusion interaction parameter k_D [163] that itself is widely-used in developability assessments to predict the viscoelastic behavior of mAb solutions [164–166]. Several case studies confirmed that these computational tools as well as their underlying principles are powerful to rationally (re-)design antibodies with decreased viscosity [137, 138, 141, 142, 167–171]. Further studies indicated that MD simulations can be conducted to additionally estimate the effect of formulation components on the viscosity [137, 170, 172–176].

14.3.7 Aggregation and Self-Association

Aggregation in biotherapeutics can nucleate from a wide array of origins ranging from reversible self-association, conformational change, chemical modification, hydrophobic and electrostatic complementarity among the interacting partners. Despite this complexity, aggregation could be directly associated to characteristics such as presence of aggregation prone regions, hydrophobicity [177], electrostatics [178], and dipole moments [179], paving the way for both sequence- and structure-based computational predictions. The in silico tools are particularly powerful in guiding the design of mAb candidates with high colloidal stability, since the effect of single and multiple amino acid exchanges on the aggregation propensity can be predicted. A selection of these predictors is summarized below, and this aspect has been described in several reviews in more details [147, 151, 180].

A vast majority of aggregation predictors are based on the sequence as they calculate aggregation propensity values for sequence stretches. These methods are based on the fact that natural amino acids differ in their physicochemical properties, and individually influence the aggregation of peptides and proteins [181]. Therefore, techniques as Zyggregator and Pafig use intrinsic properties of the amino acids such as hydrophobicity, charge and secondary-structure propensity to identify aggregation-promoting regions and predict the rate of growth [182, 183]. Similarly, a method called AGGRESCAN identifies aggregation-prone segments via the aggregation propensity of natural amino acids, whereby this method was trained on empirical data [184]. Alternative tools like TANGO, PASTA, PASTA2, FoldAmyloid, SALSA; and AggreRATE-Pred detect aggregation-prone regions and impact of mutations on aggregation kinetics via the physicochemical properties of secondary structure elements, in particular the ability to form intermolecular cross-β-structures [133, 150, 185–188]. In accordance, the frequency of β-sheets was amongst others (e.g., pI and atom-based hydrophobic moment) carved out by a neural network approach that led to the development of APPNN [189]. Another predictor named WALTZ consults a database of experimentally determined amyloid-forming hexapeptides to predict amyloid-forming sequences [190, 191]. Following a consensus prediction algorithm, 5 and later on 11 other methods (AGGRESCAN, Pafig, Tango, Waltz, and other conformational predictors) have been combined to produce AMYLPRED and AMYLPRED2 [192, 193]. Similarly, a statistical approach combined existing tools (SALSA, PAFIG, FoldAmyloid, and Waltz) to develop a meta-predictor for amyloid proteins called MetAmyl [194]. Alternatively, artificial neuronal networks (ANN) were applied to predict the aggregation temperature (T_{agg}) of antibodies based on the amino acid composition, of which charged residues and Met affected the temperature induced aggregation the most [195]. Machine-learning techniques were conducted to develop CISI, which predicts the cross-interaction or self-interaction of mAbs on the basis of tripeptide sequences [196]. Interestingly, this study indicated that tripeptides comprising Tyr, Ser and Ala residues were amongst the most prominent antibody-binding sites.

In comparison to these sequence-based methods, the structure-based ones consider the three-dimensional context of the protein fold to predict aggregation. Therefore, methods as AGGRESCAN3D (A3D) or Solubis combine distinct aggregation propensity scales (AGGRESCAN or TANGO) with the protein structure [148, 197, 198]. The A3D server further allows to include dynamic fluctuations of the protein structure in solution that might affect the aggregation tendency [197]. It was further pointed out that atomic level characteristics might be relevant for understanding aggregation, which is offered by ANuPP [145]. Another method called AggScore identifies aggregation hotspots on the protein surface via calculation of hydrophobic and electrostatic patches [199]. Orthogonal to these technologies, SAP (spatial aggregation propensity) calculates dynamically exposed hydrophobicity of protein patches [123, 200]. The Developability Index (DI) has complemented the SAP concept with additional information about the net charge of the full-length mAb to account for hydrophobicity and electrostatic interactions [201]. Recently, a machine learning approach combined SAP with other tools (Spatial negative charge map (SCM neg), Spatial positive charge map (SCM pos)) that extract molecular features from MD simulations to train predictive models for the aggregation rate at high mAb concentrations [202].

14.3.8 Prediction of Paratope

Once an experimentally determined or accurately modeled structure of the therapeutic candidate is available, the mAb can be further analyzed to identify the paratope comprising the amino acid residues that mediate the binding of its cognate antigen epitope. Oversimplified approaches assume that the paratope equals the six CDR regions comprising together 50–60 residues on average, but statistic evaluation of Ab/Ag complexes showed that the actual paratope is only formed by 18–19 residues on average [203]. Therefore, gathering a detailed understanding of the exact paratope is crucial to guide an individually tailored engineering and development strategy that avoids costly and time-consuming failures. However, especially at early research stages, the Ab/Ag complex might not have been experimentally characterized. In case information about the antigen sequence and/or structure are known, bioinformatic tools offer the prediction of the mAb interaction with the Ag structure/model that can complement experimental paratope studies and streamline conventional R&D workflows [204]. Similar to homology modelling, great advances have also been made in the field of paratope prediction [204, 205]. In the last years, several paratope predictors that pursue individual approaches were established. The most common technique considers the docking of Ab/Ag for the prediction of the paratope [206–208]. Another method identifies the paratope via circular patches of the antibody surface that exhibit certain physicochemical properties [209]. Deep neural networks are not only suitable for the paratope prediction but outperformed classical methods [205, 210]. Accurate prediction of paratopes is essential to identify potential physicochemical liabilities motifs that may overlap with antigen-binding regions. Removal of such motifs via mutations in paratope, might lead to significant modification of the biologic drug candidate's affinity toward its cognate receptor.

14.3.9 Post-Translational Modifications (PTM)

Many different approaches have been pursued to develop predictive tools for sites of post-translational modifications of which oxidation, deamidation, isomerization, and glycation are the most prominent ones.

Oxidation occurs most frequently at the side chains of Trp and Met that differ in their atomistic nature (aromatic ring; sulfur), sensitivity toward different oxidative stress (light exposure, metal-catalyzation and free radicals), and oxidized products [211]. Therefore, different in silico tools have been developed for the prediction of Trp and Met oxidation sites. For the prediction of Met oxidation, correlations between the experimental oxidation data of stressed mAbs and the solvent accessible surface area (SASA) of the Met side chain have been seen [212]. Recently, several different machine learning approaches have been conducted to train and fit different classification models such as MetODeep for the prediction of oxidation-sensitive Met sites [213–217]. Some of the resulting models used random forest algorithm and consider characteristics about sequence, secondary and tertiary structures, and dynamics to predict Met oxidation. Interestingly, again the SASA of the Met side chain was amongst the most important discriminators. Furthermore, the number of residues between the analyzed Met and the next Met toward the N-terminus and the spatial distance between the sulfur atom and the closest aromatic residue affected the oxidation propensity [215]. Typically, MD simulations are conducted to capture the dynamics of the structure and improve the prediction of critical oxidation sites [218, 219].

Like the Met oxidation sites, high solvent accessibility of the Trp residue is a prerequisite for oxidation [159, 220, 221]. However, it is hypothesized that in case of Met residues the solvent exposure of the sulfur atom is more important than that of the entire side chain [27, 215, 219]. In contrast, the entire side chain is considered for the prediction of Trp sites sensitive to oxidation. The oxidation susceptibility does not linearly correlate to the SASA of the Trp residue, indicating that other structural characteristics as side chain orientation and/or surrounding structural elements might play a role [221]. The predictive power of such tools remains to be verified for molecules that underwent different types of stress (light, reactive oxygen, metal).

Sites sensitive to deamidation and isomerization have been historically identified via sequence motifs (e.g., NG), but this approach was shown to be less accurate and prone to overprediction [222]. Therefore, the risk predictions could be improved for mAbs by not only considering the motif but also the CDR region [223]. Later algorithms as NGOME were developed under consideration of simple predictions of secondary structure elements [224]. Finally, the complex conformational environment of the reactive groups, particularly the local structure, structure flexibility and solvent accessibility, has been considered in several other studies [159, 225–228]. Furthermore, a predictor for Asn deamidation was recently trained on structural features via machine learning algorithms [229]. Thereby, Asn torsion, flexibility, solvent exposure, structural elements, and the distance between C-N for the nucleophilic attack were identified as the most relevant descriptors [229].

14.4 Conclusion and Future Directions

Computational approaches are finding increasing applications toward discovery and development of biotherapeutics in recent years. This chapter has attempted to provide an overview of this emerging field, we call as biopharmaceutical informatics. However, several major challenges remain toward increasing the contribution of biopharmaceutical informatics and thereby increasing the probability of successful translation of biologic drug candidates into medicines available in the clinic. One of the most significant hurdles for greater acceptance for biopharmaceutical informatics is the predictability and validation of the computational tools. Validation via self-consistent experimental data is important for the development of these emerging computational tools, but self-consistent experimental data sets on biologic macromolecules are rarely available in public domain. A second hurdle is that the use of antibody-based biologics as medicines is more recent when compared to the small molecule drugs. Therefore, the number of biologic drug products currently available in the market is small and our overall pharmaceutical development experience with these macromolecules is not as extensive as in the case of the small molecule drug products. This situation is further complicated by the pervasive nature of empiricism inherent to both experimental and computational tools used to study biologic drug candidates, and with the arrival of novel biotherapeutic formats such as bi-specific and multi-specific antibodies. A third hurdle for the progress of biopharmaceutical informatics is created if the computational tools are not available easily to everyone, and the underlying core programming codes are not published. A fourth hurdle is the general unavailability of the data in digital format in central repositories within and outside the biopharmaceutical companies. The need for digital transformation of biopharmaceutical industry cannot be understated. However, digital transformation of biopharmaceutical industry is hampered by the lack of digitization of the experimental data. Overcoming the digitization hurdle that makes the experimental data available to computational biophysicists, data scientists, and machine learners is the foundational step toward digital transformation. Once this is accomplished, the scientists shall then be able to use the data and connect them with sequence and structural characteristics of biologic drug candidates. Discovery of correlations among "microscopic" (properties/descriptors derived from amino acid sequence, protein structure and molecular simulations) and "macroscopic" (experimental biochemical and biophysical experiments) attributes of biologic drug candidates will prove crucial for navigating developability of biologic drugs and will truly enable biopharmaceutical informatics. As summarized in this article, these are still early days for this field as computational as well as hybrid studies containing computational and experimental data are beginning to emerge. Overall, this field is expected to grow exponentially in the coming years.

References

1. Kaplon H, Chenoweth A, Crescioli S, Reichert JM (2022) Antibodies to watch in 2022. MAbs 14:2014296. https://doi.org/10.1080/19420862.2021.2014296
2. Yamaguchi S, Kaneko M, Narukawa M (2021) Approval success rates of drug candidates based on target, action, modality, application, and their combinations. Clin Transl Sci 14:1113–1122. https://doi.org/10.1111/cts.12980
3. Ecker DM, Jones SD, Levine HL (2015) The therapeutic monoclonal antibody market. MAbs 7:9–14. https://doi.org/10.4161/19420862.2015.989042
4. Bailly M, Mieczkowski C, Juan V et al (2020) Predicting antibody developability profiles through early stage discovery screening. MAbs 12:1743053. https://doi.org/10.1080/19420862.2020.1743053
5. Saxena V, Panicucci R, Joshi Y, Garad S (2009) Developability assessment in pharmaceutical industry: an integrated group approach for selecting developable candidates. J Pharm Sci 98:1962–1979. https://doi.org/10.1002/jps.21592
6. Xu Y, Wang D, Mason B et al (2018) Structure, heterogeneity and developability assessment of therapeutic antibodies. MAbs 11:1–26. https://doi.org/10.1080/19420862.2018.1553476
7. Bui LA, Hurst S, Finch GL et al (2015) Key considerations in the preclinical development of biosimilars. Drug Discov Today 20:3–15. https://doi.org/10.1016/j.drudis.2015.03.011
8. Vulto AG, Jaquez OA (2017) The process defines the product: what really matters in biosimilar design and production? Rheumatology 56:iv14–iv29. https://doi.org/10.1093/rheumatology/kex278
9. Rathore AS, Winkle H (2009) Quality by design for biopharmaceuticals. Nat Biotechnol 27:26–34. https://doi.org/10.1038/nbt0109-26
10. Carrara SC, Ulitzka M, Grzeschik J et al (2020) From cell line development to the formulated drug product: the art of manufacturing therapeutic monoclonal antibodies. Int J Pharm 594:120164. https://doi.org/10.1016/j.ijpharm.2020.120164
11. Jarasch A, Koll H, Regula JT et al (2015) Developability assessment during the selection of novel therapeutic antibodies. J Pharm Sci 104:1885–1898. https://doi.org/10.1002/jps.24430
12. Wang Q, Chen Y, Park J et al (2019) Design and production of bispecific antibodies. Antibodies 8:43. https://doi.org/10.3390/antib8030043
13. Furtmann N, Schneider M, Spindler N et al (2021) An end-to-end automated platform process for high-throughput engineering of next-generation multi-specific antibody therapeutics. MAbs 13:1955433. https://doi.org/10.1080/19420862.2021.1955433
14. Cocco P, Ayaz-Shah A, Messenger MP et al (2020) Target Product Profiles for medical tests: a systematic review of current methods. BMC Med 18:119. https://doi.org/10.1186/s12916-020-01582-1
15. Baumann A (2006) Early development of therapeutic biologics – pharmacokinetics. Curr Drug Metab 7:15–21. https://doi.org/10.2174/138920006774832604
16. Thomas F (2019) Considering alternative dosage forms in biologics. BioPharm Int 32:16–18
17. Kumar S, Plotnikov NV, Rouse JC, Singh SK (2018) Biopharmaceutical Informatics: supporting biologic drug development via molecular modelling and informatics. J Pharm Pharmacol 70:595–608. https://doi.org/10.1111/jphp.12700
18. Chiu ML, Goulet DR, Teplyakov A, Gilliland GL (2019) Antibody structure and function: the basis for engineering therapeutics. Antibodies 8:55. https://doi.org/10.3390/antib8040055
19. Kuroda D, Tsumoto K (2020) Engineering stability, viscosity, and immunogenicity of antibodies by computational design. J Pharm Sci 109:1631–1651. https://doi.org/10.1016/j.xphs.2020.01.011
20. Tomar DS, Kumar S, Singh SK et al (2016) Molecular basis of high viscosity in concentrated antibody solutions: strategies for high concentration drug product development. MAbs 8:216–228. https://doi.org/10.1080/19420862.2015.1128606

21. Hebditch M, Warwicker J (2019) Charge and hydrophobicity are key features in sequence-trained machine learning models for predicting the biophysical properties of clinical-stage antibodies. PeerJ 7:e8199. https://doi.org/10.7717/peerj.8199
22. Jain T, Sun T, Durand S et al (2017) Biophysical properties of the clinical-stage antibody landscape. Proc Natl Acad Sci USA 114:944–949. https://doi.org/10.1073/pnas.1616408114
23. Lecerf M, Kanyavuz A, Lacroix-Desmazes S, Dimitrov JD (2019) Sequence features of variable region determining physicochemical properties and polyreactivity of therapeutic antibodies. Mol Immunol 112:338–346. https://doi.org/10.1016/j.molimm.2019.06.012
24. Raybould MIJ, Marks C, Krawczyk K et al (2019) Five computational developability guidelines for therapeutic antibody profiling. Proc Natl Acad Sci USA 116:4025–4030. https://doi.org/10.1073/pnas.1810576116
25. Starr CG, Tessier PM (2019) Selecting and engineering monoclonal antibodies with drug-like specificity. Curr Opin Biotechnol 60:119–127. https://doi.org/10.1016/j.copbio.2019.01.008
26. Zhang Y, Wu L, Gupta P et al (2020) Physicochemical rules for identifying monoclonal antibodies with drug-like specificity. Mol Pharm 17:2555–2569. https://doi.org/10.1021/acs.molpharmaceut.0c00257
27. Grassi L, Cabrele C (2019) Susceptibility of protein therapeutics to spontaneous chemical modifications by oxidation, cyclization, and elimination reactions. Amino Acids 51:1409–1431. https://doi.org/10.1007/s00726-019-02787-2
28. Alam ME, Barnett GV, Slaney TR et al (2019) Deamidation can compromise antibody colloidal stability and enhance aggregation in a pH-dependent manner. Mol Pharm 16:1939–1949. https://doi.org/10.1021/acs.molpharmaceut.8b01311
29. Aswad DW, Paranandi MV, Schurter BT (2000) Isoaspartate in peptides and proteins: formation, significance, and analysis. J Pharm Biomed Anal 21(6):1129–1136. https://doi.org/10.1016/s0731-7085(99)00230-7
30. Geiger T, Clark S (1987) Deamidation, isomerization, and racemization at asparaginyl and aspartyl residues in peptides. J Biol Chem 262:785–794
31. Füssl F, Trappe A, Cook K et al (2018) Comprehensive characterisation of the heterogeneity of adalimumab via charge variant analysis hyphenated on-line to native high resolution Orbitrap mass spectrometry. MAbs 11:116–128. https://doi.org/10.1080/19420862.2018.1531664
32. Gervais D (2016) Protein deamidation in biopharmaceutical manufacture: understanding, control and impact. J Chem Technol Biotechnol 91:569–575. https://doi.org/10.1002/jctb.4850
33. Yan B, Steen S, Hambly D et al (2009) Succinimide formation at Asn 55 in the complementarity determining region of a recombinant monoclonal antibody IgG1 heavy chain. J Pharm Sci 98:3509–3521. https://doi.org/10.1002/jps.21655
34. Rehder DS, Chelius D, McAuley A et al (2008) Isomerization of a single aspartyl residue of anti-epidermal growth factor receptor immunoglobulin γ2 antibody highlights the role avidity plays in antibody activity. Biochemistry 47:2518–2530. https://doi.org/10.1021/bi7018223
35. Wakankar AA, Borchardt RT, Eigenbrot C et al (2007) Aspartate isomerization in the complementarity-determining regions of two closely related monoclonal antibodies. Biochemistry 46:1534–1544. https://doi.org/10.1021/bi061500t
36. Cacia J, Keck R, Presta LG, Frenz J (1996) Isomerization of an aspartic acid residue in the complementarity-determining regions of a recombinant antibody to human IgE: identification and effect on binding affinity. Biochemistry 35(6):1897–1903. https://doi.org/10.1021/bi951526c
37. Harris RJ, Kabakoff B, Macchi FD et al (2001) Identification of multiple sources of charge heterogeneity in a recombinant antibody. J Chromatogr B Biomed Sci Appl 752:233–245. https://doi.org/10.1016/s0378-4347(00)00548-x
38. Sreedhara A, Cordoba A, Zhu Q et al (2012) Characterization of the isomerization products of aspartate residues at two different sites in a monoclonal antibody. Pharm Res 29:187–197. https://doi.org/10.1007/s11095-011-0534-2

39. Sakhnini LI, Greisen PJ, Wiberg C et al (2019) Improving the developability of an antigen binding fragment by aspartate substitutions. Biochemistry 58:2750–2759. https://doi.org/10.1021/acs.biochem.9b00251
40. Zhang B, Yang Y, Yuk I et al (2008) Unveiling a glycation hot spot in a recombinant humanized monoclonal antibody. Anal Chem 80:2379–2390. https://doi.org/10.1021/ac701810q
41. Wei B, Berning K, Quan C, Zhang YT (2017) Glycation of antibodies: modification, methods and potential effects on biological functions. MAbs 9:586–594. https://doi.org/10.1080/19420862.2017.1300214
42. Miller AK, Hambly DM, Kerwin BA et al (2011) Characterization of site-specific glycation during process development of a human therapeutic monoclonal antibody. J Pharm Sci 100:2543–2550. https://doi.org/10.1002/jps.22504
43. Manning MC, Liu J, Li T, Holcomb RE (2018) Rational design of liquid formulations of proteins. Adv Protein Chem Struct Biol 112:1–59. https://doi.org/10.1016/bs.apcsb.2018.01.005
44. Oyama H, Koga H, Tadokoro T et al (2019) Relation of colloidal and conformational stabilities to aggregate formation in a monoclonal antibody. J Pharm Sci 109:308–315. https://doi.org/10.1016/j.xphs.2019.10.038
45. Rizzo JM, Shi S, Li Y et al (2015) Application of a high-throughput relative chemical stability assay to screen therapeutic protein formulations by assessment of conformational stability and correlation to aggregation propensity. J Pharm Sci 104:1632–1640. https://doi.org/10.1002/jps.24408
46. Kalonia CK, Heinrich F, Curtis JE et al (2018) Protein adsorption and layer formation at the stainless steel–solution interface mediates shear-induced particle formation for an IgG1 monoclonal antibody. Mol Pharm 15:1319–1331. https://doi.org/10.1021/acs.molpharmaceut.7b01127
47. Brader ML, Estey T, Bai S et al (2015) Examination of thermal unfolding and aggregation profiles of a series of developable therapeutic monoclonal antibodies. Mol Pharm 12:1005–1017. https://doi.org/10.1021/mp400666b
48. Fleischman ML, Chung J, Paul EP, Lewus RA (2017) Shipping-induced aggregation in therapeutic antibodies: utilization of a scale-down model to assess degradation in monoclonal antibodies. J Pharm Sci 106:994–1000. https://doi.org/10.1016/j.xphs.2016.11.021
49. Nejadnik MR, Randolph TW, Volkin DB et al (2018) Post-production handling and administration of protein pharmaceuticals and potential instability issues. J Pharm Sci 107:2013–2019. https://doi.org/10.1016/j.xphs.2018.04.005
50. Cromwell MEM, Hilario E, Jacobson F (2006) Protein aggregation and bioprocessing. AAPS J 8(3):E572–E579. https://doi.org/10.1208/aapsj080366
51. Yadav S, Laue TM, Kalonia DS et al (2012) The influence of charge distribution on self-association and viscosity behavior of monoclonal antibody solutions. Mol Pharm 9:791–802. https://doi.org/10.1021/mp200566k
52. Yadav S, Sreedhara A, Kanai S et al (2011) Establishing a link between amino acid sequences and self-associating and viscoelastic behavior of two closely related monoclonal antibodies. Pharm Res 28:1750–1764. https://doi.org/10.1007/s11095-011-0410-0
53. Geng SB, Cheung JK, Narasimhan C et al (2014) Improving monoclonal antibody selection and engineering using measurements of colloidal protein interactions. J Pharm Sci 103:3356–3363. https://doi.org/10.1002/jps.24130
54. Nishi H, Miyajima M, Nakagami H et al (2010) Phase separation of an IgG1 antibody solution under a low ionic strength condition. Pharm Res 27:1348–1360. https://doi.org/10.1007/s11095-010-0125-7
55. Ghazvini S, Kalonia C, Volkin DB, Dhar P (2016) Evaluating the role of the air-solution interface on the mechanism of subvisible particle formation caused by mechanical agitation for an IgG1 mAb. J Pharm Sci 105:1643–1656. https://doi.org/10.1016/j.xphs.2016.02.027
56. Duerkop M, Berger E, Dürauer A, Jungbauer A (2018) Impact of cavitation, high shear stress and air/liquid interfaces on protein aggregation. Biotechnol J 13:1800062. https://doi.org/10.1002/biot.201800062

57. Hauptmann A, Podgoršek K, Kuzman D et al (2018) Impact of buffer, protein concentration and sucrose addition on the aggregation and particle formation during freezing and thawing. Pharm Res 35:101. https://doi.org/10.1007/s11095-018-2378-5
58. Koepf E, Eisele S, Schroeder R et al (2018) Notorious but not understood: how liquid-air interfacial stress triggers protein aggregation. Int J Pharm 537:202–212. https://doi.org/10.1016/j.ijpharm.2017.12.043
59. Fesinmeyer RM, Hogan S, Saluja A et al (2009) Effect of ions on agitation- and temperature-induced aggregation reactions of antibodies. Pharm Res 26:903–913. https://doi.org/10.1007/s11095-008-9792-z
60. Lapidoth GD, Baran D, Pszolla GM et al (2015) AbDesign: an algorithm for combinatorial backbone design guided by natural conformations and sequences. Proteins Struct Funct Bioinf 83:1385–1406. https://doi.org/10.1002/prot.24779
61. Sirin S, Apgar JR, Bennett EM, Keating AE (2016) AB-Bind: antibody binding mutational database for computational affinity predictions. Protein Sci 25:393–409. https://doi.org/10.1002/pro.2829
62. Robinson J, Barker DJ, Georgiou X et al (2019) IPD-IMGT/HLA database. Nucleic Acids Res 48:D948–D955. https://doi.org/10.1093/nar/gkz950
63. Swindells MB, Porter CT, Couch M et al (2017) abYsis: integrated antibody sequence and structure—management, analysis, and prediction. J Mol Biol 429:356–364. https://doi.org/10.1016/j.jmb.2016.08.019
64. Kovaltsuk A, Leem J, Kelm S et al (2018) Observed antibody space: a resource for data mining next-generation sequencing of antibody repertoires. J Immunol 201:2502–2509. https://doi.org/10.4049/jimmunol.1800708
65. Dunbar J, Krawczyk K, Leem J et al (2014) SAbDab: the structural antibody database. Nucleic Acids Res 42:D1140–D1146. https://doi.org/10.1093/nar/gkt1043
66. Berman HM, Westbrook J, Feng Z et al (2000) The Protein Data Bank. Nucleic Acids Res 1:235–242. https://doi.org/10.1093/nar/28.1.235
67. Ehrenmann F, Kaas Q, Lefranc M-P (2010) IMGT/3Dstructure-DB and IMGT/DomainGapAlign: a database and a tool for immunoglobulins or antibodies, T cell receptors, MHC, IgSF and MhcSF. Nucleic Acids Res 38:D301–D307. https://doi.org/10.1093/nar/gkp946
68. Tiller T, Schuster I, Deppe D et al (2013) A fully synthetic human Fab antibody library based on fixed VH/VL framework pairings with favorable biophysical properties. MAbs 5:445–470. https://doi.org/10.4161/mabs.24218
69. Pan X, Kortemme T (2021) Recent advances in de novo protein design: principles, methods, and applications. J Biol Chem 296:100558. https://doi.org/10.1016/j.jbc.2021.100558
70. Amimeur T, Shaver JM, Ketchem RR et al (2020) Designing feature-controlled humanoid antibody discovery libraries using generative adversarial networks. bioRxiv 2020.04.12.024844. https://doi.org/10.1101/2020.04.12.024844
71. Baran D, Pszolla MG, Lapidoth GD et al (2017) Principles for computational design of binding antibodies. Proc Natl Acad Sci USA 114:10900–10905. https://doi.org/10.1073/pnas.1707171114
72. Chowdhury R, Allan MF, Maranas CD (2018) OptMAVEn-2.0: de novo design of variable antibody regions against targeted antigen epitopes. Antibodies 7:23. https://doi.org/10.3390/antib7030023
73. Nimrod G, Fischman S, Austin M et al (2018) Computational design of epitope-specific functional antibodies. Cell Rep 25:2121–2131.e5. https://doi.org/10.1016/j.celrep.2018.10.081
74. Sheng Z, Bimela JS, Katsamba PS et al (2022) Structural basis of antibody conformation and stability modulation by framework somatic hypermutation. Front Immunol 12:811632. https://doi.org/10.3389/fimmu.2021.811632
75. Liu X, Taylor RD, Griffin L et al (2017) Computational design of an epitope-specific Keap1 binding antibody using hotspot residues grafting and CDR loop swapping. Sci Rep 7:41306. https://doi.org/10.1038/srep41306

76. Li W, Prabakaran P, Chen W et al (2016) Antibody aggregation: insights from sequence and structure. Antibodies 5:19. https://doi.org/10.3390/antib5030019
77. Roguska MA, Pedersen JT, Keddy CA et al (1994) Humanization of murine monoclonal antibodies through variable domain resurfacing. Proc Natl Acad Sci USA 91(3):969–973. https://doi.org/10.1073/pnas.91.3.969
78. Townsend S, Fennell BJ, Apgar JR et al (2015) Augmented binary substitution: single-pass CDR germ-lining and stabilization of therapeutic antibodies. Proc Natl Acad Sci USA 112:15354–15359. https://doi.org/10.1073/pnas.1510944112
79. Singh S, Kroe-Barrett RR, Canada KA et al (2015) Selective targeting of the IL23 pathway: generation and characterization of a novel high-affinity humanized anti-IL23A antibody. MAbs 7:778–791. https://doi.org/10.1080/19420862.2015.1032491
80. Teplyakov A, Obmolova G, Malia TJ et al (2016) Structural diversity in a human antibody germline library. MAbs 8:1–19. https://doi.org/10.1080/19420862.2016.1190060
81. Choi Y, Hua C, Sentman CL et al (2015) Antibody humanization by structure-based computational protein design. MAbs 7:1045–1057. https://doi.org/10.1080/19420862.2015.1076600
82. Teplyakov A, Obmolova G, Malia TJ et al (2018) Structural insights into humanization of anti-tissue factor antibody 10H10. MAbs 10:269–277. https://doi.org/10.1080/19420862.2017.1412026
83. ULC CCG (2021) Molecular Operating Environment (MOE), 2019.01.1010 Sherbooke St. West, Suite #910, Montreal, QC, Canada, H3A 2R7
84. Olimpieri PP, Marcatili P, Tramontano A (2015) Tabhu: tools for antibody humanization. Bioinformatics 31:434–435. https://doi.org/10.1093/bioinformatics/btu667
85. Abhinandan KR, Martin ACR (2007) Analyzing the "degree of humanness" of antibody sequences. J Mol Biol 369:852–862. https://doi.org/10.1016/j.jmb.2007.02.100
86. Gao SH, Huang K, Tu H, Adler AS (2013) Monoclonal antibody humanness score and its applications. BMC Biotechnol 13:55. https://doi.org/10.1186/1472-6750-13-55
87. Seeliger D (2013) Development of scoring functions for antibody sequence assessment and optimization. PLoS One 8:e76909. https://doi.org/10.1371/journal.pone.0076909
88. Lazar GA, Desjarlais JR, Jacinto J et al (2007) A molecular immunology approach to antibody humanization and functional optimization. Mol Immunol 44:1986–1998. https://doi.org/10.1016/j.molimm.2006.09.029
89. Choi Y, Verma D, Griswold KE, Bailey-Kellogg C (2017) EpiSweep: computationally driven reengineering of therapeutic proteins to reduce immunogenicity while maintaining function. Methods Mol Biol 1529:375–398
90. Clavero-Álvarez A, Mambro TD, Perez-Gaviro S et al (2018) Humanization of antibodies using a statistical inference approach. Sci Rep 8:14820. https://doi.org/10.1038/s41598-018-32986-y
91. van der Kant R, Bauer J, Karow-Zwick AR et al (2019) Adaption of human antibody λ and κ light chain architectures to CDR repertoires. Protein Eng Des Sel 32:109–127. https://doi.org/10.1093/protein/gzz012
92. Lehmann A, Wixted JHF, Shapovalov MV et al (2015) Stability engineering of anti-EGFR scFv antibodies by rational design of a lambda-to-kappa swap of the VL framework using a structure-guided approach. MAbs 7:1058–1071. https://doi.org/10.1080/19420862.2015.1088618
93. Mimoto F, Kuramochi T, Katada H et al (2016) Fc engineering to improve the function of therapeutic antibodies. Curr Pharm Biotechnol 17:1298–1314. https://doi.org/10.2174/1389201017666160824161854
94. Lord DM, Bird JJ, Honey DM et al (2018) Structure-based engineering to restore high affinity binding of an isoform-selective anti-TGFβ1 antibody. MAbs 10:444–452. https://doi.org/10.1080/19420862.2018.1426421
95. Somani S, Jo S, Thirumangalathu R et al (2021) Toward biotherapeutics formulation composition engineering using site-identification by ligand competitive saturation (SILCS). J Pharm Sci 110:1103–1110. https://doi.org/10.1016/j.xphs.2020.10.051

96. Fasnacht M, Butenhof K, Goupil-Lamy A et al (2014) Automated antibody structure prediction using Accelrys tools: results and best practices. Proteins Struct Funct Bioinf 82:1583–1598. https://doi.org/10.1002/prot.24604
97. Kuhlman B, Bradley P (2019) Advances in protein structure prediction and design. Nat Rev Mol Cell Biol 20:681–697. https://doi.org/10.1038/s41580-019-0163-x
98. Leman JK, Weitzner BD, Lewis SM et al (2020) Macromolecular modeling and design in Rosetta: recent methods and frameworks. Nat Methods 17:665–680. https://doi.org/10.1038/s41592-020-0848-2
99. Norman RA, Ambrosetti F, Bonvin AMJJ et al (2019) Computational approaches to therapeutic antibody design: established methods and emerging trends. Brief Bioinform 21:1549–1567. https://doi.org/10.1093/bib/bbz095
100. AlQuraishi M (2019) AlphaFold at CASP13. Bioinformatics 35:4862–4865. https://doi.org/10.1093/bioinformatics/btz422
101. Wei G-W (2019) Protein structure prediction beyond AlphaFold. Nat Mach Intell 1:336–337. https://doi.org/10.1038/s42256-019-0086-4
102. Leem J, Dunbar J, Georges G et al (2016) ABodyBuilder: automated antibody structure prediction with data–driven accuracy estimation. MAbs 8:1259–1268. https://doi.org/10.1080/19420862.2016.1205773
103. Lepore R, Olimpieri PP, Messih MA, Tramontano A (2017) PIGSPro: prediction of immunoGlobulin structures v2. Nucleic Acids Res 45:W17–W23. https://doi.org/10.1093/nar/gkx334
104. Yamashita K, Ikeda K, Amada K et al (2014) Kotai Antibody Builder: automated high-resolution structural modeling of antibodies. Bioinformatics 30:3279–3280. https://doi.org/10.1093/bioinformatics/btu510
105. Kemmish H, Fasnacht M, Yan L (2017) Fully automated antibody structure prediction using BIOVIA tools: validation study. PLoS One 12:e0177923. https://doi.org/10.1371/journal.pone.0177923
106. Teplyakov A, Luo J, Obmolova G et al (2014) Antibody modeling assessment II. Structures and models. Proteins Struct Funct Bioinf 82:1563–1582. https://doi.org/10.1002/prot.24554
107. Marks C, Deane CM (2017) Antibody H3 structure prediction. Comput Struct Biotechnol J 15:222–231. https://doi.org/10.1016/j.csbj.2017.01.010
108. Marze NA, Lyskov S, Gray JJ (2016) Improved prediction of antibody VL–VH orientation. Protein Eng Des Sel 29:409–418. https://doi.org/10.1093/protein/gzw013
109. Nishigami H, Kamiya N, Nakamura H (2016) Revisiting antibody modeling assessment for CDR-H3 loop. Protein Eng Des Sel 29:477–484. https://doi.org/10.1093/protein/gzw028
110. Regep C, Georges G, Shi J et al (2017) The H3 loop of antibodies shows unique structural characteristics. Proteins Struct Funct Bioinf 85:1311–1318. https://doi.org/10.1002/prot.25291
111. Tsuchiya Y, Mizuguchi K (2016) The diversity of H3 loops determines the antigen-binding tendencies of antibody CDR loops. Protein Sci 25:815–825. https://doi.org/10.1002/pro.2874
112. Zhu K, Day T, Warshaviak D et al (2014) Antibody structure determination using a combination of homology modeling, energy-based refinement, and loop prediction. Proteins Struct Funct Bioinf 82:1646–1655. https://doi.org/10.1002/prot.24551
113. Jay JW, Bray B, Qi Y et al (2018) IgG antibody 3D structures and dynamics. Antibodies 7:18. https://doi.org/10.3390/antib7020018
114. Yanaka S, Yogo R, Inoue R et al (2019) Dynamic views of the Fc region of immunoglobulin G provided by experimental and computational observations. Antibodies 8:39. https://doi.org/10.3390/antib8030039
115. Silva D-A, Yu S, Ulge UY et al (2019) De novo design of potent and selective mimics of IL-2 and IL-15. Nature 565:186–191. https://doi.org/10.1038/s41586-018-0830-7
116. Sormanni P, Aprile FA, Vendruscolo M (2018) Third generation antibody discovery methods: in silico rational design. Chem Soc Rev 47:9137–9157. https://doi.org/10.1039/c8cs00523k
117. Lipinski CA, Lombardo F, Dominy BW, Feeney PJ (2001) Experimental and computational approaches to estimate solubility and permeability in drug discovery and development settings. Adv Drug Deliv Rev 46:3–26. https://doi.org/10.1016/s0169-409x(00)00129-0

118. Lee HS, Qi Y, Im W (2015) Effects of N-glycosylation on protein conformation and dynamics: protein Data Bank analysis and molecular dynamics simulation study. Sci Rep 5:8926. https://doi.org/10.1038/srep08926
119. Sawant MS, Streu CN, Wu L, Tessier PM (2020) Toward drug-like multispecific antibodies by design. Int J Mol Sci 21:7496. https://doi.org/10.3390/ijms21207496
120. Jain T, Boland T, Lilov A et al (2017) Prediction of delayed retention of antibodies in hydrophobic interaction chromatography from sequence using machine learning. Bioinformatics 33:3758–3766. https://doi.org/10.1093/bioinformatics/btx519
121. Jetha A, Thorsteinson N, Jmeian Y et al (2018) Homology modeling and structure-based design improve hydrophobic interaction chromatography behavior of integrin binding antibodies. MAbs 10:1–11. https://doi.org/10.1080/19420862.2018.1475871
122. Karlberg M, de Souza JV, Fan L et al (2020) QSAR implementation for HIC retention time prediction of mAbs using fab structure: a comparison between structural representations. Int J Mol Sci 21:8037. https://doi.org/10.3390/ijms21218037
123. Chennamsetty N et al (2009) Design of therapeutic proteins with enhanced stability. Proc Natl Acad Sci USA 29:11937–11942. https://doi.org/10.1073/pnas.0904191106
124. Waibl F, Fernández-Quintero ML, Kamenik AS et al (2021) Conformational ensembles of antibodies determine their hydrophobicity. Biophys J 120(1):143–157. https://doi.org/10.1016/j.bpj.2020.11.010
125. Lai P-K, Fernando A, Cloutier TK et al (2021) Machine learning applied to determine the molecular descriptors responsible for the viscosity behavior of concentrated therapeutic antibodies. Mol Pharm 18:1167–1175. https://doi.org/10.1021/acs.molpharmaceut.0c01073
126. Magnan CN, Randall A, Baldi P (2009) SOLpro: accurate sequence-based prediction of protein solubility. Bioinformatics 25:2200–2207. https://doi.org/10.1093/bioinformatics/btp386
127. Smialowski P, Doose G, Torkler P et al (2012) PROSO II – a new method for protein solubility prediction. FEBS J 279:2192–2200. https://doi.org/10.1111/j.1742-4658.2012.08603.x
128. Hebditch M, Carballo-Amador MA, Charonis S et al (2017) Protein–Sol: a web tool for predicting protein solubility from sequence. Bioinformatics 33:3098–3100. https://doi.org/10.1093/bioinformatics/btx345
129. Sormanni P, Aprile FA, Vendruscolo M (2015) The CamSol method of rational design of protein mutants with enhanced solubility. J Mol Biol 427:478–490. https://doi.org/10.1016/j.jmb.2014.09.026
130. Sormanni P, Amery L, Ekizoglou S et al (2017) Rapid and accurate in silico solubility screening of a monoclonal antibody library. Sci Rep 7:8200. https://doi.org/10.1038/s41598-017-07800-w
131. Berner C, Menzen T, Winter G, Svilenov HL (2021) Combining unfolding reversibility studies and molecular dynamics simulations to select aggregation-resistant antibodies. Mol Pharm 18:2242–2253. https://doi.org/10.1021/acs.molpharmaceut.1c00017
132. Pérez A-MW, Sormanni P, Andersen JS et al (2018) In vitro and in silico assessment of the developability of a designed monoclonal antibody library. MAbs 11:388–400. https://doi.org/10.1080/19420862.2018.1556082
133. Walsh I, Seno F, Tosatto SCE, Trovato A (2014) PASTA 2.0: an improved server for protein aggregation prediction. Nucleic Acids Res 42:W301–W307. https://doi.org/10.1093/nar/gku399
134. Walsh I, Martin AJM, Domenico TD, Tosatto SCE (2012) ESpritz: accurate and fast prediction of protein disorder. Bioinformatics 28:503–509. https://doi.org/10.1093/bioinformatics/btr682
135. Piovesan D, Walsh I, Minervini G, Tosatto SCE (2017) FELLS: fast estimator of latent local structure. Bioinformatics 33:1889–1891. https://doi.org/10.1093/bioinformatics/btx085
136. Paladin L, Piovesan D, Tosatto SCE (2017) SODA: prediction of protein solubility from disorder and aggregation propensity. Nucleic Acids Res 45:W236–W240. https://doi.org/10.1093/nar/gkx412

137. Kuhn AB, Kube S, Karow-Zwick AR et al (2017) Improved solution-state properties of monoclonal antibodies by targeted mutations. J Phys Chem B 121:10818–10827. https://doi.org/10.1021/acs.jpcb.7b09126
138. Bauer J, Mathias S, Kube S et al (2020) Rational optimization of a monoclonal antibody improves the aggregation propensity and enhances the CMC properties along the entire pharmaceutical process chain. MAbs 12:1787121. https://doi.org/10.1080/19420862.2020.1787121
139. van der Kant R, Karow-Zwick AR, Durme JV et al (2017) Prediction and reduction of the aggregation of monoclonal antibodies. J Mol Biol 429:1244–1261. https://doi.org/10.1016/j.jmb.2017.03.014
140. Ferreira GM, Shahfar H, Sathish HA et al (2019) Identifying key residues that drive strong electrostatic attractions between therapeutic antibodies. J Phys Chem B 123:10642–10653. https://doi.org/10.1021/acs.jpcb.9b08355
141. Geoghegan JC, Fleming R, Damschroder M et al (2016) Mitigation of reversible self-association and viscosity in a human IgG1 monoclonal antibody by rational, structure-guided Fv engineering. MAbs 8:941–950. https://doi.org/10.1080/19420862.2016.1171444
142. Kumar S, Roffi K, Tomar DS et al (2018) Rational optimization of a monoclonal antibody for simultaneous improvements in its solution properties and biological activity. Protein Eng Des Sel 31:313–325. https://doi.org/10.1093/protein/gzy020
143. Prabakaran R, Goel D, Kumar S, Gromiha MM (2017) Aggregation prone regions in human proteome: insights from large-scale data analyses. Proteins Struct Funct Bioinf 85:1099–1118. https://doi.org/10.1002/prot.25276
144. Rawat P, Kumar S, Gromiha MM (2018) An in-silico method for identifying aggregation rate enhancer and mitigator mutations in proteins. Int J Biol Macromol 118:1157–1167. https://doi.org/10.1016/j.ijbiomac.2018.06.102
145. Prabakaran R, Rawat P, Kumar S, Gromiha MM (2020) ANuPP: a versatile tool to predict aggregation nucleating regions in peptides and proteins. J Mol Biol 433:166707. https://doi.org/10.1016/j.jmb.2020.11.006
146. Courtois F, Agrawal NJ, Lauer TM, Trout BL (2015) Rational design of therapeutic mAbs against aggregation through protein engineering and incorporation of glycosylation motifs applied to bevacizumab. MAbs 8:99–112. https://doi.org/10.1080/19420862.2015.1112477
147. Ebo JS, Guthertz N, Radford SE, Brockwell DJ (2020) Using protein engineering to understand and modulate aggregation. Curr Opin Struct Biol 60:157–166. https://doi.org/10.1016/j.sbi.2020.01.005
148. Baets GD, Durme JV, van der Kant R et al (2015) Solubis: optimize your protein. Bioinformatics 31:2580–2582. https://doi.org/10.1093/bioinformatics/btv162
149. Gil-Garcia M, Bañó-Polo M, Varejão N et al (2018) Combining structural aggregation propensity and stability predictions to redesign protein solubility. Mol Pharm 15:3846–3859. https://doi.org/10.1021/acs.molpharmaceut.8b00341
150. Rawat P, Prabakaran R, Kumar S, Gromiha MM (2019) AggreRATE-Pred: a mathematical model for the prediction of change in aggregation rate upon point mutation. Bioinformatics 31:409–418. https://doi.org/10.1093/bioinformatics/btz764
151. Santos J, Pujols J, Pallarès I et al (2020) Computational prediction of protein aggregation: advances in proteomics, conformation-specific algorithms and biotechnological applications. Comput Struct Biotechnol J 18:1403–1413. https://doi.org/10.1016/j.csbj.2020.05.026
152. Yang W, Tan P, Fu X, Hong L (2019) Prediction of amyloid aggregation rates by machine learning and feature selection. J Chem Phys 151:084106. https://doi.org/10.1063/1.5113848
153. Tartaglia GG, Cavalli A, Pellarin R, Caflisch A (2005) Prediction of aggregation rate and aggregation-prone segments in polypeptide sequences. Protein Sci 14:2723–2734. https://doi.org/10.1110/ps.051471205
154. Gupta P, Makowski EK, Kumar S et al (2022) Antibodies with weakly basic isoelectric points minimize trade-offs between formulation and physiological colloidal properties. Mol Pharm 19(3):775–787. https://doi.org/10.1021/acs.molpharmaceut.1c00373

155. Lu G, Moriyama EN (2004) Software review Vector NTI, a balanced all-in-one sequence analysis suite. Brief Bioinform 5(4):378–388. https://doi.org/10.1093/bib/5.4.378
156. Rice P, Longden I, Bleasby A (2000) EMBOSS: the European molecular biology open software suite. Trends Genet 16:276–277. https://doi.org/10.1016/s0168-9525(00)02024-2
157. Goyon A, Excoffier M, Janin-Bussat M-C et al (2017) Determination of isoelectric points and relative charge variants of 23 therapeutic monoclonal antibodies. J Chromatogr B 1065:119–128. https://doi.org/10.1016/j.jchromb.2017.09.033
158. Olsson MHM, Søndergaard CR, Rostkowski M, Jensen JH (2011) PROPKA3: consistent treatment of internal and surface residues in empirical pKa predictions. J Chem Theory Comput 7:525–537. https://doi.org/10.1021/ct100578z
159. Sharma VK, Patapoff TW, Kabakoff B et al (2014) In silico selection of therapeutic antibodies for development: viscosity, clearance, and chemical stability. Proc Natl Acad Sci USA 111:18601–18606. https://doi.org/10.1073/pnas.1421779112
160. Li L, Kumar S, Buck PM et al (2014) Concentration dependent viscosity of monoclonal antibody solutions: explaining experimental behavior in terms of molecular properties. Pharm Res 31:3161–3178. https://doi.org/10.1007/s11095-014-1409-0
161. Agrawal NJ, Helk B, Kumar S et al (2015) Computational tool for the early screening of monoclonal antibodies for their viscosities. MAbs 8:43–48. https://doi.org/10.1080/19420862.2015.1099773
162. Tomar DS, Li L, Broulidakis MP et al (2017) In-silico prediction of concentration-dependent viscosity curves for monoclonal antibody solutions. MAbs 9:476–489. https://doi.org/10.1080/19420862.2017.1285479
163. Tomar DS, Singh SK, Li L et al (2018) In silico prediction of diffusion interaction parameter (kD), a key indicator of antibody solution behaviors. Pharm Res 35:193. https://doi.org/10.1007/s11095-018-2466-6
164. Garidel P, Blume A, Wagner M (2014) Prediction of colloidal stability of high concentration protein formulations. Pharm Dev Technol 20:367–374. https://doi.org/10.3109/10837450.2013.871032
165. Kingsbury JS, Saini A, Auclair SM et al (2020) A single molecular descriptor to predict solution behavior of therapeutic antibodies. Sci Adv 6:eabb0372. https://doi.org/10.1126/sciadv.abb0372
166. Neergaard MS, Kalonia DS, Parshad H et al (2013) Viscosity of high concentration protein formulations of monoclonal antibodies of the IgG1 and IgG4 subclass – prediction of viscosity through protein–protein interaction measurements. Eur J Pharm Sci 49:400–410. https://doi.org/10.1016/j.ejps.2013.04.019
167. Nichols P, Li L, Kumar S et al (2015) Rational design of viscosity reducing mutants of a monoclonal antibody: hydrophobic versus electrostatic inter-molecular interactions. MAbs 7:212–230. https://doi.org/10.4161/19420862.2014.985504
168. Casaz P, Boucher E, Wollacott R et al (2014) Resolving self-association of a therapeutic antibody by formulation optimization and molecular approaches. MAbs 6:1533–1539. https://doi.org/10.4161/19420862.2014.975658
169. Apgar JR, Tam ASP, Sorm R et al (2020) Modeling and mitigation of high-concentration antibody viscosity through structure-based computer-aided protein design. PLoS One 15:e0232713. https://doi.org/10.1371/journal.pone.0232713
170. Tilegenova C, Izadi S, Yin J et al (2019) Dissecting the molecular basis of high viscosity of monospecific and bispecific IgG antibodies. MAbs 12:1692764. https://doi.org/10.1080/19420862.2019.1692764
171. Singh SN, Yadav S, Shire SJ, Kalonia DS (2014) Dipole-dipole interaction in antibody solutions: correlation with viscosity behavior at high concentration. Pharm Res 31:2549–2558. https://doi.org/10.1007/s11095-014-1352-0
172. Cloutier T, Sudrik C, Mody N et al (2019) Molecular computations of preferential interaction coefficients of IgG1 monoclonal antibodies with sorbitol, sucrose, and trehalose and the impact of these excipients on aggregation and viscosity. Mol Pharm 16:3657–3664. https://doi.org/10.1021/acs.molpharmaceut.9b00545

173. Chowdhury A, Guruprasad G, Chen AT et al (2019) Protein-protein interactions, clustering, and rheology for bovine IgG up to high concentrations characterized by small angle x-ray scattering and molecular dynamics simulations. J Pharm Sci 109:696–708. https://doi.org/10.1016/j.xphs.2019.11.001
174. Izadi S, Patapoff TW, Walters BT (2020) Multi-scale coarse-grained approach to investigate self-association of antibodies. Biophys J 118:2741–2754. https://doi.org/10.1016/j.bpj.2020.04.022
175. Chaudhri A, Zarraga IE, Yadav S et al (2013) The role of amino acid sequence in the self-association of therapeutic monoclonal antibodies: insights from coarse-grained modeling. J Phys Chem B 117:1269–1279. https://doi.org/10.1021/jp3108396
176. Lai P-K, Swan JW, Trout BL (2021) Calculation of therapeutic antibody viscosity with coarse-grained models, hydrodynamic calculations and machine learning-based parameters. MAbs 13:1907882. https://doi.org/10.1080/19420862.2021.1907882
177. Münch C, Bertolotti A (2010) Exposure of hydrophobic surfaces initiates aggregation of diverse ALS-causing superoxide dismutase-1 mutants. J Mol Biol 399:512–525. https://doi.org/10.1016/j.jmb.2010.04.019
178. Buell AK, Hung P, Salvatella X et al (2013) Electrostatic effects in filamentous protein aggregation. Biophys J 104:1116–1126. https://doi.org/10.1016/j.bpj.2013.01.031
179. Tartaglia GG, Cavalli A, Pellarin R, Caflisch A (2004) The role of aromaticity, exposed surface, and dipole moment in determining protein aggregation rates. Protein Sci 7:1939–1941. https://doi.org/10.1110/ps.04663504
180. Navarro S, Ventura S (2019) Computational re-design of protein structures to improve solubility. Expert Opin Drug Discov 14:1–12. https://doi.org/10.1080/17460441.2019.1637413
181. Pawar AP, DuBay KF, Zurdo J et al (2005) Prediction of "aggregation-prone" and "aggregation-susceptible" regions in proteins associated with neurodegenerative diseases. J Mol Biol 350:379–392. https://doi.org/10.1016/j.jmb.2005.04.016
182. Tartaglia GG, Vendruscolo M (2008) The Zyggregator method for predicting protein aggregation propensities. Chem Soc Rev 37:1395–1401. https://doi.org/10.1039/b706784b
183. Tian J, Wu N, Guo J, Fan Y (2009) Prediction of amyloid fibril-forming segments based on a support vector machine. BMC Bioinformatics 10:S45. https://doi.org/10.1186/1471-2105-10-s1-s45
184. Conchillo-Solé O, de Groot NS, Avilés FX et al (2007) AGGRESCAN: a server for the prediction and evaluation of "hot spots" of aggregation in polypeptides. BMC Bioinformatics 8:65. https://doi.org/10.1186/1471-2105-8-65
185. Fernandez-Escamilla A-M, Rousseau F, Schymkowitz J, Serrano L (2004) Prediction of sequence-dependent and mutational effects on the aggregation of peptides and proteins. Nat Biotechnol 22:1302–1306. https://doi.org/10.1038/nbt1012
186. Garbuzynskiy SO, Lobanov MY, Galzitskaya OV (2010) FoldAmyloid: a method of prediction of amyloidogenic regions from protein sequence. Bioinformatics 26:326–332. https://doi.org/10.1093/bioinformatics/btp691
187. Zibaee S, Makin OS, Goedert M, Serpell LC (2007) A simple algorithm locates β-strands in the amyloid fibril core of α-synuclein, Aβ, and tau using the amino acid sequence alone. Protein Sci 16:906–918. https://doi.org/10.1110/ps.062624507
188. Trovato A, Seno F, Tosatto SCE (2007) The PASTA server for protein aggregation prediction. Protein Eng Des Sel 20:521–523. https://doi.org/10.1093/protein/gzm042
189. Família C, Dennison SR, Quintas A, Phoenix DA (2015) Prediction of peptide and protein propensity for amyloid formation. PLoS One 10:e0134679. https://doi.org/10.1371/journal.pone.0134679
190. Louros N, Konstantoulea K, Vleeschouwer MD et al (2019) WALTZ-DB 2.0: an updated database containing structural information of experimentally determined amyloid-forming peptides. Nucleic Acids Res 48:D389–D393. https://doi.org/10.1093/nar/gkz758
191. Beerten J, Durme JV, Gallardo R et al (2015) WALTZ-DB: a benchmark database of amyloidogenic hexapeptides. Bioinformatics 31:1698–1700. https://doi.org/10.1093/bioinformatics/btv027

192. Frousios KK, Iconomidou VA, Karletidi C-M, Hamodrakas SJ (2009) Amyloidogenic determinants are usually not buried. BMC Struct Biol 9:44. https://doi.org/10.1186/1472-6807-9-44
193. Tsolis AC, Papandreou NC, Iconomidou VA, Hamodrakas SJ (2013) A consensus method for the prediction of "aggregation-prone" peptides in globular proteins. PLoS One 8:e54175. https://doi.org/10.1371/journal.pone.0054175
194. Emily M, Talvas A, Delamarche C (2013) MetAmyl: a METa-predictor for AMYLoid proteins. PLoS One 8:e79722. https://doi.org/10.1371/journal.pone.0079722
195. Gentiluomo L, Roessner D, Augustijn D et al (2019) Application of interpretable artificial neural networks to early monoclonal antibodies development. Eur J Pharm Biopharm 141:81–89. https://doi.org/10.1016/j.ejpb.2019.05.017
196. Dzisoo AM, He B, Karikari R et al (2019) CISI: a tool for predicting cross-interaction or self-interaction of monoclonal antibodies using sequences. Interdiscip Sci Comput Life Sci 11:691–697. https://doi.org/10.1007/s12539-019-00330-1
197. Zambrano R, Jamroz M, Szczasiuk A et al (2015) AGGRESCAN3D (A3D): server for prediction of aggregation properties of protein structures. Nucleic Acids Res 43:W306–W313. https://doi.org/10.1093/nar/gkv359
198. Durme JV, Baets GD, Kant RVD et al (2016) Solubis: a webserver to reduce protein aggregation through mutation. Protein Eng Des Sel 29:285–289. https://doi.org/10.1093/protein/gzw019
199. Sankar K, Krystek SR, Carl SM et al (2018) AggScore: prediction of aggregation-prone regions in proteins based on the distribution of surface patches. Proteins Struct Funct Bioinf 86:1147–1156. https://doi.org/10.1002/prot.25594
200. Chennamsetty N, Voynov V, Kayser V et al (2010) Prediction of aggregation prone regions of therapeutic proteins. J Phys Chem B 114:6614–6624. https://doi.org/10.1021/jp911706q
201. Lauer TM, Agrawal NJ, Chennamsetty N et al (2012) Developability index: a rapid in silico tool for the screening of antibody aggregation propensity. J Pharm Sci 101:102–115. https://doi.org/10.1002/jps.22758
202. Lai P-K, Fernando A, Cloutier TK et al (2021) Machine learning feature selection for predicting high concentration therapeutic antibody aggregation. J Pharm Sci 110:1583–1591. https://doi.org/10.1016/j.xphs.2020.12.014
203. Stave JW, Lindpaintner K (2013) Antibody and antigen contact residues define epitope and paratope size and structure. J Immunol 191:1428–1435. https://doi.org/10.4049/jimmunol.1203198
204. Esmaielbeiki R, Krawczyk K, Knapp B et al (2016) Progress and challenges in predicting protein interfaces. Brief Bioinform 17:117–131. https://doi.org/10.1093/bib/bbv027
205. Deac A, VeliČković P, Sormanni P (2019) Attentive cross-modal paratope prediction. J Comput Biol 26:536–545. https://doi.org/10.1089/cmb.2018.0175
206. Bourquard T, Musnier A, Puard V et al (2018) MAbTope: a method for improved epitope mapping. J Immunol 201:3096–3105. https://doi.org/10.4049/jimmunol.1701722
207. Hua CK, Gacerez AT, Sentman CL et al (2017) Computationally-driven identification of antibody epitopes. Elife 6:e29023. https://doi.org/10.7554/elife.29023
208. Weitzner BD, Jeliazkov JR, Lyskov S et al (2017) Modeling and docking of antibody structures with Rosetta. Nat Protoc 12:401–416. https://doi.org/10.1038/nprot.2016.180
209. Daberdaku S, Ferrari C (2018) Antibody interface prediction with 3D Zernike descriptors and SVM. Bioinformatics 35:1870–1876. https://doi.org/10.1093/bioinformatics/bty918
210. Liberis E, Veličković P, Sormanni P et al (2018) Parapred: antibody paratope prediction using convolutional and recurrent neural networks. Bioinformatics 34:2944–2950. https://doi.org/10.1093/bioinformatics/bty305
211. Trnková L, Dršata J, Boušová I (2015) Oxidation as an important factor of protein damage: implications for Maillard reaction. J Biosci 40:419–439. https://doi.org/10.1007/s12038-015-9523-7

212. Yang R, Jain T, Lynaugh H et al (2017) Rapid assessment of oxidation via middle-down LCMS correlates with methionine side-chain solvent-accessible surface area for 121 clinical stage monoclonal antibodies. MAbs 9:646–653. https://doi.org/10.1080/19420862.2017.1290753
213. Veredas FJ, Urda D, Subirats JL et al (2020) Combining feature engineering and feature selection to improve the prediction of methionine oxidation sites in proteins. Neural Comput & Applic 32:323–334. https://doi.org/10.1007/s00521-018-3655-2
214. Sankar K, Hoi KH, Yin Y et al (2018) Prediction of methionine oxidation risk in monoclonal antibodies using a machine learning method. MAbs 10:1–10. https://doi.org/10.1080/19420862.2018.1518887
215. Aledo JC, Cantón FR, Veredas FJ (2017) A machine learning approach for predicting methionine oxidation sites. BMC Bioinformatics 18:430. https://doi.org/10.1186/s12859-017-1848-9
216. López-García G, Jerez JM, Urda D, Veredas FJ (2019) MetODeep: a deep learning approach for prediction of methionine oxidation sites in proteins. IJCNN 1–8. https://doi.org/10.1109/ijcnn.2019.8851901
217. Delmar JA, Buehler E, Chetty AK et al (2021) Machine learning prediction of methionine and tryptophan photooxidation susceptibility. Mol Ther Methods Clin Dev 21:466–477. https://doi.org/10.1016/j.omtm.2021.03.023
218. Agrawal NJ, Dykstra A, Yang J et al (2018) Prediction of the hydrogen peroxide–induced methionine oxidation propensity in monoclonal antibodies. J Pharm Sci 107:1282–1289. https://doi.org/10.1016/j.xphs.2018.01.002
219. Chennamsetty N, Quan Y, Nashine V et al (2015) Modeling the oxidation of methionine residues by peroxides in proteins. J Pharm Sci 104:1246–1255. https://doi.org/10.1002/jps.24340
220. Barnett GV, Balakrishnan G, Chennamsetty N et al (2019) Probing the tryptophan environment in therapeutic proteins: implications for higher order structure upon tryptophan oxidation. J Pharm Sci 108:1944–1952. https://doi.org/10.1016/j.xphs.2018.12.027
221. Pavon JA, Xiao L, Li X et al (2019) Selective tryptophan oxidation of monoclonal antibodies: oxidative stress and modeling prediction. Anal Chem 91:2192–2200. https://doi.org/10.1021/acs.analchem.8b04768
222. Phillips JJ, Buchanan A, Andrews J et al (2017) Rate of asparagine deamidation in a monoclonal antibody correlating with hydrogen exchange rate at adjacent downstream residues. Anal Chem 89:2361–2368. https://doi.org/10.1021/acs.analchem.6b04158
223. Lu X, Nobrega RP, Lynaugh H et al (2018) Deamidation and isomerization liability analysis of 131 clinical-stage antibodies. MAbs 11:1–13. https://doi.org/10.1080/19420862.2018.1548233
224. Lorenzo JR, Alonso LG, Sánchez IE (2015) Prediction of spontaneous protein deamidation from sequence-derived secondary structure and intrinsic disorder. PLoS One 10:e0145186. https://doi.org/10.1371/journal.pone.0145186
225. Plotnikov NV, Singh SK, Rouse JC, Kumar S (2017) Quantifying the risks of asparagine deamidation and aspartate isomerization in biopharmaceuticals by computing reaction free-energy surfaces. J Phys Chem B 121:719–730. https://doi.org/10.1021/acs.jpcb.6b11614
226. Sydow JF, Lipsmeier F, Larraillet V et al (2014) Structure-based prediction of asparagine and aspartate degradation sites in antibody variable regions. PLoS One 9:e100736. https://doi.org/10.1371/journal.pone.0100736
227. Yan Q, Huang M, Lewis MJ, Hu P (2018) Structure based prediction of asparagine deamidation propensity in monoclonal antibodies. MAbs 10:901–912. https://doi.org/10.1080/19420862.2018.1478646
228. Kuang J, Tao Y, Song Y et al (2021) Understanding the pathway and kinetics of aspartic acid isomerization in peptide mapping methods for monoclonal antibodies. Anal Bioanal Chem 413:2113–2123. https://doi.org/10.1007/s00216-021-03176-z
229. Jia L, Sun Y (2017) Protein asparagine deamidation prediction based on structures with machine learning methods. PLoS One 12:e0181347. https://doi.org/10.1371/journal.pone.0181347

Chapter 15
Advanced Data Analytics Application in Biomanufacturing Processes

Jun Luo, Lin Qiu, Yang Tang, Grant Sumida, Sid Kundu, and Yiming Peng

Abstract Biomanufacturing drug substance processes include cell culture (thaw, seed train, inoculum, and protein production), harvest, and purification steps. Due to the living nature, the bioprocess inherently has higher variability when compared to chemical reactions. Ensuring consistent process performance and continuous process improvement are ongoing challenges for bioprocess.

Thorough characterization of process and product attributes using statistical design of experiments (DoE) during the process design (PD) stage is a proven means to establish the process and product knowledge. Recent accelerated development timelines are putting pressure and limitations on the extent of characterization during the PD stage, and, therefore, increasing the importance of a lifecycle approach to accumulate process and product knowledge during the post-approval stage of process validation (i.e., continued process verification).

Following telecoms, advertising, and insurance, the biopharma industry has started to embrace methods like Bayesian statistics and advanced data analytics (ADA) to gain additional process and product understanding, improved process control, and process performance. Using ADA may elucidate previously undetected relationships between process inputs and outputs, which hold promise as an additional tool to augment traditional DoE as a means to gain actionable insights, process and product knowledge.

An internal multidisciplinary team (e.g., ADA) comprising data engineers, data scientists, bioprocess experts, and a translator /project manager is formed.

J. Luo (✉) · G. Sumida
Manufacturing Sciences, Genentech, Vacaville, CA, USA
e-mail: luo.jun@gene.com

L. Qiu · Y. Peng
Nonclinical Biostatistics, Genentech, South San Francisco, CA, USA

Y. Tang
Nonclinical Biostatistics, Roche, Mississauga, ON, Canada

S. Kundu
Cell Culture and Bioprocess Operations, Genentech, South San Francisco, CA, USA

K. Gadamasetti, S. A. Kolodziej (eds.), *Bioprocessing, Bioengineering and Process Chemistry in the Biopharmaceutical Industry*,
https://doi.org/10.1007/978-3-031-62007-2_15

IT infrastructure to support ADA projects is established, the standard approaches to run the case studies are developed, and a library of models for bioprocess is built. An upstream process case study to improve upstream productivity and process robustness is discussed. The team is able to extract, cleanse, ingest, and process 800 GB of data from 11 disparate data sources into a cloud environment to evaluate more than 100 hypotheses. Implementing the insights leads to improved process performance. Through the case study, the strategy to sustain the internal capability is developed. The importance of building ADA capabilities to enable more efficient and reliable bioprocesses is discussed.

Keywords Advanced data analytics · Machine learning · Cell culture · Process performance variation

15.1 Introduction

Biomanufacturing drug substance processes include cell culture (thaw, seed train, inoculum and protein production), harvest, and purification (chromatography, viral filtration, ultrafiltration diafiltration, and freeze) steps. Due to the living nature, the bioprocess inherently has higher variability when compared to chemical reaction. To ensure consistent process performance and product quality, a bioprocess validation lifecycle program is needed. The lifecycle includes three stages: PD (Stage 1), process qualification (PQ, Stage 2), and continued process verification (CPV, Stage 3) [1]. The data are collected and evaluated through all three stages of the lifecycle to generate process knowledge and establish scientific evidence that a process is capable of consistently delivering quality products.

Information used to characterize the process can be split as inputs and outputs of the process. For a specific step of mammalian cell culture processes, process inputs could be raw material, equipment, operational parameters (e.g., operator, operation sequence, operating parameters), and any measurements of the initial state, etc. Process outputs could be process performance indicators (PI, or KPI: key performance indicator) from the specific process step or any downstream steps, and product quality attributes (PQA or CQA: critical quality attributes). Process knowledge could be summarized in an equation or model to describe the correlation between inputs and output. Quantitative models for optimization commercial drug substance manufacturing processes have recently attracted increasing attention. Different data analysis methods may be used to build process knowledge during different process validation stages.

Per International Council for Harmonisation of Technical Requirements for Pharmaceuticals for Human Use (ICH) quality guidelines ICH Q8 (R2), ICH Q9, and ICH Q10, the concept of quality by design is introduced. This concept includes quality-risk management systems and the implementation of Pharmaceutical Quality Systems. Key elements are critical process parameters (CPP), which affect CQAs, and DoE [2]. Thorough characterization of process and product attributes using statistical DoE during the PD stage is a proven means to establish process/product knowledge. DoE-based approaches can provide a deeper understanding of how product quality attributes and process-related impurities change when several process operating parameters vary simultaneously for a given unit operation, and thus lead to increased process knowledge with respect to the robustness of a culture step or combination of culture steps. Prior to designing a DoE study, a risk assessment is typically conducted to rank the risk of process parameters regarding the impact on process performance and product quality. This is based on the previous knowledge from the platform process, first principles, early development data, and literature review, etc. Based on the outcome of risk assessment, follow-up studies will be executed to identify the acceptable range for process parameters [3]. Since most bioprocess unit steps easily have more than four factors, the first round of experiments typically uses the screening design (e.g., Plackett-Burman method) to identify the most important factors early in the experimentation phase when complete knowledge about the system is usually unavailable [4]. Once the important factors are identified, a follow-up DoE study (e.g., response surface designs or factorial design) to understand the interaction, and to refine the model. Note: Response surface design is recommended especially if there is suspected curvature in the response surface. Once the model is refined, the model could be used to predict worst-case conditions for certain process output, and the predicted worst-case conditions could be further tested in experiments. Separately, certain univariate studies could be performed to provide knowledge for ongoing manufacturing support. For example, a pH excursion study (effect of the short-term high/low pH) is typically executed to understand the effect of the pH probe spike issue occasionally observed in commercial manufacturing.

Many bioprocess companies have a CPV program that involves univariate trending of quality attributes and process performance indicators, which are used to establish the evidence for process in control or identify process change/shift/excursion for further investigation. If process performance or capability is trending unfavorably, the ability to adaptively respond and adjust controls is contingent on the availability of documented causal relationships (i.e., models) between controllable inputs and important process/product outputs. With accelerating development timelines, the accumulation of process knowledge during the CPV stage of process validation is becoming increasingly important. Different statistical methods/models could be used to collect knowledge during the CPV phase in addition to performing experiments. For example, when the sample size is small, Bayesian statistics is a preferred method. With the increase of sample size, advanced data analytics (ADA) can be used.

Sample sizes of data in biomanufacturing drug substance processes are frequently quite small due to time, economic, and physical constraints. The effect of sample size on the interpretation of classical significance tests has been emphasized [5]. The predictive model built upon a small number of samples may highly rely on the particular method for parameter estimation [6]. Under such circumstances, to enable meaningful model prediction and results interpretation, Bayesian statistics is usually preferred [7] because Bayesian methods incorporate both the observed data information and the prior knowledge via Bayes' theorem. Bayes' theorem describes the conditional probability of an event based on data as well as prior information or beliefs about the event or conditions related to the event.

There are increasing numbers of empirical studies that have demonstrated Bayesian methods are explicitly selected to better accommodate reduced sample sizes, ranging from life science, psychology, exercise science, cross-cultural studies, and prevention science [8, 9]. Among many Bayesian methods, a framework that combines cause and effect diagrams with Bayesian belief networks (BBNs) to estimate causal relationships could be leveraged for the biomanufacturing process [10]. The framework translates qualitative insights from a cause-and-effect diagram into a closed-form relationship between inputs and outputs utilizing BBN models, simulation, and regression. The framework starts with a typical Fishbone diagram to construct a causal network, then estimate the probability of problem occurring using BBN and probabilistic elicitation, and finally identify the vulnerabilities of the network through simulation. This allows investigation practitioners to identify critical root causes of a given problem when data collection activities can often consume valuable resources or events occurring at a low frequency with a high risk of severity.

When a large enough data set is generated at a commercial scale, ADA is a tool that can complement traditional univariate monitoring programs to bolster knowledge, improve capability, and optimize bioprocess performance. Within the data science community, the use of statistical modeling to reach conclusions from data can be divided into two separate "cultures" [11]. One assumes that the data are generated by a given stochastic data model. The other uses algorithmic models and treats the data mechanism as unknown. Because data collected during commercial manufacturing (where the process runs at or close to the target as designed) may involve complex systems with unknown physical, chemical, or biological mechanisms, signals from stochastic data models may be weak and difficult to interpret. Therefore, algorithmic models using ADA may elucidate previously undetected relationships between inputs and outputs, and hold promise to gain actionable insights and knowledge to better understand, predict, and control sources of variation in a bioprocess [12].

While there are many ADA techniques available, two types of ADA models could be built based on different biomanufacturing applications, namely diagnostic (or root cause) models and predictive models. Both models provide information to gain process understanding. Specifically, the root cause model (explanatory model) focuses on actionable inputs for process troubleshooting and improvement, and makes sure the insights from the "black box" models can be explained. When the model focuses on actionable process inputs, the newly identified cause could either be verified through small-scale lab studies or large-scale process changes and subsequently monitoring. The predictive model uses all available inputs for output prediction, regardless of the model's interpretability. This could be very helpful in making early and/or real-time forward process decisions. It could also be extremely useful if there is an intrinsic correlation between process outputs. Using easy to detect process output to predict the hard-to-measure process output could significantly improve the process efficiency.

With the help from an external consultant company, an internal ADA program including a cross-functional team, IT infrastructure, data engineering technique, data visualization and modeling methodology is established. This article will describe the ADA program developed in-house and the established procedure to conduct the analysis. Specifically, a case study to gain additional insights for an intractable cell culture performance problem is conducted. The model insights, the observed and predicted benefit based on the model outcome, are discussed. The importance of building ADA capabilities to enable more efficient and reliable bioprocesses is also discussed.

15.2 ADA Program Development

With the help from McKinsey/Quantum Black (McK/QB) consultant firm, an in-house ADA program was developed and established. The program includes a cross-functional team, established environmental and tool, business process to run individual projects (use cases), and a library of models. In addition, data standardization, digitization, and data sharing are key factors to ensure the success of the program.

15.2.1 Building a Cross-Functional Team for ADA Program

The team composed of global and local cross-functional experts is formed. The team members and their corresponding responsibilities are as follows:

- *Program leader:* to set up program governance and to manage ADA engagement
- *Project leader/translator:* to set and track progress, to provide guidance/steering to the team, and to coordinate and facilitate interactions across disciplines
- *Technical subject matter experts (SMEs):* to share expertise and help with problem-solving and to help define root causes/drivers based on insights from analytics
- *Data engineer (DE):* to work with source system owners to make data available and to cleanse and link data in preparation for analysis
- *Data scientist (DS):* to define the analytic approach and execution of analysis and to interpret model results
- *Information technology (IT)/technical liaison:* to provide information on technical architecture, and to establish tools and standard, and to continuously optimize and quality check for development and deployment

15.2.2 IT Infrastructure and Tools

New environments and tools are established for the ADA program (Table 15.1). The platform is built on the Google Cloud Platform (GCP). Google Cloud Storage, BigQuery, and Dataproc are often used for data storage, data exploration, and high-performance distributed computing, respectively. Python is used as the primary programming language for data analysis. The Data Pipeline was introduced and built

Table 15.1 IT infrastructure and tools used in ADA program

Environment / tool	Description	Key capabilities built
Google Cloud Platform (GCP)	A suite of cloud computing services that runs on the same infrastructure offered by Google	Learn/practice/improve on working in GCP (cluster, codebase, notebook, etc.). Run data engineering pipeline and data science analytical modules in GCP. Master basic DevOps (Spinning up a cluster, Terminal SSH, basic Linux commands)
Dataproc	Analytics engine for large-scale data processing	Use Dataproc and Pyspark for distributed computing
Google Cloud Storage (GCS)	Worldwide cloud storage that can retrieve any amount of data at any time	Ingest data into the GCS environment
BigQuery	Cloud data warehouse	Load/Query data in BigQuery
Python	A programming language	Develop and improve the codebase, follow good coding practices, peer code review, test.
Jupyter Notebook	Open-source web application for creating and sharing documents that contain live code, equations, visualizations, and narrative text	Perform data analysis in CLI/Jupyter Notebooks
Kedro	An open-source Python framework for creating reproducible, maintainable and modular data science code	Understand the Data Pipeline concept Create raw layers, intermediate layers, primary layers, feature layers, model input table of the data engineering pipeline
GitHub	Version control development platform	Understand Git workflow. Update data engineering pipeline with Git
JIRA	Agile sprint planning tool	Manage tasks and collaborate with the virtual team in agile ways. Implement SCRUM and work in sprints as foundations
Confluence	Remote-friendly team workspace	Use as a team workspace, knowledge sharing, and collaboration tool

https://Cloud.google.com
https://spark.apache.org/
https://Python.org
https://Jupyter.org
https://kedro.readthedocs.io
https://github.com/
https://www.atlassian.com/software/confluence

using Kedro, an open-source framework for creating reproducible, maintainable, and modular data science code. All codes are version controlled in Github. Jupyter Notebook is used to support data analysis and result sharing. The project is operated in an agile paradigm running sprints using JIRA. All the knowledge and documentation are stored in Confluence as a team workspace.

15.2.3 Project Approach

The program is executed through individual use cases. Analytics problem-solving requires translating a business problem (use case) into an analytics problem and solving it through an iterative process of hypotheses generation, exploratory analysis, and feature creation. Each project could be separated into five phases: preparation, initial analysis, model iteration, translating to actionable insights, and model outcome confirmation. Table 15.2 summarized the detailed deliverables from each phase. Figure 15.1 showed a map of the project approach to data transformation processes (i.e., data pipeline). Throughout the project, translators, Des, and DSs worked together with the process SMEs to define hypotheses and select variables/features (model inputs), and to prioritize the work based on the hypothesis and data availability. Among different use cases, the continuous improvement on the overall program platform and the model methods are important, and the reusability of the pipeline built from one use case to another should be maximized.

Table 15.2 ADA project phase deliverables

Phase	Deliverable
Preparation	Select use case Align on scope and definition Develop detailed execution plan Set up the technical execution environment Define initial hypothesis Identify/prioritize data source based on initial hypotheses Align on people/capabilities and roles
Initial analysis	Define model outputs Model selection Data ingestion (data extract from data sources) Data clean/transformation & quality check (link data based on hypotheses) Feature creation Exploratory data analysis (EDA)
Model development and iteration	Build model/Analytical engine Iterate could be based on new data sources, new hypotheses, or refined features
Model insights	Build graphical representations of the Insights gained. Develop reports for a wide range of users
Insights confirmation	Confirm the insight by further lab study, additional monitoring after proposed change implementation

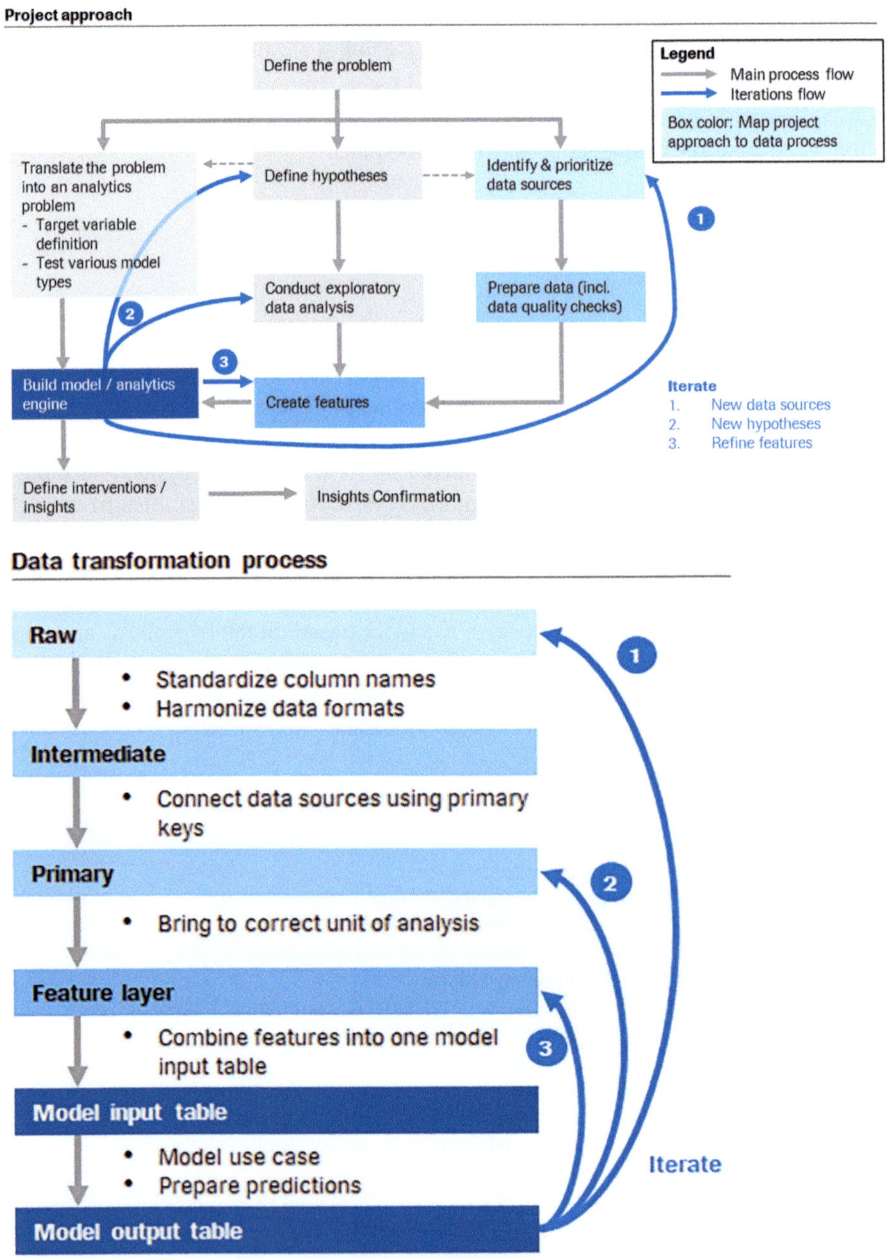

Fig. 15.1 A map of the project approach and data transformation processes (i.e., data pipeline)

15.2.4 Model Library

Machine learning is used to teach machines how to handle the data more efficiently. Simply, machine learning can be better understood as "learning from data" [13]. There are two types of machine learning algorithms, unsupervised and supervised [14]. Unsupervised learning uses machine learning algorithms (e.g., K-means, PCA) to analyze and cluster unlabeled data sets. These algorithms discover hidden patterns in data without the need for human intervention (hence, they are "unsupervised"). It is mainly used for clustering, association, and dimensionality reduction. Supervised learning is a machine learning approach that's defined by its use of labeled datasets. These datasets are designed to train or "supervise" algorithms into classifying data or predicting outcomes accurately. Using labeled inputs and outputs, the model can measure its accuracy and learn over time. Currently, supervised machine learning methods are primarily used in the ADA program, which includes Linear regression, Lasso, Decision trees, Random forest, and Gradient boosting. Lasso, Random forest, and Gradient boosting are chosen because of their feature selection properties. Linear regression and Decision trees are used as the linear and nonlinear benchmarks for model comparison. Other methods such as generalized random forest have been studied and applied as appropriate.

15.3 Case Study

15.3.1 Case Study Selection for ADA Modeling

The goal of the case study was to build a model to explain variation and identify actionable insights for a fed-batch Chinese hamster ovary (CHO)-based bioprocess process (Fig. 15.2).

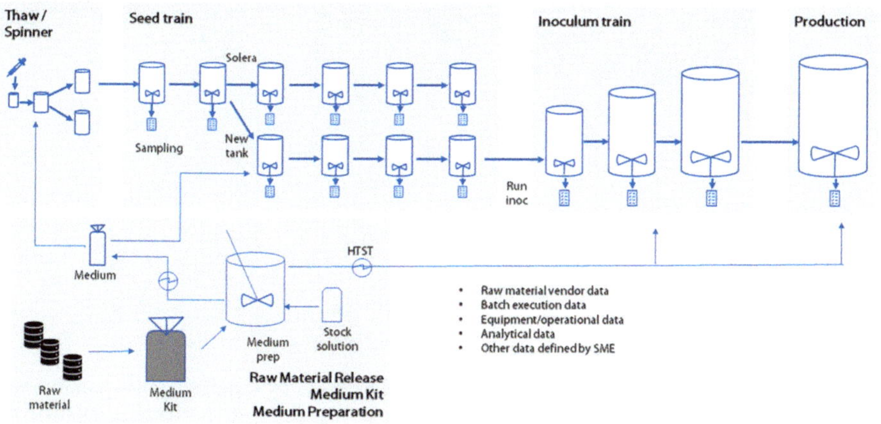

Fig. 15.2 Process flowchart from seed train to large-scale cell culture process, including raw material and medium prep. HTST refers to high temperature short time

Many factors were considered for the case study selection process, such as sample sizes, the variability of the process output, the impact on business, the variability of data sources, data accessibility, and process knowledge. For example, to develop a meaningful ADA model, it was suggested to have a manufacturing history of at least 100 large runs. Second, if the variation in interest is observed throughout the process history, it may be a more suitable use case. Since cell culture processes typically run in campaign mode, the campaign-to-campaign (raw material variability) or thaw-to-thaw variability is a common variation observed in bioprocess. Without variation observed in the historical data, there is nothing for the ADA model to optimize. In addition, where and how data were generated, captured, and stored will determine the data availability, data quality, and amount of effort for data engineering.

In this specific case, the selected process (Process A) generated more than 300 runs over the manufacturing history. As shown in Fig. 15.3, 95% of all runs have normalized titer ranging from 0.77 to 1.22. Among the total observed variation, the within-campaign variation takes about 52%, and the between-campaign variation takes 48%. Table 15.3 summarizes an example on process inputs and output for cell culture process. For process input, in addition to the typical primary process parameter for characterization, the associated control strategy and secondary process parameters should also be included in the analysis.

The total number of runs available and relatively high historical variability made the process a good candidate for the analysis. In addition, most site data sources existed in digital format except for a subset of raw material release test results and paper-based executed batch records. The data availability will shorten the data preparation phase and enable more rapid problem solving. Based on the aforementioned

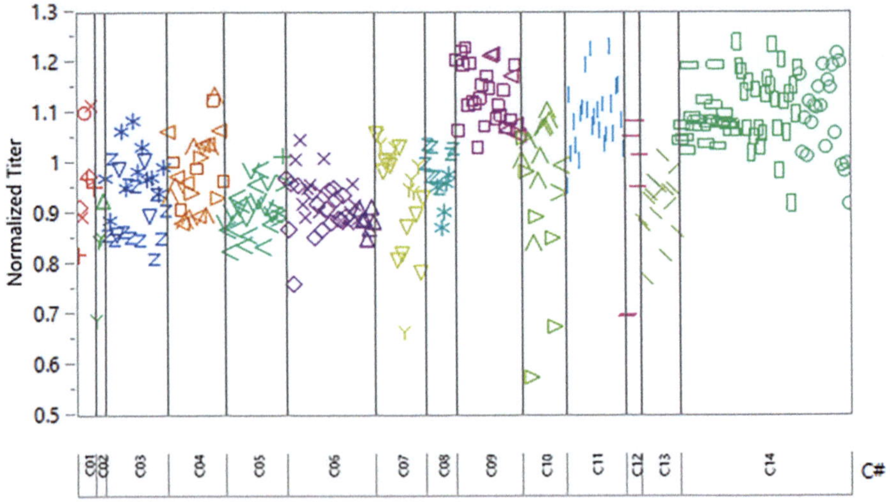

Fig. 15.3 Normalized historical titer performance of Process A by campaign (shown in different colors and separated by vertical lines) and thaw line (shown in different symbols). C# Campaign number

Table 15.3 Upstream process inputs and outputs

Process input		
Attribute	*Attribute description*	*Associated control/parameter*
Thaw	Time	Bank storage, transfer operation
Temp	Setpoint, shift, shift timing	Control loop, tank vs jacket temp
pH	Setpoint, shift, shift timing	Deadband, probe calibration
DO	Agitation, sparge, overlay flow rates	Aeration strategy
Medium	Concentration	Medium prep/transfer operation, medium composition, raw material release specification
Feed	Number, concentration, volume, timing	Medium prep/transfer operation
Inoculation density	Viable cell density	
Culture duration	Time	
Cell age	Seed train maintenance, # of nonselective passage	Solera operation, scale up strategy
Glucose	Min/Max concentration	Addition strategy
Antifoam		Addition strategy
Process output		
Category	*Attribute description*	
Cell state	Viability, integrated growth (growth rate)	
Metabolite	Lactate, ammonium, pCO_2	
Productivity	Titer	
Purity	Charge variants, size distribution, glycosylation	
Specific modification	Deamidation, glycation, oxidation	
Impurity	CHOP, DNA, sequence variants	
Potency	Potency by Anti-Proliferation, potency by Antibody-Dependent Cellular Cytotoxicity (ADCC)	
Contaminants	Bioburden, endotoxin, virus, mycoplasma	

factors, combined with a positive business value associated with understanding or reducing the performance variation, Process A was selected for the case study.

15.3.2 Data Sources Identification

A holistic review of the manufacturing process was conducted, and 11 data sources were identified that store manufacturing information. The data in 11 data sources were stored or extracted in different formats, such as Oracle databases, time-series data historian, scanned paper copy, and Microsoft Excel file.

15.3.3 Hypothesis Generation

The generation of the hypotheses is a collaborative effort between the business, the analytical team, and domain experts. Hypotheses are at the heart of every project and inform how data is ingested and how analytics is performed. In this use case, more than 100 hypotheses were generated and prioritized among process, raw material, and test method areas.

Based on the importance of the hypothesis and data availability, data ingestion process was prioritized. For example, a key hypothesis for culture performance variation was associated with raw material variation [15]. Therefore, incorporating raw material test results into the ADA model was required. However, because raw material release test information was paper based, digitizing raw material data became a top priority early in the project. In addition, the analysis is to identify factor to further improve titer/reduce titer variability; only process inputs are included in the model.

15.3.4 EDA and Feature Engineering

Once the raw data were available, data was cleaned, rearranged in smaller tables, and data from different tables were joined as a pipeline. Then data was ready for exploratory data analysis (EDA) and feature engineering. Once the feature was created, all features were joined to a master data table, and a model could be built based on the master data. Prior to running the model, Variance Inflation Factor (VIF) was used for feature selection; this is to deal with multicollinearity between features and achieve more stable explanations of models. A VIF of 5 was used as a threshold for feature selection, which means the feature will be removed from model input if it can be explained 80% by other features. Based on the EDA or the initial model outcomes, additional data might be added to the pipeline or additional features could be created for model updates. As mentioned above, the overall pipeline was built on Kedro using Python [16].

15.3.5 Feature Engineering Example

For mammalian cell culture, pH setpoint and pH control are known factors to impact cell growth and productivity [17, 18]. Therefore, it is expected to build features based on pH information. Features created from pH information described below show an example of the data ingestion process. The online pH (PI data historian) for one inoculum train stage is the raw data from the manufacturing process. In addition to the culturing phase, the pH information before (medium only) and during inoculation (transfer cells from the previous culture to medium), and transfer out (inoculation for the next stage) could be extracted as different phases from data. In addition, during the culturing phase, the pH experienced a shift, which is due to the interaction between cell growth and pH control deadband. Therefore, pH during culture could be further partitioned into before, during, and after the shift. During culture transfer out, pH control was turned off, and the probe would be exposed to air at a certain level (tank-specific information). Linking other time-series data (e.g., tank volume or weight) and equipment information is, thus, required to get the transfer-out information. With this detailed process knowledge, the features were built for pH in one inoculum train culture stage as shown in Table 15.4. As part of feature generation, EDA to review pH effect is performed to verify the correction of data ingestion and linkage and to understand if there is any effect.

Table 15.4 Data ingestion process for online pH in one inoculum train

Steps	pH example	Chart example
Raw data	pH in one inoculum train stage	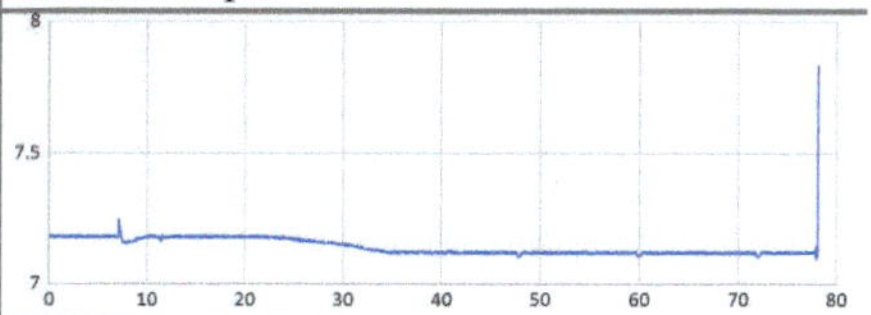
Intermediate data	Associate pH with equipment data and tank weight (kg) to identify different phases of pH	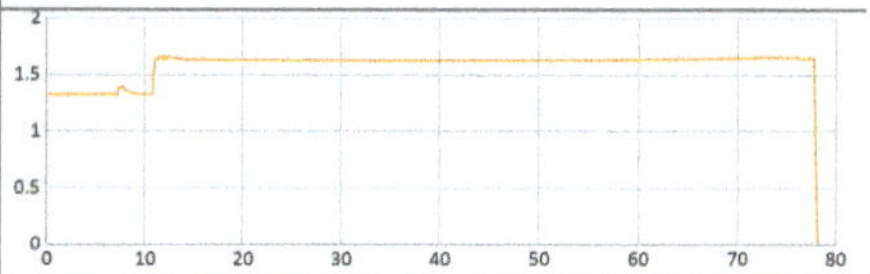
Primary layer	Based on pH and weight trend, separate pH trend in one culture stage into six different phases 1: medium batch 2: inoculation[a] 3: pH at the top of deadband 4: pH transition from top to bottom of dead band 5: pH at the bottom of deadband 6: Transfer out[a]	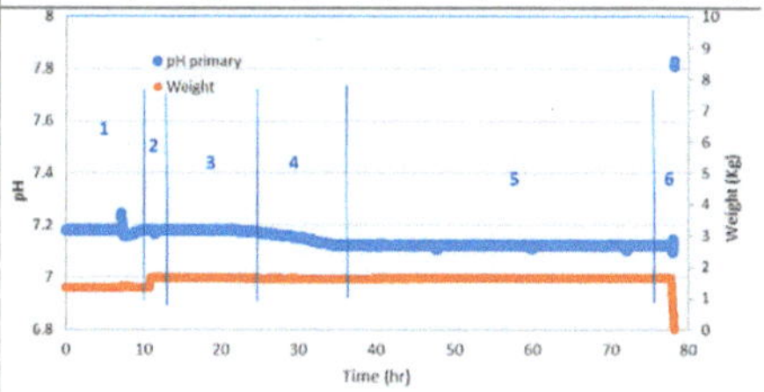
Feature	For each pH phase, explore the pH: min, max, average, 25, 50, 75 percentile of pH For each pH phase, explore the duration, the ratio of each duration to overall duration For culturing phase, explore any pH excursion: spike, change due to online/offline adjustment, etc.[a]	Max phase 3 pH for all runs in one inoculum train stage 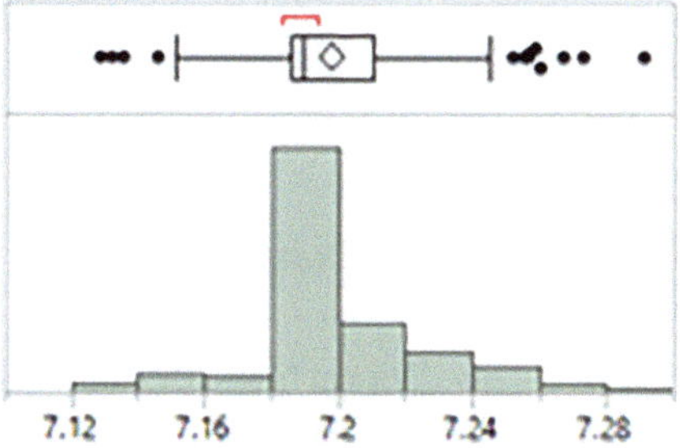

[a]Might go back to raw data based on initial data extraction frequency

15.3.6 Identify Model Output and Model Selection

Two factors to decide before modeling are to identify the model output (target variable) and identify the unit of analysis. As discussed above, the main purpose is to reduce the titer variability of the production culture to make the production culture titer a default candidate for model output. However, reviewing the production culture has shown that there are known factors that will affect titer (e.g., production culture duration). To remove the known culture duration effect, instead of titer, a growth indicator (integrated viable packed-cell volume at fixed culture duration of x hours: ivPCVx) was also used as a model output. This is because there is a linear relationship observed between titer and ivPCV at harvest and data availability (i.e., time course date of PCV measurement is available, but no time course date for titer). Other normalizations based on process characterization could be used to remove the known factor from modeling. In this specific example, the unit of analysis is each run when inoculum train culture starts. As shown in Fig. 15.2, there are many seed train cultures connected to each inoculum train culture. Therefore, a method to aggregate seed train information to run level is needed to build the model.

Once the team identified which is the target variable and the desired outcome of the model (e.g., if the intent of the model is diagnostics analytics or predictive analytics), a model was recommended by the DS team. Because the manufacturing process could benefit from both (target variable and desired outcome) to identify factors for growth/titer improvement and to predict the titer of culture for downstream processing, a random forest model with explainable artificial intelligence (AI) was used in this study. The random forest model was selected based on its strong prediction accuracy. It is good at capturing complex non-linear relationships in the data while being fairly robust to noise. To better interpret the complex and predictive models, an explainable AI (e.g., Shapley additive explanation [SHAP]) was used to generate insights [19]. When running the model, a fivefold cross-validation was used. This is to reduce overfitting of the available dataset and make the model outcome more robust to the broader dataset. The overall modeling process runs on Google cloud platform. The results were summarized in a global feature importance plot for each model iteration, with the top 20 features, and their impact on scale and direction based on the SHAP value. To further understand the feature effect and model performance, the training and test sets R^2, partial dependence plot, scatter plot, and trend stability were also reviewed.

15.3.7 Model Insights

As with many data analyses, the first model often identifies what was already known to process SMEs. Therefore, model iterations are a must-to-have for the ADA process. The review of hypotheses and the model outputs is ongoing through the model iteration phase, which helps to create new hypotheses as well as fine-tune and

Table 15.5 Example of model insights and potential actions

Outcome	Hypothesis description	Learning	Potential action
Process insight			
Confirm	Seed train status impacts production culture [18]	Evidenced by process understanding, confirmed through modeling	Select better seed train culture to source large-scale runs
Falsified	Effect of sampling volume impacts pH measurement	Sample volume variation observed but does not correlate with pH measurement	Since the model verified the variation, recommend best practice for sampling volume
New	Effect of sampling time close to the shift change time	Aggregate data in different ways through modeling	Build the knowledge into future resource planning, allocation
New	Effect of kitted medium storage time	Longer kitted medium storage correlates with reduced growth	Avoid longer storage time through planning
New	Raw material attribute	Important to include raw material test data. Different model outputs (even highly correlated) could provide different insights	Follow up with the vendor to understand raw material variability and any vendor manufacturing process change. Assess if specification change is needed
Operational insight			
Confirm	Measurement variation of the offline analyzer	Evidence by a small dataset, confirmed through modeling	Establish a monitoring program

prioritize them. In turn, it will provide direction of ongoing DE work (e.g., which data sources to link, what kind of features to create).

Table 15.5 shows a few examples of insights generated from this model exercise. This exercise provided an opportunity to verify or falsify previous hypotheses and create new insights (features not previously analyzed) on both process and operational perspectives.

15.4 Conclusion and Discussion

Leverage modeling for process and product understanding, monitoring, troubleshooting, prediction, and process control improvement is the trend for bioprocessing. Different statistical models can be used in different stages of the lifecycle. Advanced data analytics program should be considered for ongoing commercial manufacturing.

A multi-disciplinary team of bioprocess experts, data engineers, data scientists, and IT specialists, in collaboration with skilled consultants were able to employ advanced data analytics techniques to identify contributing sources of variation and actionable insights for a complex and unsolved culture performance problem.

By utilizing a structured but agile approach, the team was able to extract, cleanse, ingest, and process 800 GB of data from 11 disparate data sources into a cloud environment to evaluate more than 100 independent hypotheses. More than 500 plots to support exploratory data analysis were generated, reviewed, and a complex model used to iteratively evaluate hypotheses leading to new "actionable insights" and proposed interventions. Several of the new insights had not been previously discovered or documented during process design or from multiple years of univariate monitoring of commercial manufacturing campaigns. More than 100 runs have been executed after implementing some of the changes proposed. The normalized titer for 95% of those runs ranges from 0.97 to 1.17, which is ~6% improved on upstream productivity on Process A. Furthermore, the potential for future optimization of culture performance is bolstered by the addition of the complex model as an additional tool to complement the formal CPV program.

For complex investigations seeking root causes for process performance or product quality variation, in some cases, questions can go unasked and unanswered because of the complexity and scope of data required to evaluate a specific hypothesis. A key benefit of developing ADA capability is to accelerate the iterative cycle of asking and answering structured questions and quantifying the likelihood of potential causal factors contributing to the variation in the chosen target (response) variable. Even when evaluation of a hypothesis does not result in a "smoking gun," the documentation of "non-effects" provides evidence to rule out factors, sharpens the focus, adds to the body of process/product knowledge, and deepens the quality and thoroughness of the investigation. Establishing data pipelines and improved ADA skills holds promise for increasing hypothesis evaluation productivity, thereby bolstering institutional knowledge and problem solving speed.

In addition, the established platform and internal capability enable the building of unit operation or holistic process models, integrating with existing and new PAT devices with soft sensors, driving bioprocess digitization, and revolutionizing biomanufacturing.

Acknowledgments The authors would like to thank Mckinsey/Quantam-Black for guidance in the project execution and support in building the internal capability. The authors would also like to thank Sally Kline, Lisa Vulliet, Karen Roque, Juan Melendez, Cristen Peterson, Veronica Carvalhal, Luis Avila, Duyen Tran, Michael Siani-Rose, Ann Rea, Ulrike Strauss, Ravi Medandrao, Paul Paczuski, Jason Gu, and Daniel Coleman for their contributions to the project.

References

1. FDA Guidance for Industry (2011) Process validation: general principles and practices
2. Möller J, Pörtner R (2017) Chapter 5: Model-based design of process strategies for cell culture bioprocesses: state of the art and new perspectives. In: New insights into cell culture technology. InTech, pp 157–172

3. Horvath B, Mun M, Laird MW (2010) Characterization of a monoclonal antibody cell culture production process using a quality by design approach. Mol Biotechnol 45(3):203–206
4. Jayakumar M (2013) When and how to use Plackett-Burman experimental design. https://www.isixsigma.com/tools-templates/design-of-experiments-doe/when-and-how-to-use-plackett-burman-experimental-design/
5. Royall RM (1986) The effect of sample size on the meaning of significance tests. Am Stat 40(4):313–315
6. McNeish D (2016) On using Bayesian methods to address small sample problems. Struct Equ Model Multidiscip J 23(5):750–773
7. Melchers RE, Beck AT (2018) Structural reliability analysis and prediction. Wiley
8. Fujiwara Y, Miyawaki Y, Kamitani Y (2009) Estimating image bases for visual image reconstruction from human brain activity. Adv NIPS 22:576–584
9. Stenling A, Ivarsson A, Johnson U, Lindwall M (2015) Bayesian structural equation modeling in sport and exercise psychology. J Sport Exerc Psychol 37:410–420
10. Rodgers M, Oppenheim R (2019) Ishikawa diagrams and Bayesian belief networks for continuous improvement applications. TQM J 31(3):294–318
11. Breiman L (2001) Statistical modeling: the two cultures. Stat Sci 16(3):199–231
12. Eriksson L, McCready C (2018) Characterizing a bioprocess with advanced data analytics. Modeling at various stages of the data analytics continuum aids scale comparison of a bioreactor. BioPharm Int 31(3):18–23
13. Vamathevan J, Clark D, Czodrowski P, Dunham I, Ferran E, Lee G (2019) Application of machine learning in drug discovery and development. Nat Rev Drug Discov 18(6):463–477
14. Ayon D (2016) Machine learning algorithms: a review. IJCSIT J 7(3):1174–1179
15. Yuk IH, Russell S, Tang Y, Hsu WT, Mauger JB, Aulakh RP, Luo J, Gawlitzek M, Joly JC (2015) Effects of copper on CHO cells: cellular requirements and product quality considerations. Biotechnol Prog 31(1):226–238
16. Stichbury J (2019) Kedro: a new tool for data science, June 4. https://towardsdatascience.com/kedro-prepare-to-pimp-your-pipeline-f8f68c263466
17. Li F, Vijayasankaran N, Shen A, Kiss R, Amanullah A (2010) Cell culture processes for monoclonal antibody production. mAbs 2(5):466–477
18. Tung M, Tang D, Wang S-H, Zhan D, Kiplinger K, Pan S, Jing Y, Shen A, Ahyow P, Snedecor B, Gawlitzek M, Misaghi S (2018) High intracellular seed train BiP levels correlate with poor production culture performance in CHO cells. Biotechnol J 13:1700746
19. Molnar C (2019) Interpretable machine learning, a guide for making black box models explainable, November 17. https://christophm.github.io/interpretable-ml-book/index.html

Part VIII
Biopharmaceutical Regulatory CMC

Chapter 16
Overview of Complexities of Global CMC Regulatory Affairs

Katherine Arch-Douglas, Nathalie Dubois, Stephen Mayer, Rich Pelt, and Andrew Nelson

Abstract As the technologies within the biopharmaceutical industry evolve so does the global chemistry, manufacturing, and control (CMC) regulatory environment with ever-present complexity of gaining regulatory approvals around the world for new products for patients. From the earliest pharmaceutical laws in the 1900s to the inception of the International Council for Harmonization (ICH) to the transfer of regulatory oversight of well-characterized biologics from CBER (Center for Biologics Evaluation and Research) to CDER (Center for Drug Evaluation and Research), there has been a continuum of regulatory growth and oversight to ensure compliance and safety of all products globally. Every country and/or region has a separate health authority (HA) that oversees commercial licensure and clinical studies for all biopharmaceuticals. History, regional laws, regional cultures, and resources are many of the factors that influence the regulatory environment around the world, and all HAs must be engaged appropriately for successful global rollouts of biopharmaceuticals.

Keywords International Council for Harmonisation (ICH) · Health Authority (HA) · Biopharmaceutical · Regulatory · Commercial · Clinical

K. Arch-Douglas · S. Mayer
Global Regulatory Sciences CMC, Pfizer Inc., Pearl River, NY, USA

N. Dubois
Global Regulatory Sciences CMC, Pfizer Inc., Brussels, Belgium

R. Pelt (✉) · A. Nelson
Global Regulatory Sciences CMC, Pfizer Inc., Sanford, NC, USA
e-mail: Rich.peltiii@pfizer.com

K. Gadamasetti, S. A. Kolodziej (eds.), *Bioprocessing, Bioengineering and Process Chemistry in the Biopharmaceutical Industry*,
https://doi.org/10.1007/978-3-031-62007-2_16

16.1 Introduction

Every industry has its own semantics (e.g., set of terms, acronyms, abbreviations, and hyphenations) that facilitate workflow processes of a particular discipline. The biopharmaceutical industry is definitely no slouch when it comes to its own semantics and it is often very complex and can vary across the regions of the globe. In order to find success in the biopharmaceutical industry we need to understand the regulatory jargon of the industry, and also appreciate the complexities in how biopharmaceutical products are regulated by various health authorities (HAs) globally. This is especially true when we consider the time, cost, and risks associated with developing a biopharmaceutical product.

A biopharmaceutical product can take 10–15 years to develop and almost 90% of drug candidates in development fail before becoming a marketed product [1]. This is a huge risk for any potential drug development, especially considering that research published in 2016 found that the estimated research and development (R&D) costs associated with developing a new pharmaceutical product from discovery through development and to product launch was upward of 2.8 billion dollars [2].

Throughout the remainder of this chapter, we will begin to discuss the complexities associated with global regulatory affairs. The content discussed in the various underlying sections will only touch the surface of what you will come to see is a very diverse and agile field. As we overview the various regulatory bodies and their regulations, laws, and guidances, you will begin to understand the language that is biopharmaceutical global regulatory affairs.

16.1.1 Regulatory Bodies in Major Markets

A major challenge in global regulatory affairs is to put together a core dossier that can be leveraged in multiple markets; this adds to the complexity of what should and should not be included when you build your regulatory strategy as there is a need to consider multiple health authorities' (HAs) country regulations. Flushing out this global regulatory strategy early can reduce submission timelines, unnecessary additional authoring, and potential extra costs associated with what would otherwise be country-specific documentation for each dossier put forward. Country-specific documentation is not always avoidable and is sometimes mandatory such as in Module 1 of Common Technical Documentation (CTD) Pyramid put forward by the International Council for Harmonization (ICH) (more on ICH and CTD later in this chapter), but that does not mean that there are not places in a regulatory dossier where synergies can be in place.

So, with this in mind, as a regulatory affairs professional working for a pharmaceutical company, it is important to understand how the major markets are structured and what their regulatory requirements are: this works to your benefit in two ways. One, major markets are usually associated with the highest profit margins available for a biopharmaceutical manufacturer, and, two, many smaller or developing markets rely on approvals from reference HA markets, in this case markets such as the United States or the European Union (EU).

Below is a brief summary of some of the major HA organizations and clinical trial and license application processes.

16.1.1.1 United States

The U.S. Food and Drug Administration (FDA) is an agency within the Department of Health and Human Services and "*... is responsible for protecting the public health by assuring the safety, efficacy and security of human and veterinary drugs, biological products, medical devices, our nation's food supply, cosmetics, and products that emit radiation. FDA is also responsible for advancing the public health by helping to speed innovations that make medicines more effective, safer, and more affordable and by helping the public get the accurate, science-based information they need to use medicines and foods to maintain and improve their health....*" The FDA is subsequently organized into multiple centers and offices:

- Office of the Commissioner (OC)
- Office of Foods and Veterinary Medicine
- Office of Global Regulatory Operations and Policy
- Office of Regulatory Affairs (ORA)
- Office of Operations
- Office of Policy, Planning, Legislation, and Analysis
- Office of Medical Products and Tobacco
- Center for Biologics Evaluation and Research (CBER)
- Center for Devices and Radiological Health (CDRH)
- Center for Drug Evaluation and Research (CDER)
- Center for Tobacco Products
- Office of Special Medical Programs
- Office of Combination Products
- Oncology Center of Excellence (Operates across CDER, CBER, CDRH)

The Center for Biologics Evaluation and Research has eight offices and is responsible for regulation of biological and related products including blood, vaccines, allergenics, tissues, and cellular and gene therapies. The Office of Vaccines Research and Review (OVRR) is within CBER.

The Center for Drug Evaluation and Research has various offices and is responsible for regulation of over-the-counter and prescription drugs, including biological therapeutics and generic drugs.

Applications for clinical trials are filed under Investigational New Drug applications (IND). Typically, a single IND is opened per product and as clinical development progresses, IND amendments are filed with new information (e.g., new clinical protocols, clinical and nonclinical study reports, CMC [Chemistry, Manufacturing and Control] changes).

Biologics and Vaccine license applications are submitted as Biologics License Applications (BLA). License applications for over-the-counter and prescription drugs are submitted as New Drug Applications (NDA).

16.1.1.2 European Union

The European Union (EU) is organized quite differently. The European Commission (EC) proposes new legislation and ensures that EU law is correctly applied by member states. EU legislation concerning the regulation of medicines is adopted by the European Parliament and the Council of the European Union. The European Commission takes binding decisions on the centralized authorization of medicines based on European Medicines Agency (EMA) scientific assessments of marketing authorization applications (MAA), ensuring that medicines comply with high quality, safety, and efficacy standards.

The European Medicines Agency has seven scientific committees and a number of working parties and related groups that conduct the scientific work:

- Committee for Medicinal Products for Human Use (CHMP)
- Pharmacovigilance Risk Assessment Committee (PRAC)
- Committee for Veterinary Medicinal Products (CVMP)
- Committee for Orphan Medicinal Products (COMP)
- Committee on Herbal Medicinal Products (HMPC)
- Committee for Advanced Therapies (CAT)
- Paediatric Committee (PDCO)
- Working parties and other groups

Applications for clinical trials are filed as a Clinical Trial Application (CTA) and quality information is provided in the Investigational Medicinal Product Dossier (IMPD) component of the CTA. Typically, a new CTA is filed for each clinical study. The CTA review process was historically governed by individual National Competent Authorities; however, an optional centralized Voluntary Harmonized Procedure (VHP) is currently in place. Also, since January 2022, the Clinical Trials Regulation (CTR) came into application, harmonizing the submission, assessment, and supervision processes for clinical trials in the European Union (EU).

License applications for biologics, drugs, and vaccines are submitted as a Marketing Authorization Application (MAA).

16.1.2 Global Markets

There are numerous other markets and associated HAs in which CMC content and license review will likely differ around the globe. A few rest-of-world market/HA examples are:

- China: NMPA (National Medical Products Administration)
- Japan: PMDA (Pharmaceuticals and Medical Devices Agency)
- Canada: HC (Health Canada)
- Australia: TGA (Therapeutic Goods Administration)
- Korea: MFDS (Ministry of Food and Drug Safety)
- India: MHFW (Ministry of Health and Family Welfare)
- Hong Kong Department of Health
- Taiwan: MHW (Ministry of Health and Welfare)
- Brazil: ANVISA (Agencia Nacional de Vigilancia Sanitaria—Brazilian Health Regulatory Agency)
- Mexico Ministry of Health
- Russia Ministry of Health
- Israel Ministry of Health
- Other Latin America Countries (Argentina, Peru, etc.)

There are several key aspects about vaccines that greatly influence regulatory behaviors. First, vaccines are not "drugs"; they are vehicles to elicit a response (e.g., antibodies) to combat disease. Prophylactic vaccines aim to prevent disease and are given to healthy subjects while therapeutic vaccines aim to treat disease and are given to patients. Next, the concept of a "well-characterized product" does not apply to vaccines (i.e., "the process is the product") and each vaccine manufacturing process is unique.

16.1.3 Examples of Core Differences in Regulatory Requirements Globally

In the United States alone, the Code of Federal Regulations (CFRs) have different regulations for NDAs, which fall under 21 CFR Part 314, and BLAs, which fall under 21 CFR Part 600 (for Biologics Products: General), 21 CFR Part 601 (Licensing), 21 CFR Part 610 (General Biologics Products Standards), 21 CFR 601.2(a) Specified biological products, and Biosimilar and interchangeable products subject to licensure under section 351(k) of the Public Health Service (PHS) Act (42 U.S.C. 262(k)). There are also a number of FDA-issued guidance documents for small molecules, biologics, vaccines, and medical devices.

As new technology and understanding develops with these products, the regulations sometimes change or are updated. One recent example is the removal of the General Safety Test (GST) from the US CFR. This test was designed to test for extraneous toxic contaminants on each product lot intended for human use. The test required one-week observations for general health and weight following injection of specified volumes of product batches into guinea pigs and mice. This test was previously listed in 21 CFR 610.11 but was removed from the CFR in 2015. FDA stated in the Final Rule for removal that "FDA is finalizing this action because the existing codified GST regulations are duplicative of requirements that are also specified in BLAs or are no longer necessary or appropriate to help ensure the safety, purity, and potency of licensed biological products. FDA is taking this action as part of its retrospective review of its regulations to promote improvement and innovation, in response to the Executive order."

Similarly, the Abnormal Toxicity Test (ATT) was previously listed in the Ph. Eur. 2.6.9 and was required for testing Immunosera and Vaccines for Human Use was also previously used for testing general health of mice and guinea pigs 1 week after injection. This test was removed from Ph. Eur. in 2017 as part of European Commission Directive of Replacement, Reduction and Refinement of animal testing (3Rs) per Directive 2010/63/EU on the protection of animals used for scientific purposes.

It should be noted that while the GST and the ATT are no longer required by the FDA and EMA, respectively, regulatory authorities in other commercial markets still require these tests to be performed.

In recent years, a hot topic for CDER-regulated Biologics was concerning the inclusion of microbial control information in the NDAs and BLAs, which was driven by the formation of CDER's Office of Pharmaceutical Quality (OPQ) and increased involvement in dossier review by OPQ's Division of Microbiology Assessment.

16.2 Complexities in Navigating Global Market Drug Development and Managing Multiple Market Dossiers

Even though the major markets (e.g., the United States, EU, Japan, Canada, Australia) follow the ICH CTD format for CMC information, each market has its own set of regulations and guidance and may have their own compendial requirements for testing.

Even within these markets, the regulatory requirements are different for small molecules, biologics, and vaccines. For example, most major markets require lot release testing of vaccine lots before they are allowed into commercial distribution in that market. All lots of commercial vaccines, in both the United States (21 CFR 610.2) and EU (Directive 2001/83/EC amendment Directive 2004/27/EC (7, 8)), require lot-by-lot release and testing by government laboratories prior to commercial distribution. This lot release testing consists of the sponsor sending samples of each lot intended for commercial distribution to each country's testing lab for testing against the company's Certificate of Analysis (COAs) and releasing the lot. This lot release testing is no longer a requirement for most biologics and small molecules.

Several other markets also require in-country testing of products before they can be commercially approved (for example: Australia, Canada, China, Russia, Saudi Arabia). In addition, due to numerous climates and supply chain requirements, various shipping validation requirements would need to be met (e.g., cycling studies to support excursions during transportation).

In China, the Drug Registration Regulation became effective in July 2020. A new guideline on Manufacturing and Testing Procedures (MTP) and JS Specification templates became effective in July 2021. Drug Master Files (DMFs) or Technical Documents are required for excipients and container closures. In-country testing is a requirement for Drug Substance and Drug Product for biologic products and vaccines at commercial stage and vaccines at the clinical stage, while there is no formal process for transfer of test methods to the National Institutes for Food and Drug Control. This highlights the evolving landscape of regulatory requirements for commercial registrations in China.

Some global markets are moving toward a Reliance Model pathway (Eurasian Economic Union [EAEU], Balkans, Egypt) to reduce initial application approval timelines. However, divergence between the World Health Organization (WHO), national regulatory authorities (NRA), and industry on the definition and scope of "product sameness" is a challenge. Most global emerging markets now follow the ICH CTD template format for chemistry, manufacturing, and control (CMC) information but many markets require specific detailed regional documentation, which make it difficult to maintain a core dossier type of approach for CMC content. These regional requirements vary from country to country and include items such as:

- Country-specific normative documents
- Local application forms
- Canada specific Certified Product Information Document (CDIP)
- Certified Product Details (CPDs)
- Manufacturing and Testing Procedures (MTP)
- GMP (Good Manufacturing Practice) documents such as site/plant master files
- Country-specific declarations for CMC materials
- Ancillary documents such as stability chromatograms, cell bank COAs, certificates of source of raw materials

For some markets, such as China, Russia, and South Korea, these country-specific CMC documentation requirements require a significant amount of time and resources to prepare the documentation. Additionally, some countries, such as Turkey, require GMP pre-clearance before the commercial application for the drug or vaccine can be submitted. This entails either an onsite or paper inspection of each manufacturing site by Turkey's Ministry of Health where a GMP certificate is issued for each site following pre-clearance.

Additionally, each market has its own targeted timelines for initial review of submissions. For example, in the United States, a BLA review period is 12 months for standard review and 8 months for BLAs that receive priority review designation from FDA. Initial applications in various global markets have taken more than 2 years to obtain commercial approval. Subsequent initial applications to global markets often occur over the next several years. Any changes to manufacturing processes, shelf-life, test procedures, etc. during this timeframe may cause challenges from a global supply perspective due to the amount of time it takes to obtain approvals in all markets. Many health authorities do not allow changes to be submitted during the initial registration review period.

Timeframes to obtain approvals for post-approval CMC changes in global markets may vary from a minimum of a few months to in excess of 2 years. Management of these changes requires close communication between regulatory affairs and commercial supply colleagues and strong oversight of global commercial inventory to ensure that there is no stock-out of commercial inventory in each market.

16.3 The Role of WHO and ICH in Global Drug Development

The World Health Organization (WHO) is a United Nations agency founded in 1948 to connect nations and promote health and keep the world safe and especially serve the vulnerable. Through working with 194 Member States across 6 regions, the WHO is a decision-making body that is responsible for collaborations with governments to ensure everyone in the world can remain healthy and for coordinating the world's response to health emergencies like the recent COVID-19 pandemic.

As part of the WHO's essential global role in governance of health and disease, there are a number of technical and regulatory guidance documents drafted by the WHO to help unify approaches of regulation of medicinal products and vaccines. In response to the COVID-19 pandemic, the WHO published a number of technical guidances for the detection, prevention, and control of this virus.

The mission of the ICH "is to achieve greater harmonization worldwide to ensure that safe, effective, and high-quality medicines are developed and registered in the most resource-efficient manner. Harmonization is achieved through the development of ICH Guidelines via a process of scientific consensus with regulatory and industry experts working side-by-side. Key to the success of this process is the commitment of the ICH regulators to implement the final Guidelines" (ICH Official web site: ICH).

An example of the ICH harmonization process was the advent of the Common Technical Document (CTD) in 2003. This conceptualized the idea of global harmonization by providing a format for the assembly of quality, safety, and efficacy information and data to be included in a pharmaceutical product dossier. Global acceptance for dossier formatting by the manufacturer is important because it allows a significant ease in the burden associated with the regulatory review process by health authorities, which allows regulatory approvals to accelerate. Often overlooked and just as important, however, is the desire for acceptance of dossier formatting and required information by global health authorities, as aligning global country requirements eases the regulatory burden on the manufacturer. Encouraging the acceptance of CTD format is important for HAs and manufactures alike, as it benefits both. The ICH has provided the CTD triangle, which is a visual representation of the structure of the CTD; this is provided in Fig. 16.1.

The CTD is organized into five modules. Module 1, consists of regional administrative information. This is not considered part of the CTD, because regional information is always country specific, so it cannot be aligned across global applications. Examples of Module 1 documentation could be country-specific forms, cover letters, meeting requests, labeling or promotional materials, general correspondence with the HA, among others.

Module 2, consists of summaries of the quality, pre-clinical, and clinical modules that are all the remaining modules (3, 4 and 5).

Module 3, is the CMC module. The Quality-M4Q(R1) guidance by the ICH describes the CTD backbone structure for all quality information as it relates to manufacturing, drug substance, and drug product.

Module 4, is the pre-clinical and nonclinical module. The Safety-M4S(R2) guidance by the ICH describes the CTD backbone structure for all safety information as it relates to pharmacology, pharmacokinetics, and toxicology.

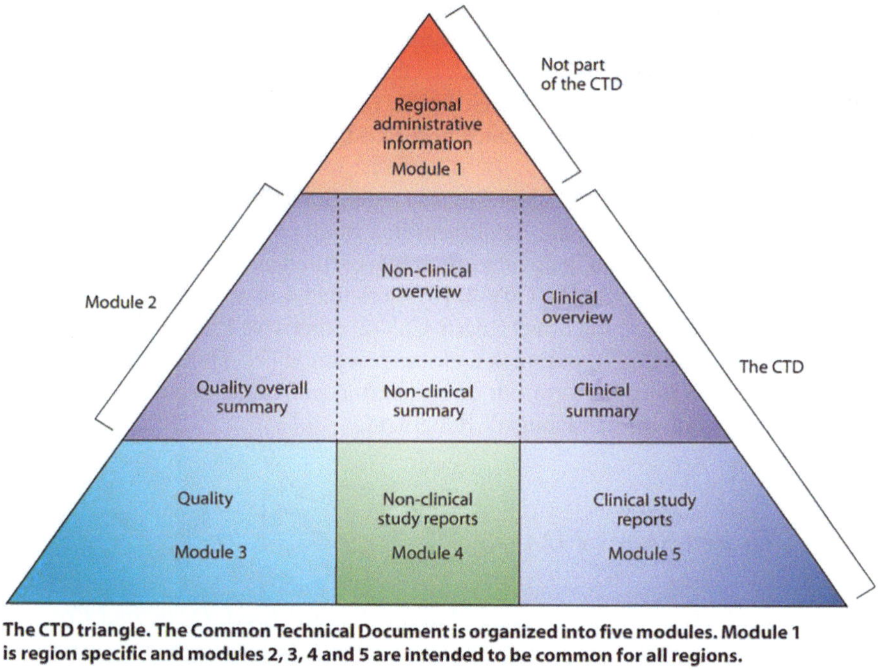

Fig. 16.1 CTD triangle. (Captured from www.ICH.org [3])

Module 5, is the clinical study module. The Efficacy-M4E(R2) guidance by the ICH describes the CTD backbone structure for all clinical pharmacology, clinical efficacy, and clinical safety studies.

Prior to the implementation of CTD, the CMC content for investigational and commercial submissions varied from country to country. Although the CTD format is now accepted globally, each market has its own regulations (i.e., laws, legal requirements) and guidance documents (i.e., regulatory agency interpretation/application of laws) that often require different submission content that results in divergence from a common technical document.

While ICH and WHO help to harmonize practices, several health authorities throughout the world maintain their own national laws and guidelines that govern regulatory activities. As with other medicinal products, there are numerous markets, global and national, in which CMC content, license review, and approval timelines will differ depending on the associated health authority. For example, some health authorities require a lot-to-lot manufacturing-consistency clinical study to be conducted during Phase 3, additional country-specific documentation to be filed, or lot-by-lot release testing to be executed by country labs prior to distribution.

16.4 Complexities of Emergency Use in a Pandemic

A pandemic is defined as a global outbreak of an infectious disease. The world has seen its share of pandemic-level events from the Bubonic plague, that killed about 30% of Europeans during the 1300s [4], to the Spanish Flu, that killed 50 million plus people worldwide in 1918 [5]. However, no global outbreak of an infectious disease has been more front and center in recent times than that of the COVID-19 pandemic.

On Tuesday, December 21, 2019, Chinese authorities alerted the WHO to a mysterious virus causing pneumonia-like illness in a small cluster of patients within the city of Wuhan. Shortly after, a novel virus was identified as severe acute respiratory syndrome coronavirus 2 (SARS-CoV-2) [6]. Less than 4 months later on March 11, 2020, the WHO declared the novel SARS-CoV-2, otherwise known as Coronavirus Disease-2019 (COVID-19), outbreak a global pandemic [7]. This declaration drove novel technologies into the forefront of biopharmaceutical science and challenged the traditional paradigm of vaccine development.

16.4.1 United States FDA

In the United States, *Under section 564 of the Federal Food, Drug, and Cosmetic Act (FD&C Act), when the Secretary of HHS declares that an emergency use authorization is appropriate, FDA may authorize unapproved medical products or unapproved uses of approved medical products to be used in an emergency to diagnose, treat, or prevent serious or life-threatening diseases or conditions caused by chemical, biological, radiological and nuclear (CBRN) threat agents when certain criteria are met, including there are no adequate, approved, and available alternatives* [8]. On January 31, 2020, the Secretary of the Health and Human Services (HHS) determined the most appropriate course of action was to do just that, and they issued a declaration of a public health emergency related to COVID-19 and mobilized the Operating Divisions of HHS [9].

The FDA and EMA began to collaborate with other countries to help facilitate COVID-19 vaccine development. They first convened on March 18, 2020, with representative experts from 17 different countries and the World Health Organization to discuss the pre-clinical data requirements that would be needed from a manufacturer for a pandemic vaccine candidate to precede to first-in-human clinical trials. The FDA and EMA would also continue to convene at future dates with other global regulators to enable sharing of clinical trial data as a means of ensuring global preparedness and streamlining vaccine development [10].

The FDA has the authority to issue an Emergency Use Authorization (EUA), which is in accordance with section 564 of the Food, Drug, and Cosmetic Act (21 U.S.C. 360bbb-3). During the initial development of a vaccine candidate for COVID-19, vaccine manufacturers only had the FDA guidance published in June of 2020, *Development and Licensure of Vaccines to Prevent COVID-19*, to use as a source guidance document to build their regulatory strategies. In this guidance the agency put forth their requirements for developmental vaccines to be considered for EUA. From a clinical trial perspective, they stated that "To ensure that a widely deployed COVID-19 vaccine is effective, the primary efficacy endpoint point estimate for a placebo-controlled efficacy trial should be at least 50%" [11], meaning that a vaccine candidate need only show 50% effectiveness, similar to the annual influenza vaccine to be considered for licensure.

The FDA later released the guidance document, *Emergency Use Authorization for Vaccines to Prevent COVID-19*, in May of 2021, which has since been superseded as of March 31, 2022. This guidance provides the requirements for key information that should be provided to the IND application prior to submitting an EUA request. Providing this key information will allow the FDA to facilitate a timely review of submission content and convene with the Vaccines and Related Biological Products Advisory Committee (VRBPAC) [12].

In addition to the FDA guidance documents that were set forth specifically to manage the COVID-19 pandemic, the FDA also allowed the manufacturers of COVID-19 vaccine candidates to meet more frequently either via virtual telecommunications or through briefing documents (since the country was on lockdown, there were no face-to-face meetings) to get their questions answered for what would be needed in a potential emergency use application.

The United States government started Operation Warp Speed that was being overseen by HHS Secretary Alex Azar and Defense Secretary Mark Esper, which aimed at speeding up vaccine development efforts to enable the delivery of safe and effective vaccine doses by January of 2021. This was done by focusing on three key phases: development, manufacturing, and distribution.

This effort from the United States government and the FDA as well as open global collaboration helped set the stage for major manufacturers like Pfizer and Moderna, during the clinical development phase, to be able to establish their developmental strategies with more certainty as everyone was operating with the same end goal, of getting lifesaving preventative vaccines for patients.

In the book *Moonshot* by Pfizer Inc. CEO Dr. Albert Bourla, he notes that Pfizer began testing four different vaccine candidates in Phase 1 trials in May of 2020 and developed a clinical trial design and manufacturing infrastructure plan that would cost upward of two billion dollars to implement. There are many ways Pfizer was able to develop a vaccine candidate in 9 months and one major reason was because of their clever trial design strategy. Dr. Bourla notes "Traditionally, those would have been tested sequentially, but we decided to test them in parallel – four different vaccine candidates, each at three different dosage levels. Such an effort might take a year under normal circumstances. We did it in a month. By late July 2020, we were prepared for a combined Phase 2 and 3 trial in which eventually over 46,000 patients would be recruited at 153 clinical sites in six countries" (*Moonshot*, p. 51) [6]. Pfizer received the clinical readout of their pivotal Phase 3 data on November 8; the readout showed to be 95.6% efficacious. They then proceeded to submit their initial EUA submission on November 20 and the U.S. FDA granted Pfizer Emergency Use Authorization on December 11, 2020.

16.4.2 European Medicines Agency (EMA)

In Europe (EU), COVID-19 pandemic has had a considerable impact on citizens, patients, and businesses, and this posed unprecedented challenges in ensuring continuity of supplies of medicines or providing medicines against the COVID-19 virus while ensuring that the high standards of quality, safety, and efficacy of medicines made available to patients are maintained. In June 2020, the European Commission (EC) announced the EU strategy (1) around COVID-19 vaccines. Taken the unprecedented crisis, the ambition was to deliver a successful vaccine within a timeframe of 12–18 months and, once a successful COVID-19 vaccine is available, to produce high amounts of doses in order to cover global needs, without compromising the production of other essential vaccines. To this end, the EU mobilized resources and joined forces with global health organizations through the Access to COVID-19 Tools (ACT) Accelerator collaborative effort. As part of the effort to help protect people everywhere and EU citizens in particular, the EC proposed an EU strategy to accelerate the development, manufacturing, and deployment of vaccines against COVID-19.

The strategy has three major objectives and is based on two pillars:

(a) Ensuring the quality, safety, and efficacy of vaccines
(b) Securing timely access to vaccines for Member States and their population while leading the global solidarity effort

(c) Ensuring equitable access for all in the EU to an affordable vaccine as early as possible
(d) Securing sufficient production of vaccines in the EU and thereby sufficient supplies for its Member States
(e) Adapting the EU regulatory framework to the current urgency and making use of existing regulatory flexibility

The EMA offers applicants a pathway to a conditional marketing authorization, whereby medicines may be granted conditional use with less comprehensive initial clinical data when the benefit of the immediate availability outweighs the inherent risk associated with the use of the therapeutic or preventative drug. Conditional marketing authorization may be granted by the Committee for Medicinal Products for Human Use (CHMP) in accordance with Article 14-a of Regulation (EC) No 726/2004 [13].

EMA's CHMP may grant a conditional marketing authorization for a medicine if it finds that all of the following criteria are met:

- *The benefit–risk balance of the medicine is positive.*
- *It is likely that the applicant will be able to provide comprehensive data post-authorization.*
- *The medicine fulfils an unmet medical need.*
- *The benefit of the medicine's immediate availability to patients is greater than the risk inherent in the fact that additional data are still required.*

16.4.3 The World Health Organization

The WHO has an Emergency Use Listing Procedure (EUL), which is used as a risk-based procedure to assess unlicensed vaccines, therapeutics, and in vitro diagnostics. The intention of this procedure is to expedite the availability of these products to people that are affected by public health emergencies [14].

On November 25, 2020, the WHO released a guidance document titled *Considerations for the Assessment of COVID-19 Vaccines for Listing by WHO*. This document was to be used by Manufacturers to understand the WHO's terms for what they would consider a viable vaccine candidate for an EUL. Like any guidance document, it is not a comprehensive step-by-step guide on vaccine development but it did put forward some important considerations.

This guidance also covers CMC considerations such as characterization of the starting material and drug substance, process controls, and process validation. The guidance states that: "As an alternative to the traditional process validation, continuous process verification can be utilized in process validation protocols for the initial commercial production and also for manufacturing process changes for the continual improvement throughout the remainder of the product lifecycle" [14].

The WHO has since issued an EUL to ten vaccine products (Comirnaty, Vaxzevria, Covishield, COVID-19 Vaccine by Janssen-Cilag Int NV, Spikevax, Inactivated COVID-19 Vaccine [Vero Cell], CoronaVac, Covaxin, Covovax, and Nuvaxovid).

16.4.4 Examples of Global Regulatory Challenges During a Pandemic

In order for a manufacturer to market their product in a country, they must ensure they are in compliance with local country laws and regulations. This is a challenge when you consider the dossier mentioning the need to alleviate concerns for a single board of health. However, when you consider the complexities of alleviating those same concerns across 50 or 80 plus countries, that regulatory challenge exacerbates ten-fold.

Every major change that a manufacturer makes (e.g., increasing batch size, extending shelf-life, or increasing the amount of manufacturing sites) to their dossier must receive approval or concurrence from every country's board of health in which the drug is registered.

There are some boards of health that require a single license per drug product manufacturing site; they may also require a new license for a different concentration or formulation of that same drug (such as dosing concentration changes for pediatric indications or changing product formulation from a liquid drug product to a lyophilized drug product as a means of increasing product shelf-life). Having to maintain five or six licenses for a single country adds to the increased regulatory burden on manufacturers. The increased burden could be having to submit different submission packages across the various licenses for the same change. There could also be parallel queries received across licenses, as well as parallel commitments across licenses, all needing to be managed, authored, and submitted within very specific timeframes.

During the COVID-19 pandemic, many countries understood this added burden and made new guidances or even changed laws to alleviate the burden so that they could receive vaccine doses at a rapid rate.

Many global boards of health rely on a reference market to leverage for submission approvals. In many cases that reference market is either the European Union (EU) or the United States. During a pandemic, the manufacturer and the country may prefer global harmonization for their dossiers and may attempt to align global market dossier packages with either the United States or the EU. This would allow the manufacturer to rapidly submit supplements, amendments, or variations, removing the need to customize a submission for specified markets on an individualized basis. The country using the EU conditional marketing authorization as a reference market dossier in turn may require the manufacturer to submit EU queries, specific obligations, and recommendations as commitments to maintain conditional use of the drug in their country. In the case of the COVID-19 pandemic, this was more prevalent as the FDA and EMA were openly collaborating together with other countries to help facilitate COVID-19 vaccine development; this included specific obligations and recommendations to manufacturers with other countries.

The COVID-19 pandemic was and continues to be the most globally collaborative therapeutic, diagnostic, and vaccine development effort in human history. It is because of this global collaboration from various governments, regulatory entities, private businesses, philanthropists, and global health organizations that we were able to see companies like Pfizer and Moderna develop and successfully launch a viable vaccine candidate with very high efficacy in under a year—something that until December of 2020 had never been done in the modern biopharmaceutical age.

References

1. https://www-sciencedirect-com.eu1.proxy.openathens.net/science/article/pii/S2211383522000521
2. DiMasi JA, Grabowski HG, Hansen RW (2016) Innovation in the pharmaceutical industry: new estimates of R&D costs. J Health Econ 47:20–33. https://doi.org/10.1016/j.jhealeco.2016.01.012. Epub 2016 Feb 12. https://pubmed.ncbi.nlm.nih.gov/26928437/
3. https://admin.ich.org/sites/default/files/2021-02/CTD_triangle_color_Proofread.pdf
4. https://www.ncbi.nlm.nih.gov/pmc/articles/PMC4013036/pdf/pone.0096513.pdf
5. https://www.ncbi.nlm.nih.gov/pmc/articles/PMC3291398/pdf/05-0979.pdf
6. Bourla A (2022) Moonshot—inside Pfizer's nine-month race to make the impossible possible. HarperCollins Publishers
7. https://pubmed.ncbi.nlm.nih.gov/32191675/
8. https://www.fda.gov/emergency-preparedness-and-response/mcm-legal-regulatory-and-policy-framework/emergency-use-authorization
9. Secretary of Health and Human Services Alex M. Azar. Determination that a public health emergency exists (January 31, 2020, renewed April 21, 2020). Available at https://www.phe.gov/emergency/news/healthactions/phe/Pages/default.aspx
10. https://www.fda.gov/news-events/fda-voices/fda-and-ema-collaborate-facilitate-sars-cov-2-vaccine-development
11. https://www.fda.gov/media/139638/download
12. https://www.fda.gov/media/142749/download
13. https://www.ema.europa.eu/en/human-regulatory/marketing-authorisation/conditional-marketing-authorisation
14. https://www.who.int/teams/regulation-prequalification/eul#:~:text=The%20WHO%20Emergency%20Use%20Listing,by%20a%20public%20health%20emergency

Chapter 17
CMC Considerations for Continuous Bioprocess Design, Development, and Manufacturing

Yanhuai (Richard) Ding and Margaret (Peggy) Marino

Abstract This chapter describes monoclonal antibody (mAb) continuous bioprocess (CBP) from design, implementation to manufacturing based on the scientific understanding of mAb physicochemical properties, proven bioprocessing principles, available technologies, chemistry, manufacturing, and control (CMC) considerations, current industrial practices, regulatory guidelines, challenges, potential solutions, and future perspectives. The discussion addresses the conventional, intensified, integral, and/or fully automated end-to-end mAb CBP manufacturing process from cell line development (CLD); cell culture process development (e.g., upstream process); protein purification process development (e.g., downstream process); analytical method development, qualification, and validation for process performance; and product quality monitoring and control perspectives. The increasing interest in the application of CBP in biopharmaceutical manufacturing is associated with increased mAb market demand, demonstrated process consistency and product quality, and potential cost of goods (COGs) reduction. Some unprecedented challenges of CBP application are discussed. Some innovative technologies are assessed with practical solutions proposed. A case study is presented and discussed regarding a flowthrough mode of cation exchange chromatography for potential CBP implementation. The goal of this chapter is to propose a design and establish a fully automated CBP platform for an end-to-end mAb production from cell culture to drug substance (DS) formulation. Finally, the CBP technology is proposed for other bioprocess and manufacturing such as adeno-associated virus (AAV) vector for gene therapy.

Y. Ding (✉)
CMC Drug Substance/Drug Product, EvolveImmune Therapeutics, Branford, CT, USA
e-mail: yanhuai.ding@evolveimmune.com

M. Marino
Downstream Process Development & Manufacturing, AnaptysBio, San Diego, CA, USA

K. Gadamasetti, S. A. Kolodziej (eds.), *Bioprocessing, Bioengineering and Process Chemistry in the Biopharmaceutical Industry*,
https://doi.org/10.1007/978-3-031-62007-2_17

Keywords Continuous bioprocess · Chemistry · manufacturing · and control · Critical quality attribute · Drug substance · Monoclonal antibody · Adeno-associated virus

17.1 Introduction

Continuous manufacturing (CM) has been used for petrochemical, chemical, pharmaceutical, and food processing for decades [1, 2]. CM can provide consistent product quality; cleaner, more flexible, and higher process efficiency; automation, low in-process hold times, and sample conditioning; smaller footprint for equipment and facility; and lower operating cost of goods (COGs) [3, 4]. The CM concept has been applied for continuous bioprocess (CBP) as well for small molecule production such as amino acids, vitamins, peptides, and antibiotics [5]. Currently, the continuous processes for small molecule manufacturing have been applied for Orkambi and Symdeko for cystic fibrosis treatment (Vertex 2015) and Prezista for HIV treatment (Janssen 2016) [6].

However, CBP application to large molecule biologics (e.g., monoclonal antibody [mAb]) manufacturing has been slow to adopt in the industry due to several challenges. The challenges include: (A) mAb biologic complexity; (B) cell line and mAb stability; (C) manufacturing raw material properties and variability; (D) lack of CBP-designated systems for effective and robust process performance monitoring and control; (E) immature CBP-designated analytical systems and methods for real-time product quality monitoring and control; (F) lack of CBP-specialized process-related hardware and software; and (G) lack of sufficient financial resources, and scientists and engineers with CBP expertise and experience.

Conventional mAb drug substance (DS) manufacturing starts from Master Cell Bank (MCB) or Working Cell Bank (WCB) vial thaw followed by seeding train expansion, cell culture production, bulk cell harvest, clarification, purification, and final DS formulation. The cell culture process is conducted in batch, fed-batch, hybrid fed-batch, or perfusion mode [7–9]. The mAb purification framework is mainly carried out in a unit-by-unit operational manner [10]. In-process sample and final DS testing is primarily conducted by offline assays, which are labor intensive, time consuming, and do not provide real-time supportive data. The conventional process has intrinsic drawbacks associated with lower productivity and efficiency, less automation and robust control, higher waste generation, higher COGs from operational to capital expenditure, and larger equipment and facility footprint. In contrast, CBP implementation is an effective strategy to overcome these drawbacks per current industrial needs.

17.2 Regulatory Guidance, Expectations, and Support for CBP

FDA has been collaborating with both industry and academia to assess new technology applications for CM [11], analytical method improvement, and microbial control (Fig. 17.1). To accelerate CM application in biopharmaceuticals, FDA supports the quality by design (QbD) and process analytical technology (PAT) approaches [12, 13].

QbD is a systematic, risk-based, holistic, proactive, and continuous improvement approach for pharmaceutical development from discovery, development to manufacture of drugs [14, 15]. The key considerations from QbD are: (1) product knowledge including quality target product profile (QTPP), dose form, formulation, and stability; (2) process design including unit operation, in-process control (IPC), and critical process parameter (CPP) assessment; and (3) product quality control strategy including critical quality attributes (CQAs) and real-time testing with predetermined acceptance criteria and specification.

PAT framework supports innovative pharmaceutical development, manufacturing, and quality assurance [16]. It includes designing, analyzing, and controlling manufacturing through timely measurements of CQAs and critical performance attributes (CPAs) from raw material, in-process material to process control (Fig. 17.2) in a timely manner (e.g., in-, on-, or at-line monitoring and control). The PAT implementation can provide the scientific basis for establishing regulatory specifications. It also enhances continuous process improvement and product quality control [16]. Regulatory authorities support an enhanced and innovative

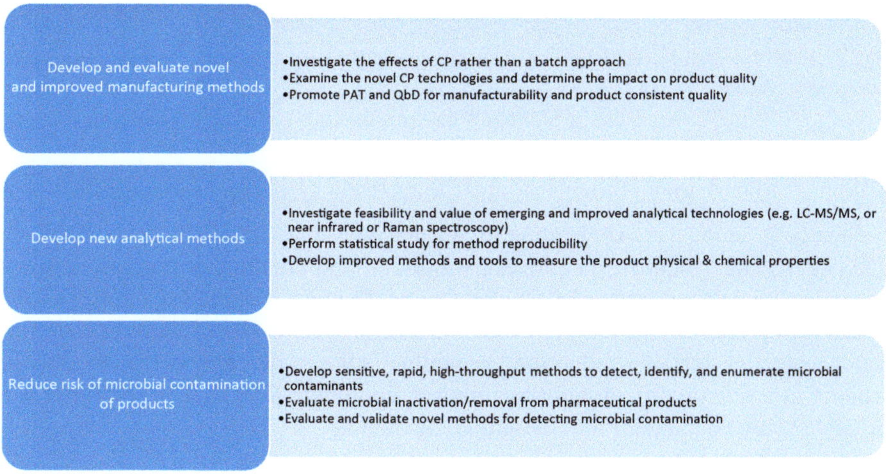

Fig. 17.1 New strategies and approaches supported by FDA for product manufacturing and quality improvement

Fig. 17.2 PAT-A framework for innovative pharmaceutical development, manufacturing, and quality assurance from FDA

bioprocessing with more flexible, compatible, and disposable technology for simpler process procedure and better product quality control [13].

Standard Guide for Application of Continuous Quality Verification (ASTM) E2537 provides guidance for process validation and continuous process verification, which is aligned with ICH Q8 and Q9. ASTM E2537 provides a guidance for continuous process monitoring, measuring, and analyzing; risk evaluation and mitigation; process variable reduction; capability adjustment; and enhancement for product lifecycle quality monitoring, control, and improvement. This guidance is a valuable source document for CBP implementation.

17.3 Current Industrial Practice for mAb Production

mAb therapeutics not only improve human quality of life but also generate billions in revenue each year for the biopharmaceuticals industry [17]. Over 80 therapeutic mAbs have been approved by global regulatory authorities [17, 18]. To bring a mAb candidate from discovery to commercialization, it is necessary for a biopharmaceutical company to spend a significant amount of money (e.g., hundreds of millions of US dollars), time (e.g., ≥10 years), and resources to complete essential activities from chemistry, manufacturing, and control (CMC), clinical trials to regulatory filings and approval.

Figure 17.3 illustrates a roadmap of mAb drug development from research to commercialization. Current mAb production is primarily carried out in a batch-mode including both upstream process (USP) and downstream process (DSP). The batch-mode has its intrinsic limitations as described previously; however, scientists and engineers from academia and biopharmaceutical industry are investigating alternative approaches from an intensified process [19], an integral process [4, 7], a hybrid process [20] to a fully automated bioprocess [21].

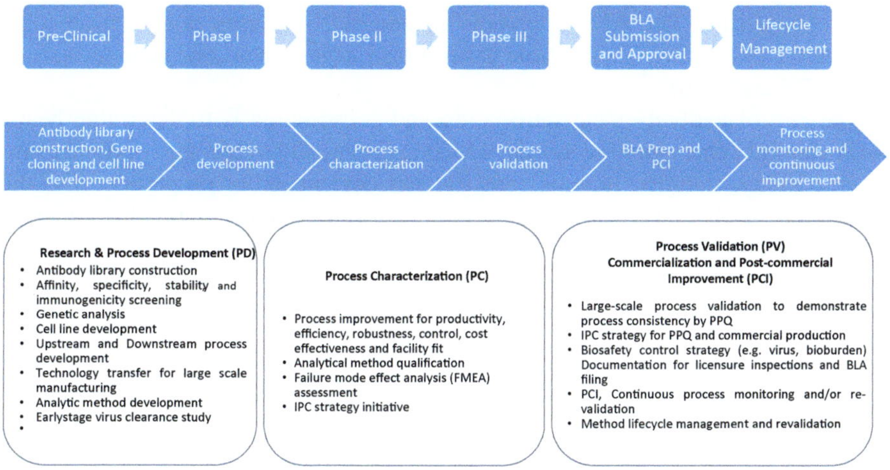

Fig. 17.3 Roadmap of mAb drug development from research to commercialization

17.3.1 Cell Line Development: The Foundation of Biologics Development

17.3.1.1 Antibody DNA Sequence and Gene Construction

Figure 17.4 illustrates a flow diagram for a typical mAb cell line development (CLD) process. The gene sequence of a mAb candidate can be identified and determined by Xenomouse hybridoma (e.g., Herceptin), phage display (e.g., Humira), transgenic mouse (e.g., Stelara), or single B-cell (e.g., HIV neutralization antibody) [18, 22]. The hybridoma technique, which has been the most widely used approach, has led to the majority of mAbs (48 out of 80 mAbs) approved on the market [18, 23]. The antibody heavy chain and light chain variable (V_H and V_L) region cDNA sequence can be generated by reverse transcription polymerase chain reaction (RT-PCR) from the isolated mRNA [24]. The determined variable region from each gene can be humanized by complementarity-determining region (CDR)-grafting and codon optimization followed by the assessment of affinity, functional potency, and stability of the mAb [24].

In addition to hybridoma technology, antibody phage display (APD) has been also used for therapeutic antibody discovery and development [25]. The APD approach is based on the pioneer work from Smith (1985) and Parmley (1988) for peptide expression followed by John McCafferty and Gregory Winter [26–28] for mAb engineering. The APD approach includes DNA library construction (e.g., mRNA, RT-PCR, cDNA), phagemid generation, phage display, and genetic analysis (Figs. 17.5 and 17.6).

The aim of antibody gene sequencing and construction is to determine heavy chain and light chain gene sequence, which is genetically stable with optimal affinity, specificity, stability, immunogenicity, and developability properties. It is the foundation for the development of all biologics and a critical step, which enables USP development and manufacturing (Fig. 17.7).

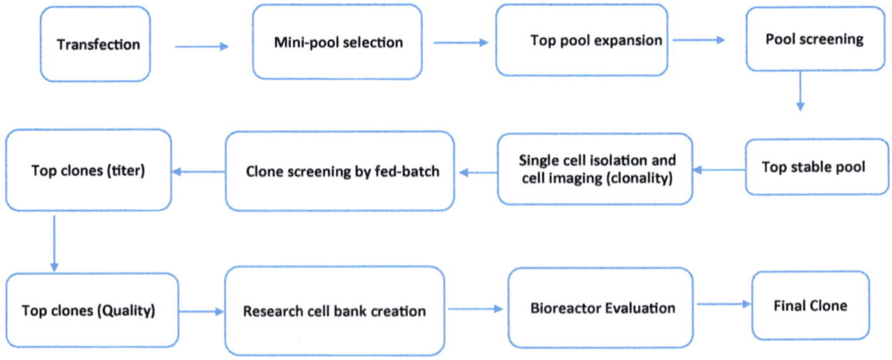

Fig. 17.4 Flow diagram for mAb cell line development

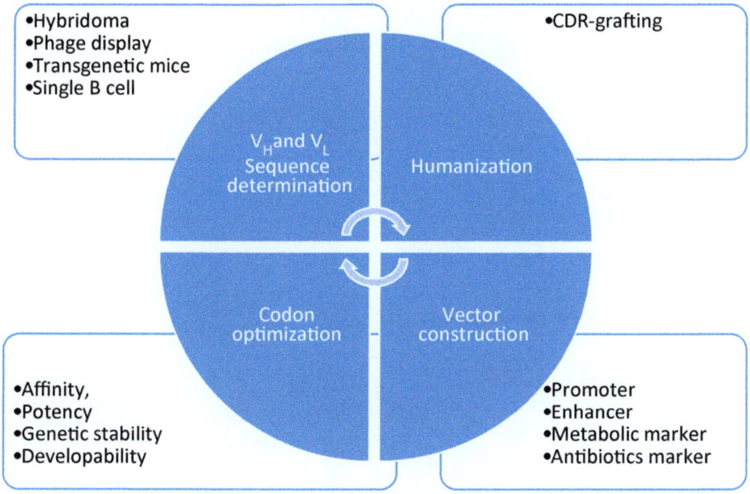

Fig. 17.5 Key considerations for mAb gene sequencing and vector construction

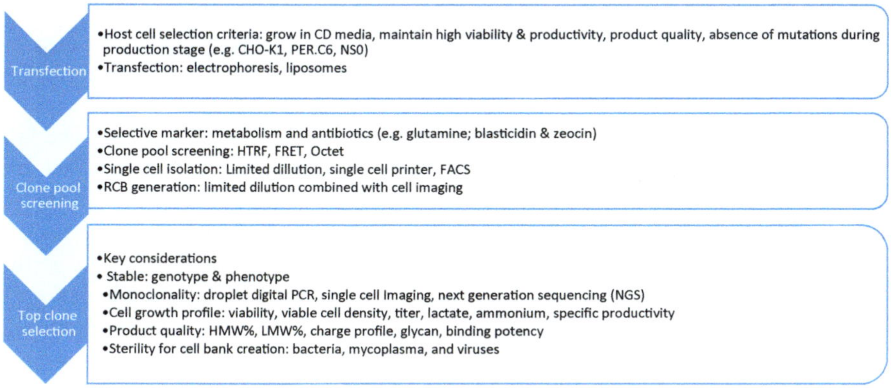

Fig. 17.6 Clone selection procedure and key considerations

17.3.1.2 Clone Selection

Once the mAb gene sequences including the signal peptide sequence for the heavy chain and light chains are determined, each sequence is synthesized and subcloned into an expression vector (Fig. 17.4). The vector contains the promoter/enhancer for mRNA transcription, translation, maturation, and selective markers (Fig. 17.5). The optimal expression vector and host cell line are crucial for the target mAb gene expression, cell culture process development, and manufacturing [29].

There are different types of expression vectors currently used in the industry. For example, CHOZN® GS−/-with ubiquitous chromatin opening elements (UCOE®) can minimize gene silencing and mutation for high productivity (e.g., 3–10 g/L titer) (Millipore Sigma). The pXC Multigene vector is designed for expressing

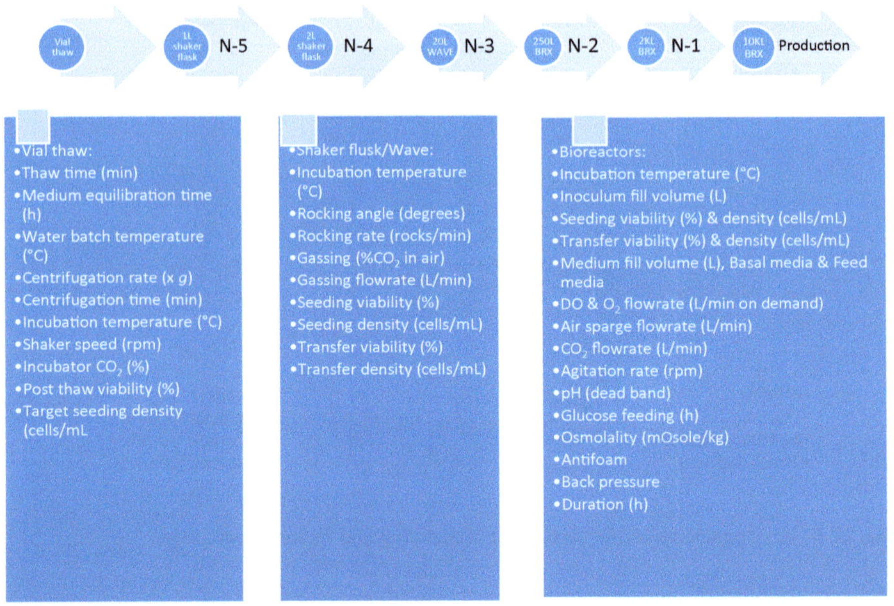

Fig. 17.7 Key process parameters of fed-batch cell culture from vial thaw to large-scale bioreactor production

heavy chain and light chain genes in one vector (e.g., Lonza Biologics). The GS piggyBac™ vector has been designed for more complex protein formats that are under investigation in the next generation, and more challenging, biologics (e.g., Lonza Biologics).

The cell line selection is vital for CBP implementation particularly for a perfusion process compared to the batch process. The selection criteria include cell productivity, viability, viable cell density, stability (e.g., genotype and phenotype), media (e.g., chemical defined media), product quality (e.g., aggregates, charge profile), post-translational modification (e.g., glycosylation), and pricing (e.g., licensing fee or royalty). Some commercial cell lines are available from murine myeloma (e.g., NS0), Chinese Hamster Ovarian (CHO) to human (e.g., PER.C6®) [8]. The CHO cell line is widely used because of robust growth in CD media, high productivity, antibiotic and metabolic selectivity, low virus susceptibility, regulatory approval (e.g., >50 mAbs), and genetic modification (e.g., gene knock-out, knock-down, post-translational silencing) (Enzo Life Sciences).

Clone selection starts with the screening of the transfected cells using Homogeneous Time Resolved Fluorescence (HTRF), Octet, or other well-based titer measurements (e.g., CisBio, ForteBio). Once the clone pool is screened (Fig. 17.4), the top clones are selected based on cell viability, viable cell density, titer, product quality, charge profile, glycan profile, and product-specific binding affinity. Single cells from the top clones are sorted or picked by different techniques such as limited dilution (0.3 cells/well), single cell printers and ClonePix 2 (molecular devices), or fluorescence-activated cell sorting (FACS) (BD FACSMelody™ Sorter) prior to research cell bank (RCB) creation. Clone imaging from Cell Metric™ is often used for monoclonality verification (Solentim Ltd).

The Key considerations of CLD should be focused on genetic stability, cell line stability, monoclonality, cell growth profile, specific productivity, and product quality (Fig. 17.6).

17.3.1.3 Cell Culture Process Development

The USP development includes media screening, feed formulation (e.g., enriched nutrient), feed strategy (e.g., timing, amount), seed train optimization, and process parameter determination (e.g., inoculation density, production temperature, pH setpoint and deadband, P_{CO2}, and dissolved oxygen [DO] setpoint with mixing and aeration strategy). Additionally, it is important to understand the nutrient and metabolite profiles (e.g., lactate accumulation/consumption, ammonium accumulation, osmolality, glucose, and amino acids), productivity (e.g., specific cell productivity and titer), and product quality (e.g., aggregates, fragments). Extensive medium optimization for media and feed combination determination is essential for fed-batch and steady-state perfusion. Media with proper redox such as glutathione or cystine could improve productivity (e.g., 10 g/L) and product quality (e.g., charge profile) [30]. Scientists have developed other novel processes such as concentrated fed-batch (CFB) or concentrated perfusion (CP) with shorter seeding process and reduced medium volume to achieve highest cell density (e.g., 200×10^6 cells/mL) and productivity (e.g., 17 g/L) [3].

Figure 17.7 illustrates a conventional fed-batch USP process from vial thaw scaled up to a 10,000 L bioreactor. Several key and/or CPPs should be controlled properly. This approach is very important for a fully automated USP process for CBP implementation (Table 17.1). These CPPs are associated with thaw time, bioreactor temperature, pH, DO, P_{CO2}, aeration strategy, feed strategy, viability, or duration (Table 17.1). The process performance monitoring and control can be conducted real-time by in-line or on-line pH, DO, and P_{CO2} probes (e.g., Hamilton, OXYPro® from PreSens Precision Sensing GmbH). The metabolite monitoring for glucose and lactate concentration can be performed by Fourier transform near-infrared (FT-NIR) spectroscopy techniques (Yokogawa Corp.).

Biopharmaceuticals have strengthened the efforts for CBP implementation for the USP. Cytiva reported a continuous mAb perfusion cell culture process with high cell density (e.g., $>7.0 \times 10^7$ cells/mL) using a 50 L single-use bioreactor (SUB) Xcellerex XDR 50 integrated with the Automated Perfusion System (APS) for media feed addition, permeate harvest, and cell bleed. The continuous cell bleed was conducted automatically based on the growth rate, reactor volume, and viable cell density. The control strategy was based on a constant XDR weight by the Xcellerex APS.

ThermoFisher Scientific (TFS) demonstrated a continuous perfusion process from 50 L to 500 L SUB [31] with a specialized precision drilled hole sparger and

Table 17.1 Control strategy for upstream process

Upstream process parameters	Vial thaw (Shaker flask)	Seeding expansion (wave, bioreactor)	Production Bioreactor
Thaw time (min)	H		
Temperature (°C)	M	M	H
Agitation rate (rpm)	L	L	M
Feed strategy			H
Cell viability (%)	M	M	H
Cell density (cells/mL)	M	M	M
Aeration strategy		L	M
CO_2 (%)	M	M	H
DO	M	M	H
pH	L	L	H
Glucose feeding strategy			M
Osmolality(mOsm/kg)	L	L	M
Duration (h)	L	L	H

H = CPPs impacting CQAs; L = no impact to CQAs or key process performance; M = key process parameters (KPPs) impacting process performance

oversized impeller for high cell density (200 × 10^6 cells/mL). The study used conventional and disposable (e.g., pH, oxygen, foam sensor) monitoring and control system (e.g., Aber's Bio-capacitance probe integrated into DeltaV for cell density prediction and bleed). TFS has also developed automated cell counters (e.g., Invitrogen Countess 3) with artificial intelligence (AI) using a deep-learning neural network algorithm to generate highly accurate cell counts. Other biopharma companies such as WuXi Biologics have announced an intensified and integrated bioprocess platform for continuous biomanufacturing [32] and Bristol Myers Squibb has demonstrated an intensified and/or integral upstream process from both lab and production scales [7].

For product quality monitoring and control, the current industry practice primarily relies on offline sample testing, which is time consuming and labor intensive. For real-time or near-real-time sampling and testing, we propose a design for an aseptic sampling system (Fig. 17.8). It is a closed system using a hollow fiber system to separate cells and cell-free media with one-way flow architecture. This system has an aseptic connector fit to both stainless steel and SUB system. The impact of the sampling amount from bioreactor would be negligible.

In short, it is feasible to adopt a CBP strategy for a multi-pack fed-batch or perfusion process based on scientific understanding of mAb properties and the availability of conventional, emerging, and cutting-edge technologies.

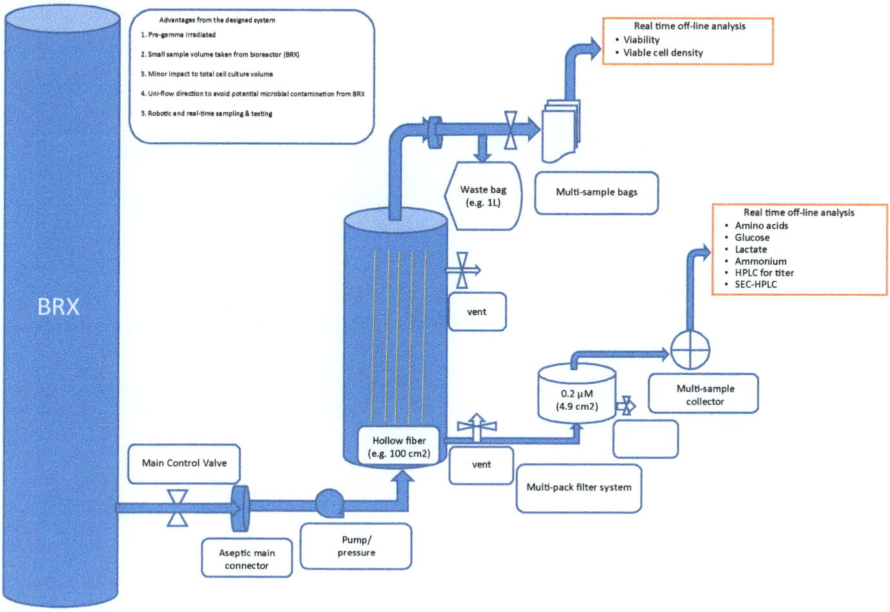

Fig. 17.8 Aseptic mini-sampling system for USP CBP implementation

17.3.1.4 Harvest and Clarification

Prior to the downstream purification, the mAb is usually harvested by depth filtration (DF) particularly when a smaller fed-batch bioreactor (e.g., ≤2000 L) is used for early clinical needs (e.g., phase I/II). DF filters with different tightness grades are used to remove viable cells and dead cell debris by size exclusive, sieving retention, and/or charge adsorption. The filters are made from cellulose or polymers with filter aids such as diatomaceous earth, charged ligands, glass, silica, or activated carbon (Millipore Merck, Pall, Sartorius, and 3 M). The DF filtrate quality (e.g., rHCP, hcDNA, aggregates, fragments), capacity, and recovery are closely related to USP, cell line, titer, viable cell density, cell viability, filter property, and feed flux and pressure. High-grade DF filters with low extractability and/or low endotoxin are available as well (3 M). It is also reported that charged depth filter can remove viruses such as retrovirus by adsorption mechanism [33].

The current industrial practice for DF is a manual mode operation. To design a continuous DF process, we proposed to use the 3 M multi-round DF system as a prototype model [21]. The 3 M system provides sufficient filter area and can be performed automatically; however, all filter-rounds are connected manually together with tubing. To upgrade the system for continuous harvest, a special filter holder was designed (Fig. 17.9) with an automatic filter assembly and re-assembly described for continuous harvest [21].

Once the drug safety and efficacy are assessed in early-stage clinical trial studies, a larger bioreactor (e.g., ≥10,000 L) is often employed to scale up manufacturing

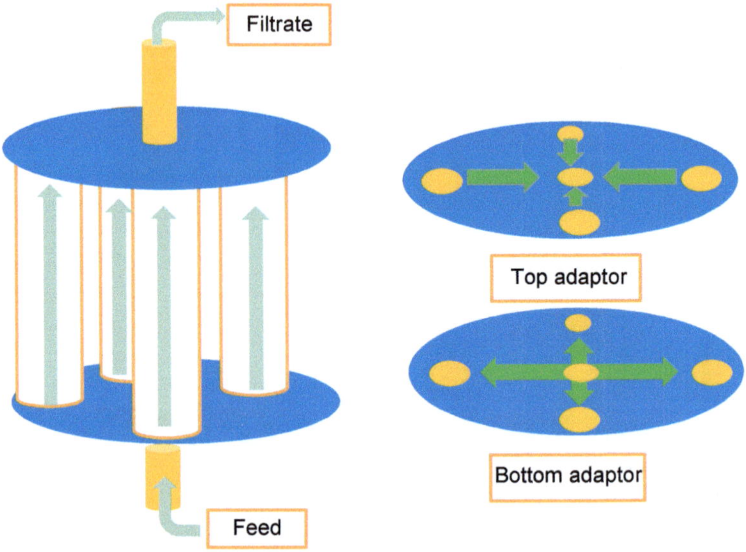

Fig. 17.9 Multi-depth filter system designed for continuous harvest

for late-stage clinical trials, process validation, and potential commercial production [34]. A continuous disk-stack centrifuge used for the mammalian bulk cell harvest is typically hydro-hermetic, where there is an air–liquid interface within the centrifuge, or fully hermetic, which allows for the centrifuge bowl to be filled with liquid. A continuous-flow centrifugation is usually used based on the established process parameters of feed rate, bowl speed, centrate back-pressure, and discharge frequency. The load cell density, cell viability, titer, and cell culture viscosity may impact the establishment of the process parameters. The centrate quality can be monitored by turbidity (Mettler Toledo) or particle size distribution (Particle Technology Labs).

Our preference is to use alternative tangential flow (ATF) for perfusion harvest. The automated multi-round DF system can be used for 2000 L bioreactor production when titer is high (e.g., >5 g/L) and/or drug product demand is low (e.g., orphan drug with lower dose administration). The continuous centrifugation should be the best approach for larger-scale harvest (e.g., >10,000 L bioreactor).

17.3.2 DSP Development for CBP Implementation

Biopharmaceutical companies have been exploring and adopting more efficient and cost-effective approaches such as an intensified and/or integrated process by continuous capture of product with the Protein A affinity column to relieve the bottleneck resulting from upstream higher titer cell cultures and the larger culture volume from perfusion processes (Fig. 17.10). To date, a few reports have been published

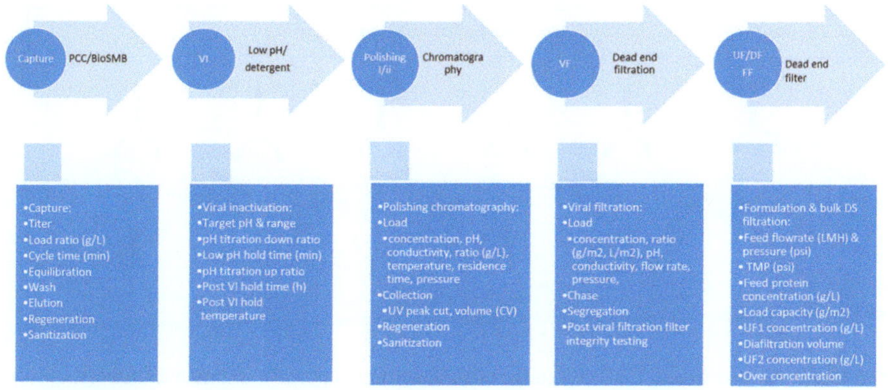

Fig. 17.10 mAb purification process parameters and control

for an end-to-end DSP from design, implementation [21] to execution at pilot scale [35].

17.3.2.1 Continuous Capture with Multi-columns

Continuous capture (e.g., BioSMB from Sartorius or periodic counter-current chromatography from Cytiva) has been successfully demonstrated at both lab and manufacturing scales [7, 19, 36]. BioSMB technology uses multi-column (up to 8 columns) capture (MCC) approach to increase process efficiency (up to 75%) and productivity (up to five-fold), maximize resin-binding capacity, and reduce column size, buffer consumption, and equipment facility footprint (Forte from Sartorius, BPI digital week 2020). One may consider exploring Expanded Bed Adsorption (EBA) Chromatographic technology (Biotechflow). This technology has also been improved and used for large-scale manufacturing to support phase I–III clinical productions (e.g., column ID: 60 cm). Its unique feature improves process efficiency by merging harvest and capture into one step without negatively impacting product purity and yield (Biotechflow).

MCC process development starts from Protein A resin screening. There are a variety of resins available in market such as MabSelect Sure PCC from Cytiva, Toyopeal AF-rProtein A HC-650F from Tosoh, and Poros ProA from TFS for screening. The screening criteria are typically focused on resin property (e.g., ligand matrix), binding capacity (e.g., >60 g/L), process efficiency (e.g., short residence time, less compression factor), product purity (e.g., rHCP, hcDNA, rProA, aggregates, fragments), resin performance (e.g., resin stability, consistency of elution profile and yield), bioburden, and endotoxin control.

Once the resin is selected, capture step parameters are assessed and established based on process design, equipment (e.g., skid, column size and number, cycle number, tank, filter, tubing, probes, sensors), and software (e.g., Unicorn, delta V).

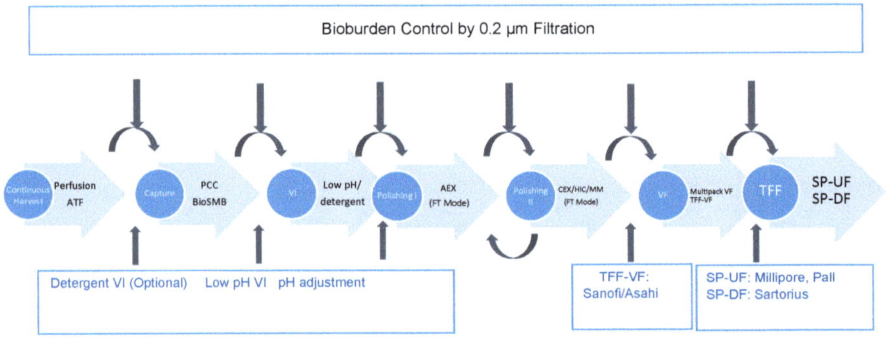

Fig. 17.11 An end-to-end CBP design for mAb downstream process and manufacturing

The process design strategy is based on the unit operational mode. Once the process is developed and optimized for process efficiency, product purity, quality, and recovery, the process can be transited into continuous mode [35]. The skid software program should be created and tested based on column size, equilibration volume/time, load titer/density, washing residence time/volume, elution peak cut and eluate volume, column post-usage regeneration, sanitization, reuse, and storage. For example, the optimization of the wash step (e.g., pH, conductivity, detergent, or amino acids addition) not only results in significant impurity and/or virus removal but also facilitates the multi-column operational programming (Fig. 17.11).

17.3.2.2 Integral of Low pH Viral Inactivation with Capture Step

Integration of capture and low pH viral inactivation (VI) into one step is an ideal CBP design strategy. The challenges and solutions for continuous VI have been discussed [37]. We have proposed to control ProA eluate pool pH by optimizing the elution buffer strength, pH, and total volume if the mAb is stable at certain pH range (e.g., 3.4–3.6) [21]. Three surge tanks (either stainless steel or disposable) with mobile mixer each can be connected sequentially for eluate mixing, pH adjustment for VI, and for the next process step if three ProA columns are used. The in-line pH adjustment for fine-tuning can be achieved by using the Cadence VI system from Pall. This system is a fully automated system with built-in software for both batch and continuous operation.

Continuous low-pH VI was evaluated based on fluid dynamic, mixing, and pressure drop by Parker et al. (2017) [38]. In addition, a coiled flow inverter (CFI) system has been successfully invented and achieved robust retrovirus inactivation (>4 log reduction) within 14.5 min [39]. The CFI application demonstrated the transition from batch VI platform into continuous mode [39]. Coolbaugh et al. (2021) used an in-line pH titration system, which consisted of a helical static in-line mixer (Koflo) and a tubular flow reactor targeting a minimum residence time of 30 min [35].

If the mAb is sensitive to low-pH treatment, detergent treatment such as Triton X-100 is an option. Triton X-100 usage is banned by the EMA due to its hazard to marine life (REACH—Registration, Evaluation, Authorization, and Restriction of Chemicals). The search for Triton X-100 alternatives is making progress. For example, Triton CG-110 with less toxicity to marine life has comparable capability for VI to Triton X-100. Nereid, another alternative, has also been evaluated by Takeda Pharmaceuticals [40]. Polysorbate 20 + 0.03% tri-*n*-butyl-phosphate (e.g., PS20 + TnBP) can inactivate enveloped virus effectively within 5 min (e.g., > 5 log reduction) [41]. The detergent can be added in the harvest tank prior to capture loading.

17.3.2.3 Continuous Polishing

It is highly recommended using a two-column process (ProA affinity capture step followed by a polishing step) or a three-column process (capture plus two flowthrough mode polishing steps) for downstream CBP. To design the process, one needs to consider USP performance, harvest methods, and harvested material quality for DSP (Fig. 17.11). For example, upstream materials with high cell viability and high titer usually have lower rHCP and hcDNA after harvest. Flocculation with soluble polyionic polymers in harvest and acid precipitation after low-pH VI treatment followed by a charged depth filtration can maximize rHCP and hcDNA reduction [10]. If the product purity does not meet the pre-determined product safety, quality, identity, potency, and efficacy (SQUIPE) requirement from the two-column process, a third column should be screened and implemented [35]. The third column (e.g., CEX, MM, or HIC resin) run in flowthrough mode is highly recommended. Our previous study using a dual flowthrough mode of AEX followed by CEX showed a much higher process efficiency (e.g., 4× faster), low cost (e.g., 90% of resin cost reduction), and high impurity removal for high molecular species (>2×) and HCP (>1.5×) from the flowthrough mode [21]. We found that the flowthrough mode CEX was effective in removing residual host cell proteins, which were evenly distributed from the CEX elution peak from the bind and elute mode. The results demonstrate a continuous process from AEX followed by CEX step with same load density, buffer pH, and conductivity can be adopted for future CBP applications. A continuous countercurrent tangential chromatography using mixed mode post-capture operations in mAb purification reduced host cell protein (HCP), leached protein A, DNA, and aggregates [42].

17.3.2.4 Design of Continuous Viral Filtration (VF)

VF is a robust viral clearance (VC) step recognized by the industry and regulatory authorities. It is routinely performed in batch-mode. Current VF technology and the clearance step is a bottleneck regarding load capacity, and filtration performance to viral leakage associated with filter integrity for CBP applications. To implement a continuous VF processing, Pall has applied the principles of QbD to assess the different design spaces for batch versus CBP and identify potential risks. Pegasus™ Prime virus filter was used for virus stability testing in different fluids and filter robustness. *Pseudomonas aeruginosa* phage PP7 was stable within 22 days; minute virus of mice (MVM) was stable within the tested time (e.g., 2 days); and Xenotropic murine leukemia virus (X-MuLV) was less stable after 2 days. CPPs identified by the Pall team were lower flux, extended process time, and process interruption by filter change-out. Asahi Kasei (Kozaili, BioProcess International 2020) assessed the feasibility of continuous VF. A recent report using tangential flow virus removal filtration from Planova 20 N (Asahi Kasei Bioprocess) demonstrated its high capacity (>3500 L/m^2, 30 kg/m^2) and long duration of up to 2 weeks without filter fouling [35]. However, the authors did not demonstrate the viral removal capability under such a high load capacity and duration.

To overcome the limitation from filter capability and change-out frequency, and prevent the viral leakage from the long process time, we proposed to invent a novel multi-pack VF system [21]. The filter property and surface area can be designed and manufactured by vendors. The filtration system can be monitored and controlled automatically per filtration flux decay, pressure drop, process capacity, time, and recovery. Each filter pack can be plug-in and change-out robotically based on its load capacity or feed pressure. The filtrate can be pooled for formulation by ultrafiltration and diafiltration (UF/DF) after post-use integrity testing pass from each individual filter pack. This practice can minimize the entire VF process failure risk by assuring individual filter integrity.

17.3.2.5 Design of Continuous UF/DF

Current UF/DF is conducted in batch-mode for final DS concentration and buffer exchange for formulation. To adopt a CBP technology for UF/DF, single pass tangential flow filtration (SPTFF) has been used for protein concentration [43]. The filter system is manufactured by either Merck Millipore or Pall (Cadence™ SPTFF). The Cadence™ system has achieved a concentration factor range of 3–25x at 10–70 L/m^2/h permeate flux within 25–65 psid feed pressure drop [44]. This technology supports in-process sample concentration prior to column or viral filter load. Multi-module SPTFF may allow the mAb bulk drug substance (BDS) to reach high concentration (e.g., 100 g/L). One study demonstrated a three-stage single pass diafiltration (SPDF) that resulted in >99.75% buffer exchange [45]. Other studies have also demonstrated SPDF systems that have been successfully applied at lab and pilot scales [35, 46], which surely will accelerate the use of continuous UF/DF at large-scale manufacturing.

As discussed above, it is feasible to design an end-to-end downstream CBP based on mAb properties, raw material control, process design, process monitoring and control, innovative software and hardware availability, product

Table 17.2 Downstream control strategies

Process parameters	ProA Capture	VI	Polishing Chromatography	VF	UF/DF, FF
Load concentration	M	M	M	M	M
Load ratio (g/L, g/m2)	M	M	H	M	M
Load or start pH	M	H	M/H	M	M
Load or start conductivity (mS/cm)	M	M	M/H	M	M
Load Pressure (psi or bar)	M		M	H	M
Load temperature (°C)	M/H		M/H	M	M
Cycle/process time (min)	M	M	M	H	M
Load flowrate (L/min)	M		M	H	M
Equilibration pH	M	M	M	M	M
Equilibration conductivity (mS/cm)	M	M	M	M	M
Wash or chase pH	M	M	M/H	M	H
Wash or chase conductivity (mS/cm)	M	M	M/H	M	H
Elution pH	M		H	M	H
Elution conductivity (mS/cm)	M		H	M	H
UV peak cut	M		M/H		
Column regeneration/sanitization	M		M		
Post usage filter integrity testing (0.2 micron or VF)	H	H	H	H	H

H = CPPs impacting CQAs; M = KPPs impacting process performance; M/H = impact associated with process mode (bind/elute or flowthrough)

quality real-time or near-real-time testing, product safety, and VC capability (Figs. 17.10, 17.11 and Table 17.2).

17.3.2.6 Control of Process Performance and Product Quality

The mAb aggregation after low-pH hold step should be closely monitored and controlled. Lot-to-lot pH variation from elution pool should be well studied to understand the negative impact from the variation of elution buffer pH and column load density. Intermediate homogeneity can be achieved by using disposable mixers (Allegro from Pall, or Mobius® Power *MIX* from Millipore). Aseptic sampling can be conducted manually by use of Sartorius Take One or Millipore Nova Septum. Ideally, an automated sampling system needs to be invented suitable for both stainless steel and disposable systems. Bioburden and endotoxin monitoring and control should be applied from raw material release through all operational steps.

To reduce microbial contamination, the use of in-line 0.2 μm filtration before load and after elution is essential (Fig. 17.11). Currently, the 0.2 μm filtration system is manually assembled for single-set usage. A multiple 0.2 μm filter assembly (similar design principle from Fig. 17.9) should be invented for CBP. Each individual filter sizing should be pre-determined based on load property (e.g., turbidity, concentration, pH, conductivity) and filter property (e.g., polyethersulfone, polyvinylidene difluoride). In-line real-time testing method and technologies for rHCP, hcDNA, CGE, glycan, and cIEF assays are not yet available.

17.3.3 Analytical Method Development, Qualification, and Validation for CBP Implementation

Analytical method development, qualification, validation, and lifecycle management are essential for batch-mode and CBP manufacturing (Table 17.3). The method development, qualification, and validation include assay specificity, identity,

Table 17.3 Analytical method development, qualification, and validation

Assay	Development/ Qualification	Validation	Purpose
SEC-HPLC	√	√	Aggregates, monomers, fragment separation
icIEF	√	√	Product charge profile assessment: deamidation, formation of N-terminal pyroglutamate, aggregation, isomerization, sialylated glycan, antibody fragmentation, and glycation at the lysine residues
CE-R	√	√	Purity, fragments, clippings, disulfide isomers, identity
CE-NR	√	√	
LC-MS/MS	√	√	Primary mAb protein sequence confirmation Identification and determination of mAb post-translational modifications (e.g. oxidation, deamidation, pyroglutamate, glycosylation) Impurity identification such as HCPs
UV Spectrometry or SoloVPE	√	√	Protein concentration
ProA-HPLC	√	√	Protein titer from cell culture and harvest
HCP-ELISA	√		Total HCP content
ELISA binding	√	√	Potency
HCP-LC-MS/MS		√	Individual HCP species identification and/or quantification
ProA-ELISA	√	√	Total residual protein A content
DNA-qPCR	√	√	Process-related impurity
Endotoxin	√	√	biosafety
Bioburden	√	√	biosafety
DP Sterility	√	√	biosafety
HVLD for CCIT	√	√	container closure integrity testing.
Amino acid analysis		√	protein concentration, identity and quality
DSC		√	protein thermostability
CD		√	protein thermostability and structure
Ellman assay		√	Free sulfhydryl
Raman spectroscope		√	The concentration of mAb, glutamine, glutamate, glucose, lactate, ammonium, cell viability and density
Fluorescence spectroscopy		√	Cell metabolism, fluorescent compounds in media
Multi-angle light scattering		√	Protein aggregation
Antifoam-HPLC	√	√	Residual antifoam
PS80 or 20	√	√	Protein stability

SE-HPLC size exclusion-high-performance liquid chromatography, *icIEF* imaged capillary isoelectric focusing, *CE-R* capillary electrophoresis-reduced, *CE-NR* capillary electrophoresis-nonreduced, *UV* ultraviolet, *VPE* variable pathlength extension, *HCP-ELISA* host cell protein enzyme-linked immunosorbent assay, *LC-MS-MS* liquid chromatography with tandem mass spectrometry, *HVLD* high voltage leak detection, *CCIT* container closure integrity test, *DSC* differential scanning calorimetry, *CD* circular dichroism, and *UPLC* ultra-performance liquid chromatography

accuracy, precision, reproducibility, linearity, absolute recovery, robustness, limit of quantitation (LOQ), and limit of detection (LOD). These activities should be conducted under the regulatory guidance such as ICH Q2 (R1), "Validation of Analytical Procedures: Text and Methodology (2005)," Q14 "Analytical Procedure Development" (expected in 2021), and "Analytical Procedures and Methods Validation" (2015) by FDA. The United States Pharmacopeia (USP) general chapters for method validation USP <1225>, method verification of compendial procedure USP <1226>, method transfer USP <1224>, instrument qualification USP <1058>, and EMA Guideline (2017) on the requirements for the chemical and pharmaceutical quality documentation concerning investigational medicinal products in clinical trials are also valuable references.

The intent of method development, qualification, and validation for CBP support should focus on real-time testing for process performance and product quality monitoring [47, 48] and control (Table 17.4). However, the challenge is that real-time testing methods for rHCP, hcDNA, and rProA are currently unavailable. To overcome the limitation from analytical methods, QbD from process design and robustness to control [49] may play critical role for an end-to-end CBP (Fig. 17.11).

Table 17.4 Essential analytical methods for CBP real-time support

Assay	Real-time on-line, in-line, at-line testing
Protein aggregates and fragment monitoring	SEC-UPLC: Waters patrol with automatic and aseptic sampling and testing
	Data collection and analysis by Supervisory Control and Data Acquisition (SCADA) software
Titer A280 measurement	ProA-UPLC: Waters or Agilent
	FlowVPE instrument: Repligen (Cygnus Technologies)
HCP content individual species identification and quantification	HCP-ELISA: Cygnus Technologies
	LC-MS/MS: HCP identity
	Real-time testing: not available
Residual protein A content	ProA-ELISA: Cygnus Technologies, Repligen
	Real-time testing: not available
Protein structure–disulfide bond formation	Raman spectroscopy by Horiba Scientific
Biosafety	Endotoxin/bioburden/virus control
	Real-time testing: not available
The concentration of cell metabolites (glutamine, glutamate, glucose, lactate, ammonium) and cell viability and density	Raman spectroscopy: metabolites (glutamine, lactate, glucose)
	Fluorescence spectroscopy

SE-HPLC size exclusion-high-performance liquid chromatography, *icIEF* imaged capillary isoelectric focusing, *CE-R* capillary electrophoresis-reduced, *CE-NR* capillary electrophoresis-nonreduced, *UV* ultraviolet, *VPE* variable pathlength extension, *HCP-ELISA* host cell protein enzyme-linked immunosorbent assay, *LC-MS-MS* liquid chromatography with tandem mass spectrometry, *HVLD* high voltage leak detection, *CCIT* container closure integrity test, *DSC* differential scanning calorimetry, *CD* circular dichroism, and *UPLC* ultra-performance liquid chromatography

17.4 Adeno-Associated Virus (AAV) Vector Production

17.4.1 AAV Application for Vaccine and Gene Therapy

Gene therapy products have potential to cure genetic deficiency, cancer, and infectious diseases [50–52]. Since the first gene therapy was initiated at the National Institutes of Health (NIH, 1989) and the first gene therapy product (Luxturna) was approved by FDA (2017), over 2600 clinical trials have been carried out using viruses from lentivirus, adenovirus (Ad) to AAV as vectors [52–54]. AAV is a safe and widely used vehicle of genes for exhausted disease treatment because of its wide range tropism, ability to infect dividing and non-dividing cells, high efficacy for long-term transgene expression, high safety with low immunogenicity, and low levels of integration into the host genome [55–57]. The diseases include Hemophilia B [56], Parkinson's disease [57, 58], and Duchenne muscular dystrophy [59].

17.4.2 AAV Plasmid Construction, Transfection, and Cell Culture

AAV genome is ~4.7 kb of single-stranded DNA (ssDNA) containing two genes, *rep* for replication and *cap* for encapsidation. The genome is flanked at both ends by the inverted terminal repeats (ITRs) allowing foreign DNA sequence to be inserted and become an AAV transfer plasmid [57, 60]. The vector gene expression needs a complementing plasmid to be co-transfected with adenovirus (Ad) helper for AAV structural and non-structural gene expression [60]. A triple-plasmid transfection method usually includes an AAV serotype-specific *rep* and *cap* plasmid, a vector plasmid with target DNA, and a third plasmid with Ad genes for transgene expression and encapsulation [60, 61].

The gene transfection is conducted by chemical-based (e.g., calcium phosphate, cationic polymers and liposomes) or non-chemical-based methods (e.g., electroporation). The calcium phosphate for adherent cells and polyethylenimine (PEI) for suspended cells have been routinely used [54]. Others such as lipofectamine reagents (Invitrogen) are also commercially available. The transfection efficiency relies on multiple factors such as the ratio of plasmid DNA to cells (e.g., 0.44 μg DNA/1E6 cells at 48 h) (Millipore, 2020). The viral titer from infected cells can be determined by quantitative real-time polymerase chain reaction (qRT-PCR), digital droplet PCR (ddPCR), Dot Blot, infectivity, and ELISA methods [54, 62].

There are different types of cells such as HEK293, 293 T, HEK Express, PER. C6, HeLa-RC32, or *Sf9* used for viral vector production (Cevec Pharmaceuticals, Millipore) [56, 63]. Cell culture process and manufacturing should be focused on media (e.g., CD media), agitation, pH, DO, sparger type, and post-transfection time for harvest (e.g., Millipore 2020) [54, 60]. Piras et al. reported [54, 60] that the AAV8 productivity increased by prolong post-transfection time from 76% at day 3 to 96% at day 7 from adherent HEK 293 T/17 cells [64]. Due to the scalability concerns from high labor intensity, high risk of contamination, large footprint of equipment and facility, and high cost from adherent cell culture process, suspended cell culture process has been investigated and implemented for large-scale production [54, 60]. HEKExpress and HEK293 suspended cell lines have been successfully cultivated in serum-free media from 20 L WAVE bioreactor continuous perfusion process for rAAV8 and rAAV9 production [54, 60].

Here we propose a novel AAV vector gene expression system for large-scale cell culture process to meet future market need. An engineered cell line (e.g., HEK294™) with AAV expression cassettes with serotype-specific *rep, cap*, and Ad genes integrated in the host chromosome should be created. These genes are constitutively replicated with cell growth. Expression of these genes is expected to be regulated by an inducer—a mechanism like the *Escherichia coli* (*E. coli*) *lac* operon. A primary master cell bank (PMCB) containing these genes could be generated and characterized for a particular serotype of AAV production. When a target gene plasmid is transfected in the PMCB, a working cell bank (WCB) could be created. The WCB can be used for different scales for AAV vector production. The *rep*/*cap*, and Ad gene expression could be induced chemically, and target gene replication could be regulated by a different mechanism. The AAV vector could be released by a similar mechanism as Phage λ from the host *E. coli* [65].

17.4.3 Harvest and Clarification

Currently the suspension cells from 48-h post-infection is lysed by detergent. The cell debris is removed by centrifugation, microfiltration, or depth filtration. Due to lower titer and higher impurities such as hcDNA and rHCP from the cell lysate, scientists often use benzonase to digest hcDNA first to minimize its negative impact to downstream resin performance from binding capacity to reuse [63]. After the DNA digestion, process-related impurities can be further reduced by flocculation followed by a charged depth filtration. The rHCPs and hcDNA can be reduced significantly from flocculation and charged depth filtration. The filtrate can be concentrated by SPTFF or conditioned by a conventional TFF to increase load concentration, shorten capture load time, and avoid potential product degradation associated with long loading time due to lower titer and larger load volume.

17.4.4 AAV Purification

The goals for the AAV vector purification process and manufacturing should focus on process scalability, high recovery, low COGs, and high product safety and quality. There are several challenges for viral vector purification. The first challenge is related to the unique viral properties such as size, shape, serotype, charge variants, stability, and binding to receptors. The second challenge is due to process-related impurities (e.g., hcDNA, rHCP, cell culture additives, helper virus, detergent, bio-burden, endotoxin, and other potentially contaminated viruses) and product-related impurities (e.g., empty/partial/aggregated capsids), which are difficult to remove. The third challenge is the lack of profound knowledge, experience, and universal platform to remove empty capsid from variable serotypes of the AAV vector.

17.4.4.1 AAV Capture

AAV vectors can be captured by affinity resins such as POROS™ CaptureSelect™ AAV8, AAV9, AAVX 1–9 (TFS), AVB Sepharose (Cytiva), or heparin affinity resin (ViraKit™). The AAV affinity capture should be the best approach to reach high recovery, product purity, and process efficiency. Besides the affinity capture, CEX resins have also been selected for AAV capture [66]. Potter et al. (2014) used a bind and elute mode of CEX chromatography (HiPrep SP HP) for rAAV serotypes 2, 8, and 9 purifications from both HEK293 and Sf9 cell culture [63]. Adamson-Small et al. (2016) achieved 40–50% overall recovery from an SP FF chromatography purification followed by UF/DF step [62].

17.4.4.2 Intermediate and Polishing Steps

The goals for the intermediate and polishing process steps are two-fold, one is to remove product-related impurities and the other is to remove process-related impurities to acceptable levels. The product-related impurities are primarily from empty capsid removal. Due to similar size, shape, and charge between empty and full capsids, CsCl gradient ultracentrifugation is the most effective method for removal of empty capsids. It was reported that the non-functional empty vector could reduce the gene transduction efficiency by competing with the fully packaged therapeutic vector particles for receptor uptake, internalization, and intracellular trafficking [67]. In contrast, some reports demonstrated that the empty AAV2 vector could decoy the human immune system and enhanced gene transfer efficacy [68, 69]. Mingozzi et al. (2013) suggested a proper ratio of full/empty capsids in the final formulation could maximize the efficacy of gene transfer [68]. This report shed lights on a flexible purification process and manufacturing with chromatographic approach to purify different serotypes of AAV vectors without using gradient ultracentrifugation.

Wang et al. (2019) reported AEX resin effectively separate the empty capsids from the full capsids based on their isoelectric point (*pI*) of 6.3 and 5.9, respectively [69, 70]. Tomono (2016) reported the comparable purity from rAAV1 obtained from ultracentrifugation and chromatography methods [61]. Qu et al. (2007) reported that empty capsid to vector ratios could vary widely from 3:1 to 30:1 when the triple-transfection production method is used [71]. They used POROS 50HQ to remove empty capsids up to 86-fold and achieve a 74% yield for AAV2 vector purification.

Abrecht et al. (2019) compared AEX membrane with resin [72]. They described membrane chromatography and its advantages in larger pore size for larger molecular accessory, higher binding capacity (up to 20×), higher process efficiency (e.g., >10× flowrate), smaller footprint, disposable-minimizing cross contamination, and ready-to-use without column packing and unpacking requirement. The authors reported that the higher salt in the load allowed increased removal of HCP (LOQ) and DNA (>4 LRV). Although the advantages from membrane chromatography have been discussed above, we would recommend the use of an AEX resin (e.g., Poros 50HQ) for empty vector removal to avoid the limitation of membrane-binding capacity and separating resolution from the gradient elution.

The process-related impurities are primarily removed from the capture step followed by an intermediate chromatographic step. The remaining residual impurities can be further reduced by use of hydrophobic interaction chromatograph (HIC) or size exclusive chromatography (SEC) [68, 73]. Chahal et al. (2007) reported an overall 75% yield was achieved from AAV2 extraction from cell pellets by detergent (e.g., 0.5% sodium deoxycholate), capture by CEX (e.g., Fractogel SO_3^-), and HIC (e.g., Toyopearl butyl 650 M) process [73]. SEC is effective to separate AAV and other impurities, but this approach has its intrinsic limitations from load capacity, process efficiency, and scalability to product quality. Other novel resins such as monoliths (BIA Separations) based could be employed [74]. We recommend using Capto Core 700, a multi-modal resin with IEX, HIC, and SEC properties. The Capto Core 700 bead has an outer layer for SEC function and a core with octyl amine with both hydrophobic and positively charged functions (at pH < 10). The core has highly efficient multi-modal binding of various impurities over a wide range of pH and salt concentrations (Cytiva). This resin was designed for viruses, DNA, large protein, or protein complexes (native size ≥700 kDa) purification in flowthrough mode with its high loader capacity and process efficiency (e.g., 500 cm/h) compared with conventional SEC technology. Per our previous experience for Collagen 7 purification, Capto Core 700 resin was extremely effective to remove both product- and process-related impurities such as HCPs (data not published).

Multiple methods have been reported to characterize empty AAV capsid levels [54], including transmission electron microscopy (TEM), analytical ultracentrifugation (AUC), charge detection mass spectrometry (CDMS), UV spectrophotometry, capsid and genome copies by ELISA and qPCR [70], and endotoxin content measurement [75].

17.4.5 *Future AAV Manufacturing*

It is well known that AAV vector processes and manufacturing are currently in its infant stage. It is premature to develop a CBP for AAV manufacturing based on current knowledge, experience, technologies, and regulatory guidelines. However, we proposed a prototype CBP here for AAV purification and manufacturing from design to development for future clinical and/or market needs (Fig. 17.12).

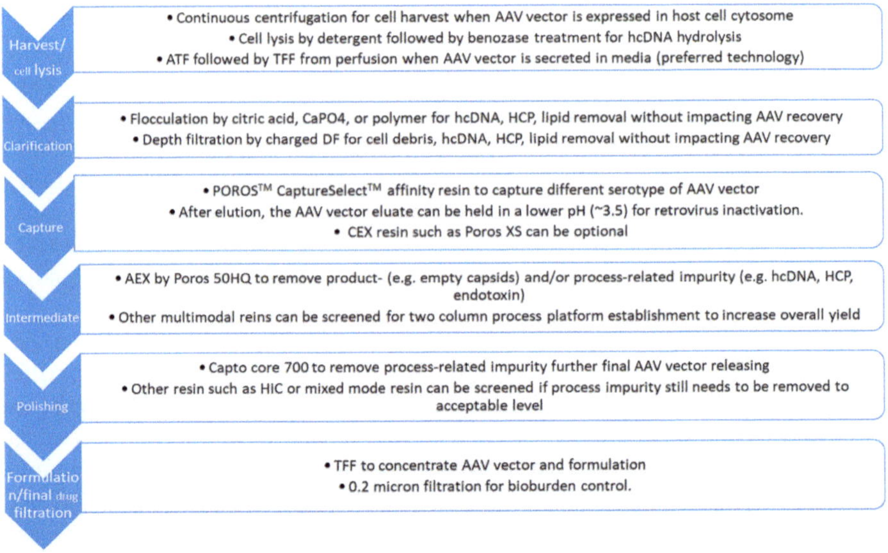

Fig. 17.12 CMC considerations for an intensified or a continuous process design at large scale of AAV vector purification

17.5 Summary

Continuous process operated and controlled by automation has been successfully adapted in car assembly, petro refining, food manufacturing, and pharmaceuticals for small molecules for decades. As we discussed in this chapter, the CBP for biologics manufacturing has been slow to adopt due to several challenges associated with process design, process monitoring and control, product SQUIPE monitoring and control, real-time in-process testing, batch defining, and equipment and software availability. However, it is encouraging to see more and more scientists and engineers from both academia and biopharmaceutical companies have turned their interest and efforts from batch-mode production to CBP. Several successful studies from intensified, integrated, and/or end-to-end process for mAb production have shed light on the future implementation of CBP technology at commercial scale production. The CBP approach has tremendous potential to meet the increasing market demand for patient healthcare by increasing productivity, enhancing process consistency, assuring product quality, and reducing COGs and facility footprint. Longer term, the CBP technology may offer benefits for the manufacturing of vaccine, enzyme, and gene therapy therapeutics.

Acknowledgments The authors would like to thank Charles Winter for critical review of this chapter and Michael Hippach for review of the upstream sections.

References

1. Webb D, Jamison TF (2010) Continuous flow multi-step organic synthesis. Chem Sci 1:675–680. https://doi.org/10.1039/c0sc00381f
2. FDA (2019) Quality considerations for continuous manufacturing guidance for industry, pp 1–24
3. Shevitz J, Bonham-Carter J, Sinclair A et al (2011) An economic comparison of three cell culture techniques. BioPharm Int 24(2):54–60
4. Walther J, Godawat R, Hwang C et al (2015) The business impact of an integrated continuous biomanufacturing platform for recombinant protein production. J Biotechnol 213:3–12. https://doi.org/10.1016/j.jbiotec.2015.05.010
5. Brown MR, Collier PJ, Giolbert P (1990) Influence of growth rate on susceptibility to antimicrobial agents: modification of the cell envelope and batch and continuous culture studies. Antimicrob Agents Chemother 34(9):1623–1628. https://doi.org/10.1128/aac.34.9.1623
6. Maruthamuthu MK, Rudge SR, Ardekani AM et al (2020) Process analytical technologies and data analytics for the manufacture of monoclonal antibodies. Trends Biotechnol 38(10):1169–1186. https://doi.org/10.1016/j.tibtech.2020.07.004
7. Xu J, Rehmann MS, Xu M (2020) Development of an intensified fed-batch production platform with doubled titers using N-1 persusion seed for cell culture manufacturing. Bioresour Bioprocess 7(17):1–16. https://doi.org/10.1186/s40643-020-00304-y
8. Li F, Vijayasankaran N, Shen AY et al (2010) Cell culture processes for monoclonal antibody production. mAbs 2(5):466-79:466. https://doi.org/10.4161/mabs.2.5.12720
9. Yang WC, Lu J, Kwiatkowski C et al (2014) Perfusion seed cultures improve biopharmaceutical fed-batch production capacity and product quality. Biotech Progress 30(3):616–625. https://doi.org/10.1002/btpr.1884
10. Ding Y-HR, Marino MP, Kumar H (2019) Antibody purification process development and manufacturing. BioPharm Int 32(12):24–29
11. Lee S (2016) Current FDA perspective for continuous manufacturing. In: MIT-CMAC 2nd international symposium on continuous manufacturing of pharmaceuticals, pp 26–27
12. Juran JM (1992) Quality by design: the new steps for planning quality into goods and services. The Free Press, New York
13. Chatterjee S (2012) FDA perspective on continuous manufacturing. In: IFPAC annual meeting, pp 1–22
14. Yu LX (2008) Pharmaceutical quality by design: product and process development, understanding, and control. Pharm Res 25(4):781–791. https://doi.org/10.1007/s11095-007-9511-1
15. Rathore AS (2009) Roadmap for implementation of quality by design (QbD) for biotechnology products. Trends Biotechnol 27(9):546–553. https://doi.org/10.1016/j.tibtech.2009.06.006
16. FDA (2004) Guidance for industry PAT – a framework for innovative pharmaceutical development, manufacturing, and quality assurance, pp 1–16
17. Walsh G (2018) Pharmaceutical benchmarks. Nat Biotechnol 36(12):1136–1145. https://doi.org/10.1038/nbt.4305
18. Lu RM, Hwang YC, Liu IJ et al (2020) Development of therapeutic antibodies for the treatment of diseases. J Biomed Sci 27(1):1–30. https://doi.org/10.1186/s12929-019-0592-z
19. Warikoo V, Godawat R, Brower K et al (2012) Integrated continuous production of recombinant therapeutic proteins. Biotechnol Boieng 109(12):3018–3029. https://doi.org/10.1002/bit.24584
20. Hiller GW, Ovalle AM, Gagnon MP et al (2017) Cell-controlled hybrid perfusion fed-batch CHO cell process provides significant productivity improvement over conventional fed-batch cultures. Biotechnol Bioeng 114(7):1438–1447. https://doi.org/10.1002/bit.26259
21. Ding Y-HR, Marino MP (2020) Points to consider for continuous downstream bioprocess and manufacturing. BioPharm Int 33(8):22–26

22. Lei L, Tran K, Wang Y et al (2019) Antigen-specific single B cell sorting and monoclonal antibody cloning in Guinea pigs. Front Microbiol 10(672):1–19. https://doi.org/10.3389/fmicb.2019.00672
23. Parray HA, Shukla S, Samal S et al (2020) Hybridoma technology a versatile method for isolation of monoclonal antibodies, its applicability across species, limitations, advancement and future perspectives. Int Immunopharmacol 85:106639. https://doi.org/10.1016/j.intimp.2020.106639
24. Meyer L, López T, Espinosa R et al (2019) A simplified workflow for monoclonal antibody sequencing. PLoS One 14(6):1–19. https://doi.org/10.1371/journal.pone.0218717
25. Almagro JC, Pedraza-Escalona M, Iván Arrieta H et al (2019) Phage display libraries for antibody therapeutic discovery and development. Antibodies 8(3):44. https://doi.org/10.3390/antib8030044
26. Smith GP (1985) Filamentous fusion phage: novel expression vectors that display cloned antigens on the virion surface. Science 228(4705):1315–1317. https://doi.org/10.1126/science.4001944
27. Parmley SF, Smith GP (1988) Antibody selectable filamentous fd phage vectors: affinity purification target genes. Gene 73(2):305–318. https://doi.org/10.1016/0378-1119(88)90495-7
28. Alfaleh MA, Alsaab HO, Mahmoud AB et al (2020) Phage display derived monoclonal antibodies: from bench to bedside. Front Immunol 11(1986):1–37. https://doi.org/10.3389/fimmu.2020.01986
29. Haryadi R, Ho S, Kok YJ et al (2015) Optimization of heavy chain and light chain signal peptides for high level expression of therapeutic antibodies in CHO cells. PLoS One 10(2):1–16. https://doi.org/10.1371/journal.pone.0116878
30. Handlogten MW, Lee-O'Brien A, Roy G et al (2018) Intracellular response to process optimization and impact on productivity and product aggregates for a high-titer CHO cell process. Biotechnol Boieng 115:126–138. https://doi.org/10.1002/bit.26460
31. Madsen B, Cobia J, Jones N (2018) Continuous process performance enhancements for 50 L to 500 L single-use bioreactors. In: Proceedings (conference poster)
32. Zhou WC, Zhou H, Fang MY et al (2020) Continuous biomanufacturing implementation using an intensified and integrated bioprocess platform. BioProcess Int Ind Innov:16–17
33. Shukla AA, Aranha H (2015) Viral clearance for biopharmaceutical downstream processes. Pharm Bioprocess 3(2):127–138
34. Ding YH, Marino MP, Zen K et al (2020) Considerations for MAb bioprocess and manufacturing validation. Pharm Technol 44(8):31–36
35. Coolbaugh MJ, Varner CT, Vetter TA et al (2021) Pilot-scale demonstration of an end-to-end integrated and continuous biomanufacturing process. Biotechnol Boieng:1–15. https://doi.org/10.1002/bit.27670
36. Jungbauer A (2013) Continuous downstream processing of biopharmaceuticals. Trends Biotechnol 31(8):479–492. https://doi.org/10.1016/j.tibtech.2013.05.011
37. Schofield M (2018) Continuous low-pH virus inactivation: challenges and practical solutions. GENS 38(15)
38. Parker SA, Amarikwa L, Vehar K et al (2018) Design of a novel continuous flow reactor for low pH viral inactivation. Biotechnol Boieng 115(3):606–616. https://doi.org/10.1002/bit.26497
39. David L, Maiser B, Lobedann M et al (2019) Virus study for continuous low pH viral inactivation inside a coiled flow inverter. Biotechnol Boieng 116(4):857–869. https://doi.org/10.1002/bit.26872
40. Farcet JB, Kindermann J, Karbiener M et al (2019) Development of a Triton X-100 replacement for effective virus inactivation in biotechnology processes. Eng Rep 1:e12078. 1–10. https://doi.org/10.1002/eng2.12078
41. Roberts PL, Lloyd D, Marshall PJ (2009) Virus inactivation in a factor VIII/VWF concentrate treated using a solvent/detergent procedure based on Polysorbate 20. Biologicals 37:26–31. https://doi.org/10.1016/j.biologicals.2008.08.003

42. Dutta AK, Fedorenko D, Tan J et al (2017) Continuous countercurrent tangential chromatography for mixed mode post-capture operations in monoclonal antibody purification. J Chromatogr A 1511:37–44. https://doi.org/10.1016/j.chroma.2017.06.018
43. Teske CA, Lebreton B, Reis R (2010) Inline ultrafiltration. Biotech Progress 26(4):1068–1072. https://doi.org/10.1002/btpr.404
44. Casey C, Gallos T, Alekseev Y et al (2011) Protein concentration with single-pass tangential flow filtration (SPTFF). J Membr Sci 384(1–2):82–88. https://doi.org/10.1016/j.memsci.2011.09.004
45. Rucker-Pezzini J, Arnold L, Hill-Byrne K et al (2018) Single pass diafiltration integrated into a fully continuous MAb purification process. Biotechnol Boieng 115(8):1949–1957. https://doi.org/10.1002/bit.26608
46. Tan R, Hezel F, Franzreb M (2021) Continuous single pass diafiltration with alternating permeate flow direction for high efficiency buffer exchange. J Membr Sci 619:118695. https://doi.org/10.1016/j.memsci.2020.118695
47. Amand MM, Ogunnaike BA, Robinson AS (2014) Development of at-line assay to monitor charge variants of MAbs during production. Biotechnol Prog 30:249–255. https://doi.org/10.1016/j.memsci.2020.118695
48. Vargas JV, Nielsen S, Cárdenas V et al (2018) Process analytical technology in continuous manufacturing of a commercial pharmaceutical product. Int J Pharm 538(1–2):167–178. https://doi.org/10.1016/j.ijpharm.2018.01.003
49. Schweitzer M, Pohl M, Hanna-Brown M (2010) Implications and opportunities of applying QbD principles to analytical measurements. Pharm Technol 34(2):52–59
50. FDA (2019) Statement from FDA Commissioner Scott Gottlieb, M.D. and Peter Marks, M.D., Ph.D., Director of the Center for Biologics Evaluation and Research on New Policies to Advance Development of Safe and Effective Cell and Gene Therapies
51. Nienhuis AW, Nathwani AC, Davidoff AM (2017) Gene therapy for Hemophilia. Mol Ther 25(5):1663–1667. https://doi.org/10.1016/j.ymthe.2017.03.033
52. Burova E, Loffe E (2005) Chromatographic purification of recombinant adenoviral and adeno-associated viral vectors: methods and implications. Gene Ther 12(S1):S5–S17. https://doi.org/10.1038/sj.gt.3302611
53. van der Loo JCM, Wright JF (2016) Progress and challenges in viral vector manufacturing. Hum Mol Genet 25(R1):1–11. https://doi.org/10.1093/hmg/ddv451
54. Blessing D, Vachey G, Pythoud C (2019) Scalable production of AAV vectors in orbitally shaken HEK293 cells. Mol Ther Methods Clin Dev 13:14–26. https://doi.org/10.1016/j.omtm.2018.11.004
55. Dong B, Nakai H, Xiao WD (2010) Characterization of genome integrity for oversized recombinant AAV vector. Mol Ther 18(1):87–92. https://doi.org/10.1038/mt.2009.258
56. Allay JA, Sleep S, Long S (2011) Good manufacturing practice production of self-complementary serotype 8 adeno-associated viral vector for a Hemophilia B Clinical Trial. Hum Gene Ther 22:595–604. https://doi.org/10.1089/hum.2010.202
57. Maurer AC, Weitzman MD (2020) Adeno-associated virus genome interactions important for vector production and transduction. Hum Gene Ther 31(9 &10):499–511. https://doi.org/10.1089/hum.2020.069
58. Wills PW (2020) Unprecedented growth – and challenges in gene therapy. Life Science Leader Newsletter
59. Arimura S, Okada T, Tezuka T et al (2014) DOK7 gene therapy benefits mouse models of diseases characterized by defects in the neuromuscular junction. Science 345(6203):1505–1508. https://doi.org/10.1126/science.1250744
60. Grieger JC, Soltys SM, Samulski RJ (2016) Production of recombinant adeno-associated virus vectors using suspension HEK293 cells and continuous harvest of vector from the culture media for GMP FIX and FLT1 clinical vector. Mol Ther 24(2):287–297. https://doi.org/10.1038/mt.2015.187

61. Tomono T, Hirai Y, Okada H (2016) Ultracentrifugation-free chromatography-mediated large-scale, purification of recombinant adeno-associated virus serotype 1 (rAAV1). Mol Ther Methods Clin Dev 3(15058):1–8. https://doi.org/10.1038/mtm.2015.58
62. Adamson-Small L, Potter M, Falk DJ et al (2016) A scalable method for the production of high-titer and high-quality adeno-associated type 9 vectors using the HSV platform. Mol Ther Methods Clin Dev 3(16031):1–11. https://doi.org/10.1038/mtm.2016.31
63. Potter M, Lins B, Mietzsch M (2014) A simplified purification protocol for recombinant adeno-associated virus vectors. Mol Ther Methods Clin Dev 1(14034):1–8. https://doi.org/10.1038/mtm.2014.34
64. Piras BA, Drury JE, Morton CL et al (2016) Distribution of AAV8 particles in cell lysates and culture media changes with time and is dependent on the recombinant vector. Mol Ther Methods Clin Dev 3(16015):1–5. https://doi.org/10.1038/mtm.2016.15
65. Ryan GL, Rutenberg AD (2007) Clocking out: modeling phage-induced lysis of *Escherichia coli*. J Bacteriol 189(13):4749–4755. https://doi.org/10.1128/JB.00392-07
66. Qu WH, Wang MX, Wu YQ et al (2015) Scalable downstream strategies for purification of recombinant adeno-associated virus vectors in light of the properties. Curr Pharm Biotechnol 16(8):684–695. https://doi.org/10.2174/1389201016666150505122228
67. Gao K, Li MX, Zhong L et al (2014) Empty virions in AAV8 vector preparations reduce transduction efficiency and may cause total viral particle dose-limiting side effects. Mol Ther Methods Clin Dev 1(9):1–7. https://doi.org/10.1038/mtm.2013.9
68. Mingozzi F, Anguela XM, Pavani G et al (2013) Overcoming preexisting humoral immunity to AAV using capsid decoys. Sci Transl Med 5(194):194ra92. https://doi.org/10.1126/scitranslmed.3005795
69. Wright JF (2014) AAV empty capsids: for better or for worse? Mol Ther 22(1):1–2. https://doi.org/10.1038/mt.2013.268
70. Wang CL, Mulagapati HR, Chen ZY et al (2019) Developing an anion exchange chromatography assay for determining empty and full capsid contents in AAV6.2. Mol Ther Methods Clin Dev 15:257–263. https://doi.org/10.1016/j.omtm.2019.09.006
71. Qu G, Bahr-Davidson J, Prado J et al (2007) Separation of adeno-associated virus type 2 empty particles from genome containing vectors by anion-exchange column chromatography. J Virol Methods 140(1–2):183–192. https://doi.org/10.1016/j.jviromet.2006.11.019
72. Abrecht H, Pressac G, Boulais A et al (2019) Adenovirus downstream process intensification implementation of a membrane adsorber. BioProcess Int 17(10):38–44
73. Chahal PS, Aucoin MG, Kamen A (2007) Primary recovery and chromatographic purification of adeno-associated virus type 2 produced by Baculovirus/insect cell system. J Virol Methods 139(1):61–70. https://doi.org/10.1016/j.jviromet.2006.09.011
74. Gagnon P, Leskovec M, Goričar B (2020) Streamlining industrial purification of adeno-associated virus. BioProcess International
75. Kondratova L, Kondratov O, Ragheb R et al (2019) Removal of endotoxin from rAAV samples using a simple detergent-based protocol. Mol Ther Methods Clin Dev 15:112–119. https://doi.org/10.1016/j.omtm.2019.08.013

Part IX
Technology Transfer

Chapter 18
Three Decades of Advancements in Technical Transfer of Biologics: A Blueprint for Advanced Therapeutics

Mairead Looby, Ciaran Brady, Melissa Bentley, Amanda M. Lewis, Conor Layden, Barak Barnoon, Erin Abbott, and Brendan Hughes

Abstract Technology transfer of the manufacturing process and associated analytical methods is a critical process in the lifecycle of biotech products. It enables the production of clinical material to test a hypothesis developed in the drug discovery environment, expansion of clinical supplies, and establishment of a robust platform for commercial production. It can also provide flexibility in the product lifecycle by adding additional manufacturing nodes to supplement supply or provide business continuity.

In the early history of the development and manufacturing of biotech products, there was reticence to locate commercial manufacturing facilities too far from the locations where the processes were developed since the technology was viewed as complex and not easy to transfer successfully. Over time, as technology matured, and in some cases driven by a need to acquire capacity not available within the innovator, companies began to transfer processes around internal networks and to newly constructed purpose-built sites, as well as to business partners and contract manufacturing organizations (CMOs) that sprung up to meet the expanding needs of successful pipelines.

M. Looby (✉) · C. Layden
Global Product Development and Supply, Bristol Myers Squibb, Dublin, Ireland
e-mail: mairead.looby@bms.com; conor.layden@bms.com

C. Brady · A. M. Lewis · B. Barnoon · B. Hughes
Global Product Development and Supply, Bristol Myers Squibb, Devens, MA, USA
e-mail: barak.barnoon@bms.com

M. Bentley
Global Product Development and Supply, Bristol Myers Squibb, Phoenix, AZ, USA
e-mail: Melissa.bentley@bms.com

E. Abbott
Global Product Development and Supply, Bristol-Myers Squibb Company, East Syracuse, NY, USA

K. Gadamasetti, S. A. Kolodziej (eds.), *Bioprocessing, Bioengineering and Process Chemistry in the Biopharmaceutical Industry*,
https://doi.org/10.1007/978-3-031-62007-2_18

Technology transfer is a complex multi-disciplinary activity involving process and analytical development, engineering, operations, quality, and increasing automation and IT, as facilities are designed with high levels of automation and control. Factors influencing success include the level of process understanding and the understanding of the relationship between the process, equipment, and automation. Effective project management is a critical enabler that ensures successful orchestration of multiple functions and effective decision-making in a very complex and dynamic environment.

The typical pathway to a successful transfer will be through a facility fit and gap assessment between the donor and acceptor facility, design and execution of any facility and equipment modifications, raw material procurement, process trial lots, and ultimately process validation lot manufacture. All of these activities will also be dependent on the transfer and qualification of appropriate test methods.

Technology transfer, while not explicitly regulated by health authorities, is described in International Council for Harmonisation (ICH) Q10 as "the transfer of product and process knowledge between development and manufacturing and within manufacturing sites to achieve product realization. This knowledge forms the basis for the manufacturing process, control strategy, process validation approach and ongoing continual improvement." Additionally, it is widely appreciated that the outcome of validation and comparability studies is critically reviewed as part of the regulatory approval of the process in the new location. To this end, care must be taken to ensure data integrity and continued compliance with U.S. 21 code of federal regulations (CFR) Part 11 and Eudralex Volume 4, Annex 11 across all of the steps from the donor to acceptor site. Clear definition of protocols and success criteria is critical to ensure effective transfer and decision-making, especially when projects are under time pressure.

The relationships between the donor and acceptor teams are sometimes overlooked. Generation of close, collaborative relationships with shared goals are major enablers of success. These cultural and behavioral enablers can be built into the program, and on-site presence and visits to facilities can enhance these relationships.

In the past 30 or so years, the transfer of robust processes like those used to produce monoclonal antibodies has become extremely effective and can be achieved with relatively modest investment of expert resources and can be completed in a matter of months. Mature biopharma companies have made significant advances in the development of platform processes and, with the associated maturity of expertise and capability and knowledge management within these companies, can perform effective technology transfer in a matter of months in a clinical setting and less than one year in a larger-scale commercial setting. Leveraging modeling software and other digital tools has also enabled speed and robustness. Despite these advances, it should be acknowledged that the timeline and investment may be much longer and costly for tech transfers to new facilities (e.g., biotech to external contract manufacturing organizations [CMOs] or between external CMOs) (i.e., where an effective technology transfer platform process is not in place). Interestingly, in the emerging and advanced therapies environment, the situation is more like the early days of recombinant protein production where process knowledge is still under development and the expertise is often localized within a small number of companies. Nevertheless, the same principles of technology transfer can be applied and over time will result in a broader proliferation of skills and the same flexibility that applies now for monoclonal antibody production.

Keywords Facility fit modeling · Drug product · Drug substance · Clinical tech transfers · Single-use facilities · Commercial tech transfers · CMOs

18.1 Introduction

Technology transfer as applied to biopharmaceutical manufacturing can broadly be defined as the transfer of information, technology, and methodology for the production of a biotherapeutic substance or product from a sending party (typically Process Development [PD] or an existing manufacturing site) to a receiving party (a new or additional manufacturing site). Technology transfer of the manufacturing process and associated analytical methods is a critical process in the lifecycle of biotech products. It enables the production of clinical material to test a hypothesis developed in the drug discovery environment, expansion of clinical supplies, and establishment of a robust platform for commercial production. It can also provide flexibility in the product lifecycle by adding additional manufacturing nodes to supplement supply or provide business continuity.

In the early history of the development and manufacturing of biotech products, there was reticence to locate commercial manufacturing facilities too far from the locations where the processes were developed since the technology was viewed as complex and not easy to transfer successfully. Over time, as technology matured, and in some cases driven by a need to acquire capacity not available within the innovator, companies began to transfer processes around internal networks and to newly constructed purpose-built sites, as well as to business partners and contract manufacturing organizations (CMOs) that sprung up to meet the expanding needs of successful pipelines.

Technology transfer is a complex multi-disciplinary activity involving process and analytical development, engineering, operations, quality, and increasing automation and IT, as facilities are designed with high levels of automation and control. Factors influencing success include the level of process understanding and the understanding of the relationship between the process, equipment, and automation. Effective project management is a critical enabler that ensures successful orchestration of multiple functions and effective decision-making in a very complex and dynamic environment. Typically, a project manager is assigned who builds a technical transfer project plan that would integrate the deliverables and resource plan needed to successfully execute the transfer including the overall timeline and milestones in the transfer. This project manager or tech transfer leader could also run the tech transfer team's meetings; this team is typically comprised of technical experts across multiple functions, both on the sending and receiving end, who are tasked with executing the transfer. The project manager/tech transfer lead would manage the team and ensure that the team members work together collaboratively to execute the transfer and manage timelines.

The relationships between the donor and acceptor teams are sometimes overlooked. Generation of close, collaborative relationships with shared goals are major enablers of success. These cultural and behavioral enablers can be built into the program, and on-site presence and visits to facilities can enhance these relationships.

The typical pathway to a successful transfer will be through a facility fit and gap assessment between the donor and acceptor facility, design and execution of any facility and equipment modifications, raw material procurement, process trial lots, and ultimately process validation lot manufacture. In parallel, a thorough gap assessment of analytical equipment, capabilities, status of analytical methods, and other potential gaps to support transfer of analytical method is also executed. This analytical gap assessment is a critical activity in the tech transfer of the product to the new site or location, and risks highlighted during the analytical gap assessment can sometimes emerge as critical path activities for the overall tech transfer.

In the past 30 or so years, the transfer of robust processes like those used to produce monoclonal antibodies has become extremely effective and can be achieved with relatively modest investment of expert resources and can be completed in a matter of months. Mature biopharma companies have made significant advances in the development of platform processes and, with the associated maturity of expertise and capability and knowledge management within these companies, can perform effective technology transfer in a matter of months in a clinical setting and less than one year in a larger-scale commercial setting. Leveraging modeling software and other digital tools has also enabled speed and robustness. Figure 18.1 illustrates the typical process steps in a drug substance manufacturing process and Fig. 18.2 illustrates a timeline and sequence of activities of a typical tech transfer projects demonstrating the differences between an early-phase clinical tech transfer and a scale-up of a late-phase/commercial process to a large-scale commercial facility. Clinical tech transfers, that rely heavily on the use of platform technologies and raw materials, as well as highly flexible/disposable systems, can now be effectively completed in less than 3 months in mature organizations. Larger-scale tech transfers, which typically have more significant requirements in terms of the need to scale-up and perform process validation, in addition to complex automation updates, can take up to one year or longer depending on the maturity of the facility and systems and whether a platform process is being employed. Non-platform processes can take significantly longer as such a transfer may include the need to make expensive and time-consuming facility engineering modifications, and also order long lead time raw materials (cell culture media or purification resins, for example). More established companies, that have developed core expertise and established business processes for developing and manufacturing biopharmaceuticals, have invested heavily in shortening these timelines through employing platform technologies to avoid facility modifications and the need to purchase long lead time

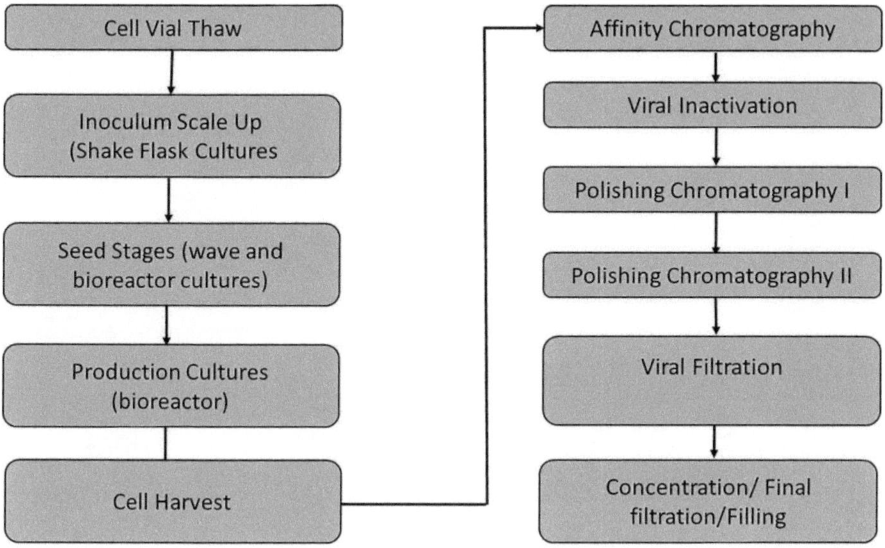

Fig. 18.1 Key bioprocessing steps in a typical drug substance manufacturing process

materials. The added benefits of utilizing a platform do not stop there however, as such platforms reduce the need to change complex automation recipes, allow the use of platform documents (such as control strategies and batch records), reduce training needs for operators, and allow for the development of deep technical expertise within the company on the platform technologies.

In the emerging and advanced therapies environment, the situation is more like the early days of recombinant protein production where process knowledge is still under development and the expertise is often localized within a small number of companies. Nevertheless, the same principles of technology transfer can be applied and over time will result in a broader proliferation of skills and the same flexibility that applies now for monoclonal antibody production.

In this chapter, various examples and case studies are presented that provide some historical context on how the process for tech transfer for biopharmaceutical development and manufacturing evolved as organizations became more mature in the field. A process for facility fit assessments, which is typically performed at the front end of the tech transfer, is presented to determine if constraints exist that may necessitate either changes to the facility or the process; these analyses can also be used to determine what facility is the best fit for the product of interest. A case study is also presented on the transfer of a process into a clinical pilot-scale facility, which highlights the value of speed to manufacture, utilizing platform technologies and highly flexible systems and equipment in the facility. Additionally, case studies of transfers of late-phase biologics drug substance and drug product processes, both to internal large-scale commercial facilities and external contract manufacturing organizations (CMOs), are presented.

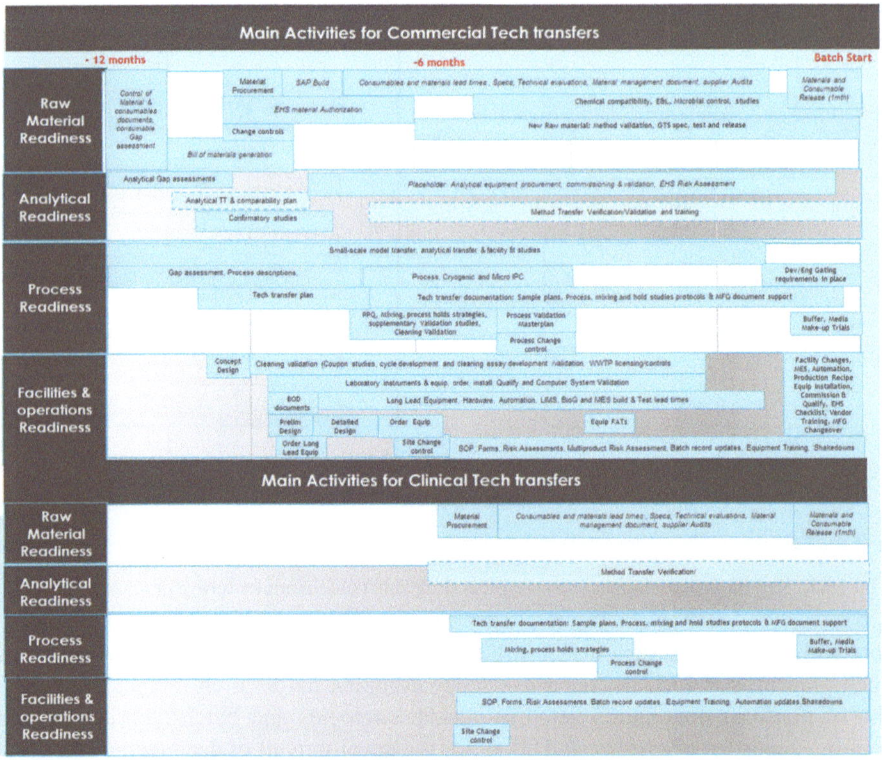

Fig. 18.2 Main activities and associated timelines for a drug substance tech transfer into a commercial and clinical facility

18.2 Facility Fit Modeling and Engineering Considerations

At the outset of any tech transfer, some fundamental questions relating to the overall program scope need to be addressed and aligned before embarking on any tech transfer activities to ensure that the tech transfer project meets organizational needs and product requirements. One question that is often asked at the start relates to product lifecycle and clinical needs of the product: *Is the tech transfer required to generate clinical material for phase I/II clinical studies or indeed another clinical study?* In this type of tech transfer, where the process is being transferred to generate clinical material at sufficient scale, speed is often the primary goal and often the critical path activity is how quickly can the clinical material be generated. In these tech transfers, the tech transfer team may forgo some facility modifications that would ordinarily have enhanced operational flexibility, just to reduce tech transfer scope and ultimately reduce overall tech transfer timelines. Similarly, the batch output requirements for the particular product are also important, i.e., the extent of facility changes (and associated automation change) that might be undertaken for a short clinical campaign to deliver a finite number of batches might be very different to the facility and automation changes undertaken to enable the transfer of a proven commercial product into the facility. Another question that might be asked is whether the process needs to be scaled up and out for current or future needs as this will drive the final process scale. This may involve a number of different tech transfers or one main tech transfer to the final manufacturing facility; often these types of tech transfer involve scaling up from smaller single-use bioreactors to larger stainless steel bioreactors. Once these initial questions are addressed, the focus then shifts to facility selection: *where to transfer the product to*? The scope, duration, and technical risk of a tech transfer is highly dependent on the capabilities of a receiving site's equipment, infrastructure, personnel, and quality systems, as well as available capacity and macro network strategy. Implementing a robust site selection process can ensure that the desired business outcomes for the product transfer will be achieved.

Facility fit and engineering analysis tools are key enablers of a robust site selection process. A robust site selection analysis will compare relative scope of work to implement the new manufacturing process; it will allow comparison of expected project timelines, resources, costs, and risks. The analysis will identify changes required for equipment, facility, and process automation, as well as scope for development studies and process changes. The level of detailed analysis required should increase with level of project commitment. Initial manufacturing facility selection should be based on feasibility study utilizing the mass balance model to identify any major scope items resulting from fit gaps with the existing manufacturing equipment and services. A rough order of magnitude cost estimate along with high-level project timeline can provide a rapid evaluation framework to decide between alternative manufacturing site choices, including both internal and external options.

The engineering toolset relies on inputs such as process description and expected or actual process data to inform sizing of key unit operations. The workhorse of the engineering toolset is the mass balance model, ideally with Process Flow Diagrams. The mass balance model should allow scaling the process within expected commercial boundaries. The mass balance tool enables scenario analysis via input of key manufacturing scale parameters (such as final production bioreactor volume, or size of available chromatography columns) and deliver as outputs, the resultant required equipment sizing at each step of the process. Other tools that can be utilized include: (1) Facility capacity models (built utilizing discrete event simulation tools such as BioG, Simio, ProModel, or other similar platforms) to understand facility batch cadence and throughput capabilities; and (2) Mechanistic models of key equipment for streamlined process scaling and selection of operational parameters. Computational fluid dynamics (CFD) models can reduce the burden of at-scale development trials or engineering runs by identifying the appropriate operating parameters for the new process on existing equipment.

Once a decision on site selection is made, further detailed analysis is required to develop the full scope of the tech transfer project. A detailed facility fit study informs the full scope of enabling work to implement the new manufacturing process. This should focus on both the main process train and all the support functions, i.e., warehousing, storage (ambient and cold), laboratories, etc. It should identify any changes required to equipment, procedures, process automation, qualified materials, recipes, quality control (QC), data analytics, etc. This detailed facility fit should be a unit operation by unit operation assessment of the equipment suitability and sizing. A key part of this assessment should be agreed ranges for product titer and process yields, to enable a robust and realistic facility fit. Monte Carlo analysis can be used at this stage to assess the distribution of probable operating scenarios.

A tech transfer project risk register will identify both technical and business risks to the overall tech transfer program and should be maintained through the full duration of the product transfer. It can help identify where additional mitigations are required in the form of development and scale-up studies.

Platforming of processes, components, and operational approaches can lead to acceleration of technology transfers and reduced risk. It can reduce the number and severity of process fit gaps as well as reduce the scope of work in supporting functions (e.g., qualification of new raw materials and consumables). A key variable here is the source of the product being transferred. Products from within the company may align well with existing processes and facilities, but externally sourced products may be based on very different process flows, and present a larger facility fit challenge.

This can be an excellent time for meaningful engagement with Process Development to develop mutual understanding of the specific process requirements, and also to develop an understanding of the main facility fit challenges. While the main process requirements will always exist, there is often scope within ranges to modify setpoints, possibility of increased filter loadings or scope in some of the "preparation" activities to achieve an easier facility fit, and reduce the amount of change required to facilitate the new process, i.e., increased filter loadings may be possible but just has not been investigated by PD, preparation of filters using water for injection (WFI) versus buffer, how recovery flushes are operated, etc.

The tables below represent a real-life example of the "before" and "after" facility fit summary for a process being transferred to an internal site. Early engagement with Process Development, and a mutual understanding of where the main facility constraints lay, and options to alleviate them, led to a significant reduction in facility changes required (Tables 18.1 and 18.2).

Figure 18.3 shows another real-life example of "platform conversion" changes made during the transfer of a clinical molecule into an internal manufacturing site. The extent to which changes can be made to suit the facility will obviously depend on the lifecycle stage of the product, and the specifics of the process, with more significant "platform conversion" changes perhaps being possible earlier in the lifecycle. In addition to yielding facility fit advantages, this platform conversion can also have benefits in terms of consumables cost per batch (reduced filter numbers), raw material savings (higher utilization of expensive chromatography resin), and cadence benefits from rationalized unit operations.

Another consideration during the "engagement" phase with Process Development is the process critical consumables required. Increasingly, tech transfer timelines are being driven by consumable and raw material deliveries, and these can often represent the most significant risk to successfully delivering on a tech transfer schedule.

Table 18.1 Initial facility fit assessment

Unit Operation	Facility Target	Bioreactor Titre (g/L)			Notes
		6 g/L	6.5 g/L	7 g/L	
Chrom 1 Cycles (2.0m column)	< 3 cycles	2	2	3	
Chrom 2 Cycles (1.6m column)	< 2 cycles	1	1	1	
Chrom 3 Cycles (1.4m column)	< 2 cycles	1	1	1	
Virus Filtration (No. of Filters)	< 10 filters	9	10	11	Current Virus Filter skid has max capacity for 10 filters. Current loading limit of 200L/m2, mean this is a faciity constraint at higher titres.
Final UF Skid Membrane Capacity	≤ 32 cassettes	40	44	48	UF Skid max capacity is 32 cassettes. Current membrane loading range of 1500 g/m^2 means this is a constraint at all titres.
Final UF Vessel Capacity	≤ 2000L	1811	1962	2113	Final UF Tank volume is 2000L. Current concentration target of 35 g/L means tank is not large enough to handle process titres of 7 g/L, and titre of 6.5 g/L is a borderline vessel fit issue.

Table 18.2 Final facility fit assessment

Unit Operation	Facility Target	Bioreactor Titre (g/L)			Notes
		6 g/L	6.5 g/L	7 g/L	
Chrom 1 Cycles (2.0m column)	< 3 cycles	2	2	3	
Chrom 2 Cycles (1.6m column)	< 2 cycles	1	1	1	
Chrom 3 Cycles (1.4m column)	< 2 cycles	1	1	1	
Virus Filtration (No. of Filters)	< 10 filters	7	8	8	Current Virus Filter skid has max capacity for 10 filters. Development study executed to increase loading from 200L/m^2 to 300L/m2 to avoid this constraint. Also reduced numbers of expensive consumables required.
Final UF Skid Membrane Capacity	≤ 32 cassettes	32	32	32	UF Skid max capacity is 32 cassettes. Skid sizing now ok based on successful process development study to test membrane loadings up to 2100 g/m2.
Final UF Vessel Capacity	≤ 2000L	1585	1717	1849	Final UF Tank volume is 2000L. Agreement with Process Development that diafiltration can run at a concentration setpoint of 40g/L, which eliminates this tank sizing constraint.

Where possible, existing site consumables with established supply chains should be used. A review of new process critical consumables (typically filters) should be undertaken with the assistance of Supply Chain and Procurement to establish if there are any particularly high-risk items. Engaging with Process Development at this stage, and discussing which of these high-risk filters might be suitable for secondary filter qualification, can significantly reduce the risk to the overall tech transfer program.

Automation can either be the bane or boon of a streamlined technology transfer—it really depends on implementation of a scalable and thoughtful automation architecture as well as on the level of platforming and standardization within the organization.

- Design for agility and flexibility with parameterization of recipe and operational variables. When used with a sound modular S88 approach can greatly reduce the effort required to implement a new product recipe. It can be the difference between code changes at multiple layers of the automation versus a change to parameter values in a recipe table.

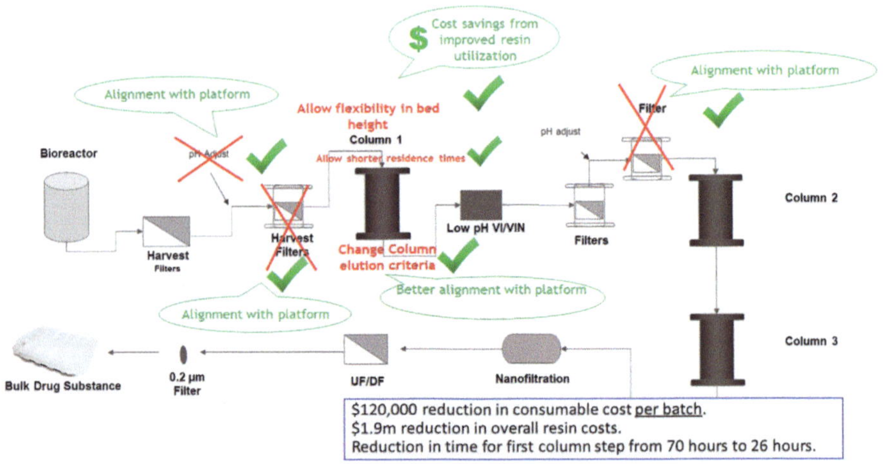

Fig. 18.3 "Platform conversion" of a clinical molecule as part of tech transfer

- Consistency in how equipment is operated across sites can result in fewer changes required during a new product introduction as well as reduced risk.
- Automation can lead to quicker changeovers between products and with aid of the above elements can lead to quicker implementation of new recipes.
- Automated processes typically have more real-time data and greater control, which can offer greater insights into process performance and tighter process capability.

Process Data consisting of online data trends (e.g., bioreactor pH, temperature, and dissolved oxygen) are required to monitor process performance—arguably the truest measure of the success of a tech transfer. Data analytics must be built into the technology transfer project plan to ensure that data hierarchies, data pipelines, and analytics are established to aid in the transfer; to ensure that the process performance is meeting acceptance criteria; and product quality is maintained appropriately in the process and release of product quality specifications. Building the data analytics hierarchies for the new product is often an after-thought since regulatory expectation around continuous process verification (CPV) only manifests into published outputs at the annual product quality report (APQR) stage, long after the original tech transfer project is closed-out. The monitoring plan for the product should be developed as part of the tech transfer and the enabling data analytics configuration of data repositories should be completed as part of the enabling works of the tech transfer, if possible. In this way the systems can be leveraged throughout the development and PPQ (process performance qualification) activities. Manufacturing and quality data availability provides the ability to understand process performance and a mechanism to compare to historical performance from the donating site.

18.3 IND Enabling Technology Transfer of a Clinical Asset to a Single-Use Facility

The generation of high-quality, kilogram-scale protein therapeutics is necessary to enable an Investigational New Drug (IND) filing and the start of clinical trials in humans. This technology transfer is typically the first time the process is being manufactured in a good manufacturing practice (GMP) facility, and often little to no at-scale operations have been performed to date. The focus throughout the transfer is on speed, while maintaining quality.

In order to facilitate the speed of the transfer, the process development work typically utilizes a standard platform. Where possible, this platform is an established framework for the manufacture of the product. By leveraging platform raw materials that the company has experience with, the complexity of the Bill of Materials (BOM) and sourcing is minimized. Ideally, no new vendor relationships are required, and materials are already established in the inventory system. For raw materials that have expiries (e.g., media powders, salts, resin), the expiry limits will be well understood and previously established. For consumable components (e.g., filters, bags, probes), operations will have direct experience, reducing human error. Additionally, previous operational limits from other programs can be leveraged to establish operational limits and best practices. Additionally, the use of common materials can be a critical advantage during periods when the global supply chain is not robust. Critical materials are more likely to be at hand if they are used across multiple processes.

A platform process also enables speed of the transfer. Operations require a process description or definition in order to develop the operational instructions required to execute manufacturing tasks. This process description defines in what order and how each step of the process is executed, including acceptable operational ranges, target values, and which parameters are most critical to product quality. By leveraging a platform process, the operational ranges, targets, and critical parameters for each step are largely defined in advance and optimized based on development data and scale-up runs. By combining program-specific data with process experience gained through other programs that use the same materials and operational steps and sequence, better and more informed risk-based decisions can be made quickly. FAILURE MODE and effects analysis (FMEA) and other risk assessment tools are typically utilized throughout the tech transfer process to make a risk-based decision that balances data and historical experience with identified gaps. In some cases, the gaps are significant enough that additional experimental work is required, and then a decision can be made to perform it in advance of the transfer or concurrent with the manufacturing activities.

The use of platform analytical assays will also speed up technology transfers. Just as with the process platform, an analytical platform has advantages of common materials, historical experience from personnel, established limits, and previous method descriptions as a starting point. In most tech transfers, there is still a need to tech transfer non-platform product-specific assays namely Enzyme Linked Immunosorbent Assay (ELISA) and cell-based potency assays that will still need to be transferred to support the specific process and product being transferred. For an IND program, there will be limited or no regulatory requirements for comparability data. However, the analytical profile for clinical batches will set the foundation for future, commercial expectations. Therefore, it is important to balance speed with quality in making decisions around analytical specifications.

The process description is the starting material for the generation of the manufacturing batch records (MBRs), or detailed instructions that operators use to manufacture the batch. As part of the tech transfer activities, process-specific MBRs need to be authored and approved for each unit operation, inclusive of media and buffer preparations. A single manufacturing process will generate in excess of 20 MBRs. Adherence to a platform process can greatly accelerate speed in this step, as common media and buffer formulations would require little to no MBR modification, and only minor modifications to major unit operations. Additionally, the use of paper batch records can provide a speed advantage compared to fully or highly automated instructions. In general, the more manual the operational process is, the faster the tech transfer activities can be completed. This will need to be balanced with quality, as automation or engineering initiatives in manufacturing can greatly reduce the overall deviation rate and mitigate human error. Therefore, the extra time to automate select operational steps may be a worthwhile tradeoff.

Finally, the design of the facility itself can enable the speed of the transfer. A single-use manufacturing facility makes use of disposable materials and single-use consumables for the majority of the product contact steps. This provides the advantage of rapid product changeover and reduced cleaning. Limited equipment will require cleaning validation studies, and this can be further reduced by the use of platform resins, buffers, and cleaning agents. Extractables, leachables, and container closure studies can be performed on platform consumables at one point in time and then leveraged for future programs. Additionally, a single-use facility inherently introduces operational flexibility. Unit operations can be performed in different sequences with minimal impact to facility setup. The use of single-use mixers and totes enables flexibility for material preparation and storage between operations. One potential drawback of a single-use facility is the scale of operations. To date, within the industry, throughput is limited by upstream operations at a maximum scale of 2000 L working volume. This physical limitation translates to <10 kg/batch for a typical monoclonal antibody (mAb), resulting in high cost of goods. However, the volumes needed for clinical studies is typically low, therefore these volumes are appropriate for this stage of process lifecycle.

18.4 Tech Transfer of Biologic Drug Substance (DS) Process to External CMO for Commercial Manufacturing

Technology transfers of commercial processes from manufacturing sites to external CMOs pose many challenges for the sending and receiving sites that require attention to detail, collaboration, communication, and transparency to data and information to be successful. Innovative ways of supporting need to be adopted to ensure the successful transfer of the process.

One such example was the technology transfer of a commercial monoclonal antibody process to a new facility at an external CMO. This transfer involved the transfer of a next-generation process from an existing stainless steel commercial facility to a new facility at a larger scale containing a mixture of stainless steel and single-use bioreactors. This case study focuses specifically on the upstream portion of the process, with the same procedure followed for the downstream process. Prior to the initiation of the technology transfer itself, the team met through video conferencing and face-to-face meetings to conduct introductions to the core team, define the scope of the project, and outline the starting points.

The team consisted of subject matter experts (SMEs) from manufacturing sciences and technology, manufacturing support, analytical, and project managers from the sending site and process development and manufacturing support from the receiving site. The project managers were assigned to help facilitate the transfer and be the overall accountable parties that ensured the successful transfer. These meetings proved to be instrumental to forming the close working relationship needed to ensure success. Additionally, it allowed the members of the transfer team to identify what information was going to be needed and outline the steps that would be followed to transfer that information. Depending on the stage of the tech transfer and information being exchanged, the tech transfer core team membership flexed to include other functional teams on ad hoc basis, namely global regulatory science and quality organizations, to garner their input on specific topics throughout the tech transfer.

In terms of tech transfer scope, a key aspect for the team to align on initially was the scope of the analytical tech transfer, i.e., which assays should be transferred (in-process versus release assays) and qualification status needed for these methods at various stages of the program, e.g., Development batches require a lesser state of readiness in terms of analytical methods compared to PPQ batches where fully validated methods are required.

A key component of tech transfer is information and knowledge exchange and as such it was important to set up a secure repository for the information to flow from one site to another. In this instance, a secure SharePoint site was utilized to place necessary source documents and data openly between the sites. Additionally, a new idea project management tool was implemented during this process, a request log. Previously, requests were being made via emails and could be overlooked and there was little to no accountability for actions. The request log allowed for tracking of the responses, accountability, and visibility to all parties to ensure that needed information was being seamlessly communicated.

During the initial meetings, the quantity of raw materials needed to run the process at-scale was calculated and ordered. Raw materials often have long lead times for large GMP batches, especially if the order requires international shipping as in this case. International shipping requires stricter documentation and temperature monitoring. These raw materials included materials to prepare media, solutions, buffers, and cryovials with the cells. Certain raw materials cannot be shipped internationally, or indeed there may be additional environmental considerations in one location versus another, which means that alternative sources of those raw materials needed to be identified. The qualification of alternative vendors can be a lengthy process and should be factored in tech transfer timelines as early as possible to mitigate overall program delays. In some cases, small-scale studies may be required to verify the suitability of the alternatives, which can significantly lengthen the overall qualification timelines. Samples of the raw materials were sent from the sending site, as well as samples of alternative materials and cryovials to the receiving site to initiate lab-scale studies.

During the weeks that followed the meetings, the teams initiated the sharing of important process and facility-related information. The documents shared from the sending site included process flow diagrams containing volumes and bioreactor dimensions, in-process control documents, manufacturing batch records, process characterization reports, media preparation records, and historical manufacturing and lab-scale data. A process gap analysis was performed by the receiving team based on data provided and the knowledge of the facility limitations. The gap analysis was essential for understanding how the process will fit into the facility and what modifications needed to be done. As part of the gap analysis, mass balance calculations were performed to evaluate how the process would look at each step and identify areas that might be of concern or need modifications to fit the facility. The mass balance calculations identified the need for an additional seed bioreactor step due to volume constraints in the facility and different aspect ratios of the bioreactors that could lead to mixing and gassing differences. A risk assessment was then performed by both teams to review these concerns as well as others and assign the likelihood of them having impact on process performance (viable cell density, viability, and cell productivity) and also subsequent product quality (such as glycans and charge variants). Any concerns that were assigned a medium or high risk with respect to these performance parameters and critical quality attributes were identified for characterization work in the process development labs to understand the impact of the changes/gaps prior to initiating at-scale runs. It was identified that the bioreactor and medium batching vessels aspect ratios were different, and studies would be needed to ensure proper mixing speeds and gas flowrates. Additional equipment/procedural differences were identified. The sites were utilizing different instruments to monitor the offline gasses and metabolites throughout the process. At that time, the receiving site was performing hand counting of the cells throughout the expansion process to the production bioreactors. This differed from the sending site that was using an automated instrument to perform the cell counting.

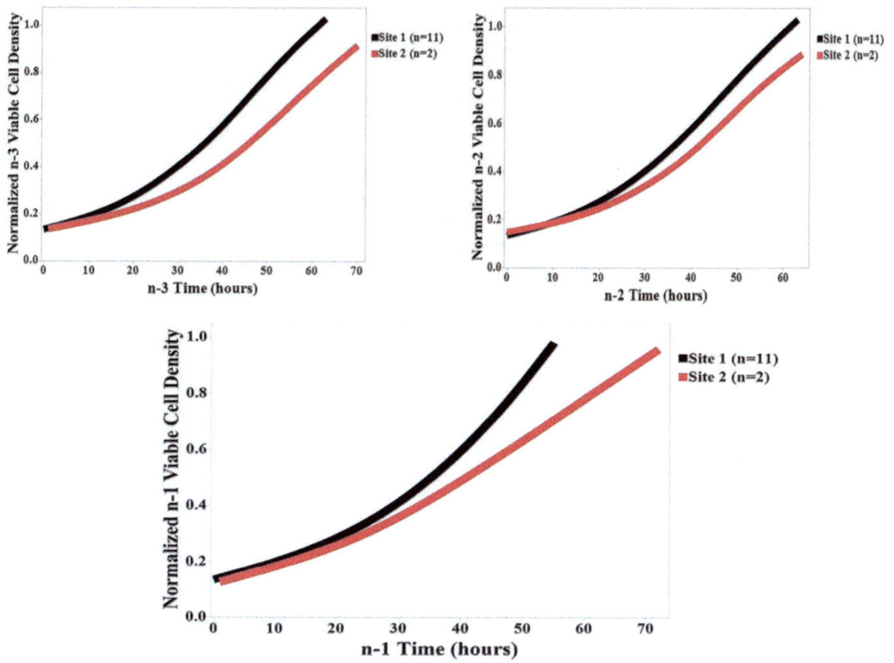

Fig. 18.4 Seed bioreactor viable cell density trends

As it was identified that additional studies would need to be conducted in the process development lab to support changes to the process and gaps, the transfer of the lab-scale model from the sending site laboratory to the receiving site laboratory was imperative. The lab-scale models would also be utilized for satellites from at-scale runs to provide potential troubleshooting support. The process development lab planned to conduct shake-down runs to verify the comparable performance of the process in the lab-scale bioreactors. The shake-down runs were also conducted to familiarize the on-site staff with the process before it was conducted at-scale. During the shake-down runs, slower growth was observed in the seed bioreactors compared to the historical sending site as shown in Fig. 18.4. An investigation into the growth differences identified that preparation differences in the seed bioreactor media caused the reduction in growth. Further data comparison showed that the seed bioreactor performance was similar to those observed at a previous CMO utilized for the process. The final cell densities achieved were also within the acceptable final cell density range, which ensures that enough cells are present to meet volume requirements at the next process step. Despite growth profiles being similar to historical trends, lower final titer values were observed in the production bioreactor as shown in Fig. 18.5. A cross-site investigation into the final titer differences was initiated. Feed medium raw materials were shipped back to the sending site for side-by-side comparisons of the performance with the receiving site. It was identified that the final titer performance differences were linked to a specific lot of a

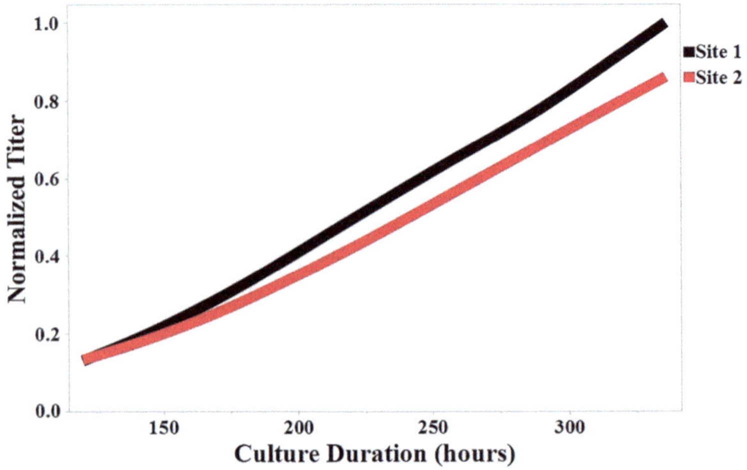

Fig. 18.5 Production bioreactor titer trends

highly variable raw material in the feed medium, an issue seen before at the sending site but not expected at the CMO and that did not become apparent until the initial large-scale runs were executed at the CMO.

After the completion of the lab-scale gap analysis studies and demonstration of comparable performance, several engineering runs were performed to evaluate the performance of the modified process at-scale. Samples were removed from the production bioreactor after inoculation to perform lab-scale satellites as a troubleshooting tool should any differences at-scale be observed. The satellites allow for identification of a process problem, media problem, or a facility fit problem. In addition, employees from the sending site (also called person-in-plant) were present on-site during the runs to observe the process, assist with decisions, and analyze the data for comparability. The data from the engineering runs and satellites demonstrated approximately a 10% reduction in titer compared to historical data. An investigation into the root cause determined that the same raw material responsible for the reduced titers in the shake-down runs was responsible for the reduction in titer at-scale. The trends are presented in Fig. 18.6.

As the health authority agencies require a comparison of the performance of all steps and product quality when filing a new facility, a process performance qualification (referred to as PPQ) campaign was conducted at the CMO. These data are important in demonstrating that the product produced is the same as the originating site, despite any modifications that were needed for facility fit related reasons. Data from this campaign were compared to historical data from the originating site. The PPQ performance data were compared to historical performance data as shown in Fig. 18.7. Despite subtle differences in some of the final process performance data, the process performance data overall at the receiving site were similar to the sending site, within acceptable ranges and deemed well controlled.

The technology transfer of this commercial monoclonal antibody process was a major success for both companies with both process performance (viability, titer)

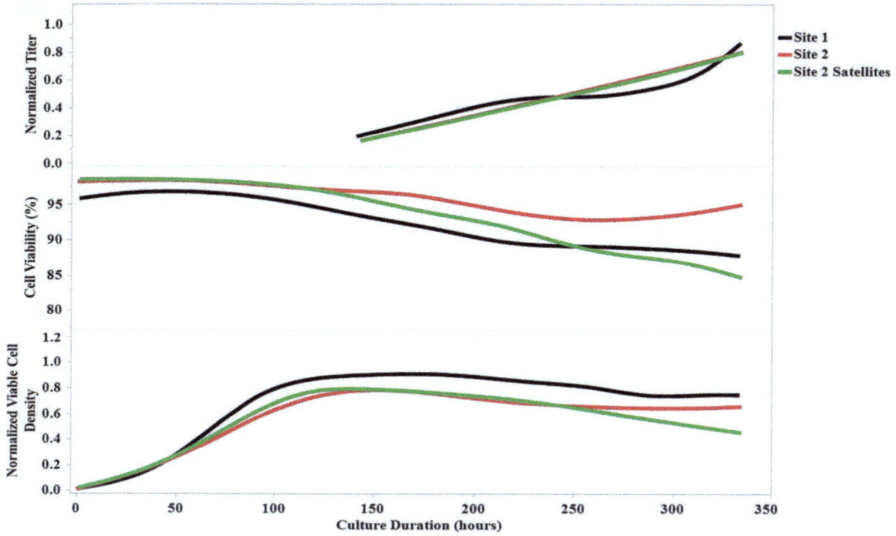

Fig. 18.6 Engineering production bioreactor trends. In the normalised titre figure, the titre results for Site 2 (Red line) completely overlap with site 1 (Green line) results

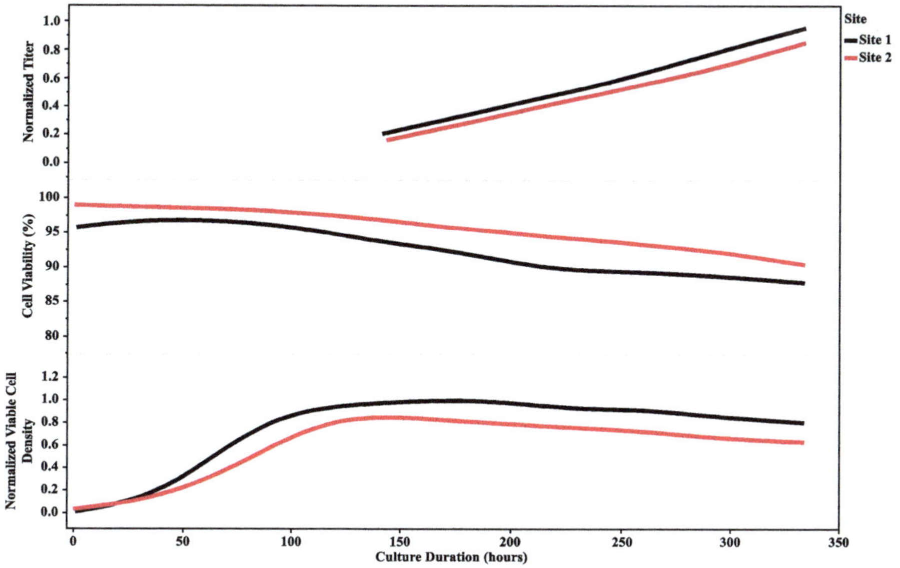

Fig. 18.7 PPQ production bioreactor trends

and product quality (release and extended characterization analytical data) results comparable between the sites, and ultimately resulted in the approval of the process at the CMO for multiple regulatory agencies. The transfer utilized many different tools: face-to-face meetings, gap analysis, risk assessments, lab-scale and at-scale studies to assess performance, and person in plant support. The steps taken in this transfer have subsequently been utilized as a template for additional process transfers to the same CMO based on the working relationships that were established and the overall success of the transfer.

18.5 Technology Transfer of Clinical Asset to a Large-Scale Commercial Facility

Introduction of late-stage clinical programs into large-scale commercial facilities (>15,000 L) can pose many challenges during tech transfer at a time when speed, agility, and cost minimization are key priorities within a tech transfer program. Often, the process development team that has developed the late-stage clinical molecule may still be utilizing non-platform upstream and downstream technologies and continuing to develop and optimize the process as it evolves and moves through the clinical stages. In this case study, the tech transfer of a late-stage clinical program involving the use of non-platform technologies from process development into commercial manufacturing is described. In this instance, the process was originally developed to generate clinical material in single-use bioreactors but the process was subsequently modified and trailed at 500 L single-use bioreactor stage and then scaled up to 15,000 L stainless steel bioreactor stage to generate material for long-term stability studies, subsequent drug product studies, and also generate additional material for later-stage clinical bridging studies. In this instance, the distinct absence of platform technology in the late-stage program presented many challenges for the tech transfer as the process was being moved into a large-scale (15,000 L), heavily automated facility containing mostly stainless steel technology that would require significant intrusive modification to cater for the needs of this late-stage non-platform molecule. Such facility modifications involving use of new equipment or complex modifications to existing equipment often require long lead times and not only increase the tech transfer timelines but also require plant down time to make the necessary changes, hence driving increased costs. During such tech transfer, often the process areas that presented the most complex and ultimately time-consuming and expensive modifications were associated with media formulations, scale-up train, sequencing of unit operations, and additional filtration requirements in term of new filtration capacity or new technologies such as a switch from filtration at harvest to centrifugation to cater for the increased volumes at the larger scales.

Upstream, in mammalian cell culture processes, the cells are fed to sustain growth and productivity. In the commercial setting, this feeding regime is typically heavily automated based on defined time intervals, viable cell densities, or using digitalized feedback control loops using Raman or Biomass technology to predict and control cell culture feeds at-scale. In this tech transfer, the final process definition was still evolving, and optimal feeding strategy was not fully defined resulting in a feed medium strategy that was akin to that used in a process development laboratory, which required multiple components to be added at regular intervals throughout a day. Such an intensive and demanding control strategy is not sustainable at manufacture scale and presented a conundrum for the tech transfer into the commercial facility: Do we build an expensive, complex feed control strategy for a late-stage clinical asset that has yet to complete clinical trials and may not move forward into commercial plant long term or sacrifice some process efficiency at-scale to enable a more rapid initial introduction of the process, but may impact batch cadence and demand more resources in the commercial facility in the interim. To add to the complexities, the media used in the upstream process contained large numbers of media components, with unproven solubility profiles and media compositions that yet had to be fully consolidated into a more manageable media for manufacturing scale. Many components have to be individually prepared as stock solutions, which drives additional equipment requirements around storage of multiple media components each with their own specific storage requirements in terms of designated cold storage conditions and environmental, health, and safety (HSE) storage considerations. Furthermore, the large volumes for multiple components required at-scale do not lend themselves to straightforward integration into the commercial facility and careful consideration had to be given to the space constraints associated with storing large quantities of powder media at controlled temperatures, and handling and dissolution of large volumes of different components to enable robust formulation of media in large-scale preparation vessel (10,000 L). A key lesson learned from this tech transfer, particularly for the early development teams, was to consider the complexity and consequences of the culture media/feeding strategy that is used in anticipation of large-scale manufacturing as failure to pay attention to this can lead to some very complex technical challenges on plant that are not readily resolved without significant investment (e.g., additional cold rooms to store heat labile media components) or impact to facility cadence and batch throughout.

Challenges in the downstream purification train were not dissimilar to those in upstream due to the non-platform nature of the late-stage molecule. Unit operations downstream called for multiple filtration steps using differing technologies that were simply not in place at the manufacturing facility thus posing a number of challenges. The scope and complexity were dramatically enhanced as the new filtration carts and associated filters were required but lead time and uncertainty around supply of these filters meant that technologies had to be ordered well in advance to guarantee supply and delivery times. Due to ongoing manufacture in the commercial plant at the time of the tech transfer, the seamless integration of these technologies within the existing fabric of the facility from an automation and hardware perspective was a necessity. In other unit operations, the sequencing and pooling of unit operations particularly post the affinity capture step appeared relatively straightforward. However, despite having all the main equipment in terms of vessels and columns, subtle differences in sequencing of operations significantly increased facility modifications requirements. For example, in the downstream process, the viral inactivation step using low pH is performed after an affinity capture step. Traditionally, the virus inactivation is performed once on a combined pool post the column step. Conversely, in this scenario, the inactivation steps are performed multiple times on each eluate pool that comes off the column. Post each inactivation, the product is then subsequently filtered and pooled. Sequencing of operations of this step was not aligned with the platform process in the commercial plant from a hardware or automation perspective. Changing the sequence of operations in the facility was identified as a process requirement during the tech transfer gap assessment and would require intrusive facility modifications to existing stainless steel pipework but additionally require new pipework to be fabricated to create new process routes to facilitate the specific sequence of these unit operations.

To streamline the scope of the tech transfer activities, the process development team worked in tandem with the manufacturing site team, identifying all facility constraints and then executing appropriate small-scale studies to move the process into a more platform direction. Sequence of individual unit operations was a key focus area downstream particularly around column steps and associated filtration unit operations. The original affinity capture step had a high and very narrow bed height range, which meant that large and costly volumes of resin would be required in the large 2M columns in the commercial facility. Process development teams conducted a number of small-scale studies that supported the use of smaller bed height and shorter residence times. In addition, they also modified some of the affinity capture conditions to enable material to be held for longer period, which allowed product pools to be combined for viral inactivation. Downstream processing previously demanded a lot more filtration capacity; process development teams executed small-scale filter fractionation studies focused on impurity removal together with filter capacity studies to support increased filter loading, reducing the requirement for additional membrane area or alternative filters. Upstream, process development teams executed small-scale studies that focused on simplifying media preparation stock solutions and developing more streamlined feeding regimes that eliminated the need to build bespoke automated recipes but meant that some steps were more manual for initial process introduction with an opportunity to automate at a later point. Given the many changes in the process, primarily changes that focused on enabling better facility fit in the new manufacturing facility, clinical material was generated at-scale with the new modified processes. The clinical material was introduced into a phase III bridging study to generate clinical comparability data to support the subsequent submission of the product. Enabling this opportunity to garner clinical data provided an opportunity to bridge between the original clinical batches and subsequent PPQ studies that will be executed at-scale and reduce any risk to commercialization.

A key enabler to the successful tech transfer of this late-stage clinical molecule was the excellent partnership between the process development and the commercial manufacturing site, each side sharing a common and aligned goal throughout the program. The partnership enabled transparent, open discussions on everything from facility fit, platform processes, process critical consumables and supply risks, and overall program risk tolerance to contingency plans and operational agility. Frank discussions enabled early investment in critical path items, and continuous and

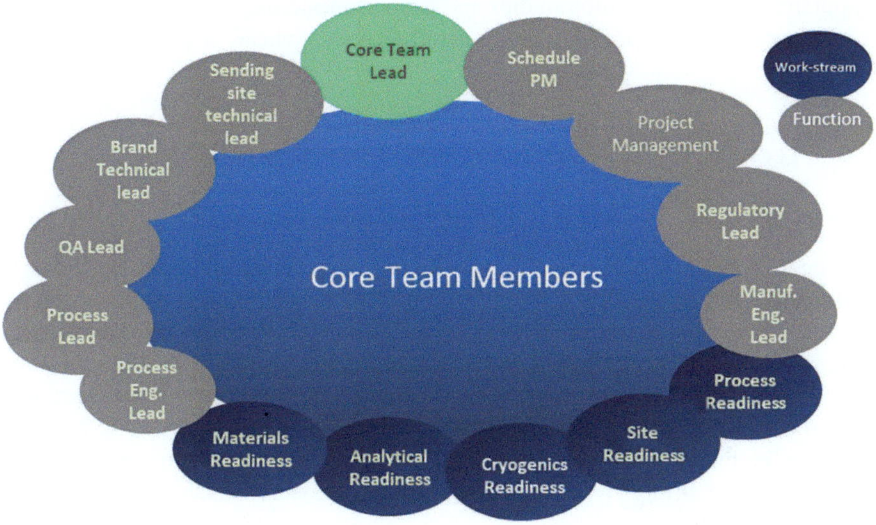

Fig. 18.8 Tech transfer core team structure

shared risk management enabled critical path tech transfer activities to be identified and prioritized between the teams without sacrificing quality demands. Throughout the program, a structured tech transfer framework was in place to support the cross-functional partnership. The structure incorporated a core tech transfer team (Fig. 18.8) and associated governance committee with defined roles and responsibility and support/escalation pathways for the tech transfer. Included in the tech transfer core team, was a diverse team of experts, which paid particular attention to ensuring process robustness and focused on achieving analytical comparability of the product from the new facility. The team also paid a lot of attention to regulatory requirements and constraints but also balanced this with the need for speed, agility, and appropriate risk. From the process development side, there was a single point of contact on the core tech transfer team that not only exercised oversight on technical project elements, but also served as clear decision-makers to ensure the programs consistently moved forward. Due to the non-platform nature of the process that was being transferred, the governance committee provided oversight, direction, and resourcing to enable process development studies to be executed to mitigate the need for intrusive and costly facility modifications. This support enabled teams to prioritize studies by focusing on those components that presented the greatest challenge in terms of facility operations such as increasing filter loadings that meet commercial facility hardware constraints and also reducing buffer flush volumes requirements to meet commercial facility capability. The process development team also focused studies on the key cost driving unit operations, such as the affinity column chromatography, and the modification to that step alone resulted in a 6% saving in overall program costs. A quality risk assessment as per ICH Q10 was executed to assess any residual risks associated with the process introduction into the commercial facility. To mitigate these risks, additional studies namely computational fluid dynamics studies and process modeling were executed to optimize fluid dynamics in process and preparation vessels ahead of any campaigns.

18.6 Tech Transfer of Drug Product Process to CMO Case Study

Working with external partners presents a unique challenge of adapting processes to fit into facilities that are meant to manufacture a wide variety of products. External transfers require a finesse to establish a balance between the donor site's knowledge of the product and the acceptor site's knowledge of their facility. Communication and collaboration between both parties are critical to the success of the project. Key considerations for drug product transfers focus on the unique drug product unit operations including component selection, aseptic processing, formulation, sterile filtration, aseptic filling, lyophilization, container closure, visual inspection, and bulk packaging.

The following case study focuses on a fusion protein transferred from an internal (donor) site to an external (acceptor) site. This is a commercial parenteral product in a lyophilized presentation. The transfer of a parenteral product from an internal donor site to an external partner starts in much the same way as a drug substance tech transfer. Core and extended teams were formed and the tech transfer was kicked off. The internal team consisted of technology transfer lead (TTL), subject matter experts (SMEs), manufacturing technology product owner, analytical, and project managers. The external team mirrored the internal team to ensure seamless information flow. A series of tech transfer meetings were initiated with both internal and external team members. The first step was to define the tech transfer strategy and perform a comprehensive gap assessment. The gap assessment identified clear areas of focus to define the direction for the project including media fill strategy, container closure system (CCS), sterile filter validation, filling technology, lyophilization cycle optimization, and visual inspection.

The container closure system (vials, stoppers, and seals) is an integral part of the parenteral product because it is qualified to maintain sterility, therefore machinability, equipment fit, and component availability are critical path items for a drug product technology transfer. For this transfer the primary container was changed from a custom-sized vial to the International Organization for Standardization (ISO) standard vial to enable faster timelines by utilizing change parts already available at our selected CMO. To qualify the container closure system, process parameters were challenged during the capping process and samples were tested for residual seal force (RSF) and container closure integrity (CCI). RSF is an indirect measure of the compression force exerted by the stopper on the vial land surface post crimping and gives an indication of capping parameter appropriateness. In this case, RSF indicated the parameters were not optimized as they were below the target force (indicated by the red dashed line in Fig. 18.9). A design of experiment was performed to

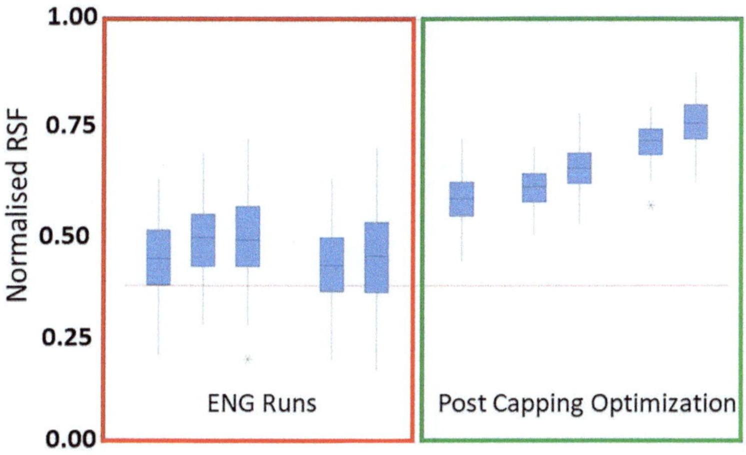

Fig. 18.9 The normalized residual seal force (RSF) profiles during engineering batches (prior to cap changes) and then RSF results post capping optimization work

adjust the capping parameters and RSF showed improvement (refer to Fig. 18.8) with all values above the target force, demonstrating successful parameter optimization and decreased risk of CCI failures.

Several programs make up sterility assurance for manufacturing parenteral products including cleaning, environmental monitoring, equipment/parts sterilization, and media fill/aseptic process simulation. CMOs typically have clearly defined bracketing approaches for qualifying a wide range of container closure systems (CCS) during media fill challenges. When using a system outside of the bracket, or new to the CMO, it may be beneficial to perform container closure specific media fill simulations prior to GMP manufacturing. For this case study, one of the components of the CCS (stoppers) was new to the CMO, therefore container closure specific media fills were performed to ensure the CMO was able to aseptically manufacture product using the defined CCS for this product.

The development phase of the technology transfer is important as it provides the opportunity to define the process capability and identify challenges at-scale prior to advancing into qualification phase. For this case study, challenges were encountered during the engineering runs while attempting to dissolve one of the excipients added during the compounding step. As a result of the challenge encountered during the engineering run, a step was added to pre-dissolve the excipient into a stock solution to optimize the drug product formulation step at the CMO. Figure 18.10 illustrates the process flow and highlights the step that was optimized.

Sterile filtration is one of the most important processes for parenteral manufacturing, and the filter validation package is a critical deliverable for the tech transfer. Key studies for filter validation include filter sizing, bacterial retention, compatibility, extractables/leachables, and filter integrity testing. Additionally, the potential for dilution during filtration, or adsorption of protein/excipients to the filter membrane must be evaluated with filter flush studies. In our case study, the filter flush

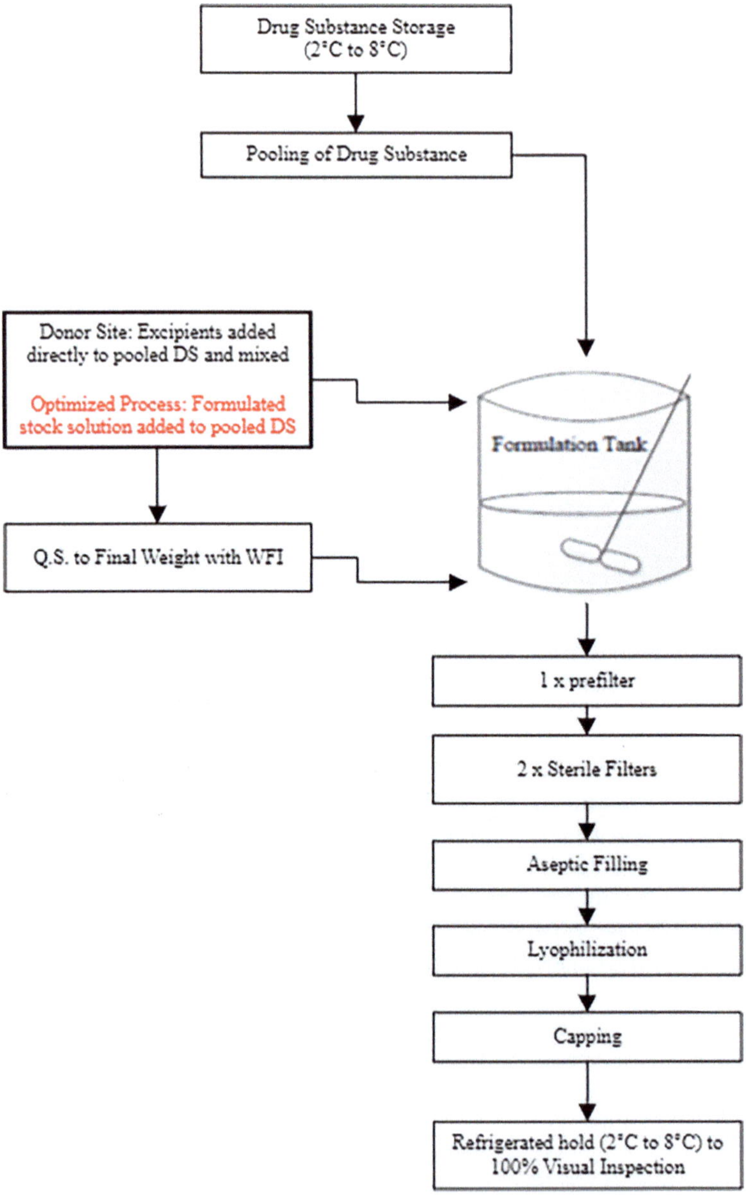

Fig. 18.10 High-level process flow for a drug product manufacturing process

characterization involved pulling sample fractions throughout filtration to monitor key parameters such as protein concentration, osmolality, and excipient concentration. Filter flush characterization for this process indicated an appropriate flush volume of 750 mL to ensure homogeneity (refer to Fig. 18.11).

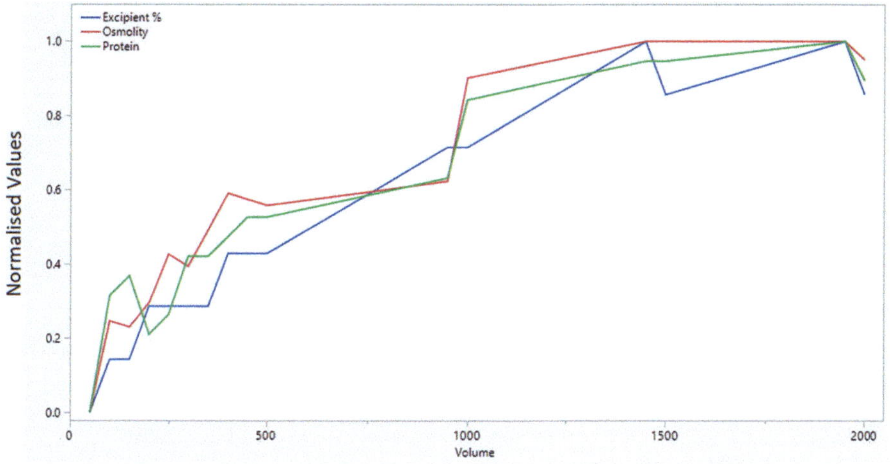

Fig. 18.11 Change in osmolality, protein, and %excipient as a function of filter flush volume during filter flush studies

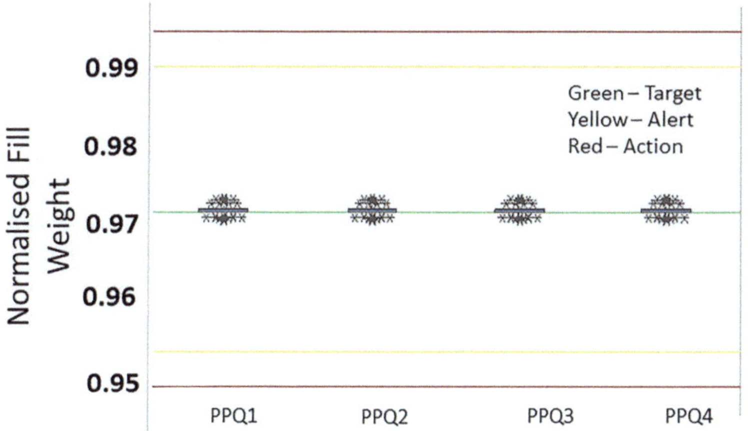

Fig. 18.12 Normalized fill weights for the four different PPQ batches

It is critical to characterize the filling process in order to have tight control over the fill weight delivered and ensure each container receives the appropriate dosage. Acceptable ranges are defined and tightly controlled during filling, which include overages to account for vial/syringe losses. Fill weight checks are performed at defined intervals during the filling process and the capability is monitored during characterization. Newer technology allows for 100% fill weight checks during routine filling. For this case study, though the filling mechanism at the receiving site was different from the sending site, the process was capable of robustly delivering the appropriate quantity of drug product (refer to Fig. 18.12).

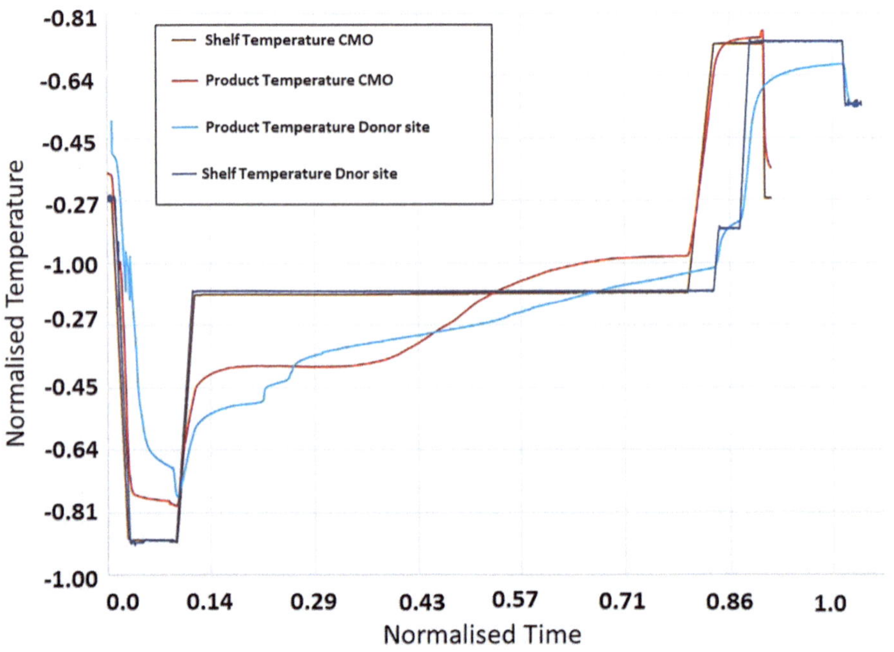

Fig. 18.13 Lyophilization cycle profiles at the company's internal drug product (DP) manufacturing site (donor site) at the CMO (receiver site)

When transferring a lyophilized product, it is key to understand the equipment capabilities of the acceptor site along with process knowledge from the donor site. In this case, the team used the technology transfer to a CMO as an opportunity to optimize the lyophilization cycle in order to increase efficiency and robustness of the parameters. Extensive development studies were performed both at lab and commercial scale to ensure the optimized parameters did not impact product quality (Fig. 18.13). Ultimately the optimized lyophilization cycle was successfully implemented at the CMO, reducing cycle time and increasing parameter robustness.

Parenteral products are 100% visually inspected. During the gap assessment a difference was identified between the donor and acceptor facility visual inspection program defect classifications. A key defect for the donor site was not captured in the receiving site process. The donor site TTL worked with internal SMEs to clearly define the defect and provide instructions to the receiver site. The receiver site then worked with their internal visual inspection SMEs to incorporate the new defect into their process.

Ultimately, the technology transfer of this fusion protein was a success. Drug product samples were tested with the previously established assays at a centralized analytical lab at the donor site and analytical comparability demonstrated that both the donor and acceptor site processes produced comparable drug product.

Additionally, the transfer resulted in approval of the new process with multiple regulatory agencies and the CMO is now successfully producing commercial product for patients.

18.7 Conclusion

Technology transfer is a complex multi-disciplinary activity that strives to balance speed, cost, quality, and regulatory compliance. At the outset of the tech transfer, implementation of a robust site selection process utilizing facility fit and engineering tools is the key enabler to ensure that the desired business outcome for any product tech transfer is achieved. Throughout the tech transfers many challenges can present themselves that require close collaboration, communication, and transparency to data and information to be successful. Here, effective project management is a critical enabler that ensures successful orchestration of multiple functions, risk assessment, and effective decision-making in a very complex and dynamic environment, which is especially important in transfers to external organizations.

Over the last three decades, organizations such as BMS have evolved and streamlined tech transfers by leveraging modeling software and other digital tools enabling more rapid and robust tech transfers. Adherence to platform processes, analytics, materials, and application of disposable technologies that negate the need for lengthy and often troublesome cleaning validation studies has enabled transfers in a matter of months in many clinical settings. However, in late-stage clinical programs moving into commercial facilities for the first time, such platform technologies are often absent in the manufacturing processes, presenting many costly facility fit considerations. Strong partnership between process development and the tech transfer teams and a focus on the technical development work have supported program acceleration. Keeping a strong focus on the individual process steps as highlighted in the drug product tech transfer in this chapter, understanding equipment capabilities in the manufacturing facility, and also understanding the supporting materials and analytic strategy are key considerations that can be applied in any tech transfer, particularly in emerging therapeutic areas that are relatively new to tech transfer.

Part X
Emerging Trends and Future of Biopharmaceuticals

Chapter 19
Emerging Biopharmaceutical Technologies and Trends

David W. Wood and Izabela Gierach

Abstract The biopharmaceutical industry is driven to innovate by the discovery of new therapeutics and modalities, as well as shifts in regulatory guidelines, market forces, and intellectual property law. Most recently, gene therapy, drug conjugates, chimeric antigen receptor (CAR)-T therapies, and a variety of fusion proteins have demanded effective manufacturing antibody methods that do not conform to established platforms. In addition, patent expiration on historic blockbuster drugs has driven a gold rush for biosimilars, which will also require more efficient and inexpensive production methods in a newly competitive market. At the same time, the need for process mobility, flexibility, and miniaturization has increased, with the goal of serving new patient populations under ambitious business growth models. Thus, it is useful to examine the new technologies that are being developed to address these needs. In this chapter, we will review several emerging trends and technologies, with some attention to the philosophical and economic drivers for their development.

Keywords Cell and gene therapy · CRISPR · Continuous biopharmaceutical manufacturing · Portable manufacturing · mRNA therapeutics · Accelerated process development

---

The original version of the chapter has been revised. A correction to this chapter can be found at https://doi.org/10.1007/978-3-031-62007-2_20

D. W. Wood (✉)
Department of Chemical and Biomolecular Engineering, Ohio State University, Columbus, OH, USA
e-mail: Wood.750@osu.edu

I. Gierach
Protein Capture Science, Columbus, OH, USA
e-mail: igierach@proteincapturescience.com

K. Gadamasetti, S. A. Kolodziej (eds.), *Bioprocessing, Bioengineering and Process Chemistry in the Biopharmaceutical Industry*,
https://doi.org/10.1007/978-3-031-62007-2_19

19.1 Introduction

Advances in biopharmaceutical technologies are closely tied to public health needs and are driven and enabled by new scientific discoveries and innovations in process and product development. Financial resources are also critical for the development of innovative medications, manufacturing platforms, and supporting technologies, and therefore inevitably drive innovation toward profitability. At the regulatory level, the Food and Drug Administration (FDA), European Medicines Agency (EMA), and other international agencies establish requirements to assure clinical safety and efficacy, where regulatory successes and failures provide precedents to inform the development of new approaches. Additional, albeit less obvious industry drivers include increased public access to scientific and clinical information, ease of scientific communication across the globe, and the expansion of biotech-relevant databases. Implementation of artificial intelligence, the Internet of Things, process simulation, and integrated digital processing has further enabled rapid prototyping and cross-team communication.

The drugs that came out of the early biopharmaceutical industry were typically human or human-like globular proteins that could be produced using established platform methods. Also observed were high development costs, where dedicated production facilities were often being built as each drug worked its way through the approval process. A bad clinical trial during late development could spell disaster for even the most established multinational companies, as in the more recent case of Biogen's aducanumab treatment for Alzheimer's disease. In this case, Biogen halted to two Phase 3 clinical trials in March 2019, sending their stock price down by over 33% practically overnight. These inherent development risks produced a highly conservative industry, where commonly large patient populations, high selling prices, and established development and approval pathways became the norm. Innovation was mainly limited to identifying validated drug targets for the development of new treatments, which typically arose from federally supported basic research in academic institutions and startup companies. Although production processes over the last 30 years have become more efficient, most biologics on the market are still expressed in one of only a handful of established hosts and are purified by a series of well-established chromatographic steps in dedicated facilities. While new drug targets and disruptive therapies have been discovered and developed over this time, truly disruptive innovations in biopharmaceutical production processes have been rare, despite changing pressures on the industry.

The discovery of new approaches to medicine, including cell and gene therapy, as well as engineered fusion proteins and complex drug carriers have challenged the status quo. Today, modified cells can be used with high precision to treat genetic disorders, kill cancer cells, or possibly rebuild damaged organs. The direct administration of mRNA that encodes a desired therapeutic protein has now been established with the Coronavirus disease of 2019 (COVID-19) vaccine. As with early biopharmaceuticals, the potential rewards for successfully developing these therapies are boundless, yet these vehicles and modalities are often not amenable to current production platforms. At the same time, patent expirations on early blockbuster drugs have led to dozens of new and established players to enter the biosimilars market, where development costs and risks are greatly reduced, but production costs are now critical in a newly competitive market. Societal pressures, including greater access to medicines for underserved populations, rapid development of new drugs for smaller patient populations, and even green and sustainable bioprocessing are having a greater influence on the industry as well. Taken together, these trends have significantly increased the need for new platforms and paradigms in bioprocessing, which has resulted in new and disruptive processes for making new and existing classes of products. This chapter introduces a few of the high-profile changes taking place in the industry, both in terms of new treatments and modalities and new manufacturing approaches. This list is far from being comprehensive, and additional chapters in this volume provide a higher-resolution view of many new technologies. Our primary goal here is to introduce these advances and provide some context for their development.

19.2 Novel Gene and Cell Therapies

19.2.1 Gene Therapy

Gene therapy as a possibility to treat inherited diseases surfaced as a concept about 50 years ago [1]. Today, gene therapies refer to in vivo and ex vivo methods that result in a modification of cellular DNA to prevent, treat, or cure infectious or inherited diseases and cancer. There are several techniques used today to introduce a new/modified gene, inactivate a defective gene, or replace genes. Among them are ex vivo techniques, such as genetically modified cell-based immunotherapies, where a patient's cells are harvested, grown, activated/genetically modified externally, and infused back to the patient's body (e.g., chimeric antigen receptor (CAR)-T cell therapies, T-cell receptor (TCR) therapies, natural killer (NK) cell therapies, and sickle cell disease therapies) [2]. There are also in vivo methods, which rely on delivery of a therapeutic gene to humans via bacterial vectors, non-viral vectors (e.g., nanoparticles), and viral vectors used to "encapsulate" therapeutic genes and transfer them into human cells (e.g., adeno-associated viruses [AAVs], adenovirus, lentivirus) [3–7]. At the same time, gene-editing methods can be used to repair or mutate existing human genes using CRISPR (clustered regularly interspaced short palindromic repeats) technologies to treat patients [8, 9].

As is obvious, the ability to modify genomic DNA for therapeutic effect has enormous potential, and a large number of companies are developing therapeutics in this space [10]. Several challenges remain, however, which largely center on difficulties in manufacturing and remaining concerns about safety. In the manufacturing space, the vehicles to effectively transfer therapeutic DNA to the cell nucleus are large and complex [11, 12]. Many approaches involve viral vectors, with lentivirus being preferred for ex vivo applications [3, 13], and either adeno-associated virus (AAV) or nanoparticles being preferred for in vivo applications [14–16]. Among the main challenges that remain are the ability to purify viral vectors away from other cellular products, including exosomes and other vesicles [17], as well as the empty/full problem of selectively purifying only viral particles that contain the expected therapeutic DNA or RNA [18, 19].

The safety of gene therapy approaches has a mixed past, where some significant adverse events have slowed progress in this area [20–22]. In the case of Jesse Gelsinger, a volunteer with mild case of ornithine transcarbamylase deficiency, the patient died shortly after receiving a gene therapy treatment, likely due to an unexpectedly acute immune response to a recombinant adenoviral vector containing the corrected gene. The publicity surrounding this case left a shadow on gene therapy that would last for decades. Perhaps more distressing was the case of the retroviral treatment of children for X-linked severe combined immunodeficiency (X-SCID), where the treatment was rated a success before all five pediatric patients later developed T-cell acute lymphoblastic leukemia (ALL). Although subsequent research identified the causes for these events, it is clear that gene therapy still presents some significant safety concerns. Even within the last 2 years, adverse events have taken the lives of additional children and have halted otherwise promising clinical trials [23].

Despite these risks, the potential positive impact of gene therapies continues to drive their development, and early successes have been achieved. After tremendous medical, clinical, and research effort, the first gene therapies were FDA approved for ocular and muscular diseases. Currently, the largest obstacle for most of the patients is still the cost of gene therapy. For example, a one-time treatment using Zolgensma® costs $2,125,000, which was the most expensive single treatment ever seen back in 2019, although it could be financed over a 5-year period [24]. Indeed, as with most biopharmaceuticals, it may be the willingness of insurance companies and the federal government to cover patient costs that ultimately determines what drugs are commercialized.

19.2.2 CRISPR Technology

The revolutionary Clustered Regularly Interspaced Short Palindromic Repeats (CRISPR) technology is often referred to as genetic "scissors," which allows high-precision in vivo and ex vivo genome edits. The CRISPR-Cas9 system contains two main components, a CRISPR-associated protein known as Cas (e.g., Cas9 and Cas12 enzymes) and a guide RNA, which are typically delivered ex vivo using lentiviral vectors. Among the new CRISPR-based companies are CRISPR Therapeutics and ERS Genomics. CRISPR Therapeutics' mission is to develop gene-based therapies for a variety of diseases such as cancer and diabetes. Additionally, CRISPR Therapeutics is working with Vertex Pharmaceuticals on CTX001, designed to be used for patients with severe hemoglobinopathies [25]. ERS Genomics is focused on bringing global access to the CRISPR-Cas9 technology by providing non-exclusive licenses for research, and sales of products and services to other companies and researchers. The advantage of CRISPR technology is that it can be used for a wide range of applications such as editing the human germline, development of animal therapeutics, creation of modified strains for agricultural applications (e.g., food and feed for livestock), and strain optimization for commercial manufacturing of other products [9, 26–28]. There are also those who argue that the increased precision of CRISPR methods can lead to greater safety in gene therapy applications.

Mammoth Biosciences is focused on the development of diagnostic tests using CRISPR-Cas12, including tests for COVID-19 (in collaboration with Sherlock Biosciences) [29–32]. In 2020, the FDA issued an Emergency Use Authorization (EUA) for Mammoth Biosciences' SARS-Cov-2 DETECTR BOOST™ test that can be used only in authorized labs. Mammoth Biosciences partnered with MilliporeSigma (a contract manufacturer of the kits) and Agilent Technologies (a co-marketer) to implement Agilent's automated liquid handling systems into lab workflows for rapid testing for severe acute respiratory syndrome coronavirus 2 (SARS-Cov-2).

Currently, CRISPR technologies are being developed and utilized by other small companies, which are partnering and licensing the technology with well-established biopharmaceutical companies. As an example, Precision BioSciences is working in collaboration with Eli Lilly to use the genome editing platform ARCUS® to develop new therapies for diseases such as Duchenne muscular dystrophy [33]. Additionally, Intellia Therapeutics is collaborating closely with Regeneron and Novartis to apply CRISPR-Cas9 technology to develop new medicines to treat immuno-oncology and autoimmune diseases (e.g., sickle cell disease and acute myeloid leukemia). It is also focused on the treatment of genetic diseases such as transthyretin amyloidosis caused by accumulation of misfolded transthyretin protein. One of these, NTLA-2001, is in Phase I clinical trials now [34, 35]. Finally, Beam Therapeutics is working on new medicine development utilizing several gene-editing technologies, including CRISPR-Cas [36, 37]. A more extensive review of CRISPR-Cas technology itself is included in an additional chapter of this book.

19.2.3 Cell Therapy

Cell therapy combines a designated cell line, which can include T-cells, dendritic cells, or natural killer cells, with engineered receptors to boost their effectiveness and selectivity for killing cancer cells. Currently, there are several therapies under development, including tumor-infiltrating lymphocyte therapy, engineered T-cell receptor therapy, CAR-T cell therapy, and natural killer cell therapy. Tumor-infiltrating lymphocyte therapy relies on harvesting naturally occurring T-cells from the patient's blood, which are actively fighting cancer cells, growing them in an external bioreactor, and activating them ex vivo. Large volumes of T-cells are then re-infused back into the patient as part of therapy to enhance the patient's existing cancer-killing capability.

If natural T-cells require a boost, a patient's T-cells can be modified via ex vivo gene therapy to add an engineered chimeric antigen receptor, or CAR, to target specific cancer antigens. The resulting CAR-T cells are highly effective in treating blood cancers, and there are many re-engineered CAR designs that have been tested in various clinical and preclinical research programs. Over time their structure expanded from three (see domains 1, 3, and 4) to five main components, depending on their developmental generation: (1) *binding domain/extracellular antigen recognition domain* (e.g., single-chain variable fragments [scFvs], cytokines, ligands, nanobodies, and other peptides), (2) *hinge* (e.g., CD8, IgG1), (3) *transmembrane domain* (e.g., CD3ζ, CD4, CD8α, CD28, ICOS), (4) *intracellular signaling domain(s)* containing co-stimulation (e.g., 4-1BB, CD27, CD28, ICOS, OX40, KIR2DS2, MYD88-CD40), and activation fragments (e.g., CD3ζ), and (5) *armor* (e.g., bispecific T-cell engagers [BiTEs], cytokines and chemokines, monoclonal antibodies [mAbs], scFv, and other peptides or enzymes). To better understand the evolution of CAR technology, the generational differences are shown in Fig. 19.1 [38, 39].

The engineered cells are able to express specific surface receptors that play a role in a "disease-fighting" mechanism [38, 40, 41]. Since only small amounts of CAR-T cells can be modified, they must be allowed to multiply and grow in a lab using an external bioreactor. The CAR-T cells are then washed and formulated (e.g., with dimethyl sulfoxide, albumin human and/or dextran 40) [42]. Finally, the CAR-T cells can be infused back to the patient so they can recognize and attack cancer cells. The rapid response of the CAR-T cells may result in cytokine release syndrome, which is a large release of cytokines into the bloodstream, resulting in overreaction

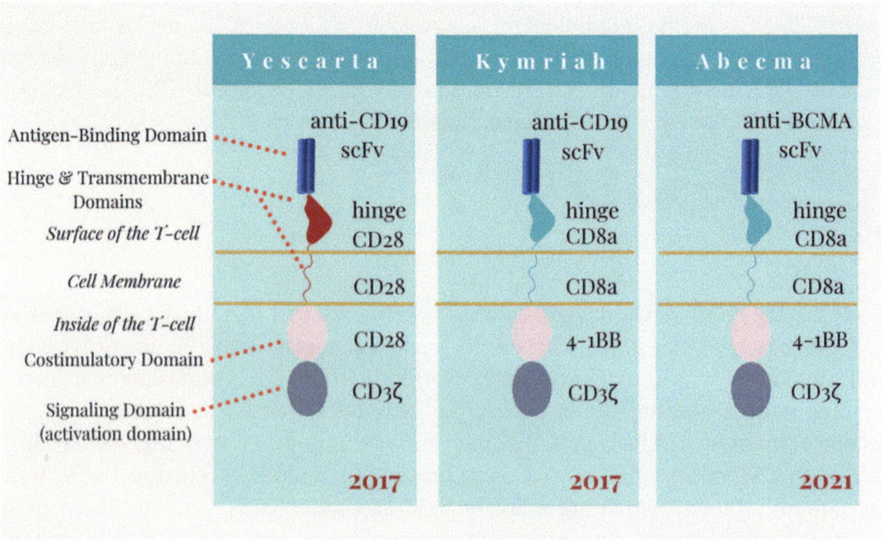

Fig. 19.1 Evolution of FDA-approved CAR-T technology over past several years [38, 39]

of the immune system, acute systemic inflammation, and multiple organ dysfunction. These potentially lethal side effects are in addition to the literally pounds of dead cancer cells that the body needs to gradually clear, requiring careful monitoring of patient progress throughout the treatment. Currently, the preparation of engineered CAR-T cells takes a couple of weeks for each patient, where the parent cells are derived from the individual patient receiving the therapy.

Originally, Dr. Carl June led a research team to develop CAR-T therapy at University of Pennsylvania medical school, followed by clinical trials initiated at Penn Medicine Hospital in 2010. Thanks to an extensive collaboration with Novartis, the first CAR-T therapy, Kymriah (tisagenlecleucel) was approved by the FDA in 2017 [43]. Kymriah is manufactured by Novartis, and is a treatment for relapsed/refractory (R/R) B-cell acute lymphoblastic leukemia (ALL). Kymriah therapy uses lentiviral vectors to insert CAR genetic material into isolated patient T-cells, causing the CAR to be expressed on the surface of the T-cells. The CAR function is specifically to recognize CD19 receptors on the surface of B-cells and bind to them, resulting in initiation of cancer cell death via the action of the expressing T-cell.

The CAR-T development story illustrates the importance of financial support in developing disruptive breakthrough technologies. In this case, several previous attempts had been made to develop successful CAR-T therapies for solid tumor diseases such as ovarian cancer and neuroblastoma. These attempts unfortunately failed due to low efficacy of CAR-T against solid tumors, and proposals to support Dr. June's work with blood cancers were therefore rejected by the National Cancer Institute. Therefore, Dr. June's first clinical trial was financially supported by a private foundation, the Alliance for Cancer Gene Therapy. In this case, the patient had chronic lymphoblastic leukemia (CLL) and was not responding to other treatments, but showed complete cancer remission after one round of CAR-T therapy. He passed away nearly 11 years later due to COVID-19 infection in 2021. Since FDA approval, Kymriah is available to children and young adults (25 years old or younger) with ALL cancers in 27 countries (mainly Europe, the USA, and Japan) across 290 dedicated treatment centers, and its worldwide sales reached $141 million by the end of 2020 [44]. Its success rate is significantly higher for ALL cancers compared to CLL, and it is also approved to treat adult R/R diffuse large B-cell lymphoma (DLBCL) [45].

A second therapy, Yescarta (axicabtagene ciloleucel, Kite Pharma), was also approved by the FDA in 2017 for large B-cell lymphoma, which was quickly followed by Kite Pharma's acquisition by Gilead Sciences the same year [46]. Currently the FDA has approved several CAR-T therapies, mainly for lymphoma and leukemia. In March 2021, the FDA also approved Abecma (idecabtagene vicleucel, BMS) [47]. It is the first cell-based gene therapy for adult patients with R/R multiple myeloma. Abecma can recognize and selectively bind to the B-cell maturation antigen (BCMA) expressed on multiple melanoma cancer cells, which results in cell death.

It is important to note that cell therapies under development are designed for increased safety through several different approaches, including the implementation of a safety "off" switch controlled by a small molecule. In one case, a chemical inducer of dimerization (CID) such as rimiducid would bind to inducible Caspase-9, resulting in dimerization of iCasp9, followed by Caspase 3, 6, 7 activation and cell apoptosis (e.g., CaspaCIDe® technology by Bellicum Pharmaceuticals) [48–50]. In contrast, the GoCAR-T® technology is based on the ability to switch "on" T-cell activation by using rimiducid [2]. In this case rimiducid triggers dimerization of inducible MyD88/CD40 (iMC), while a simultaneously expressed extracellular single-chain variable fragment antibody domain (linked to a single T-cell receptor signaling domain, CD3ζ) would specifically bind to an antigen of cancer cells, resulting in cancer cell death [48, 51–54]. This type of technology is now being tested against solid cancers.

19.3 Novel RNA-Based Therapies and Vaccines

19.3.1 mRNA Therapies and Modes of Action

Today, several RNA therapies have been approved by the FDA. These include the first ever antisense RNA (single-stranded RNA; ssRNA) drug Fomivirsen® (Vitravene) approved in 1998, the RNA aptamer drug Pegaptanib® (Macugen) approved in 2004, and the small interfering RNA (siRNA; double-stranded RNA, dsRNA) drug Patisiran® (Onpattro) approved in 2018 [55]. In late 2021, one of the mRNA vaccines for COVID-19, BNT162b2 by Pfizer-BioNTech, became FDA approved after an initial EUA granted by the FDA and EMA [56].

RNA is a nucleic acid, which as an RNA aptamer can block protein function or as an mRNA can be translated into a protein, triggering expression of that protein. RNA may present other different functions as well. For example, ssRNA can specifically bind to cellular mRNA transcripts to modulate splicing or induce degradation of that specific mRNA. Similarly, siRNA can bind to mRNA and result in mRNA cleavage and effective gene silencing. Existing RNA therapies are used to treat many different medical conditions, including cytomegalovirus (CMV) retinitis, an infection of the retina caused by CMV (Vitravene®) [57]. Some RNA therapies are used to treat genetic diseases. Among them are Duchenne muscular dystrophy, an inherited progressive weakness and degradation of muscles caused by mutations of gene encoding dystrophin (Exondys 51® and Vyondys 53®), and hereditary transthyretin amyloidosis, a build-up of amyloid in human tissue (Onpattro® and Tegsedi®) [58–60]. RNA therapy can also be used to treat a metabolic disorder known as acute hepatic porphyria caused by a deficiency of hydroxymethylbilane synthase production in the body (Givlaari®) [61].

19.3.2 mRNA Vaccines

In the last 2 years, a leap in vaccine innovation was triggered by the COVID-19 pandemic, leading to the development and approval of the world's first mRNA vaccines. These vaccines deliver directly to human cells a "short-living" mRNA that encodes a fragment of the COVID-19 spike protein, resulting in the production of the antigen and an induced immune response. The "prototype" of the mRNA-1273 used in the Moderna COVID-19 vaccine was made in 42 days from selected viral sequences, while the development time for the creation of a useful vaccine was shortened from 10–15 years to a couple of months [62]. During the pandemic, manufacturing capabilities of vaccine and antibody development companies were also accelerated in a way that has never been seen before.

As a result of these heroic efforts, the first mRNA vaccines received an EUA at the end of 2020. The BNT162b2 vaccine, also known as Comirnaty® by Pfizer-BioNTech, showed 95% primary efficacy in phase 3 compared to 94.1% for the mRNA-1273 by Moderna [63–65]. The EUAs were granted on December 11 and 18 of 2020, respectively, with full approval of the Pfizer vaccine in August 2021 and the Moderna vaccine, Spikevax™, in January 2022.

Typical clinical development phases for vaccines take 5–7 years [66]. However, once the World Health Organization declared COVID-19 a pandemic on March 11, 2020, an emergency authorization for Pfizer-BioNTech's mRNA vaccine was granted in the UK after only 266 days (see Fig. 19.2) [67, 68]. In phase 3 clinical trials, over 160 sites in 6 countries agreed to participate in the NCT04368728 clinical trial, including Argentina, Brazil, Germany, South Africa, Turkey, and the USA [69]. As a result, the clinical trials for these vaccines were completed in record time, allowing their authorization in time to save countless lives.

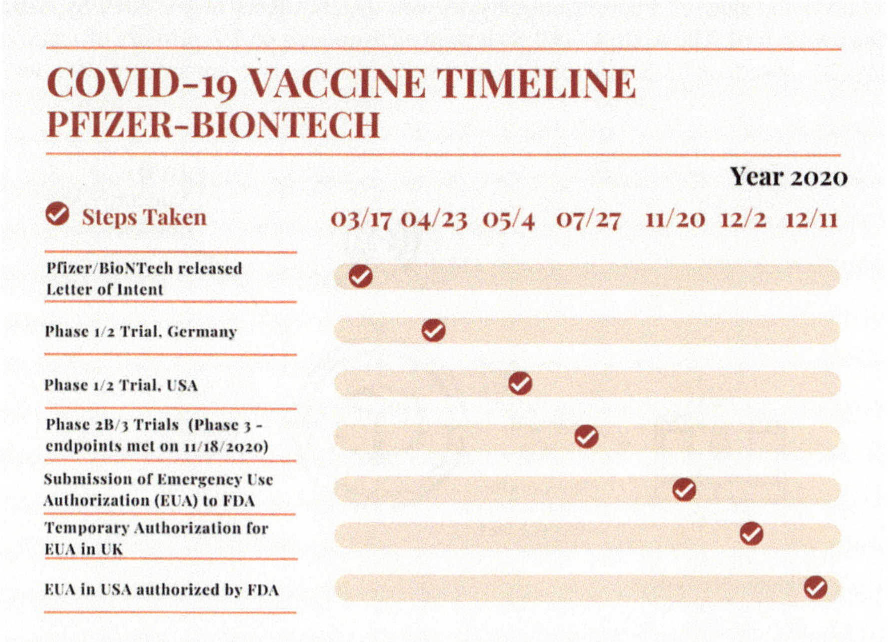

Fig. 19.2 Breaking world records in vaccine production and its delivery to patients by Pfizer-BioNTech [67, 68]

The successful speed into the market relied heavily on the agility of the supply chain, which allowed rapid manufacturing, multidose-vial packaging and labeling, and distribution of vaccines to clinical sites and patients. Moreover, it was critical that BioNTech partnered with Pfizer, which is not only the top biopharmaceutical company in the world, but also one of the largest dry ice manufacturers (~10 tons/day at the production peak). Storage and shipping conditions were also critical to prolong vaccine activity. Based on initial data, frozen vials prior to use have a six-month shelf-life when stored at −80 °C to −60 °C (−112 °F to −76 °F). However, during transportation, undiluted and thawed vials have up to only 12 hours storage time at temperatures between 2 and 8 °C (35–46 °F). The vials with vaccines are recommended to be used within 2 h after dilution when they are stored between 2 and 25 °C (35–77 °F) [70].

The tremendous success of the vaccine launch would not be possible without operational excellence and the emergence of Industrial Internet of Things (IIoT) solutions, supported by monitoring of vaccine shipment and temperature using GPS devices and radio-frequency identification (RFID)-based monitoring for tracking and authentication of vaccines during shipment (Fig. 19.3) [71, 72]. At the same time, parallel activities had to be initiated to accelerate COVID-19 vaccine development and fillings.

It is important to note that other vaccines for COVID-19 were also selected from a short list of six recommended options. Among them were non-replicating viral vector-based vaccines, Ad26.COV2.S (other name: JNJ-78436735) by Johnson & Johnson and Janssen Pharmaceuticals, as well as CHAdOx1 nCoV-2019 by Astra Zeneca-Oxford. The Ad26.COV2.S showed an estimated 66.1% primary efficacy at day 28, based on infection of 66 vaccinated volunteers compared to 193 who

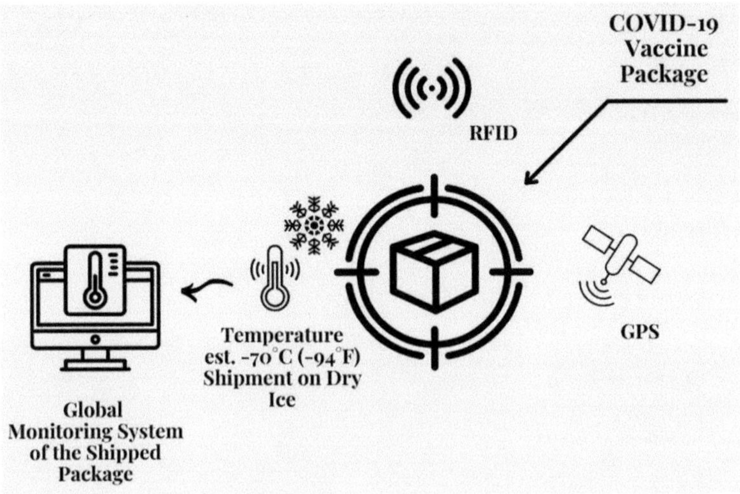

Fig. 19.3 Modern monitoring of transported vaccines in trackable and temperature-controlled packages [71, 72]

received a placebo in one study (after 14 days the primary efficacy was 66.9% based on 116 infections among vaccinated participants versus 348 for placebo) [73]. The EUA for this vaccine was granted on February 27, 2021. The CHAdOx1 vaccine is a Chimpanzee Adenovirus encoding the SARS-CoV-2 spike glycoprotein, also known as AZD1222 [74]. This vaccine showed that after two doses, the efficacy at preventing symptomatic COVID-19 after at least 15 days was 76% [75]. Two other vaccines, NVX-COV2373 (Novavax), made using nanoparticle technology and Matrix-M™ adjuvant, and the mRNA vaccine MRT5500 (Sanofi-Translate Bio), were in clinical trials as of summer 2021 [76–78].

19.4 Trends in Biotech Manufacturing Modalities

An ideal path for accelerated drug development and production is possible if unified platform(s) are used early on with processes that are flexible, cost-effective, agile, automated, and to some extent parallel. At the same time, there is a strong interest in developing semi-continuous, continuous, and non-chromatographic methods for manufacturing. To this day, most processes for downstream operations are chromatography-based, and it is important to note that chromatographic methods are specific to the type of protein(s) that are purified. For example, a wide variety of Protein A (ProA) resins can be used to purify monoclonal antibodies (mAbs), while proteins that do not contain the Fc-domain are typically purified via various complex methods optimized for each protein of interest. Today, mAbs are purified using different types of ProA resins, which are packed in pre-packed chromatographic columns of widely varying size from well-recognized companies like Repligen (OPUS®, column volume 50 μL up to 150 L) or other vendors like MilliporeSigma and ThermoFisher. Since mAbs can be engaged to fight many different diseases [79, 80], they attract high interest among biopharmaceutical companies, which is strengthened by well-established manufacturing processes as production titers are still pushed to even higher limits. These advantages have led to over 100 mAb-based medications out of a total of ~400 FDA-approved protein-based biopharmaceuticals by the end of 2020 [81].

19.4.1 The Rise of Continuous Processing

In the past, semi-continuous unit operations gradually replaced portions of established batch and fed-batch processes, and there is high interest in the further development of continuous processes for biopharmaceutical production [82]. Automation and an increase in single-use process technology are now key factors in the push toward process standardization and adaptability. A big advantage of continuous manufacturing is lower cost and a smaller footprint of the production site.

As an example, in previous years, the innovative iSKID™ was introduced as a next-generation continuous manufacturing platform for mAbs by Pfizer and Boehringer Ingelheim, with the goal of accelerating the manufacturing of protein therapeutics [83]. The iSKID™ is a fully automated, controlled continuous/periodic stage/batch hybrid, disposable system, and concentrated buffer and media. It addresses various disadvantages of conventional processes, including poor Protein A resin utilization during clinical manufacturing, long in-process holds, large media and buffer preparation areas, isolated production stages and teams, and overall a large process footprint in the manufacturing facility. The iSKID™ allows controllable, integrated manufacturing that scales from 100 L (prototype) to planned 2000 L with a smaller footprint (initially estimated to be 5 m × 9 m for smaller scale) than previous facility layouts, while preserving high efficiency, and results in maximized productivity. The process itself utilizes a non-steady state perfusion bioreactor. Medium containing expressed mAbs is passed onto two Protein A columns working in tandem to continuously purify the target, followed by elution, continuous viral inactivation at low pH, and an anion-exchange polishing step run in flow-through mode. The purified biopharmaceutical then undergoes additional steps that include single-pass tangential flow filtration followed by a single-use mixer, viral filtration, and ultrafiltration/diafiltration, formulation, and the fill/finish step. Initially tested mAbs showed volumetric productivity of 1–4 g/L/day, which is a 3–4× increase in productivity over a typical conventionally optimized fed-batch process.

An end-to-end integrated and continuous bioprocessing (ICB) proof-of-concept for mAb production was closely analyzed by development teams from Sanofi, Moderna, Novartis, and Pfizer [84]. The main idea was to highly automate the process itself with minimal hold-up time in the process and to scale down overall unit operation volume. An estimated ~20-fold reduction in overall vessel volume was achieved compared to a typical fed-batch process. The initial tests of the process showed a total of 4.9 kg of drug substance production from a 100 L perfusion bioreactor (BRX) in 25 days. The overall product quality was within acceptable limits, although the ~60-day campaign results that compared six test batches showed some disturbances of the process (e.g., related to product concentration and osmolality variations), which would need to be further investigated and optimized in the future. However, drug substance purity (over 95%), pH (between 5 and 5.4), charge isoform profile, host cell protein level (~2 ng/mg), and high molecular weight species levels (below 0.4%) were relatively steady.

19.4.2 *Rapid Cell Line and Process Development Approaches*

WuXi Biologics has recently developed a process known as WuXiUP™ (Ultra-high Productivity). It is based on an optimized 20-day continuous cell culture bioreactor process that results in a very high titer of 51 g/L compared to a typical fed-batch process with ~6 to 11 g/L. This process can be scaled up as well [85].

Since WuXi Biologics is working on bispecific antibodies for various diseases as well as neutralizing antibodies against COVID-19, their chemistry, manufacturing, and control (CMC) work includes cell line development, formulation, purification process development, and toxicology studies followed by Good Manufacturing Practice (GMP) studies (stability study and manufacturing). It is important to note that their development time is significantly shortened from typical 12–18 months to 3–6 months [86]. To achieve this, early processes simply overlap and run in parallel. Some of the processes, such as tech transfer, are planned and performed earlier than in a conventional development campaign. WuXi Biologics also implements high-throughput systems and platform processes to accelerate overall development and testing. To optimize clone selection, WuXi Biologics uses artificial intelligence to analyze genomic data. As a result, new biopharmaceuticals are approved faster while the development cost is lower. For example, during the pandemic, the development of anti-COVID-19 antibodies from DNA to IND (Investigational New Drug) was finalized in 4.5 months (instead of 12–18 months). At the end of 2020, WuXi Biologics was working on 10 anti-COVID-19 antibodies [87].

19.4.3 Portable Manufacturing

Today, GMP mini-manufacturing processes can also fit into portable pods to assure flexibility and adaptability of biopharmaceuticals development, which can literally be dropped into locations that are closer to patients [88]. At the extreme side in the push for miniaturization, the U.S. Defense Advanced Research Project Agency funded prototype development for a briefcase-sized biomanufacturing platform for military use, which included collaborations with industry and academic institutions [89, 90]. Besides miniaturization, disposable single-use perfusion bioreactors and multicolumn continuous chromatography systems with disposable flow paths have gained interest. Single-use bioreactors for mammalian cell culture are also becoming more popular due to their advantages, such as flexibility, portability, reduction of cross-contamination, overall lower cost, lower energy consumption, and lack of cleaning requirement, resulting in less overall chemical usage. For the last few years, innovation in fill and finish processes led to the introduction of smaller and flexible filling machines that are automated with standardized aseptic enclosure with minimal human interaction (e.g., Vanrx's robotic SA25 Aseptic Filling Workcells) [91].

19.5 Discussion and Future Trends

It is important to note that innovation in biopharmaceutical technology is also influenced by a well-developed biotech ecosystem with vendor support, external investors, governmental financial support, and public health needs in each particular country. Over time, innovative platforms and novel therapies coming out of startup accelerators or academic labs commonly feed into larger corporations like Pfizer or Merck via M&A, licensing agreements, co-partnerships, or joint ventures after successful clinical trials or FDA and/or other regulatory agency approvals. In 2020 alone, the total number of deals taking place through partnerships, joint ventures, and licensing agreements increased 86% compared to 2019 and reached $170.6 billion (based solely on 26% of disclosed deals; the real number is likely higher). The number of companies taking the initial public offering (IPO) route for fundraising also increased dramatically (186% higher in 2020 compared to 2019), equating to $34.3 billion in raised capital. Globally, venture capital (VC) companies funded 45% more biotech companies over the last year with a cumulative funding equal to $36.6 billion [92]. Due to higher investor interest in financing biotech companies, three particular areas became dominant within pre-commercial biopharmaceutical development. These are mAbs (13% increase in 2020 over 2019), cell therapy, and gene therapy (~24% and 20% increase, respectively). Four other growing areas of the biopharmaceuticals market are therapies based on oligonucleotides, RNAi, mRNA, and antisense drugs. As expected, the record increase in interest was predominantly in the mRNA therapy development space (47% increase). There were also nearly double the number of deals performed within the gene and cell therapy space, showing increasing trust in those technologies compared to the previous 4 years. These have mainly been licensing deals, corporate acquisitions, and strategic R&D collaborations [93]. A record 121 corporate acquisitions valued below 30 *billion*, 1*mega* – merger (*over* 30 billion), and 367 product acquisitions were performed within biopharma in 2020 alone, a level that has not been seen over the past 10 years. Among these were Gilead's acquisition of Immunomedics ($21B), AstraZeneca's acquisition of Alexion ($39B), and J&J's acquisition of Momenta ($6.5B).

Platforms for making new antibodies "on demand" are also very attractive, especially due to the rapid development of new biopharmaceuticals with high specificity, precision, and accuracy. However, even during the pandemic, we have seen only a handful of companies that were able to put forth an agile response to the public health emergency. Among those companies that put forth a good initial response were Regeneron (Casirivimab/Imdevimab), Gilead (Remdesivir or combination of Baricitinib and Remdesivir), and Eli Lilly (Bamlanivimab and Etesevimab).

The impact of innovative biopharmaceutical technologies cannot be better measured than through outcomes seen in patients and the ability to select them with more accuracy and precision for particular clinical trials. Therefore, an additional trend that is growing, especially in the USA, is a great effort from many organizations (e.g., FDA, National Institutes of Health (NIH), National Cancer Institute (NCI), and Centers for Disease Control and Prevention (CDC)) overseeing biopharmaceutical development and innovation, and tracking public health to better communicate their findings, progress, and clinical trials' opportunities to the public. Since 1993, MedWatch, FDA's Safety Information and Adverse Event Reporting Program, was introduced to voluntarily communicate any safety issues and adverse effects to the public [94, 95]. In 1996, PubMed was launched as a free search engine of scientific and medical research information, providing access to knowledge of current innovations and exploratory research findings, including biopharmaceutical development and biotech processes [96]. In 2006, DailyMed was launched online. It is updated daily and as of August 2021 it contains 140,261 drug entries provided by FDA, ensuring prescription data are up-to-date and publicly available [97]. Moreover, there is a trend to decentralize clinical trials and to use digital technology for remote monitoring of patients.

To conclude, there is no doubt that there will be more interest in gene-based therapies and RNA-based vaccines due their rapid development and ability to cover wide-range indications, but mAb-based and conventional therapeutics will continue to be developed. However, due to a large number of proteins in human body, there is still a gap in fully understanding many proteins, proteins-interactions, and its involvement in diseases development or their potential as a treatment on a basic molecular biology level. Therefore, it is expected that in near future there will be many opportunities to develop new protein-based biopharmaceuticals and multiple-cells-targeted therapies. Many medical conditions still need better treatments, including neurogenerative diseases, psychiatric therapies, and aging-associated diseases. At the same time, there is a need to reduce the cost of biopharmaceuticals so they can be made available to low-income and developing countries. These challenges will be met with innovation at all stages of drug discovery, development, and production, which will continue to provide new treatments for all human and animal diseases.

References

1. Friedmann T, Roblin R (1972) Gene therapy for human genetic disease? Science (New York, NY) 175(4025):949–955. https://doi.org/10.1126/science.175.4025.949
2. Becerra CR, Manji GA, Kim DW, Gardner O, Malankar A, Shaw J et al (2019) Ligand-inducible, prostate stem cell antigen (PSCA)-directed GoCAR-T cells in advanced solid tumors: preliminary results with cyclophosphamide (Cy) ± fludarabine (Flu) lymphodepletion (LD). J Clin Oncol 37(15_suppl):2536. https://doi.org/10.1200/JCO.2019.37.15_suppl.2536
3. Bulcha JT, Wang Y, Ma H, Tai PWL, Gao G (2021) Viral vector platforms within the gene therapy landscape. Signal Transduct Target Therapy 6. https://doi.org/10.1038/s41392-021-00487-6
4. Celec P, Gardlik R (2017) Gene therapy using bacterial vectors. Front Bioscience-Landmark 22(1):81–95. https://doi.org/10.2741/4473
5. Zu HG, Gao D (2021) Non-viral vectors in gene therapy: recent development, challenges, and prospects. AAPS J:23. https://doi.org/10.1208/s12248-021-00608-7
6. Patil S, Gao Y-G, Lin X, Li Y, Dang K, Tian Y, Zhang W-J, Jiang S-F, Qadir A, Qian A-R (2019) The development of functional non-viral vectors for gene delivery. Int J Mol Sci 20(21):5491. https://doi.org/10.3390/ijms20215491
7. Jensen TL, Gøtzsche CR, Woldbye DPD (2021) Current and future prospects for gene therapy for rare genetic diseases affecting the brain and spinal cord. Front Mol Neurosci 14(212). https://doi.org/10.3389/fnmol.2021.695937
8. Knott GJ, Doudna JA (2018) CRISPR-Cas guides the future of genetic engineering. Science (New York, NY) 361(6405):866–869. https://doi.org/10.1126/science.aat5011

9. Doudna JA (2020) The promise and challenge of therapeutic genome editing. Nature 578(7794):229–236. https://doi.org/10.1038/s41586-020-1978-5
10. Mendell JR, Al-Zaidy SA, Rodino-Klapac LR, Goodspeed K, Gray SJ, Kay CN et al (2021) Current clinical applications of in vivo gene therapy with AAVs. Mol Ther 29(2):464–488. https://doi.org/10.1016/j.ymthe.2020.12.007
11. Jiao Y, Xia ZL, Ze LJ, Jing H, Xin B, Fu S (2020) Research progress of nucleic acid delivery vectors for gene therapy. Biomed Microdevices 22(1):16. https://doi.org/10.1007/s10544-020-0469-7
12. van Haasteren J, Li J, Scheideler OJ, Murthy N, Schaffer DV (2020) The delivery challenge: fulfilling the promise of therapeutic genome editing. Nat Biotechnol 38(7):845–855. https://doi.org/10.1038/s41587-020-0565-5
13. Dong W, Kantor B (2021) Lentiviral vectors for delivery of gene-editing systems based on CRISPR/Cas: current state and perspectives. Viruses 13(7). https://doi.org/10.3390/v13071288
14. Tang H, Zhao X, Jiang X (2021) Synthetic multi-layer nanoparticles for CRISPR-Cas9 genome editing. Adv Drug Deliv Rev 168:55–78. https://doi.org/10.1016/j.addr.2020.03.001
15. Karlsson J, Rhodes KR, Green JJ, Tzeng SY (2020) Poly(beta-amino ester)s as gene delivery vehicles: challenges and opportunities. Expert Opin Drug Deliv 17(10):1395–1410. https://doi.org/10.1080/17425247.2020.1796628
16. Tarvirdipour S, Huang X, Mihali V, Schoenenberger CA, Palivan CG (2020) Peptide-based nanoassemblies in gene therapy and diagnosis: paving the way for clinical application. Molecules 25(15). https://doi.org/10.3390/molecules25153482
17. Minh AD, Kamen AA (2021) Critical assessment of purification and analytical technologies for enveloped viral vector and vaccine processing and their current limitations in resolving co-expressed extracellular vesicles. Vaccines (Basel) 9(8). https://doi.org/10.3390/vaccines9080823
18. Schnodt M, Buning H (2017) Improving the quality of adeno-associated viral vector preparations: the challenge of product-related impurities. Hum Gene Ther Methods 28(3):101–108. https://doi.org/10.1089/hgtb.2016.188
19. Hajba L, Guttman A (2020) Recent advances in the analysis full/empty capsid ratio and genome integrity of adeno-associated virus (AAV) gene delivery vectors. Curr Mol Med 20(10):806–813. https://doi.org/10.2174/1566524020999200730181042
20. Staal FJ, Pike-Overzet K, Ng YY, van Dongen JJ (2008) Sola dosis facit venenum. Leukemia in gene therapy trials: a question of vectors, inserts and dosage? Leukemia 22(10):1849–1852. https://doi.org/10.1038/leu.2008.219
21. Yarborough M, Sharp RR (2009) Public trust and research a decade later: what have we learned since Jesse Gelsinger's death? Mol Genet Metab 97(1):4–5. https://doi.org/10.1016/j.ymgme.2009.02.002
22. Smith L, Byers JF (2002) Gene therapy in the post-Gelsinger era. JONAS Healthc Law Ethics Regul 4(4):104–110. https://doi.org/10.1097/00128488-200212000-00004
23. Third patient dies in halted study of Audentes gene therapy (2020). https://www.biopharmadive.com/news/audentes-gene-therapy-patient-deaths/580670/. Accessed 01/10/2021
24. Cross R (2019) FDA approves second gene therapy, Zolgensma, to treat spinal muscular atrophy in infants. Chem Eng News 97(22)
25. CRISPR Therapeutics and Vertex Announce New Clinical Data for Investigational Gene-Editing Therapy CTX001™ in Severe Hemoglobinopathies at the 25th Annual European Hematology Association (EHA) Congress (2020). https://investors.vrtx.com/news-releases/news-release-details/crispr-therapeutics-and-vertex-announce-new-clinical-data. Accessed 07/27/2021
26. Nidhi S, Anand U, Oleksak P, Tripathi P, Lal JA, Thomas G et al (2021) Novel CRISPR-Cas systems: an updated review of the current achievements, applications, and future research perspectives. Int J Mol Sci 22(7). https://doi.org/10.3390/ijms22073327

27. Zhao Z, Li C, Tong F, Deng J, Huang G, Sang Y (2021) Review of applications of CRISPR-Cas9 gene-editing technology in cancer research. Biol Proced 23(1):14. https://doi.org/10.1186/s12575-021-00151-x
28. Zhang S, Guo F, Yan W, Dai Z, Dong W, Zhou J et al (2019) Recent advances of CRISPR/Cas9-based genetic engineering and transcriptional regulation in industrial biology. Front Bioeng Biotechnol 7:459. https://doi.org/10.3389/fbioe.2019.00459
29. Broughton JP, Deng X, Yu G, Fasching CL, Servellita V, Singh J et al (2020) CRISPR-Cas12-based detection of SARS-CoV-2. Nat Biotechnol 38(7):870–874. https://doi.org/10.1038/s41587-020-0513-4
30. Rahman MR, Hossain MA, Mozibullah M, Mujib FA, Afrose A, Shahed-Al-Mahmud M et al (2021) CRISPR is a useful biological tool for detecting nucleic acid of SARS-CoV-2 in human clinical samples. Biomed Pharmacother 140:111772. https://doi.org/10.1016/j.biopha.2021.111772
31. Liu TY, Knott GJ, Smock DCJ, Desmarais JJ, Son S, Bhuiya A et al (2021) Accelerated RNA detection using tandem CRISPR nucleases. Nat Chem Biol. https://doi.org/10.1038/s41589-021-00842-2
32. Broughton JP, Deng X, Yu G, Fasching CL, Singh J, Streithorst J et al (2020) Rapid detection of 2019 novel coronavirus SARS-CoV-2 using a CRISPR-based DETECTR lateral flow assay. medRxiv. https://doi.org/10.1101/2020.03.06.20032334
33. Crispi V, Matsakas A (2018) Duchenne muscular dystrophy: genome editing gives new hope for treatment. Postgrad Med J 94(1111):296–304. https://doi.org/10.1136/postgradmedj-2017-135377
34. Gillmore JD, Gane E, Taubel J, Kao J, Fontana M, Maitland ML et al (2021) CRISPR-Cas9 in vivo gene editing for transthyretin amyloidosis. N Engl J Med 385(6):493–502. https://doi.org/10.1056/NEJMoa2107454
35. Mullard A (2015) Novartis secures first CRISPR pharma collaborations. Nat Rev Drug Discov 14(82). https://doi.org/10.1038/nrd4546
36. Slaymaker IM, Gaudelli NM (2021) Engineering Cas9 for human genome editing. Curr Opin Struct Biol 69:86–98. https://doi.org/10.1016/j.sbi.2021.03.004
37. Gaudelli NM, Lam DK, Rees HA, Solá-Esteves NM, Barrera LA, Born DA et al (2020) Directed evolution of adenine base editors with increased activity and therapeutic application. Nat Biotechnol 38(7):892–900. https://doi.org/10.1038/s41587-020-0491-6
38. Rafiq S, Hackett CS, Brentjens RJ (2020) Engineering strategies to overcome the current roadblocks in CAR T cell therapy. Nat Rev Clin Oncol 17(3):147–167. https://doi.org/10.1038/s41571-019-0297-y
39. Tokarew N, Ogonek J, Endres S, von Bergwelt-Baildon M, Kobold S (2019) Teaching an old dog new tricks: next-generation CAR T cells. Br J Cancer 120:26–37. https://doi.org/10.1038/s41416-018-0325-1
40. Sterner RC, Sterner RM (2021) CAR-T cell therapy: current limitations and potential strategies. Blood Cancer J 11(4):69. https://doi.org/10.1038/s41408-021-00459-7
41. Dressel R, Greinix HT, Holler E, Dickinson AM (2018) Editorial: cellular therapies: past, present and future. Front Immunol 9:1966. https://doi.org/10.3389/fimmu.2018.01966
42. Li R, Johnson R, Yu G, McKenna DH, Hubel A (2019) Preservation of cell-based immunotherapies for clinical trials. Cytotherapy 21(9):943–957. https://doi.org/10.1016/j.jcyt.2019.07.004
43. FDA. Kymriah (tisagenlecleucel). https://www.fda.gov/vaccines-blood-biologics/cellular-gene-therapy-products/kymriah-tisagenlecleucel. Accessed 08/09/2021
44. Novartis. Financial Results, Q4 2020 Media Release. https://www.novartis.com/sites/www.novartis.com/files/q4-2020-media-release-en.pdf. Accessed 08/09/2021
45. Novartis. Kymriah® (tisagenlecleucel), first-in-class CAR-T therapy from Novartis, receives second FDA approval to treat appropriate r/r patients with large B-cell lymphoma. https://www.novartis.com/news/media-releases/kymriah-tisagenlecleucel-first-class-car-t-therapy-from-novartis-receives-second-fda-approval-treat-appropriate-rr-patients-large-b-cell-lymphoma. Accessed 12/01/2021

46. FDA. Yescarta (axicabtagene ciloleucel). https://www.fda.gov/vaccines-blood-biologics/cellular-gene-therapy-products/yescarta-axicabtagene-ciloleucel. Accessed 08/09/2021
47. FDA. Abecma (idecabtagene vicleucel). https://www.fda.gov/vaccines-blood-biologics/abecma-idecabtagene-vicleucel. Accessed 08/10/2021
48. Duong MTC-PMR, Foster AE, Bayle JH, Spencer DM (2016) Abstract A057: Uni-CIDeCAR-T cells: MyD88/CD40-enhanced, Ab-directed CAR incorporating the CaspaCIDe® safety switch. Cancer Immunol Res 4(1 Supplement):A057
49. Gargett T, Brown MP (2014) The inducible caspase-9 suicide gene system as a "safety switch" to limit on-target, off-tumor toxicities of chimeric antigen receptor T cells. Front Pharmacol 5:235. https://doi.org/10.3389/fphar.2014.00235
50. Zhang J, Wang L (2019) The emerging world of TCR-T cell trials against cancer: a systematic review. Technol Cancer Res Treat 18:1533033819831068. https://doi.org/10.1177/1533033819831068
51. Wang X, Cabrera FG, Sharp KL, Spencer DM, Foster AE, Bayle JH (2021) Engineering tolerance toward allogeneic CAR-T cells by regulation of MHC surface expression with human herpes Virus-8 proteins. Mol Ther 29(2):718–733. https://doi.org/10.1016/j.ymthe.2020.10.019
52. Foster AE, Mahendravada A, Shinners NP, Chang WC, Crisostomo J, Lu A et al (2017) Regulated expansion and survival of chimeric antigen receptor-modified T cells using small molecule-dependent inducible MyD88/CD40. Mol Ther 25(9):2176–2188. https://doi.org/10.1016/j.ymthe.2017.06.014
53. Collinson-Pautz MR, Slawin KM, Levitt JM, Spencer DM (2016) MyD88/CD40 genetic adjuvant function in cutaneous atypical antigen-presenting cells contributes to DNA vaccine immunogenicity. PLoS One 11(10):e0164547. https://doi.org/10.1371/journal.pone.0164547
54. Collinson-Pautz MR, Chang WC, Lu A, Khalil M, Crisostomo JW, Lin PY et al (2019) Constitutively active MyD88/CD40 costimulation enhances expansion and efficacy of chimeric antigen receptor T cells targeting hematological malignancies. Leukemia 33(9):2195–2207. https://doi.org/10.1038/s41375-019-0417-9
55. Damase TR, Sukhovershin R, Boada C, Taraballi F, Pettigrew RI, Cooke JP (2021) The limitless future of RNA therapeutics. Front Bioeng Biotechnol 9:628137. https://doi.org/10.3389/fbioe.2021.628137
56. LaFraniere S, Weiland N (2021) F.D.A. Aims to Give Final Approval to Pfizer Vaccine by Early Next MonthF.D.A. Aims to Give Final Approval to Pfizer Vaccine by Early Next Month. https://www.nytimes.com/2021/08/03/us/politics/pfizer-vaccine-approval.html. Accessed 08/09/2021
57. Advances in CMV management: fomivirsen (Vitravene) approved (1998) Proj Inf Perspect 26:7
58. Iftikhar M, Frey J, Shohan MJ, Malek S, Mousa SA (2021) Current and emerging therapies for Duchenne muscular dystrophy and spinal muscular atrophy. Pharmacol Ther 220:107719. https://doi.org/10.1016/j.pharmthera.2020.107719
59. Gales L (2019) Tegsedi (Inotersen): an antisense oligonucleotide approved for the treatment of adult patients with hereditary transthyretin amyloidosis. Pharmaceuticals (Basel, Switzerland) 12(2). https://doi.org/10.3390/ph12020078
60. Urits I, Swanson D, Swett MC, Patel A, Berardino K, Amgalan A et al (2020) A review of Patisiran (ONPATTRO®) for the treatment of polyneuropathy in people with hereditary transthyretin amyloidosis. Neurol Ther 9(2):301–315. https://doi.org/10.1007/s40120-020-00208-1
61. Scott LJ (2020) Givosiran: first approval. Drugs 80(3):335–339. https://doi.org/10.1007/s40265-020-01269-0
62. Moderna (2019) Moderna Ships mRNA Vaccine Against Novel Coronavirus (mRNA-1273) for Phase 1 Study. https://investors.modernatx.com/news-releases/news-release-details/moderna-ships-mrna-vaccine-against-novel-coronavirus-mrna-1273. Accessed 07/27/2021
63. Sahin U, Muik A, Derhovanessian E, Vogler I, Kranz LM, Vormehr M et al (2020) COVID-19 vaccine BNT162b1 elicits human antibody and T(H)1 T cell responses. Nature 586(7830):594–599. https://doi.org/10.1038/s41586-020-2814-7

64. BioNTech&Pfizer (2020) Study to describe the safety, tolerability, immunogenicity, and efficacy of RNA vaccine candidates against COVID-19 in healthy individuals. Identifier: NCT04368728. https://clinicaltrials.gov/ct2/show/NCT04368728. Accessed 08/09/2021
65. Lamb YN (2021) BNT162b2 mRNA COVID-19 vaccine: first approval. Drugs 81(4):495–501. https://doi.org/10.1007/s40265-021-01480-7
66. Krammer F (2020) SARS-CoV-2 vaccines in development. Nature 586(7830):516–527. https://doi.org/10.1038/s41586-020-2798-3
67. Thomas SJ, Moreira ED, Kitchin N, Absalon J, Gurtman A, Lockhart S et al (2021) Safety and efficacy of the BNT162b2 mRNA Covid-19 vaccine through 6 months. N Engl J Med 385(19):1761–1773. https://doi.org/10.1056/NEJMoa2110345
68. Pfizer-BioNTech. https://www.pfizer.com/news/press-release/press-release-detail/pfizer-and-biontech-celebrate-historic-first-authorization. Accessed 12/11/2020
69. National Library of Medicine (U.S.). https://ClinicalTrials.gov/show/NCT04368728. Accessed 07/01/2021
70. FDA. Pfizer-BioNTech COVID-19 Vaccine, Fact Sheet for Healthcare Providers Administrating Vaccine (Vaccination Providers). https://www.fda.gov/media/144413/download. Accessed 06/25/2021
71. Pfizer (2021) COVID-19 vaccine U.S. distribution fact sheet. https://www.pfizer.com/news/hot-topics/covid_19_vaccine_u_s_distribution_fact_sheet. Accessed 08/09/2021
72. Kaplan DA (2020) 3 applications for RFID in the fight against COVID-19. https://www.supplychaindive.com/news/coronavirus-rfid-sensor-tracking-hospital-supply-chain/581066/. Accessed 08/09/2021
73. Sadoff J, Gray G, Vandebosch A, Cárdenas V, Shukarev G, Grinsztejn B et al (2021) Safety and efficacy of single-dose Ad26.COV2.S vaccine against Covid-19. N Engl J Med 384(23):2187–2201. https://doi.org/10.1056/NEJMoa2101544
74. Folegatti PM, Ewer KJ, Aley PK, Angus B, Becker S, Belij-Rammerstorfer S et al (2020) Safety and immunogenicity of the ChAdOx1 nCoV-19 vaccine against SARS-CoV-2: a preliminary report of a phase 1/2, single-blind, randomised controlled trial. Lancet 396(10249):467–478. https://doi.org/10.1016/s0140-6736(20)31604-4
75. AstraZeneca. AZD1222 US Phase III primary analysis confirms safety and efficacy. https://www.astrazeneca.com/media-centre/press-releases/2021/azd1222-us-phase-iii-primary-analysis-confirms-safety-and-efficacy.html. Accessed 08/09/2021
76. Shinde V, Bhikha S, Hoosain Z, Archary M, Bhorat Q, Fairlie L et al (2021) Efficacy of NVX-CoV2373 Covid-19 vaccine against the B.1.351 variant. N Engl J Med 384(20):1899–1909. https://doi.org/10.1056/NEJMoa2103055
77. Kalnin KV, Plitnik T, Kishko M, Zhang J, Zhang D, Beauvais A et al (2021) Immunogenicity and efficacy of mRNA COVID-19 vaccine MRT5500 in preclinical animal models. NPJ Vaccines 6(1):61. https://doi.org/10.1038/s41541-021-00324-5
78. Sanofi and Translate Bio initiate Phase 1/2 clinical trial of mRNA COVID-19 vaccine candidate. https://www.sanofi.com/en/media-room/press-releases/2021/2021-03-12-07-00-00-2191846. Accessed 08/10/2021
79. Lu R-M, Hwang Y-C, Liu IJ, Lee C-C, Tsai H-Z, Li H-J et al (2020) Development of therapeutic antibodies for the treatment of diseases. J Biomed Sci 27(1):1. https://doi.org/10.1186/s12929-019-0592-z
80. Kaplon H, Muralidharan M, Schneider Z, Reichert JM (2020) Antibodies to watch in 2020. MAbs 12(1):1703531. https://doi.org/10.1080/19420862.2019.1703531
81. Mullard A (2021) FDA approves 100th monoclonal antibody product. Nat Rev Drug Discov 20(7):491–495. https://doi.org/10.1038/d41573-021-00079-7
82. Zydney AL (2016) Continuous downstream processing for high value biological products: a review. Biotechnol Bioeng 113(3):465–475. https://doi.org/10.1002/bit.25695
83. Jankowski M (2020) Development of an integrated manufacturing process: the iSKID. In: 24th symposium on the interface of regulatory & analytical sciences for biotechnology health products. WCBP, Washington, DC

84. Coolbaugh MJ, Varner CT, Vetter TA, Davenport EK, Bouchard B, Fiadeiro M et al (2021) Pilot-scale demonstration of an end-to-end integrated and continuous biomanufacturing process. Biotechnol Bioeng. https://doi.org/10.1002/bit.27670
85. Stanton D (2019) Up titer: WuXi breaks 50g/L with continuous CHO process. https://bioprocessintl.com/bioprocess-insider/upstream-downstream-processing/up-titer-wuxi-breaks-50g-l-with-continuous-cho-process/. Accessed 07/28/2021
86. Zhou W (2022) Secure the future of bioprocessing. Genet Eng Biotechnol News 42(1)
87. Bala S (2021) WuXi Biologics CEO expects approval for Covid-19 antibody treatment 'late this year or early next year'. https://www.cnbc.com/2020/11/19/wuxi-biologics-ceo-predicts-approval-for-covid-19-antibody-treatment-.html. Accessed 08/09/2021
88. Pardee K, Slomovic S, Nguyen PQ, Lee JW, Donghia N, Burrill D et al (2016) Portable, on-demand biomolecular manufacturing. Cell 167(1):248–59.e12. https://doi.org/10.1016/j.cell.2016.09.013
89. Arnold C (2019) Who shrank the drug factory? Briefcase-sized labs could transform medicine. Nature 575:274–277. https://doi.org/10.1038/d41586-019-03455-x
90. Arnaud CH (2018) Making biologics on demand. Chem Eng News 96(45)
91. Vanrx (2021) Meet the SA25 aseptic filling workcell. https://vanrx.com/products/sa25-aseptic-filling-workcell/. Accessed 07/02/2021
92. Cancherini LL, Santos da Silva J, Zemp A (2021) What's ahead for biotech: Another wave or low tide? https://www.mckinsey.com/industries/pharmaceuticals-and-medical-products/our-insights/whats-ahead-for-biotech-another-wave-or-low-tide. Accessed 04/30/2021
93. Stoll J, Pothier KC, Sapletal S, Milton A, Stephenson AM, Katz J (2021) Biopharmaceuticals deal trends: what to expect in 2021. https://assets.kpmg.com/content/dam/kpmg/ie/pdf/2021/04/ie-2021-kpmg-biopharmaceuticals-deal-trends-stoll-040521.pdf. Accessed 09/08/2024
94. MedWatch VC (2007) The FDA safety information and adverse event reporting program. J Med Libr Assoc 95(2):224–225. https://doi.org/10.3163/1536-5050.95.2.224
95. FDA. MedWatch: The FDA Safety Information and Adverse Event Reporting Program. https://www.fda.gov/safety/medwatch-fda-safety-information-and-adverse-event-reporting-program. Accessed 08/09/2021
96. PubMed. https://pubmed.ncbi.nlm.nih.gov/. Accessed 08/01/2021
97. DailyMed. https://dailymed.nlm.nih.gov/dailymed/. Accessed 08/08/2021

Correction to: Bioprocessing, Bioengineering and Process Chemistry in the Biopharmaceutical Industry

Kumar Gadamasetti and Stephen A. Kolodziej

Correction to:
Chapters 1 and 19 in: K. Gadamasetti, S. A. Kolodziej (eds.), *Bioprocessing, Bioengineering and Process Chemistry in the Biopharmaceutical Industry*, https://doi.org/10.1007/978-3-031-62007-2

The book was inadvertently published with incorrect author information for one of the chapter authors in the initially published versions of Chapters 1 and 19, listing "David W. Woods" instead of the correct "David W. Wood." This has now been corrected.

The updated version of these chapters can be found at
https://doi.org/10.1007/978-3-031-62007-2_1
https://doi.org/10.1007/978-3-031-62007-2_19

K. Gadamasetti, S. A. Kolodziej (eds.), *Bioprocessing, Bioengineering and Process Chemistry in the Biopharmaceutical Industry*,
https://doi.org/10.1007/978-3-031-62007-2_20

Index

K. Gadamasetti, S. A. Kolodziej (eds.), *Bioprocessing, Bioengineering and Process Chemistry in the Biopharmaceutical Industry*,
https://doi.org/10.1007/978-3-031-62007-2

P

Q

R

S

MIX
Papier aus verantwortungsvollen Quellen
Paper from responsible sources
FSC® C105338

If you have any concerns about our products,
you can contact us on
ProductSafety@springernature.com

In case Publisher is established outside the EU,
the EU authorized representative is:
Springer Nature Customer Service Center GmbH
Europaplatz 3, 69115 Heidelberg, Germany

Printed by Libri Plureos GmbH
in Hamburg, Germany